FUSION TECHNOLOGY 1980

Volume 2

AF901495

SYMPOSIUM COMMITTEE

ORGANIZING COMMITTEE

R Hancox	— Culham Laboratory (Chairman)
P A Davenport	— Culham Laboratory (Scientific Secretary)
H Conrads	— KFA Jülich, FRG
A M Van Ingen	— FOM Jutphaas, Netherlands
A Malein	— EURATOM Brussels, Belgium
G Rostagni	— Padua University, Italy
P Schiller	— JRC Ispra, Italy
K H Schmitter	— IPP Garching, FRG
B Taquet	— CEN Grenoble, France
A Torossian	— CEN Fontenay-aux-Roses, France

SPONSORS

Culham Laboratory, UKAEA
European Atomic Energy Community — EURATOM

LOCAL COMMITTEE	SECRETARIAT
R Hancox	Mrs R Bullough
D Bayes	Miss N Clargo
P A Davenport	Mrs J Hitchon
B J Green	Mr I E Pollard
J H C Maple	Miss H A V Smith

LEGAL NOTICE

Neither the Commission of the European Communities, nor any person acting on behalf of the Commission, is responsible for the use which might be made of the following information.

Association
European Atomic Energy Community — EURATOM
UKAEA — Culham Laboratory

FUSION TECHNOLOGY 1980

Proceedings of the Eleventh Symposium
The Examination Schools, Oxford, UK
15—19 September 1980

In Two Volumes

Volume 2

11th
SOFT
1980

Published for the
COMMISSION OF THE EUROPEAN COMMUNITIES
by
PERGAMON PRESS

OXFORD · NEW YORK · TORONTO · SYDNEY · PARIS · FRANKFURT

U.K.	Pergamon Press Ltd., Headington Hill Hall, Oxford OX3 0BW, England
U.S.A.	Pergamon Press Inc., Maxwell House, Fairview Park, Elmsford, New York 10523, U.S.A.
CANADA	Pergamon of Canada, Suite 104, 150 Consumers Road, Willowdale, Ontario M2J 1P9, Canada
AUSTRALIA	Pergamon Press (Aust.) Pty. Ltd., P.O. Box 544, Potts Point, N.S.W. 2011, Australia
FRANCE	Pergamon Press SARL, 24 rue des Ecoles, 75240 Paris, Cedex 05, France
FEDERAL REPUBLIC OF GERMANY	Pergamon Press GmbH, 6242 Kronberg-Taunus, Hammerweg 6, Federal Republic of Germany

Published for the Commission of the European Communities,
Directorate General
Information Market and Innovation —
Scientific and Technical Communication, Luxembourg

Copyright © 1981 ECSC, EEC, EAEC Luxembourg,

All Rights Reserved. No part of this publication may be reproduced, stored in a retrieval system or transmitted in any form or by any means: electronic, electrostatic, magnetic tape, mechanical, photocopying, recording or otherwise, without permission in writing from the copyright holders.

First edition 1981

EUR 7035 EN

British Library Cataloguing in Publication Data

Symposium on Fusion Technology,
11th, Oxford, 1980
Fusion technology, 1980
1. Nuclear fusion - Congresses
I. Title II. Commission of the
European Communities
621.48'4 TK9204 80-41607

ISBN 0 08 025697 X

In order to make this volume available as economically and as rapidly as possible the authors' typescripts have been reproduced in their original forms. This method unfortunately has its typographical limitations but it is hoped that they in no way distract the reader.

CONTENTS

VOLUME I

Opening Addresses

Invited Lectures

Contributed Papers

Data Acquisition & Control

REAL-TIME PROCESSOR FOR A TOKAMAK

A. Ogata, T. Kumahara, T. Matoba, A. Sawai* and Y. Ohtake#
Japan Atomic Energy Research Institute, Tokai-Mura, Ibaraki-Ken 319-11, Japan
* Central Research Laboratories, Nippon Electric Co. Ltd.,
Takatsu-ku, Kawasaki-City 213, Japan
Systems Integration Division, Nippon Electric Co. Ltd.,
Minato-ku, Tokyo 108, Japan

ABSTRACT

A real-time processor is designed for a tokamak, which not only gathers but processes data during a shot. It has a structure of pipelined-microprocessors. Integrated signal processors with instruction cycle of 250 nsec are adopted as its main elements. A method based on a directed graph is employed to define computational requirement given to the real-time processor.

1. INTRODUCTION

This paper describes design of a real-time processor for tokamak experiments. Conventional scheme has been gathering data during a shot and processing them during a shot interval. However, this real-time processor not only gathers but processes data during a shot, so that all its processing procedures finish at the moment when a shot finishes.

Requirement of such real-time processing has arisen from the following facts;

- Feedback control: Several proposals have been made to control plasma parameters such as temperature and density by feedback[1]. To realize such proposals, diagnostic information has to be supplied to a controller in real-time.
- Long discharge: Shots with more than 10 sec duration are planned in some large tokamak experiments[2]. If the data processing relies entirely upon the conventional inter-shot processing, we need huge memory capacity to store raw data. We have only to store the processed data, if the real-time processing is successfully carried out.
- High duty ratio: Future tokamaks such as INTOR will have higher duty ratio than the large tokamaks now in gestation[3]. The real-time processor will be essential for such operation, because little time will be available between two successive shots to exclude possibility of the inter-shot processing.

This paper proposes an example of pipelined multiprocessor structure of the real-time processing. It has following advantages:

- High throughput.
- Cost effectiveness resulting from the exploitation of many low cost microprocessors.
- Expandibility due to modular concept.

In section 2 of this paper we summarize the computational requirements for the tokamak real-time data processing by a directed graph[4]. The directed graph clarifies which tasks can be carried out in parallel and which should be pipelined. We can easily construct a real-time processor, assigning one processor module for each task, or each node, and pipelining processor module to connect tasks which are interconnected by branches in the graph. This pieplined multi-processor structure is described in section 3 in detail. Section 4 estimates the number of the microprocessor modules necessary in the real-time processor to fulfil the requirement of section 2. Section 6 contains discussion.

Table 1. Required computations

task number	objective parameter	required computations
11	emissivity of visible bremsstrahlung	Abel inversion: $y(r) = \sum_x C_{rx} y_L(x)$, $r = 1,---,10$; $x = 1,---,10$
12	electron density	$y = \varepsilon$ or n_e, subscript L denotes line integral
14	current density	$T_e^{3/2}$ law: $j(r) = I_p T_e^{3/2}(r) / \sum_r T_e^{3/2}(r)$, $r = 1,---,10$
15	Ohmic power	Ohm's law: $P_{ohm} = V_L I_p$
16	power lost to first wall	heat conduction equation[6]:
17	power lost to first wall at nbi region	$x(t,r) = \sum_j C_{jr}\ T(t-j,r)$, $r = 1,---,10$; $t > j$ $\bar{x} = \sum_r C_r x(r)$, $x = P_{loss}$ or $P_{loss} + P_p$
21	effective ion charge	bremsstrahlung dependence[7]: $\bar{Z}(r) = C\varepsilon(r) T_e^{1/2}(r) / [n_e^2(r) \exp\{T_e(r)\}]$, $r = 1,---,10$
22	second momentum of space distribution	$\mu(x) = \sum_r r x(r)$, $V(x) = \sum_r r^2 x(r) - \mu^2(x)$, $r = 1,---,10$
24		$x = T_e$, n_e or j
25		
23	electron pressure	$W_e = \sum_r C_r n_e(r) T_e(r)$, $r = 1,---,10$
26	safety factor	$q(r) = r^2 / \{\sum_{r'} j(r')r'\}$, $r' < r$; $r = 1,---,10$
28	power from nbi to plasma	$\overline{P_{nbi}} = P_{inj} - \overline{\{P_{loss} + P_p\}} + \overline{P_{loss}}$
31	energy confinement time	power balance equation: $\tau_E = W_e / \{P_{ohm} + \overline{P_{nbi}} - \overline{P_{loss}} - \frac{W_e(t) - W_e(t-1)}{\Delta t}\}$

2. DIRECTED GRAPH OF THE COMPUTATIONAL PROCEDURE

Figure 1 shows the directed graph for an example of the tokamak real-time processing. The graph consists of directed branches interconnected at nodes. Each node corresponds to a computational task, and branches which grow from each node correspond to plasma parameters obtainable from the corresponding task.

The processing given in fig.1 is proposed for the large tokamak JT-60 experiments. It is required for the real-time processor to carry out all the tasks every one millisecond. Objective plasma parameters and their respective computation methods are given in table 1.

Some computations required at the tasks in fig.1 are not popular in the conventional processing. Specifically, tasks 22, 24 and 25 compute parameters characterizing spatial distributions of density, temperature and current respectively; i.e., they give variances or second momenta of the spatial distributions. These parameters will be fed to the feedback controller. Of course we can feed data of raw distribution, but simpler parameters are more welcomed by the controller because they lighten its load.

The tasks 11 to 18 are executable in parallel. Tasks 21 to 29, most of them have more than two incoming branches, are also executable in parallel, only if the results of their preceding tasks are synchronously available. In principle, we can construct a real-time processor, assigning one processor module for each node provided that the computational time of each tasks is less than 1 msec. Our real-time processor is based on this idea. However, the computation time differs among tasks. To make pipelining successful, we have to partition

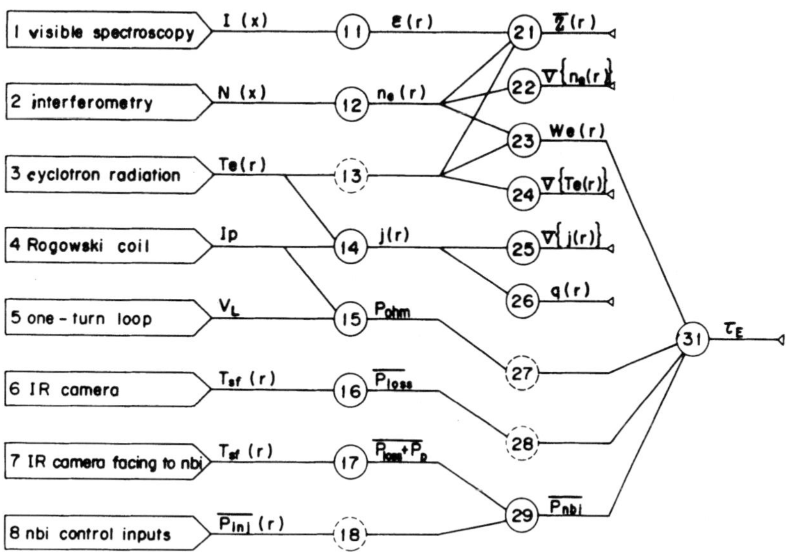

Fig. 1 Directed graph for an example of tokamak real-time data processing.

a time-consuming task onto plural processors. On the contrary, result of a non-time-consuming task has to wait until others finish. We regard this waitings also as tasks and show them by circles of broken line in fig.1.

We will come back to fig.1 and table 1 again in section 4, where we estimate the number of processor modules to carry out all the task every one millisecond.

3. REAL-TIME PROCESSOR STRUCTURE

Structure of the real-time processor based on pipelined processor modules is given in fig.2. It consists of a processing part and a controlling part. The processing part consists of serially connected stages. Each stage has several signal processor modules (SPMs). SPMs in the i-th and (i+1)-st stages are pipelined, while all in the same stage operate in parallel to attain high throughput.

The controlling part consists of the control processor (CP), main memory, peripherals, a disk, buffer memories (BMs) and stage controllers (SCs). The CP controls the whole system and, in addition, provides program development utility. An SC controls SPMs belonging to its stage and enables data transmission. The processed data in SPMs are stored temporalily in BMs and transferred to the disk. Some processed data are available in real-time at the monitor. Charge-coupled devices are adopted as BM elements.

Data transmission between stages is realized by the packet switching, which provides flexibility and expandibility. The transmission rate is 2.048 $MWsec^{-1}$. The number of bus lines is 17 of which 16 are used for data and one for control. The packet length is made variable to attain efficient transmission.

Time chart of computation and data transmission is shown in fig.3. SPMs with less time-consuming tasks are arranged ahead of those with more time consuming tasks in a stage so as to obtain high-speed pipeline processing. Time delay at a stage is given by the most time-consuming task. Proposed packet format is given in fig.4, where a packet is formed with a header and data. A header consists of one or more words which define talker-listener relations among SPMs and the corresponding number of data words.

Structure of an SPM is given in fig.5. The input selector receives the addressed data and transfers them to the signal processor through FIFO memory. The processed data are fed to the output control through another FIFO. The output control generates packets and outputs them under the specified timing. The module controller, which has a general purpose

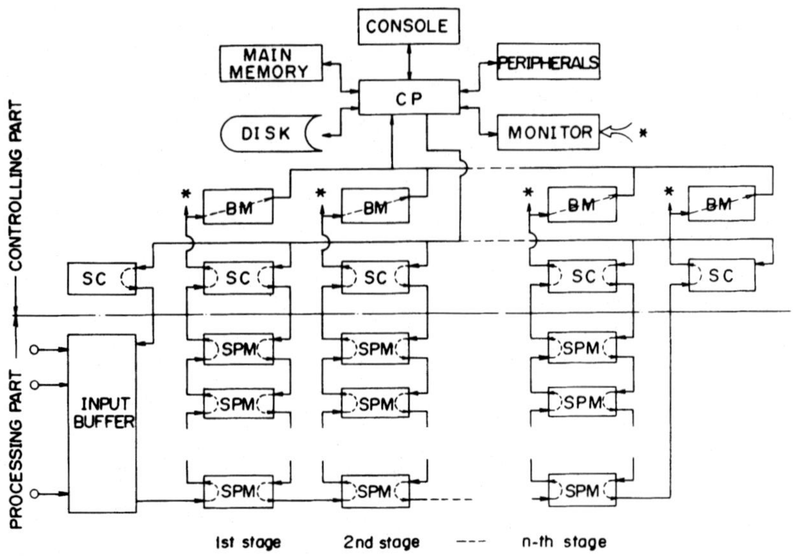

Fig. 2 Structure of the real-time processor.

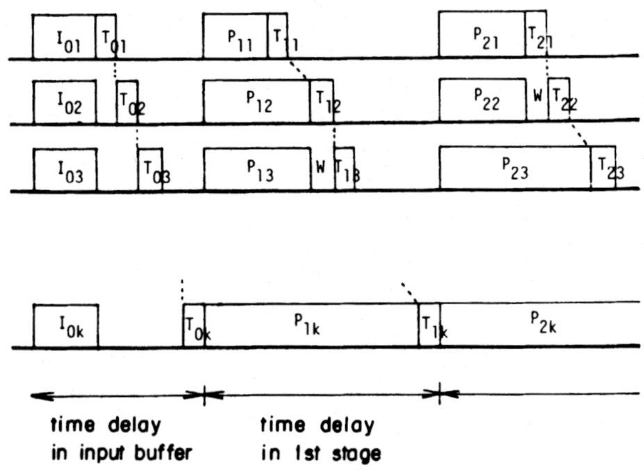

Fig. 3 Time chart.
W:waiting time, T:transmission time.

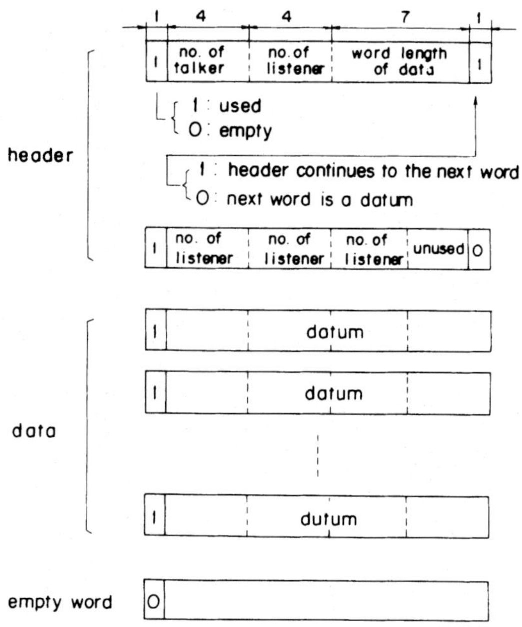

Fig. 4 Packet format.

microcomputer Z80, supervises the whole module through the 8 bit bus. The number of integrated circuits to constitute each part of the SPM is also given in the figure.

As the signal processor, μPD7720D is employed. Its performance is summarized in table 2. The signal processor chip has a built-in multiplier and executes in parallel such functions as move, add and multiply during one micro-instruction cycle of 250 nsec, or equivallently 4 million steps per second[5].

An SPM is designed on a 250 x 290 mm printed card board with four sets of 17 way connector plug. An SC and a BM are also implemented on cards with the same size. Each stage consisting of an SC, a BM and SPMs is housed in a crate. A data transmission bus is installed on the mother board behind the crate. There piled-up are as many crates as the number of stages of the real-time processor. On the top of the piled-up crates are mounted the CP, the main memory and the monitor.

4. DESIGN EXAMPLE

In this section, we first examine the number of steps for basic computations; square rooting and division. We then estimate the steps for the computations given in section 2.

Table 2. Performance of μPD7720D

PROGRAM ROM	512 x 23 bits
DATA ROM	512 x 13 bits
RAM	128 x 16 bits
MULTIPLIER	16 x 16 parallel
DATA WORD LENGTH	16 bits
I/O	8 bit parallel serial
CLOCK	8 MHz
CYCLE TIME	250 nsec
POWER SUPPLY	+ 5 V single
DEVICE TECHNOLOGY	3 μm N channel MOS
NUMBER OF ELEMENTS	about 40,000
DIE SIZE	5.47 x 5.20 mm
PACKAGE	28 pin DIP

Finally, the number of SPMs to realize the given computations is estimated.

Square rooting: We first normalize the given value x into the range between 1/4 and 1. Then the square root is approximated by $Ax + B$ (x [1/2,1]) or $2Bx + A/2$ (x [1/4,1/2]), where $A = 151/256$ and $B = 427/1024$. Using this approximated value as the initial value, we apply Newton-Raphson method, which reduces to the computation $y_{k+1} = (3y_k - y_k^3 x)/2$, where y_k is the approximation of $x^{1/2}$ obtained by k-th iteration. Just one iteration gives enough accuracy for 16 bit word length. Denormalizing the result, we finally accomplish the square rooting. The number of dynamic steps are 75 for normalization and linear approximation, 11 for Newton-Raphson iteration, 8 for denormalization and 94 in total.

Division: Similar estimation for division results in approximately 300 steps. Because it is time-consuming, algorithm to reduce the number of inevitable divisions is to be studyed.

Input/output: Approximately 5 μsec is needed for an input datum to be taken into RAM area of μPD7720D from the input. Another 5 μsec is needed for outputing.

Steps for computations given in section 2: We now estimate the computational steps and the number of SPMs for each task of table 1. Abel inversion (task No.11 and No.12) reduces to multiplication between 10 x 10 matrix and 10 dimensional vector. Not only the vector but all the matrix elements can be stored in RAM of the signal processor. The number of required dynamic steps is 221. One SPM is enough for each task.

Current density computation (No.14) includes square rooting of 10 data, which requires about 1000 steps, and one division. Though ten divisions are included, nine of them are reduced to multiplications, once the reciprocal is computed. The total number of steps then becomes 1600; i.e., 1000 for square rooting, 300 for division and 300 for others. One SPM can carry out the whole computation within given time interval. Because the calculated square roots are again used in task No.21, $T_e^{1/2}(r)$ values are also forwarded to the next stage.

Ohmic power computation (No.15) is very simple. Do-nothing tasks (No.13 and No.18) are also loaded on the SPM for this task.

Straightforward programming of heat conduction equation solution in tasks No.16 and 17 requires extravagant memory capacity and program steps. We are going to develop some approximation program, but tentatively allot two SPMs for each task.

Computation of effective ion charge (No.21) include exponential computation and division. Exponentials are computed by Maclaurin expansion. The total time mainly depends on the division time. Estimating that computation for one datum requires 300 dynamic steps, we partition this task onto two SPMs.

Computation of variances (No.22, 24, and 25) are relatively simple. One SPM for each is more than enough. Computation of

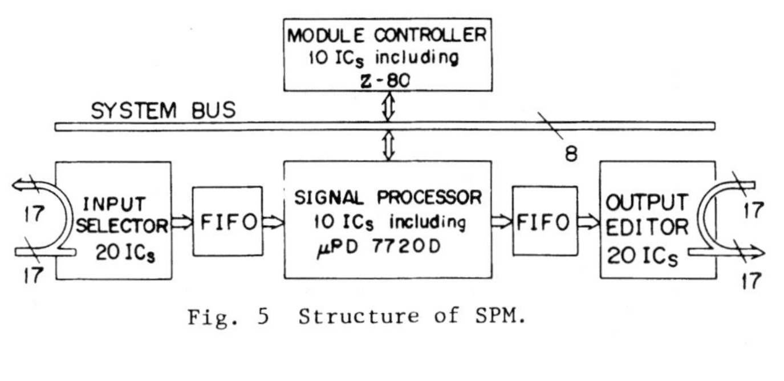

Fig. 5 Structure of SPM.

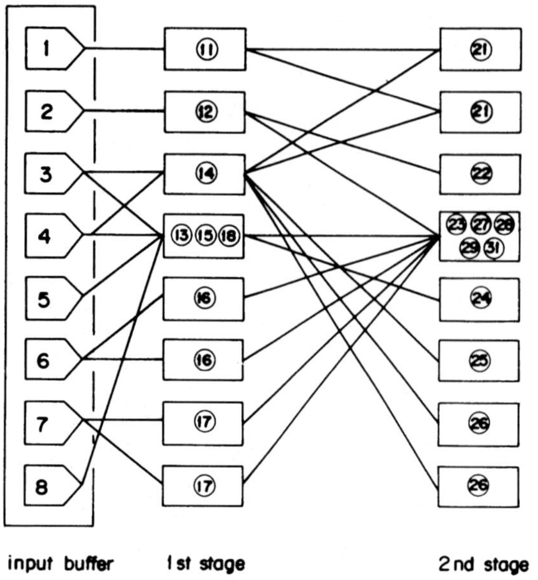

Fig. 6 Configuration of processing part to realize the given requirement. Rectangle : SPM. number in circle : task number.

safety factor (No.26) includes division. This task is also partitioned onto two SPMs. All other tasks (No.23, 27, 28, 29 and 31) are carried out on one SPM. The numbers of steps for No.23 and 29 are so small that most dynamic steps of this SPM are dedicated to the division to derive the confinement time. No.27 and No.28 are do-nothing tasks.

Figure 6 shows configuration of processing part of the real-time processor to fulfil the requirement given in fig.1. It has only two stages, each of which has 8 SPMs.

5. DISCUSSION

The design of real-time processor described in this paper was started as a part of the data processing system for the large tokamak JT-60. Deisgn of the whole system is now proceeding, and some changes will be made in the real-time processor.

First, most SPMs will be CAMACized and housed in the crates directly interfacing diagnostic devices. As in shown in fig.1, most computations in the first stage are characteristic to the diagnostic devices, and hence, are computable in parallel. It is natural to carry out them in the CAMAC crates.

Second, more SPMs than described in this paper will be adopted. We should note the fact that the more real-time processing is adopted, the lighter the load of the conventional inter-shot processor becomes. The inter-shot processor will then be liberated from simple but time-consuming computations, and will be used for those which are hitherto batch-processed. One example is calculation of the energy transfer from injected neutral beams to ions and electrons.

Though the real-time processor was designed for the JT-60, design principle is applicable not only for data processing of other fusion devices but for applications of other field.

REFERENCES

1. OGATA, A. and NINOMIYA, H., Proc. 8th Symp. Engng. Problems Fusion Research (San Francisco, Calif., 1979) IEEE Pub. No.79CH1441-5, p.1879.
2. OGATA, A., Large Tokamak Experiments, to be published in Nuclear Fusion.
3. INTOR Group, Nuclear Fusion 20 (1980) 349.
4. OGATA, A. and YUASA, K., Comput. Phys. Commun. 19 (1980) 35.
5. NISHITANI, T., KAWAKAMI, Y., MARUTA, R. and SAWAI, A., Proc. ICASSP '80 (Denver, Collorado, 1979) IEEE Pub. No.80CH1559-4, p.386. POSA, G., Electronics, Feb.14 issue (1980) p.100.
6. CARSLAW, A.N. and JAEGER, J.C., Conduction of Heat in Solid (Oxford Press, Oxford, 1959) 2nd ed. p.361.
7. KADOTA, K., OTSUKA, M. and FUJITA, J., Nuclear Fusion 20 (1980) 209.

CONTROL, ACQUISITION AND DATA RETRIEVAL FOR THE TCA TOKAMAK

J. B. Lister and A. Simik
Centre de Recherches en Physique des Plasmas
Association Euratom - Confédération Suisse
Ecole Polytechnique Fédérale de Lausanne
CH-1007 Lausanne / Switzerland

ABSTRACT

A Modular Control and Acquisition System based on the extensive use of distributed processing power using microprocessors has been designed and implemented in Lausanne for use on the TCA Tokamak experiment {1}. The control system is entirely software-driven. Standardization of the main components has allowed the rapid realisation of a variety of functions for both control and data acquisition. Although there has been a considerable initial investment of design effort, we are now in a position to gain from the simplicity inherent in the modularity of both hardware and software, allowing the rapid realisation of new units tailored to the exact requirements of the physicist or engineer. The communication between these units and the PDP 11/60 analysis computer is via the RS232 terminal interface. This solution has demonstrated considerable advantages during its implementation.

1. INTRODUCTION

The control and acquisition units for the TCA Tokamak are based on many autonomous microprocessor-controlled units which are individually simple in function. Standardization of basic parts for both control and acquisition functions has greatly simplified the system construction and allows us to develop new units with minimal effort. The standardization has not been based on CAMAC hardware as recommended by the Euratom Data Acquisition Coordinating Committee {2} since, at the time of system conception, Switzerland was not a member of Euratom, and CAMAC was considered to be far too expensive for use in a distributed system. It is interesting therefore to consider a system which has not been designed around the CAMAC requirements, but around the use of a microprocessor.

The requirements of the integrated control and acquisition system can be summarized as:

- control of many power supplies either locally for commissioning and maintenance or from the central control unit during operation;
- provision of a safety interlock chain;
- display of the machine status and comparison with the status requested by the physicist;
- provision of a timing sequence for the precise control of a Tokamak pulse and the logging of the timing values in the sequence;
- acquisition of diagnostic and machine data by autonomous units during a Tokamak pulse;
- retrieval of acquired data (typically 100 kBytes) into the analysis computer;

- routine analysis of data without operator intervention;
- storage of up to 250 full shots on disk and backup storage on magnetic tape;
- automated production of a machine log.

2. CONTROL SYSTEM DESCRIPTION

The layout of the complete system is shown in Figure 1. We see that the central control is a single unit which contains a program which communicates with all other units. There is one additional hardware circuit between all units which acts as a safety interlock and which remains valid when any of the control system units is unpowered. The central control unit polls or commands all other units in a fixed sequence, interrogating the various active units to determine their status, transfers data to the display system and finally checks whether the whole system is ready. All remote active units are autonomous (controllable locally or remotely), providing their own fault detection and status monitoring. There is, therefore, no need for remote units to interrupt the central control unit, which greatly simplifies the organisation of the software. The total cycle time is below one second. The central control panel button array is identical to the one on the active units, removing any difference between remote and local operation as far as the operator is concerned.

3. BASIC UNITS

The basic unit common to all modules is centred around the Motorola 6800 8-bit-microprocessor (to be updated later to the 6809) programmed at present in assembler language. An editor plus diskette facility exists with a cross-assembler which should reduce software development time in the future. The processor card has 128 Bytes of RAM and 1 KByte of ROM (to be updated to 2K). There is a 6.144 MHz clock divided down to 1.024 MHz which provides timing for the processor and also for I/O transfers. Program initialisation is automatic on power-up. The bus used is the "Mubus" {3} proposed as a European standard but slightly downgraded for our use. It is a printed circuit bus mounted as the backplane of a commercial ELMA crate powered with +5V/5A regulated and filtered and with +12, - 5V for the interfaces and memory. Additional memory is provided on cards of 8K RAM (of which only 4K need be present for the card to function) with a write-protect switch for debugging. As on the processor card, the memory and data bus lines are buffered on the card. A general purpose I/O board provides a serial interface by ACIA, satisfying the requirements of EIA-RS232C transmission and reception which is used throughout for inter-module communication and module-PDP communication, all at 9600 Baud. The same board provides parallel I/O through PIA with 2 Bytes plus control lines. This I/O is used for interfacing the panel buttons and indicators, for example. The structure of one unit is illustrated in Figure 2.

4. CONTROL SYSTEM UNITS

The active units control the charging of eight separate capacitor banks, the status of the toroidal field supply, the operation of the preionisation/discharge-cleaning supply and the operation of the vertical field supply. They examine local interlocks (e.g. overvoltage protection, dump positions) and the general safety interlock, all of which are optically decoupled at their inputs to the unit. The parameters required by the physicist are read from local thumbwheels or from the central control system. Actual operation is also either local or remote and a change from local to remote or vice versa always returns a unit to idle. The local control panel status and voltage indication are valid under both local and remote control.

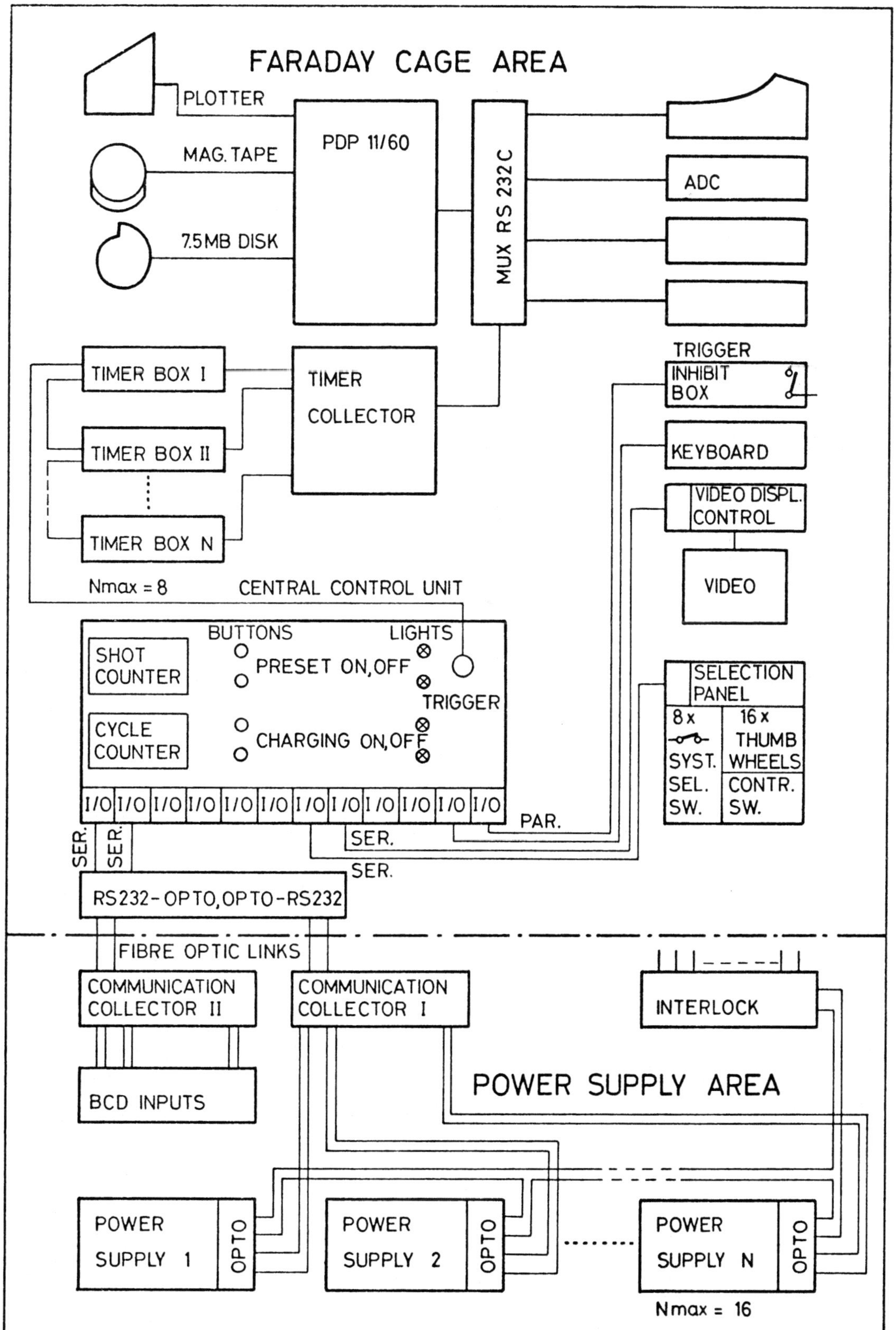

COMPLETE SYSTEM

FIG. 1

The central display is a standard TV driven by a Motorola 6845 CRT control chip producing normal video output. 8 KBytes of RAM are used to produce the picture using 3K of ROM containing fixed descriptions and program together with the status values returned from the remote units. Up to 4 pages can be produced enabling us to display machine status and also information of a general nature inserted by hand for subsequent logging in the PDP.

The system control panel contains a select switch for each subsystem, and thumbwheel control for values required to be preset values. For a small system this is more convenient for the physicist than keyboard control. Two mechanical counters are installed to log the machine cycles and Tokamak shots.

In order to reduce the cabling, communication between central and remote units is made via a remote microprocessor-controlled multiplexer which handles eight channels of RS232 communication. The link between multiplexer and control room is by fibre-optic link (one cable for each direction) using PIFAX 120 fibre and TTL compatible Burr-Brown transmitters and receivers.

The experiment timer is based on the Motorola 6840 programmable triple-timer, using TTL falling-edge timing. Four digits and an exponent are set by thumbwheels for each channel, giving microsecond to second resolution and 5 microsecond to 2½ hour timing capability. Each unit of 12 such channels is microprocessor driven and can communicate with the PDP via a microprocessor multiplexer for reading out the timing values. 32 of the timing signals can pass through an inhibit system which blocks a trigger when that particular subsystem to which the trigger belongs has not been selected. The timing sequence is started from the firing button on the control panel, which also informs the PDP that a shot has started. All timing signals which leave the screened control room are sent by fibre-optic cable, in the same manner as that described above.

5. ACQUISITION UNITS

At present most acquisition units are under development. The main unit already working is an ADC based on the DATEL DAS250 Module with inputs multiplexed at 250 kHz. A control panel allows selection of 1-16 channels and up to 500 samples/channel. The sampling interval is variable from 4 microseconds x No of channels up to 10 msec. We developed the card to control this unit. In addition, a 1 MByte/second DMA interface was constructed to facilitate the transfer to the local memory at a rate compatible with the 4 microsecond conversion rate. The data are resequenced in the microprocessor before transmission to the PDP, saving access time to the data during analysis. This unit is intended for most of the analogue diagnostic signals. A fast acquisition unit is planned for the soft-X-ray signals. A unit has been built for controlling the 2mm microwave interferometer and calculating the total phase shift {4}. Units are under development for an eight-channel scaler to count NPA signals, a software-driven ADC to control an ece interferometer and a unit to control an eight-channel Far Infrared Laser Interferometer. All acquisition units can communicate with the PDP via the standard I/O interface. The single CAMAC crate for diagnostics will also interface to the PDP by RS232.

6. DATA RETRIEVAL AND ANALYSIS

With full use of autonomous acquisition and control modules the idea of a central computer disappears. The PDP 11/60 is now only a retrieval and analysis computer with no real-time function in either of the acquisition or control processes. As mentioned above, all communication with remote units uses the RS232 protocol, which is well established. The interface to the PDP bus is the DZ11 terminal multiplexer which provides us with 16 lines for terminals or microprocessors. The driver used for remote communication is provided by DEC and is the standard terminal driver (TTDRV) which we have not had to modify, thus saving considerable time and effort. Our modules can be quickly connected to any computer, as few computers do not have a terminal I/O facility available. In order to reduce further the retrieval task, we have imposed a common software protocol on all PDP communication, regardless of module type. The PDP interrogates the module initially by sending two Bytes (DC3, Ø) and the module returns the number of logical blocks of 128 Bytes which it wishes to transfer. The PDP requests these blocks one at a time (DC3, N) and they are sent together with a 129th Byte as a checksum. The PDP validates the checksum, or times out if no answer is received. The data rate for each line is 9600 Baud and so we multiplex the lines to reduce retrieval time. At the same time as we asynchronously read up to 8 lines, we write data to disk. In this way, we can retrieve and dump data at a rate of 3.3 KBytes/sec. The reduction from a theoretical maximum of 7.5 KBytes/sec is due to the CPU loading by the character-oriented driver. This disadvantage is outweighed by the following advantages: we can interchange a terminal, remote module or PDP at will, enabling a complete debugging of the software and hardware, we can write all the retrieval software in FORTRAN, saving an enormous programming effort, and we can run on almost any computer. The standardization of communication protocol, both hardware and software, considerably reduced the complexity of the data base used to drive the retrieval software, which can be modified simply by any of the physicists.

For data analysis, we have several graphics-terminals and a printer-plotter for which a common set of graphics routines has been written, merging the manufacturers' software. This enables the physicist to switch from one medium to another during program execution. The main analysis program for the analogue signals can be run manually or driven from a disk file which enables an automatic analysis to be executed after each Tokamak shot. To support the analysis the PDP has 25 MByte of fixed disk on which many shots can be stored and a magnetic tape drive on which shots may be archived using a protected book-keeping system. This solution to the data retrieval problem required a minimum of effort compared with other systems, and the slow I/O speed is nonetheless reasonable for a small-sized experiment, or for a subsystem of a large experiment.

7. CONCLUSION

A system making extensive use of microprocessors in control and acquisition has been commissioned. Avoiding CAMAC was essential to us in a system using over thirty crates, for reasons of cost. Complete standardization of components allowed hardware and software to be executed by one engineer in under three years. Simplification of data retrieval by using a standard protocol, and programming only in FORTRAN allowed this problem to be solved in less than one man year, while achieving an acceptable data transfer rate.

8. ACKNOWLEDGEMENTS

This work was supported by the Swiss National Science Foundation, by the Ecole Polytechnique Fédérale de Lausanne and by Euratom. It owes a lot to the encouragement and support of Dr. A. Heym, R. Schreiber did much of the FORTRAN programming.

9. REFERENCES

{1} CHEETHAM, A.D. et al., 11th Symposium on Fusion Technology, Oxford, these proceedings.

{2} READ, A.H., 10th Symposium on Fusion Technology, Padua, 1978. Proceedings p. 451.

{3} NICOUD, J.D. and DEL CORSO, D., Microscope, 1 No 8 (1977).

{4} LISTER, J.B. and MEANS, R.W., CRPP Report, LRP 171/80.

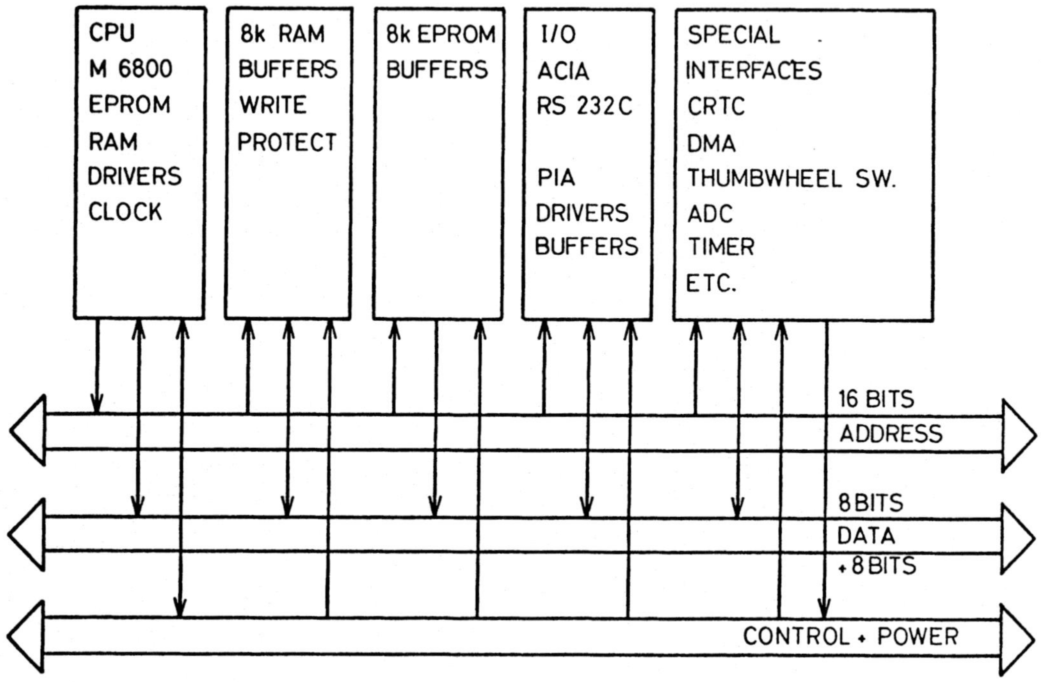

CPU AND I/O ARE IN EVERY CONFIGURATION,
THE OTHER BOARDS ARE OPTIONAL

GENERAL CONFIGURATION OF ONE BOX

FIG. 2

DESIGN OF THE CONTROL AND DATA ACQUISITION SYSTEM FOR SPICA II

W. Kooijman
Association Euratom-FOM, FOM-Instituut voor Plasmafysica
Rijnhuizen, Nieuwegein, The Netherlands

ABSTRACT

The main features of the computerized control and data acquisition system that will be used in the new screw-pinch experiment SPICA II are described. The system is based on a PDP 11/10 computer, interfaced to the experiment via CAMAC equipment and linked to a remote PDP 11/70. This provides long-term archiving facilities and allows several data analysis programs to be run simultaneously.

1. INTRODUCTION

The SPICA II device, currently under construction, is a modification of the SPICA device. The main characteristics of SPICA II are given in [1]. The most important constraints on the design of the control and data acquisition system are:

a. The timescale of interest ranges from a tenth of a microsecond to several minutes.
b. In the experiment pulsed, strong, electric and magnetic fields are used.
c. The characteristics of the discharge may differ from shot to shot.

The data acquisition system described in section 2 is based substantially on experience with the system now in use for the SPICA device [2]. Much of the hardware has already undergone the test of every day operation, which gives confidence that constraints (a) and (b) will impose no major problems. Constraint (c) asks for the storage of as many data as possible from each discharge.

With respect to the control system, described in section 3, the most important constraint is (c). Unwanted variations in the initial conditions must be kept to a minimum. Therefore, a large number of control channels to monitor to a rather high degree of accuracy the time history of the external parameters such as filling pressure and firing of the spark gaps is necessary.

The distinction between the data acquisition system and the control system is more or less artificial. Some signals from specific diagnostics (like Rogowski coils) are also used for the control of the experiment.

2. THE DATA ACQUISITION SYSTEM

2a. The hardware part

The lay-out of the hardware part of the data acquisition system is given in fig. 1. The system is based on a PDP 11/10 computer, located in the screened room. The main characteristics of this computer are given in table I.

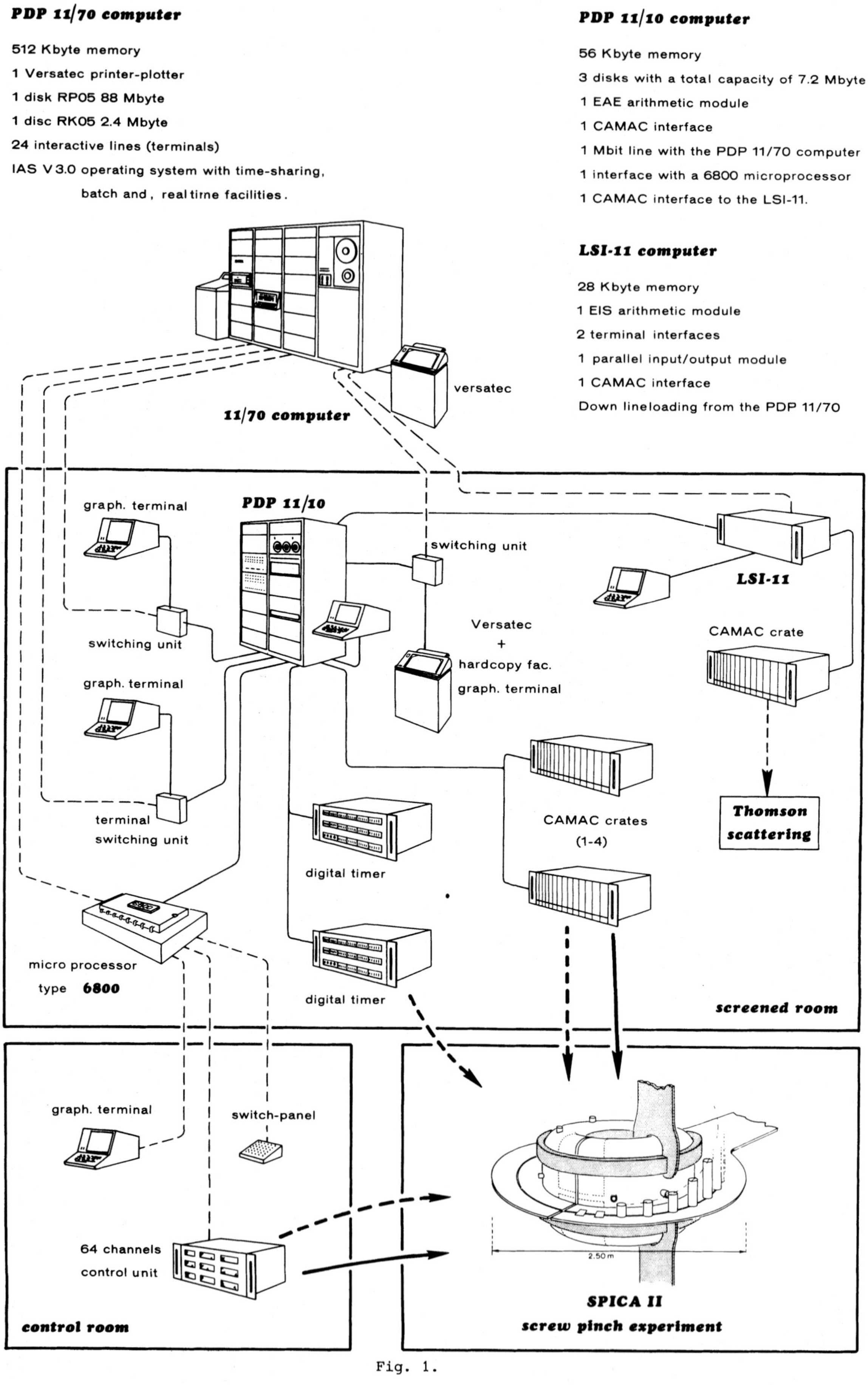
PDP 11/70 computer
512 Kbyte memory
1 Versatec printer-plotter
1 disk RP05 88 Mbyte
1 disc RK05 2.4 Mbyte
24 interactive lines (terminals)
IAS V3.0 operating system with time-sharing, batch and, realtime facilities.
PDP 11/10 computer
56 Kbyte memory
3 disks with a total capacity of 7.2 Mbyte
1 EAE arithmetic module
1 CAMAC interface
1 Mbit line with the PDP 11/70 computer
1 interface with a 6800 microprocessor
1 CAMAC interface to the LSI-11.
LSI-11 computer
28 Kbyte memory
1 EIS arithmetic module
2 terminal interfaces
1 parallel input/output module
1 CAMAC interface
Down lineloading from the PDP 11/70
versatec
11/70 computer
graph. terminal
PDP 11/10
switching unit
LSI-11
switching unit
Versatec
+
hardcopy fac.
graph. terminal
CAMAC crate
graph. terminal
Thomson scattering
terminal
switching unit
CAMAC crates
(1-4)
digital timer
micro processor
type 6800
digital timer
screened room
graph. terminal
switch-panel
64 channels
control unit
2.50 m
SPICA II
screw pinch experiment
control room

Fig. 1.

TABLE I

Characteristics of the PDP 11/10 computer

56	kbyte memory
3	disks with a total capacity of 7.2 Mbyte for the storage of the operating system and data
1	EAE arithmetic module
1	CAMAC interface linked to 5 CAMAC crates
1	Mbit line with the PDP 11/70 computer
1	interface with a 6800 microprocessor
1	CAMAC interface to the LSI-11 mini-computer

The main task of the local computer is the performance of the first checks on the initial conditions and on the data from the diagnostics. The data are then transmitted via an optically coupled 1 Mbit line to the central computer. The main characteristics of the PDP 11/70 computer are given in table II.

TABLE II

Characteristics of the PDP 11/70 computer

512	kbyte memory
1	disk RP05 88 Mbyte
1	disk RK05 2.4 Mbyte
24	interactive lines (terminals) coupled via an optical network
An IAS V3.0 operating system with time-sharing, batch, and real time facilities	

The PDP 11/10 is also coupled to an LSI-11 mini-computer, which is used for the data handling of the Thomson-scattering diagnostics.

With the help of the LSI-11 the calibration and the alignment of the diagnostic can be done independently from the local computer. The characteristics of the mini-computer are given in table III.

TABLE III

Characteristics of the LSI-11 mini-computer

28	kbyte memory
1	EIS arithmetic module
2	interfaces for terminals
1	parallel input/output module of 16 bits
1	CAMAC interface
Down line loading from the PDP 11/70 is applied	

Most of the data from the other diagnostics will be collected by means of fast ADC's with local memories. The total number of channels is 58, the total number of data to be stored during each discharge amounts to ~ 0.1 Mbyte. The various types of ADC's are:

- 36 Culham AD convertors (Culham 95/8190).
  These units, built in CAMAC modules, have a sample frequency of 5 MHz and a memory of 1024 words of 7 bits. It is possible to operate all units with a single time-sequence generator. Interweaving of a maximum of 5 ADC's can be performed by means of a pre-set delay unit (Culham 95/8252). Thus, a sample frequency of 25 MHz can be obtained.
- 12 Biomation 610 B AD convertors. The sample frequency of these units is 10 MHz. The local memory consists of 256 words of 6 bits. The convertors are modified in order to make it possible to use only one clock and, furthermore, to store the settings of the recorders in the last words of their memories.
- 10 in-house developed AD convertors. These CAMAC units have a sample frequency of 16 MHz and a memory of 1024 words of 8 bits. It is possible to operate these units with one external time-sequence generator. The units also have an internal, computer-controlled, variable attenuator.

Processing of the signals before storage will be minimized and carefully recorded in order to enable correction in the processing whenever necessary. Special CAMAC modules (Rijnhuizen EL 994) have been developed in our laboratory to be used together with the Culham convertors. These computer-controlled units contain an amplifier, an attenuator, a filter, an integrator and a DC balance adjustment (see also fig. 2).

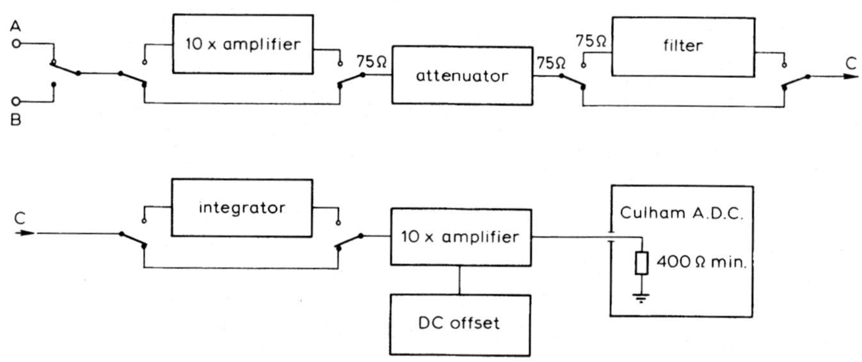

Fig. 2. Block scheme of the signal-processing units.

2b. Transport

2b.1 Digital signals

The digital signals out of or into the screened room are transported via fiber optics. Units to convert electrical signals into optical signals and vice versa have been developed. Throughout the laboratory a network of optical lines exists between terminals and the central computer. The largest distance bridged in this way is 120 m. By means of a specially developed unit it is possible to connect two lines of different computers onto one terminal. In this way it is possible to work interactively on one computer and to get simultaneously the output of programs running in the other computer.

2b.2 Analog signals

Most of the analog signals from the experiment to the screened room will still be transported by coaxial cables lying in copper tubes, which also provide the grounding of the experiment. For those diagnostics which require an insulation between plasma and the screened room, like the magnetic probe, a multi-channel electro-optical convertor with a bandwidth of 3 MHz is being developed.

2c. The software part

The two major components of the software system are:

a. The program which stores the data in the computer. All the data from the CAMAC units described previously are transferred to the memory of the PDP 11/10 computer. The information necessary for an immediate check of the convertors (e.g., underflow/overflow) is then given a data sheet, together with the standard information used in the administration of the discharges. The program is also used for transmission of the data to be stored in the standard data files [4] of the PDP 11/70.
b. The program for the initial data analysis. After each discharge some of the diagnostic signals as well as some of the control signals will be processed and displayed automatically in order to decide on the conditions for the next shot. To do this in a fast and flexible way an updated version of the INTRA program [3] will be used. To avoid long intervals between discharges, large parts of this monitor program will be switch-controlled.

3. THE CONTROL SYSTEM

Two control phases can be distinguished:

a. one in which the variations are slow (seconds). During this phase the capacitor banks are charged, the torus is filled with gas, the system is made ready for a discharge;
b. one in which the variations are fast (micro-seconds). During this phase the banks are fired and crowbarred and diagnostics are triggered.

ad a. During the first phase most of the control (including all safety precautions) will be hand-wired into the system. Like in SPICA, this will largely be a pneumatic system, which has proven to operate satisfactorily. It will be extended, however, with a microprocessor coupled to the PDP 11/10 computer. The microprocessor, type 6800, continuously compares the actual conditions with the requested conditions. If large deviations occur this will be reported to the operator. Therefore, the microprocessor has been built into an in-house developed module with the possibilities to have a 4 k ROM or RAM memory, 2 interfaces for terminals, a parallel input/output unit of 16 bits and a variable number of optical input/output channels. Most of the time the microprocessor will have a fixed program. Down line loading will be possible. The actual conditions of the experiment will be monitored on 64 channels. The most crucial ones will be on display in the control room. The output of all 64 channels will be stored into the computers with a sample frequency of 3 Hz.

ad b. In the second phase the programming of the experiment will be governed by digital timers with a total of 44 channels. Each timer can be used independently during tests and calibrations, but during a discharge they will share one clock to ensure proper synchronization. The channels will provide optical as well as electrical pulses with time delays adjustable from 0.1 μs to 10 ms in 32 channels and from 0.01 μs to 10 ms in 12 channels. The settings of the channels are stored in the computer by means of a CAMAC interface. The control during this phase of the discharge will be governed by an interrupt unit, very similar to the one described previously [2]. Its main action will be to stop the digital timer in case of malfunctioning of one of the capacitor banks. Several spark-gap monitors will be used to store the moment of firing of the 58 spark gaps in use. These CAMAC modules all use a time-to-voltage convertor. The output is transferred to the computer. Whether the time of firing fell within a pre-set time interval is also made visible with LED's.

4. ACKNOWLEDGEMENTS

The author wishes to express his gratitude for the cooperation and assistance he has received from the members of the SPICA II team, the Design Group Electronics and the team of the central computer facility.

This work was performed under the Euratom-FOM association agreement with financial support from ZWO and Euratom.

5. REFERENCES

[1] SPICA II TEAM, "The design of SPICA II", this conference.

[2] BOBELDIJK, C. et al., "Proposal for a modification of SPICA", Rijnhuizen Report 78-112 (1978).

[3] KOOIJMAN, W., Proc. 8th Symp. on Fusion Technology, Noordwijkerhout (1974) p. 477.

[4] VAN DER HEIDEN, T.C. et al., "Standardization of data organization and management", this conference.

STANDARDIZATION OF DATA ORGANIZATION AND MANAGEMENT

Rijnhuizen Users Software Team*. Presentation by G.M.D. Hogeweij
Association Euratom-FOM, FOM-Instituut voor Plasmafysica
Rijnhuizen, Nieuwegein, The Netherlands

ABSTRACT

In this paper a standardized data handling software system is described. This system has been developed at the Rijnhuizen laboratory, to be used with all experiments.

1. INTRODUCTION

At the Rijnhuizen laboratory for plasma physics three experimental lines exist: the RINGBOOG (gas-blanket research), SPICA (screw pinch) and the TORTUR (turbulent heating) line. Each one has its own mini-computer for real time data acquisition; see [1] for a description.

The central computer, a DEC PDP 11/70 system under the IAS operating system, is operational since 1st January 1978. There is a central data team managing this system. The philosophy behind the Rijnhuizen data acquisition system is to centralize the data manipulation, analysis, storage and retrieval, in order to provide optimal computer facilities for all experimental groups. This philosophy led to the need of standardized data handling software, so a good cooperation between the programmers of the experimental groups and the central data team became necessary.

For this purpose a coordinating software team was formed. It is named RUST, the Rijnhuizen Users Software Team. The main task of the 11/70 system is the data analysis in the widest sense; this is the main subject for RUST to deal with.

In the following, we shall describe aims and method of work of RUST in short. Thereafter we shall discuss the software needed for data analysis, and the work that RUST has done until now.

2. RUST

The software team RUST was formed in the autumn of 1979. Anyone who uses or plans to use one of the computers of the laboratory, can become a member of this team. At this moment, the team has 11 members; the programmers of the experimental groups, as well as the central data team, participate in RUST.

The goals of RUST are:

- to be a platform where computer workers from the various groups in the laboratory exchange experiences, wishes and problems;
- to provide a set of library routines and complete programs to all users of the 11/70 computer and its satellites.

Ideas for program libraries or complete programs can be brought in by each member of the team. These will be discussed by RUST. If an idea is useful, it will become a project of the complete team; all active members of RUST will participate in the development of such a project.

There are meetings of the group at least each month, but usually each week. Subjects that are discussed at these meeting are:

- projects in progress;
- ideas for new projects;

- management of the central computer;
- various experiences.

3. DATA ANALYSIS

Let us consider a general data analysis program, as it might run on the Rijnhuizen PDP 11/70 computer, or in another plasma-physical institute. What facilities are needed to write such a program with as little effort as possible?

We come to the following list; see fig. 1. Most of these facilities have a more general scope than data analysis only.

1. A common data file definition is needed to be used for all experimental data; routines to read, write and update such files must be available.
2. There must be the possibility to catalog data files.
3. Almost every program interacts with its user. It is important to have uniform conventions for the way in which such interactions take place; one possibility to achieve this, is to use a small number of standard routines for all interaction between program and user.
   This easens the use of developed programs by non-experienced users.
   Moreover, such a standardization is very important for maintenance people.
4. It should be possible to use a user profile (with information, specific for one user).
5. There must be print and plot facilities.
6. Routines must be available for mathematical, physical and statistical functions.
7. A display utility is needed.

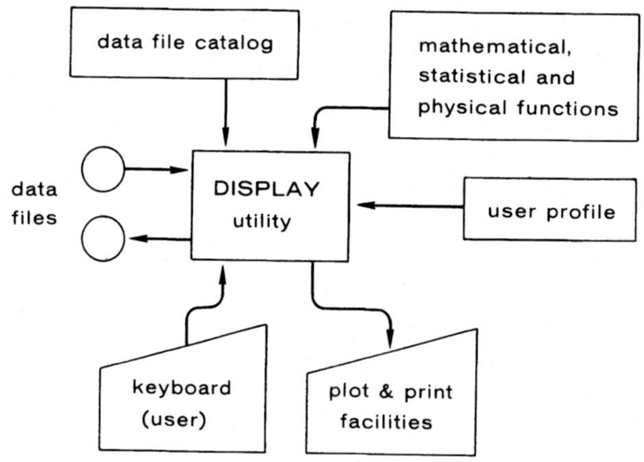

Fig. 1.
Facilities.

So far, RUST has mainly payed attention to the first point; in this report we shall restrict ourselves to that point.

4. DATA ORGANIZATION AND MANAGEMENT

The first concern of RUST was the organization and management of data. First of all, a general data file definition was needed. We shall describe here the definition adapted by RUST. For ease of reference, a data file produced by a plasma-physical experiment is called a 'shot'.

We must draw a distinction between the structure of a data file as it presents itself to the user and the invisible, internal structure.

For the user, a shot is a collection of 4 kinds of tables:

- one shot information table (abbreviated SHT), containing general information about the shot;
- condition tables (COT's), containing information about static conditions (parameters); one COT for each condition;
- signal information tables (SIT's) and signal value tables (SIV's): for each signal a table with information about the signal, and a table with signal values.

The user is not concerned with the order of the tables in the file.

Each table consists of a number of fields. Each field is identified by a mnemonic code, so the user is not concerned with the actual placement of data in a table. Moreover, this has the advantage that programs are compatible over maintenance releases.

The internal structure of a data file is as illustrated in fig. 2. There are two kinds of file information tables:

- the file pointer area (FPA), which contains pointers to the SHT, the COT's and the SPA's;
- the signal pointer areas (SPA's): for each signal there is a SPA, containing pointers to the corresponding SIT and SIV.

The ordering of the tables is unimportant, except for the first one, which must be the FPA.

The user is completely unaware of the existence of these file information tables; they only serve the internal organization.

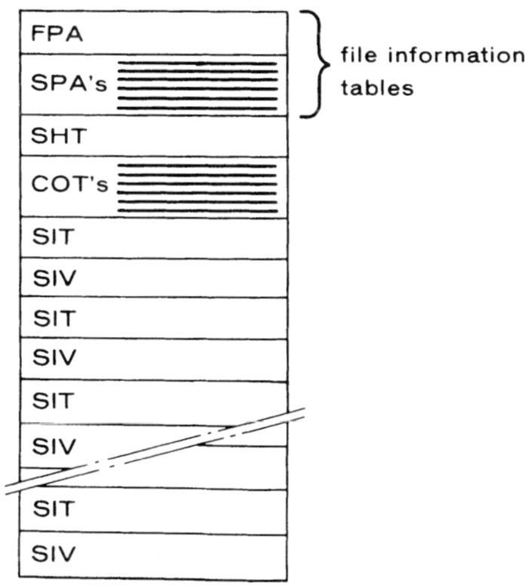

Fig. 2.

Internal structure of a DOM data file.

The whole data organization and management system (briefly DOM system) is built on this data file definition. The system consists of a set of routines, enabling people to write their own data analysis program with great ease, and a number of utility programs.

The DOM system is set up with great modularity. The routines can be divided in a number of levels; see fig. 3. Only the upper two levels are user callable; the other levels are internal.

Next we shall devote a few words about the programming languages used. The routines which are closest to the machine, the block I/O and open/close routines, are written in MACRO (the assembly language of the PDP 11/70).

For the rest of the DOM system, the choice was between PASCAL and FORTRAN. We decided to write the first version in FORTRAN. There was little experience with PASCAL in the laboratory; almost all existing programs have been written in FORTRAN. Moreover, it is possible to call FORTRAN routines from a PASCAL main program, which is not the case vice versa.

If the use of PASCAL increases in the laboratory, we might co-write the DOM system in PASCAL.

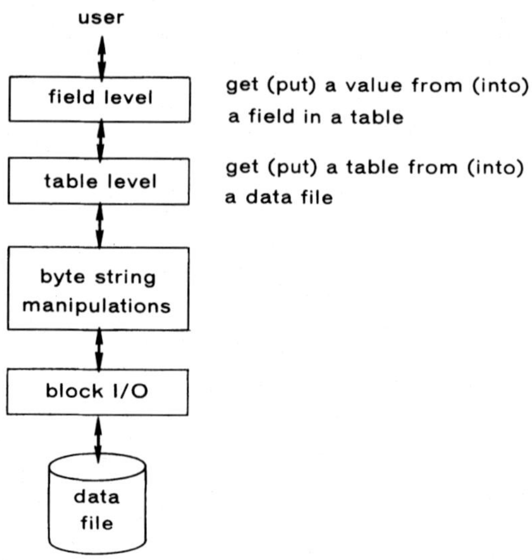

Fig. 3.

Levels of the DOM system.

By July 1980 the DOM routines were written, and all routines were tested separately.

At this momemt the DOM system is tested as a whole: programmers from the experimental groups are developing programs based on DOM, and are creating DOM data files. We have set up a scheme, leading to the acceptance of the whole DOM system before September 1st.

So far, experiences with DOM are good. Problems can be located quickly, because of the high modularity of the system. The speed of the system will prove in practice; ideas exist how to speed up eventual bottlenecks at the cost of some extra programming or some extra computer memory.

5. CONCLUSIONS

Since the existence of RUST many things have changed:

- there is a much better contact between the computer workers;
- programmers learn from each other;
- in the experimental groups the (im)possibilities of the central computer are better understood.

The first project of RUST, the DOM system, runs successfully. The good experiences so far guarantee the survival of RUST after the finishing of this first project.

New projects that have been proposed and which are partially in progress already are:

- a stripped version of DOM for the PDP 11/10;
- a PASCAL version of DOM;
- a data file catalog system;
- a display utility;
- a CAMAC library;
- a library of routines for general use (with a wider scope than data files only).

6. REFERENCES

[1] VAN DER HEIDEN, T.C., VAN INGEN, A.M., VERDOES, J.A., Proc. 10th Symp. on Fusion Technology, Padova (1978), Vol. I, pp. 521-523.

* Members of the Rijnhuizen Users Software Team are:
H.J.B.M. Brocken, J.J. Busser, B. de Groot, T.C. van der Heiden, G.M.D. Hogeweij, T.L.M. Hoogenboom, A.M. van Ingen, W. Kooijman, P.A. van Kuyk, D. Oepts, J.A. Verdoes.

ACQUISITION, ARCHIVING, AND ANALYSIS OF DOUBLET III DIAGNOSTIC DATA ON A DISTRIBUTED COMPUTER SYSTEM*

A. Glad, P. Henline, B. McHarg, D. Drobnis, S. Karin, G. Shephard
General Atomic Company
10955 John J. Hopkins Drive, San Diego, CA 92121, USA

ABSTRACT

Computer support of a Tokamak fusion experiment for the purpose of data acquisition, analysis, and archiving required handling the large amounts of data generated each shot, and presenting the data for both individual diagnostic experiments and shot summary in a timely and useful manner. A distributed computer system with processing at three levels supports General Atomic's Doublet III Tokamak by providing: (1) acquisition and permanent archiving of all diagnostic data, (2) analysis and control for each physics diagnostic, (3) overall shot summary after each shot, and (4) detailed analysis of current and past shot data.

1. INTRODUCTION

This paper describes the diagnostic data acquisition, archiving and processing system for Doublet III, with emphasis on management of the large mass of diagnostic data through a distributed computing system. The number of diagnostics currently operating and planned required a system which provides concurrent acquisition and processing, and allows the addition of new diagnostics without unduly perturbing the existing system. In addition, generalized applications software tools were necessary for use by a variety of diagnostics. These are transferable from microprocessor to minicomputer to large mainframe, allowing consistency and adaptability of software.

2. CENTRALIZED MINICOMPUTER SYSTEM

The centralized minicomputer system was designed and implemented to be the host computer, with responsiblities for acquisition and management of all Doublet III data. The minicomputer system design requirements were demanding and included the acquisition, processing and archiving of approximately two million samples of data over a shot cycle 4-7 minutes in length. In addition, flexibility was necessary to enable modifications reflecting rapidly changing physics diagnostics.

Three objectives were defined. The first was to acquire and manage all data for every physics diagnostic on the Doublet device. To do this, a centralized data acquisition system was designed using one set of services to acquire and manage all diagnostic data. The data acquired was of two types: 1) raw data from hardware devices such as waveform digitizers, analog-to-digital converters (ADCs), etc., which were acquired and managed directly by the minicomputer system, and 2) preprocessed data from microprocessors used to control and interface individual physics experiments.

The second objective was to archive all data for permanent storage and retrieval. An existing DECsystem-10 computer, which is part of the National Magnetic Fusion Energy (NMFE) computer network, was used. This allowed for post-shot analysis too elaborate to perform on the mini- or microcomputers within the shot time frame, as well as access to it through time-shared terminals and the NMFE computer network. The data archived includes both the raw or processed data acquired by the minicomputer system and all data base parameters necessary to interpret and process the data.

The third objective was to provide limited real-time analysis for each physics diagnostic. This analysis provides an indication of whether the diagnostic is functioning successfully and of overall plasma behavior. The analysis is distributed between the mini- and micro- systems to allow a large amount of output within the shot time.

2.1 Hardware configuration

The diagnostic minicomputer hardware configuration is illustrated in fig. 1. The computer system consists of a ModComp Classic computer with 512K 16-bit words of memory. Devices connected to the computer system include four moving head disks. One 2.5 megabyte cartridge disk is used as the system disk containing fixed partitions of all operating system libraries, tasks, etc. The second 2.5 megabyte cartridge is organized with flexible file structure, using ModComp's File Manager, containing auxiliary files for data analysis tasks. The two large 50 megabyte disks are both file managed disks, containing the current data files for all physics diagnostics for the past 10-20 shots and all system, application and user source and object programs. A high speed bulk core memory contains a portion of the data base accessed frequently and all non-resident services used to manage both the data base and all experimental data. A 6250 cfi tape drive is currently being added to the system, replacing an 800 bpi unit, as the medium for archiving data. The 6250 will reduce the number of tapes generated and increase the speed for archiving the data. A high speed synchronous link to DECsystem-10 is currently being implemented; tapes will then be used only as backup to recover when transmissions between the DECsystem-10 and the ModComp fail.

A communications interface connects 64 devices, both asynchronous and synchronous, to the ModComp. Currently, eight Tektronix Series 4000 graphics terminals are used. In addition, the soft X-ray diagnostic microprocessor system, controlling a commercial multichannel analyzer (MCA) system, interfaces through an RS-232 asynchronous channel.

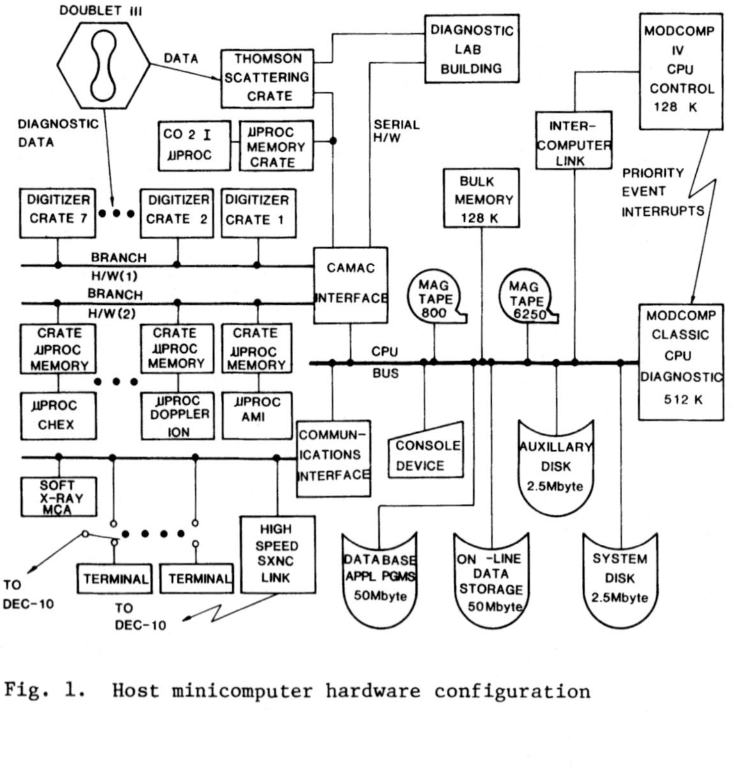

Fig. 1. Host minicomputer hardware configuration

Finally the high speed synchronous link to the DECsystem-10 uses a channel on the communications interface.

Most of the diagnostic data acquisition for the Doublet device is performed through a CAMAC system. The Doublet III diagnostic CAMAC system consists of two branch highways and a serial highway. The branch highways, because of their higher transfer rate, contain a majority of the devices producing large data blocks. Two major types of data modules are connected onto the branch highway. LeCroy waveform digitizers, which convert analog signals into digital arrays, are a majority of the data acquisition hardware on the CAMAC system. Currently, a little over 250 channels are being acquired with a future expansion of up to about 500 channels. Most of the digitizers are used to acquire magnetic waveforms, although some diagnostics also utilize this type of data acquisition. The second type of modules predominating on the branch highway are microprocessors with dual port memories. Present practice is for each new diagnostic to use a CAMAC microprocessor system to perform local acquisition and processing. The microprocessor then writes data into its dual port memory with the ModComp acquiring the data over the CAMAC. This distributed processing allows the diagnostic to run independently of the ModComp computer system and also to remove some application processing from the ModComp. The CAMAC serial highway contains all physics diagnostics which require remote capability from the control room. Currently, the Thomson scattering and the CO_2 laser interferometer experiments use the serial highway in order to have their interfaces close to their other electronics.

Logging printers log messages indicating normal operation and abnormal conditions. Error messages to indicate problems in acquisition, processing or archiving of diagnostic data are printed with shot and diagnostic experiment identification.

Finally, the diagnostic computer system interfaces the control computer system through an intercomputer link and a series of priority interrupts. The control computer is used to control and monitor the operating of the Doublet device. The intercomputer link is used to transfer the data base of the control computer each shot and archive it along with the diagnostic data. The priority interrupts are used by the diagnostic system to synchronize data acquisition, processing and archiving with Tokamak operation.

2.2 Software configuration

The software system is driven by information organized into a series of files and tables called the data base. The data base is a complete grouping of all information related to the data collected from experiments on the Doublet device. All system tasks are driven from the data base and include the centralized data acquisition tasks, the data manager responsible for dynamically managing all data locally, the data archiver responsible for transmitting all data to the DECsystem-10, and the task manager responsible for scheduling all system and application tasks. In addition, the experiment application software accesses the data base indirectly in requesting shot data for analysis through the common data file services. Finally, the data base generator and associated interactive programs enable operators to build, view, and modify the data base files and tables.

2.2.1 Shot timing

The shot cycle on Doublet III from shot initialization through completion is approximately 3-5 minutes. This time span is divided up into a series of time slices, termed phases. Three event times in the shot cycle initiate sequences of phases causing specific data acquisition tasks to be performed.

The diagnostic system utilizes these three events to time and initiate all data acquisition, processing and archiving tasks. The first event of a shot cycle, Start Phase (approximately two minutes before a shot), causes the data base files and tables to be initialized, performs calibration of acquisition hardware , initiates the data manager to allocate space for the new data, and performs any acquisition and archiving non-shot dependent data, e.g., switch settings on acquisition hardware, etc. The second event (occurring ten seconds before a shot) causes all preshot control and acquisition to be performed including initializing all transient digitizers and dual port memories, acquiring the control computer's data base, etc. The third event, indicating shot completion, initiates the post shot (data acquisition phases). Since all the Doublet diagnostic data is stored locally in hardware, i.e., transient digitizers, dual port memories, etc., the computer system simply reads all of the data for a shot and stores it on the data files. To accommodate the large amount of data input per shot, which requires up to two minutes to acquire, the post shot time is divided into ten phases (nine synchronous sequential and one asynchronous). This enables data required for immediate viewing and analysis to be acquired immediately after a shot, while allowing data used for off-line analysis to be read later in post shot phases.

2.2.2 Data acquisition/archiving

The data acquisition and archiving for the host computer (ModComp) incorporates three decisions. First, all information such as calibration, gain settings, etc., necessary for analysis must be appended to the data, thus making data structures and programs transferable between computers. Second, all experiments without microprocessors have their data acquired and processed by the host computer. Third, all microprocessors will be polled rather than interrupting the host when data is available. Thus, when a microprocessor has loaded its "mailbox" memory with data, a flag is set and the polling will cause the data to be acquired.

3. SATELLITE MICROCOMPUTER SYSTEMS

The microprocessor computer systems remove some acquisition and processing work load from the host, providing output soon after a shot is complete. In addition, each diagnostic can be developed, tested, and if necessary operated as an independent system without relying on the host.

3.1 Microprocessor hardware configuration

The microprocessor hardware configurations, although different for each application, have a number of common features. Each microprocessor is a CAMAC based LSI-11 processor with 32K bytes memory. Floppy disks with 200K bytes of storage are used for mass storage. Each micro contains a hardware floating point chip to speed data analysis. Finally, each micro system contains a graphics terminal used to provide application graphics output each shot.

The hardware necessary to interface the micro system to the host computer is done through CAMAC. Each micro system is configured as two CAMAC crates (fig. 2). One crate contains the microprocessor and all CAMAC modules necessary to control and acquire data from the experiment and interface the floppy disk, terminal, etc. This crate also contains the physics diagnostic interface which can be either commercially purchased or constructed locally. The second crate is on the host computer's CAMAC. Contained in this crate is a dual port memory used as a mailbox where micro writes all its data (processed and/or raw) for the host to acquire.

3.2 Software configurations

The primary software functions of the microprocessors are standard for all and include: 1) providing control of the experiment and acquiring all of the diagnostic data,

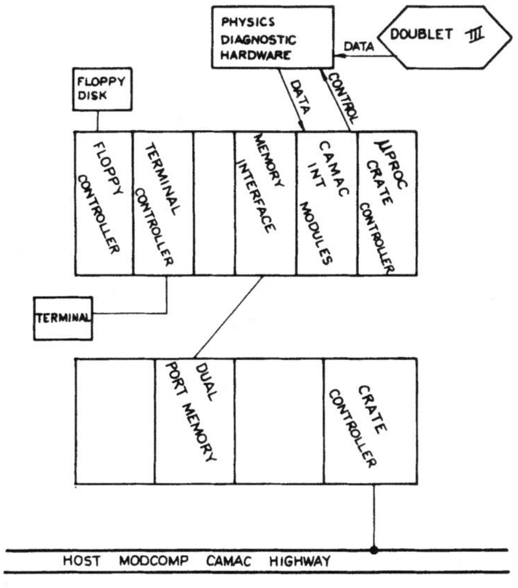

Fig. 2. Microprocessor hardware configuration

2) archiving the data to the host ModComp computer and locally on the floppy disk, and
3) providing initial data reduction, analysis and output thus reducing the work load in the host.

One motive for standardizing on the LSI-11 was the early availability of an adequate operating system for software development in a higher level language. The basic microprocessor configuration supports RT-11, and a somewhat larger LSI-11 work station supports development and maintenance for active diagnostics. FORTRAN, BASIC, MACRO, and PASCAL are available, with FORTRAN predominating.

The microprocessor software systems are designed to operate in an unattended mode requiring operator intervention to change full floppies and for physicists to examine the shot output to verify correct operation of the diagnostic.

4. DECSYSTEM-10 COMPUTER

The purpose of the General Atomic User Service Center (USC) DECsystem-10 computer is threefold. First, the USC provides access to the NMFE large scale central computers (CDC-7600 and Cray-1) for both the theoretical and experimental investigators at General Atomic requiring such resources. Such large scale requirements arise from the great mathematical complexity of the analysis of the plasma physics problems of interest. Figure 3 shows the overall DECsystem-10 and Doublet III interface to the NMFE computer network.

The second major task of the USC is to provide data archival, retrieval, and analysis facilities for the Doublet III experiment. Use of the DECsystem-10 for this removes unfavorable interaction of later analysis with data collection and allows use of the proven DECsystem-10 file handling and time-sharing software. The problems of data retrieval and analysis are more concerned with the central processing power of the computer system. A planned upgrade of the USC to dual KL-10 processors is intended to satisfy this requirement so that the analysis of Doublet III data may proceed at a pace determined by the progress of the experiment.

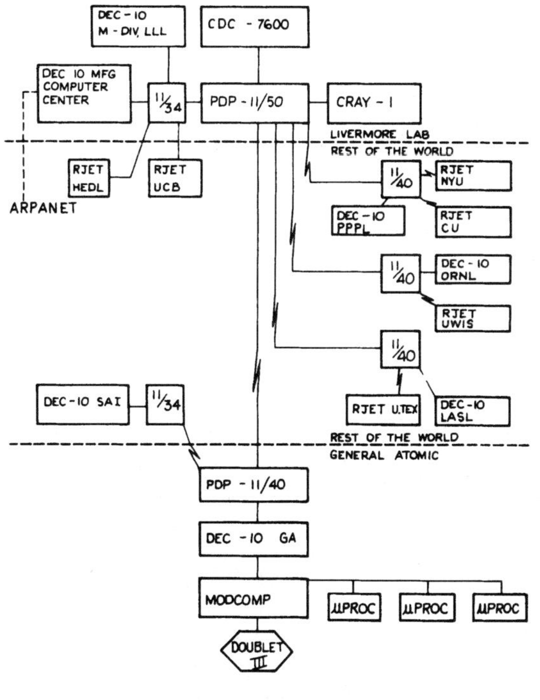

Fig. 3. DECsystem-10 hardware configuration

The third task that the USC is used for is local computing unrelated to the Doublet III program. This includes some theoretical plasma physics investigations, engineering and physics calculations related to the Engineering Test Facility program and other DOE Development and Technology tasks.

4.1 Hardware configuration

The DECsystem-10 hardware configuration is based upon a model KL-10D central processor. Five 200 megabyte disk drives and one 100 megabyte drive are used to contain approximately 100-200 shots of Doublet III data currently being accessed along with all the operating system libraries, source, etc., and system and application programs. Three tape drives are currently in use, processing current Doublet III data, retrieving data already processed and saving systems and Doublet III data for backup. A PDP 11/40 minicomputer provides the synchronous 50K baud link to the NMFE computer center at Lawrence Livermore Laboratory and a 50K baud link to another DECsystem-10 at Science Applications, Inc. A PDP 11/40 with data link hardware provides the 100K baud link to the ModComp computer. Two PDP minicomputers (an 11/34 and 11/40) provide 80 asynchronous (300-9600 baud) TTY lines to the DECsystem-10 system to support the high usage of terminals accessing the Doublet III data from throughout the GA facility. An array processor is used by the DECsystem-10 to provide computing power necessary to run some of the elaborate analysis codes processing Doublet III data. Finally, the DECsystem-10 has all the normal high speed peripherals including printers, plotters, card readers, etc.

4.2 Software configuration

The DECsystem-10 is responsible for archival and retrieval of all data generated by Doublet III. Data arrives at the DECsystem-10 either on magnetic tape or, in the near future, via a hardwired communications link in a form convenient to the ModComp in order to alleviate disk throughput constraints. The data is translated to DECsystem-10 format and archived in a form that may be both conveniently accessed by experimentalists and archived on magnetic tape.

By default, as soon as the data from a shot has been received by the DECsystem-10, it is archived on magnetic tape in an easily transportable and retrievable format and deleted from the disk. Only shots which have been explicitly requested by users remain on disk.

Experimentalists may request that shots be restored to the disk interactively. When a shot request is made, a message containing the shot number, magnetic tape reel and the user name is printed on the operator's personal terminal. Response is a few minutes to a few hours depending on the number of shots to be restored. Procedures exist to archive all unarchived shots, restore a shot to disk, bring the number of shots on the disk to a set level and to query the status of any shot.

Users may access data on the DECsystem-10 in one of several ways. ODDBAL, a program developed at the Oak Ridge National Laboratory, emulates an HP45 calculator with data base access capabilities. PLOTER, a code developed at GA on the ModComp, plots data in a variety of ways. MDAT, which does a quick broad spectrum analysis of a given Doublet shot, runs on both the ModComp and the DECsystem-10. Lastly, an MHD code, which is available on the DECsystem-10, the CRAY, the DECsystem-10 array processor and the LLL 7600, is used extensively to investigate plasma current and flux profiles.

*Work supported by the U. S. Department of Energy, Contract No. DE-AT03-76ET51011

CONTROL SYSTEMS FOR SUSTAINING NEUTRAL BEAM POWER SUPPLIES

by James D. Cummins
Aydin Energy Division
Palo Alto, California USA

ABSTRACT

A system is described which was designed for the MFTF system at the Lawrence Livermore Laboratory, but the approach has applications wherever high-power systems must be controlled.

Microprocessor based logic was used for flexibility in the initial design process, and to permit easy modification of operation as required.

An important consideration was the ability of the system to function reliably in the presence of electromagnetic interference produced by large switching currents, high switching voltages, and arcing in the neutral beam injection system. Extensive use of fiberoptics in this system provided noise immunity as well as voltage standoff between modules. Electrical links between modules utilize transmission lines isolated at both ends by means of optical couplers or relays. Power supplies are isolated from power line transients by shielded transformers, line filters, and by transient suppressors.

The system is also required to operate in the fringe magnetic field of the superconducting magnets in the MFTF system. This restricted the choice of components and led to the selection of alphanumeric plasma display panels in the local control consoles.

Data communication with the MFTF central control system is provided by an interface to CAMAC equipment, utilizing an internal 24 bit bus for optimum data transfer to and from the CAMAC dataway.

1. INTRODUCTION

The System controls the power supplies required to operate a neutral beam injector. A high voltage dc power supply, accelerator power modulator, filament power supply, arc power supply, and a suppressor power supply make up the principal modules which are controlled. The power modulator consists of the following modules: A series switch tetrode regulates the voltage of a neutral beam accelerator. A shunt switch tetrode controls the current when the load pulse must be interrupted. A crowbar consisting of a series string of ignitrons provides protection for the switch tube and neutral beam accelerator (see Figure 1).

The series modulator tetrode is installed on a "floating deck" which is insulated from ground and follows the tetrode output voltage. Some of the auxiliary power supplies and controls for the series modulator are mounted on the floating deck. The rectifiers and some controls associated with the arc and filament power supplies are also located on the floating deck.

Each module of the power supply contains its own protection circuits so that the modules are self-protecting to the extent that this is possible. Several control modules provide additional protection for personnel and equipment. The center of the control system

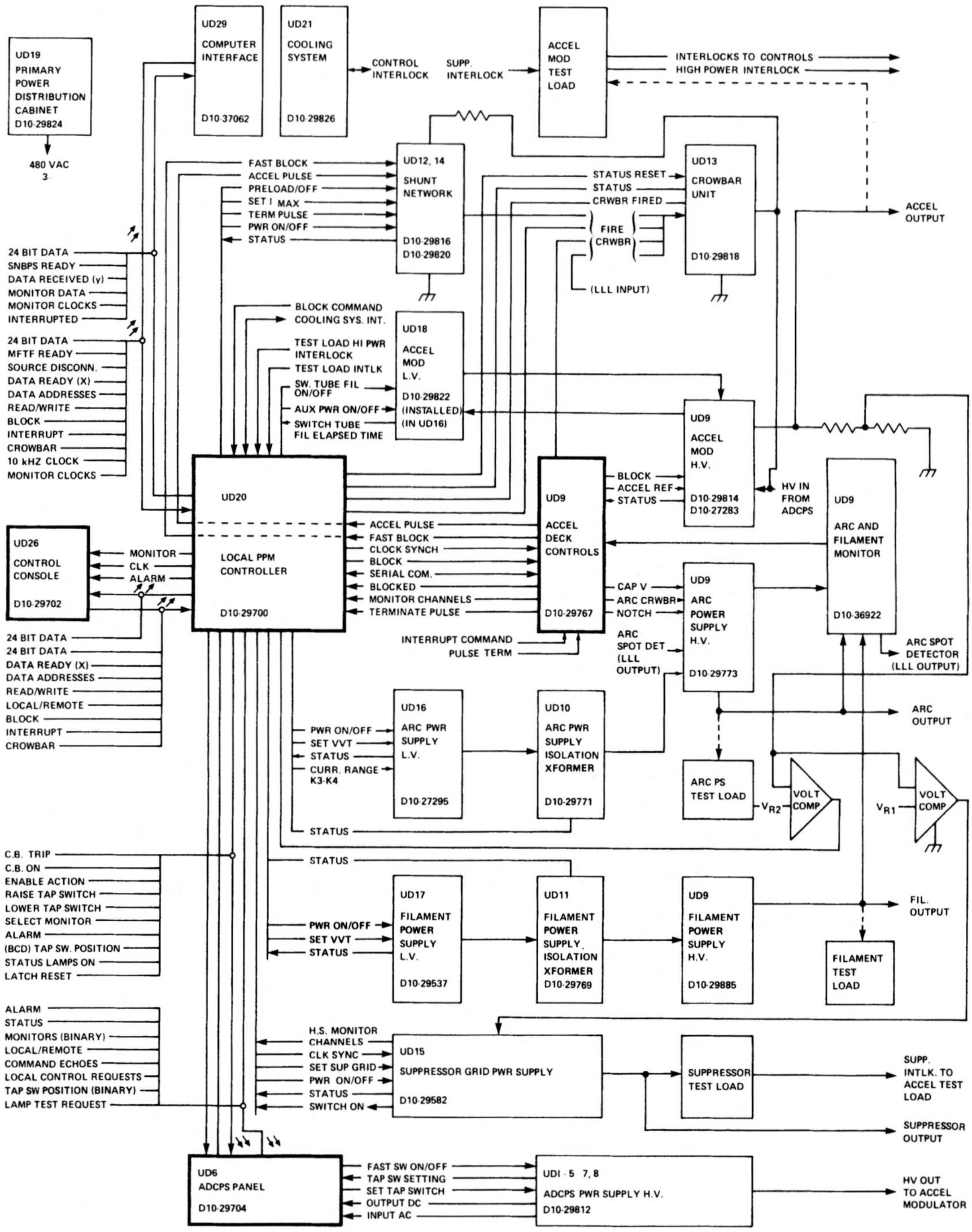

MFTF SUSTAINING NEUTRAL BEAM POWER SUPPLY BLOCK DIAGRAM

FIGURE 1

is the pulse power module control (PPMC). It is located in the modulator area and is housed in a steel cabinet. An adjacent cabinet houses the customer's CAMAC interface equipment.

The PPMC communicates with all ground level modules over fiberoptic links or over conductive links to be described later. It communicates with all floating deck controls over fiberoptic links for voltage isolation and noise immunity.

The accelerator dc power supply (ADCPS) equipment is located outside the building at some distance, so that a local control panel is required for maintenance purposes. An outdoor cabinet houses the local control panel and protection system for the ADCPS. The PPMC communicates with the ADCPS control cabinet over fiberoptic links for noise immunity. Additional conductive relay links are provided for safety and reliability.

In normal operation each neutral beam injector system is operated by the customer's central computer complex, over a CAMAC system which is interfaced to the PPMC with conductive links and optical isolators. Commissioning and maintenance operation is provided by a local control console (LCC) - each LCC is connected to six PPMCs. One or two of the neutral beam injector power systems may be operated manually from the console, which can duplicate all of the control and monitoring functions of the computer interface. Two systems may be operated at the same time, so that any interaction between systems can be detected.

2. MICROPROCESSOR LOGIC

The logic design for the control system is based on the Z80 microprocessor. The tasks assigned to the microprocessor are many and varied, but no high speed control loops are required and all critical operations are provided by hardware circuits. The microprocessor provides a means of change in function within hours instead of days. This is important in the design stage, but may also prove useful in operation.

Some of the tasks performed by the microprocessor are: data communication control, status checking, parameter scaling, closed loop control of the ADCPS except during a pulse, calculation of filament power supply output current, translation of required output voltages into parameter settings for the hardware and redundant time-out protection.

3. NOISE IMMUNITY

The large voltages and currents provided by the power supplies presents a problem in electrical interference with the control circuits. Power levels to 10 megawatts may switch in less than 1 microsecond under some conditions. The control circuits may respond to as little as 10 milliwatts. This requires a coupling coefficient of 10^{-9} or less to prevent interference.

Some of the noise immunity measures are: fiberoptic links, optical isolators, shielding, power line filters and transient suppressors.

Three types of fiberoptic links are used in this system (see Table 1). Plastic optical fibers are used for the 5 meter links between the PPMC and the floating deck controls. This fiber provides voltage isolation greater than 200 kV in addition to avoiding any electromagnetic coupling. The electronics for this fiberoptic link were designed by Aydin engineers, using commercially available components. The fiberoptic links to the group control consoles, at distances up to 70 meters, require glass fibers for lower losses. The principal reason for the use of fiberoptics in this application is noise immunity, since some of these cables will pass near 6 PPMs. The transmitters and receivers are manufactured by the Hewlett-Packard Corporation, and are mounted on PC boards designed by Aydin engineers.

The fiberoptic links to the ADCPS, at distances up to 300 meters, require low loss glass fibers and high output transmitter diodes. These links run in the same trench with the high voltage dc power lines so that noise immunity and voltage standoff are both required. These links were purchased from the Canoga Data Systems, Inc.

FIBEROPTIC LINKS

FIBER MATERIAL	LENGTH (METERS)	APPLICATION	FREQUENCIES USED
Plastic	5	Floating Deck	0-3MHz
Glass	10 - 70	Console	500KHz-3MHz
Glass	350	ADCPS	500KHz

TABLE 1

Optical isolators are used to prevent common mode coupling into input circuits from conductive links. The many short control links needed for this system would not be economical to implement with fiberoptics. The solution is to use shielded twisted pairs to minimize transverse (unbalanced) noise pickup. Optical isolators, and shields that are terminated only at one end, minimize common mode noise pickup. Links which run between modules of the PPM use optical isolators at both ends to avoid common mode noise coupling back into the transmitting end electronics. This is made possible by the use of separate link power supplies. The only ground return for these links is the link power supply.

Shielding for the control circuits includes both enclosure shielding and cable shielding. Control cabinets are made of steel sheet to provide shielding down to power line frequencies. This is necessary to protect sensitive electronic circuits from the magnetic fields around transformers and cables carrying large currents.

Control cables passing through the cabinet walls have their shields interrupted at the bulkhead connector, and the cable shields inside the cabinet are grounded only at the end near the bulkhead connector. This shield inside the cabinet prevents radiation of common mode noise into internal circuits.

Further shielding against high frequency noise is provided by internal enclosures made of aluminum sheet. The enclosures are usually card cages for the printed wiring cards.

The large printed wiring cards used for most control logic circuits have provision for added shielding on the card. A wide plated metal band down each side of the card, with holes drilled at intervals for mounting screws, provides a convenient means of attaching shields. The shields may be used for prevention of noise radiation from incoming lines, or they may be used for the prevention of noise pick-up in internal circuits. The circuits connected to external wiring are grouped on one section of the printed wiring card for better isolation and easier shielding. The plated metal bands may be cut to provide independent shields on the same board.

The electrical links are current loop circuits. The receiver is a diode with light output proportional to current. The high speed signals (up to 3MHz) required in some circuits make it necessary to use a high speed isolator such as the 6N137 or 6N139. These devices operate at 5 volts and level shifters convert 5 volt signals into 15 volt CMOS signals. CMOS is used for internal logic because of the higher threshold (larger noise voltages are required to interfere).

4. MAGNETIC FIELDS

The superconducting magnets used in the MFTF system produce a magnetic field, at some control module locations, in the order of 50 GAUSS. Test results indicated that the relays used in this system would function properly. Tests were conducted on a CRT from a commercially available terminal with less encouraging results. Even within a mu-metal shield it was necessary to orient the CRT in a preferred direction to avoid distortion.

Two plasma display panels, providing a total of 24 lines 40 characters wide, were used in place of a CRT to display information at the group control console. The display panels manufactured by IEE were found to be unaffected by the magnetic field and capable of adequate updating of the display. The viewing angle, both in the horizontal and vertical direction compares favorably with a CRT. The number of alphanumeric characters available with the two panels used is less than the number provided by a standard CRT terminal, but is adequate for this application.

5. DATA COMMUNICATION

Communication with the customer's computer is via CAMAC equipment (see Figure 2). The customer in this case provides the CAMAC equipment. The interface to CAMAC is via a 24 bit transmitter and a 24 bit receiver. The receiver also accepts 6 address bits and a read/write signal. Other special control lines and monitor data lines are also provided. All signals are transmitted across the interface on shielded twisted pairs. The receivers are optical couplers to provide isolation between the two systems.

A bidirectional 24 bit bus is utilized internally where this is practical. The low speed requirements in this system allow much of the internal data handling to be done using 8 bit busses which match the Z80 microprocessor used. Conversion betweeen the 24 bit bus and the 8 bit busses is provided where required.

Data communication within the control system is aided by a versatile set of printed wiring cards. These are the serial data multiplexer, and the serial data demultiplexer. These cards provide 48 bit serial transmission in 200 microseconds. Parallel inputs and outputs may be 48 bits, 2 words of 24 bits each, or 6 bytes. This provides the versatility required to use these cards in all of the major control modules.

Simulators were built to aid in testing the control system without using the high power equipment. The group control console (GCC) provides a means of testing the controls without the computer. An additional computer interface simulator provides a means of testing those circuits which are not utilized by the GCC. The power supply simulators are modular so that one section of the system may be tested with a simulator while other sections are not disturbed.

6. FLOATING DECK CONTROLS

The floating deck controls are pre-set prior to the pulse. During a pulse the only input required is a 10KHz clock to drive the timers. Noise pulses on this link can only shift the pulse timing by a small amount.

The accel voltage rise curve is required to match the arc current rise curve. This control of accel voltage rise is provided by a voltage reference which is derived from successive 8 bit digital values stored in a programmable read only memory (PROM). A 1MHz oscillator drives a counter circuit to provide successive addresses to the PROM. The 256 digital

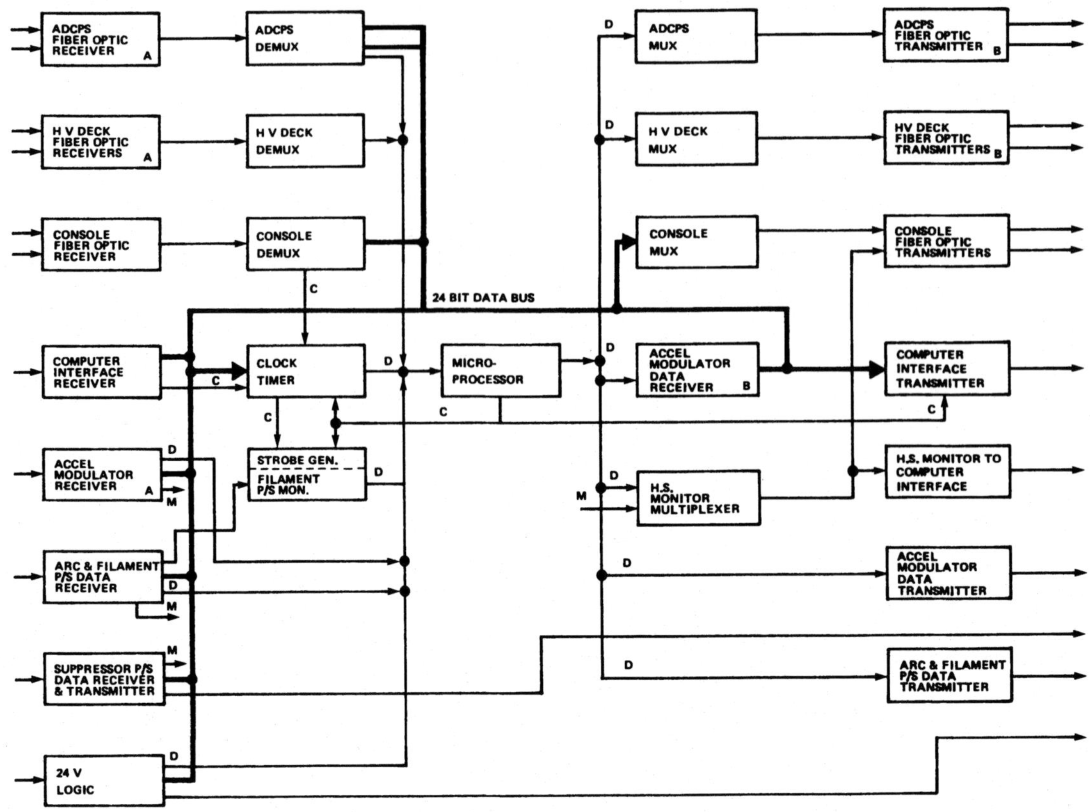

BLOCK DIAGRAM PULSE POWER MODULE CONTROLLER

FIGURE 2

values stored in the PROM are converted to points on the reference curve by a digital to analog converter (DAC). All of this circuitry is on the accel deck so it is not affected by external link noise.

7. CONCLUSION

Each major module of the power supply system is internally self-protecting. The control system then provides communication among modules, accepts command inputs from the central computer, and provides for the monitoring of performance. The power supply system is internally self-protecting because the controls will reject any inputs which would cause unsafe or destructive operation.

A set of simulators were built to permit a complete test of the control system without operating any of the high power modules in the power supply system. This prevents the possibility of damage to the power supply system while testing the control system.

The ADCPS panel controls have been tested and other controls have been operated with simulators.

The arc and filament power supplies have been operated into test loads and into an experimental arc chamber with control test boxes using the conductive links. Complete system operation is expected this year.

A DISTRIBUTED INTELLIGENCE CONTROL SYSTEM USING FIBRE OPTICS

A J Vickers, E G Murphy, S E Dorling, S E Hughes, E M Jones,
P Nicholson, D Richardson, C A Steed,* D L Trotman, I Young**
Culham Laboratory, Abingdon, Oxon, OX14 3DB, UK.
(Euratom/UKAEA Fusion Association)

ABSTRACT

A distributed intelligence control system has been developed using fibre optics to interconnect the individual modules. This has been built initially for use in the development programme for RFX. In this application it operates under the control of a microcomputer which communicates with the operator via a colour display and special purpose keyboard. The control system has been implemented and is undergoing commissioning on an existing capacitor bank test facility which uses capacitor modules of the type which might be used for RFX. This test facility is primarily for components of experimental assemblies and provides a suitable working environment for gaining operational experience with newly developed modules intended for control of fusion experiments.

As a result of experience gained during commissioning, this paper gives an assessment of the validity of this approach to control systems required for fusion experiments and discusses other areas in which such a system would find application.

*Now with JET
**Attached Student from North Staffordshire Polytechnic

1. INTRODUCTION

Fusion experiments present a hostile environment to control systems. The switching of large currents produces heavy electrical interference and high voltage capacitor banks necessitate stringent isolation precautions being taken to separate individual sections of an experiment. Furthermore these sections frequently require at least some degree of autonomous and local control if the central system is not to be impractically complex. Last but not least is the problem of system integrity which necessitates extensive error checking facilities being built into the system. This paper describes a solution to these problems based on an earlier system [1] and briefly outlines its implementation on a capacitor bank test facility.

2. SYSTEM CONCEPTS

The main concepts on which the system is based are:

- Distributed intelligence
- System integrity
- Flexibility
- Fibre optics

2.1 Distributed intelligence

The main advantages of distributed intelligence are that it reduces the load on the control computer and allows independent commissioning and fault finding of individual sections of the experiment. It can also enable operation in a reduced mode under some failure conditions. A further benefit is that intelligence in peripheral units enables them to incorporate self checking facilities. A key question is how far to distribute the intelligence of the system; as currently designed, there are three levels of intelligence:

1. The controlling computer
2. The local controllers
3. The peripheral units

Although all three levels contain processors, only level 1 initiates any actions or indeed does any co-ordination or correlation of signals. Level 2 is told what the individual signals from level 3 should be and it checks that these are correct, notifying level 1 if this is not the case. Level 3 polls the signals from the experiment and passes them on transparently to level 2. Both levels 2 and 3 transparently pass on output signals to the experiment. The intelligence in levels 2 and 3 therefore is used largely in handling the link protocol and system error checking.

2.2 System integrity

It is essential that such a system as this not only operates correctly, but that it can be seen to be so doing. This is particularly true of the peripheral units which operate in the hostile environment of the experiment. In addition to parity checks, all messages to and from peripheral units are address checked. These units also give error information on their own status and that of the message to which they are replying.

2.3 Flexibility

This system is intended for use in a wide range of applications which differ significantly in the degree of autonomy required by the local controllers and peripheral units. The control of the majority of the systems operation of these items has therefore been put into their software, so that although as stated in 2.1 no signal co-ordination is done by them, this could easily be altered. This could be particularly useful both to further reduce the load on the control computer and also to give better response times than can be obtained by the normal hierarchical system. The flexibility of the system is further increased by the use of multiplexers which increase the number of peripheral units which can be driven by one local controller. The system can therefore readily be extended or re-configured to adapt to changing experimental requirements.

2.4 Fibre optics

The main advantages of fibre optics are:

1. High voltage isolation
2. Noise immunity - they neither pick up nor radiate interference

Again system integrity is important and so an inbuilt facility for testing the operation of installed units at reduced power is incorporated as a means of detecting any marginal operation. A bi-directional single cable is used to reduce cabling costs.

3. SYSTEM IMPLEMENTATION

In order to obtain some operating experience of the system it is being integrated into a complete control system currently being commissioned on a capacitor bank test facility [2]. Two banks (1MJ, 8kV start bank and a 100kJ, 250/500V clamp bank) and a load assembly are being operated under conditions close to that expected from an experiment. The computer in this case is an LSI-11 microcomputer and the operator interface is a special purpose keyboard with a colour video display which provides mimic diagrams, parameter lists and results. The control system schematic is shown in fig. 1.

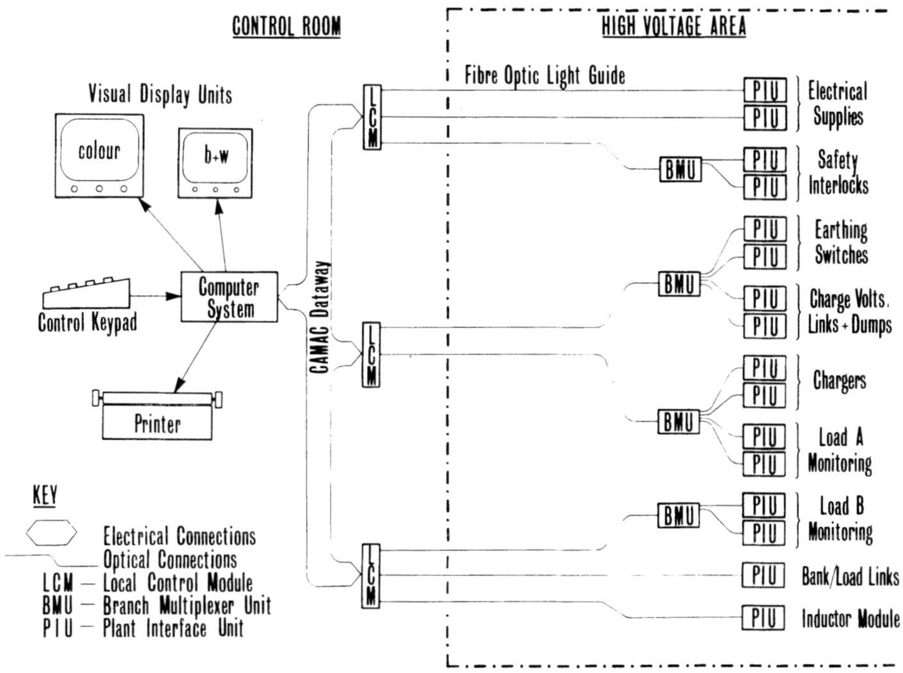

Fig.1. Control system schematic

4. SYSTEM COMPONENTS

In addition to the controlling computer and operators console which will not be described further, the main components of the system are:

1. Plant Interface Units (PIU)
2. Branch Multiplexer Units (BMU)
3. Local Control Modules (LCM)
4. Optical Links

4.1 Plant Interface Unit (PIU)

There are built in free standing screened boxes and are intended to be capable of operation close to the experiment. They have two sections, one for control and the other for signal input/output. All PIUs have the same control hardware which is based on an Intel 8085 microprocessor, but each type of PIU has its own program and input/output section. The types of PIU currently in production are:

PIU 3	24 digital outputs, 24 digital inputs
	4 analogue outputs, 8 analogue inputs
PIU 4	72 digital inputs

All the digital signals are optically isolated from one another. These inputs can handle up to 24V with a minimum current of 5mA and the outputs can drive up to 100mA from a 24V supply. The analogue inputs are 100k impedance differential, $\pm$5V range with 10 bit resolution and 8 bit accuracy. The analogue outputs have the same range resolution and accuracy and can supply up to 10mA drive current. All outputs are latched. In the quiescent state, the microprocessor is continually polling all its signal inputs and alternately updating a pair of input data buffers. On receipt of a data command from the optical link, the microprocessor updates the signal outputs and returns the contents of the full input data buffer down the link. There are three levels of error checking done by the PIU; each message must have correct horizontal and vertical parity, the message header byte must match the PIU's hardware address (set on a coding plug) and the command must be a valid one. If any of these checks fail, the PIU will not update its output signals but will send back its input data. The header byte of a message from a PIU contains its hardware address and the command it was asked to implement. The second byte contains status information on the PIU's voltage, current and temperature levels and also error information associated with the previous command (parity, address, invalid command etc). In addition to the data command described above, further commands can initialize the PIU or cause it to echo a message (for test purposes). It can also be put in half power mode, when the current to the optical transmitting diode is halved. This again is intended for test purposes as an early warning of any degradation of the optical system.

4.2 Branch Multiplexer Unit (BMU)

This multiplexes the signal from one light guide to up to 15 light guides, the object being to increase the fan out of an LCM. The BMUs can be cascaded so that with two layers for instance, the hardware allows for the driving of up to 225 PIUs from one channel of an LCM. The BMU switches the link messages transparently without any error checking, this being left to the LCM and PIU.

4.3 Local Control Module (LCM)

This is an Intel 8080 microprocessor based CAMAC module which controls and monitors PIUs either directly or via BMUs. The LCM is initially loaded by the controlling computer with the parameter information on all the input and output signals of the PIUs under its control. It then continually polls all its PIUs transmitting the state of all the PIU output signals and receiving the state of all the PIU input signals. At the end of each complete scan, each input signal is checked for errors. For digital signals this means looking for an erroneous logic 0 to 1 or 1 to 0 transition, and for analogue signals it means checking if the signal has exceeded either the upper or lower, warning or failure limits. An analogue signal can therefore be in any of 5 bands, ie. upper failure, upper warning, normal, lower warning or lower failure. All the limits for each individual signal are under the control of the user thereby providing a very flexible system. The LCM remembers the state of the signal on the previous scan, and in the event of the signal being in a different error band to the previous scan, then an error status bit is set and if the user has enabled that particular interrupt, then the computer is interrupted. Although the LCM can work autonomously, the computer can intervene at any time, requesting status or data information or updating output information. Indeed the whole operation of the LCM can be controlled on a single step basis if required. In addition to the error checks mentioned above, the LCM also checks all messages for parity, address and status errors. As a further safeguard of system integrity there is also a watchdog timer in the LCM which has to be repeatedly reset by the LCM's microprocessor otherwise an interrupt is generated.

4.4 Optical links

These interconnect all the PIUs, BMUs and LCMs. Single rather than dual links are used since this halves cabling costs with minimal degradation of system performance. The optical element used is a gallium arsenide emitter photodiode which functions alternately both as a transmitter and as a receiver thereby dispensing with the need for a beam splitter and separate transmitting and receiving diodes.

5. OPERATIONAL EXPERIENCE

After testing that individual LCMs and PIUs conformed to specification, including cycling tests, a 72-input digital PIU was installed in the capacitor bank high voltage area and connected to its controlling LCM and the rest of the control system outside the area. None of the equipment was specially screened. As part of the commissioning procedure for the complete system a series of tests were carried out under normal operating conditions, at high magnetic fields and under abnormal (unscheduled) conditions.

5.1 Normal operational tests

Though problems with improperly made light guide connectors had been experienced, once these had been overcome the installation and the limited operational experience so far have been trouble free. The single PIU system was subjected to 150 operational shots under various conditions up to the full power of the capacitor banks delivering 50 kA at 8 kV into the inductor module. No transmission errors and no cases of incorrect information from the PIU were detected. All the tests with bank voltages above 3 kV were observed to have a transient effect on the visual displays probably through a rise in earth potential. Though more operational experience is needed it would appear that these modules would work well under normal operating conditions which produce peak magnetic fields of 0.003T at the PIU and 0.001T at the LCM and computer system.

5.2 High magnetic field tests

By placing the PIU in various positions in the inductor module its behaviour in high magnetic fields was assessed. The PIU was found to work correctly in any orientation at fields up to 0.25T (2 orders of magnitude above normal operating conditions) and, when favourably aligned with the magnetic field, up to 0.35T. When these fields were exceeded the module would usually stop working, which was detected by the control system, but sometimes it would work erroneously and therefore it would be unacceptable to use the PIU where magnetic fields above 0.25T might occur. The problem is ascribed to saturation of the power supply transformer and could be remedied if necessary.

5.3 Unscheduled tests

A number of abnormal shots were fired due to premature firing of some ignitrons. These caused overcurrents, severe electrical interference and local overheating of some parts of the circuit and one such fault generated sufficient magnetic field to magnetise the colour visual display unit so that it required degaussing. No other ill effects were observed on the control system.

A flashover which damaged the mains wiring in the area caused power supply fuses to blow in the control computer and its colour display but caused no other damage to the control system.

5.4 Future Tests

Tests with PIUs containing analogue inputs and outputs are to follow and there is a programme of operating PIUs in all possible locations in the high voltage area to verify the ability of these modules to withstand all the conditions prevailing in the capacitor bank. Attempts will be made to quantify the amount of electrical interference so that limits can be set on the environment in which such equipment would be safe to use.

6. CONCLUSIONS

Preliminary tests have shown that the system performs satisfactorily in the hostile environment of a capacitor bank power supply. The emphasis placed on system integrity appears justified though to some extent this has been obtained at the expense of size and power consumption of the peripheral units, features which it should be possible to improve in future designs.

7. REFERENCES

[1] FULLARD, K., VICKERS, A. J., 10th Symposium on Fusion Technology, Padova, 4-9 September, 1978. Proceedings pp 539-545.

[2] RICHARDSON, D., BURDEN, R.A., FULLARD, K., MURPHY E.G. and WHITTLE H. R., 10th Symposium on Fusion Technology, Padova, 4-9 September, 1978. Proceedings pp 505-510.

AN AUTOMATIC METHOD FOR STORING AND RETRIEVING RAW DATA AND ANALYSED RESULTS

V J Calderbank, J P Hatfield, E G Murphy, J Pimpernell*
Culham Laboratory, Abingdon, Oxon, OX14 3DB, UK
(Euratom/UKAEA Fusion Association)
*International Computers Limited

ABSTRACT

The installation of a new central computer system at Culham Laboratory provided an opportunity to review the methods by which both raw experimental data and analysed data from fusion experiments are stored and retrieved.

During the last 10 years, many information systems software packages have become available and these can be broadly classified into two types - pointer based systems and inverted file systems. Both of these approaches have advantages and disadvantages, and an investigation was made into the applicability of both methods to plasma physics data.

The standard pointer-type database management system on the ICL 2976 is IDMS. This is well suited to holding information about raw experimental data and accordingly a data system was designed for the information collected by the data acquisition computers at Culham. The IDMS database set up for this purpose contains the time and date of creation of the data, notes on its validity, the type of analyses carried out on it and notes about the analyses, as well as essential data management information.

Pointer based systems often lack an efficient on-line query language, yet analysed data from experiments consists of information which users wish to search according to some particular criterion. For this type of application the STATUS II free text information retrieval system has been installed on the PRIME computer at Culham. This is an inverted file system and does not have the rapid on-line updating features of IDMS but does have a powerful query language which allows information to be retrieved according to the occurrence of a particular word or phrase. Information may be selected on some particular parameter and sections may be viewed, printed or transferred to a file for subsequent input to a further analysis program. Databases of technical information may be set up very rapidly by users who do not have a high degree of specialist knowledge.

The use of these two systems in a complementary way seems to satisfy most of our requirements for a scientific data handling system at Culham and has provided us with more powerful access to information than has been possible before.

1. INTRODUCTION

The development of automatic systems for data acquisition from fusion experiments has led to the production of large quantities of data which can be collected, stored and processed by computers quite efficiently. Comparatively little effort has gone into making the data available to the experimenter or theoretician in a readily usable form; each experimenter is presented with graphs and printout and is left to communicate his findings to his colleagues.

The next stage in the development of scientific information systems is to allow users of such systems to:

a) record an opinion on the quality of the raw data so that, for example, time is not spent analysing data that is known to be suspect

b) record the results of computations together with details of the methods used so that other people, with permission to do so, can have access to the information

c) record comments on the validity of the results obtained so that, for example, restrictions on the range of validity of the results may be expressed

d) search all the above records according to some selection criteria so that knowledge accumulated from previous experiments is more easily accessible.

Steuerwald *[1]* has examined the whole problem of handling scientific information and comes out in favour of a highly integrated system encompassing data compression, standard data reduction libraries and data access via data descriptors. A design for such a system is also reported *[2]* and this incorporates a data mangement system with a query processor designed specifically to cope with scientific data and an algorithm library.

This paper describes a loosely integrated system which is being implemented at present and attempts to provide the extra facilities enumerated above by:

a) storing information of direct interest to the user in a more generalised index than that required for administrative purposes alone and allowing comments and information about analyses to be stored there;

b) storing the results of data analysis runs in such a way that quick searching of those results will be possible.

2. SYSTEM DESIGN PRINCIPLES

There were two major problems to be solved. The first arises out of the conviction that analysis programs should be written by experimenters or theoreticians who should not be restricted to specific algorithms or in the parameters they wish to calculate. This leads to the need for a database whose contents cannot be fully specified in advance and yet one which must be efficiently searched for any information that the user puts into it. This was largely solved by choosing a suitable database for the storage and retrieval of processed data. The second problem stems from this choice because the type of database that is suitable for rapid searching is generally not one that is amenable to the sort of rapid and continuous updating that is required of a raw data index where second-by-second changes must be recorded. This problem was avoided by using a separate database system for the raw data index and providing facilities, to be called by the user's analysis program, which will enter cross-references between the two databases at the time that processed data is recorded. This forms the basis of the loosely coupled data handling system shown in fig. 1, which shows the activities of raw data storage and retrieval, the analysis of raw data and the storage of selected results. The figure also shows the information that passes from one activity to another and the access that the user of this information system has at various stages. Raw data is supplied by a data acquisition activity which has been described elsewhere *[3,4,5]* and will not be discussed further in this paper.

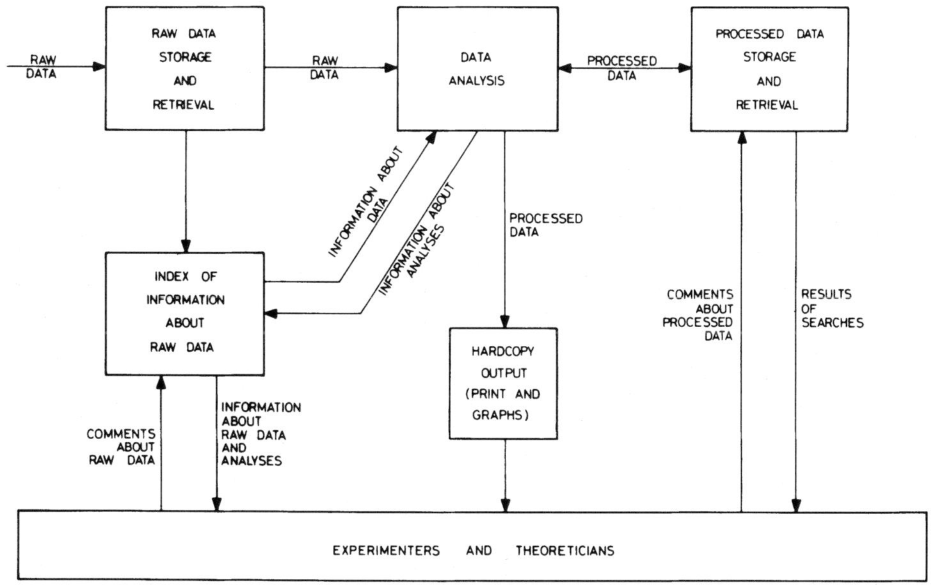

Fig. 1. Schematic of data handling system.

3. RAW DATA STORAGE

Fusion experiments are pulsed and all the data from a pulse is, for convenience, stored as one data set. The storage and retrieval of these data sets is under the control of the raw data index and the notion of this index has been expanded to encompass user information about the data and analyses which may have been carried out on that data. A data model of the user-accessible entities in the index is shown in fig. 2. The diagram shows that an experiment has many pulse data sets and each such set has many data items and pulse notes. In addition it shows that a user, who has a number of analysis programs which he can arrange in a number of analysis sequences, can, if he has access rights, run any of those analysis sequences against any selection of pulse data sets to produce named results; each analysis run can have a number of notes or comments associated with it. The implementation of this data model is such that a user will be able to:

a) find out how much data has been filed for a pulse and when the pulse occurred

b) append and read textual comments about a pulse

c) find out what analyses have been done on the raw data, by whom, using which analysis program, when and with what results. (The result could be a cross-reference to another database)

d) append and read textual comments about the analysis run.

The information comprising the raw data index is accurately definable, highly structured and has to be held in a database which must be easily and frequently updated because it is anticipated that new data can be created every few minutes and it is necessary to keep pace with the experiment producing the data so that the appropriate analysis can be done as quickly as possible. The most suitable kind of database for this purpose is a standard pointer-based one and the ICL supported IDMS (Integrated Database Management System) *[6]* has been chosen to support the raw data index. Its advantages include the ability to update the information rapidly on-line (around 10 transactions per second), the provision of automatic

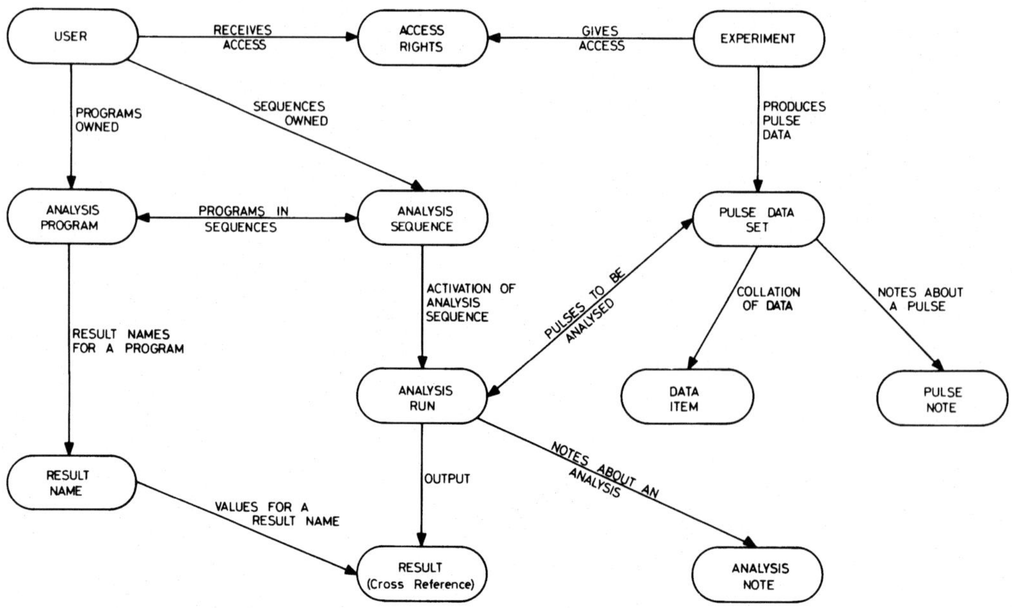

Fig. 2. Data model of user-accessible entities
(arrows show 1:N relationships with N at the arrow head,
M:N relationships are shown by double headed arrows).

recovery and warm restart facilities in the event of system failures, the elimination of data redundancy and the provision of multiple access paths to a central pool of consistent information.

As well as the user information the index contains a great deal of information to help with operating the system, the management of the filestore and the security of the data against loss.

4. PROCESSED DATA STORAGE

The information in the raw data index has to be restricted in order to limit the size of the database so that it can be rapidly updated. It is therefore not intended to hold large amounts of processed results in that index but rather to record a cross-reference to a database of processed results. As stated earlier, it is difficult to predict what analyses will be carried out on the raw data or what quantities or arrays of quantities will be calculated in the future. In order to retain as much freedom as possible to record results a free text information retrieval system, STATUS *[7]*, has been chosen to hold large volumes of analysed results in textual form. STATUS is an inverted file system which provides access to the data via an index (inverted file or concordance) in which every reference to every word is maintained. Whilst this may be expensive on disc space, such systems have the advantages of rapid on-line retrieval by means of a simple query language, and ease of use including the ability to create a database rapidly without the need for much specialist knowledge. This database does not have the advantages of rapid updating or ease of maintenance but information is only entered here when considered suitable for publication and therefore remains unchanged.

The data from many different experiments may be held in separate areas of one large STATUS database. These areas are further subdivided into shots of information (corresponding to a pulse) and STATUS will retrieve all those shots which satisfy a particular question. Users may select, display and sort information, transfer selected parts of it to a file or print it. Part of a typical on-line session is shown in fig. 3 which illustrates the sort of information which is stored. Shots of information may be further subdivided into named sections which may be accessed and displayed independently.

```
Q #BETAP>0.1?                                                      SELECT ALL SHOTS
  QUESTION IS SATISFIED BY    8  ARTICLES                          WITH BETA GREATER
>                                                                  THAN 0·1
SORT #BETAP DESCEND
>
DISSEC TBASE 1                                                     SORT INTO DESCENDING
TBASE                                                              ORDER. THE SHOT WITH
 &TBASE                                                            THE HIGHEST VALUE OF
  #TIME    = 0.180 ,  #BT     = 2.69   ,  #IGAS    = 1.62E05,      BETA IS NOW AT THE TOP
  #VLOOP   = 2.4   ,  #QLIM   = 5.2    ,  #BETAPE  = 0.174  ,      OF THE LIST. DISPLAY
  #BETAPI  = 0.143 ,  #BETAP  = 0.317  ,  #TAUE    = 0.0118 ,      THE TIME BASE SECTION
  #ZEFF    = 1.18  ,  #NEBAR  = 4.12E19,  #LI      = 1.60   ,      FOR THIS SHOT.
 &END

>
DISSEC FILE PROFILES 1                                             TRANSFER THE PROFILES
  ENTER FILE NAME                                                  TO THE FILE
HIGHBETA                                                           HIGHBETA
>
```

Fig. 3. Part of a STATUS on-line session.

5. DATA ANALYSIS

In this activity the user has complete freedom to apply any processing to any set of data. This does not mean that the user is left without support, but rather that the only support that can be offered is in the form of interface routines to other parts of the system and a library of generally useful data reduction routines which he may integrate into his analysis program. Routines are provided to access the raw data and information about the raw data and similarly a routine for an analysis program to enter processed data (or a cross-reference) into the raw data index is provided. Entering information by hand into a STATUS database is very tedious and the work involved is sufficient to deter a busy experimentalist. For this reason, a preprocessor program is being designed to automate this process. The program can form part of an analysis sequence run as the experiment is operating. Analysed results from other programs are taken and structured into a STATUS temporary file which may subsequently be used to update the database in a batch run overnight.

6. DISCUSSION AND CONCLUSIONS

In contrast to the highly integrated system proposed by Steuerwald the system described here consists of two standard general purpose databases loosely bound into one compatible system. It has a number of disadvantages:

a) the user of such a system has to be cognizant of three interfaces - to the raw data, to the raw data index and to the processed data - rather than one unified interface;

b) the storage of processed data relies on the conscientiousness of its users;

c) it allows the storage of subjective information which could be misleading.

These disadvantages are offset by the following advantages:

a) the system retains the freedom to develop in the way that its users dictate as circumstances change;

b) it makes use of developments in database technology and is supported on standard databases;

c) it takes advantage of particularly suitable, and often mutually exclusive, features of database packages to give a system which is capable of rapid updating and efficient searching where necessary.

At the present stage of the evolution of automatic methods of dealing with raw and processed data it is felt that the advantages of flexibility and freedom to develop outweigh the disadvantages.

In conclusion, modern technology has made it possible to collect and store more experimental data than ever before and the need for an automated method of retrieving selected parts of that data has become pressing. By applying advances in software technology to this problem it has been possible to produce a data handling system with the ability to store more information as the experiment is progressing, to select and retrieve pulses of raw data rapidly and to make analysed results from experiments easily accessible on-line. The resulting system is sufficiently flexible to permit further expansion and its development will lead to a better understanding of what is required of a scientific data handling system.

7. REFERENCES

[1] STEUERWALD, J., Sonderdrück aus Kleinheubacker Berichte, Vol 21, 1978, pp 183-193.

[2] STEUERWALD, J., EDDAR: Experimental Data Documentation and Retrieval, Max-Planck-Institut (Garching) report IPP R/20, July 1977 (in German).

[3] MURPHY, E.G., FULLARD, K. and STEED, C.A., Proc. 7th S.O.F.T. Grenoble (1972) pp 77-82.

[4] STEED, C.A. and MURPHY, E.G. On-line Computing in the Laboratory, International Tracts in Computer Studies. Advance Publications Ltd, London, 1976. pp 333-343.

[5] STEED, C.A. and MURPHY, E.G. Proc. 9th S.O.F.T. Garmisch (1976). pp 493-498.

[6] IDMS: Data Establishment. ICL Technical Publication, TP 6924.

[7] STATUS User Manual ed. CROALL, I.F., Computer Science and Systems Div. Harwell, Oxon January 1979.

DATA ACQUISITION FOR THE DITE SURFACE PHYSICS EXPERIMENTS.

by E K Sinton and J M Weaver

Culham Laboratory, Abingdon, Oxon OX14 3DB, UK.
(Euratom/UKAEA Fusion Association)

ABSTRACT

A system has been implemented to gather and process data from a group of surface physics diagnostics on the DITE experiment. A description of the system hardware and software is given as well as details of a satisfactory operating procedure which has been evolved. The methods adopted to overcome problems of electrical interference are outlined.

1. INTRODUCTION

Surface physics diagnostics on Tokamak experiments are directed towards studying plasma-wall interactions, impurity levels and ion energies. Much data is gathered from these diagnostics during each of the DITE Tokamak shots at Culham, and with a short time interval between shots much data is accumulated during an experimental run. Manual data analysis methods are time consuming and do not provide the fast feedback of information to the experimenter that is required for optimising discharge conditions.

To alleviate these problems it was decided to implement an electronic data acquisition and processing system that would serve four surface physics diagnostics. These are the two temperature measurement diagnostics using an infra-red (IR) camera and an array of thermocouples, the arc detector using probes giving a direct reading of the intensity and the duration of arcing, and fourthly the hydrogen flux measurement using a carbon thermal desorption probe and mass spectrometer.

2. SYSTEM HARDWARE

The present data acquisition system is shown in Fig.1 and is the culmination of improvements both in performance and experimenter acceptance of earlier surface physics data systems. Operation of the system hardware is best illustrated by describing each diagnostic sub-system and in the order in which they were started.

2.1 Temperature Measurement Data Systems

A scanning IR camera (AGA) is used to observe the transient surface temperature rise of a moveable limiter in DITE [1]. It uses a cooled InSb detector and an optical/mechanical system with rotating octagonal prisms and produces a raster display on an oscilloscope. By stopping the vertical scanning prism at pre-selected times the object can be scanned continuously for 600μs. Each selected line profile is displaced in turn on the raster display and then photographed for subsequent manual measurement. To extract the necessary detail, both for waveform amplitude and shape, only 10 line profile waveforms per DITE pulse were possible.

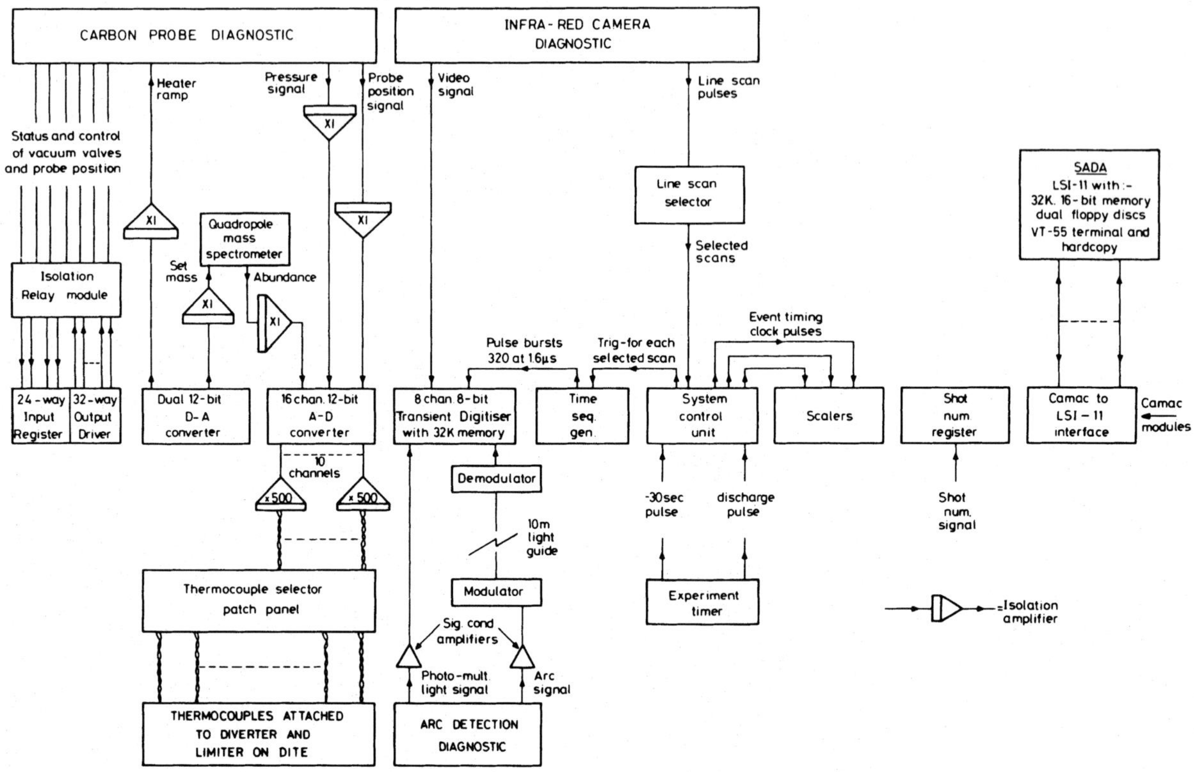

Fig.1. Block schematic of the data system for the four Surface Physics diagnostics on DITE

To automate and expand this data collection, eventually to 32 channels, and include measurements of bulk temperature of the limiter and diverter using thermocouple arrays, a CAMAC-based data system was assembled and initially linked to the DITE data computer. However, competition between different experiment diagnostics for computer data channels and limited data storage for the IR camera caused overload problems. To overcome this data "bottleneck" as well as meet the demand for other surface physics data requirements on DITE it was decided to link the CAMAC and signal conditioning hardware to an LSI-11 based Stand-Alone Data Acquisition (SADA) system [2].

In this part of the data system, as shown in Fig.1, the set of thermocouples are scanned at different times during the DITE operational cycle. The first scan is 30sec before the discharge pulse and the remaining scans are determined by the experimenter via the program. However, the period of the discharge is not available for thermocouple scans as the IR Camera measurements are taken during this time. In the earlier system occasional interference signals triggered the thermocouple scanner, which was operating in the DITE area using a digital thermometer. Its digitised data was being transmitted to the "grounded" CAMAC equipment through opto-coupling circuits as the torus could reach potentials of a kilovolt under fault conditions. To overcome this false triggering and retain safe operation, each incoming thermocouple signal is fed into an isolation amplifier before input to the CAMAC multi-channel ADC. Ten channels of voltage isolation amplifiers were designed and installed, based on the transformer-coupled 3656BG (Burr Brown) module with a x500 gain input amplifier. These were powered from the CAMAC supply and provided isolation to 3kV.

Initially the video signals from the IR Camera were fed into one channel of the ADC with a 2k word memory, but its limiting speed (10μs) and insufficient memory for 32 line profiles led to using the transient digitizer 2264 (LeCroy). Each line profile is digitised and stored autonomously with a total of 320 samples each taken at 1.6μs intervals to enable sufficient detailed information to be extracted from the incoming signals. The System Control Unit was designed to interface control signals from both the experiment Timer and the IR Camera to the CAMAC equipment, and also to organise clock pulses which initiate the time for each thermocouple scan and the time intervals between each IR line profile scan.

2.2 <u>Arc Studies Data System</u>

Arcing in DITE has been observed on probes and torus components using a time-resolved arc detection system [3]. Arcs are detected by measuring the current flowing to a probe, made of molybdenum or titanium, and correlating these current pulses with the optical emission from neutral atoms emitted from the probe material as a measure of impurity atoms into the plasma. The probe arc signal is coupled to the SADA by means of linear opto transmitter/receiver modules L2260A and L2261A (Belling Lee) which are linked by a 10 metre light guide. The system bandwidth is 20Hz to 5 MHz with a signal/noise ratio better than 40db. This signal and the optical emission signal, from a photomultiplier, are amplified before input to two channels of the transient digitiser 2264 (LeCroy). There are 8k conversions for each signal an accuracy of 8-bits, each conversion being triggered by the internally selectable 2264 clock. The conversion start time and sampling period can be set either to cover the whole DITE discharge (200ms) or can be selected at particular times during the discharge (typically every 5μs) to resolve more detailed information on the intensity and duration of the arcs.

2.3 Carbon Probe Data System

Hydrogen and deuterium flux measurements to the walls of DITE are determined using a carbon thermal desorption probe [4]. The probe is exposed during the plasma discharge, then removed into a separate vacuum system and heated linearly for 30 seconds when time flux measurements of the different hydrogen species are taken using a quadropole mass spectrometer.

In addition to its signal gathering and processing role, this part of the data system is required to control many of the operational aspects of this diagnostic - e.g. set and confirm positions of vacuum valves and the probe, produce the heater ramp and set the mass so that 100 flux measurements of 4 different masses can be recorded during the ramp. The buffer isolation amplifiers, again based on the 3656BG (Burr Brown) module, are needed for voltage isolation on the probe signals and to provide galvanic isolation between the mass spectrometer and the CAMAC equipment in order to eliminate the effect of ground loop interference which obliterated the diagnostic signals.

3. SOFTWARE

Since it was not clear at the start of the project which of the four diagnostic experiments (carbon probe, infra-red camera, arc detection and thermocouples) would be required to work in conjunction, it was thought desirable to adopt a modular construction for the software. That is to say, a number of individual subroutines were written associated with either a particular CAMAC module or a facet of the data handling. These routines could then be linked in various combinations depending on the particular pattern of diagnostics required at any one time. The routines are written in FORTRAN except where speed of execution, as in the sampling of infra-red camera signals, necessitated an assembly language procedure.

3.1 System Programs

Three main programs were developed dealing with respectively:

1. The Infra-red camera, Carbon Probe and Thermocouples
2. Arc detection, Carbon Probe and Thermocouples
3. The Thermocouples only.

Each of these programs acquire data, via CAMAC, from the appropriate experimental hardware. The data is then processed, displayed on the Video Display Unit (VDU) for immediate assessment and stored on floppy disc.

The first two programs also exercise control functions. They generate a linear heating ramp for the carbon probe desorption and simultaneously set the Mass Spectrometer control volts repetitively for four selected mass numbers.

Interaction with the user is allowed for at several stages in the program. Initially, if required, to change various user-defined parameters if they need to be altered since the last sequence of shots. Secondly to call up a specified block of data (a particular scan of the infra-red camera for example) for display on the VDU and finally to modify the graphical presentation or read-off data from it.

The display is controlled by a standard graphics package developed at the Culham Laboratory [2]. This provides automatic scaling and labelling of graphs and by light pen interaction allows the user to expand selected portions of the graph and to read-off values at positions denoted by a moveable cursor.

The three main programs have each an associated program which allows the user to call up, from the disc, archived data from past shots and display or printout as required. Figure 2 shows an outline flow chart for the two main programs.

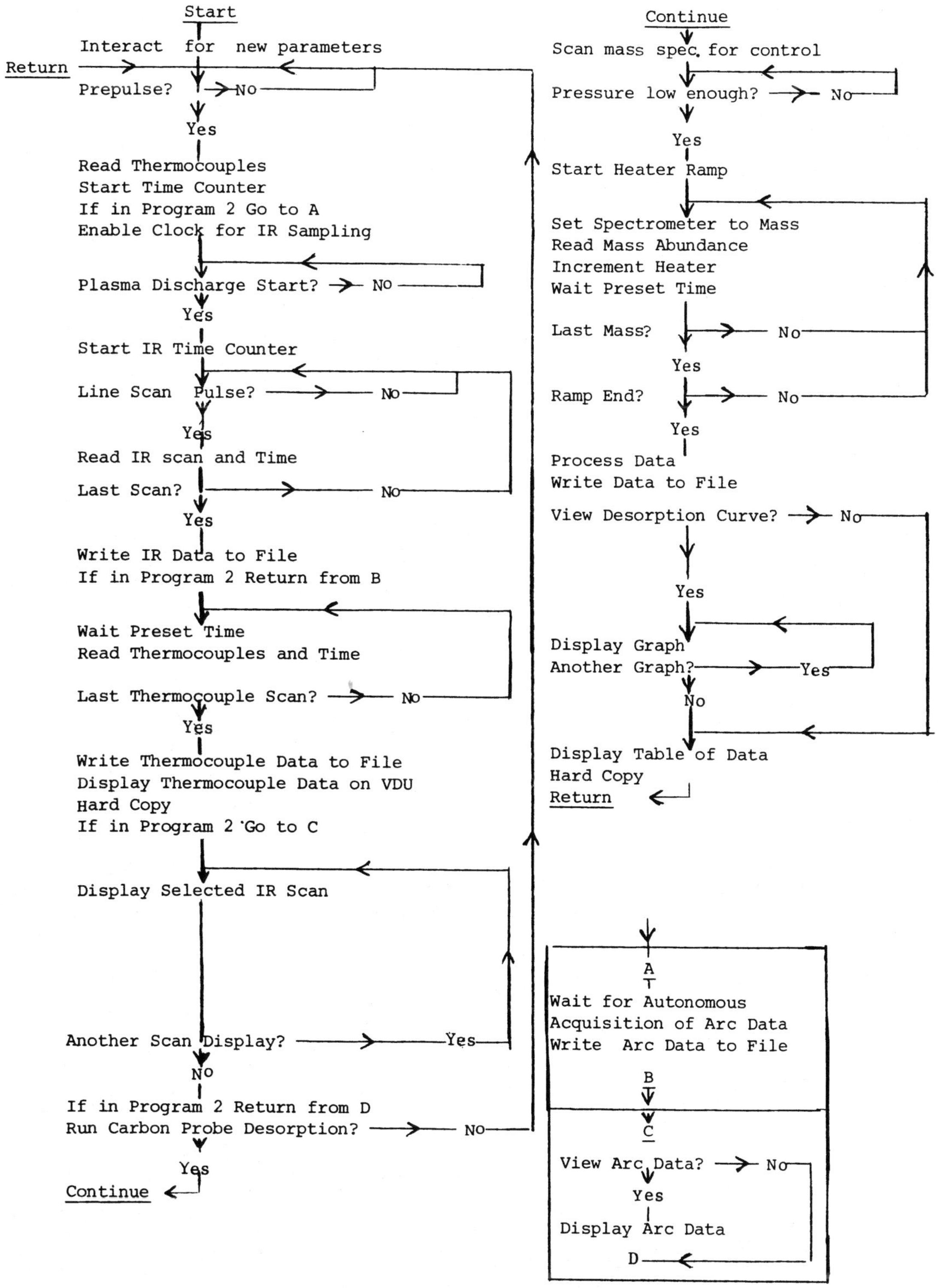

Fig.2. Outline flowchart for the two main programs

3.2 Future Extensions to Software

Additional control functions are to be provided for the Carbon Probe program which will insert the probe to a specified position in the plasma discharge vessel, hold it there for a preset number of shots and then withdraw it to the pump chamber. Then, when adequate base pressure is reached it will initiate the desorption and mass spectrometer routines. These functions are at present performed manually.

The infra-red camera results will be additionally processed to convert the present radiation intensity data to surface temperatures, which is a non-linear relationship. Then for a number of sets of line scans at different positions in the frame, contours of surface temperature are to be derived and displayed graphically.

4. SYSTEM PERFORMANCE

This data system has enabled the Surface Physics experimenters to collect and process large quantities of data from each DITE shot and provide results quickly for evaluation, which would have been impossible by manual means with limited resources. The data storage medium provides a method of filing and recalling data that can be handled with more speed and ease than rolls of chart paper. For the Carbon Probe, control of the mass spectrometer results in higher accuracy and reproducibility of measurements and will lead to more automatic operation of this diagnostic by freeing physicists from the time consuming aspects of their work.

5. ACKNOWLEDGEMENTS

Close collaboration between the authors and members of the Surface Physics Group at Culham, particularly Dr G. M. McCracken, Mr D. H. J. Goodall and Dr S. K. Erents, has contributed much to the successful progress of this work.

6. REFERENCES

[1] GOODALL D.H.J. 9th Symposium on Fusion Technology, 1976. Garmisch-Partenkirchen. Proceedings. pp 523-528.

[2] CROSS, G S., FULLARD, K., READ, A.H. and REED, K. 10th Symposium on Fusion Technology, Padova, 1978. Proceedings. pp 477-483.

[3] GOODALL, D.H.J., McCRACKEN, G.M., Nucl. Fus. 19, 1396-1401 (1979).

[4] ERENTS, S.K., McCRACKEN, G.M., VINCE, J., J.Phys. D Appl Phys., 11, 227-236 1978.

THE STRUCTURE AND SUPPORT OF A LABORATORY CONTROL AND DATA ACQUISITION SYSTEM.

by A H Read, P Nicholson, F A Raynor, D L Trotman, A J Vickers and B M White.

Culham Laboratory, Abingdon, Oxon, OX14 3DB, UK.
(Euratom/UKAEA Fusion Association)

ABSTRACT

This paper traces the development of the data acquisition systems which have been developed and brought into use within the Culham Laboratory over the last 12 years. It highlights the strong influence of the CAMAC standard on the later developments and additions to these systems and on the maintenance and support arrangements provided. The support and maintenance schemes currently in use in the Laboratory are briefly described.

1. INTRODUCTION

For a period of about 8 years until 1977, data acquisition at the Culham Laboratory was based solely upon the IDA (Integrated Data Acquisition) System [1]. In essence, the IDA system used a "star-connected" array of minicomputers, one located at each major experiment, with the site's Central Computer at the hub. The minicomputers selected for this system were CTL Modular One machines, while the Central Computer consisted of a pair of ICL 4-70s front-ended by another Modular One. Peripheral devices, including the actual data acquisition equipment, were connected to the computers by means of either a BS interface (a byte parallel path through which the flow of data was controlled by a "handshake" technique) or by the then recently introduced Harwell 7000 Series of equipment which was destined to form the basis of the current CAMAC system [2,3]. The local Modular One computers, which were used to control the data gathering and the presentation of results to the experimenters, were connected to the Central Computing System by means of a full duplex, byte parallel data link incorporating fibre-optic isolators. The Central Computer was used principally for carrying out data analysis and for data archiving purposes.

During this period, the most significant hardware change to the IDA system was the increasing use made of the CAMAC standard for the data acquisition equipment. There were two reasons for this; firstly, some Fusion laboratories began to develop modules specifically aimed at meeting their own diagnostic requirements and secondly, the Euratom Data Acquisition Co-ordinating Committee (DACC) endorsed the use of CAMAC in those laboratories with Contracts of Association. Furthermore, at this time a range of CAMAC based transient recorder modules began to appear on the market and were to form the basis of many of the diagnostic systems set up on IDA. However, as the use of the IDA system became more widespread, certain limitations and disadvantages became apparent which were mainly related to the use of a single minicomputer to serve each large experiment.

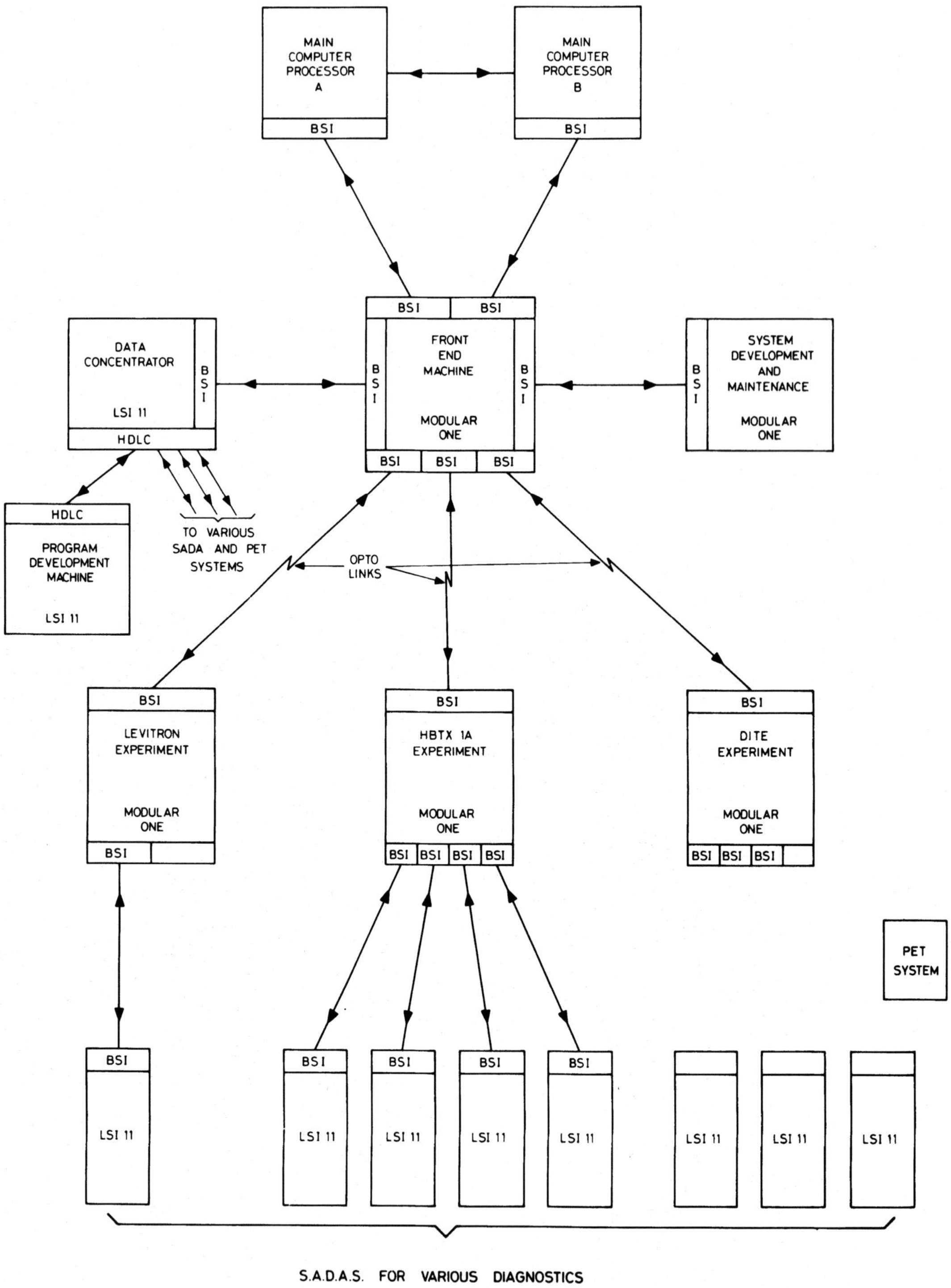

Fig.1. Data Acquisition Network at Culham

AUG 1980

2. LIMITATIONS OF SINGLE MINICOMPUTER SYSTEMS

The principal limitations encountered with the IDA arrangement were:

(i) the restrictions placed on real-time data acquisition during the experimental cycle.

(ii) the difficulty of preparing new software or modifying and testing existing software when the system was being operated in its data acquisition mode.

(iii) the difficulty of supporting those small experiments which were not related to the larger experiments without incurring great expense.

(iv) the problems of commissioning a new diagnostic system for a large experiment without interfering with existing data gathering and being subjected to the disciplines of the main experimental programme.

Users of the IDA system were able to overcome many of these limitations individually but no significant impact was made on the problem as a whole until the introduction in 1976/77 of the single board microcomputer of which the DEC LSI-11 was an early example.

3. THE USE OF MICROCOMPUTERS

These devices were relatively cheap, had many of the capabilities of the previous generation of minicomputers and had the advantage of being able to utilise most of the software which had been prepared for earlier machines. The LSI-11 microcomputer was used as the basis of the Culham Stand-Alone Data Acquisition Systems of which 36 have currently been ordered at an average price about 20% that of an IDA node [4]. All the data acquisition units of a SADA are CAMAC modules, thus the developments carried out for the IDA system were immediately compatible with SADA systems.

The simple philosophy governing the use of these systems was that a SADA should be dedicated to an individual diagnostic experiment and that a group of SADAs could be employed, where necessary, to support a Fusion experiment. For those experimenters using SADA systems, the limitations set out in 2(ii),(iii) and (iv) are removed and the restriction mentioned in 2(i) can be eased by coupling additional SADA systems to an IDA node. This has been accomplished with the design of a CAMAC Communications Module using the HDLC protocol. In order to serve the requirements of individual SADA users who have no access to the IDA system, these communications modules may be used to link their machines to the site's Central Computer through a SADA based Data Concentrator, fig. 1.

SADA systems are also finding their place in control systems and it expected that control will become a major application area of these machines in future.[5]

The recent introduction of the LSI-11/23 microcomputer which, by its use of the Q-bus makes it electrically compatible with the LSI-11, provides an opportunity to upgrade existing SADA systems and update the IDA system.

4. MEETING THE NEEDS OF THE SMALLER USER

Although the SADA systems meet the needs of many of the small experiments and individual diagnosticians, there remain two classes of user whose requirements are unfulfilled, they are:

(i) those who could usefully automate a simple set of repeated measurements if the cost were sufficiently low or,

(ii) those who wish to carry out a preliminary trial which might lead to the use of a larger system if successful.

Their needs may now be met due to the advent of the "personal computer". Furthermore, due to the introduction of a special CAMAC Controller utilising the IEEE 488 bus, systems may be built which are, in many ways, compatible with the IDA and SADA systems.

A small number of such systems are currently being set up based upon the Commodore PET Personal Computer, which has been selected due to its use of the IEEE bus. The CAMAC Controller, a Lecroy 8901 GPIB Interface module, may be directly connected to the PET with a cable. By using a relatively cheap, half-width, powered crate, an economical but effective data acquisition system may be constructed. The PET computer is programmed using BASIC and at present a BASIC subroutine is used to drive the GPIB bus and consequently the CAMAC system. Should a faster response be required, a machine coded CAMAC handler could be "poked" into memory by a BASIC program. A small data acquisition system using the PET computer can be assembled for about 5% of the real cost of one of the early IDA nodes.

5. INTERCHANGEABILITY OF HARDWARE

The decision to use the 7000 Series, later CAMAC, as the basis of the Culham Data Acquisition System has clearly influenced the course of recent developments. One reason for this choice, which is still valid today, was that CAMAC was essentially computer independent thus providing the opportunity of updating the computers without the necessity of redesigning and rebuilding the peripheral hardware. It was probably not foreseen in 1968 that the choice would provide the means of supporting three levels of system while maintaining considerable interchangeability of hardware.

The position of software is less satisfactory, the Modular One computers were programmed mainly in CLSD[6]. The SADA systems, which currently work under the RT-11 operating system, are programmed in FORTRAN with suitable assembly coded handlers where necessary. As stated before the PET machines use BASIC.

Should the decision be taken to upgrade the IDA system by the use of LSI-11/23s, the position regarding software for both the IDA and SADA systems will eventually become more satisfactory since the software should be more easily interchangeable between systems. The use of larger, faster machines from the PDP-11 Unibus range would also be attractive but the level of hardware and software compatibility between systems is likely to be less advantageous compared with an "all Q-bus" system.

6. MAINTENANCE AND THE USE OF THE CAMAC STANDARD

By concentrating purchases and developments to the CAMAC standard it has been possible to hold a wider range of equipment available for use and has facilitated the operation of an effective and responsive maintenance system. Where possible, usage of CAMAC units is limited to a subset of those available. This choice is not too difficult to make since much of the CAMAC equipment range is directed at the High Energy Physics field and only a small part is suitable for Fusion applications. The selected items are held in small quantities in an Electronics Store and are sold to users at cost price. In many cases, this gives the user the advantage of immediate availability. To ensure that modules are used as fully as possible, users are encouraged to return unused modules to the store by a rebate arrangement which restores to the user the full value of the item, less one sixtieth of the current value for every month it has been held by the user. The returned items are tested then placed in the store for reissue.

When more than two or three modules of a particular type are in the field it is normal practice for the Maintenance Section to hold "service spares" which can be used to replace faulty units on an exchange basis. The same repair arrangement exists for the LSI-11, and LSI-11/23 computers of SADA systems which are repaired on a board exchange basis, sufficient serviceable boards being held in stock to permit the faulty ones to be sent to the manufacturer for repair. Faulty CAMAC modules are brought to a central workshop for repair, which is free of charge to the owner; the workshop is equipped with SADA test systems using mainly CATY [7] as the main test language. When a module is repaired, it is returned to the owner and the "service spare" is recovered.

It is the objective of the Maintenance Section to keep downtime to a minimum and in practice most systems are operational within 30 minutes of a distress call.

A maintenance and updating service also operates for SADA system software and the handlers associated with some standard modules. Basic service with consultancy is purchased from DEC for the RT-11 and FORTRAN packages. A software fault and simulation service is provided by Electronics Group to separate user faults from operating system defects.

7. CONCLUSION

This paper indicates problems which have arisen in providing an extensive data acquisition and control coverage at Culham. In providing solutions to these problems an effective system support facility has been created with the aim of providing a service for a group of satisfied customers by keeping system downtime to a minimum.

8. REFERENCES

[1] CALDERBANK, M. and READ, A. H., 7th Symposium on Fusion Technology, Grenoble, 24-27 October, 1972. Proceedings pp83-88.

[2] ESONE, CAMAC - a Modular Instrumentation System for Data Handling. CEC Report EUR4100 (1972).

[3] ESONE, CAMAC - Organisation of Multi-Crate Systems. CEC Report EUR4600 (1972).

[4] CROSS, G. S. et al, 10th Symposium on Fusion Technology, Padova, 4-9 September, 1978. Proceedings pp477-483.

[5] FULLARD, K. and VICKERS, A J., 10th Symposium on Fusion Technology, Padova, 4-9 September, 1978. Proceedings pp539-545.

[6] CALDERBANK, V.J. and CALDERBANK, M., Culham Laboratory Program Documentation Report CLM-PDN 9/71 (1971)

[7] U K CAMAC ASSOCIATION, Report UKCA/CATY 1/01 (1977)

THE ENGINEERING DATA SYSTEM FOR THE HBTX1A EXPERIMENT

by M E U Smith and E K Sinton.

Culham Laboratory, Abingdon, Oxon, OX14 3DB, UK.
(Euratom/UKAEA Fusion Association)

ABSTRACT

This paper describes an electronic data system for measuring, displaying and recording engineering signals and machine operating conditions on HBTX1A. The modular system has been designed using fibre optic light guides for data transmission and includes microprocessors to provide distributed data processing.

1. INTRODUCTION

The High Beta Toroidal Experiment (HBTX1A) is being built to extend the research into reversed field pinches. The increasing size and complexity of such experiments has led to a growing trend towards the use of minicomputer and microprocessor-based systems for monitoring the numerous machine parameters [1].

For the HBTX1A system it is necessary to perform the tasks of collecting, transmitting, displaying and logging the data originating from or arriving at the engineering control desk. These data correspond to the various machine parameters, which may be either analogue or digital, and include voltages, temperatures, switch settings and status signals, which all need to be safely and accurately monitored.

2. SYSTEM REQUIREMENTS

It was desirable that the electronic system should be modular in construction. This was dictated both by safety factors and the layout of the plant area, control desk and computer screened room of the experiment. Electrical isolation between various parts of the equipment was needed because the "earth" potential of the different areas may vary with respect to one another, so presenting problems of personnel safety and equipment malfunction.

When the experiment is operating normally a minicomputer will be available for processing and controlling the engineering data and will be linked to the main HBTX1A control and diagnostic computer for machine operation. The data modules in the HBTX1A computer use the CAMAC interface and for hardware compatibility it would be sensible to use the same interface for communicating with the engineering data system. However, it is important that the engineering data system should operate independently of its minicomputer so that engineers could commission the experiment apparatus using manual methods. It should perform the tasks of collecting data from the plant area and displaying meaningful information about this data at the control desk.

The data to be processed by the engineering data system may be divided into two types:

(i) that originating at the plant, which needs to be transmitted to the control desk for processing and display. Selected items of data would be transmitted to the minicomputer for logging on request. Examples of this data are:

  (a) The 8 capacitor bank voltages such as the primary, crowbar and programming banks with a range from 350V to 40 kV.
  (b) The voltages associated with the power supplies for the laser diagnostics, up to 20kV.
  (c) The temperature profiles of the torus components and vacuum system during machine "bake-out", using up to 108 thermocouple inputs.

(ii) the data being communicated solely between the control desk and the computer and having no direct involvement with the plant. Examples of this data are:

  (a) The selected switch settings that determine both the capacity and operating voltage for each bank.
  (b) The status of the power supplies, such as whether they are switched on, set positive or negative and whether charging is complete.
  (c) Commands from the control desk such as alert, fire, override and abort; signals showing the progress of the machine operation and status of the computer such as idle, busy, ready, accept and reject.

3. DESCRIPTION OF ENGINEERING DATA SYSTEM

The requirement for electrical isolation has been met by using fibre optic light guides for the serial transmission of data from one area to another. This form of communication has the added advantage of reducing the amount of interference normally experienced with long runs of cable in electrically and magnetically hostile environments - like fusion machines. Due to the relatively slow nature of the parameters of interest it has been possible to use a system which both transmits and receives along the same light guide. This is achieved by using an emitter diode for both transmitting and receiving [2],[3].

The modular implementation of the system is best illustrated in the schematic layout in Figure 1. To meet the requirement that the system should operate independently of its minicomputer each of the Controllers contains a single-chip microcomputer thus providing distributed intelligence within the system. The different tasks to be performed by each Controller are determined by software resident in the micro-memory. Thus a number of identical modules have been designed for this system to perform a range of different functions.

The capacitor bank voltages are accurately attenuated before input to the Analogue Interface Module (AIM), which is situated in the plant area. The AIM multiplexes and digitises each input and transmits its data sequentially and serially to the Controller. Each voltage is then scaled to its actual decimal value and all 8 are displayed at the control desk. The two laser voltage measurements are similarly digitised and scaled, but only one voltage from each group of 8 inputs need be displayed at any one time. This is determined by a digiswitch setting on the Selector connected to the Controller, which signals the appropriate AIM to transmit information of the desired channel to be displayed.

Each thermocouple output is scanned and fed to a digital thermometer as a 4-digit BCD code, which is continuously updating a data store. Only one display, set by a Selector, is required at the control desk. The Controller signals a Digital Interface Module (DIM) to transmit the temperature information together the with requested channel, as confirmation, for display as degrees Celsius. The switch settings and status information, being digital in form and requiring no control or processing, are selected and transmitted transparently via DIM units situated in the control desk.

To connect the Engineering data system to its minicomputer by means of the 4 Controllers and the two Switch and Status DIMs are coupled through light guides to two Local Control Modules (LCM) fitted in its CAMAC crate [4]. Requests from the minicomputer for particular data, for instance the temperature of thermocouple 27, are relayed by the LCM to the appropriate Controller or DIM, which collects and sends the required information. In the case of the Controller, linking software in the micro breaks into to the independent operating sequence to initiate the request.

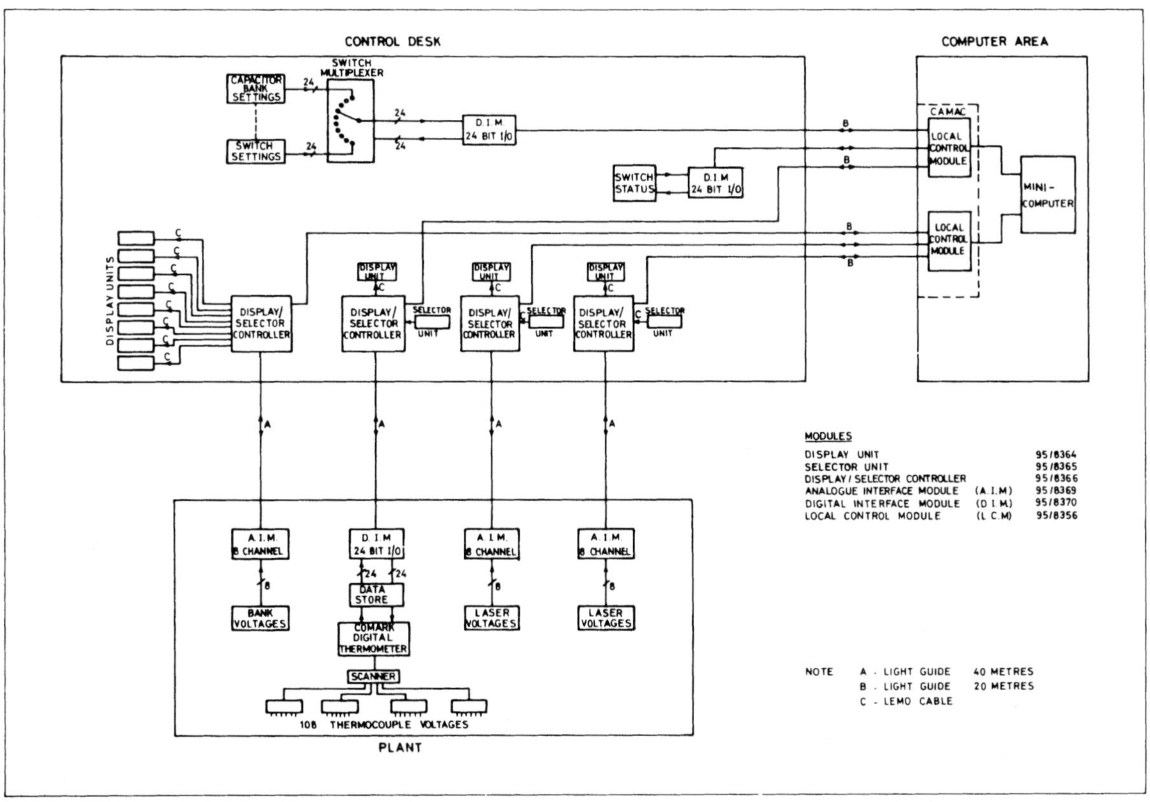

Fig.1. Block schematic of engineering data system for HBTX1A

4. MODULE DESCRIPTION

4.1 Display/Selector Controller (see Fig.2)

This module uses a single chip microcomputer (INTEL 8748) to process and/or control the transmission of serial data between the various modules in the system. The precise function of this module is determined by the software stored in the UV erasable PROM of the microcomputer. The three Universal Asynchronous Receive/Transmit (UART) microcircuits used for communication with all the other modules are connected to the microcomputer via a common bus and controlled by one of its I/O ports.

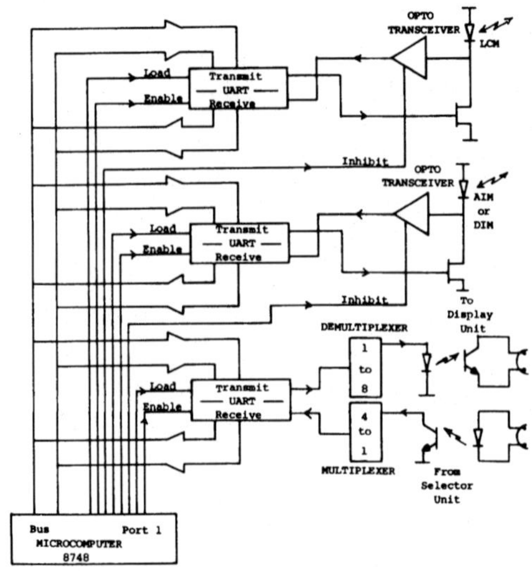

Fig.2. Display/Selector Controller

4.2 Display unit (see Fig.3)

This unit consists of a five character, seven segment LED display and has been designed to accept optically isolated serial data from the Controller. The display driver contains an 8 x 8 static memory for storage of the received data for consequent multiplexing to the display. A "time-out" circuit is incorporated in order to blank the display if data is not received to update the display within a preselected period.

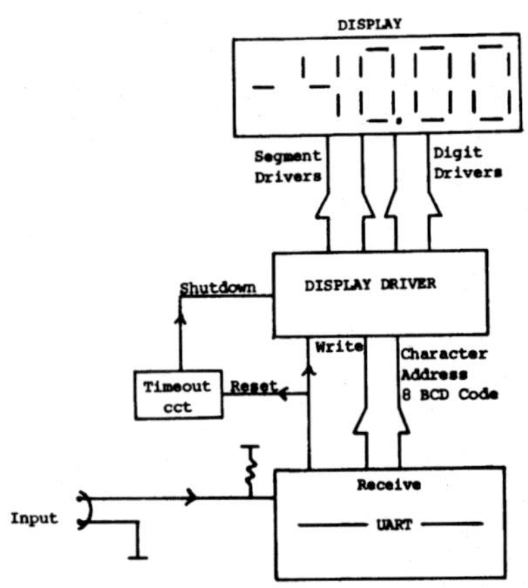

Fig.3. Display unit

4.3 Selector Unit (Fig. 4)

This unit is designed to output, to the Controller, serial data encoded to give the positional address together with the BCD code of up to 8 thumbwheel switches mounted on the unit.

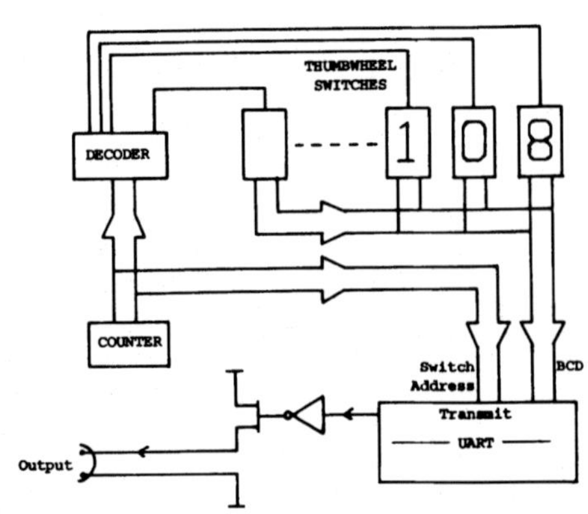

Fig.4. Selector unit

4.4 Analogue Interface Module (Fig.5)

This module is designed to transmit, in response to a request from the Controller, a digital representation of the analogue signal present at any one or all of its inputs. The eight differential-input amplifiers are fitted with protection circuits and accept signals of $\pm 5V$ amplitude. These signals are multiplexed and fed to a single sample and hold for subsequent digitisation by a 10-bit analogue to digital converter. Data Transmission is in the form of two 8-bit bytes, allowing the inclusion of the data and its corresponding channel address.

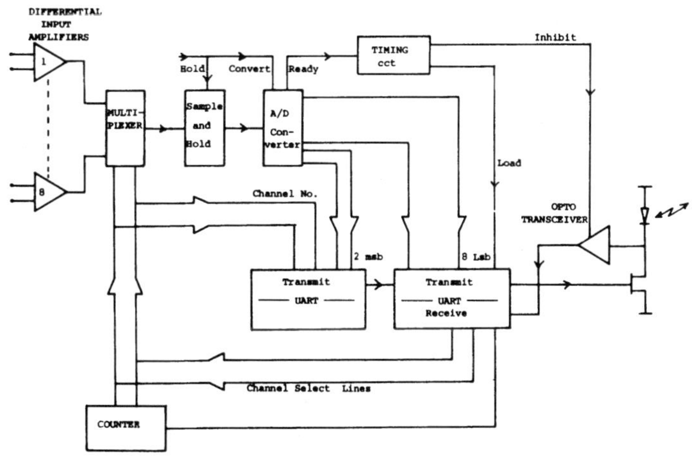

Fig.5. Analogue interface module

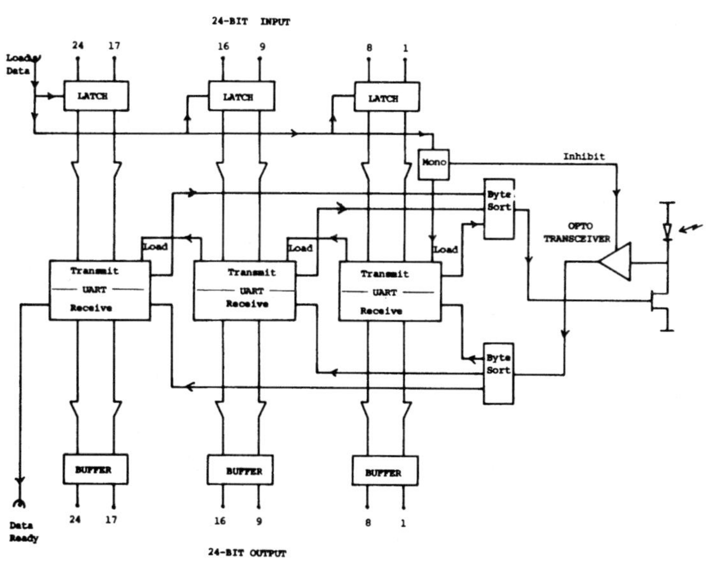

Fig.6. Digital interface module

4.5 Digital Interface Module (Fig. 6)

The module is designed to receive and transmit 24 bits of information along a single light guide. Serial data is received in the form of three 8-bit bytes. This data is presented as a buffered 24-bit word. Data present at the 24-bit input is latched and serialised prior to transmission.

5. SYSTEM PERFORMANCE

The design, development and manufacture of the hardware together with the production of the software has been completed and the various sub-systems have been bench tested.

Before installation on HBTX1A the capacitor bank sub-system was tested on the DITE experiment neutral injector supplies for several weeks. Although subjected to various high voltage transients and numerous injector electrical breakdowns, the sub-system continuously digitised and transmited the correct data. The complete system is now at HBTX1A ready for commissioning and should provide an economic and reliable method of logging and processing engineering signals in hazardous environments.

6. ACKNOWLEDGEMENTS

We gratefully acknowledge the contributions made by our colleagues in the development of the system outlined in this paper. In particular we thank Messrs. P J Roberts, P Nicholson and B J Ward.

7. REFERENCES

[1] WHITTLE, H.R., FULLARD, K., MURPHY, E.G., and RICHARDSON, D. 7th Symposium on Engineering Problems in Fusion Research, Knoxville, Tennessee, Oct. 1977. Proceedings pp 1223-1226.

[2] FULLARD, K., VICKERS, A.J., 10th Symposium on Fusion Technology Padova, Italy. 1978. pp 539-545.

[3] TROTMAN, D.L., FULLARD, K. Two way Optical Transmission System. Patent Application. P6023/78.

[4] VICKERS A J et al. 11th Symposium on Fusion Technology, Oxford, 1980.

IMPROVED DATA ACQUISITION SYSTEM FOR T.F.R.

Mme J. Breton, A. Cohen, F. Hennion, J. Idmtal, J. Touche

ASSOCIATION EURATOM-CEA SUR LA FUSION
Département de Recherches sur la Fusion Contrôlée
Centre d'Etudes Nucléaires
Boîte Postale n° 6. 92260 FONTENAY-AUX-ROSES (FRANCE)

I. INTRODUCTION

The data acquisition system for TFR was untill recently composed of an old T 2000 minicomputer connected to a Solar 16/65 both from the SEMS. The task of the T 2000 was limited to data acquisition, storage of raw data of each shot on magnetic tape and several simple processings with display of signal output.

The use of a large Solar minicomputer allows to store on disk a few number of shots, to perform new and more sophisticated processings, and to prevent the overloading of system. Furthermore it allows the use of advanced languages such as FORTRAN. This system is described in reference 1.

After important breakdowns on the T 2000, this computer has been eliminated and the whole system has been modified and is described below.

II. HARDWARE

The different type of acquisition modules are : binary switches (256 bits), digital oscilloscopes with local buffers for fast measurements (such as plasma current, voltages ...), 4 K words memories with digitalization modules adapted on each diagnostics (fast neutral particle analyser), special diagnostics (Thomon scattering, Infra-red interferometry) and Lecroy. All this modules are connected to the CAMAC branches. The acquisition systems are connected to the minicomputer by opto-electronical serial link for electronic insulation.

A Solar 16/65 of 96 K word memories and its 10 megabyte disk is connected to a Solar 16/75 of 256 K word memories and its 50 megabyte disk. Two magnetic tape drivers can be linked on either one of the two minicomputers. Figure 1 shows the hardware configuration.

III. SOFTWARE

The large storage space of the memory and the capacity of the disks allows to use in better conditions the software SEMS system : RTESD (Real Time Executive System) on the Solar 16/65 and MPES (Multi Programming executive System) on the Solar 16/75, and FMS (File Management System) for both computers.

On line processing

The organisation of the system for a line processing is shown in figure 2.

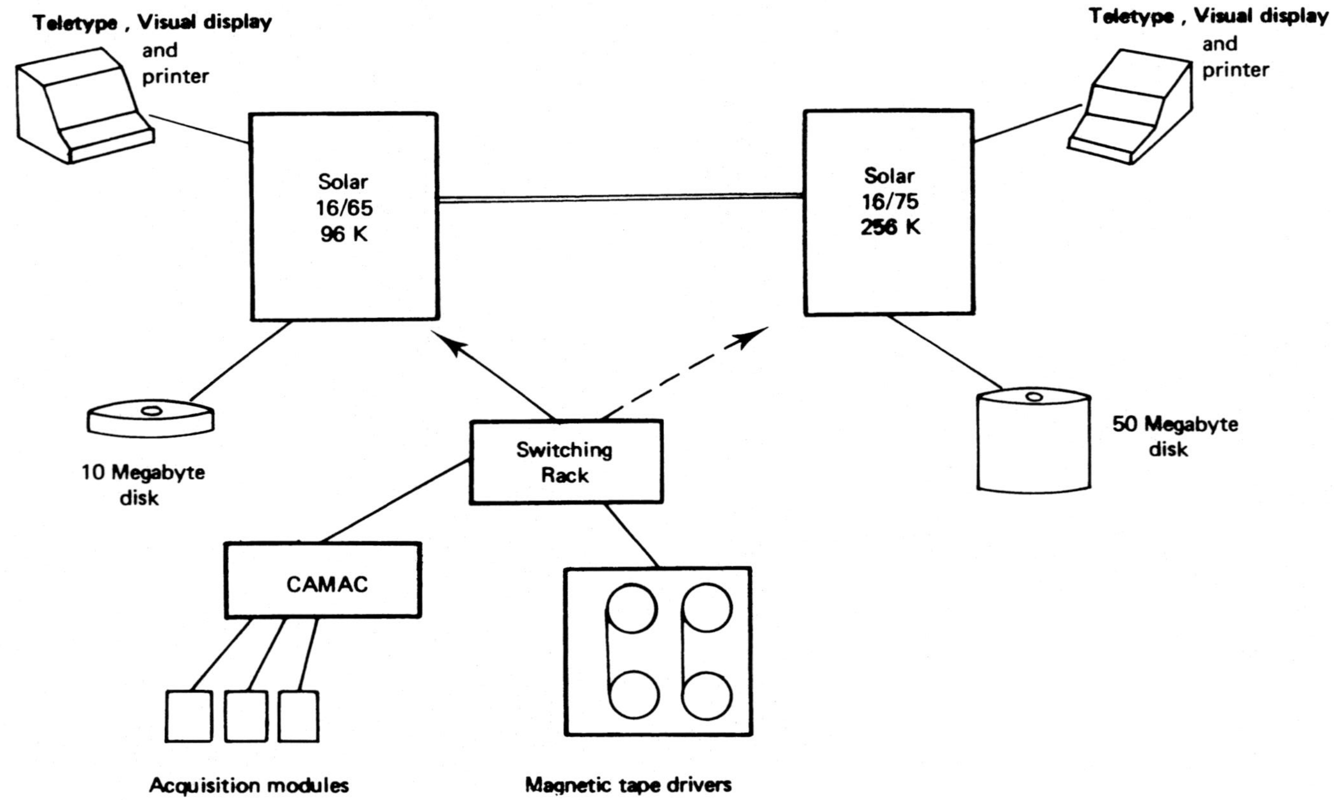

HARDWARE CONFIGURATION

FIGURE 1

a) Raw data storage

The raw data coming from the CAMAC are stored on disk and on a magnetic tape.

b) Processings :

We distinguish between three classes of processings :

- First level of processings ((1) on the figure 2) which read on disk the raw data and after calculations store on disk the different signals in a very standard tabular form.
- Second level of processings ((2) on the figure 2), which read the precedent standardized signals and create new data blocks, after many elaborated calculations (such as plasma displacements). This new results are stored on disk in the same standardized form as the former signals.
- Taskes for particular processings and displays ((3) on the figure 2) : Particular forms of display cannot be taken care of by the usual display programs.

c) Display :

The display programs have a great versatility, variable format, fixed scales or scales adapted to the display signal, function (smooth, integration, logarithm) and optional hard copy. Processings (2) and (3) can be written by the physicists themselves because they use FORTRAN. Moreover there is an interface between this language and the programs for writting and reading on disk. During the shots, acquisitions, processings and displays can be modified both in nature and number thanks to some "file menus" (list of options available). Furthermore, the elaborated data for each shot (in standardized form) can be stored on a second magnetic tape.

Off line processing :

All the previous processings and the display programs can be performed as well either from the raw data stored on the first magnetic tape (processings (1), (2), (3), or from the elaborated data coming from the second magnetic tape (one part of processing (2) and (3). In addition, very sophisticated processings are performed off line after the shots (statistical processings, Abel inversions, elaborated smoothings .. i.. e..).

IV. MULTICOMPUTER

The link between the two Solars allows to share out processings (classes (2) and (3) and display programs). The raw and elaborated data are stored simultaneously in both disks through the two computers. In addition of the real time processings, the software from the SEMS allows to use on the Solar 16/75 : batch, time sharing options and, in the futur, remote batch.

V. RELIABILITY

Hardware : The similary of the two computers emphaze the reliability of the system. If one breakdown occurs on one of the computers, the other one can takes care of the main part of the processings and displays. So the system can run in a degraded mode.

Software : Furthermore each acquisition task, processing and even display programs gives an account on a special display of possible failures to control easily the system and ensure its good working condition.

VI. EXTENSION OF THE SYSTEM

The implantation of another Solar 16/05 provides the data acquisition of residual gas analyser. In the futur it is intended to implant new Solar 16/40 minicomputers for R.F. and neutral injectors. All these system are independant of the data acquisition on TFR, but the link with the TFR system allows to store the data and perform advanced processings.

Reference :

[1] Multi computer system for the acquisition and the processings of the data of TFR Tokamak experiment.
Mme J. Breton, J. Idmtal, F. Hennion, J. Touche - Fusion Technologie 1978.

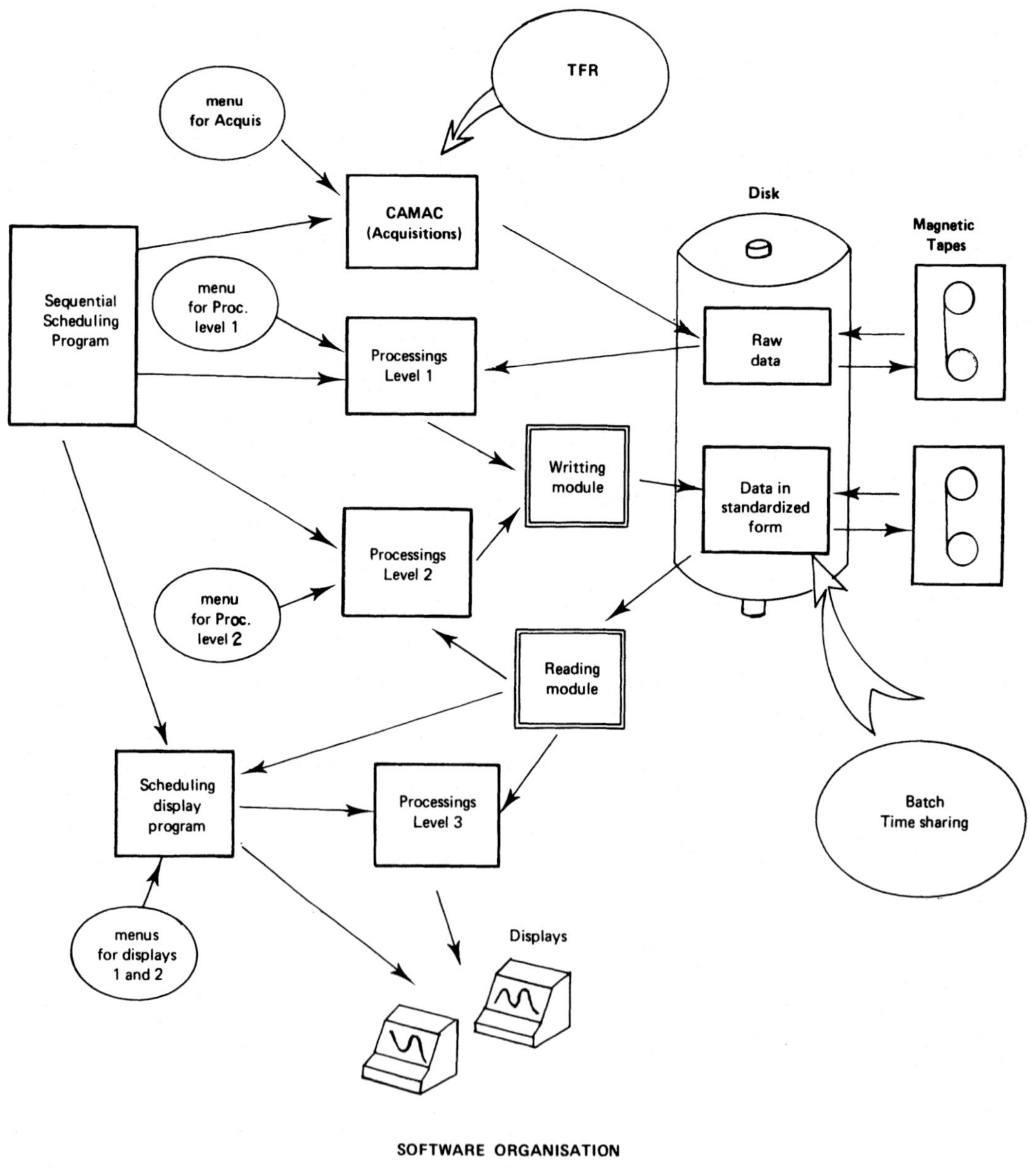

SOFTWARE ORGANISATION

FIGURE 2

PROGRESS IN COMPUTER-ASSISTED DIAGNOSIS AND CONTROL OF NEUTRAL BEAM LINES

E. Theil, V. Elischer, J. Fiddler, N.J.D. Jacobs, V. Jacobson, R. Lawhorn, D. Uber, D. Wilner

Real Time Systems Group
Lawrence Berkeley Laboratory

University of California
Berkeley, CA 94720

ABSTRACT

This paper discusses the principles that have guided the development of a computerized diagnostic and control system for both the Neutral Beam Systems Test Facility at Lawrence Berkeley Laboratory and the Doublet III neutral beams at the General Atomic Company. The emphasis is not on the particular details of the implementation, but on general considerations which have influenced the design criteria for the system. Foremost among these are the requirements of an appropriate human interface to the system, and effective use of a relational data base. Examples are used to illustrate how these principles are carried out in practice. A systems view of diagnostic programs is suggested in the light of our experience.

1. INTRODUCTION

Prior to the start of the Neutral Beam Test Facility at LBL in the mid-1970's, the Real Time Systems Group had been engaged in implementing computer control and diagnostic systems for several of the particle accelerators at the Laboratory. From experience gained there, two principles emerged which have helped shape the present neutral beam computer system. They are

1. The need for an appropriate human interface to the computer system. It was recognized that the operators of particle beam accelerators have to deal with large amounts of complex data in real time and frequently need to access several different kinds of data quickly and easily. Moreover, they are generally well acquainted with the more conventional forms of data displays, notably oscilloscope traces, and have neither the time nor the inclination to sit at a computer terminal and learn even a very high level set of protocols while they are busy doing their primary tasks. Thus, one of the prime objectives of our group was the development of a graphics system that was both natural to use and "friendly". This system is discussed further in Section 2.

2. The importance of a relational data base as an integral part of the system design. Much has been written about data bases, their theory and implementation. Nevertheless, their use is not as common in a scientific environment as elsewhere. We discuss the advantages of a data base system in the context of neutral beams in Section 3.

In Section 4 we suggest that the traditional view of diagnostic programs as a collection of clever algorithms is not sufficient, in view of the need for an integrated systems approach.

2. THE OPERATOR INTERFACE

In an effort to create a simple and convenient communication medium between the human operator of the beamline and the computer, LBL designers are using three large CRT screens (one color, two black/white), with one screen equipped with a Mylar touch panel [1, 2]. The touch panel, combined with logical (software) buttons, constitutes an effective way for the operator to generate an external interrupt to the computer. Its most obvious use is in selecting displays from a "menu" of choices, but there are more interesting uses, especially when the touch panel is combined with input knobs via a microcomputer, as in the following example.

2.1 The Timing System -- An Example of the Use of the Touch Panel

Figure 1 shows one of the seven timing displays that can be called up on the touch panel by pressing a menu button. Once the display appears, any of the timing signals named on it can be referenced by first pressing the button surrounding the name (e.g., "Gas" in Fig. 1). Two knobs then control the leading and trailing edges of the pulse. As the knobs are turned, the pulse will move left or right and widen or narrow accordingly. At the same time, the appropriate numbers will appear in the box immediately to the left of the pulse name. When the operator has set the timing pulses to his satisfaction, "Change Timers" is pressed and the modifications are entered into the computer's data base. As may be seen in the figure, the displays are also equipped with a "zoom" feature (Expand/Shrink Display). Misplaced pulses (e.g., arc before filament) are avoided by software interlocks that automatically move related pulses as needed or prohibit the adjustment of others. Virtually all timing signals at NBSTF are now under this type of computer control.

2.2 Deleting Sensors -- Another Touch Panel Example

Another interesting use of the touch panel/logical button interface is the ability to delete temperature sensors during calorimetry analyses. Fig. 2 shows a sensor control display, a Doublet III calorimeter. By touching any of the sensor buttons and then updating the status control, the temperature measured at that position is removed from (or brought into) consideration in data fitting and smoothed when plotting isothermal contours. Since thermal sensing devices may fail for a variety of reasons, this ability to "remove" them immediately can be and is very useful.

The buttons acknowledge a hit by brightening their borders and carry whatever legend the programmer desires to place there. A set of Fortran-callable subroutines that interface with the graphics system is available for manipulating the buttons. All changes to display layouts are therefore done in a high-level language.

3. THE IMPORTANCE OF THE DATA BASE

This concept has been discussed at length in [3]. For our purposes, two particular characteristics of the data base are worth noting:

1. It contains a description of all the physical characteristics of the neutral beam lines needed for computer analysis and display.

2. It permits a "decoupling" of the analysis and display programs from the details of the instrumentation hardware and data acquisition tasks.

There are three important advantages to be gained from these features:

First, the data base helps to create a natural modularization of the system, since it is possible to think of single purpose tasks which need not carry along explicit assumptions about the data inputs or beamline devices with which they are dealing.

Second, it provides a central store for all data, permanent and transient (shot-to-shot). This improves debugging and maintenance efficiency, since changes in the physical description of the system (which are frequent) occur in only one place.

Third, several different beam lines, perhaps with quite different physical characteristics, can be described in the data base, and the individual programs need have no "wired-in" knowledge of their differences, or of which one is currently active.

An example seems appropriate here to illustrate these three points:

3.1 The Data Base In Use: Many Devices on Many Beam Lines

As one of our diagnostics, there is a task that fits the temperature data, read from any of several calorimeters, to a theoretical model of temperature distribution [4], and another task that displays the results of that analysis. Let us look more closely at the input data needed by the display task. It requires the size and shape of the calorimeter and the relative positions of the thermocouples. It needs the parameters obtained from the data fitting process, the maximum temperature, and a host of other data items that change from shot to shot. (Fig. 3 shows this display). It also needs some text that does not change from shot to shot in order to make the information meaningful. It does not need to know, however, what all this information is, only where to access it (just the opposite of the human who reads it).

Our system uses a message service. When a task is activated, it receives a start-up message which contains the beam line number, source number and a two character device code. If, for example, the system is the present NBSTF beamline at LBL, the start-up message will contain the ASCII string 0002BD, which translates to beamline 0, source 2, beamline calorimeter. This provides the only specific information which the task will need in order to form a unique data base name. That name will, in turn, refer the task to pointers and files that contain the specific instances of the general data types described above. Moreover, much of this data is shared by various tasks. For instance, the locations of the thermocouples on a variety of instrumented devices are stored in the data base and used by a number of programs, each time accessed by the technique just described. Thus, to continue the examples, the display task will display the acquired data in positions which appear on the screen in the same relative positions as the thermocouples on the device. Figures 4 and 5 show two such displays, for the NBSTF neutralizer and ion dump calorimeter, respectively. Exactly the same task is used for all such displays (presently more than twenty) with the identifying information contained in the message received from the message service.

Since similar message services are generally available in real time operating systems now, these techniques can be implemented in any such system with a relational data base.

These ideas can be carried a step further. From the computer's point of view, the positions on the CRT screens and the colors of the pictures and text constitute part of the description of the display. Thus, these too are included in the data base and a set of simple Fortran-callable subroutines are used first to get this information and then automatically display it. The considerable advantage of this approach is that the same information can be displayed in different arrangements on the screen, as a function of the geometry of the associated device (see Figs. 4 and 5), again without explicit assumptions built into the display program.

4. DIAGNOSTICS: A CHANGE IN VIEWPOINT

A wide variety of diagnostics is presently available on our system, with more under development. Included are:

1. A full range of calorimetry tasks (temperature displays, isothermal contouring, temperature data fitting and analysis), all of which respond to operator control of the sensor status buttons, as described in Section 3.2.
2. Power supply waveform displays.
3. Plasma probe waveform displays and profiles.
4. Photodiode analysis of the beam.

5. Perveance line and tuning curve displays for source conditioning.

This list is not exhaustive. Some of these diagnostics are used only occasionally, while others, such as the calorimetry analyses and the plasma probe profile, relieve the operators and physicists of an immense amount of data gathering and reduction and speed up beamline operations immeasurably.

Traditionally, in scientific laboratories, the emphasis on diagnostics has been almost totally algorithmic. How to derive the information from raw data naturally takes a great deal of thought and effort. However, with an emphasis on graphical output and a data driven system, the presentation of that information and the integration of the diagnostic tool into the system play an equally important role. No one should think that this change in viewpoint constitutes a denigration of the importance of first rate computational algorithms. They require special knowledge and often are an art in themselves. However, we recognize that in a well-designed real time system, which must provide flexibility, ease of maintenance and comprehension to the operator, these other considerations must also enter. Thus, when a diagnostic is designed, thought must be given to the system tools which will be useful to the task.

4.1 The Plasma Probe Profile: An Example of an Integrated Diagnostic

The plasma probe profile display relieves the operator of a tedious chore, while providing important information on the plasma density in the source's arc chamber.

The waveforms of eight Langmuir probes, located in the chamber, are read, digitized and acquired by the computer. Ideally, all probes should have the same average value and significant deviations from a central value are carefully monitored.

The display is seen in Fig. 6, with idealized data. A line graph of the average values is drawn (the vertical lines are a measure of the variation in each waveform). The average values themselves appear in a table at the right and the ratio of largest to smallest is shown. (This is the key value; a ratio greater than about 1.25 is cause for concern.)

A typical version of such a diagnostic tool might first get the probe waveforms, average them in a subroutine and then display the results, with a good deal of in-line assumptions and decision making. While the process is conceptually and computationally simple, the job can become difficult because of the large number of input variables that have to be specified: e.g., the number of digitized samples, the sampling rate, the gains and offsets for conversion to engineering units, the waveforms themselves, the locations on the screen and scaling constants for displays, etc. Many of these are subject to frequent change.

In our system, waveform averaging is accomplished by a separate task which, given a list of the waveform names, does the averaging and computation of the variation, taking waveform interrupts into account. Needless to add, this task finds its inputs by forming a unique database name, as described in Section 3. For displaying the information, the probe profile task uses our standard graphical routines which allow the formats and screen location of the information to be specified in a database file. Thus, the integrated display task reduces to identification of the inputs, design of the data structures and graphical format, effective use of system aids and, in this case, a relatively small amount of algorithmic processing.

5. SUMMARY

We have tried to show, in this brief exposition, that a systems approach to a diagnostic and control design of the type required for neutral beam operations is desirable for several reasons. It leads to a unified view of the various diagnostic programs, encourages modularity and helps programs to avoid the dangers of wired-in assumptions. In addition, we believe the attention given to the human interface is worthwhile from the point of view of operator acceptance and comprehension.

This approach is not without its price. Writing programs will sometimes require more work to assimilate the program into the system than would otherwise be the case, and a certain amount of discipline is necessary to guarantee that standard interfaces and programming conventions are maintained. But our experience suggests that this is a price worth paying.

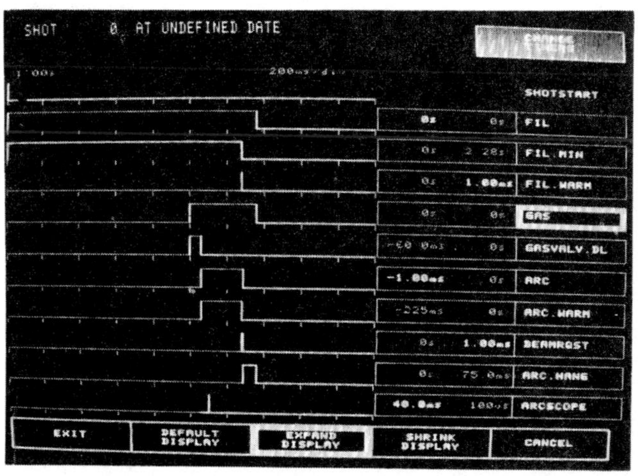

Figure 1.

NBSTF Control Room

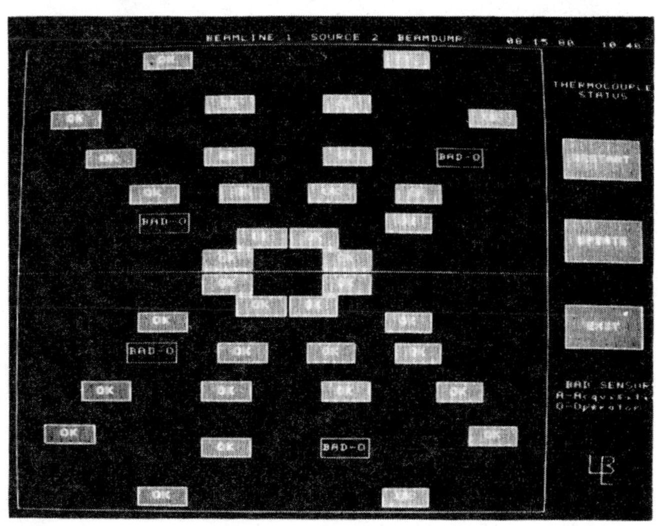

Figure 2.

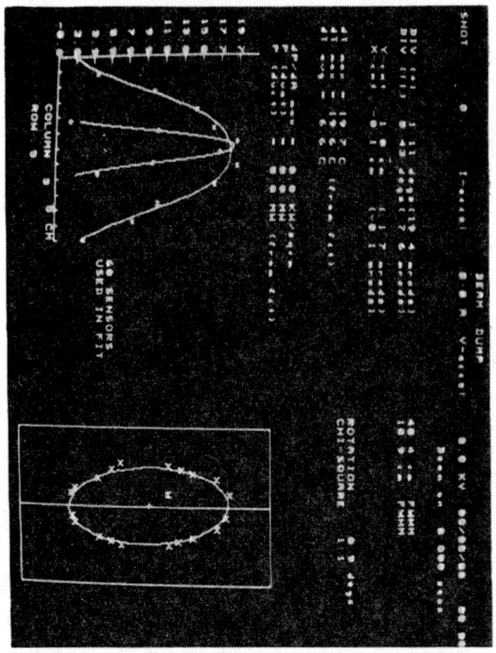

Figure 3.

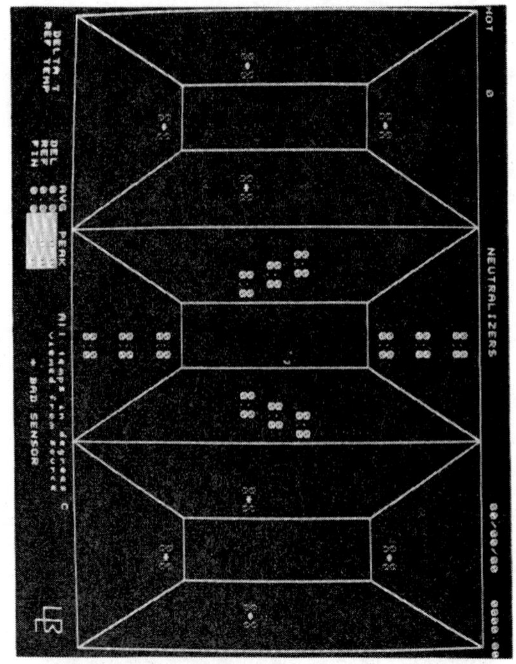

Figure 4.

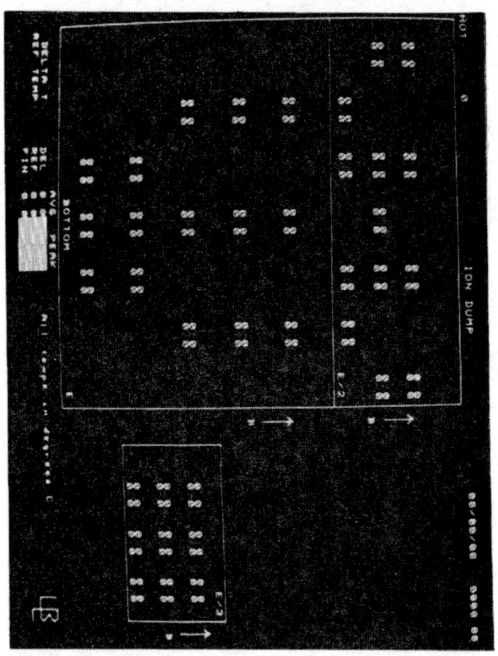

Figure 5.

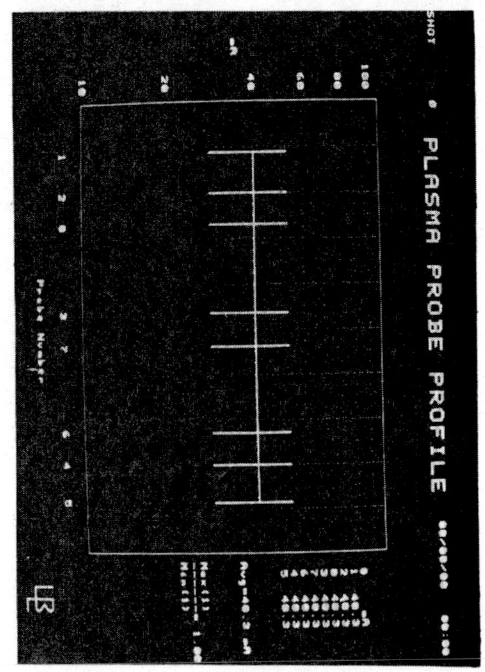

Figure 6.

We thank Kate Hughes for her expert preparation of the typescript of this paper.

6. REFERENCES

[1] ELISCHER, V. and FIDDLER, J., Graphical Human Interface to a Neutral Beam System, Lawrence Berkeley Laboratory Report, LBL 9146.

[2] CROWLEY-MILLING, M.C., Proc. of the Tenth Symposium on Fusion Technology, 1978.

[3] ELISCHER, V., FEHR, S., JACOBSON, V., WILEY, K., Proceedings of the Sixth Symposium on Engineering Problems of Fusion Research, San Diego, California, 1975.

[4] KIM, J., and WHEALTON, J.H., Nucl. Instrum. and Meth. 141, 187 (1977).

SYSTEM DESIGN OF ZENKEI, THE CENTRAL CONTROL SYSTEM FOR JT-60

Y. Suzuki, I. Kondo, A. Ogata, T. Kumahara
Japan Atomic Energy Research Institute, Tokai, Japan

T. Hatakeyama, K. Murai, D. Iba
Hitachi Ltd.
Hitachi, Japan

ABSTRACT

ZENKEI, the central control system for JT-60, is a multi-computerized system to supervise various control subsystems and to maintain harmonious operations of sybsystems under plasma dischargings. It provides also machine monitoring and feed-back control of plasma shape and position.

1. INTRODUCTION

JT-60 is an experimental equipment whose system size is comparably large as ordinal industrial plantations. Therefore ZENKEI is required to be extremely reliable but also flexible and expansible.

Design philosophy of JT-60 control system will be realized in the following category.

(1) It is hierarchical system so that ZENKEI supervises subcontrol systems.

(2) It applies the distributing control method with many computers so that test run of individual subsystems may be performed in easier situations.

(3) It adopted CAMAC interfaces between ZENKEI and the subsystems so that data transfer will be uniformalized.

(4) Safety interlock of JT-60 is also hierarchical without hardwired signal transformation.

2. CONSTRUCTION OF JT-60 CONTROL SYSTEM

The construction of JT-60 control system is shown in fig. 1. JT-60 control system is composed of ZENKEI and control subsystems.

ZENKEI consists of control panels, computer system, CAMAC system, timing system, plasma monitor devices and protective interlock system. These will be explained in later sections.

Here we classify the control subsystems as follows. Machine I, Machine II, Toroidal field coil P.S., Poloidal field coil P.S., 2ndary Coolings, Powersupply facilities, NBI heating, RF heating where Machine I includes Primary Cooling, Vacuum Pumping, Baking P.S., Gate valves, Vacuum Vessel Instrumentation and Vacuum Vessel Protection systems,and Machine II includes Gas Injection, Preionization, Movable Limiter and E & M measurement devices.

Each subsystem has local control panels at the places where the control objects are installed and system console in the control building to have individual control over the system.

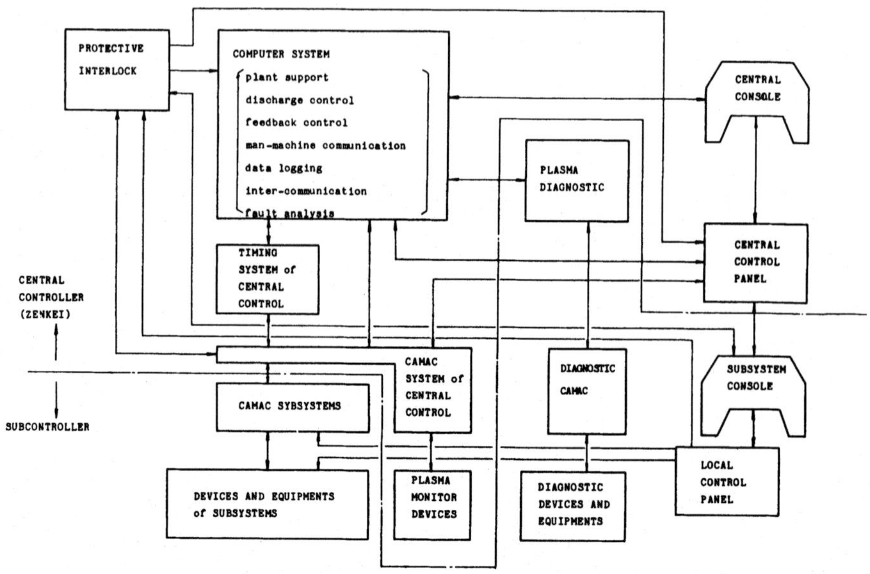

FIG. 1. CONSTRUCTION OF JT-60 CONTROL SYSTEM

3. CONTROL PANELS OF ZENKEI

There are six kinds of panels under the name of ZENKEI control panels, the Central Console, Supervisor's Console Auxiliary Graphic Panel, ZENKEI Auxiliary Relay Circuit Panels and Diagnostic Room Display Panels and ZENKEI Alarm Display Panel.

Central Console is nothing but on input and output devices of ZENKEI computer system. It is a desk type console with 12 CRT Display units (8 of character displays and 4 of graphics) so that JT-60 control operators may supervise all subsystems, prepare and execute the discharge sequences.

They may also refer and set the data, obtained and required at the 2ndary heating systems, and check and print out the results of various informations of JT-60 control systems.

Supervisor's Console is a desk type console where one may refer any information appears on the central console CRT but no executing input. One may also control ITV from this console to monitor many important points of JT-60 sites.

Auxiliary Graphic panels will display constantly overall monitor of JT-60 such as control status, alarms, and principal monitoring values of the sub-systems. This would help the control operators to grasp JT-60 status at once. This panel is designed so that in case of black out of CRT caused by ZENKEI computer system down. The JT-60 control operators may be informed indis-pensable control parameter for the manual control from this panel as well as from the graphic panels on the subsystem console.

Pen-Recorders which monitor radiations will be mounted on this panels.

Diagnostic Room Display Panels will help to inform the JT-60 control status to the experimentalists in Diagnostic Rooms of the Experimental Building.

ZENKEI Auxiliary Relay Circuit Panels contains hardwire circuit of Relays and Power Supply Units for the Central Console and Auxiliary Graphic Panel.

ZENKEI Alarm display panels will display of alarms of ZENKEI Computer System, and ZENKEI CAMAC system.

4. ZENKEI COMPUTER SYSTEM

This computer system is the heart of ZENKEI entire system. It is multi-computerized system of seven HIDIC 80E minicomputers with buble memory of discs many kinds of I/Os and peripheral I/Os. Fig.2 shows the construction of the system. This system has the following performances.

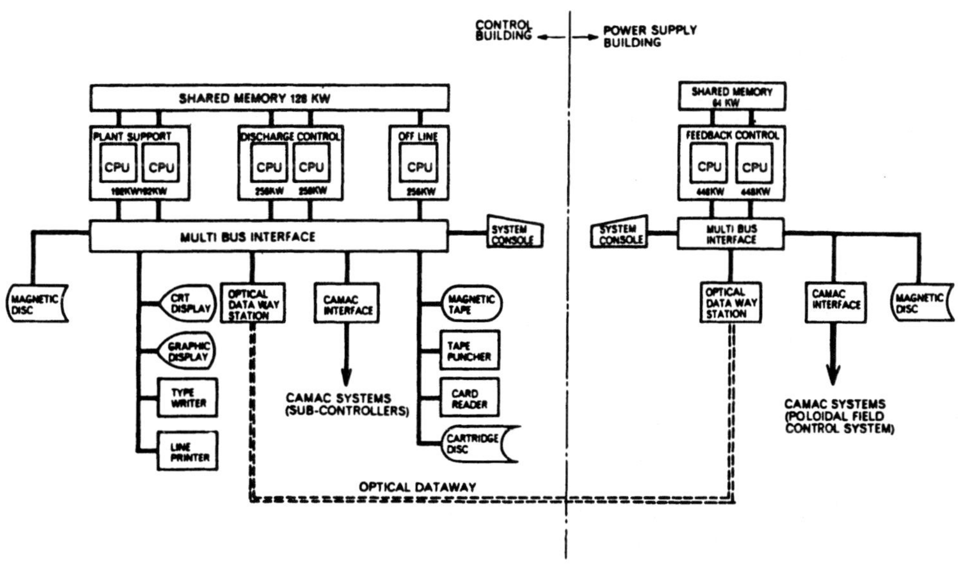

FIG. 2. ZENKEI COMPUTER SYSTEM

(1) Plant Supporting Activities

This system gathers plant data from the subsystems through CAMACs, logs processes, checks and examines them. In case of JT-60 mode transition, these data will be used to judge whether the transition will be acceptable or not. ZENKEI computer system will supervise the subsystems in the sense of overall control of JT-60 plant supportings.

(2) Discharge Control Activities

This system conducts entire JT-60 discharging sequences (includes discharge cleanings) with Toroidal Field Coil P.S. and Poloidal Field Coil P.S., Preionization, Gas Injection, Movable Limiter and Heating Systems. Successful Generation, Holding and Ceasing of Plasma in the main purpose of this conditions. This function also includes superintendence of discharging status and plasma, and checkings of JT-60 prior to the discharging.

(3) Feedback Control Activities

This system will do the feedback control of plasma position and shape with Poloidal Field Coil P.S., plasma control system, using the data obtained from Electro-Magnetic Plasma Monitor devices. This control is therefore real time control of discharge sequence.

(4) Man-Machine Communications

This system will display many kinds of figure and informations on CRT display on the Central Console and Supervisor's Console and will output the status and alarms on the Auxiliary Graphic Panels and Diagnostic Room Display Panels and so on.

This system will accept various kinds of input and setting from the Central Console so that operator may control overall JT-60 Plant Supports and Dischargings.

(5) Recordings

This system may record following major items.

- o Conditions and results of discharges.
- o Status of operations of the subsystems, apparatus and devices.
- o Monthly report.
- o Automatical output in case of specified occurances.
- o Operational input.

(6) Data transfer and intercommunications

This system will intercommunicate with the computers of NBI heating, RF heating, Plasma Diagnostic and Joint Tokamak Systems.

(7) Fault Tree Analysis

In case of alarms and unexpected conducts of JT-60, this system collects the transition data from subsystems and analyze them to inform the results and remarks to the operators so that they might be also resume the normal operation.

5. ZENKEI CAMAC SYSTEM

ZENKEI CAMAC system is formed of three groups, plant support CAMAC system, Discharge Control CAMAC system and Feedback Control CAMAC system.

ZENKEI CAMAC system connects ZENKEI computer system to ZENKEI timing system, the subsystems and Central Power Conversion station so that ZENKEI may have smooth and supervisory computerized control. CAMAC system also connects the computer system to the Protective Interlock Panels, and Auxiliary Graphic Panels so that it may coordinates with the hardwire circuits. Fig.3 shows the construction of ZENKEI CAMAC system.

6. ZENKEI TIMING SYSTEM

ZENKEI timing system is devided to two groups. One is in the Control Building and the other is in the Power Supply Building. These control discharging sequences of entire JT-60 control system following to the delay and logics set by the computer systems. Fig.4 shows the construction of ZENKEI timing system.

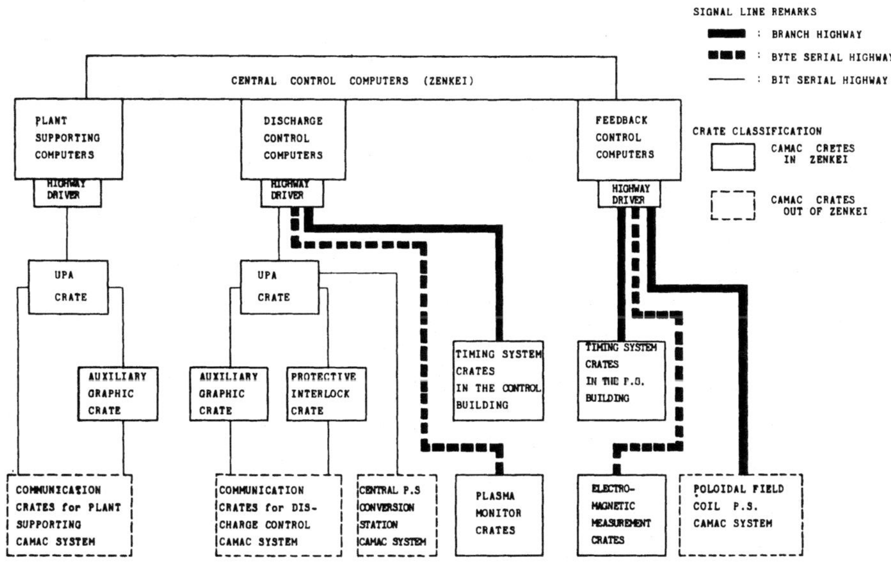

FIG. 3. CONSTRUCTION OF "ZENKEI" CAMAC SYSTEM

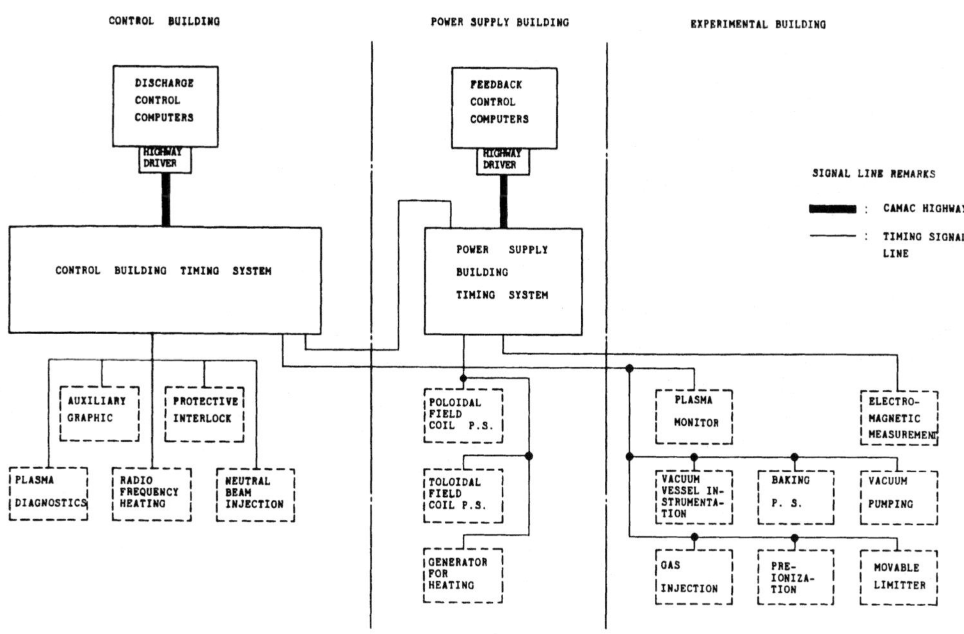

FIG. 4. CONSTRUCTION OF "ZENKEI" TIMING SYSTEM

7. PLASMA MONITOR DEVICES

Among the plasma monitor devices, we have following kinds of devices in ZENKEI control system

- o Electro-magnetic signal conditioning and transform circuits, CAMAC modules for the electron temperature monitor, electron density monitor and pre-ionization density monitor and the devices for Radiation Monitors.

Here the radiation monitors are the monitors of hard X-Ray and neutron in the vacuum vessel room of JT-60 Experimental Building so that Protective Interlock may activate to the runaway electron mode and may suppress the mode in short time as possible as it may.

8. ZENKEI PROTECTIVE INTERLOCK SYSTEM

The system is formed of Protective Interlock Panels and detective devices. It will correlate to the interlock of the subsystems in order to protect human and devices beyond the subsystem, when accidents occur. This system is all hardwired.

9. ACKNOWLEDGEMENT

Although we discussed here principal property of ZENKEI constituents, more detailed designings are now under the process. Specially more detaled design of the software on the ZENKEI computer system and the handlings of the JT-60 control are our recent main obligation as well as hardware designings.

Here we thank to the many others of JT-60 members and wish good luck!

DATA ACQUISITION AND PROCESSING SYSTEM FOR THE HCN INTERFEROMETER DIAGNOSTIC AT THE TEXTOR EXPERIMENT

M. Korten, H. Soltwisch
Institut für Plasmaphysik der Kernforschungsanlage Jülich GmbH
Association EURATOM-KFA, Postfach 1913, 5170 Jülich, FRG

ABSTRACT

A computerized data acquisition system for collection, storage and analysis of the experimental data from the Tokamak experiment TEXTOR is under development and implementation.

A nine-channel, high sensitivity HCN Laser interferometer is under construction and will be used to measure the plasma electron density of TEXTOR with spatial resolution. On line data collection and analysis will be supplied for this Diagnostic by the TEXTOR Diagnostic data acquisition system.

A one channel pilot system has been set up and tested successfully by measuring the electron density of a test microwave discharge plasma.

1. INTRODUCTION

At the TEXTOR experiment, a data acquisition system is set up to support the handling of the Diagnostic instrumentation and the processing of the experimental data. It is to perform the acquisition, storage retrieval and display of the data from the diagnostic systems connected to it; i.e.

- the TEXTOR control an monitoring diagnostics
- the diagnostics of the plasma core
- the diagnostics of the plasma wall interaction.

High modularity, hardware and software standardisation is aimed at for the system. Data will be filed centralized and be accessable by all users, whereas mode of operations for the various diagnostics will be decentralized to enable autonomous control.

For several diagnostic systems envisaged, test systems have already been set up providing an interface to the TEXTOR Diagnostic Data Acquistition System during the build up phase of the experiment. We will describe here the planned data acquisition and processing system for the HCN-Interferometer diagnostic of TEXTOR, which is under construction as a multichannel system for time and space resolved measurement of the plasma electron density.

2. THE TEXTOR DIAGNOSTIC DATA ACQUISITION SYSTEM

A dual Computer configuration is planned to provide dedicated hardware for

- the acquisition, storage and communication (ASC)
- the analysis and display of the experimental data (AD)

The ASC Computer will have to perform data acquisition and control of the diagnostic local units by interfacing to two serial highways, central filing of the collected data and communication to the other connected computers, e.g. the control computer, the analysis and display computer and the central timesharing computer IBM 370/168 of the KFA Jülich.

The AD computer will interface to the experimentalists by interactive display terminals and graphic monitors. After the TEXTOR pulse the acquired data will be written here and open to further processing by user application software.

A first stage of the system has been set up and will be described here as shown in fig. 1.

The ASC - computer comprises a PDP 11/45 CPU, floating point processor, cache memory 256 kbyte main store and 2 176 Mbyte disk drives.

The AD computer will be a powerful 32 bit computer with 4 Mbytes main store to serve as the interface to the users by TEXTRONIX 4010/4025 graphic terminals located near the site of the diagnostic systems. As the AD computer has not yet been installed, the terminals are connected to the ASC Computer for the present development phase of the system.

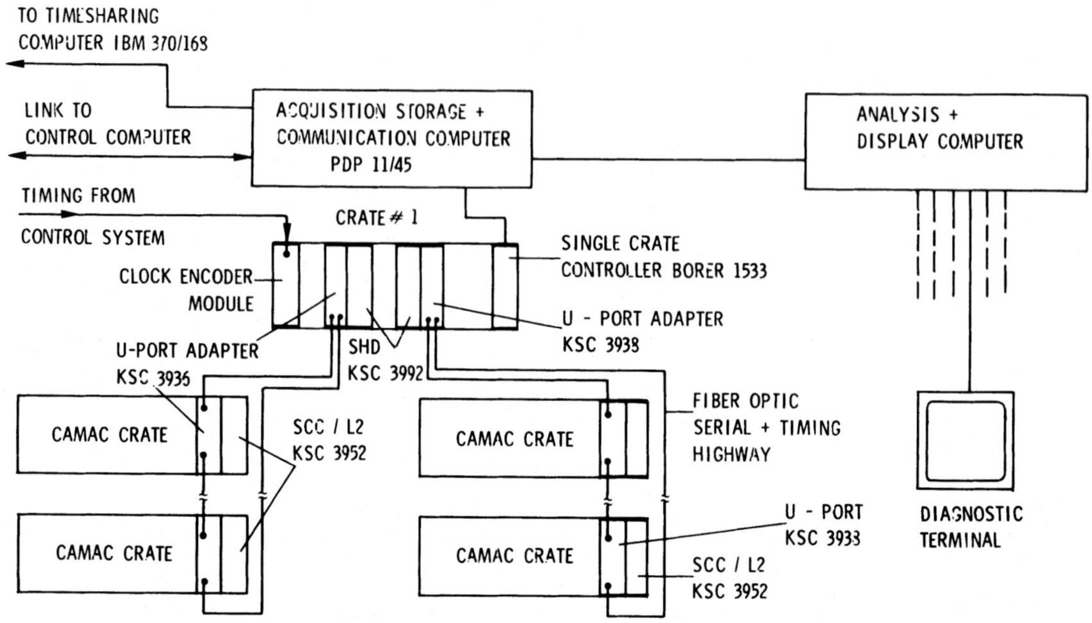

Fig. 1 TEXTOR diagnostic data acquisition system configuration

Interfaced to the ASC Computer, two serial highways are being established to service the various diagnostic systems:

- A 100 KHz serial highway connecting remote diagnostics and test systems by telephone lines,
- a 5 MHz fiber optic highway to be extended for final TEXTOR operation.

The serial highways are driven by two Kinetics Systems Type 3992 serial highway drivers contained in a crate which is interfaced to the PDP11/45 by a Borer Type 1533 crate controller. The CAMAC Crates are accessed through U-Port adapters (Kinetic Systems Type 3936) and controlled by L2 type serial crate controllers (Kinetic systems type 3952).

Software for control, data acquisition and display has been adopted from the Max-Planck-Institut für Plasmaphysik Garching [1, 2] and extended for our CAMAC interface and modules applied to the diagnostic systems. The system allows for autonoms operation of each diagnostic as well as control by the central timing system of TEXTOR.

The TEXTOR timing system will use modules adopted from the TFTR computerized instrumentation control and data acquisition system at Princeton. The timing is accomplished by distributing a coded clock signal through a fibre optic highway to the various remote receivers of the system which utilize this clock to achieve real time synchronism and drive a common reference clock signal.

3. THE HCN INTERFEROMETER DIAGNOSTIC METHOD

The interferometer system for time and space resolved measurement of plasma electron density consists of nine vertical probing beams traversing a poloidal plasma cross section and a reference beam travelling outside the plasma. Due to changes of plasma refractivity caused by variations of electron density each of the probing waves experience a phase shift relative to the reference wave:

$$\Delta\varphi = -2.82 \cdot 10^{-17} \lambda \int N_e \, ds$$

where λ denotes the probing wavelength in μm, Ne is the electron density in cm^{-3}, and the line integral refers to the path length inside the plasma taken in cm. By simultaneously measuring these phase shifts - and hence the electron line densities - along different chords and processing these data by appropriate inversion procedures (e.g. Able inversion in case of axisymmetric plasmas) the electron density profile is obtained.

The optical arrangement conforms largely to the multichannel interferometer developed by Dr. Vĕron for the TFR tokamak [3]. It is supplied by a 150 mW HCN laser [4] operating at a wavelength of 337 μm. Phase modulation is introduced by slightly shifting the frequency of the reference beam by diffracting it off a rotating cylindrical grating [3]. In that way low-frequency beat signals are produced when recombining the shifted reference wave with the unshifted probing waves on square-law detectors. Consequently, the plasma induced phase shifts $\Delta\varphi$ are transformed to convenient low-frequency signals (typ. 10 kHz) which may simply be compared with a fixed phase reference signal (the latter being provided by a reference interferometer without any plasma in its probing arm but using the same rotating grating). In practice $\Delta\varphi$ is deduced from the time lag Δt between corresponding zero crossings of each probe signal and the reference signal using the relation

$$\Delta\varphi = 2\pi \frac{\Delta t}{\mathcal{T}} = \omega_o \Delta t$$

wherein $\mathcal{T}$ is the period and ω_o the beat signal frequency.

Phase shift measurement and data acquisition will be done in two ways:

For system control and monitoring purposes each of the probing channels is equipped with a simple phase comparator circuit [5]. This circuit essentially consists of a flip-flop which is set by a zero-crossing transistion of the reference signal and reset by the next probe signal transistion. So, the flip-flop yields a pulse train with period $\mathcal{T}$ and pulse-width Δt whose time average is proportional to $\Delta t/\mathcal{T} = \Delta\varphi/2\pi$. By putting this pulse train through a low pass filter onto a storage oscilloscope the phase shift - or the electron line density - will be displayed as a function of time. However, the circuit is only capable of detecting fractional phase shifts since it automatically resets to zero when $\Delta\varphi$ passes through multiples of 2π. The resolution is determined by the filter time constant $\mathcal{T}_F$ which has to be chosen larger than the signal period $\mathcal{T}$ ($\mathcal{T}_F = 10\mathcal{T}$; typ. 1 ms).

More accurate measurements of total phase shifts even beyond multiples of 2π with better time resolution will be done by means of the computerized acquisition system. Its lay-out is schematically shown in fig. 2. Both the reference and the probe signals pass through signal converters which form a series of voltage spikes occuring at the zero-crossing transitions. These spikes serve to read an electronic clock formed by a 10 MHz oscillator and a 16 bit binary counter. Thus, for each zero crossing the time elapsed from a common starting point (i.e. deblocking of the oscillator pulses) is measured with a resolution of 0.1 μsec which is small compared to the signal period ($\mathcal{T}$ = 100 μs).

By comparing the times elapsed up to corresponding zero crossing events of the plasma penetrating and reference signals respectively, the correct time lags can be calculated even for phase shifts larger than 2π.

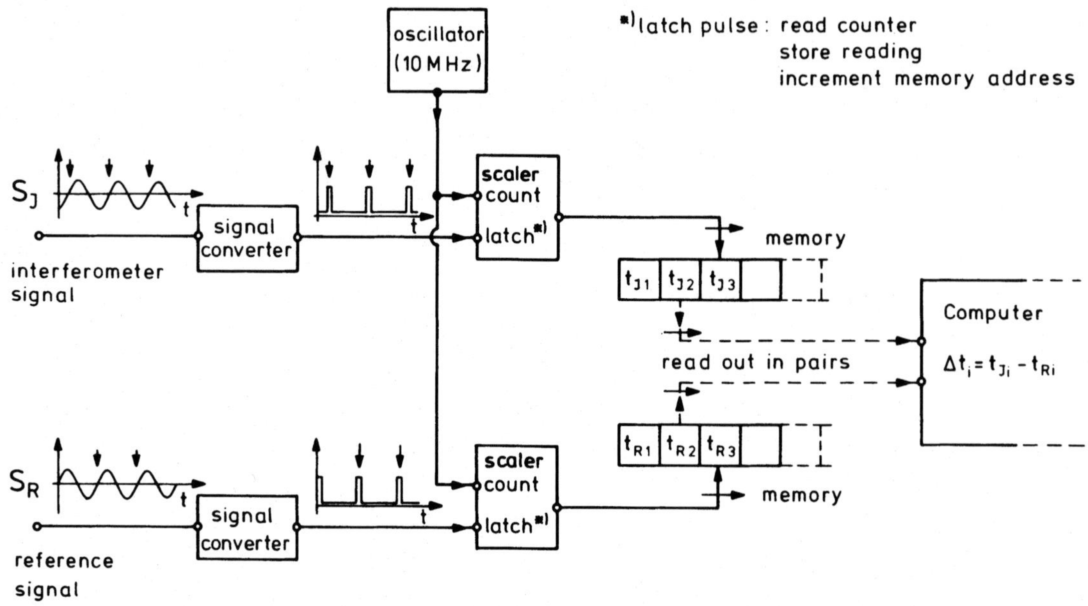

Fig. 2   HCN-Interferometer data acquisition method

4. THE TEXTOR HCN-INTERFEROMETER DATA ACQUISITION SUBSYSTEM

The configuration of the complete 9-channel subsystem is shown in fig. 3. Except of the signal conditioning electronics it is set up by commercially available CAMAC modules.

The system is controlled by the TM404 Timing and DTG 305 digital timed gate modules procured from the Princeton Plasma Physics Laboratory. Run-down will be started under remote control of the TEXTOR central timing or locally under manual control from the user's computer terminal. The centralized clock encoder module, triggered by software or by an external pulse, will send one encoded bi-phase pulse train to all connected crates.

The TM 404 modul in the Crate monitors the "Facility clock" signal and causes the pulse output channels to go into an active state upon recognition of a software asigned code. Pulses from this system, after preprogrammed delays, are used to control the experimental system (i.e. trigger monitor) and to trigger the digital timed gate, which provides transformer coupled pulses to start and stop the 10 MHz reference clock and an open collector optically isolated output to be used to gate the interferometer beat signal latch pulses. The reference clock signal will be generated by a Jorway 217 Gated Clock module, multiplied by Joerger FT fan outs to drive the 9 (+1 reference) channel scaler inputs. As scalers the Le Croy 100 MHz Latching scalers (CAMAC Model 8590) are used. The zero crossing pulses from the experiment connected to the scaler latch inputs cause data to be strobed from the counter into the storage latches and then transfered into subsequent locations of the dual port CAMAC Memory Type Le Croy 8801 (16 bit, 16 k). Data acquisition will be stopped on address overflow of the memory or by closing the latch pulse gate.

After the discharge pulse of the experiment, the data resident in the CAMAC memories are subject to be read from the computer. Data values in corresponding memory addresses are subtracted to show the net charge of phase shift from the starting point. The resulting values are displayed on the terminal screen as curves being proportional to the line density of the discharge.

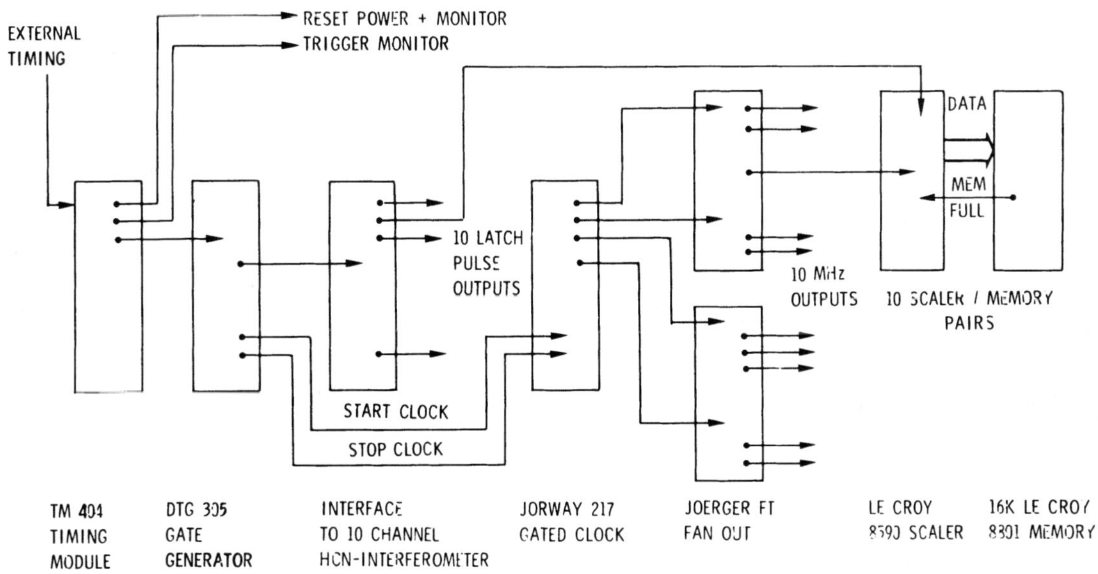

Fig. 3 HCN-Interferometer data acquisition instrumentation

5. EXPERIMENTAL TEST

A single-channel pilot system has been set up and tested by measuring the electron density of a pulsed microwave discharge plasma. Both the optical arrangement and the data acquisition closely correspond to the final TEXTOR system as described before. The test plasma has a line density of $\int N_e\, ds \leq 10^{15}$ cm^{-2} and a pulse duration of about 5 msec. It is produced by feeding 25 kW microwave power at 2.45 GHz into a 150 cm long quartz tube filled with hydrogen (for further details see e.g. [6]).

Figs. 4 and 5 show the results of density measurement for two different shots. Subfigs. 4a and 5a represent scope traces yielded by the monitor system; the ripple does not indicate periodic density fluctuations but is due to imperfect behaviour of the phase comparator circuit. Subfigs. 4b and 5b show corresponding hardcopies obtained from the computerized acquisition system. Scope traces and hardcopies are in good qualitative agreement, but the hardcopies exhibit much more details in consequence of better resolution. Small fluctuations imposed on the base line indicate the accuracy of measurement. They are mainly caused by detector noise and insufficient mechanical stability of the interferometric set up.

Fig. 5 gives an example for a high-density plasma producing phase shifts larger than 2π. Whereas the monitor signal resets to zero when $\Delta\varphi$ passes through 2π the computerized system acquires the total phase shift without any detriment of accuracy.

6. REFERENCES

1 Engelhardt, K. Lathe, R., Müller, E., Report IPP Garching R/26 (1977)

2 Lathe, R., Müller, E., Report IPP Garching R/27 (1978)

3 Vêron, D., Certain, J., Crenn, J.P., 7. Opt. Soc. Am 67, 964 (1977)

4 Belland, P., Vêron, D., Withbourn, L.B.,
Appl. Optics 15, 3047 (1976)
5 Wolfe, S.M., Button, K.J., Waldman, J., Cohn, D.R.,
Appl. Optics 15, 2645 (1976)
6 Beerwald, H., Böhm, G., Krämer, M.,
Z. Physik 268, 165 (1974)

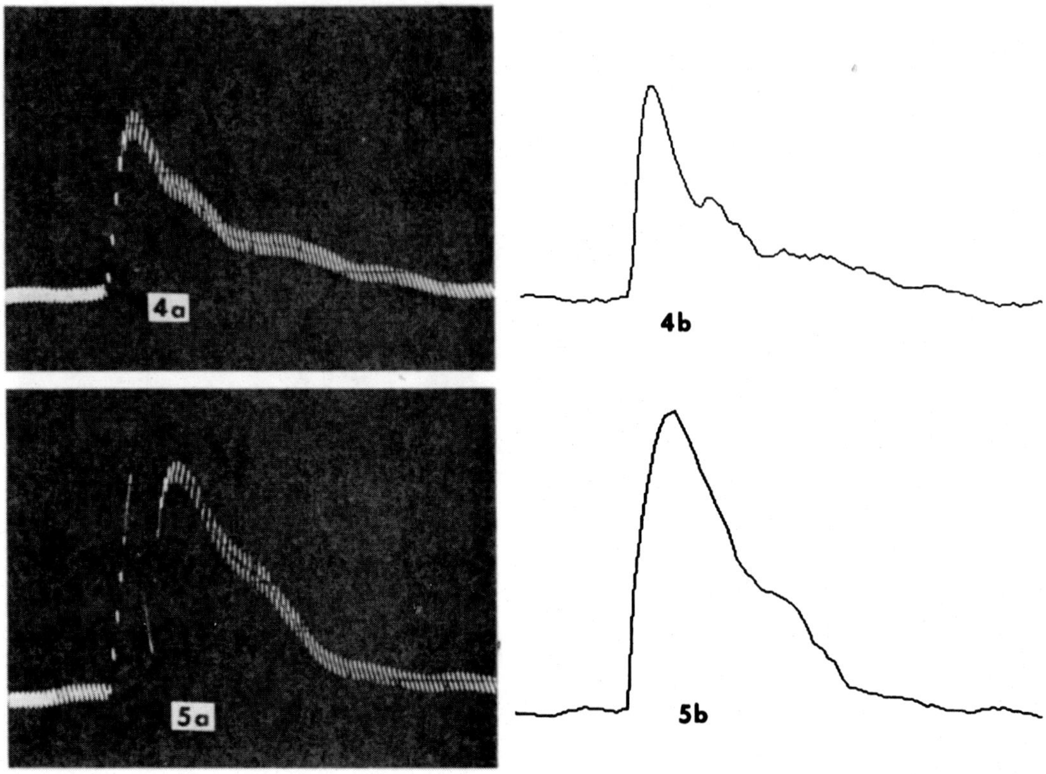

Phase shift V.S. time, 1 ms/DIV.

Fig. 4: Lower electron density plasma

Fig. 5: Higher electron density plasma

a: Oscilloscope Traces

b: Terminal Hardcopies

MICRO-COMPUTER CONTROL OF VACUUM AND GAS FILLING SYSTEMS

L Firth
UKAEA, Culham Laboratory, Abingdon, Oxfordshire OX14 3DB, UK
(*Euratom/UKAEA Fusion Association*)

ABSTRACT

A complete system based on an inexpensive micro-computer has been built and is currently in use to control all functions of the two main vacuum systems for the RFP experiment, HBTXIA, together with two diagnostic pumping systems and the gas filling unit. At the control desk a VDU shows the status of the four systems, chosen gas filling mode and previous filling pressure. Push buttons allow the operator an over-ride facility but apart from the selection of the gas filling mode and pressure, operator intervention is not required. Vacuum system faults are diagnosed and displayed and the appropriate actions taken. Software is written in BASIC and can be modified via the computer keyboard. Programs can contain up to 16K bytes with storage facility on magnetic tape or disc. Output and input controls communicate via pneumatic lines with the vacuum systems to permit high voltage isolation. The computer's 5V D.C. output signals are converted at the electrical/pneumatic interface unit, first to 240V A.C. and then to air signals by solenoid valves. Pneumatic micro-switches are used to change air signals to electrical ones. Simple procedures such as leak rate calculations and leak hunting are easily included. By the addition of an analogue output/input interface an auger surface analyser or mass spectrometer may be completely computerised.

1. INTRODUCTION

HBTXIA is a toroidal RFP experiment under construction at Culham.[1] The principles of the vacuum pumping systems were discussed previously.[2] This paper describes the construction and implementation of the micro-computer based control for the vacuum systems.

The commercial availability of inexpensive micro-computers makes it possible to improve control and data acquisition functions. For a vacuum system in a complex experiment there are several advantages:

(a) the operator's tasks are simplified;

(b) routine procedures such as shut down are predetermined, minimising the risk of error;

(c) gas filling and pumping procedures are integrated.

The time (and hence cost) of programming in a high level language such as BASIC can be reduced and the program readily modified to suit changing requirements. This has advantages over the dedicated micro-processor with its complex machine language and its initial requirement of an expensive supporting development system.

2. THE VACUUM SYSTEMS

The toroidal plasma vessel (liner) is pumped by two turbomolecular based systems.[2] Two smaller systems are used to pump diagnostic chambers which communicate directly with the liner. Each system can be isolated from the liner by a gate valve. A gas filling system with four filling modes is situated equidistant from the main pumps. Since each system could be at the liner potential associated with its position, electrical isolation to 15kV is required.

3. CONTROL AND INDICATION

3.1 Control

Five basic services provided are as follows:

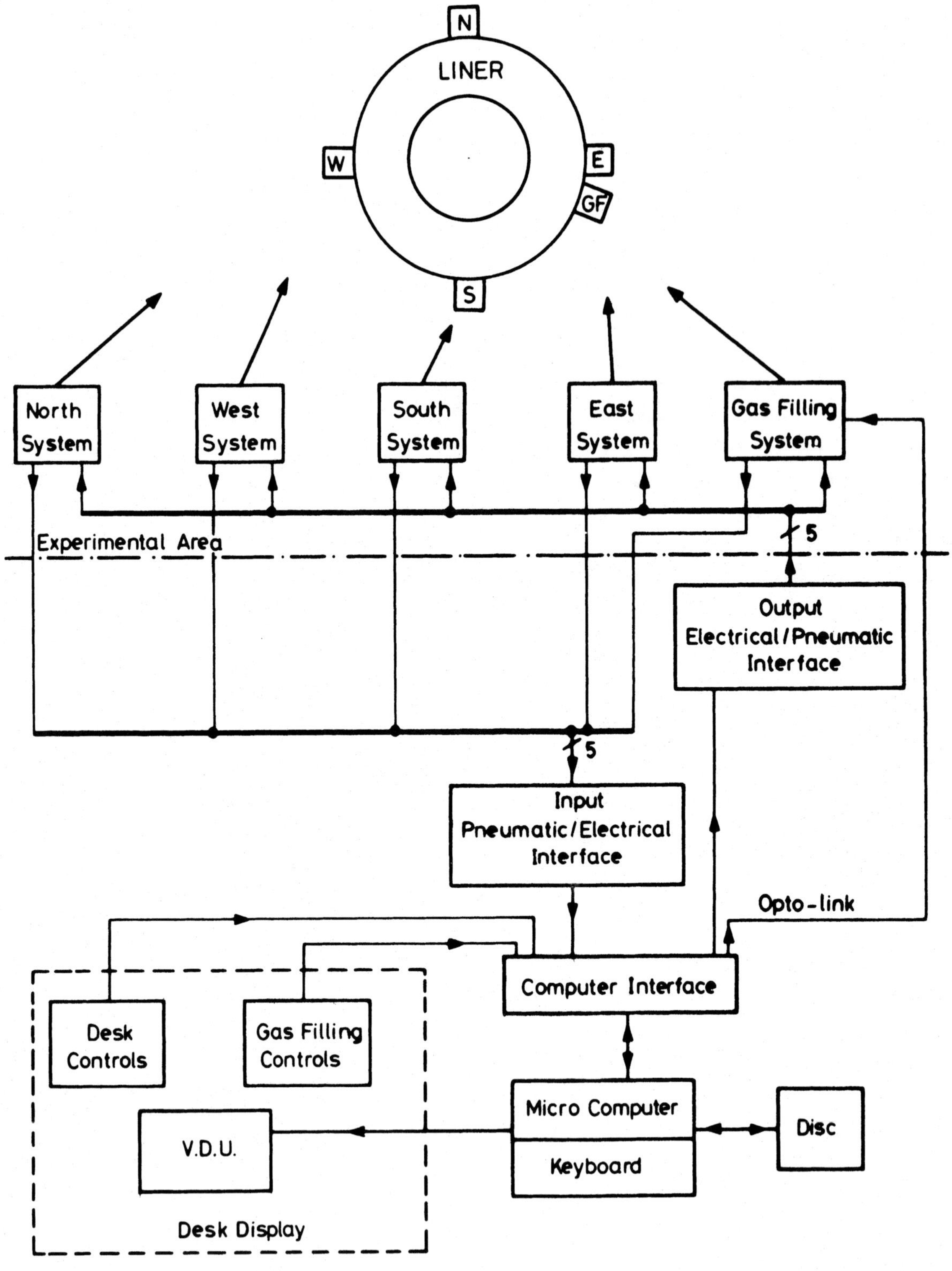

Fig 1   Schematic diagram of the complete system.

3.1.1 To evacuate the liner from atmosphere, checking at each stage for correct operation and bring all systems on line.

3.1.2 To start up an individual vacuum system finally opening the main isolating valve to the liner.

3.1.3 To continuously monitor the performance of all operational systems.

3.1.4 To shut down a faulty system in the correct order.

3.1.5 To interpret the predetermined gas filling requirement set by the operator and set the gas filling system. Prepare the vacuum system for a filling sequence.

3.2 Manual control

Local manual control is provided at the pumping modules for testing and maintenance.

3.3 Indication

A VDU is used to indicate the status of the systems and main isolating valves, base pressure and previous filling pressure. During start up and shut down the current state is indicated; for example, main valve closed, TMP on.

4. OVERALL SYSTEM

The main components of the system are identified in the schematic Fig 1. A brief description of the system components follows.

4.1 Micro-computer, external program store and VDU

A Tandy radio shack* TR80 level II micro-computer with 16K RAM is used. This proved satisfactory for the system described but since other program development was taking place concurrently with vacuum control the cassette recorder was replaced with a minidisk system and its associated expansion interface providing 48K bytes of random access memory and storage capacity of 59K bytes per diskette.

4.2 Computer interface

A sub-rack houses the buffered RS232 interface with communicates with the TRS80 via its expansion port edge connector. On board power supplies are used to drive up to 15 input/output modules.

4.3 Output unit

This unit decodes the instructions from the microcomputer and provides up to 16 +5V D.C. output signals for control or indication. The address of each unit can be pre-selected by a bank of 7 DIL switches on the p.c.b.

4.4 Input unit

This unit accepts 15 off 0 - 5V + D.C. signals from the vacuum system and routes them via the bus to the micro-computer. These inputs are used to check for correct operation.

4.5 Analogue input/output unit

This module is used to output a D.C. voltage between -10V and +10V to control the gas filling PSUs. The input half accepts a ± 4V D.C. signal, converts them into binary and inputs them to the bus. It is used to input the vacuum base pressure of the experiment from an ion gauge via a Culham 95-8337/8 opto-link. The analogue input/output unit will later be used to output a program to a mass spectrometer or auger surface analyser and input the ion current etc back to the computer.

* registered trade mark

```
10200 '
10201 ' * * * * * * * * * * SHUTDOWN (NORTH) * * * * * * * * * *
10202 '
10300 GOSUB 11900
11810 PRINT@271,"SHUTDOWN";:OUT132,1
11811 A=A AND NOT NM:GOSUB999
11812 FOR T=1 TO 600:NEXT:GOSUB888
11813 IF (IN AND NC)=NC THENPRINT@400,"CLOSED"; ELSE GOTO11812
11814 IF (IN AND NL)=NL THEN 11817
11816 A=A AND NOT NT AND NOT G2 AND NOT G3:GOSUB999
11817 A=A AND NOT NT AND NOT G2:GOSUB999
11818 PRINT@271,"TMP OFF";
11820 FOR T=1 TO 1000:NEXT
11822 A=A AND NOT NR:GOSUB999
11824 PRINT@271,"B.P.OFF";
11826 FOR T=1 TO 1000:NEXT
11828 A=A AND NOT NV:GOSUB999
11830 PRINT@271,"VENT OPEN";
11832 FOR T=1 TO 2000:NEXT
11834 A=A AND NOT NB AND NOT ND:GOSUB999
11836 PRINT@271,"SHUTDOWN";:SN=0
11838 RETURN
11840 '
```

Fig 2 Sample sub-routine.

4.6 Output electrical/pneumatic interface

5V D.C. signals from the output unit are converted via opto-coupled solid state relays to 240V A.C. which is used to operate air solenoids. Air signals for 15kV isolation are fed into the high voltage area to service the pumping modules.

4.7 Pumping modules

Some of the incoming air signals are converted back to 240V A.C. by pneumatically operated micro-switches and used to control the pump and ion gauge supplies. Others are used directly to open and close valves. Manual control is coupled by "OR" gates with the incoming signals to facilitate local control. Signals returning to the computer are derived from pneumatic switches on the valves, water flow relays etc. Sixteen channels of air signals return to the input pneumatic/electrical interface for processing. Isolated 3 phase A.C. supplies are used to power the pumping modules.

4.8 Input pneumatic/electrical interface

A bank of 48 pneumatically operated micro-switches and a 5V PSU convert the air signals to electrical ones to feed the input units at the computer interface.

4.9 Desk controls

A total of 19 momentary push button switches connected to an input unit allows vacuum personnel some desk control of the four systems.

4.10 Gas filling controls

Four self latching push button switches allow selection of the filling mode and two thumb wheel switches are used for selection of filling pressures. A manual fill button is also provided. These switches are connected to an input unit, decoded by the computer and used to control the gas filling pressure.

4.11 Gas filling system

Two types of valves are used:

(a) a magnetically driven pulsed gas valve of in house manufacture;

(b) a Veeco* Peizzo electric valve.

The pulsed gas valve is driven by an SCR fired capacitor charged by a computer controlled PSU. The Veeco valve can be used in a similar way (pulsed input) or computer controlled by a D.C. supply for continuous flow.

4.12 Operation of the system

The operator selects the filling mode and pressure at the desk. The micro-computer decodes the requirement and directs the analogue output (see Section 4.5) to the appropriate PSU via an opto-link. The computer then waits for an external signal to arrive from the experiment's main timing circuits and prepares all the pumping systems for filling. Gas valves are triggered from the main timer. After a preset delay the computer initiates pump out of the gas and returns to normal operation.

5. SOFTWARE

BASIC is used to program the TRS80. Flow diagrams were prepared for each part of the program. Typical sub-routines are: start up, normal operation and shut down. A sample subroutine in shown in Fig 2. A set of sub-routines was written for one vacuum system control and duplicated with minor modifications for the other three systems. The main program flow diagram is shown in Fig 3. The cycling time with all systems operational is 4.5s. No attempt has been made to speed up execution, for example deleting all REM statements. POKE statements are, however, used for video display formatting. To reduce the response time to the control

* registered trade mark

panel software interrupts, the desk controls are interrogated four times per program cycle.

6. ENHANCEMENTS

Since the complexity of hardware design is effectively transferred into the software program, it is now very easy to incorporate a changed or new experimental requirement. Enhancements in the present program include fault diagnosis and a leak rate test sub-routine with the appropriate calculations. The system will also interface with the engineering data system[3] for the logging of vacuum parameters.

7. CONCLUSION

An inexpensive micro-computer has been used to provide multi-system control for the vacuum and gas filling systems of the HBTXIA experiment, requiring minimal supervision by the machine operator and affording greater flexibility than the conventional hard-wired system.

8. ACKNOWLEDGEMENTS

The author gratefully acknowledges the work of J A Fessey on interface designs and J Wyatt for prototype manufacture.

9. REFERENCES

[1] P D Wilcock: Proc 10th SOFT Conference.

[2] L Firth, F C Jones and A A Newton: Proc 10th SOFT Conference.

[3] M E U Smith and E K Sinton: this Conference.

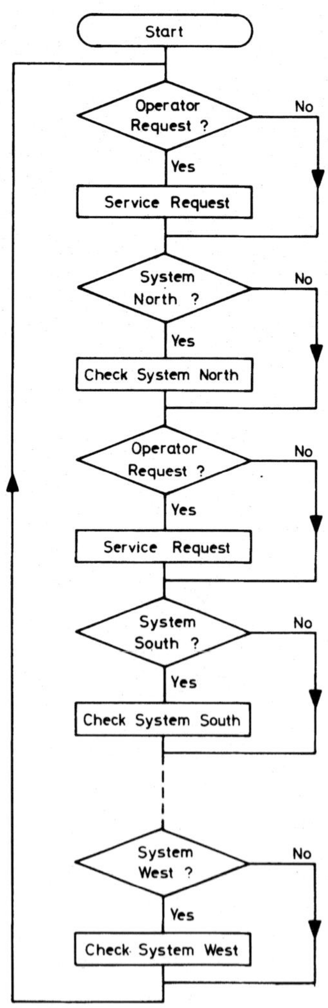

Fig 3 Main program flow diagram.

REMOTE CONTROL AND DISPLAY FOR TEXTOR
WITH A CAMAC-COMPUTER-SYSTEM

W. Tenten, H. Huppertz, W. Janßen, J. Lerch,
K.D. Müller, K. Pütz*, F. Rongen, M. Sauer
Zentrallabor für Elektronik
KFA Jülich GmbH
Postfach 1913, 5170 Jülich, W-Germany

* Ingenieurbüro K. Pütz
Wiedfeldtstr. 60B, 4300 Essen-Bredeney, W-Germany

ABSTRACT

The revised control and data acquisition scheme for the TEXTOR tokamak, which is in the final stage of construction, contains a CAMAC computer system for remote control and display. The major components of this system, including fibre optic data transmission and the central control console incorporating advanced control concepts, are described.

1. INTRODUCTION

TEXTOR is a tokamak experiment for technology oriented research in the field of plasma wall interaction which is in the final stage of construction /1/. The objectives of TEXTOR demand a powerful control and data acquisition concept. It has proved to be a reasonable approach to make a clear distinction between the control area and the acquisition of diagnostic data. Commissioning of the TEXTOR apparatus is envisaged for mid 1981. Long before this date an operable system for remote control and display is needed. This system is computerized using CAMAC standards for solving all occuring interface problems including the central control console for the TEXTOR operation.

2. TEXTOR CONTROL AND DATA ACQUISTION CONCEPT

In comparison with the original design of a computerized control system for TEXTOR /2/ the concept has been revised. Due to organisational difficulties and manpower limitations a redesign took place which strongly emphasized the aspects of modularity with respect both to functionality and the time scale of implementation. The actual control and data acquisition system for TEXTOR (fig. 1) consists of two virtually independent computer installations which serve different tasks. One computer system /3/ is intended to fulfil all requirements concerning data acquisition and display of diagnostic data. The data acquisition is assisted by locally distributed diagnostic subsystems which are connected to the central data acquisition computer. The other system is responsible for remote control and display in order to facilitate the operation of the TEXTOR machine. The locally distributed TEXTOR

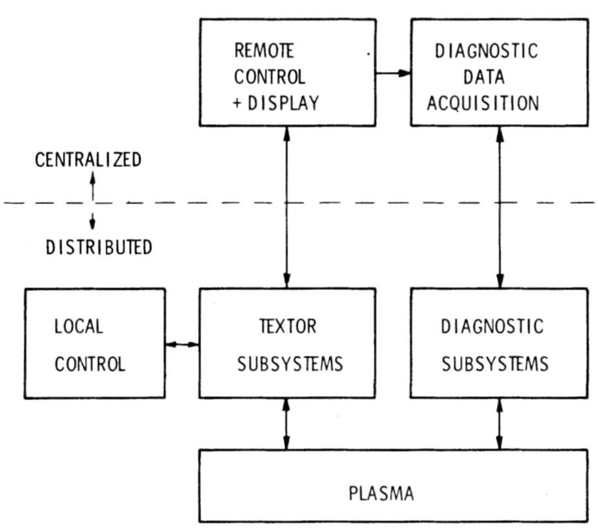

Fig. 1. Control and data acquisition concept for TEXTOR experiment

subsystems have an additional local control panel to alleviate commissioning and testing of single devices and components without the assistance of the central remote control and display system. This means in consequence that all interlocking is done in the subsystems themselves. Looking at the aspect of the time schedule for the implementation of TEXTOR this structure moves the central control system out of the critical path.

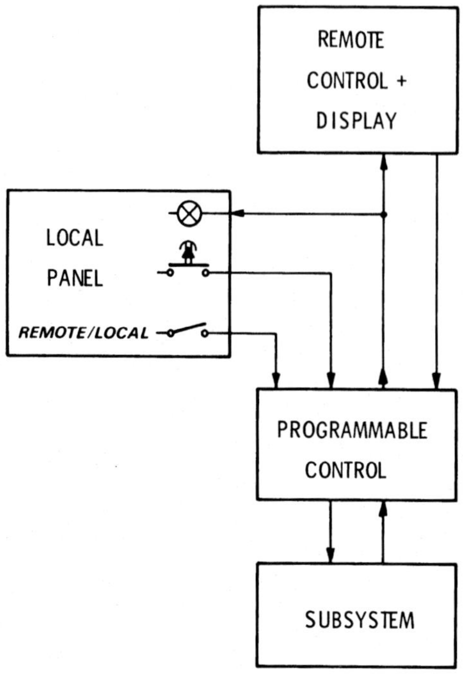

Fig. 2. Local and remote control

The basic hardware design and many of the software features were adopted from the control system for the VICKSI-accelerator project /4/ at HMI Berlin and extended to fulfil the general and specific requirements of TEXTOR. The local control feature (fig. 2) gave reason to the principal approach between the individual subsystems and the central control system. It was considered to be a good practice to duplicate the capabilities of local display and control at the central console. In order to avoid dubious situations when using both modes of control simultaneously an appropriate interlock switch is used regulating the command access to a subsystem. The position of this switch is fed to the programmable control which is used at every subsystem for local interlocking purposes.

3. THE CONTROL SYSTEM HARDWARE

The control system for TEXTOR is composed from hardware and software building blocks. The major hardware components comprise the central control console for man-machine interaction, the computer system including appropriate peripherals and finally the process control interface.

Great emphasis was put to the fact that all equipment is commcercially available. Employing an international standard like CAMAC and a member of a widely used computer family this has been achieved with a few exceptions.

3.1 The computer system

Unlike in its outer appearance in its functional design the control computer plays a central role because it is an important factor for both the versatility and the performance of the remote control and display concept. Whereas the present version is running on a PDP-11/40 computer the final installation will have a PDP-11/44 as central computer. This latest member of the widely known PDP-11 family has some outstanding characteristics which led to its selection for TEXTOR. Because it is the latest addition to its family, it has good price/performance ratio. A feature which makes it a favourite for a complex system with lots of software to be run concurrently is the expandability of its main memory which can be extended to 1 MBytes at present.

The performance of the control computer is complemented by a set of standard peripherals. Two different types of disc storage are used. A RP06 compatible disc controller and drive is used to yield a 200 MByte capacity per drive as a system device. For individual use and portability two RL02 disc drives with a capacity of 10 MBytes each have been added. These mass storage devices are supplemented by a console printer, a line printer and several alphanumeric and graphic display terminals. In addition to the line time clock of the central computer an eternal calender plus clock is used which retains its information even in the case of a power failure. During the boot and startup procedure of the computer system the information of the eternal clock is accepted and inserted into the operating system. Furthermore provision has been made to synchronize the mentioned methods of time counting with the radio controlled central clock system of the KFA research center.

3.2 The control console

The most impressing component of the TEXTOR remote control and display system is the control console which is located in the central control room adjacent to the diagnostic room. This console consists of ten bays (fig. 3) in a curved arrangement. The different bays contain various equipment in 19-inch-format. The vertical cross section of the console can be seen in fig. 3. Only the center six bays have a connection to the control system computer. The two leftmost bays contain CCTV equipment for room monitoring and visual inspection of the TEXTOR device by means of remote controlled TV-cameras connected to black/white monitors. Sections of these two bays are also reserved for use by the safety system which is self-contained and has no direct connection to the remote control and display system. A telephone set and a public address facility is also built into this section.

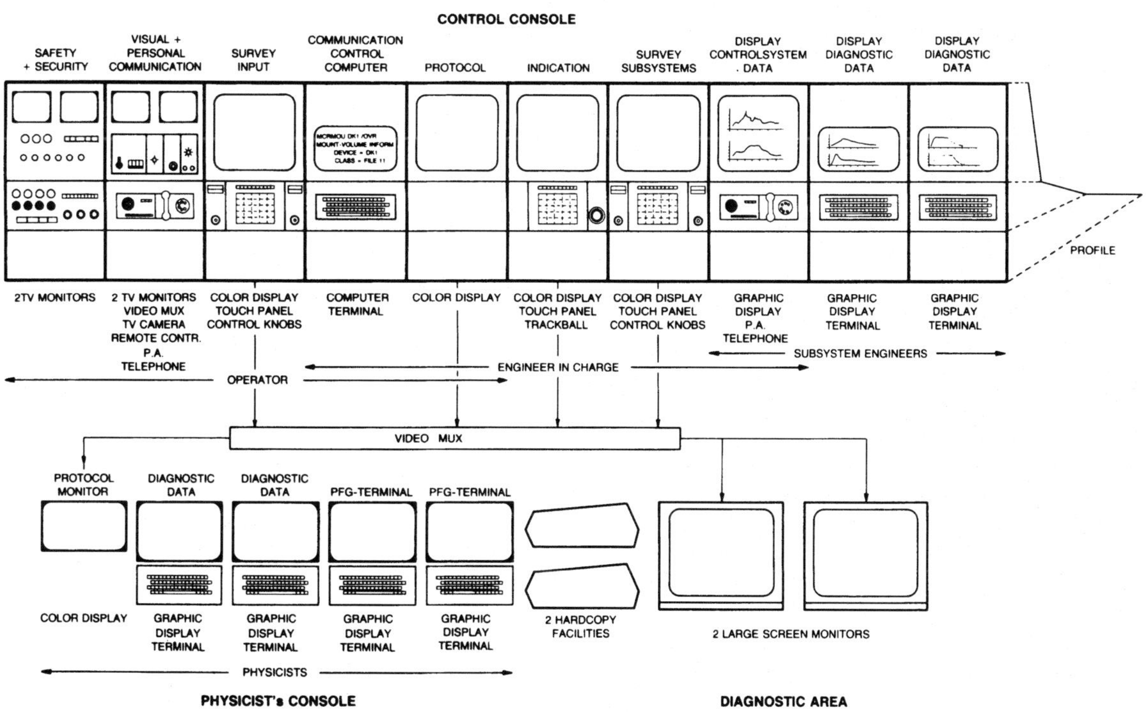

Fig. 3. Central control room instrumentation

The rightmost two sections contain two graphic display terminals which are connected to the diagnostic data acqusition system. Diagnostic data relevant to the operation of the machine can be observed here during shot intervals. Four of the inner six bays contain a "standard man-machine interface" (fig. 4) or parts thereof.

3.2.1 The standard man-machine interface

A one person working station comprises at most 3 different devices. A large color television screen is used for display of machine parameters in various formats. This information is continuously updated. Important forms of representation are binary display of status information, numerical display of analog values and analog display of analog values.

These options can be mixed on the screen. An additional way of representing the state of TEXTOR or one of its subsystems is a symbolic display e.g. in form of a relay network where switches and other elements change color according to their actual state. A high degree of familiarity is achieved by choosing forms of symbolic display which resemble strongly the layout of the local panels at the different subsystems. Space on the screen has been reserved for other purposes like display of date and time and the logging of error messages.

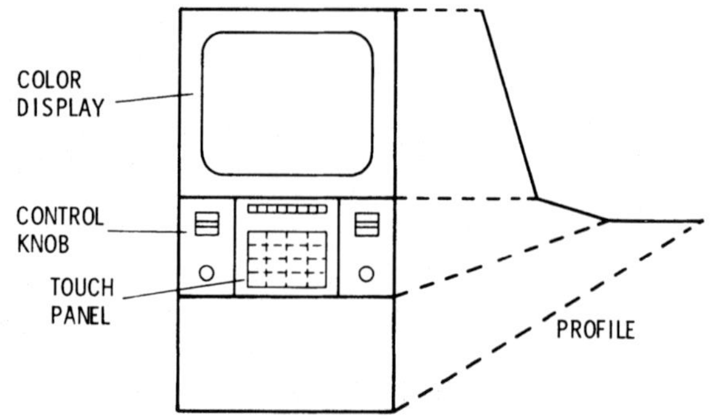

Fig. 4. The standard man-machine interface

The main medium for operator action within the standard working station is the touch panel display. This is a combination of a pressure sensitive 4 x 4 matrix which acts as push-buttons located in front of a black/white television screen used to label the push-buttons. The labelling and with it the function can be exchanged by the computer depending on the intended use of the touch panel. Thus it is possible to perform program requests, program option selection, device to display assignment, display selection or just simple turning over between pages. Another application of the touch panel is the option to use it as a numeric key pad e.g. to control a process variable. This method is complemented by two analog control knobs on either side of the touch panel which allow to control process variables in a more conventional fashion like with a potentiometer.

The described standard man-machine interface has been built into the central control console four times. Two stations have a complete outfit, one is lacking the control knobs and the fourth consists only out of the color television screen.

3.2.2 Additional console equipment

One of the four color television terminals is equipped with a trackball device. It is primarily intended to serve as designing aid in composing the layout of symbolic display pictures.

A large storage tube display with high resolution is used for displaying data of the control system in form of curves. This can be done either in point plot or vector mode. For labelling of the curves and the related axes a character generator is available.

Finally an alphanumeric display terminal of the control system computer is integrated into the control console. This device can be used for various forms of communication with the operating system, with the application software package and with an interpreter which has access to all parameters of the control system.

With the exception of this terminal all devices in the console are connected to the central computer via the CAMAC branch highway according to EUR 4600. The branch was preferred because several crates have to be controlled in distances of several ten meters from the computer. The interrupts which origin from some of the console devices are handled efficiently by the branch concept and by the CAMAC branch driver which connects directly to the PDP-11 Unibus. The block transfer capabilities implemented in the Kinetic Systems 2053 branch driver are used to enhance the reaction upon an operator action.

3.3 <u>The process control interface</u>

Two conditions dictated the choice of the interface system for the process control equipment. At first the distributed location of the TEXTOR subsystems requested an interconnection which could handle distances of up to several hundred meters at reasonable speed. The concentration of interconnection points at the different subsystems led to a configuration with distributed clusters.

The second constraint is related to the hazard of high voltages which may be present at specific subsystem in the case of a plasma disruption. This criterion led to the requirement of an appropriate isolation between both the subsystems and the central control room with all its instrumentation.

These combined demands could ideally be fulfilled by the introduction of the CAMAC serial highway, as specified in the EUR 6100 document, together with a fibre optic transmission between crates or clusters thereof. The total loop has a length of 300 m with 14 crates at 12 different locations.

Similar to the branch driver the Kinetic Systems serial highway driver with various block transfer modes is connected directly to the PDP-11 Unibus in order to avoid the bottleneck involved in other known solutions.

3.3.1 The interface modules

A limited amount of standardization was possible with respect to the types of CAMAC modules which perform the link between the CAMAC dataway and the outer world in form of the TEXTOR subsystems. According to the nature of the signals a general distinction can be made according to input and output, analog and digital signals. The indicated direction is related to the control system. The field of digital input and output is covered by two types of CAMAC modules which perform functions of the type 'pulse out', 'word in' and 'word out' in different numerical combinations. Depending on the interpretation the function 'word in' can also act as 'status in'. In this manner approximately 920 bits of status information and 4 digital words are input to the control system. 400 different standard pulses and 16 digital words origin from the control system. The interface logic was selected as high-level-logic (HLL) to yield sufficient noise immunity. The principal isolation being done by the fibre optic serial highway, any additional decoupling is within the responsibility of the subsystem itself.

The signal standard for analog signals was chosen to be 4 - 20 mA. The input to the control system is performed by 16-channel dual-slope ADC modules with differential inputs and a resolution of 10 bits. To minimize delays due to the time consuming integration technique a self-scanning type of ADC was selected which has an internal memory containing the newest updates of all 16 input channels. The ADC in principle being voltage sensitive the necessary conversion from current to voltage is performed by precision resistors mounted external to the CAMAC-modules. This allows the input to be disconnected without disrupting the current loop.

For analog output from the control system 8-channel D/A converter CAMAC modules with a resolution of 10 bits were chosen. Their voltage outputs are converted to the required current signals in an adjacent module which at the same time performs an optical isolation of every current output.

The present signal statistics of TEXTOR show 210 analog signals to be measured by the control system and 30 output channels to be continuously delivered to the subsystems. All these CAMAC modules used as interface to the TEXTOR subsystems utilize the same front panel connector type.

3.3.2 Other CAMAC modules

For display and monitoring purposes every crate, including the parallel ones in the console, is equipped with a system monitor and dataway display module. The standard modes of monitoring dataway signals are greatly enhanced by the addition of a control module which allows selective display during specific CAMAC commands.

The D-port of the standard L-2 serial crate controller, according to EUR 6100, is converted to a fibre-optic U-port by means of a U-port adapter. This adapter has an additional set of optical transmitters and receivers for the transmission of timing system signals.

3.4 The timing system

As there are a lot of time critical sequences during a shot of TEXTOR which must be synchronized a very efficient timing system is added to the control system. In spite of the resemblance to the control system in topological and functional respect the timing system has to operate autonomously due to speed considerations. The different modules are dedicated to the locally distributed TEXTOR subsystems and are interconnected by means of an own transmission medium. As the requirements of speed and isolation are comparable to the serial highway the fibre optic transmission was duplicated.

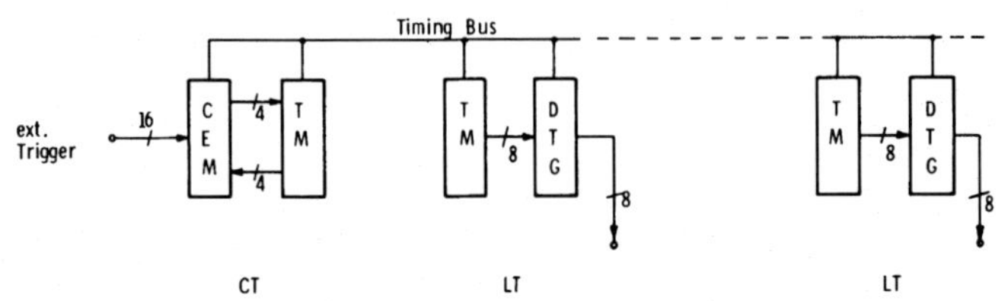

Fig. 5. The TEXTOR timing system

The CAMAC timing modules developed for TFTR /5/ could be used for TEXTOR. The central timer CT (fig. 4) generates a sequence of coded time signals which modulate a continuously running 1 MHz-clock. The modules of the local timer assemblys LT receive the transmitted clock and generate timing signals with individual code reference, delay and duration with a granularity of 1 μsec. All modules plug into present crates of the control system. Thus programming of pulse properties can be achieved from the central console using all resources of the control system.

4. THE CONTROL SYSTEM SOFTWARE

The features and characteristics of the hardware building blocks for man-machine interaction and process control are linked together in the central computer of the control system. The linkage is performed by the computer hardware and mainly by the software that is running on it. In spite of the generally very complex software interaction three basic levels can be clearly distinguished. The operating system is provided by the computer manufacturer. The system software is necessary to fulfil the requirements of remote control and display. Finally any user can add specialized tasks of local importance using the resources provided for this purpose.

4.1 The operating system

The control system was originally developed to run under the operating system RSX-11D. Unfortunately this system is no longer maintained by the manufacturer. On the other hand the PDP-11/44 computer is not sustained by RSX-11D. The designated successor is IAS which was built from RSX-11D. The present version 3.0 of IAS does not support the PDP-11/44. This will be amended in the next version available in early 1981.

The necessary software changes are prepared now. In the meantime the control system can operate with reduced performance using the PDP-11/40 under RSX-11D.

4.2 The system software

An extensive software package was received from HMI Berlin which they use for the control of the VICKSI accelerator. This system produces the linkage between the operator using the devices of the central console to achieve remote control and display of components connected to the process control interfaces located at locally distributed subsystems. This software package turned out to be an extremely valuable tool for the remote control and display of TEXTOR, too. All real time software modules have been written in assembler language. Only minor portions which are used in the startup or rundown procedure could be formulated in FORTRAN. A set of FORTRAN written utility programs for maintaining the various collections of data about CAMAC modules, signals connected to these and about touch panel page descriptions is available.

4.3 User software facilities

All operator actions can be duplicated by software. For this purpose the appropriate interface either for assembler or a FORTRAN environment is provided.

For the ease of programming an interpreter /6/ is available, which combines the known advantages of interpreters with the access to the control system variables.

With these tools a user of the system has the opportunity of writing special tasks that fulfil his specific requirements which are not covered by the standard operation features.

5. CONCLUSION

Using hardware products in form of a member of a widely applied digital computer family with standard peripherals and interface techniques which comply with well known international standards the remote control and display system for the operation of TEXTOR was implemented. In order to minimize the often underestimated software expenses an existing software package was taken over from an acquainted research institution and complemented for TEXTOR purposes. Thus it was possible to have an operating control system before the TEXTOR subsystems started their autonomous operation. This is of special importance regarding the usual manpower limitations.

6. ACKNOWLEDGEMENT

All authors have the urgent need of emphasizing their thanks to the control system group of HMI Berlin who developed and implemented the control system for the VICKSI-accelerator. Without their experience and advice during the design phase and without their software system TEXTOR would probable lack an operable system for remote control and display at present. Special mentioning should be done to the helpful discussions with Dr. W. Busse and to the assistance of Frl. T. Kiehne during the installation of the Berlin software package in Jülich.

7. REFERENCES

/1/ TEXTOR, Association KFA-EURATOM, Technical Data I, KFA publication

/2/ HOPMANN, W., HÜTTEMANN, P.W., KRAFT, J., The design of a computerized control system for TEXTOR, 10th Symposium on Fusion Technology, Padua, Italy, Sept. 4-9, 1978. Proceedings pp. E4

/3/ KORTEN, M., KRAFT, J., LORENZEN, J., Data acquisition for magnetic diagnostic and monitoring at the TEXTOR experiment, 8th Symposium on Engineering Problems of Fusion Research, San Francisco, CA., USA, Nov. 13-16, 1979. Proceedings p.914

/4/ BUSSE, W. et al., The VICKSI control system, 15th European Cyclotron Progress Meeting, Berlin, Germany, April 5-7, 1978

/5/ CICADA engineering specifications, TFTR-10A2-H56 Facility Clock Characteristics
TFTR-10B4-H401 Facility Clock Subsystem

/6/ BUSSE, E. et al., MUMTI, a Multi-User-Multi-Task-Interpreter for process-control applications with CAMAC, HMI-Report B-228, Berlin 1977

Power Supplies

THE SUSTAINING NEUTRAL BEAM POWER SUPPLY SYSTEM
FOR THE MIRROR FUSION TEST FACILITY*

R. D. Eckard, J. H. Wilson, H. W. Van Ness
Lawrence Livermore National Laboratory, University of California
Livermore, California, USA 94550

ABSTRACT

In late August, 1978, a fixed price procurement contract for $25,000,000 was awarded to Aydin Energy Division, Palo Alto, California, for the design, manufacture, installation and acceptance testing of the Lawrence Livermore National Laboratory Mirror Fusion Test Facility (MFTF) Sustaining Neutral Beam Power Supply System (SNBPSS). This system of 24 power supply sets will provide the conditioned power for the 24 neutral beam source modules. Each set will provide the accel potential, the arc power, the filament power, and the suppressor power for its associated neutral beam source module.

The design and development of the SNBPSS has progressed through the final design phase and is now in production. Testing of the major sub-assembly power supply is proceeding at Aydin and the final acceptance testing of the first two power supplies at LLNL is expected to be completed this year.

A special design review was held on the accel dc rectifier transformer. The rectifier transformer was subjected to full short circuit current tests at the McGraw-Edison testing laboratory, brought back to the vendor, opened, and examined for damage. An independent analysis of the short circuit current stress, plus the results of the actual short circuit current tests confirmed the adequacy of the transformer construction.

The arc power supply is one of the more complex subsystems of the sustaining neutral beam power supply set. It must insert enough impedance to stabilize the arc; be able to reduce the 4000 A output current to 600 amperes and return to full output in about 100 microseconds; and be able to reduce the current to zero in case of a metal ion arc in the ion chamber. The power supply has been tested into an MFTF 80 kV prototype arc chamber and the results tabulated.

1. MFTF SUSTAINING NEUTRAL BEAM POWER SUPPLY SYSTEM

1.1 Introduction

The Mirror Fusion Test Facility requires three separate injector systems to establish, heat, and maintain a plasma at fusion temperatures. A target plasma of deuterium

*Performed by LLNL for USDOE under contract W-7405-Eng-48.

ions is established by first streaming electrons and ions into a confining magnetic field followed by a 20 millisec burst of 20 keV deuterium atoms. Higher energy deuterium atoms (average of 50 keV) are then injected to increase the ion density, heat and sustain the plasma for the remainder of the 1/2 sec experiment. The sustaining injector system includes twenty four 80 kV ion accelerators each with their own power supply set.

Two hundred and fifty megawatts of power are required to drive the injector system. Due to the large size of the power supply equipment, the ac power control and the dc rectifier systems are located outside Building 431 in an area the size of two football fields, fig. 1. The pulse power modulators that switch and condition the dc power into the neutral beam injectors are located in Building 431 and occupy about 40 percent of the space on the north and south 2nd and 3rd floors, approximately 13 thousand square feet.

Fig. 1

The major subsystems of the sustaining neutral beam power supply system, SNBPSS, fig. 2, are the Accel dc Power Supply (ADCPS), the Pulse Power Modulator (PPM), the Arc Power Supply, the Suppressor Power Supply and the Control System.

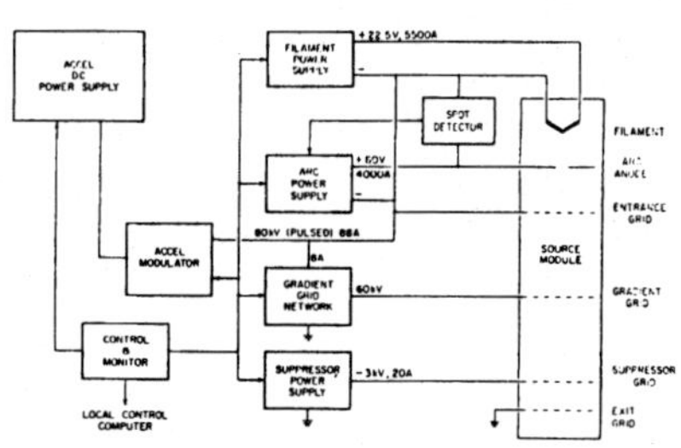

FIG. 2 FUNCTIONAL BLOCK DIAGRAM OF SUSTAINING NEUTRAL BEAM POWER SUPPLY SET

Commensurate with MFTF project goals of maximum industrial involvement a performance specification was written and a contract awarded to Aydin Energy Division to Design, Manufacture, install and test the total power supply system for $25,000,000. The system is scheduled for completion in April, 1981.

2. SNBPSS SYSTEM CONFIGURATION

2.1 Accel dc power supply (ADCPS)

A dedicated 230 kV 250 megawatt pulse power substation provides the 13.8 kV power for the ADCPS. The power supplies convert the 13.8 kV ac power into 85 kV dc and can deliver up 88 amperes to the pulse power modulator. Each power supply set consists of a switch gear assembly, a step voltage regulator, a rectifier transformer, series power factor correction assemblies, and a rectifier. The series power factor correction assembles correct the power factor to within 0.96 at full load to meet the requirement of the local utility power company. Each rectifier transformer has two secondaries, one delta and one wye, to provide for twelve pulse rectification. The primary is wound as an extended delta with 12 of the power supplies phase shifted plus 7-1/2 degrees and 12 power supplies phase shifted minus 7-1/2 degrees. When all power supplies are operating, the apparent load to the utility power line is that of a 24 pulse power supply system, and all harmonic interference below the 23 harmonic is reduced to nearly zero.

2.2 Accel pulse power modulator (PPM)

The accel pulse power modulator conditions and switches the high voltage from the ADCPS to operate its associated Neutral Beam Source Module. The PPM consists of a resistor-capacitor compensation network, a crowbar for protection, a series CQK 200-4 vacuum tube and controls, and a shunt CQK 200-4 and controls. The shunt tube requires about 50 millisec to establish the initial current from the ADCPS and then transfers it to the series switch in about 100 microseconds.

When the series tube turns on to start the accelerating pulse, the shunt tube turns off at a coordinated rate to smoothly transfer the ADCPS current to the output neutral beam source load. The RC network acts as an energy reservoir to reduce transients due to small descrepancies during the transfer. The load current is transferred back to the shunt tube during neutral beam source module spark downs and at the end of pulse.

It is required that the accel voltage and arc current follow the relationship:

$$I_{arc} = kV_{acc}{}^{n}$$

where V_{acc} is expressed in kV and I_{arc} is in amperes. The factors k and n are approximately 4 and 3/2, respectively, and will be tailored to the specific source design. The voltage/current relationship is established by monitoring the arc current during set up and pre-programming the accel voltage control signal in a Programmable Read Only Memory.

2.3 Filament power supply

The Filament Power Supply is an unregulated 12 pulse rectifier that will deliver up to 5500 A with an adjustable voltage range of 12 to 20 volts to heat the tungsten filament assembly in the source module.

2.4 Arc power supply

The Arc Power Supply is also an unregulated 12 pulse rectifier, and it is designed to deliver from 600 to 4,000 A with an adjustable voltage range of 25 to 71 volts.

The power supply consists of an impedance network, a 12 pulse rectifier, an arc notcher, and a crowbar. At currents above 1000 A, the power supply internal impedance is designed to be equal to or greater than the source module arc load impedance in order to stabilize the arc. This is accomplished by adding inductors in the ac power input section.

It is necessary to switch the arc current off for a few hundred microseconds when the source module sparks down. This is accomplished by switching a partially charged capacitor across the arc power supply output through an SCR switch. Occasionally, the arc chamber will develop a metal ion discharge which requires switching off the arc power supply for several milliseconds. The "arc spot" is detected both by a Langmuir probe and by monitoring the arc voltage and current. The two signals are logically weighted and used to trigger an SCR crowbar on the output of the arc power supply.

2.5 Suppressor Grid Power Supply

The suppressor grid power supply provides 2 kV or 3 kV to the suppressor grid nominally at 20 amperes with the capability to provide up to 50 amperes for a few microseconds. The purpose of the suppressor grid is to prevent electrons from the neutralizer from being accelerated into the high voltage region of the arc chamber. The power supply consists of an SCR regulated 12 pulse rectifier with capacitive storage and a transistor switched output. Since the Neutral Beam Source Module is subject to frequent sparkdowns, the Suppressor Grid Power Supply output circuit is protected by a series inductor and a reverse biased diode assembly to prevent damage due to the 80 kV transients.

2.6 Controls

The SNBPSS controls is basically a three tier hierarchical system that ties the power supply control into the MFTF central control room. The highest level is an Interdata 8/32 mini computer and color CRT consoles. At the next level is the Local Control and Instrumentation System (LCIS) which consists of an LSI-11 microcomputer, fiber-optic communication to a CAMAC crate, and a master timing communication link. The third tier of controls is part of the vendor supplied equipment and consists of a Z80 microcomputer and fiber optic communication links to the various power supply subsystems. A control console for each set of 6 power supplies is included that can control any 2 of the six power supplies simultaneously.

3. RECTIFIER TRANSFORMER

3.1 Description

The Accel Rectifier Transformer takes incoming power at nominal voltages ranging from 6900 to 13.8 kV, 3 phase, 60 Hz, and steps the voltage up to 35 kV on each of two secondary windings. The two 35 kV AC circuits drive series-connected rectifier bridges, providing nominally 96 kV DC output. The rated input current is 390 A at 13,100 V on a 10% duty cycle, with a pulse length of 30 seconds. See Figure 1 for the single line diagram illustrating the rectifier transformer application.

4. SHORT CIRCUIT TESTS

4.1 Specification requirements

The SNBPSS specification required that the accel rectifier transformers be designed to withstand 10,000 crowbar short circuits. Since it would obviously be impossible to test the transformer for this number of short circuits, it was decided to require a more severe current for a reasonable number of short circuit tests. The number originally specified was 10. The specification required that rated voltage be maintained on the primary terminals of the transformer during the short circuit. This is a more severe situation than any which would occur during operation because the high current causes the source voltage to sag. In addition, the short circuit duration was longer; the short circuit was maintained for 9 cycles whereas the ADCPS circuit breaker would normally interrupt the short circuit in 3 cycles or less. If the control signal failed to trip the circuit breaker, then the rectifier transformer would be protected only by the current-limiting fuse, which would blow in about 9 cycles. All these factors combine to make a test which, if the transformer passed it, would give great confidence in its ability to survive the rigors of SNBPSS service.

4.2 Test description

Preliminary tests were performed, consisting of: Dielectric tests; Power loss measurements under no-load conditions; Exciting current at rated primary voltage; Impedance measurements; and low voltage impulse measurement.

Dielectric tests were performed first, including two full-wave impulses, at the specified test voltages. The other preliminary data were measured to provide baseline values, and baseline LVI measurements were made on all windings.

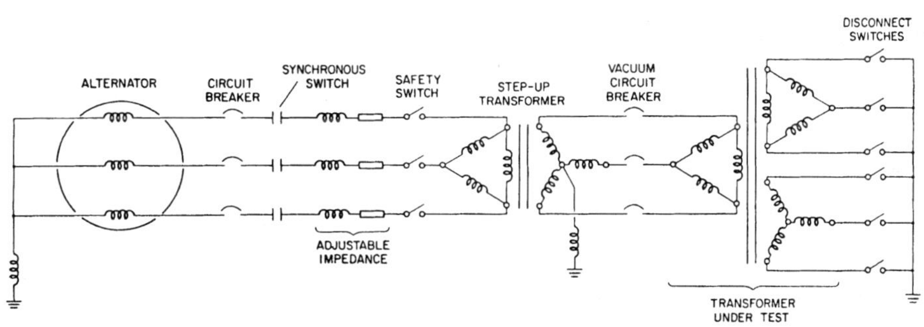

FIG. 3. RECTIFIER TRANSFORMER SHORT CIRCUIT TEST—SIMPLIFIED SCHEMATIC.

The short circuit tests proper consisted of 24 short circuits. Figure 3 shows the schematic of the short circuit test. The intent was to close the synchronous switch at such a time as to produce the calculated maximum offset current for 8 shots in each phase. However, the primary current was excessive when the circuit closure was timed for maximum offset current, due to magnetizing inrush current added to the short circuit current. Note that magnetizing inrush current is not present during a crowbar, because the transformer is already excited and the fault occurs across the secondary. By independently adjusting the closing times of the synchronous switch, the combined inrush and short circuit current was reduced to equal the calculated maximum short circuit current. Thus, a timing setup was developed which met the specification requirement for rated primary voltage to be maintained during the short circuit.

After closing times were determined, 24 short circuits were applied, eight timed for calculated offset current in each phase. The diagnostic measurements comprising LVI, impedance, and exciting current, were made after the third, sixth, and twelfth short circuits. Final tests, including the preliminary tests cited above and the full ANSI impulse tests, were then made.

4.3 Diagnostics

The principal diagnostic measurement is the transformer's impedance. In our tests, the impedance was measured after the third, sixth, twelfth, and twenty-fourth short circuits. Low-voltage impulse measurements were performed at the same points in the test sequence as impedance measurements. See fig. 4 for the "differential" LVI test scheme we used. The LVI waveform is a signature of the winding's inductance and capacitance, hence its physical arrangement.[5]

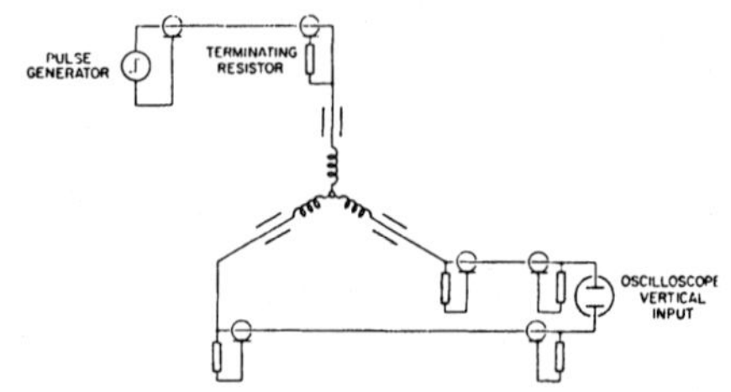

FIG. 4. DIFFERENTIAL LVI TEST.

The transformer's magnetizing current was measured at the same points in the test sequence as the impedance and LVI measurements.

4.4 Results

The transformer passed with exceptionally good performance. There were only minor changes in the LVI waveforms, and changes in impedance during the test were on the same order of magnitude as the overall accuracy of the instrumentation (under .4% of the initial value). After return to the Aydin Plant in Palo Alto, the transformer was opened for visual inspection. There was no visible damage.

5. ARC AND FILAMENT POWER SUPPLY TESTING

Initial tests are in progress at Aydin Energy Division utilizing the first production arc and filament power supplies. These units are being used to power either resistive test loads or an LLNL supplied prototype arc chamber protected by a developmental spot detector.

Pulse tests involving the arc chamber are limited (by the chamber) to a maximum of 3000 A arc current for 0.5 seconds. Tests on the test loads may be run up to the full 30 second, 4000 A specified pulse magnitude.

To date, these tests have revealed two problems requiring correction. Both of these problems stem from the fact that the arc supply is switched on and off in a non-synchronous manner by means of an electromechanical contactor in the 480 volt, 3 phase, supply lines.

On two occasions, one of the three metal oxide varistors that are designed to suppress voltage transients on the 3 phase a.c. line input to the power supply, failed with explosive force. It was determined that a damped oscillatory transient involving the 60 KVAR shunt power factor correcting capacitors resonating with the inductance of the a.c. input supply occurred when the main input contactor closed. This transient, which had a ringing frequency of approximately 700 Hz, decayed in less than one period of the 60 Hz supply. Its magnitude varied according to the time of contact closure in respect to the line a.c. waveform. Occasional voltage surges in excess of 1500 volts with an energy content greater than the 1000 joule capacity of the surge suppressors accounted for the two failures observed.

A suitable solution for this problem was obtained by the addition of another contactor in series with the shunt power factor correction capacitors. This added contactor was programmed to close approximately two cycles in advance of the main power supply contactor. By this means, when the main contactor closes, the transient has damped out and is not impressed on the surge suppressors or the remainder of the power supply components.

The second problem due to the non-synchronous switching relates to the transient performance of the rectifier isolation transformer. Since the contactor opens and closes at random times relative to the a.c. line waveform, the magnetic state of the transformer at the time of contactor closing will vary from pulse to pulse. As a consequence, an initial transient, due to transformer saturation, will occur on some pulses. This transient appears in the power supply output in the form of a distorted initial pulse front with an excess ripple magnitude along with 60 Hz and higher frequency components added to the ripple due to the 60 Hz waveform distortion. This transient output decays in about 300 milliseconds. Since turn-on of the main accel high voltage pulse may be delayed until this transient has subsided, the source output may not be seriously affected. However, this transient makes effective operation of the arc-spot detector difficult and may also contribute to erratic performance of the arc chamber.

One possible solution among several being considered for this problem is to synchronize the excitation of the 480 volt contactor coil with the 60 Hz line supply. By this means, the contactor contacts may be caused to always close and open at a predictable region on the 60 Hz cycle. As a result, an acceptable reduction in the transient may be achieved.

Aside from the foregoing problems, testing to date has indicated that these supplies are capable of meeting the specification requirements. The arc chamber has been operated up to its full rating of 3000 A for 0.5 second. The arc supply has also demonstrated the capability to meet the 4000 A, 30 second requirement when tested into the resistive test load.

6. CONCLUSIONS

A single contract was awarded to design, fabricate, install and perform the acceptance tests of the total SNBPSS system. With a single contractor having full system responsibility it has been possible to have close liaison and frequent workshops that have resulted in design optimization.

The fabrication of the first two power supply sets has progressed very well. The engineering tests have been completed on the suppressor, filament and arc power supplies. Low voltage tests have been successfully completed on the ADCPS. High voltage tests on the pulse power modulator are scheduled to be started prior to this conference. Final acceptance of the first two power supplies is expected to be completed this year.

Short circuit tests were conducted on the ADCPS rectifier transformer at the McGraw Edison Research Facility. The short circuit duration was 150 ms at full rated primary voltage. The transformer passed all the short circuit tests with less than 0.4% change in the transformer impedance. The transformer was returned to Aydin Energy Division, at Palo Alto, opened, and there was no visable evidence of damage or degradation.

Non-synchronous closing of the arc power supply contactor onto the ac power line caused two problems: A transient that had a ringing frequency of 700 Hz caused excessive power dissapation on the input voltage transient suppressors, and secondly, transformer saturation occurred on some pulses causing excessive ripple during the first 300 milliseconds of the pulse. A possible solution being evaluated, is to close and open the ac contactor at nearly the same phase of the ac sine wave each time of operation. Aside from the above two problems the arc power supply has met the intended specification when operated into both a dummy load and an actual sustaining neutral beam arc chamber.

7. REFERENCES

[1] Technical Performance Specification MF-PS-1.5-E-100 for MFTF Sustaining Neutral Beam Power Supply System, Volume II.

[2] ECKARD, R.D., MAYHALL, D.J., SHIMER, D.W., Sustaining Neutral Beam Power Supply System, Magnetic Fusion Energy Quarterly Report, LLL, January through March (1979)

[3] ECKARD, R.D., VAN NESS, H.W., HESSELGRAVE, D., Test and Evaluation of The BBC-CQK-200-4 Modulator Tube For The MFTF Sustaining Neutral Beam Power Supply System, 8th Symposium On Engineering Problems Of Fusion Research, November 1979, Paper No. 13-02.

[4] ECKARD, R.D., VAN NESS, H.W., The Sustaining Neutral Beam Power Supply System for The Mirror Fusion Test Facility, 8th Symposium on Engineering Problems of Fusion Research, November 1979, pp 337-340.

[5] VANUCCI, D.A., Medium Power Transformer Design and Short Circuit Testing Considerations.

NOTICE

This report was prepared as an account of work sponsored by the United States Government. Neither the United States nor the United States Department of Energy, nor any of their employees, nor any of their contractors, subcontractors, or their employees, makes any warranty, express or implied, or assumes any legal liability or responsibility for the accuracy, completeness or usefulness of any information, apparatus, product or process disclosed, or represents that its use would not infringe privately-owned rights.

Reference to a company or product name does not imply approval or recommendation of the product by the University of California or the U.S. Department of Energy to the exclusion of others that may be suitable.

MFTF NEUTRAL BEAM PULSE POWER MODULATOR

R. D. Eckard, H. W. Van Ness
Lawrence Livermore National Laboratory, University of California
Livermore, California 94550

D. H. Hesselgrave
Aydin Energy Division
Palo Alto, California 94302

ABSTRACT

The NBIS (Neutral Beam Injection Source) of a fusion energy reactor requires pulsed, high-energy dc input power. An electron tube pulse modulator, together with control, monitoring and protection equipment, comprise a system to regulate and switch this power. The design and development of salient components for this system are described in this paper.

1. INTRODUCTION

The Mirror Fusion Test Facility (MFTF) at the Lawrence Livermore National Laboratory, Livermore, California, requires 24 separate, independent sustaining neutral beam sources to heat the plasma. Each of these 24 sources has its own power supply system consisting of two main sub-systems: the unregulated HVDC (high-voltage direct-current) subsystem including ac switchgear, rectifier transformer, and described in a companion paper [1] is located outdoors; and the pulse modulator sub-system is located indoors. For ease in equipment maintenance and control, each of the latter 24 sub-systems is in a separate PPM (Pulse Power Module) grouping. The equipment in any one of these 24 sub-systems is shown in Figure 1. This equipment delivers regulated HVDC pulse energy to the source for pulse periods varying between 10.0 milliseconds and 30 seconds. The pulses have a normal rise and fall time of 50 to 100 microseconds, a fall time of 1-2 microseconds under load fault conditions, and an amplitude stability of 1.0 percent. The pulse power modulator equipment operates in a static magnetic field of up to 100 gauss. High speed monitors, control circuits and analog reference voltages bridge dc voltage differences of 80 kV or more. The support equipment for the pulse modulator system includes a regulated screen grid (or grid 2) power supply, a control grid (or grid 1) amplifier, and low-capacitance gas-insulated transformers.

The pulse modulator uses a single tetrode electron tube operating as a series voltage-control regulator to provide HVDC power. This switch tube is the final stage of a closed-loop dc amplifier that has a controlled response extending from dc to 1.0 MHz. An analog signal, transmitted from ground level and referenced to the switch tube cathode, establishes the output voltage at which the dc amplifier regulates. To initiate and terminate current flow in the slow-response rectifier transformer, a second tetrode tube operates as a shunt switch.

2. SALIENT COMPONENTS

2.1 Tetrode Electron Tubes Used in the Series and Shunt Switches

The BBC (Brown Boveri & Co. Ltd.) CQK 200-4 tetrode tube used in this system is derived from a family of type CQ tetrodes that were designed as RF amplifiers for radio broadcast service. The existing design was modified to optimize: (a) anode voltage hold-off rating, (b) anode dissipation capability, (c) current distribution and (d) lifetime under long-pulse (1 to 30 seconds) operation. Experimental testing to confirm satisfactory operation as a high-voltage switch tube is discussed in Section 3.

Since the new tube is derived from an existing family, there is prior production and quality control experience, as well as complete familiarity with the high-frequency behavior of the design. This latter knowledge minimized the likelihood of parasitic oscillations or resonances. Further assurance of stable operation was the tube design as a complete assembly including: (a) a tube socket with decoupling, damping, and over-voltage protection elements, and (b) an SF_6 hood assembled on an insulating base plate.

2.2 Terminating Network

This unit serves to stabilize the HVDC on a short-term basis, and as a termination for the 300-meter length of (nominally) 50-ohm transmission line from the outdoors power source. Working in conjunction with the shunt switch, the network minimizes voltage changes (i.e. from pulse ON to pulse OFF) during HVDC current transitions.

The network consists of: (a) five capacitors (each 30 microfarad, operating in series), (b) a set of noninductive resistors, totalling 50 ohms, (c) a high-impedance shunt resistive circuit (which acts to equalize the dc voltage across each of the capacitors, and to discharge them during OFF periods) and (d) a set of neon lamps (with their individual current-limiting resistors) that indicate presence of voltage across each of the capacitors.

2.3 Ignitron Crowbar

If there is a shunt or series switch tube malfunction the ignitron crowbar is backup protection. It is rated for energy delivered both by the terminating network, which is insignificant, and the follow-on current, delivered by the HVDC supply before the 2-cycle circuit breaker opens. The crowbar consists of five identical modules, each containing an EEV (English Electric Valve) BK-514 ignitron, an isolated ignitor driver transformer, shunt resistor-capacitor voltage-equalizing networks, and anode and cathode water circuits. To optimize the voltage hold-off, the latter circuits warm the anode, and cool the cathode of each ignitron.

The five crowbar ignitor isolating transformers have their primary windings connected in series, and are driven by a pulse-forming network discharged through a hydrogen thyratron. These components are in the Ignitor Driver Assembly, which includes five conductively-coupled, and two optically-coupled, crowbar-firing circuits. These are driven by a variety of fault-sensing circuits, including one originating at the nominal +80 kV dc level.

2.4 Shunt Switch

The shunt switch represents the active circuit alternative to a large, passive capacitor bank network [2]. To maintain adequate upper and lower limits on voltage, the passive network would require approximately 60 microfarad of capacity. Such a capacitor bank requires more space than available indoors; its use outdoors is equally undesirable, since it must be weather-proofed, and does not terminate the HVDC cable.

The alternative system is embodied in the shunt switch, using a CQK 200-4 tetrode tube as the high-power switch element. To illustrate the system function, a graph of shunt switch current versus time appears in Figure 2. The slow turn-on (1) and the slow turn-off (8) illustrate shunt switch load current with di/dt established by the response time of the HVDC rectifier transformer leakage reactance. Time slots (2), (6) and (7) on the figure illustrate the system response to pulse on and pulse off. Time slots (3), (4) and (5) show

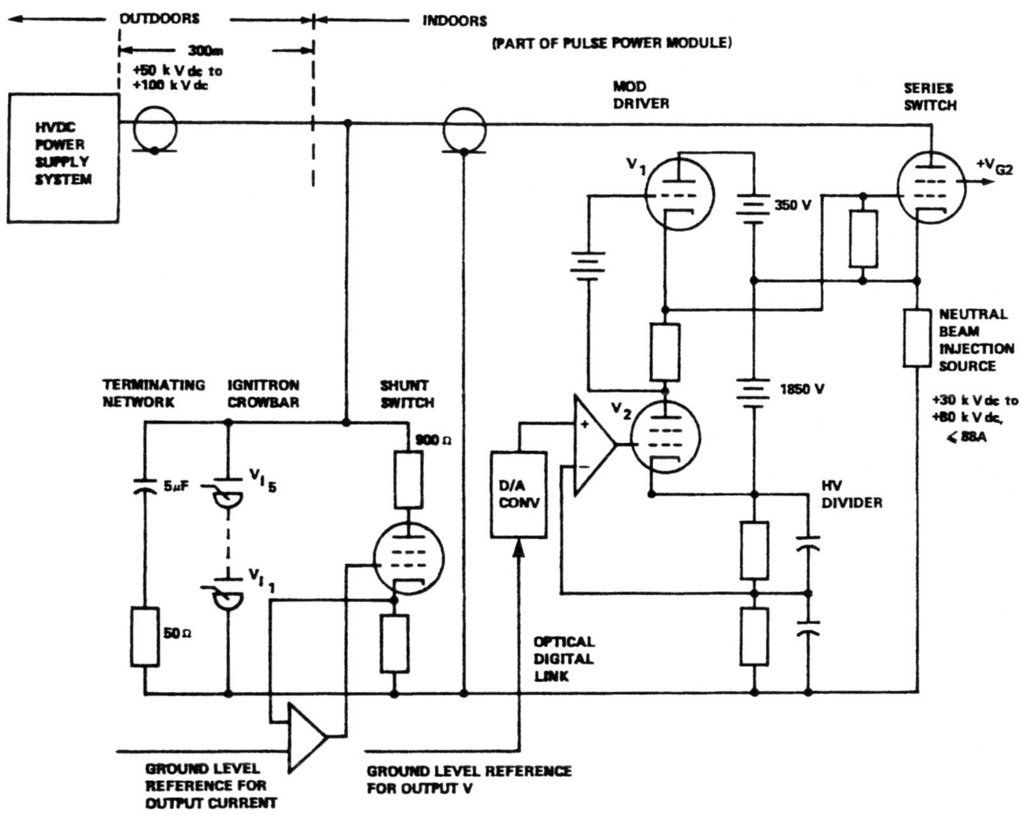

SYSTEM SIMPLIFIED SCHEMATIC
FIGURE 1

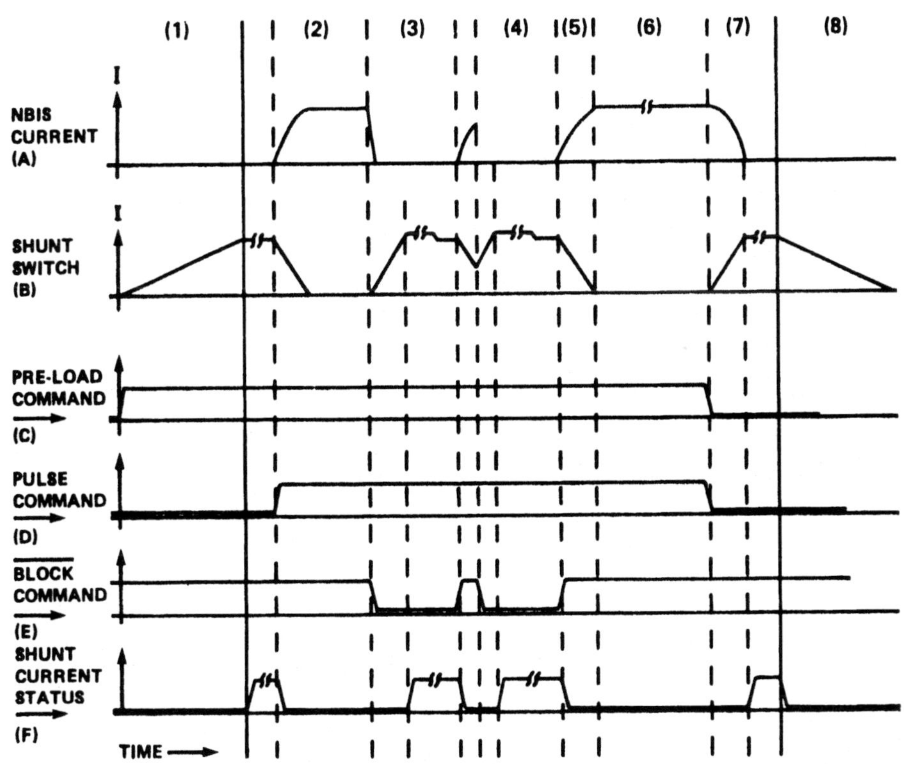

TIMING DIAGRAM
FIGURE 2

how the system responds to sudden interruptions initiated by arcs in the NBIS, requiring fast turn-off and turn-on, with a short delay for arc extinguishing.

2.5 The Floating Deck

The floating deck is a mechanical support system to carry all parts of the pulse modulator system that are operating at high voltage. In addition to the series switch equipment, the deck also supports the arc supply and the NBIS filament supply equipment, all related control equipment, power transformer and distribution circuits and water manifolds. The entire deck is about 5 meters in length, and weighs 2.5 metric tonnes. It is supported and isolated from ground by a set of six fiberglass cylinders, each 130 mm in diameter and about 1500 mm in height. To establish a uniform voltage plane, the deck has removable panels that cover all vertical surfaces. The deck is constructed of aluminum channel, and has a top surface and corona ring towering about 2900 mm above the floor. The capacitance of the deck, and all of the associated high-voltage isolation transformers connected to it, is less than 3000 picofarads.

2.6 HV Isolation Transformers and Enclosures

The pulse modulator equipment on the floating deck requires a variety of mains supply voltages. In addition, the NBIS arc and filament dc supplies are fed by isolating transformers. In all, four isolating transformers are required: (a) a 500-kVA arc supply, (b) a 100-kVA NBIS filament supply, (c) a 16 kVA 480V:480V mains isolation supply, and (d) a 10 kVA 480V:21V tetrode tube filament supply. Each of these transformers must pass a 150 kV dc hipot. To meet this requirement, the transformers are mounted in sealed fiberglass enclosures filled with SF_6 at a nominal pressure of 1.05-1.1 atmospheres. The use of SF_6 maintains a relatively low capacitance from primary to secondary, and precludes the use of flammable oil for dielectric (prohibited by specification on the MFTF project).

Two fiberglass enclosures are used for each floating deck: one enclosure contains the arc transformer only, while the other houses the NBIS filament, the mains isolation and the tetrode tube filament transformers. Each enclosure has internal radiators through which cooling water flows. Connections from the transformer secondaries to the floating deck are carried through separately molded, copper buss feedthrough fittings; these assemblies are gasketed and bolted into the fiberglass enclosures. The primary connections to the transformers, as well as the low-level sensing circuits, come through individually-sealed bushings at the rear of the housing, close to floor level.

Because of limited available space in the PPM, the enclosure permits transformer loading and unloading from the side, without the use of forklifts or other power machinery. To accomplish this loading, the side of the transformer enclosure is removable, and the transformers are mounted on tracks, with wheels and guide rails.

2.7 The Series Switch

2.7.1 General

By maintaining a predetermined level of output voltage, the series switch controls the current flow to the NBIS. A CQK 200-4 tetrode serves as the high power switch element that is driven by a two-stage amplifier. A frequency-compensated high-voltage divider on the output feeds a signal back to a comparator at the input; the series switch regulates at a point determined by the ground-level originated reference voltage signal. Figure 1 illustrates the main components.

2.7.2 Ground Level Reference

The output voltage control [3] regulates both the shape of the dV/dt curve and its maximum amplitude. The shape of this curve is generated by a digital synthesizer circuit consisting of a counter, a ROM (read only memory) and a DAC (digital to analog converter). The eight-bit counter steps the ROM through a programmed data table. To generate the analog reference waveform, the output of the ROM drives the DAC. The absolute magnitude is derived from a frequency-to-voltage converter operating from ground level through a fiberoptic link that provides the reference voltage required by the DAC. This multiplying DAC arrangement allows the MFTF central computer control to remotely program the scale factor (K). The net result is: $V_{out} = K \cdot f(t)$;f(t) is fixed by the ROM and the clock rate which is driving the DAC.

2.7.3 Grid Drive Amplifier (Mod Driver)

The Mod Driver has a solid-state hybrid amplifier for a first stage and a triode-tetrode electron tube pair in a "totem-pole" configuration for a second stage. The first stage has over 100db available gain at dc, and a gain frequency product of 5 MHz. The second stage available gain is +40db.

The quiescent output load for this second stage is relatively modest: it consists of a 12-kilohm load resistor as the grid return resistor for the BBC CQK 200-4 tube. However, during either the turn-on or the turn-off transient operation, the second stage must drive a capacitance totaling about 2500 pF. Charging current for this capacitance during positive slew conditions is supplied by V1; the positive bias raises the available current. The negative charging current is supplied by tetrode V2; at this time the current is drawn through the common resistor in the totem-pole configuration and supplies cutoff bias for tube V1. The tetrode can deliver about 50 percent more current than the triode,which results in a faster turn-off than turn-on.

2.7.4 Series Switch Tube Deck

This unit supports the series switch tube socket and the related electronics. All low-level active circuits are on two PC cards which are also mounted to the deck. Related circuits include arc-over detectors for grid 1-to-anode, and grid 2-to-anode; grid 2-to-common crowbar; filament current timer and presence detector; grid 1 voltage, or anode current by remote selection.

Because the electron tube operates in a high static magnetic field (worst-case condition is 50 to 100 gauss) it must be adequately shielded to prevent internal electron-flow distortion. Theoretical calculations and experimental tests showed that the above magnetic field could be reduced to manageable proportions (less than 10 gauss) by the use of heavy steel shielding. The steel is combined with lead to protect against X-ray radiation. The result is a tube deck supporting the tetrode tube, identical for both the shunt and series switch. It is stoutly constructed of a 6.4 mm thick steel sheet and a 3.2 mm thick lead sheet resulting in a quasi open-faced sandwich. This same construction is used in an open cylinder that surrounds the tube.

The two shields are combined with the SF_6 gas enclosure. This creates a need for a flexible "boot" or gasket at the top end of the tetrode tube anode.

3. High Power Electron Switch Tube Tests

3.1 Test Objectives

Tetrode tube BBC CQK 200-4 was tested to confirm that it would switch 80 kV after conducting 88 A for 30 seconds. There were two concerns: (a) during the 30-second high-current load, the screen grid might have excessive dissipation causing a loss of control of the tube; (b) the cyclic temperature-induced stresses in the anode might cause fatigue failures (stress cracking) that would reduce the tube's useful life. The first concern was addressed by performing full-scale tests at the LLNL test facility. The tests demonstrated that the CQK 200-4 would switch 80 kV in less than 5 μS under the specified load conditions. The second concern was initially explored by a paper study conducted at LLNL [4]followed by tests conducted in Switzerland by the Brown Boveri Co.

3.2 Test Results

The sequences of test listed below were conducted on two CQK 200-4 tubes. Both tubes yielded essentially identical test results.

3.2.1 Cold Hipot

Both tubes were hipot tested with the filament and grids deenergized. This test can detect the presence of residual gas after a period of inactivity of the tube. Tube number one was tested after three days of non-use and tube number two was tested directly after installation at LLNL. No indication of residual gas was detected.

3.2.2 Hot Hipot

This is a hipot test with the filament fully energized, the control grid set to cutoff bias, the screen grid set for full operating voltage (typically 1800 V), and the hipot connected from tube anode to cathode. It verifies hold-off and cathode current cutoff capability. The hipot test was used to condition the first tube up to 135 kV (limited by external circuit arc downs) and the second tube up to 150 kV.

3.2.3 Micro-Pulse Tests

Pulses, in the 10 to 100 microsecond region, are used to check the turn-on and turn-off characteristics of the tube. The ON and OFF slew rates of the CQK 200-4 tube are of great importance for any pulse modulator/regulator work. Test set drive amplifier equipment had significant limitations in performance. Consequently, the final results are not a true assessment of the CQK 200-4 tube capability. Thus, the rise and fall time will be equal to or better than illustrated in Figure 3.

3.2.4 Long Pulse Tests

These pulses proved the power capability of the tube. An initial series of pulses having a duration of 50 to 200 milliseconds verified the operating point of the tube. This period of time permits the HVDC supply to reach a new stable operating condition and gives a true picture of anode voltage and current. Once the operating point was verified, the test set was ready for the final pulse series that extended to the specified 30-second maximum length. During this period, data was observed on panel meters rather than as oscilloscope displays. This series of test enabled us to read the screen current; impossible to read at shorter pulses. The tests showed that the tubes could handle the current and come back to full stand-off voltage. The screen current was 0.9 A during all pulses. These tests verified that both tubes could conduct full rated anode current for 30 second pulses repeated at 5 minute intervals with no change in turn-off capability at the end of each pulse.

3.3 Anode Thermal Fatique

Calculations [4] indicate that the anode thermal stresses developed during 30 second,

1.1 MW dissipation pulsing can be as great as 155 MPa (22,500 psi). The resulting thermal fatigue life prediction indicate that for long-life performance, (10^6 pulses) the dissipation should not exceed 50 percent of the maximum design limit. Because of uncertainties introduced by the modeling approximations, coupled with the flatness of the fatigue life curve, it was recommended that physical testing be performed to assure the desired results. Consequently, Brown, Boveri & Co., Ltd. conducted a series of tests to confirm tube capability in meeting the desired SNBPSS operational life goal. The mechanical stress fatigue flexure tests were conducted on properly processed samples of anode material. These test measurements, compensated to account for anode operating temperature and water passage grooves, indicated an expected pulse life of 2×10^6 pulses at 1 MW dissipation.

To further verify this prediction, operating tests were conducted on a smaller tube with parameters scaled to simulate the 1 MW anode loading of the CQK 200-4. The pulse duration (3 seconds) and the interpulse interval (12 seconds) were sufficiently great to permit the anode to reach a stabilized temperature at each extreme. After 300,000 pulses, the experiment was terminated and the tube opened for metallographic analysis of the anode. The visual inspection revealed no fatigue cracks at the inner or at the outer anode surface. Further theoretical study is continuing.

4. CONCLUSIONS

The results obtained from the prototype testing and experimental work substantiate the design goals of the MFTF Neutral Beam Power Supply System. Components and subsystems are working within their theoretical performance margins; control instrumentation and transient frequency response of analog circuits is more than adequate for the task. The pulse modulator system is currently undergoing system tests to verify that the combined subsystems are compatible.

5. REFERENCES

[1] ECKARD, R. D., Eleventh Symposium on Fusion Technology, Oxford, 13-19 September 1980. "The Sustaining Neutral Beam Power Supply System."

[2] ECKARD, R. D., "Engineering Problems of Fusion Research," San Francisco. November 13-16, 1979. Proceedings pp 337-340.

[3] CUMMINS, J. D., Eleventh Symposium on Fusion Technology, Oxford, 13-19 September 1980. "Control Systems for Sustaining Neutral Beam Power Supplies."

[4] SUTTON, S. B., UCID-18049, "Temperature and Thermal Stress Analysis of a Switching Tube Anode." (January 1979)

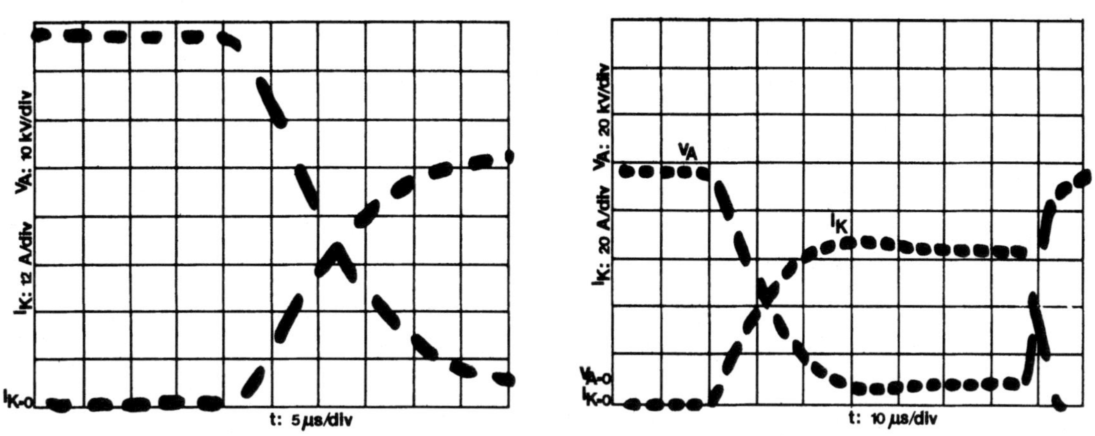

FIGURE 3

HIGH POWER VACUUM TUBE TECHNOLOGY APPLIED TO FUSION SYSTEM COMPONENTS

D. R. Carter, R. E. Reed, J. D. Stabley
RCA Corporation
Lancaster, Pennsylvania USA

ABSTRACT

RCA has developed a high voltage switch tube for use in the Neutral Beam Power Systems (NBPS) on the Tokamak Fusion Test Reactor (TFTR) being built in Princeton, New Jersey, U.S.A. Furthermore, high power vacuum tube technology is being applied by RCA to also manufacture neutral beam sources for the Mirror Fusion Test Facility (MFTF) at Livermore, California.

This paper will describe the present test results on the high power switch tube. A reliability improvement program will be briefly described which was implemented when the device was transferred from engineering development to factory production.

This paper will also review the status of the neutral beam source program. Design reviews for new power products that are being placed in manufacturing are discussed. In addition to the basic design, the design review also includes materials, mechanical construction, assembly procedures, testing, and reliability. The status of the program accomplishment to date will be highlighted.

1. INTRODUCTION

The Power Devices section of RCA's Electro Optics and Power Devices Division has been actively engaged in the fusion program since 1976. Contributions have been in the form of a 2000 Kw regulator tube for TFTR and more recently in the start of a manufacturing program to produce 80Kv, 80 amp Neutral Beam Sources for the MFTF program at Livermore, California. Both these undertakings are an extention of the Large Power Tube activity that began in the 1940's at RCA's Princeton Laboratories.

2. GENERAL TUBE DESCRIPTION

The S94000E shown in Fig. 1 is a liquid cooled ceramic to metal beam power tetrode. It is 36 inches high with a maximum diameter of 22¼ inches and weighs approximately 320 pounds.

Fig. 1 - RCA S94000E Tetrode

The tube is arranged with the anode and cathode terminal separated by an anode insulator. The lower end of the tube has appendages for electrical and water connections to the various elements of the tube. Also included with each tube is a 8 liter vac-ion pump and its associated magnet. The insides of the tube shown in Fig. 2 utilize sixty-six individual electron guns each employing a directly heated thoriated tungsten ribbon filament.

The control and screen grids are comprised of small (.0066" diameter wires) tungsten wires embedded into water cooled copper blocks. The unique anode structure is centrally located and is comprised of sixty-six individually cooled petals. These petals are set at an oblique angle to the electron beam axis as shown in Fig. 3.

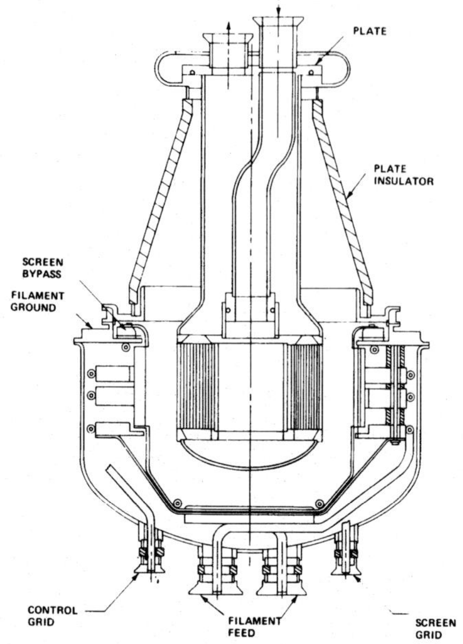

Fig. 2 - RCA HVST Internal Arrangement

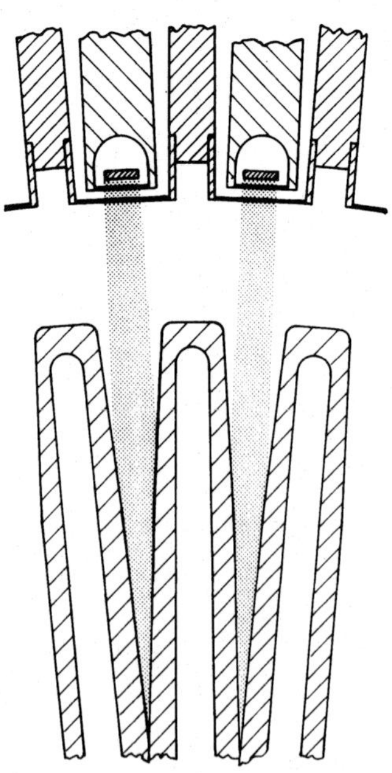
Fig. 3 - Electron Gun - Anode Cross Section

The angle is very effective in enhancing the bombarded anode area. Design of the anode petals was accomplished by the use of a Finite Element Analysis [1]. This method of analysis was very useful in the solution of transient and steady state thermal problems that were encountered in handling the $14Kw/cm^2$ of beam flux available from each electron gun.

3. PRESENT REGULATOR TUBE REQUIREMENTS FOR TFTR

Type of Tube.........Tetrode
D.C. Anode Voltage...200 Kv
Anode Current........70 Amps
Anode Dissipation....2.0 Megawatts
Instantaneous Grid No. 1
Voltage...........Negative
Pulse Length.........1.5 Seconds*

*With an upgraded ion source

4. TUBE EVALUATION

A complete tube evaluation to the above conditions was not possible at the Lancaster location. A search of prospective very high power test facilities resulted in the selection of the high power test lab at the Rome Air Defense Center (RADC) in New York. The RADC facility is capable of providing 25 megawatts of peak power at 200Kv. For the S94000E tests, the power supplies were connected in a series parallel arrangement to yield 200Kv at eighteen amperes of continuous current. RADC Engineers determined, after consultation with the power supply designers, that the rating could be nearly tripled (50 amp) if the pulse length did not exceed .5 seconds. Consequently, the tube evaluation was tailored to the RADC equipment. Shown in Fig. 4 and Fig. 5 is a simplified schematic of the test circuit and a picture of the S94000E in the lead shielded oil container.

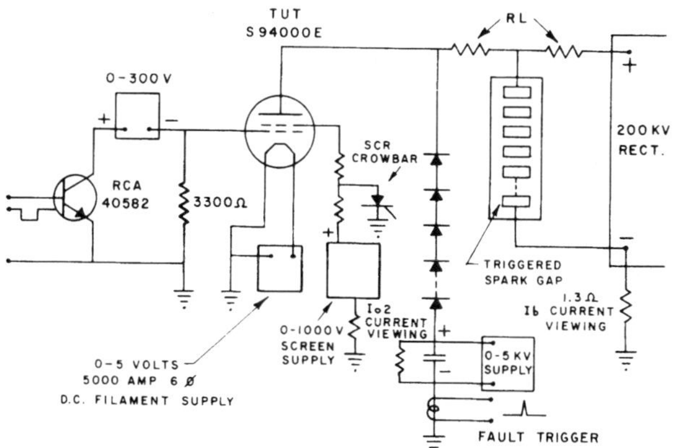

Fig. 4 - Simplified Schematic of Test Circuit

Fig. 5 - RCA S94000E in Lead Shielded Oil Container

5. RADC TEST SUMMARY

After many delays as a result of overstressing the RADC equipment, tests at the 200Kv level were successfully accomplished. The results of those tests along with the maximum ratings of the tube are shown in Fig. 6. This data shows that a new higher level of performance has been obtained with gridded tubes in combining pulse width, voltage hold-off and anode dissipation capabilities. Although it is desirable to operate the screen at much higher voltages on the tube, this was not possible due to power supply limitations.

TEST DATA SUMMARY

FILAMENT CURRENT = 4300 AMPS

PULSE LENGTH = 0.5 SECONDS AT 3 PER MINUTE

200 MICROSECONDS RISE TIME

ANODE CERAMIC IMMERSED IN OIL

E_{BB} KV	E_P KV	I_B AMPS	E_{G2} VOLTS	E_G VOLTS	ANODE DISP. KW
185	62	27.3	520	-30	1690
190	42	47.0	490	-10	1980
193	53	43.0	500	-15	2080
193	50	46.0	600	-25	2150
200	36	50.0	590	-18	1800
ANODE CERAMIC IN AIR					
107	36	52.0	590	-14	1872
107	26	60.0	740	-14	1560

*TUBE PRESSURE DURING PULSE AND INTERPULSE PERIOD.

MAXIMUM RATINGS

DC PLATE VOLTAGE	200 KV
PULSED DC PLATE CURRENT	125 AMPS
DC GRID NO. 2 VOLTAGE	1800 VOLTS
GRID NO. 2 CURRENT	7.5 AMPS
DC GRID NO. 1 BIAS VOLTAGE	1000 VOLTS
GRID NO. 1 DISSIPATION	10 KW
GRID NO. 2 DISSIPATION	10 KW
ANODE DISSIPATION	2000 KW
DC FILAMENT CURRENT	4700 AMPS

Fig. 6 - Test Data Summary and Maximum Ratings

6. LIVERMORE TEST STAND

The S94000E production tube #J105 is presently being employed as a series modulator in the test stand at the Lawrence Livermore Laboratories in Livermore, California. The tube is part of an open deck high voltage modulator that supplies the accelerating voltage for the Neutral Beam Ion sources employed in the MFTF and Doublet III programs. The MFTF source has been switched at a load voltage and current of 80Kv and 70 amperes at .3 seconds pulse length.

The Doublet III source is presently at test and is operating at 80Kv and 70 amperes at .5 second pulse length. During the .5 second pulse from 1 to 5 faults occur in the source requiring the S94000E to block and then reapply full voltage to the source after approximately 20 milli seconds. During the off time, the power supply no load voltage reaches 135Kv.

7. RELIABILITY IMPROVEMENT PROGRAMS

During the course of the tests at Lancaster, Rome, and Livermore using two Engineering tubes, several areas of design change within the tube and external to the tube became apparent. A brief discussion of these areas of concern and the changes that have occurred in the production tubes to improve reliability follows:

7.1 Screen By-pass Capacitor

The high capacity capacitor shown in Fig. 2 was originally formed by using mica between two copper plates. This high capacity immediately at the screen electrode is desirable to electrically put the screen at an RF ground. However, the capacitor is inside the vacuum enclosure and vulnerable to puncture if overvoltage occurs. As a result of this condition and the puncture of a mica during the high voltage processing, the capacitor at the screen terminal now consists of two closely spaced plates that form a vacuum capacitor which is self healing. This change makes a marked improvement in the tube reliability.

7.2 RF Absorber (Eccosorb Ring)

The physical size of the S94000E is such that high frequency (L band) resonances exist between the anode and screen of the tube. These tube associated resonances are typically circumferential in nature and have unloaded Q's that are high and capable of building up very high voltages if not loaded properly. To present a load for these resonances an RF absorbing ring is placed on the filament ground connection of the tube (shown in Fig. 2) extending upwards and following the contour of the ceramic for three inches. This ring has found to be very effective and results in a tube that is completely free of parasitic oscillations over its entire operating range.

7.3 Grid No. 1 to Filament Spacing Changes

The S94000's now being fabricated for TFTR have Grid No. 1 to filament spacings that are wider than were on the two engineering tubes. This increase in spacing has a two fold purpose. First, it lowers the mutual conductance which reduces the gain for parasitic oscillations. Second, the beam current density variation is improved.

8. S94000E SUMMARY

The S94000E has been switched at 200Kv and has operated at two million watts of anode dissipation at RADC. The tube has also successfully switched and regulated the power to Neutral Beam sources. The tube design engineers are pleased with the performance to date and look forward to gaining more operational information.

9. NEUTRAL BEAM SOURCE PROCUREMENT STATUS:

9.1 General

The U.S. Department of Energy through the University of California (U of C) Lawrence Livermore Laboratory awarded an industrialization and manufacturing sub-contract to RCA for the fabrication of 48 Neutral Beam Source Modules (NBSM's) in February, 1980. The award was based on a Department of Energy policy of involving private sector industry in the acquisition of components for fusion programs and is predicated on a source design developed by the U of C Laboratories.

Each NBSM consists of an arc chamber source for deuterium ions plus an accelerator assembly for energizing and focusing these ions into beams. Both 80Kv sustaining and 20Kv start-up modules are included in the sub-contract.

RCA's entry into the development and manufacturing of Neutral Beam Sources for fusion research is an extension of its large power vacuum tube design and manufacturing experience as used for the S94000E switch tube. Both technologies require the fabrication of large, vacuum tight envelopes encasing high voltage and focusing electrodes in a region where charged particles are accelerated to high kinetic velocities. Both technologies incorporate identical low vapor pressure materials, high voltage insulation and require dimensions with precision alignment tolerancing as tight as ±0.0003 inch. Cleanliness controls are identical since care must be taken to avoid residual films or particulate contamination. These controls are essential for reduction of excessive arcing and high voltage seasoning time.

9.2 NBSM Fabrication

Specific requirements that are being addressed in the industrialization program for the Mirror Fusion Test Facility NBSM's are described below. The manufacturer has reviewed the University's design and recommended design changes based on value engineering criteria for improved manufacturability, reliability, maintenance and cost. Extensive make-or-buy studies were performed in conjunction with vendor surveys to guide the procurement of parts. Vendor selection criteria on critical components in the beam optics assemblies included availability of modern CNS equipment capable of achieving the dimensional requirements and interchangeability on components with toler- of ±.001 or less, and the Management/ Quality staffing to assure compliance.

In order to provide a dedicated and concentrated effort, a separate activity was established combining personnel and facilities required for the design and manufacture of NBSM's. Management, engineering, administrative control, drafting, hourly and technician skills are all represented. Cognizant quality assurance engineers and technicians are functionally located in the area but organizationally report into a separate level of the plant management.

RCA plans to adapt to the NBSM the skills and techniques utilized successfully in the fabrication of power tubes. Specifically, these will include the following:

a. High voltage electrode finishing, contouring and cleaning,
b. Tungsten, directly-heated thermionic filaments,
c. Vacuum sealing techniques utilizing low vapor pressure materials,
d. Ceramic insulator technology,
e. Precision alignment techniques for brazing, welding, assembly and inspection,
f. Brazing, welding, plating and other processes suitable for high voltage, high vacuum electronic/ionic structures.

Material certification, interchangeability and traceability methods are being utilized for all raw material, parts and assemblies in the NBSM.

RCA is currently completing special facilities which include a 3500 sq. ft. temperature and humidity controlled staging area for preassembly parts inspection, parts storage, and minor part alterations; a separate cleaning area containing vapor degreasing, acid, solvent, ultra-sonic, and demineralized water facilities; and a class 100,000 clean room for the final assembly, inspection, and sealing of source modules in a precisely controlled temperature and humidity environment.

A specialized ultra-precision Computerized Optical Comparator and Coordinate Measuring Machine capable of ±.0003 accuracy will be used for establishing accelerator grid alignments and spacings. Capability exists for measuring vertical wire to wire spacings between the four layer grid assemblies.

A rigorous quality control plan has been instituted to monitor and control the revision of part, assembly and process specifications, assure compliance with material traceability, and manufacturing specifications. Travel tickets providing part and assembly history will accompany components throughout the fabrication operations. In process inspection stations monitored by Quality staffing will be utilized to control compliance with requirements. Design reviews have been held on the form, fit, and function of all parts and sub-assemblies in the source module mount, arc chamber, and accelerator assemblies. Additional reviews are scheduled to assure continued design integrity.

Certain design modifications have been incorporated into the original LLL design to accomodate proven reliable fabrication processes that are commonly used in the tube industry to improve compliance with the module design objectives. For example, brazing has replaced welding in various assemblies to eliminate warpage. Progress is being made on acquisition of components for fabrication of the first four pre-production 80Kv modules scheduled for delivery in the 1st half of 1981. Engineering efforts are thus now being redirected toward the design and fabrication of assembly fixtures and brazing jigs as well as to the detailed write-up of processing, plating, cleaning and assembly procedures.

9.3 Schedule - CPM

The time frame of performance design complexity, and sourcing of short run parts necessitates a constant awareness of deliverable components and potential vendor problems to assure schedule compliance. Overall project planning and tracking is being accomplished via a developed Critical Path Method (CPM) program.

An existing computer code or program is being used to schedule and control the various tasks involved in the NBSM program. This schedule provides critical path information in time/date sequence, Gantt milestone charts with earliest and latest dates for each work element or task, slack or float times printed-out for non-critical path tasks and the capability of sorting by task or by responsible individual. The CPM program provides an early warning system which aids program management in effectively re-assigning resources to reduce the impact of potential problem areas on cost and delivery. The use of control systems, such as CPM, is deemed essential in highly complex programs with critical time schedules.

9.4 NBSM Summary

While no sources have been delivered, the combination of prior high voltage, vacuum and manufacturing of large structure experience is avialable to assure the success of the NBSM industrialization program. The existing personnel and facilities are adaptable to other neutral beam source designs and procurements with minimum modifications.

REFERENCES:

[1] BAUDER, R.C., Finite Element Analysis of a Twenty-Five Megawatt Power Tube and a High-Energy Water-Cooled Heat Sink for Fusion Research (1978).........RCA Review

DITE MK 2 BUNDLE DIVERTOR POWER SUPPLIES AND CONTROL SYSTEM

D V Bayes, D Bell, J W Gray, J L Browning, E G Poole

Culham Laboratory, Abingdon, Oxon. OX14 3DB, UK
(Euratom/UKAEA Fusion Association)

1. ABSTRACT

The power supplies and control system currently being constructed for the DITE Tokamak MK 2 Bundle Divertor Field System [1,2] meet the following specifications:

Maximum current	20 kA
Rise time	25 ms
Flat top time	200 ms
Load parameters	2.0 mH, 21 mΩ

This is accomplished by a 2.5 kV, 800 kJ electrolytic "start" bank of dimensions 2.9 m x 1.8 m x 4.4 m high, and a 1080 V, 2.8 MJ electrolytic "sustain" bank of dimensions 3.4 m x 1.8 m x 5.7 m high, Fig. 1. The start bank is discharged into the load to establish the current and the 8 "sustain" bank sections are fired sequentially to maintain the "flat top".

The power supply and load assembly is controlled by a microprocessor based Programmable Logic System, which is a logical progression from the modular hard-wired, DITE system using solid state elements [3].

2. CHOICE OF POWER SUPPLY

The divertor power requirements are approximately 12 MW during the 200 ms flat top period and 40 MW peak during the current rise. It was not possible to obtain this directly from the site mains supply owing to the present limitation on maximum demand (20 MVA) for the whole of the Laboratory Site. In principle the demand during current rise could be reduced by extending the rise time, but since the coil temperature rise per pulse was already 80°C this was not considered a viable option. In addition the B_ϕ power supply would need to be turned off during this period to keep within the mains supply restrictions. It was therefore decided to design a power supply using capacitor banks. The optimum voltage for the start and sustain banks was found to be 2.5 kV and 1 kV respectively and the use of paper dielectric capacitors was found to be uneconomic at these voltages. The use of high energy density electrolytic capacitors has produced an economical and compact power supply. Capacitors are arranged in a series-parallel connection to produce the required output voltage.

3. POWER SUPPLY ARRANGEMENT

The power supply is constructed as ten 350/400 kJ capacitor banks (Fig. 2). Each bank is a complete subsystem with charging, switching, dumping, shorting and earthing facilities. Each bank is switched by a type E ignitron (9" diameter unit) through a current limiting inductor and diode assembly to a parallel connected collector plate module, and then cabled to the divertor coils. Each bank also has two type "E" ignitrons to rapidly discharge the capacitors (typically 5 ms) in the event of a detected fault.

The modular approach to the capacitor bank design is based on a stacking frame in which two opposite faces can each contain up to 36 capacitor trays and the two end faces can be used for switching components. Normally the capacitor tray (0.8 m x 0.25 m x 0.16 m high) can contain 5-6 kJ of electrolytic capacitors and a reverse charge protection diode. However the use of high energy density capacitors of double anode foil construction has resulted in 10 kJ per tray and more than 700 kJ in a switched module of overall dimensions 3.4 m x 1.8 m x 1.3 m high.

Maintenance is achieved by removing and replacing trays of capacitors in up to 10 kJ units. Each capacitor is protected by an HRC fuse and enables maintenance to be done at convenient times.

4. FAULT PROTECTION AND IGNITRON MONITORING SYSTEM

The bank energy will be dumped and all output switching inhibited on instruction from the DITE Microprocessor or on a malfunction of one of the 30 installed ignitrons (10 load switching, 20 dump ignitrons). The complete ignitron monitoring and protection unit is housed in a Camac crate and controlled by a small microcomputer. In the event of fault protection operating, the microcomputer will display details of the cause of the shutdown.

Each ignitron is fitted with a monitor optically coupled to the Camac crate and indicates the time conduction started and whether or not a preset current value was exceeded. The receiver units in the Camac crate interpret this information as to whether there was a prefire, late fire or overcurrent situation and respond accordingly.

Each component of the fast dump system is duplicated to increase the confidence that the protection system will operate reliably.

5. LOCAL CONTROL OF THE POWER SUPPLY

The power supply installation area is remote from the DITE machine and construction has proceeded without disrupting the machine operation. In order that most of the power supply commissioning might similarly proceed independently, the power supply is fitted with its own local control system and dummy load.

From the local charger control room capacitor banks may be selected, charged and switched in sequence into the dummy load. Thus most of the power supply can be commissioned before integrating into the DITE machine. The local diagnostics system is also Camac based. It is controlled by the same microcomputer as the fault protection system and includes transient recorders and digital voltmeter.

6. CONTROL REQUIREMENTS

A microprocessor based logic system has been chosen because of the system complexity, high costs of hard-wired systems and the need to be compatible with the existing DITE instrumentation. The system is a compromise between hard wire logic control and a computer based system and offers improved reliability. The use of 16 parallel programming facilities and a 12 μs instruction time ensures that the system will have an overall scanning response time of better than 4 ms during the 200 ms pulse.

6.1 The control system has four separate control functions:

(a) Area and personnel safety

(b) Power supply charging sequence and interlocks

(c) Divertor coil high current pulse protection and cooling cycle

(d) Vacuum control and alarm system

The control functions required for (a) and (d) are relatively slow in their operational response times, e.g. the systems operational times are measured in minutes, the charging and cooling sequence has operational times of seconds, whilst the pulse sequence is only 200 milliseconds. During the 200 millisecond pulse sequence the Divertor coil protection circuit must have an overall response time of not greater than 5 milliseconds.

In order to maintain the safety margin at least two protective and interlock circuits must fail before the system fails to danger, e.g. two sensors measuring the Divertor coil earth leakage current will instigate a fast shutdown of the power supplies during the fault conditions. The duplication is maintained into all the control functions, e.g. two independent guard lines are used, Fig. 3.

6.2 Area and Personnel Safety

The independent interlock systems are used for the area and personnel safety. A mechanically interlocked system using Castell Keys and separate pneumatic system using air sensors interlocked with pneumatic logic circuits. Both the mechanical and pneumatic systems are supervised by the microprocessor control system. The pneumatic control system monitors and interlocks the state of the High Voltage Area for the Bundle Divertor Power Supply, e.g. access doors, earth sticks, earth links and emergency stop buttons. Failure of the control air or interruption due to an emergency stop action will cause or maintain the earths and earth dumps to be in a safe state. Similarly a signal from the Microprocessor Monitoring control will return the power supply area to a safe state.

6.3 Power Supply Charging Sequence

The ten chargers are selected and controlled from the DITE control room through the input output channels of the microprocessor. Control push buttons and interlocks give 46 input commands to the processor with an associated 908 software instructions. Each charger is controlled through the processor, text outputs are displayed on a VDU (Visual Display Unit) giving the state of the charger. The completion of charge signal is transmitted from the microprocessor to the DITE Central Control logic system prior to the initiation of the plasma pulse.

6.4 Divertor Coil High Current Pulse Protection and Cooling System

There are four fault protection circuits monitoring the Divertor coils during the 200 millisecond high current pulse.

(a)	Coil Earth Leakage Resistance:	Two separate systems for the detection of a low resistance between the coils and their housing.
(b)	Voltage Balance between Divertor Coils:	Comparison of the potential across each coil detects earthed coils and shorted turns with a sensitivity of one turn.
(c)	Excess Pulse Current:	This is sensed in both the 'go' and 'return' bus bars.
(d)	Current Balance in Coil Bus Bars:	Detection of a low resistance to earth on the Power Supply, sensitivity 0.01 volt per amp difference.

The inherent time delay between the inception of a load fault and the initiation of the Protection ignitrons is 3.4 milliseconds for each of the above protection circuits and their respective Guard Lines; this allows a maximum Microprocessor Software execution time of 1.6 milliseconds for its additional protection control.

During the normal or abnormal pulse sequence the completion of the cycle is controlled by the Microprocessor by the operation of the power supply dump and earth switches before the commencement of the cooling cycle. The cooling sequence is interlocked such that the initial coolant water temperature is identical to that of the coils prior to cooling. These interlocks and the subsequent gradual reduction in temperature of the coils by adding cold water are controlled by the software programme of the control system. The cooling time period is ten minutes when the mean temperature of the coils is below their set interlock temperature. The cooling program requires 340 software instructions which includes text messages given to the operator as to the current state of the cooling cycle.

6.4 Vacuum Control and Alarm

Alarms are indicated on the Visual Display Unit by test statements showing the state of the Divertor vacuum and cryogenic pumping system.

7. MICROPROCESSOR SEQUENCE CONTROL

The four separate programme control functions required by the Mark II Bundle Divertor have differing execution times. The most stringent cycle time required is for the protection sequence during the 200 millisecond pulse sequence. It was evident that a Microprocessor Control System having only one sequence programme would not be suitable. A number of commercially available Microprocessor based Programmable logic systems were considered and the SESTEP 500 system, manufactured by Sprecker and Schuh Limited, matched the control specifications required for the Divertor System. The Sestep 500 gives the following important facilities which are required for the four control functions.

(a) There are sixteen parallel programmes available. These are independent programmes and each of the four control sequences is therefore allocated its own parallel programme set routine.

(b) The Sestep 500 has the facility of a fast central Processing Unit (CPU) giving a software instruction time of 12 microseconds.

(c) Non isolated input and output modules can be used for enhanced response times of the order of 1 millisecond, and

(d) The Programming Language is simple and takes the form of a Mnemonic code which permits programming from flow charts, ladder diagrams or Boolian algebraic expressions.

7.1 Programme Structure

The master programme (P0) controls the main sequence for the Bundle Divertor control system; this programme also contains the fast protection control function. The later part of the programme is designed such that its cycle time in software is less than 1 millisecond.

The Programme P0 is extended by using the alternating mode instruction when the additional four parallel programmes are used. Each programme has control over all the standard modules of the microprocessor control logic system, e.g. input, output modules etc., and thus the programmes are completely independent, each with their return address registers and four bit working register. Fig. 4 shows the organisation of the control programme for the Divertor Control System. The programme P0 is always the master programme. The appropriate start "parallel programme" instruction will initiate the independent control function of each parallel programme whilst the P0 programme continues in its own vertical scan. Interlock and alarm markers set up by the parallel programmes can be interrogated and acted upon by the master programme P0. This latter programme has control of the complete Divertor control functions.

Vacuum alarms and the Basic Safety programmes P1 and P2 are run continuously through the master programme P0 whilst the charge and cooling sequence programmes operate only when requested by the DITE Central Control. During the 200 millisecond pulse period the control functions are performed by the P0 programme with a limited number of instructions at 12 microsecond each to maintain the fast response time required. The overall time response of the system is less than 5 milliseconds including 3.4 milliseconds due to the monitoring and trip analogue signal sensors.

REFERENCES

[1] Coil and Structural System Design of the DITE MK II Bundle Divertor.
Thompson V K et al. Proc.10th Symp. Fusion Tech. (Padua 1978) Vol. 1. p573.

[2] DITE Status and Upgrade Proposals.
Bayes D V, Dobbing A J, Crawley H J, Cowlin M L, Hay J H, Hayward R J, Sanders B C, Thompson V K. Proc. 11th Symp. Fusion Tech. (Oxford 1980).

[3] Design and Operation of the DITE Power Supplies and Control Systems.
Bayes D V, Bell D, Galloway G, Plummer K M, Skelton D E.
Proc. 9th Symp. Fusion Tech. (Garmisch 1976) p685.

FIGURE 1

DITE MK II Bundle Divertor
Power Supply
Under Construction

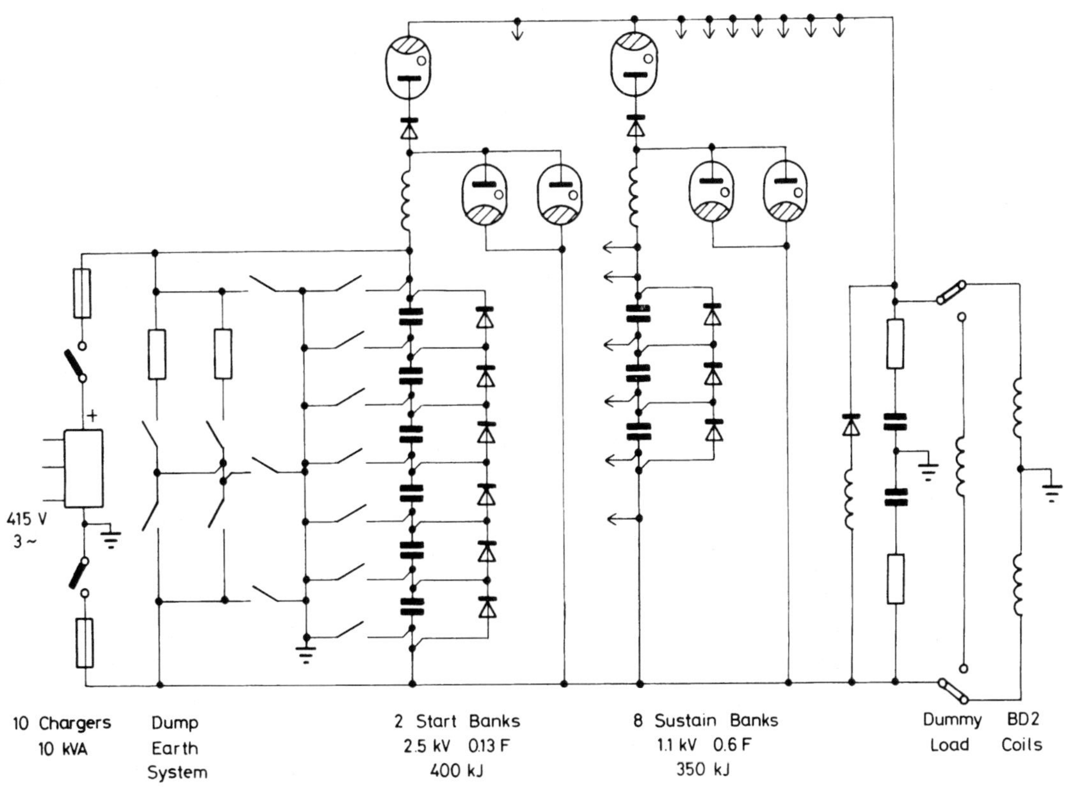

Fig. 2 DITE Mk 2 Bundle Divertor Power Supply Schematic

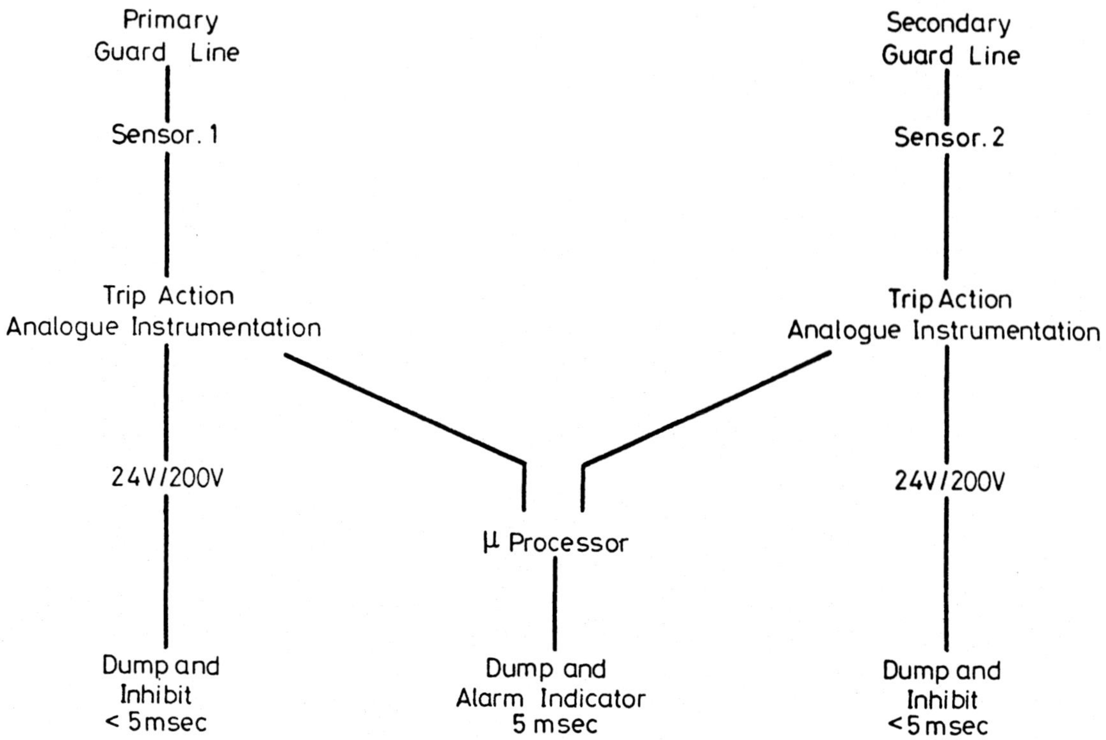

Figure 3. Trip Protection Circuits

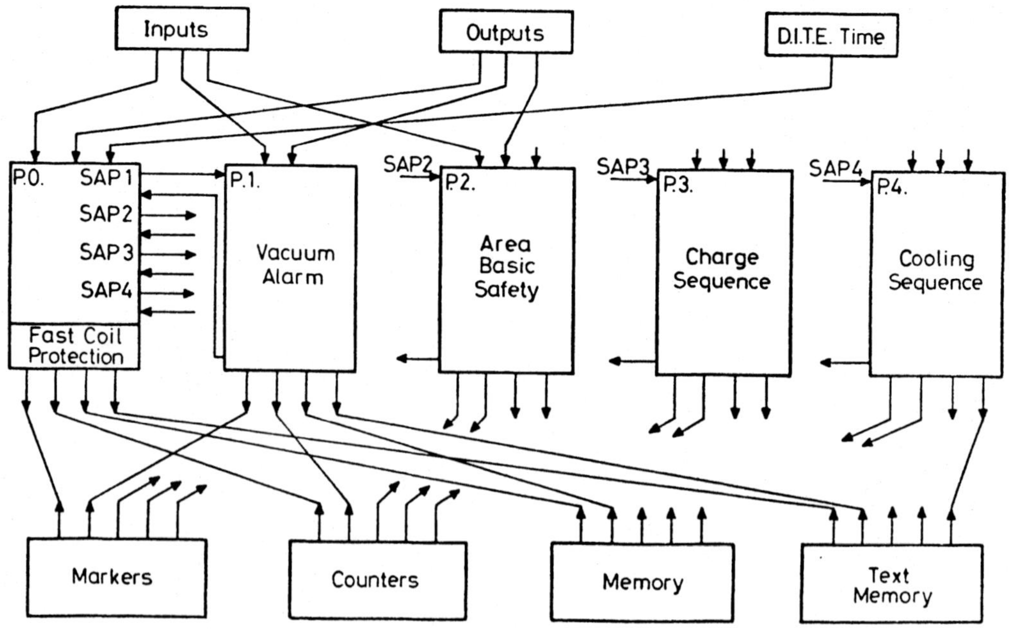

Figure 4. Program Structure

PLASMA CURRENT CONTROL SYSTEM IN JIPP T-II

S. Tanahashi, K. Matsuura, J. Fujita, A. Miyahara and JIPP T-II Group
Institute of Plasma Physics, Nagoya University, Nagoya 464, Japan

ABSTRACT

The Control of plasma current has been adopted in JIPP T-II tokamak/stellarator. The power supply for the control is a combined system of transistor and thyristor sections which are connected in series. A role of the transistor section is quick and fine control. This power supply is coupled to the primary circuit of current transformer with a stack of 24 semiconductor diodes. This stack also plays a role of crowbarring ohmic heating current.

We found experimentally that the plasma current can be kept constant within an error less than few percent. The one-turn loop voltage, ion temperature and electron temperature are found to be almost constant during the control.

1. INTRODUCTION

The constant operation of plasma current leads to considerable simplification in the analysis of experimental data, such as calculations of ohmic power input to the plasma. This operation also improves the positional stability when vertical magnetic field control is applied. The constant current operation has been widely adopted in toroidal devices by the use of transistors, thyristors, hard tubes or some pulse-forming-network techniques [1].

As it is well known, a transistor has a quick response although the rated power is generally small, while thyristor can control a high power though the response is rather slow. We have developed a power supply based on the combination of transistors and thyristors to realize a quick response at a high power level. The system has been applied to JIPP T-II for the control of plasma current.

2. OHMIC HEATING CIRCUIT OF JIPP T-II

JIPP T-II is a hybrid device of tokamak and stellarator. Basic parameters of the device are listed in Table I. A schematic diagram of ohmic heating circuit for JIPP T-II is shown in fig.1. In the figure, C_1 is the 1st capacitor bank (20 kV, 500 μF, 100 kJ) for producing plasma, C_2 is the 2nd capacitor bank (5 kV, 12800 μF, 160 kJ) for ohmic heating, C_3 is the 3rd capacitor bank (5 kV, 7200 μF, 90 kJ) additionally used to superpose a rapid increase or decrease in plasma current, S_1, S_2 and S_3 are ignitrons for firing the capacitors, and S_{C1} and S_{C3} are crowbar ignitrons for maintaining plasma current.

We have two modes of operations. One is the operation with the 3rd capacitor bank and without current control system. We can select the polarity of the capacitor bank either to increase or decrease the plasma current. The other is the operation without the 3rd capacitor bank and with control system.

In JIPP T-II, loop voltage is typically about 2 V; 360 V in terms of primary voltage of the current transformer. Plasma current up to 150 kA has been achieved; 0.8 kA in terms of primary current. The required power to control the plasma current will be of the order of ohmic input power, that is 300 kW. Details of the control circuit will be described in section 3.

The function of the ohmic heating circuit in the case of the control is as follows: When

C_1 and C_2 are fired, there appears a high voltage in the primary circuit of the current transformer as is shown in fig.2. Therefore, diodes D_1-D_{24} are back biased to be cut off (period A in fig.2). When the voltage decrease below the no-load voltage of the control system, the current flows from the control system to the current transformer. Then, we can control plasma current by driving the transistor/thyristor control system (period B in fig.2). When the reference signal to the controller is turned off after programing, the primary voltage of the current transformer decreases to be negative. In this period, diodes play a role of crowbar (period C in fig.2).

3. CURRENT CONTROL SYSTEM

3.1 Transistor section

In the transistor section, 1600 transistors of the type 2SD 520 (collector loss 100 W at 25°C) are used. Typical ratings of this transistor are listed in Table II. Here 8 transistors are connected in series. These stacks are connected in parallel; 200 stacks in total. With this arrangement, we can regulate the current more than 1 kA and voltage as high as 300 V for 1 second. The response time was found to be less than 50 μs.

There is leakage current of transistor section and it consumes the flux of the current transformer. As shown in fig.1, a thyristor switch SCR 1 is used at the output of transistor section to prevent the flux consumption. This SCR 1 is turned on at the time when C_1 is fired.

3.2 Thyristor section

Thyristor section is a conventional 3-phase SCR power supply; firing interval is 2.8 ms. The maximum current is 1.2 kA and the maximum voltage is 400 V. The output of this section is controlled so as to keep the ratio $(V_{TH}/(V_{TR} + V_{TH})$ roughly constant where V_{TH} and V_{TR} are output voltage of thyristor and transistor section respectively as shown in fig.1.

3.3 Crowbar diodes

In fig.1, D_1-D_{24} are the diodes FD1600-50. The maximum rating of the voltage of the diode is 2.5 kV (repetitive peak reverse voltage) and that of the current is 1.6 kA (average forward current). The diodes are selected so as to have the same reverse recovery times. In order to protect from break down, resistors, capacitors and Zeners are added in parallel to each diode to balance the reverse voltage. Beside this, we added a reactor L and a capacitor C as shown in fig.3 to prevent voltage surges.

4. ANALYSIS OF THE CONTROL SYSTEM

A block diagram of the plasma current control system is shown in fig.4. A reference signal is generated in Block (1). The plasma current obtained with Rogowski coil (7) and an integrator (8) is compared with this reference. The difference is amplified by pre-amplifier (2). This output is fed through insulation amplifier (3) to drive the transistor section and hence the thyristor section(5). Thus feedback loop is formed and we can control the plasma current (period B).

In fig.5, we show an equivalent circuit for plasma-transformer-transistor system in terms of primary circuit. From this equivalent circuit, plasma current i_p may be written as follows:

$$i_p = \frac{1}{R} \cdot \frac{sT_{CT}v}{1 + s\{T_{CT} + T_p(1 + M/L)\} + s^2T_pT_{CT}} , \qquad (1)$$

where $T_{CT} = M/\gamma = 300$ sec

$T_p = L/R = 75$ ms

s = complex parameter in Laplace transformation

$\gamma = 0.35\ \Omega = 0.25 + r_1$

$M = 90\ H = aM_1$

$L = 75\ mH = a^2L_p$

$R = 1\ \Omega = a^2R_p$

r_1 = resistance of primary winding in current transformer = 0.1 Ω

a = turn ratio of windings = 180

L_p = self-inductance of plasma = 2.2×10^{-6} H

R_ℓ = liner resistance = 4 mΩ

R_p = plasma resistance ≅ 30 μΩ

M_1 = mutual inductance of current transformer = 0.5 H

A simplification can be made by omitting negligible terms in the frequency range which depends on the control duration. Thus we have

$$i_p = \frac{1}{(R + \gamma)(1 + sT'_p)} V \quad , \tag{2}$$

where

$$T'_p = T_p\left(\frac{R}{R + \gamma}\right) \simeq 56\ \text{ms} \quad . \tag{3}$$

The simplified block diagram is shown in fig.6. On the basis of this block diagram, we can calculate the Bode diagram. The results are shown in fig.7. We see that there is a gain margin g_m of about 50 dB or a phase margin θ_m of about 90°. Thus we can expect that the control system is very stable. The accuracy of the control may be estimated to be $1/(1 + g_m)$= 0.03 (3 %) for slow change in plasma current.

Response of the plasma current to the stepwise change is limited due to the time constant of plasma T_p. In this case response of plasma current I_p is an exponential waveform with time constant T'_p = 56 ms. If we want more rapid change, we must use a techniques of "forcing". For this purpose the third capacitor bank C_3 can be used.

5. EXPERIMENT ON CONSTANT CURRENT OPERATION

Operation of constant plasma current has been carried out in JIPP T-II stellarator. Examples are shown in fig.8 for time variations of plasma parameters, where the control is done for the period from 90 ms to 160 ms. In the figure, evolutions of the plasma parameters without current control are also shown to demonstrate the effect of the current control. We see that the following plasma parameters are kept almost constant with this control: loop voltage V_p, plasma current I_p, electron temperature T_e at the center of the plasma column, half width of electron temperature profile and ion temperature T_i at the center of the plasma column. In this case, experimental conditions are as follows; toroidal magnetic field 22 kG, rotational transform of helical field 0.14, filling pressure of hydrogen 0.75×10^{-4} Torr and setting value of the plasma current 52 kA.

6. CONCLUSION

We have designed and constructed a power supply for plasma current control in JIPP T-II. Transistors are employed for quick and fine control in addition to thyristors. Diodes are used in order to couple the control power supply to the current transformer. These diodes also play a role of crowbar action for the decaying phase of the plasma current. The analysis of the stability of the control system shows that the system is very stable.

Experiments on constant current control show that many plasma parameters are kept almost constant for the period of the constant current operation.

Acknowledgements

The authors wish to express their thanks to Messrs. S. Hirokura, T. Tomita, J. Nishikimi, Y. Ono and H. Ishiguro for their assistance of the experiments.

REFERENCES

[1] MAENO, M., et al., Japan Atomic Energy Research Institute Report, JAERI-M6762 (1976).

TABLE I Basic parameters of JIPP T-II

Major radius	R = 91 cm
Plasma radius	a = 17 cm
Toroidal field	B_t = 30 kG
Helical field	ℓ = 2, m = 4
	$\iota/2\pi$ = 0.25 when B_t = 20 kG
Feedback-controlled vertical field	B_{vf} = 0.5 kG
Magnetic flux of iron core	ϕ = 0.58 v-s (without premagnetization)

TABLE II Maximum ratings of transistor 2SD-520

Maximum collector loss	P_c = 100 W
Maximum collector current	I_c = 7 A
Maximum voltage between collector and emitter	V_{CE} = 400 V

(made by Mitsubishi Co. Ltd.)

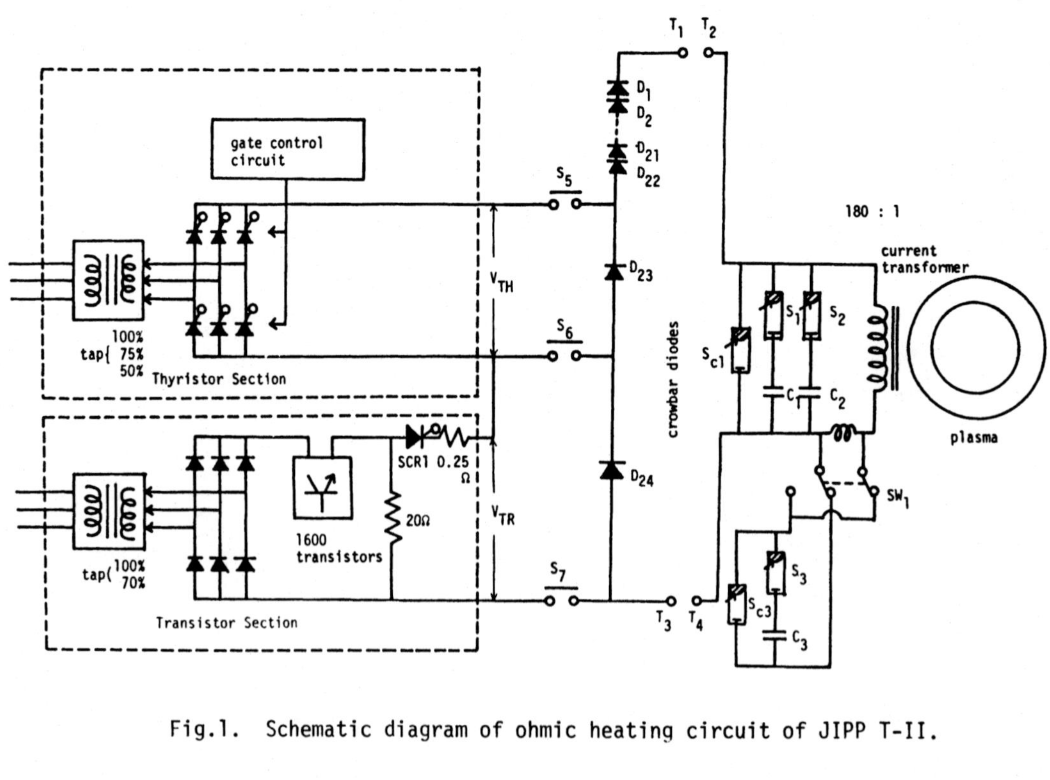

Fig.1. Schematic diagram of ohmic heating circuit of JIPP T-II.

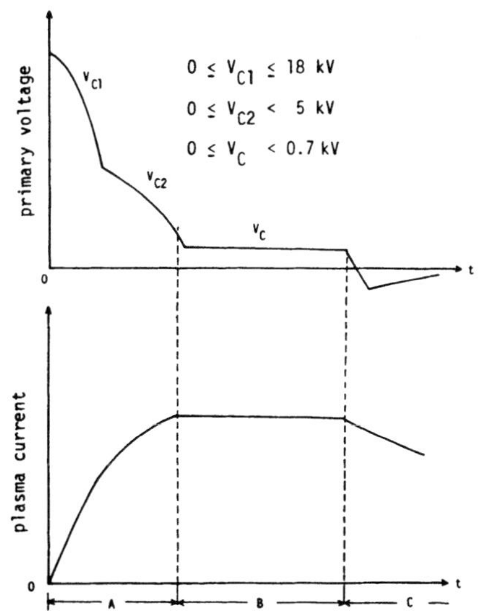

Fig.2. Schematic waveforms of primary voltage and plasma current with constant current control. V_{C1} and V_{C2} are voltages of capacitor C_1 and C_2 respectively. V_C is output voltage of the control supply.

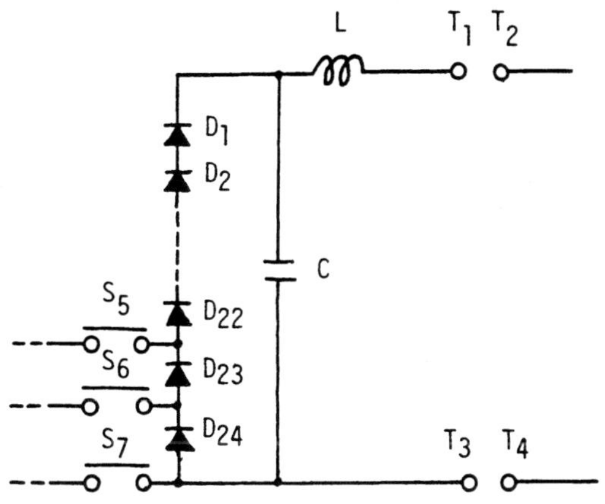

Fig.3. Reactor L and capacitor C are used in series to prevent voltage surges.

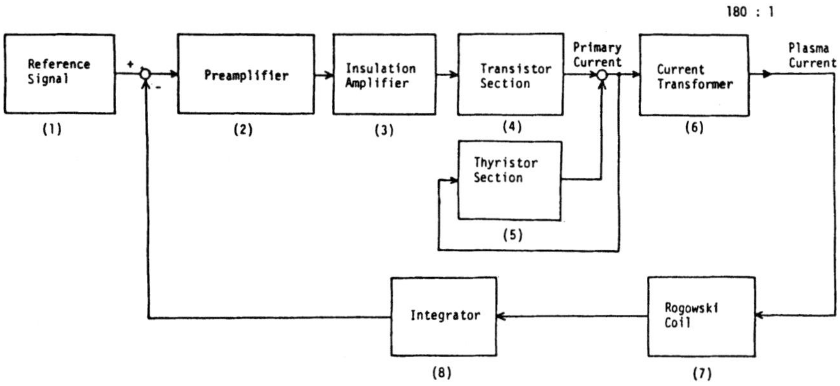

Fig.4. Block diagram of plasma current control system.

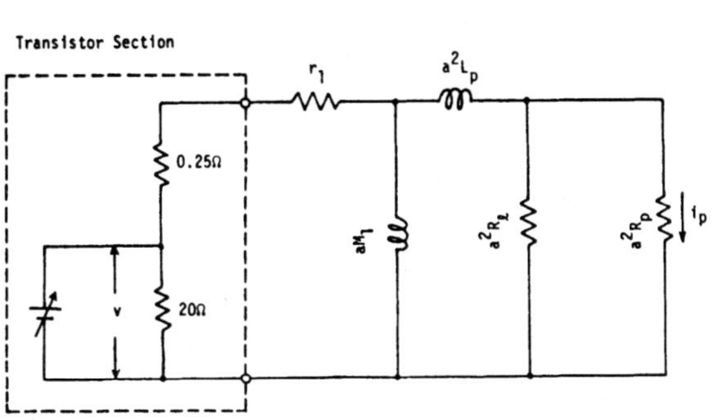

Fig.5. Equivalent circuit converted into primary circuit of current transformer.

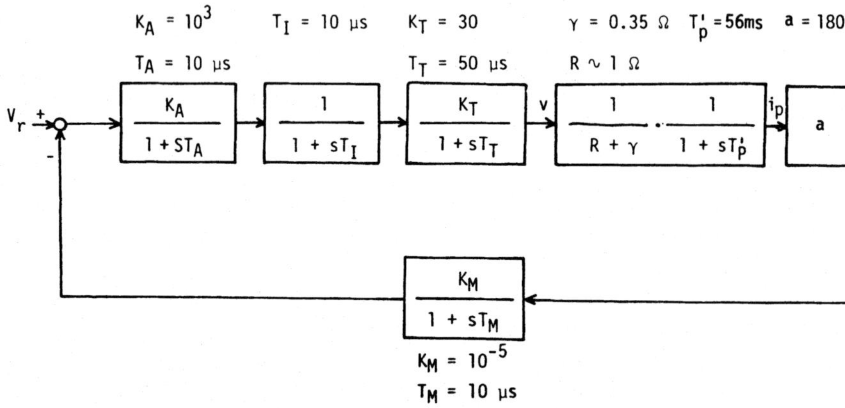

Fig.6. Block diagram of plasma current control system.

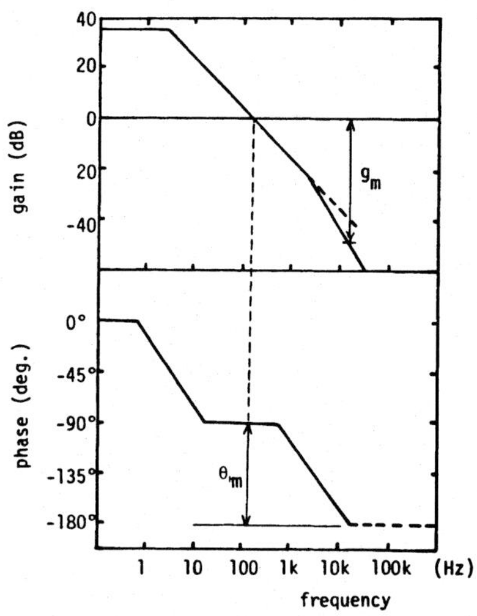

Fig.7.
Bode diagram of plasma current control system. g_m is gain margin, θ_m is phase margin.

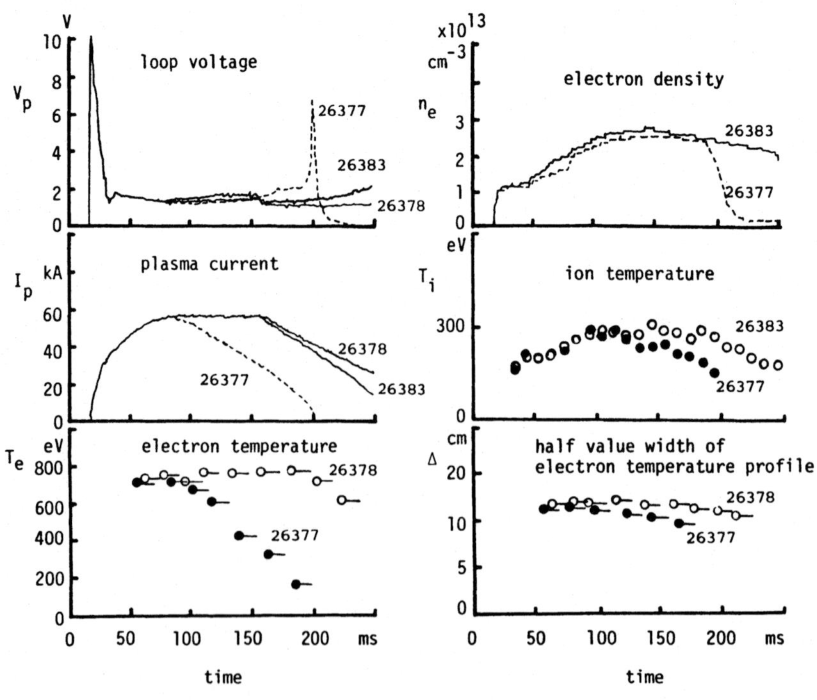

Fig.8.
Time variation of plasma parameters: loop voltage V_p, plasma current I_p, electron temperature T_e, electron density n_e, ion temperature T_i, half value width of electron temperature profile Δ. Solid lines and open circles show the case with constant plasma current operation (shot nos.26378 and 26383), broken lines and closed circles show the case without the control (shot no. 26377).

POWER SUPPLY SYSTEM OF NEUTRAL BEAM INJECTOR FOR JFT-2

S. Matsuda, T. Shibata H. Nagamura, O. Higa, S. Yoshizawa
Y. Shimomura, S. Yamamoto N. Fujiwara, S. Ioka
Japan Atomic Energy Research Institute Toshiba Corporation
Tokai Japan Tokyo Japan

ABSTRACT

The neutral beam injection system for JAERI Fusion Torus-2 (JFT-2) was developed and is now in operation. This is one of the largest injection systems in Japan and was de signed, fabricated and tested within about one year. In this paper, the design and test results of power supply system is described.

1. INTRODUCTION

The system has injected 1.5MW of 35kV neutral hydrogen atoms for 50ms through two beam lines [1] connected to a single port of JFT-2 Tokamak. Two sets of duoPIGatron ion source [2] have been mounted in each beam line. In order to supply power to these ion sources, the power supply system consists of two sets of accel and decel power supplies and four sets of PIG power supplies. The voltage, current and pulse duration of this ion source are 35kV, 28A, 100ms respectively. However, the rated voltage, current and pulse duration of the accel power supply are 40kV, 65A, 300ms for the future system.

2. POWER SUPPLY

The schematic diagram of this neutral beam power supply is shown in Fig.1.

Ac-power is delivered from the commercial power line, 3.3kV-3ϕ-50Hz to two set of accel power supplies and to the step-down transformer for PIG and decel power supplies. An ac-harmonic filter is also provided in the ac 3.3kV line.

Our design feature of the accel power supply is to use an ac-thyristor controller, a thyristor crowbar switch, a series regulator tube (Eimac X-2062J) and a surge blocking core.

A series regulator tube, E-3068 (by Toshiba corporation, 25kV-12A) is used for the decel power supply.

A set of PIG power supply consists of an arc power supply, a filament power supply, a source coil power supply and auxiliary power supplies. The dc-circuits of the PIG power supply is operated at the potential of the accel power supply. We immersed the insulation transformer for two sets of PIG power supplies in an oil tank to make it's size compact.

Fundamental specifications of each power supply are listed in Table I.

TABLE I

Fundamental specifications of each power supply

	Voltage	Current	Duration	Duty	Ripple
Accel power supply	10 ∿40 kV	65 A	300 ms	1/30	≦±1.5%
Decel power supply	-(1 ∿5) kV	10 A	300 ms	1/30	≦±1.5%
Arc power supply	180 V	800 A	300 ms	1/30	≦±1.5%
Filament power supply	10 V	1,200 A	3 sec.	1/30	≦±1.5%
Source coil power supply	15 V	150 A	cont.	-	≦±1.5%

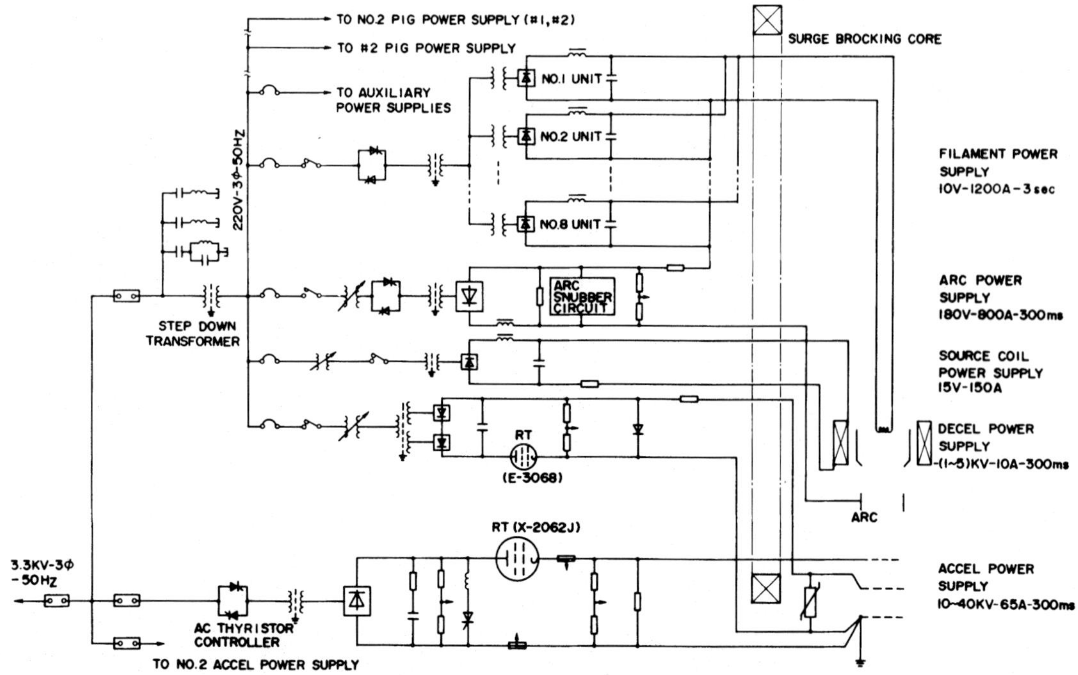

Fig.1 SCHEMATIC DIAGRAM OF NEUTRAL BEAM POWER SUPPLY SYSTEM

2.1 <u>Accel power supply</u>

The simplified schematic diagram of the accel power supply is shown in Fig.2. The ac thyristor controller is operated on utility power line, 3.3kV-3ϕ. Each phase of the ac thyristor controller is composed of 3 in series and in antiparallel connection of 4000V-1500A thyristors. This thyristor controller has following functions:

1) To regulate the voltage of dc-filter capacitors to a higher value than the output voltage of the regulator tube: The average anode dissipation of this tube depends on the reference value of the tube voltage drop. This reference value is determined considering the capaciters' voltage dip that is caused when the tube is switched on, and the minimum voltage drop at which the tube can be operated stably.

2) To interrupt the primary ac current within one cycle after ignition of the thyristor crowbar switch in order to protect the rectifier and the thyristor crowbar switch.

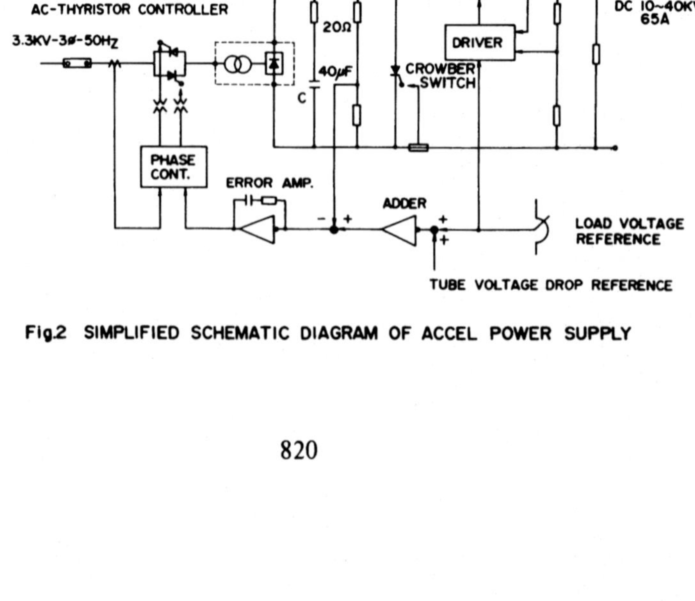

Fig.2 SIMPLIFIED SCHEMATIC DIAGRAM OF ACCEL POWER SUPPLY

The output voltage of the ac thyristor controller is delivered to the transformer and converted into dc voltage by the diode rectifier. The transformer and the rectifier are immersed in an oil tank. The rectifier is composed of bridge circuit (6 pulses). Where, each arm of this rectifier consists of 2 in parallel and 54 in series connection of 3000V-25A avallanche diode.

The regulator tube (x-2062J) has the following functions:

1) To regulate the load voltage
2) To supply the load voltage with a short rise time (within 100μsec)
3) To shut off the current within 20 μsec when break downs occur in the ion sources

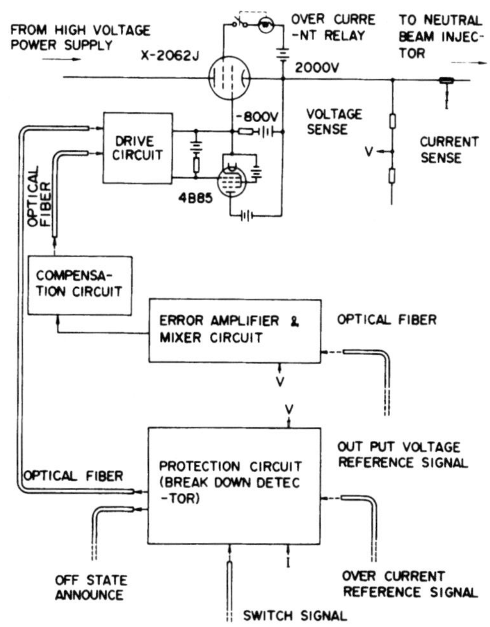

Fig.3 CONTROL AND PROTECTION SYSTEM OF ACCEL POWER SUPPLY

The control and protection system of the accel power supply is shown in Fig.3. The load voltage reference is transmitted from the ground level V/F converter (Voltage to frequency converter) to the hot deck of the regulator tube. The F/V converter demodulates the frequency to the reference voltage, and the load voltage is regulated by the regulator tube through an error amplifier. When the voltage value of the control grid is set below about -500V, the regulator tube shuts off the load current, while this tube can deliver the rated current (65A) under the condition of the control grid voltage above -50V at a screen voltage of 2 kV. The anode current is not proportional to the control grid voltage at the whole negative voltage range from-50V to-500V. Therefore a compensation circuit must have a function for linearlizing the non-linear charactaristic of the tube and this function is useful to stabilize the feed back control. The voltage break down in ion sources is detected by overcurrent and/or undervoltage senser of the load. The shut off signal is transmitted to the drive circuit through an independent optical fiber other than the voltage control loop.

To reduce the inflow of stored energy from the power supply to the electrodes of the ion source, and to suppress the surge voltage in the whole power supply system, the surge blocking core device is provided. This device consists of a ring core, resistors and a bias circuit which compensates core saturation phenomena caused by the nomal current of the accel power supply. The schematic diagram of this device is shown in Fig.4. The specifications and design features are listed in Table II.

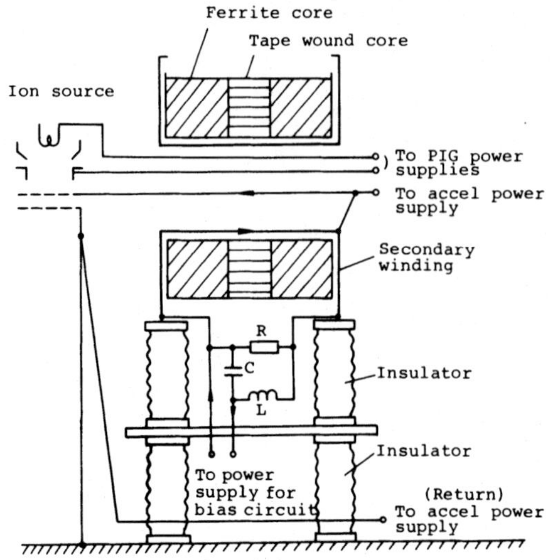

Fig. 4 Schematic diagram and configuration of surge bloking core device

TABLE II

Specifications and design features of surge blocking core

Estimated stray capacitance of accel and PIG power supply	0.07 μF
Allowable energy inflow into ion source's electrodes from stray capacitance	10 J
Testing voltage (impulse)	80 kV
Frequency characteristics (ferrite ring core and tape wound ring core were used together)	∿1 MHz
Volt - sec. of core device	0.02 V·s
Magnetizing inductance (at 1 MHz and small flux density)	>15 μH
Resistance of secondary circuit (non-inductive high voltage resistor)	7 Ω
Inductance of bias circuit's air core reactor	1 mH

2.2 Arc power supply

The schematic diagram of the arc power supply is shown in Fig.5. A dc reactor reduces the current ripple and maintains the rectifier current to a constant value while the arc snubber circuit operates.

The arc snubber circuit has functions for shortly-limiting the arc current to about 1/10 of the rated value and increasing it again to the rated value in accordance with the turn on of the accel voltage. This function is useful to decrease the probability of break downs in the beam initiation phase. The current limiting level and the rise time of the arc current are adjustable.

GTOs (gate turn off thyristor SG600R21) are used for the arc snubber circuit to realize these requirements. The whole GTO assembly consists of 2 in series and 2 in parallel connection of 1300V-600A GTOs. Two series connection has a capability of withstanding the voltage induced by the cable inductance in the case of very small rise time of the arc current. The snubbing of the arc current can be done by triggering GTOs. When the arc current is reduced to the required value, GTOs are turned off by negative gate pulses. Thus the current of the dc reactor is delivered to charge the rise-time adjusting capacitors and then to the arc load. The rise time can be adjusted easily by changing the parallel number of capacitors. The test result of this circuit is shown in Fig.6.

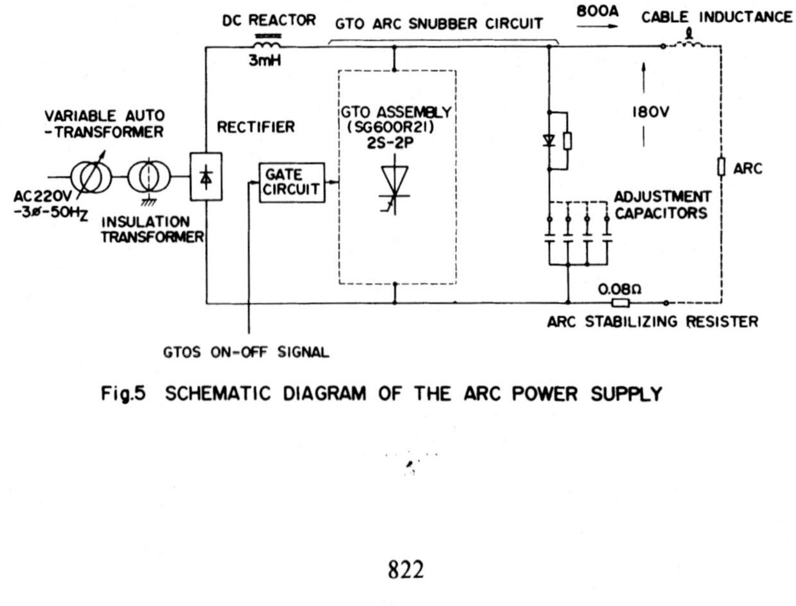

Fig.5 SCHEMATIC DIAGRAM OF THE ARC POWER SUPPLY

2.3 Grounding system

The grounding system of the power supply is separated into two sections, an ac-grounding system and a dc-grounding system. In the dc-grounding system, main components in the dc circuit such as a thyristor crowber switch, a series regulator tube of accel power supply, PIG insulating transformers are put on a common dc-grounding conducting plate which is insulated from the floor of the building.

This grounding system is expected to reduce malfunctions of the control circuits associated with break downs of the ion souces. The measured surge voltage between this conducting plate and the floor of the building was about $\pm$200V when the break down occured.

3. CONTROL SYSTEM

Control system has functions for control, monitor, and protection of the power supply system and the beamline system. There are four modes of operation. The typical mode change is illustrated in Fig.7. In the emergent case such as malfunction of equipment and/or failure, the injection is stopped immediately to protect the beamline components and the others. And the operation mode is changed to a stop mode or a preparation mode according to fault conditions.

The power supply system is controlled by a sequential controller that consists of the IC preset timer and the IC logic circuits. The signal interface between the sequential controller and the power supply is insulated by the optical device to avoid malfunctions of this controller. The indicators, test terminals and digital counters that count shut-off time of the regulator tubes are also provided. The timing chart of the power supply and the sequential controller is shown in Fig.8.

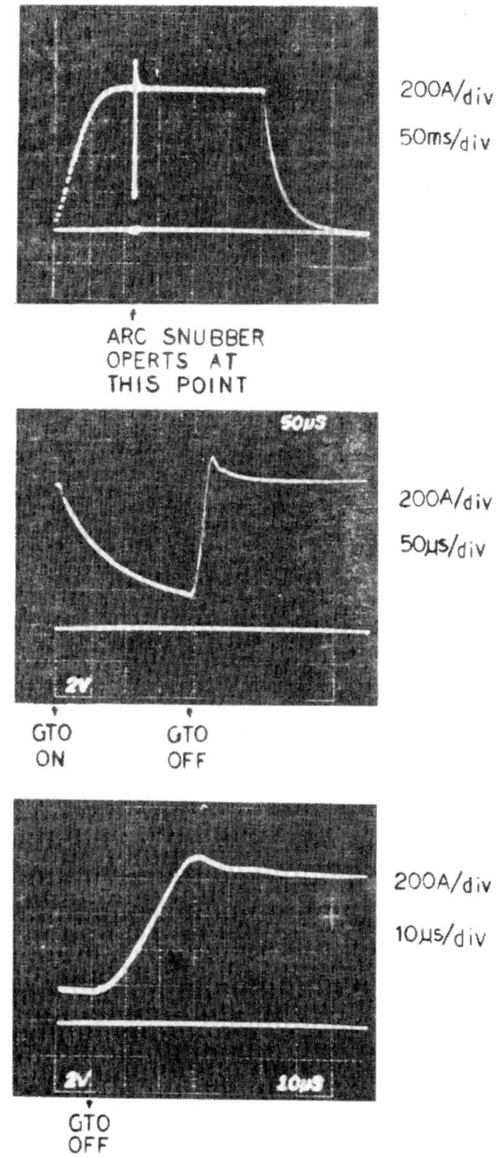

Fig.6 ARC CURRENT WHILE ARC SNUBBER OPERATS

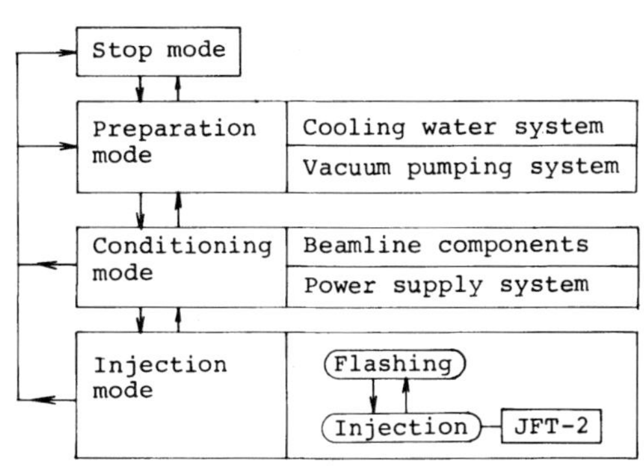

Fig. 7 Typical mode change of operation

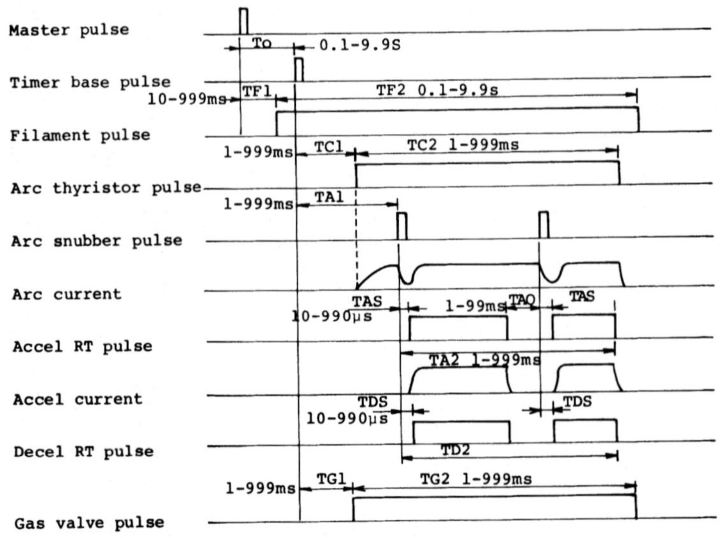

Fig. 8 Timing chart of sequential controller and power supply

4. CONCLUSIONS

Power supply system of neutral beam injector for JFT-2 was fabricated and tested both at factories and on site. This power supply system has been operating for five months at JAERI, and plasma heating experiments has been made successfully [3] [4] . As a result of these tests and experiments, authors believe that this system will make great contributions for further NBI system development.

5. REFERENCES

[1] K. Anno et al.; Sym. on Fusion Technology (1980) .... conference papers

[2] Y. Ohara et al.; Proc. 8th Sym. Eng. problems on Fusion Reserch, 198. (1979) .... conference papers

[3] N. Suzuki et al.; IAEA-CN-38/T-2-3 .... conference papers

[4] Y. Shimomura et al.; JAERI M 9065 (1980) .... for reports

A HALF MEGAWATT RF SYSTEM FOR ION-CYCLOTRON HEATING OF THE ERASMUS TOKAMAK

A.M. MESSIAEN+, G. BOSIA++, V.P. BHATNAGAR, R.R. WEYNANTS+++

Laboratoire de Physique des Plasmas - Laboratorium voor Plasmafysica
Association "Euratom-Etat belge" - Associatie "Euratom-Belgische Staat"
Ecole Royale Militaire - B 1040 Brussels - Koninklijke Militaire School

ABSTRACT

We discuss the design and operation of an economic rf heating system for ion-cyclotron resonance heating (ICRH) or Alfven wave heating experiments. It consists of a power Colpittsoscillator (4-15 MHz) delivering a pulsed power output that exceeds half a megawatt at 5 MHz, an associated strip line antenna system and a matching network. This system has been successfully operated for ICRH studies on the Erasmus tokamak and can be easily upgraded to an economic MJ-Unit.

1. INTRODUCTION

The considerable amount of additional heating power necessary for the heating of the next generation of tokamaks will require a vigorous development of the technology of additional heating system and economics will play a dominant role in the selection of such systems. For the case of RF heating, where recently very successful experimental results have been obtained, especially in the ICRH range, the future development will mainly be focused on the flexibility and power handling capability of the couplers and on the economics and reliability of the power transmission system.

For the ion-cyclotron heating experiments currently underway on the ERASMUS tokamak we have developed and successfully operated an RF generator and coupling system which can be upgraded into an economic, MJ-unit for the heating of a low field machine (< 1.5 T) at the ion-cyclotron resonance. The same technology permits however, by a straightforward extrapolation, to build viable MW modules for the Alfven wave heating of a reactor plasma at costs less than 0.3 $/W.

The main features of the RF system are the following :

a) A Colpitts power oscillator is constructed with a coupling system such that the RF power level is not very sensitive to the exact value of the loading resistance or change of the reactance of the antenna.

+ Maître de Recherches FNRS
++ On leave from Lab. di Cosmogeofisica del CNR (Torino, Italy).
+++ Bevoegdverklaard Navorser bij het NFWO.

b) This oscillator uses a reliable industrial induction heating triode (Philips YD 1342).

c) Oscillator grid keying is used which allows an accurate timing of the rf pulse.

d) The antenna is of the "all metal" shielded strip line type. The stainless steel electrostatic shield is also used as a limiter of the tokamak plasma in order to maximize the coupling of rf energy to the plasma.

e) The matching between the oscillator and the antenna is made by means of a resonating transmission line. This system is simple and allows one to achieve, even at maximum power output, a high Q anode circuit for the oscillator.

f) The oscillator and the coupling circuits are tunable.

g) The oscillator presently uses a condenser bank anode power supply but can also be energized by a high voltage, high current power supply with appropriate modifications to deliver long pulse lengths.

II. CLASS-C OSCILLATOR DESIGN.

II.a. Numerical simutation of the oscillator.

We use the basic theory of class C amplifiers and oscillator as given in reference [1]. We consider the situation and notation of fig.1 for total grid and anode voltage and current of the triode. In a (V_g, V_A) diagram of the tube characteristics, for the case shown in fig. 1 the operation takes place along the line AA', as shown on fig. 2. Point A is characterized by V_{gM}, V_{Amin} and I_{AM}, I_{gM} i.e. the maximum values of anode and grid current flowing through the tube during each period of oscillation. Let us remark that I_A flows only when the operating point on the line AA' is above the tube cut-off ($I_A = 0$) and that I_g flows when $V_g > 0$. The conduction angle for anode and grid current is represented by θ_A and θ_g respectively.

In an oscillator, such as that corresponding to the circuit of fig. 3, once the tube chosen, one has to specify the dc plate voltage supply V_{Adc}, the feedback ratio $h = V_g/V_A = (V_{g\,bias} - V_{gM})/(V_{Adc} - V_{Amin})$ and the dc grid current leak resistor R_g and then the complete operation will be determined by the value R_A of the equivalent rf resistive load seen by the anode. Referring to fig. 3, h is obtained from the ratio of C_1/C_2 of the capacitors in the tank circuit of the Colpitts oscillator. The value of the dc grid bias V_{gbias} is equal to $R_g\ I_{gdc}$ where I_{gdc} is the dc Fourier component of the I_g current pulse. The amplitude of the rf plate voltage $V_A = R_A\ I_{A1}$ where I_{A1} is the fundamental Fourier component of the I_A current pulse. The Fourier components I_{gdc}, I_{g1} and I_{Adc}, I_{A1} are respectively related to θ_g and I_{gM} or to θ_A and I_{AM} by the solutions of the Fourier analysis of the I_g or I_A current pulses as given in reference [1] (see fig. 13-7 p. 461). An approximate analytical representation of the solutions is given below. Denoting μ, the amplification factor of the tube, the class C oscillator is described by the following system of equations :

$$I_{gM} + I_{AM} = C_4 \left(V_{gM} + V_{Amin}/\mu\right)^{E_4} \quad \text{(Langmuir-Child Law)} \tag{1}$$

$$\theta_g = 2 \cos^{-1} \left(\frac{V_{g\ bias}}{V_{g\ bias} + V_{gM}}\right) \quad (2) \ ; \quad V_{g\ bias} = R_g\ I_{gdc} \tag{3}$$

$$h = \frac{V_{g\ bias} + V_{gM}}{V_{Adc} - V_{A\ min}} \quad (4) \ ; \quad \frac{I_{AM}}{I_{gM}} = C_3 \left(\frac{V_{A\ min}}{V_{gM}}\right)^{E_3} \tag{5}$$

$$\theta_A = 2 \cos^{-1} \{(1 + [(\mu V_{gM} + V_{Amin})/(\mu V_{g\ bias} - V_{Adc})])^{-1}\} \tag{6}$$

$$V_{Amin} = V_{Adc} - R_A\ I_{A1} \quad (7) \ ; \quad I_{Cdc} = C_1\ (I_{CM})^{E_1},\ C = g \text{ or } A \tag{8}$$

$$I_{C1} = C_2\ (I_{CM})^{E_2},\ C = g \text{ or } A \tag{9}$$

Eq (5) is an approximate law expressing the ratio of I_A and I_g when $V_g > 0$ for triode of cylindrical coaxial geometry [2]. The coefficients C_3, E_3, C_4, E_4 are chosen in order to have the best fit with the published characteristics of the tube in the chosen range of V_{gM} and V_{Amin}. The coefficient C_1, E_1, C_2, E_2 in eqs.(8) and (9) are chosen to give the representation of the aforementionned Fourier analysis of I_g and I_A.

For a given tube (i.e. values of μ, C_{1-4}, E_{1-4}) and a chosen operating condition (i.e. values of h, R_g and $V_{A\ dc}$) we solve numerically the non linear system (1) - (9) as a function of the rf anodic load. From this solution, we derive the following power relations [1] : input dc power to the tube $P_{in} = I_{Adc}\ V_{Adc}$; power loss in the grid resistor $P_{Rg} = (V_{g\ bias})^2/R_g$; power loss in the grid $P_g \simeq 0.9\ V_{gM}\ V_{bias}/R_g$; output rf power $P_{out} = 1/2\ R_A\ I_{A1}^2$; available power for an external load $P_A = P_{out} - P_{Rg} - P_g$; oscillator efficiency $\eta = P_A/P_{in}$.

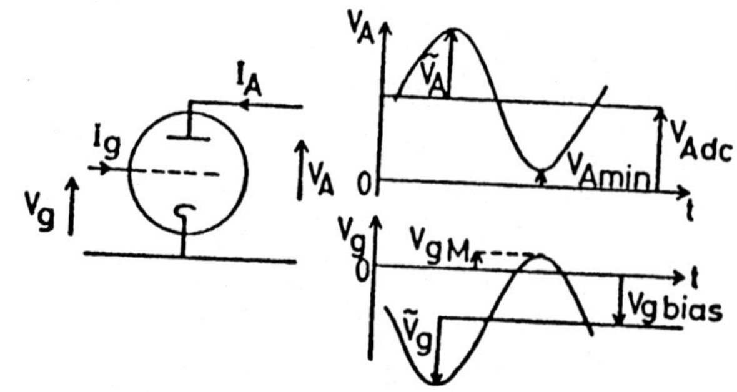

Fig. 1. Anode and grid voltage in class C operation

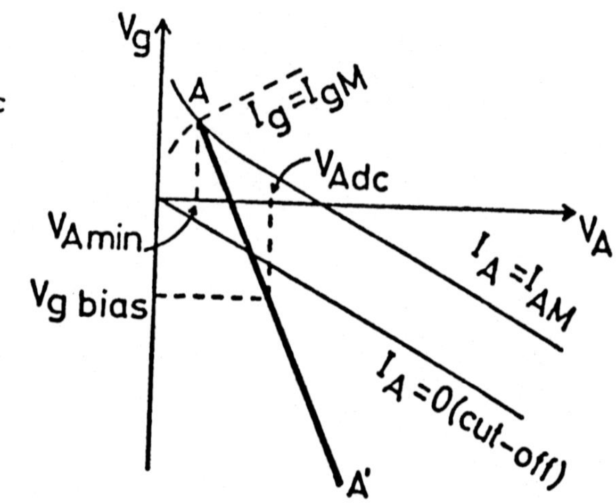

Fig. 2. Operating line AA' of the oscillator on the V_A, V_G tube characteristics

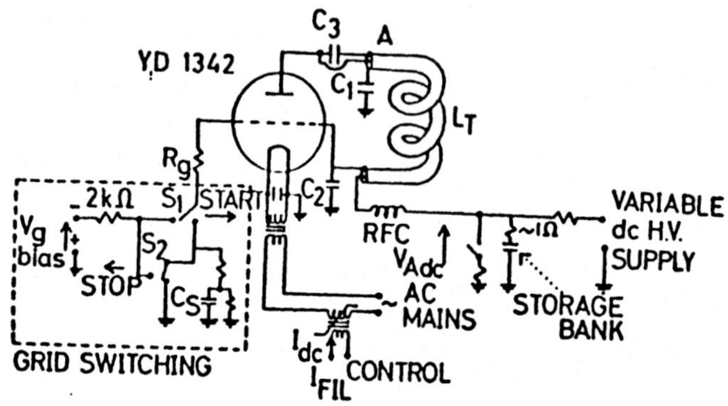

Fig. 3. Simplified schematic of the oscillator

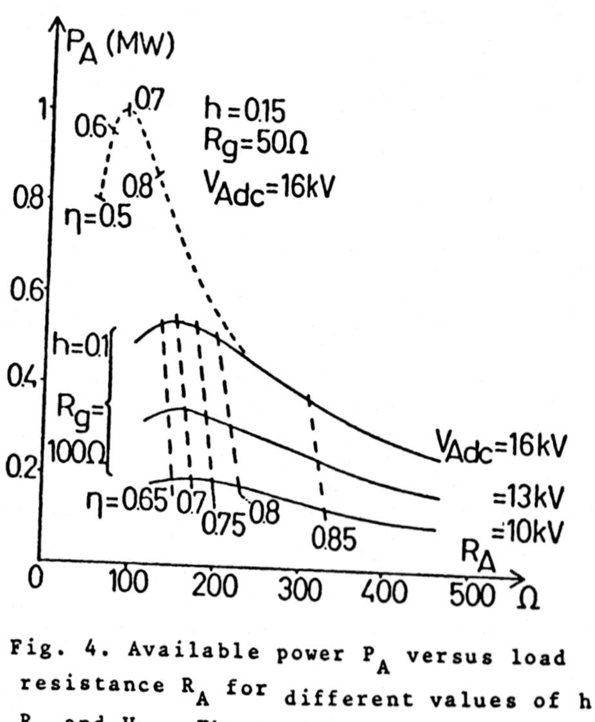

Fig. 4. Available power P_A versus load resistance R_A for different values of h, R_g and V_{dc}. The oscillator efficiency η is also shown

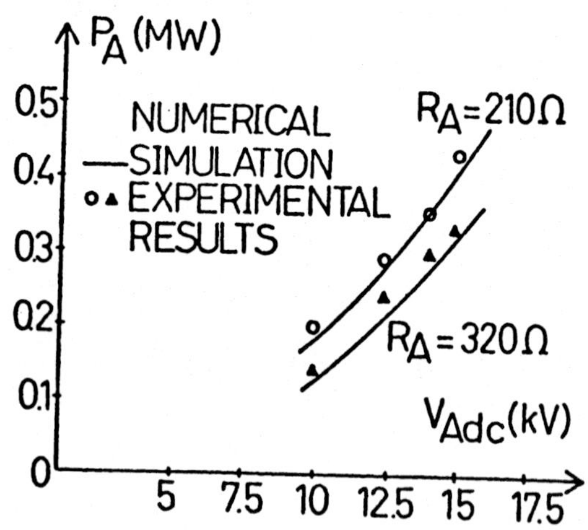

Fig. 5. Comparison between numerical simulation and experimental results for P_A versus V_{Adc} for two values of R_A.

II.b. Results of the numerical simulation.

The operation of the oscillator has been solved for different values of V_{Adc}, h and R_g as a function of R_A in the case of the Philips triode YD 1342. In fig. 4, results showing P_A versus R_A are given for two different set of values of h and R_g. In the first set (h = 0.1, R_g = 100 Ω) the peak cathode current is limited to 250 A (limit value recommended by the manufacturer). In this case, for a voltage V_{Adc} = 16 kV, a value of P_A = 540 kW can be reached. The top curve of fig. 4 corresponds to a peak cathode current of 514 A and a value of maximum P_A = 1 MW. Such a peak cathode current can be obtained for short periods of time (∿ 0.1 s) by overheating the triode filament with an increase of filament voltage of ± 10 % [3] during a few seconds before and during the rf pulse. Table I shows the optimum solution (largest P_A) for the two sets of values of R_g and h, and the values of C_{1-4}, E_{1-4} chosen.

R_g	h	V_g bias	I_{gM}	V_{gM}	I_{AM}	V_A min	θ_g	θ_p	I_{A1}	I_{Adc}	P_{out}	P_{Rg}	P_{in}	P_g	P_A	η	R_A	V_{Adc}
Ω		V	A	kV	A	kV	radian		A	A	kW	kW	kW	kW	kW	%	Ω	kV
100	0.1	710	40	0.56	212	3.3	1.95	2.62	86	49	547	5.0	785	3.6	539	69	147	16
50	0.15	1010	107	0.994	397	264	2.1	2.5	156	87	1039	20.4	1393	18	1000	72	86	16

TABLE I

TUBE	C_1	E_1	C_2	E_2	C_3	E_3	C_4	E_4	μ
YD 1342	9.67×10^{-2}	0.906	0.213	0.674	2.4	0.447	2.32×10^{-2}	1.43	33

From the numerical simulation, one can conclude :

(1) There is a relatively large range of R_A values around which P_A remains close to its maximal value $P_{A\,max}$ (see fig. 4). (2) For given h and R_g, η is roughly independent of V_{Adc}. (3) P_A is roughly proportional to V_{Adc} for given R_g, h and R_A; which implies $\hat{V}_A \propto V_{Adc}$. (4) parametrical dependance of P_{Amax} versus h and R_g for a given V_{Adc} is found to behave as $P_{Amax} \propto (h)^{0.71}(R_g)^{-0.53}$ for the YD 1342 tube; this relation holds, of course, if P_A is not limited by peak cathode current saturation.

II.c. Oscillator construction.

The grounded cathode Colpitts oscillator has been designed to operate, with only slight changes in the tank circuit, in the frequency range 4 - 15 MHz (see fig. 3). Components values quoted below corresponds to an operation near 5 MHz. The oscillator tank circuit (L_T, C_1, C_2) is constructed with short connections to the tube following the rules described in ref. [4]. C_1 is a "JENNING" vacuum tunable condenser ($C_1 \simeq 1000$ pF), C_2 is made from disk ceramic capacitors and a vacuum tunable condenser placed symmetrically around the grid connector ($C_2 \simeq 10.000$ pF), the tank coil is made from 3.5 turns of a copper tube ($L \simeq 1$ µH). The anode dc supply, made by a condenser bank, charged at the operating voltage, is connected through a RF choke and a 100 kV dc insulated wire passing through the tube of L_T. C_3 is a 40 kV, 0.01 nF rf bypass condenser. The oscillator is keyed in by the grid. In the standby mode, the grid is biased below cut-off and in the operation mode the grid switching system connect R_g to the grounded cathode, and the grid bias is then automatically obtained as described in IIa. The necessary time constant for the grid bias is obtained from C_2R_g. The grid switching system is made from two specially selected "Kilovac" vacuum relays (HC-1/S30 type) in order to have fairly precise ON and OFF timing (Jitter < 0.3 ms). To start the rf pulse a first switch "start" is activated by the discharge of a capacitor in its coil, and similarly to stop the oscillations another switch "stop" is activated. Arcing in the switch is avoided by charging C_S during the switching operation. The filament heating current is controlled by a saturable reactance for slow turn on as well as to allow filament overheating during short periods of time. The oscillator is enclosed in an aluminium housing, all supplies are fed through RFI filters, the grid switching system is optically controlled from outside.

Fig. 5. shows comparison between numerical simulation and obtained results with an external carbon resistor used as load between anode and ground (with appropriate dc blocking capacitor).

III. ANTENNA AND COUPLING CIRCUIT.

III.a. Antenna system.

The rf launching system in the Erasmus tokamak is constituted by two half-loop antennae symmetrically placed with respect to the equatorial plane (see fig. 6). Each antenna is of the "all metal" type [5] and consists of a short circuited section of a copper strip line; it is electrostatically shielded by a double array of stainless steel metal strips at the side and by a single array on the inner side which faces the plasma. This inner part of the shield in fact serves as the limiter of the plasma discharge (Ø = 39 cm). The design follows the guidelines discussed in ref [6] : 1°) The electrostatical (ES) shield avoid direct contact between edge plasma and the antenna and basically prevent the low density plasma from short-circuiting the line input rf voltage 2°) The ES shield provides the right wave polarization to excite the magnetoacoustic (M.A.) waves 3°) The ES shield is used as the limiter in order to provide the maximum possible coupling to the MA waves. 4°) The line return path is grounded to the vessel at several points which avoids parasitic loading due to rf potential difference between the wall and the ES shield. The strip line is characterized by its capacitance per unit length C (i.e. capacitance between the central strip and the return path with its ES shield), its inductance L and resistance R per unit lenght. R takes into account the losses of the line and the loading due to the plasma, L is roughly independent of the ES shield, but is influenced by the plasma. The way to estimate R and L in presence of plasma is discussed in ref. [6]. The strip line input impedance for a short-circuited section of length l, represented by Z_s is as follows :

$$Z_s = \left(\frac{R + i\omega L}{i\omega c}\right)^{1/2} \text{th}\left(\sqrt{(R + i\omega L)\, i\omega c}\; l\right) \qquad (10)$$

and the power coupled to the antenna P_S is given by

$$P_S = 1/2 \text{ Re } (1/Z_s)\, |V_s|^2 = 1/2 \text{ Re } (Z_s)\, |I_S|^2 \qquad (11)$$

where I_S and V_S are the strip line input current and voltage. The length l has to be chosen smaller than $\lambda/4$ (λ is the current wavelength on the strip line) in order that the antenna operates around a current antinode for maximum coupling to the MA waves [6]. For the case of Erasmus, we have $l \ll \lambda/4$ and $R \ll \omega L$, then $Z_S = R_S + i\omega L_S \simeq Rl + i\omega Ll$.

III.b. Coupling circuit.

The matching between the launching system impedance $Z_S/2$ and the anode resistance R_A required by the oscillator for operation at a power level near P_{Amax} is made by a variable condensor system C_B (disc ceramic + vacuum variable capacitors) placed at the input of the strip lines and a length l_1 of a number of n coaxial cables connected in parallel. The length l_1 is taken $< \lambda_1/4$ (λ_1 : wavelength on the cable). The resulting characteristic impedance Z_o is chosen by the number n (for Erasmus $Z_o = 10\ \Omega$ with n = 5 for the cable RG 218/u). The impedance seen by the line at its output (point B on fig. 6) is $Z_B = R_B + iX_B$ where $X_B = \omega L_B - 1/\omega C_B$, $L_B = L_S/2 + \Delta L$ (ΔL is due to the connections) and $R_B = R_S/2 + R_{SL}$ where R_{SL} is an equivalent serie resistance which takes into account the losses in the coupling circuit. We chose the oscillator operation at a frequency $\omega = (L_T\, C_T)^{-1/2}$ which requires the imput impedance of the line l_1 to be purely resistive, i.e. $X_L = 0$; this is obtained by an adjustment of C_B such that $X_B \simeq Z_o \,\text{cotg}(\beta\, l_1)$ with $\beta = 2\pi/\lambda_1$. The equivalent anode resistance R_A is given by

$$R_A \cong R_L = Z_o^2/(R_B \sin^2 \beta l) \qquad (12)$$

and the Q of the anode circuit (tank + coupling circuit) is given by

$$Q = \frac{1}{R_B} \left[\frac{Z_o}{\sin^2\beta l} \left(Z_o \omega C_T + \frac{\pi\omega}{4\omega_L}\right) + \frac{\omega L_B}{2}\left(1 + \frac{\omega_B^2}{\omega^2}\right)\right] \qquad (13)$$

with $\omega_L = \pi \, \omega/(2 \, \beta l)$, $\omega_B^2 = 1/(L_B C_B)$

For the case of Erasmus we have $\omega/_{2\pi} = 4{,}6$ MHz, $L_S = 0.37$ µH $C_B \simeq 6000$ pF, $l_1 = 8.8$ m, $R_A = 157/R_B$, $R_{SL} = 0.28\ \Omega$ and $Q = 28/R_B$ with R_A and R_B expressed in Ω. Note that $Q > 30$ for the oscillator operation provides a very good oscillator stability.

The value of R_S is measured from the current $I_B = 2I_S$ (by means of a current transformator of "Pearson" type) and from the output power of the generator $P_A \sim 1/2\ R_B\ I_B^2 = 1/2\ I_L\ V_L$ where I_L is the input line current and $V_L \sim V_A$ the input line voltage. These are measured respectively by a current transformer and a capacitive voltage divider which feed (via carefully adjusted lengths of coaxial lines) a "double balanced mixer". It gives a rectified output proportional to P_A. This system is calibrated by means of carbon resistors used as dummy load and placed across the antennae input. The frequency is also monitored by means of a Phase Lock Loop integrated circuit which provides a DC signal proportional to ω. The connection of the cables to the tank circuit in L (which includes a current transformer and a capacitive voltage divider) has to be carefully designed to avoid arcing problems.

IV. TYPICAL RESULTS.

The oscillator allowed us to couple to the tokamak plasma a rf power $P_{HF} = P_A\ R_S/(R_S + 2\ R_{SL})$ more than four times the OH power [7] With an equivalent serie resistance of the 2 antennae $0.15\ \Omega < R'_S = R_S/2 < 0.3\ \Omega$ we have $P_{RF} \sim 0.5\ P_A$. The experiments were performed with $P_A \leqslant 600$kW, $P_{HF} \leqslant 250$ kW, $I_B \leqslant 1500$ A (amplitude). The oscillator characteristics are such that for $I_B = 1000$ A, $I_T \simeq 470$ A, $V_L \simeq 12.5$ kV and $V_s \simeq 9$ kV (see fig. 6).

Fig. 7 shows typical rf pulses. In fig. 7a the plasma conditions are such that toroidal eigenmodes exist and their presence may produce an increase in R'_S up to 0.25 Ω.

In fig. 7 b, a typical result corresponding to the high power output operation is shown.

From rf power experiments carried out during 6.000 shots we can conclude that the operation of this oscillator was very reliable and flexible.Once good vacuum and clean wall conditions were obtained in the machine, no breakdown problems have been observed on the antennae

(V_S < 13 kV amplitude).

The trouble free operation of the oscillator at 600 KW level gives us confidence to upgrade the system successfully to the megawatt level. The economics of such a system looks to be very competitive (< 0.3 $/Watt) when compared with the price (1 $/Watt) usually quoted by the industry for a power amplifier.

ACKNOWLEDGEMENTS.

We thank A. Leskens, J. Neefs, C. Pleurdeau, F. Van Goethem and F. Van Thillo for their skilled technical assistance.

REFERENCES

[1] F.E. Terman "Electronic and Radio Engineering" Mc. Graw-Hill, New-York 1955.

[2] K. Henney "Radio Engineering Handbook" Mc. Graw-Hill, New-York 1950. Chapter 13.

[3] H.F. Dittrich, private communication (1979)

[4] H.F. Dittrich, "Tube for R.F. Heating" Philips Application book, Eindhoven 1977.

[5] TFR Group, Proc. Int. Conf. on Plasma Physics, Nagoya 1980 (invited paper)

[6] R.R. Weynants et al., Proc. 2d. Grenoble-Varenna Int. Sympos. on Heating in Toroidal Plasmas, Como 1980 (invited paper)

[7] V.P. Bhatnagar et al., Proc. 8th Int. Conf. Plasma Physics and Contr. Nucl. Fusion Research, Brussels 1980 (paper IAEA - CN-38/D-4)

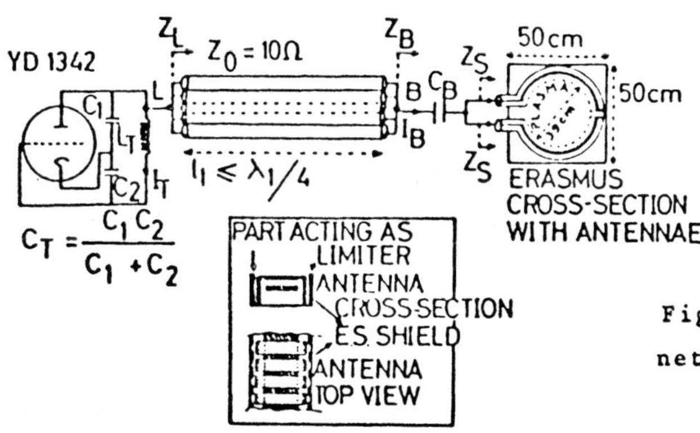

Fig. 6. Antenna system and coupling network used on Erasmus.

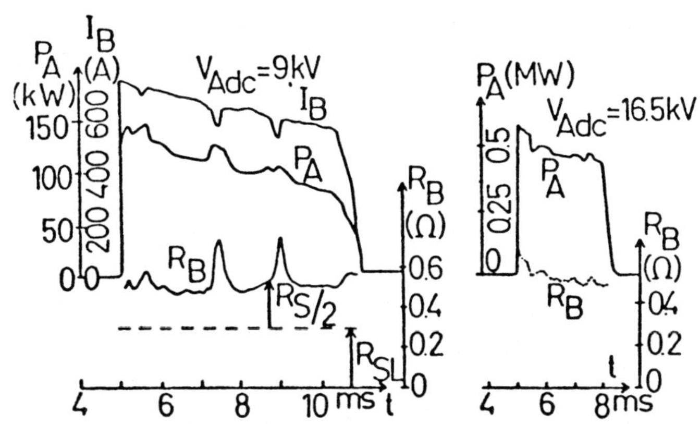

Fig. 7. Typical oscillator pulse and equivalent serie resistance R_B in presence of plasma a) in presence of toroidal resonances b) at high P_A with no toroidal resonances.

TOROIDAL FIELD CIRCUIT FOR A LARGE RFP EXPERIMENT

A. Maschio
Centro di Studio sui Gas Ionizzati,(Associazione EURATOM-CNR)
'Centro G.Levi Cases' - Università di Padova (Italy)

ABSTRACT

The characteristics of a multiturn toroidal field inductor for a large RFP device are examined; in particular an attempt is made to optimize the number of the turns and the winding radius as regards the amount of flux outside the plasma vessel and the distortion of the toroidal field lines compared with a uniform pattern of the field.

The ability of different inductors to operate in the self reversal and aided reversal modes is investigated on the basis of two models of the interaction between toroidal and poloidal circuits. Finally the parameters of two types of power supply (capacitor banks or thyristor rectifiers), which are comparable in terms of performances, are given.

1. INTRODUCTION

The recent results of the existing RFP devices [1,2] have emphasized the usefulness of a large RFP experiment, with a plasma current of the order of some MA, rise times of some tens of ms and a flat top time for both poloidal and toroidal circuits of some hundred of ms.

In such a device the toroidal inductor produces a bias magnetic field B_{zo} at the beginning of the discharge; the field at the wall B_{zw} is then reversed (spontaneously or by the external circuit) during the plasma current setting up; the device must also be able to sustain and control the reversed field for a time as long as possible, because the plasma remains confined only as long as B_{zw} remains negative [1].

To recognize the problems connected with the design of such an inductor and the characteristics of an optimized power supply system is the aim of the present work; in particular reference is made to the design of RFX as developed in [3], whose main parameters are shown in table I.

TABLE I

Parameters of a large RFP device

Minor radius	a	0.6 m
Major radius	R	1.8 m
Shell thickness	s	65 mm
Plasma current	I_z	2 MA
Mean toroidal field	B_{zo}	0.56 T
Rise time	t_r	10÷50ms
Flat top time	t_{FT}	$\geq$120 ms

2. TOROIDAL INDUCTOR

In a toroidal inductor the occurrence of the self reversal of B_{zw} is strongly connected with as low as possible values of the parameter λ, which is the ratio between the external flux (from the vessel to the 'equivalent crowbar radius') to the flux inside the plasma vessel; on the other hand, the deviation δ of the magnetic field lines from the toroidal shape has to be considered: the distorted field lines, that intersect the plasma vessel, produce 'scrape off' layers near the plasma edge, increasing the plasma losses; the largest deviations, δ_{MAX}, result on the largest equatorial circumference into the plasma vessel.

Two solutions have been considered for the toroidal inductor: single turn (directly fed shell), with transformers for matching with the power supply system, or a multiturn winding external to the shell.

The first solution has been proposed in [4] and subsequently developed in detail in [5];

the value of λ is comparatively small (0.25), while the transformers put an upper limit to the field sustainement time, due to the resistance of the transformers, connections and inductor system; the field line deviation, only due to the pumping ports and to an uneven feeding of the shell, requires a careful evaluation.

A multiturn winding, considered here as an alternative, presents a higher flexibility and considerably reduces the encumbrance around the torus; on the other hand, the value of λ and δ_{MAX} can increase, involving a careful disposition of the turns.

3. MULTITURN WINDING

The multiturn winding must take into account some constraints: first is the need of a conducting shell inside the winding in order to stabilize the plasma which increases the minimum distance of the winding from the plasma vessel and therefore the λ value. Moreover with a multiturn winding a radial component B_r of the magnetic field arises, whose value is related, with an even distribution of the turns, to the distance between two adjacent turns.

The presence of vacuum and diagnostics portholes generally requires an uneven distribution of the turns giving rise, with the same number of turns, to higher B_r and, correspondingly, δ values. An improvement is obtained by assembling the turns in a convenient number of equal and equispaced groups p.

In the following, for instance, the circumference of the device has been divided in 144 parts, with 2.5° angular width; the turns have been differently distributed in these spaces, varying the gap angular width α between two adjacent groups and keeping constant the total number of turns N and vice versa, as indicated in table II.

TABLE II

Possible displacement of the turns

p	12	24	48	12	24	36
N	96	96	96	120	96	72
total free ang. space p x α	12x10°= 120°	24x5° = 120°	48x2.5°= 120°	12x5°= 60°	24x5°= 120°	36x5°= 180°

In fig. 1 the amplitude of δ_{MAX} is shown as a function of p for the different cases of table II; the winding radius r_w is 0.7 m, and the other geometrical parameters are as in table I; the calculations have been made by utilizing a code developed at CNEN /6/. Two solutions seem in particular interesting: the first one (A; p=24, N=96) requires a primary current of about 50 kA, with 24 sections of reasonable dimensions for vacuum and diagnostics portholes; the second one (B; p=36, N=72) requires a primary current some higher than the previous case (≃70 kA), but δ_{MAX} is quite smaller and the number of porthole sections is greater (36), with the same dimensions.

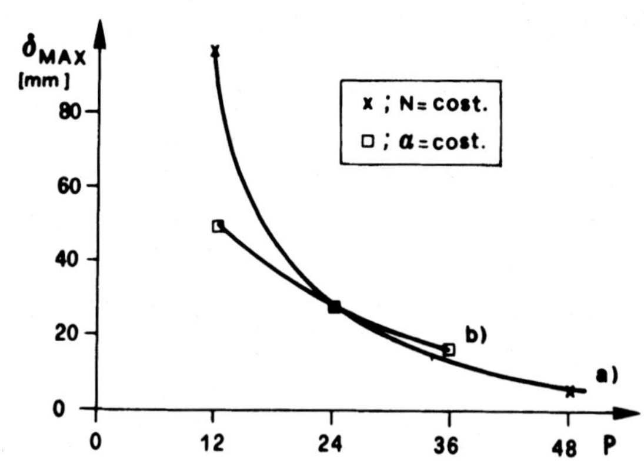

Fig. 1 - δ_{MAX} as a function of p.

It appears clear that the two requirements previously stated for a multiturn winding, i.e. to have values of δ_{MAX} and λ as low as possible are in some way each other contrasting. In fig. 2 the pattern of δ_{MAX} and λ are given as a function of $x = r_w/a$: curves a) and b) refer to the solutions A and B respectively.

Generally $\delta = 10$ mm is considered the maximum allowable field line deviation, as assumed in /4/ for the deviation produced by the presence of gaps in the shell; so different solutions relative to the cases A and B have been examined, which minimize the values of λ or δ_{MAX}, as shown in table III.

TABLE III

Values of λ and δ_{MAX}

	N = 96		N = 72	
	A1	A2	B1	B2
r_w (m)	0.7	0.82	0.7	0.74
λ	0.37	0.88	0.37	0.52
δ_{MAX} (mm)	29	10	17	10

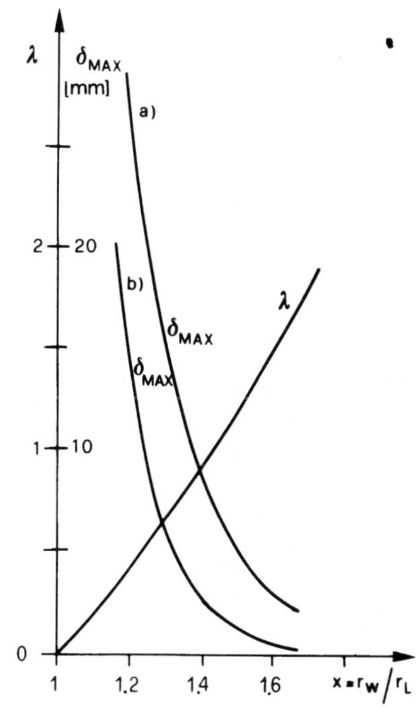

Fig.2 - Pattern of δ_{MAX} and λ as functions of $x=r_W/r_L$

4. PLASMA MODELS OF THE CIRCUIT INTERACTIONS

The evaluation of the toroidal circuit performances requires to use an 'electrical' model of the plasma behaviour that takes into account the interaction of the poloidal and toroidal circuits. The plasma is to our purpose represented by a non linear two port network specified by I_z, the toroidal plasma current, V_z, the related liner voltage, I_θ, the current corresponding to B_{zw} and V_θ, the liner voltage in the poloidal direction.

With this purpose two models have been developed by different authors [7,8] in a preliminar way; basic hypothesis for such models is that the plasma evolves through a series of equilibrium states, described by the universal curve F - θ [9], where $F = B_{zw}/B_{zo}$ and $\theta = B_{\theta w}/B_{zo}$, which gives a first link among the previous quantities.

A second link is obtained in the 'energy model' [7] by hypothesizing that the plasma internal energy is zero and neglecting the plasma losses. From the consideration of the energy conservation it is possible to work out the following expressions:

$$I_z = \int \frac{V_z - K\,V_\theta}{L_1}\,dt \qquad (1)$$

$$I_\theta = \frac{F}{L_o}\int V_\theta\,dt \qquad (2)$$

where K and L_1 are functions of θ and of parameters depending on the device dimensions [7]; L_o is the vacuum toroidal inductance.

In the 'invariant model' [8] the magnetic helicity K_o, which is supposed invariant in the plasma volume [9], can be related to θ, to parameters depending on the device dimensions and to the toroidal flux ϕ; it is then possible to derive θ as a function of time and so:

$$I_z = \frac{a\,\theta\,\phi}{R\,L_o} \qquad (3)$$

$$I_\theta = \frac{F}{L_o}\,\phi \qquad (4)$$

These two models have been used here in the simulations of the circuit behaviour, and the obtained results show an excellent agreement between them.

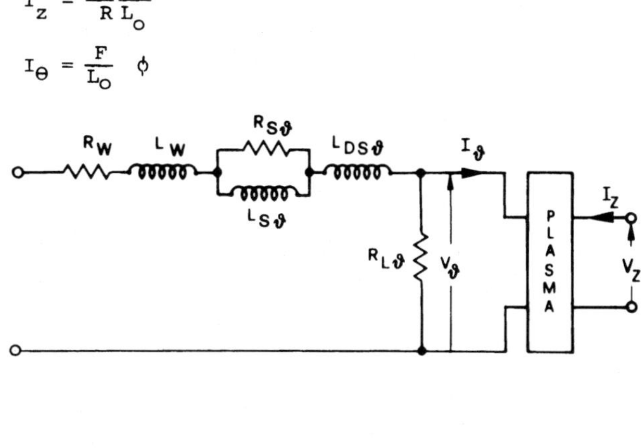

Fig. 3 - Electric scheme of the toroidal inductor

5. CIRCUIT PARAMETERS

In fig. 3 the equivalent electric scheme of the toroidal inductor is shown. $R_{L\theta}$ is the liner toroidal resistance (215 μΩ); $L_{DS\theta}$ is the dead space toroidal inductance (15 nH); $L_{S\theta}$ is the inductance equivalent to the shell volume while $R_{S\theta}$ has such a value as to give a time constant $L_{S\theta}/R_{S\theta}$ equal to the field diffusion time into the shell (15ms); R_w is the total resistance of the winding referred to the plasma side and L_w is the leakage inductance between the winding and the shell.

TABLE IV

Values of R_w, L_w and $L_{S\theta}$

	A1	A2	B1	B2
R_w μΩ	1.45	1.64	1.94	2.06
L_w nH	4	64	3	20
$L_{S\theta}$ nH	26	30	27	30

Table IV shows the values of these parameters depending on the different solutions examined; for the poloidal circuit parameters reference is made to /‾10_/.

6. CIRCUIT PERFORMANCES

6.1 Programming modes

In an RFP the reversal of B_{zw} can be obtained by exploiting the self reversal mechanism , i.e. isolating the inductor from the power supply by means of a passive crowbar and allowing B_{zw} to evolve spontaneously; the obtained reversal is low and the maximum θ obtained are slightly higher than 1.2.

The reversal has to be driven by means of the external circuit to reach higher levels (up to 100% of B_{zo}); the values of θ can be increased up to 1.5 ÷ 1.8. Maximum B_{zo} values of 0.56 T for the self reversal mode and 0.37 T for aided reversal one have been taken /‾3_/.

6.2 Self reversal

In this mode of operation any kind of power supply can be used to establish the initial field B_{zo}, as long as the crowbar branch can be considered as a perfect short circuit.

With reference to the solutions proposed in Table III, a capacitor bank requires stored energies between 2.8 and 4.0 MJ, depending on λ values (its energy is connected with $1+\lambda$). With a thyristor rectifier system, on the other hand, 4 or 6 of the 12.5 kA moduli described in /‾11_/ are required, depending on the current values (50 or 70 kA respectively).

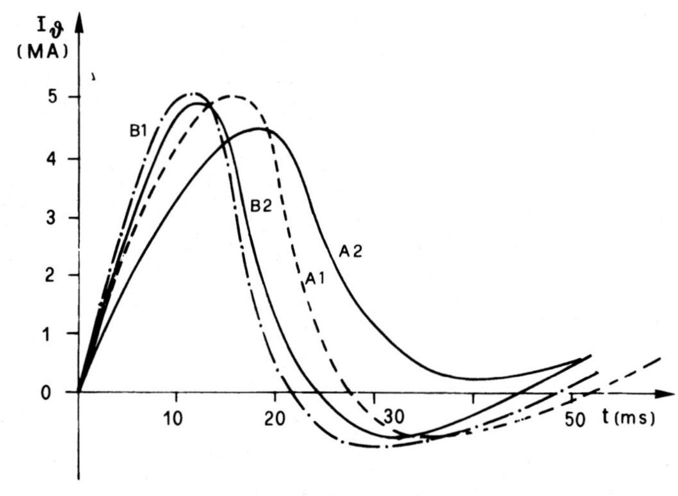

Fig. 4 - Waveforms of I_θ in the self-reversal mode

Fig. 4 shows the results of a comparison made among the different solutions of Table III, in terms of I_θ waveform, the energy stored in the capacitor bank being equal.

In case A2 the self reversal of B_{zw} is not reached, confirming that there is a critical value of λ, over which the self reversal becomes impossible; but the results of the other cases shows that this value is probably higher than previously evaluated.

Anyway in this programming mode the reversal of B_{zw} terminates within few tens of ms (fig. 4); so a further power supply, in series with the passive crowbar, is required if the confinement time should be increased.

A further consideration to be made, for the converter power supply, is the temperature-rise in the charging phase, in particular for case B, that is the more unfavourable from this point of view; anyway, if the cross section of the turns is $S = 630\ mm^2$, as taken in [3], and the charging time about 0.5 s, the temperature-rise remains below 12°C, that doesn't seem a problem.

6.3 Aided reversal

Different techniques can be used to drive the reversal of the magnetic field, utilizing either capacitor banks or thyristor rectifiers [3,12]. Two solutions have been examined here: the first one, shown in fig. 5, utilizes three capacitor banks, C_F, C_R and C_{EL}; C_F supplies the initial magnetic field in the inductor; then the voltage across it becomes negative and the current correspondingly goes to zero; when the current is zero C_R can be inserted, precharged in the opposite direction compared with C_F; finally the crowbar S_3 is fired when the current in the inductor reaches the desired negative level; the electrolytic bank C_{EL} is inserted in series with the crowbar to sustain the current for the due time (but without controlling it).

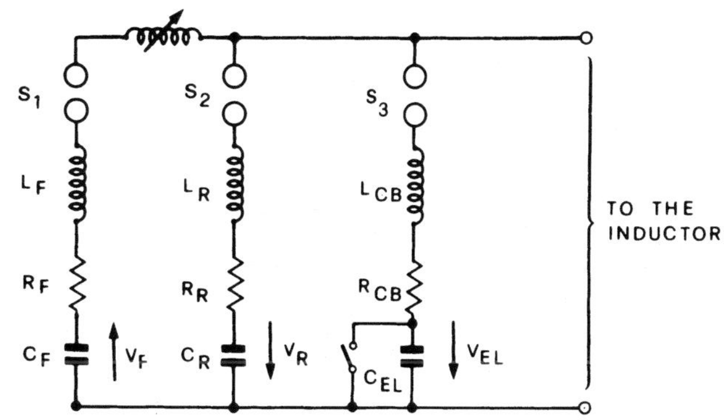

Fig. 5 - Capacitor power supply for aided reversal

An alternative solution is schematically shown in fig.6: the thyristor rectifier charges the inductor up to the desired current; then the switch S_5 closes and, immediatly afterwards, the switches S_1 and S_2 open, allowing the inductor to swing with the bank C_R. When the current in the inductor is reversed the switches S_3 and S_4 close, and insert the thyristor rectifier which terminates the reversal of the I_θ current and sustains and controls it for the desired time. In the scheme also a crowbar branch is indicated (S_6), which allows to operate also in the self reversal mode.

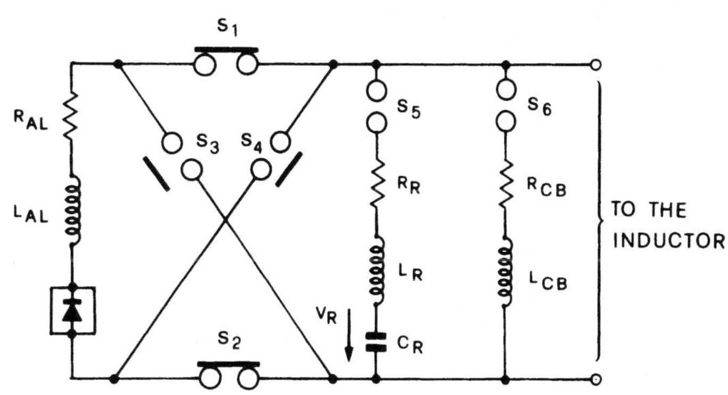

Fig. 6 - Thyristor rectifier power supply for aided reversal

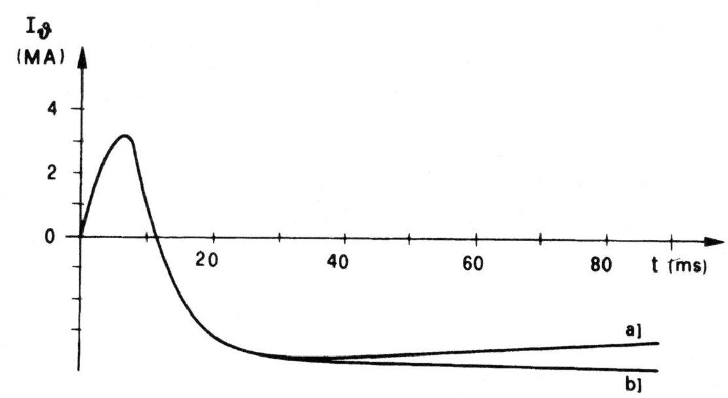

Fig. 7 - Waveforms of I_θ in the aided reversal mode

In fig. 7 the waveforms of I_θ are shown for case B2 of table III, fed either by a capacitor bank (a) or by a thyristor rectifier (b); the values of the power supply parameters have been chosen to give a level of the field reversal equal to 100% of B_{zo}, a reversal time equal to the setting up time of the plasma current and to maintain V_θ quite closely near to zero; the difference between the two curves stresses the better performances of the thyristor rectifier system which allows to control very closely the B_{zw} level.

The energy required for the banks of fig. 5 are 1.6 MJ for the bank C_F, 0.7 MJ for C_R one and 6 MJ for C_{EL} in order to give about the same performances of the thyristor rectifier power sup ply. The design energy of the bank C_R of fig. 6 is about 2 MJ; for the highest values of field reversal this bank has also to be precharged.

7. CONCLUSIONS

A multiturn inductor can be designed, that fulfils the requirements for a maximum field line deviation δ_{max} less than 10 mm and for a value of λ that allows to support the self reversal of the magnetic field at the wall B_{zw}, giving at the same time a large space for vacuum and diagnostic portholes (case B2 of table III).

For similar performances, the total energy required in a purely capacitive power supply is about 9MJ, while a thyristor rectifier system requires 4 moduli /‾11_7, rated at 1400V 12.5 kA each, and a 3 MJ capacitor bank as charge supply in the self reversal mode and as reverse bank in the aided reversal.

8. ACKNOWLEDGEMENTS

The author wishes to thank the people of the Padova RFX Team for the useful discussions on the various interacting aspects of the device and in particular G. Rostagni for his contributions during the development of this work.

9. REFERENCES

/‾1_7 Buffa A. et al., 8th Int. Conf. on Plasma Physics and Controlled Nuclear Fusion Research, Bruxelles, 1-10 July, 1980. Paper IAEA-CN-38/L-1.

/‾2_7 Baker D.A. et al.,ibidem. Paper IAEA-CN-38/L-2-1.

/‾3_7 Rostagni G. et al., University of Padova Report. UPee 80/04 (1980).

/‾4_7 Phillpott J. et al., 7th Symp. on Engineering Problems of Fusion Research, Knoxville, 25-28 October, 1977. Proceedings pp. 129-133.

/‾5_7 Quinn W.E. et al., LASL Contribution to the RFX Design, unpublished note (1980).

/‾6_7 Righetti G.B., private communication.

/‾7_7 Thompson V.K., Gray J.W., Culham note. RFX/CP/P3 (1977).

/‾8_7 Johnston J.W., private communication.

/‾9_7 Taylor J.B. 3rd Top. Conf. on Pulsed High Beta Plasmas, Culham, 9-12 September, 1975. Proceedings pp. 59-67.

/‾10_7 Guarnieri M., Rubinacci G., Stella A., these proceedings.

/‾11_7 Bolognani S., Ciscato D., Tenti P., these proceedings.

/‾12_7 Mondino P.L., 9th Symp. on Fusion Technology, Garmisch, 14-18 June, 1976. Proceedings pp. 149-155.

A MODULAR APPROACH FOR RFX D.C. POWER SUPPLIES

S.Bolognani, D.Ciscato*, P.Tenti
"Centro di studio sui gas ionizzati" (Associazione Euratom-CNR)
and "Centro G. Levi Cases" - Università di Padova - ITALY

ABSTRACT

A modular approach to the realization of D.C. Power Supplies of the Reverse Field Pinch Experiment (RFX) is presented. A basic convertor module configuration is proposed which enables a high flexibility of the power supply system to be obtained, as required to cope with the plasma behaviour uncertainties. The basic convertor module rating has been selected to meet all the expected load requirements and a preliminary design of the convertor has been carried out for evaluating the technical and economical feasibility of the system. Different fault conditions have been analyzed and protective devices have been devised accordingly. Finally simulated profiles of the active and reactive power drawn from the AC mains are reported.

1. INTRODUCTION

The Reverse Field Pinch Experiment, RFX, requires the use of five main D.C. power supplies [1,2,3] namely:

- the magnetizing (PS_M), field-shaping (PS_F) and flat-top (PS_{FT}) P.S. for the poloidal circuit,
- the charging (PS_{TM}) and the flat-top (PS_T) P.S. for the toroidal circuit.

The analysis of the various loads shows that, for most of them, the energy to be delivered and the control requirements lead to the use of thyristor AC-DC static convertors. In fact energies in the range of 100 MJ can be economically drawn from the main through thyristor convertors capable furthermore of a DC voltage/current accuracy better than 1%. As regards PS_{TM}, the use of capacitor banks seems to be more adequate because of the high power / energy ratio involved (35MW/2MJ). As regards the other P.S., for which thyristor convertors are envisaged, a modular system is proposed to obtain, at a reasonable cost, a very flexible P.S. system, as required to cope with the uncertainties of the plasma behaviour. The choice of a 4-section configuration for the poloidal circuit (fig. 1) involves voltage and current requirements for the various P.S., which enable the basic convertor modules (PS_B) to be designed fully utilizing the capability of modern thyristors, so that the additional cost of the modular solution turns out to be marginal with respect to ad hoc P.S. and it is well balanced by several advantages.

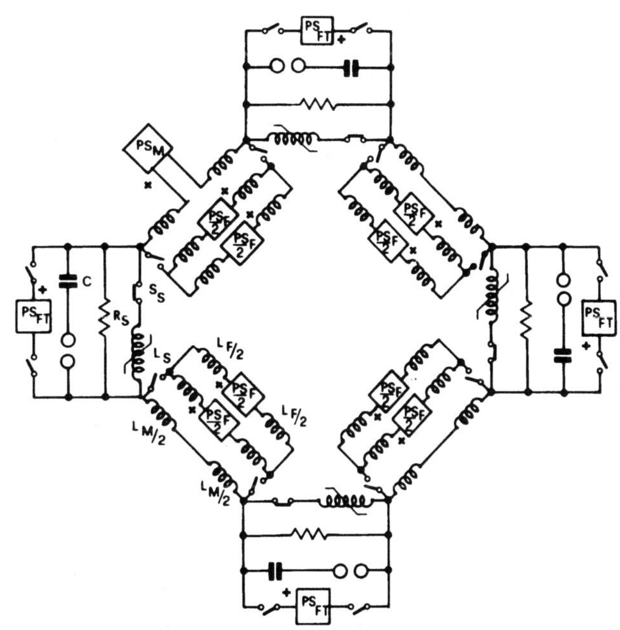

Fig. 1. Four section configuration of the poloidal circuit

2. RATING AND CONFIGURATION OF THE BASIC CONVERTOR MODULE

The expected time behaviours [3] of the various load currents at 2MA plasma current are reported in fig. 2. The maximum current demands are:

PS_M : 50 kA (6s); PS_T : 37.5 kA (150 ms); PS_F : 8 kA (200 ms); PS_{FT} : 10 kA (150 ms)

* Temporary Address: JET JOINT UNDERTAKING, Abingdon, U.K.

Moreover, to provide vertical field control, each section of the primary winding is subdivided (fig. 1) into two parallel paths separately fed by two independent half convertor module, each one capable of 6 kA. The resistive voltage drops in the loads for 2MA plasma current are of about 800 V. This value has to be increased in order to take into account the inductive drops. Moreover an additional increase of the PS_{FT} output voltage is expected to be required at low plasma current and at the beginning of the flat top, when the plasma is cold.

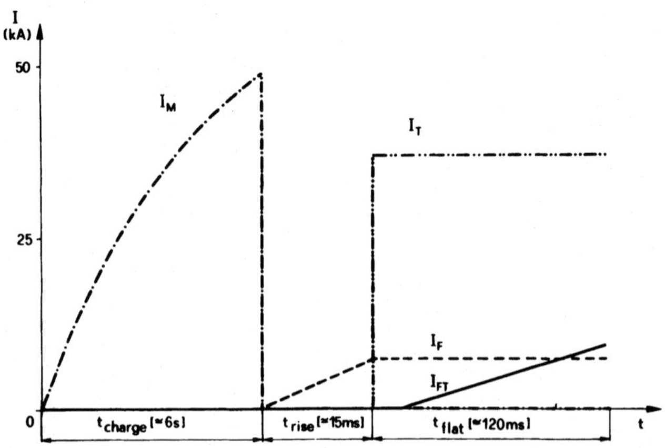

Fig. 2. Profiles of the load currents at 2MA plasma current

On the basis of the afore considerations a 12-pulse configuration, made up of two 6-pulse thyristor convertors in parallel, has been proposed for the basic convertor module PS_B (see fig. 3). The current rating of the module has been fixed to 12.5 kA and the no-load voltage to 1.4 kV. This voltage can be obtained using modern available 3.6 kV thyristors with a safety factor of 2.45. The corresponding full-load output voltage has been evaluated to 1.1 kV, which fulfils all the expected voltage requirements of the loads. The configuration chosen for PS_B enables PS_F to be divided into two independent convertors. Moreover the two half-modules may be connected in series for increasing the PS_{FT} output voltage at low plasma current. In Table I the configuration and the rating of the various P.S. are shown.

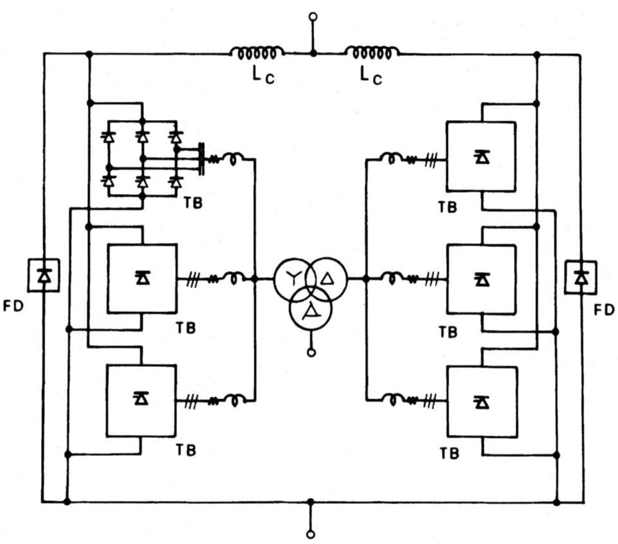

Fig. 3. Basic Convertor Module Configuration

TABLE I

Configuration and rating of RFX Power Supplies

Power supply	number of PS_B	PS_B modules connection	Full-load DC output voltage	No-load DC output voltage	DC output current	Reference Power [MVA]	Overall Data	
							PS_B numb.	Power [MVA]
PS_M	4	Parallel	1.1 kV	1.4 kV	50 kA	70	4	70
PS_T	3(4)	Parallel	1.1 kV	1.4 kV	37.5(50)kA	52.5(70)	3(4)	52.5(70)
PS_F·4	1	--	1.1 kV	1.4 kV	12.5 kA	17.5	4	70
PS_{FT}·4	2	Series	2.2 kV	2.8 kV	12.5 kA	35	8	140
						Total	19(20)	332.5(350)

The P.S. system configuration given in Table 1 is to be considered for reference only at this stage. In fact the modular approach enables the P.S. system to be easily rearranged. (This especially regards PS_{FT}, whose voltage requirement depends strongly on the plasma resistance. The DC voltage of the reference PS_{FT} configuration has been designed to assure satisfactory control of the plasma current during most of the flat-top). During the first set of experiments at 1MA plasma current, the PS_B number can be smaller than the reference one and the need of additional modules will be dictated by the results from the actual experiments.

3. DESIGN CONSIDERATIONS

A 12-pulse configuration has been chosen for PS_B to limit the harmonic contents of the A.C. line current and to improve the dynamic characteristics of the P.S.. Each half module is obtained by three parallel thyristor bridges T.B. and has its own free-wheeling diode F.D. (in the experiments with controlled plasma current decreasing, which require PS_{FT} output voltage inversion during the rigenerating phase, the FD can be disconnected). The F.D. current rating requires many components in parallel, so that the use of two separate free-wheeling branches involves only a marginal additional cost (cabinets, ecc.) with the advantage of half modules being identical and capable of independent operation. The use of three T.B. for each half module increases the flexibility, since they can be separately inserted or removed. In this way the current rating of the various P.S. can be stepped up according to the plasma current in each stage of the experiment, thus involving a distribution of the P.S. cost over the time; moreover the nominal output power of the P.S. can be tailored according to their duty cycle. Other advantages are the capability of operating at reduced power, if one of the bridges is out of service, and simpler servicing. The main PS_B design parameters have been evaluated and are given in Table II. As far as the transformer is concerned the primary winding can be of the extended - Δ or dog-leg type to allow the A.C. line current harmonic contents (24 pulse) to be further reduced if two modules operate at the same D.C. voltage and current; K_t and $v_{sc}\%$ have been chosen to limit the reactive drop of the D.C. output voltage to 18% of the no-load voltage (the residual 2-3% corresponds to the typical resistive voltage drop, including transformers). The choice has been done assuming the step-down transformer (400/33kV) be similar to those used in the JET (300 MVA, 24%) and fed by the same 400 kV line [4]. Thermic considerations would allow a higher value of K_t to be taken, but mechanical (short-circuit stresses) and economical considerations suggest to choose $v_{sc}\%$ not less than 4÷5%, thus limiting K_t to less than 4. L_c inductances have been chosen to avoid discontinuous operation of the two half-modules for the D.C. output current ranging from 40% to 100% of the nominal value; the resulting value also avoid that the fault current, in case of short-circuit at the D.C. terminals, exceed the capability of the selected thyristors. The current sharing among the three parallel T.B. of each half module can be assured by individual impedances on the A.C. side of the convertors, as shown in fig. 3. If the convertor layout is carefully designed and if the thyristors are chosen in the same class of forward voltage drop, the current sharing could be obtained separately connecting each bridge to the convertor transformer through individual cables.

TABLE II

Basic Convertor Module Rating

Bridges	: 1.1 kV at 2.2 kA; 1.4 kV no-load
Thyristors	: 0.7 kA (average); 1.2 kA (rms) 3.6 kV (V_{RRM}); safety factor K_s=2.45
Transformer:	33 kV/2·1.04 kV; $v_{sc}\% \cong 4.5\%$ 4.6 MVA (nominal); 18.4 MVA (peak) primary connection: Z or extended - Δ secondary " : Y and Δ $K_t = \frac{\text{peak power}}{\text{nominal power}} = 4;$
Inductances:	L_c = 120 μH

4. PROTECTION DEVICES

As described in [1,3] during the fast-rise phase, in normal operation, a voltage up to 50 kV would appear across the D.C. terminals of PS_{FT}. In order to protect the thyristor convertors, ignitrons have been proposed to be connected in series to the convertors according to fig. 4. Since overvoltage of either polarity are expected, back-to-back ignitron pairs have been used. Moreover in order to reduce the ignitrons rating a make-switch is connected in parallel to them.

A preliminary analysis of the fault conditions, performed in /‾5_/, has shown that the worst conditions are a fault of ignitrons and a plasma disruption. Both faults stress especially PS_{FT}. Ignitron fault occurs when both of the two ignitrons in series fail at the same time. The number of faults of this type is not expected to be very high during the life time of the experiment. The plasma disruption occurs when plasma current suddenly disappears during the flat-top. In this case the poloidal magnetizing current flows into the transfer resistors R_S (fig. 1) developing an overvoltage across the convertors. The worst case results when this fault occurs at the beginning of the flat-top, but it is in any case less severe than the ignitron fault. The plasma disruption is expected to occur fairly often.

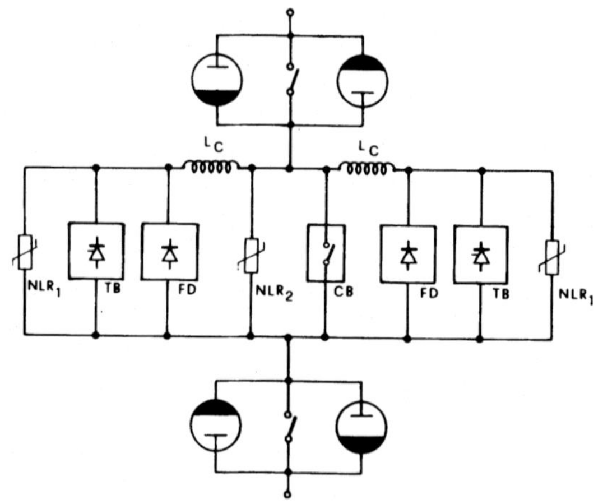

Fig. 4. PS_B Overvoltage protection circuit

In order to prevent overvoltages under fault conditions the use of nonlinear resistors and crow-bars is proposed, as shown in fig. 4. As far as the ignitron fault is concerned NLR_1 and NLR_2 varistors have to be chosen to limit the overvoltage across the convertors until the crow-bar is turned on. The energies W_1 and W_2 absorbed respectively by NLR_1 and NLR_2 can be estimated considering that during the fault the external circuit acts as a current generator and that NLR's can be considered as constant voltage devices. Therefore the current through L_c increases linearly versus time and the energies W_1 and W_2 can be estimated by

$$W_1 = (V_2-V_1)V_2t_D^2/(2L_c) \quad \text{for} \quad V_2 \geq V_1 \quad (W_1 = 0 \quad \text{for} \quad V_2 < V_1) \tag{1a}$$

$$W_2 = (I_M-V_2/R_s)V_2t_D-W_1 \quad \text{for} \quad I_M > (V_2/R_s) \tag{1b}$$

where t_D is the delay time of the crow-bar, I_M the magnetizing current (up to 50 kA in case of ignitrons fault) V_1 and V_2 the threshold voltages of NLR_1 and NLR_2. V_1 coincides necessarily with the maximum voltage which can be taken safely by the PS_{FT} convertors, and can be fixed to 9.5 kV taking into account the characteristics of the thyristors chosen. For L_c the value of 120 μH has been assumed. As far as t_D is concerned, eqs. (1) show that the losses in the NLR's increase with t_D. If metal-to-metal crow-bars are used, a value of t_D less than 50 μs seems obtainable, taking into account the typical delay times of the fault detection and switching. Substituting V_1, L_c and t_D values in eqs. (1) it can be easily obtained that W_1 and W_2 increase according to V_2 (in the field $0 \leq V_2 \leq 20$ kV). Therefore a low value of V_2 is desiderable. However the ratio between V_2 and the working voltage of NLR_2 has to be sufficiently high in order to make use of standard nonlinear resistor. Consequently a lower limit of V_2 exists which depends on the NLR type. Taking into account the characteristics of the ZnO NLR a good value for V_2 seems to be 13 kV, so that 60 cm^2/40 mm disks can be used for both NLR_2 (two parallel resistors) and NLR_1. The corresponding specific energy absorbed during the fault results to less than 50 J/cm^3 for NLR_2 and smaller for NLR_1, as evaluated through eqs. (1).

A computer simulation of the faults has been performed and the resulting profiles are shown in fig. 5. An equivalent circuit of the poloidal circuit shown in fig. 1 has been considered for the computer simulation carried out by means of the SPICE program /‾6_/. The nonlinear resistors have been simulated by adjusting the parameters of the available diode model of the SPICE in order to obtain the V-I characteristic of the ZnO nonlinear resistors /‾7,8_/. The specific energy losses of NLR_1 during the faults result 3 J/cm^3 for ignitron fault and 75 J/cm^3

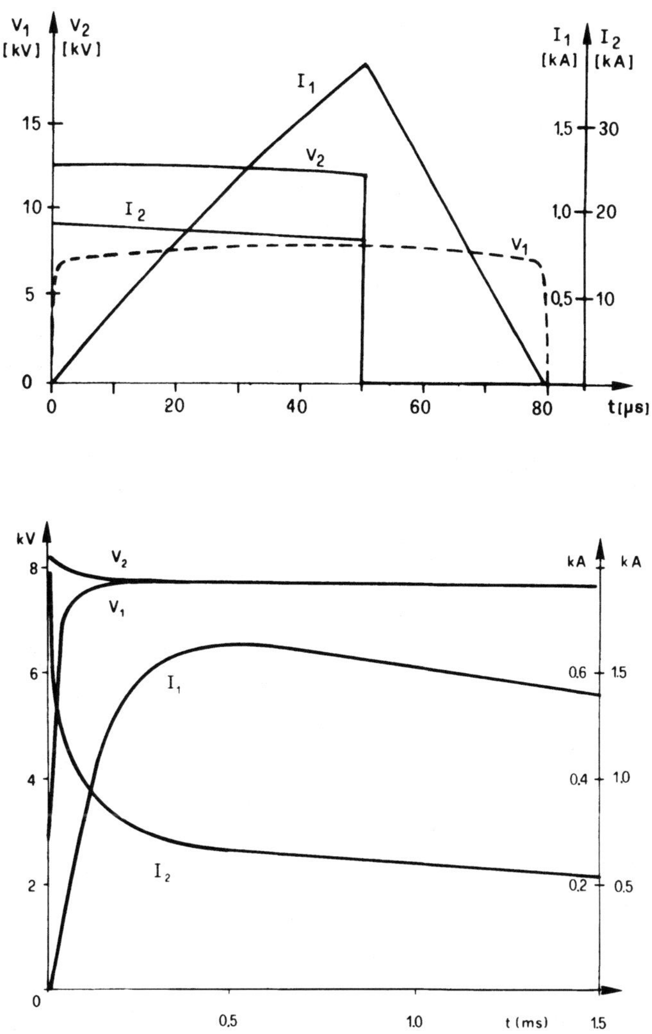

Fig. 5. Simulated behaviour of the varistors in case of fault; up: ignitron fault; down: plasma disruption

for plasma disruption (without crow-bar intervention). For NLR_2 the energy losses result 45 J/cm^3 and 80 J/cm^3 respectively. The simulation shows that in case of plasma disruption no crow-bar intervention is required as the energy absorbed by NLR's is within the capability of the devices.

As far as the overcurrent protection devices the faults to be taken into account are the short-circuit at the D.C. terminals of the converters (e.g. crow-bar operation), the thyristor failure, and the short-circuit on the A.C. side of the convertors. The fault current due to a short-circuit can be limited by a proper control of the delay angle of the thyristors with no intervention of the A.C. circuit breakers or fuses as the steady-state short-circuit current is about 5 times the nominal value. In case of thyristor failure, protection is assured by fuses in series with each thyristor; the overcurrent is in the order of 5.5 times the nominal value. The A.C. side short-circuit can be handled by the A.C. circuit breakers on the primary side of the PS_B transformer.

5. A.C. POWER

An approximate evaluation of the active and reactive power drawn from the A.C. mains by the RFX Power Supplies is required for a preliminary design of the A.C. Supply. For this purpose tentative profiles of the D.C. output voltage and current of each P.S. have been assumed (see fig. 2) as described in /‾3‾7. The overall active and reactive power consumption of the power sypply system are shown in fig. 6 and 7.

The maximum active power is about 90 MW and occurs during the flat-top, whereas the charging phase shows an active power peak of about 55 MW. The reactive power, which has been evaluated taking into account the commutation in the convertors and the

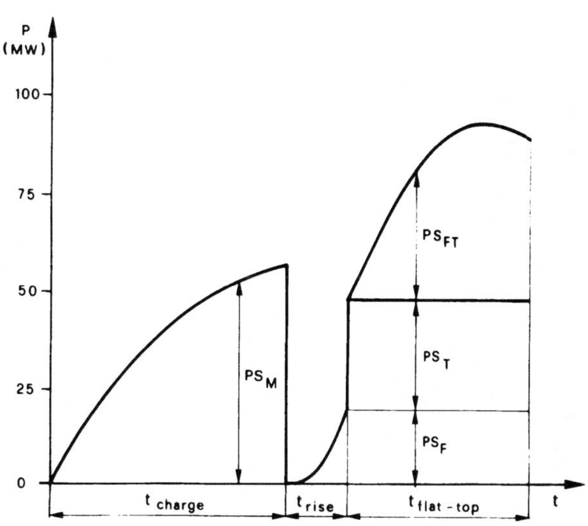

Fig. 6. RFX Active Power Consumption at 2 MA plasma current

free-wheeling diode effect, depends on the control technique adopted for the PS_{FT} power Supply. In fig. 7 the effect of sequential and nonsequential control of PS_{FT} are shown. In the former case a peak of about 140 MVAR has been found whereas in the latter case the peak is of about 215 MVAR.

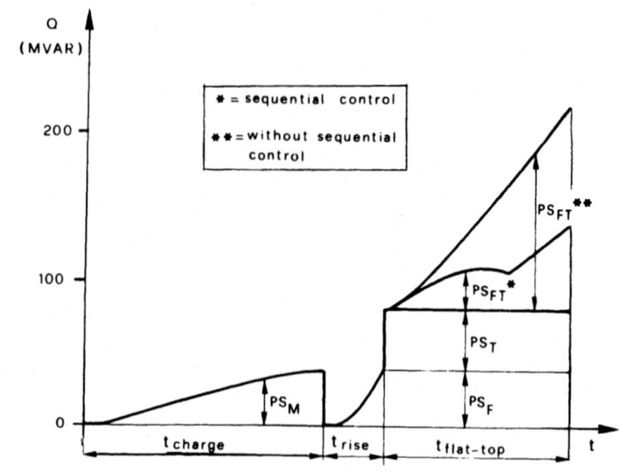

Fig. 7. RFX Reactive Power Consumption at 2 MA plasma current

6. ACKNOWLEDGEMENTS

The authors would like to thank Prof. G. Rostagni for many useful discussions and advices.

7. REFERENCES

/¯1_7 GUARNIERI, M., RUBINACCI, G., STELLA, A., these proceedings.

/¯2_7 MASCHIO, A., these proceedings.

/¯3_7 ROSTAGNI, G. et al., UPee Padova University Internal Report 80/04.

/¯4_7 CISCATO, D., COCCORESE, E., Symposium on Fusion Technology, Padova, 4-9 September, 1978, Proceedings pp. 365-370.

/¯5_7 GNESOTTO, F., MARTONE, R., PIOVAN, R., these proceedings.

/¯6_7 NAGEL, W., Berkeley Research Laboratory Memorandum ERL (1975).

/¯7_7 GENERAL ELECTRIC, Tranquell Station Surge Arresters (1978/79).

/¯8_7 SIEMENS, SIOV Metal Oxide Varistors (1978/79).

SOME ELECTRO-TECHNOLOGY ASPECTS OF HBTXIA

G C H Heywood, J H Anderson,* G E S Harding and R J Rigley
UKAEA, Culham Laboratory, Abingdon, Oxfordshire OX14 3DB, UK
(*Euratom/UKAEA Fusion Association*)
*MOD (PE), AWRE, Aldermaston, Berkshire, UK

ABSTRACT

Two parts of the electrical system for HBTXIA[1] are desribed. Firstly, the switching system for the capacitor banks which uses 39 type BK488 and 10 type BK496 ignitrons for connecting the two main banks (1MJ at 40kV and 1.25MJ at 8kV) to the I_ϕ and I_θ load coils, providing the poloidal currents and fields required. The trigger system for these ignitrons is solid state with isolation which allows the various parts of the circuit to operate satisfactorily over a wide voltage range and at various times with no spurious breakdown even with 40kV transients across some of the ignitron stacks.

Also described is the audio frequency heating system for baking out the main part of the stainless steel vacuum vessel (the liner). This system can feed up to 25kW at a frequency of about 2.5kHz into the liner.

1. CAPACITOR BANK SWITCHING SYSTEM

1.1 Introduction

The 40kV bank is split into four parts, each being connected to a multiple ignitron switch. Three of these are used for the primary I_ϕ current, as shown in Fig 1, and the fourth supplies the programming I_θ current (Fig 2).

The 8kV bank is divided into four parts, 0.25MJ is used for the I_θ initial bias current. The remaining 1MJ is divided into three and called the floating bank. These are fed to the multiple ignitron switches such that they can be switched between the ϕ and θ circuits, as required, to optimise the power crowbar and field control phases within the 1MJ of available energy, see Figs 3 and 4.

1.2 Multiple ignitron switch (S_1, S_2 and S_3)

Various combinations of ignitrons are used for the switching functions shown in Figs 1 and 2. The seven ignitron switches used for S_1, S_2 and S_3 in Fig 1 will be described, the other switches being of a similar form.

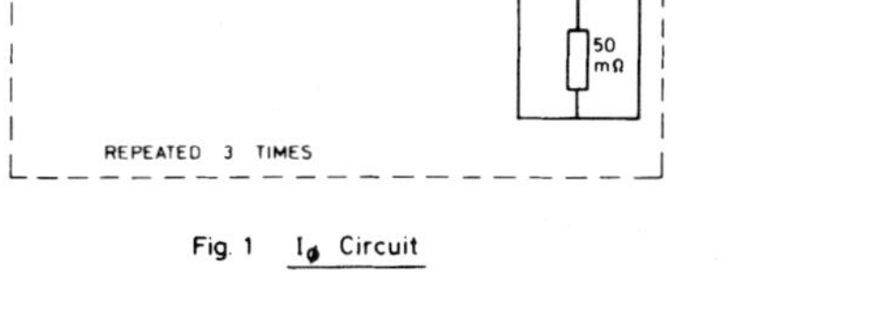

Fig. 1 I_ϕ Circuit

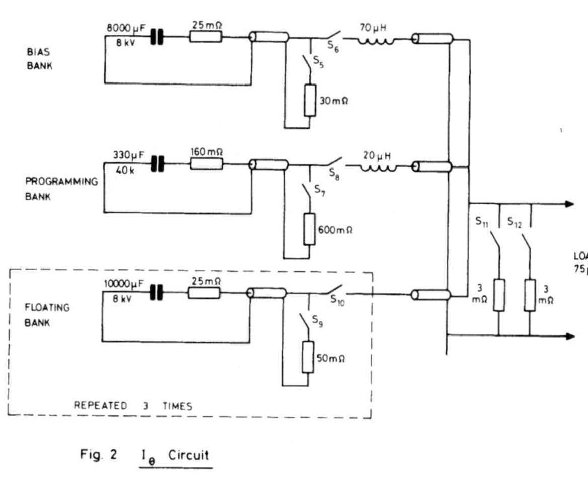

Fig. 2 I_θ Circuit

Ignitrons type BK488 often require conditioning to reach the 25kV maximum rated voltage, but both S_2 and S_3 have to be capable of holding off up to 40kV and supplying 'on' currents of up to 55kA. By using three in series instead of two, a nominal voltage of 13.3kV for each ignitron is low enough to require only trivial conditioning to reach 40kV operation and obtain good reliability.

Since S_3 is always to be closed first, to supply the primary current phase (Fig 3), S_2 can be connected in series with S_3 for the power crowbar phase. This arrangement has the advantage that the fast transient voltage swings which would occur on S_2 if it were connected directly to the load are avoided.

The switch S_1 and the 50mΩ damping resistor are fitted to keep the overswing of the 8kV bank within its rated limit.

An equal resistance of about 500kΩ and capacitance of 30pF is formed across each ignitron by the following simple means. The cooling water circuit connects the ignitrons in series and produces equal resistance between them which has been shown to give adequate sharing of voltage. Sacrificial anodes are used in each case at the junctions of the ignitron cooling jacket and the water circuit to prevent damage to the cooling jackets. The power supply transformers for the trigger units (see para 2.5.2) are designed to be run in series and give a constant feed through capacitance of 30pF. They are, therefore, both used to supply the trigger power and the A.C. voltage equalisation between ignitrons.

1.3 Layout of switch assembly

A total of seven ignitrons is, therefore, needed for one switch assembly S_1, S_2 and S_3. The ignitrons are mounted vertically at intervals below a horizontal busbar connecting system which has an earth return conductor on each side of the central high voltage conductor. A second system is mounted immediately alongside and used for switches S_9 and S_{10} in the θ circuit. This enables the floating bank, with a total capacitance of 30,000μF, to be easily transferred as required between the ϕ and θ circuits. This total assembly is 4 metres long, 1.5 metres high and 0.9 metres wide, and contains all the water resistors, trigger units and anode heaters.

1.4 Triggering sequence

The required load current waveform for I_ϕ is shown in Fig 3. It is initiated by triggering the three ignitrons forming S_3 to produce the primary rising current waveform. Normally, the ignitrons forming S_1 and S_2 are triggered at the same time, but since they are initially reversed biased, they do not pass anode current. They remain waiting, with a discharge on the holding anodes only, until the primary bank has discharged to the voltage of the floating bank, then there is a smooth transition into the power crowbar phase.

2. TRIGGER SYSTEM

2.1 Requirements

A pulse of at least 100A at 2kV must be applied to the ignitor for reliable operation of each ignitron. The trigger system has to be capable of producing separate pulses of this amplitude at the required time for each ignitron over the full range of bank voltages used (from maximum down to 5% of maximum).

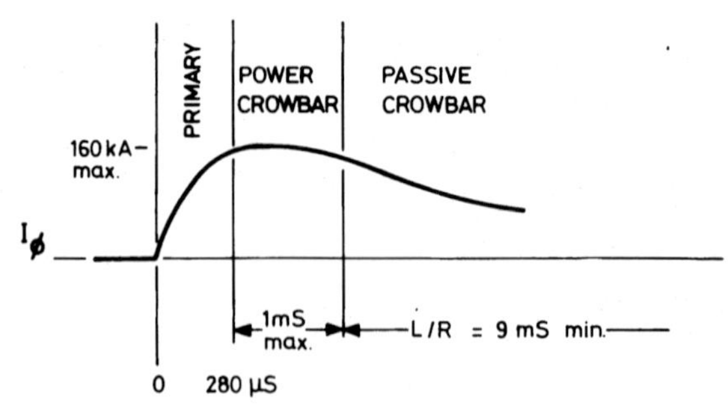

Fig. 3 I_ϕ Waveform

An additional requirement is that the trigger generator must be isolated so that it can swing with its associated ignitron cathode to follow early parts of the waveforms without triggering the ignitron or affecting its ability to operate when required. Note that in Fig 1 the switch S_4 has to withstand a pulse voltage of up to 40kV when S_3 operates.

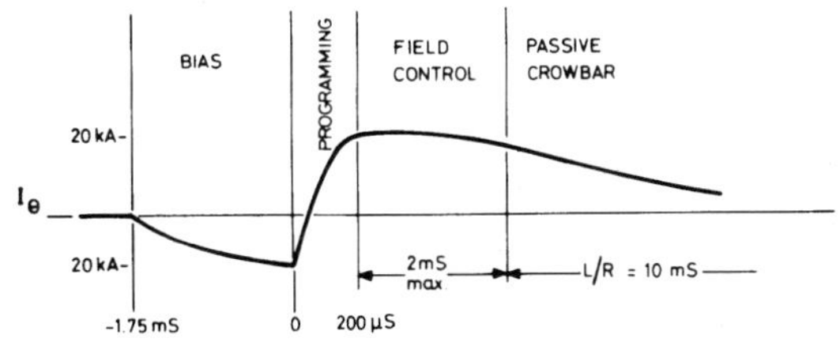

Fig. 4 I_θ Waveform

2.2 Operating sequence

Pulses of 24V amplitude are produced by the digital timer for each switch operating time needed for S_1 through to S_{12} (see Figs 1 and 2). These pulses are first amplified to 300V by thyristor circuits adjacent to the digital timer. Cables then feed these 300V pulses to a position adjacent to each ignitron switch assembly, where each pulse is converted into three simultaneous infra-red pulses. Optical cables feed on to each of the ignitron trigger units in one switch assembly. Each ignitron trigger unit generates ignitor and holding anode pulses for one ignitron. The optical cables are 2 metres long and easily guarantee the required voltage isolation of 40kV through the ignitron stack.

2.3 Thyristor amplifier

This is a conventional thyristor circuit using a 2N4444. When triggered from the digital timer it generates a 300V pulse which is applied to the output cable via a capacitor. A voltage of +150V D.C. is added to the output cable via a resistor of 470kΩ to provide the power for the infra-red pulse generator.

2.4 Infra-red pulse generator

A simple filter separates the 150V D.C. which is then used both to charge the thyristor trigger circuit and to bias off a diode gate in series with the input pulse. A degree of interference suppression is thereby obtained as only pulses in excess of 150V pass through the diode gate to trigger the 2N4444 thyristor. Only the 'front edge' of the thyristor pulse is used. A coupling capacitor of 0.05μF feeding the 150V pulse from the thyristor cathode to each R.C.A. lasing diode type SG3001, defines the laser pulse at 35A for 0.2μs. The three lasing diodes each provide an output pulse of 25W minimum at a centre wavelength of 904nm. The output window of each lasing diode is positioned about 2mm from the end of each coupling optical cable. No precision assembly is needed since the light pulse is much larger than the minimum needed.

2.5 Ignitron trigger unit

2.5.1 General description

An isolating power supply transformer and a screened box containing a D.C. power supply and a multiple thyristor trigger unit form the ignitron trigger unit. Table 1 lists the characteristics.

Table 1

Outputs	1 - ignitor pulse > 2.2kV @ 100A
	2 - holding anode pulse 190V @ 20A for 10ms
Isolation	- 60kV and 30pF
Trigger input	- infra-red light pulse 0.5W @ 904nm
Delay	- 1μs maximum
Power input	- 20V R.M.S. @ 28kHz
Power output	- 20V R.M.S. @ 28kHz for next trigger unit
Repetition period	- 60s minimum

2.5.2 Power supply

A transformer was designed by Savage Transformers, Devizes, Wiltshire for operation at 20V, 28Hz with a ratio 1:1 and a power rating of 40VA. It has an isolation of 60kV and capacity between the double screening of 30pF. These transformers can be connected up to three in series, giving a power available in each trigger unit of 13VA, sufficient to charge up the trigger circuit for operation every 60 seconds.

A voltage multiplier circuit is used to convert the 20V, 28kHz to two D.C. voltages of 190V and 380V. The 190V is used to charge up a 600μF capacitor for the ignitron holding anode output and to energise the input thyristor circuit, whilst the 380V is used for the ignitor pulse generator.

2.5.3 Input and holding anode circuit

The infra-red input pulse is applied directly to the input window of an L9B thyristor and is large enough to trigger the device in less than 1μs. A thyristor type 2N4444 is then triggered to connect the 600μF capacitor, charged to 190V, to the ignitron holding anode via a 10Ω resistor.

2.5.4 Ignitor pulse circuit

A seven stage Marx generator ladder type circuit is used for the ignitor pulse. Seven 1μF capacitors are connected in parallel via resistors and charged to 390V. Seven thyristors type 2N4444 are connected diagonally across this capacitor ladder. When the first 2N4444 is triggered, the rest of the 2N4444 thyristors also trigger causing all the 1μF capacitors to appear in series. A pulse in excess of 2.2kV is thus generated and applied to the ignitron ignitor via a 10Ω resistor.

2.6 Results achieved

The system has all been assembled. It has been tested and used as follows:

(1) One of the three sections of the I_ϕ circuit has been operated at full voltage with no trigger problems and has produced the required full energy waveforms.

(2) The whole ϕ and θ circuits have been operated, at voltages 100 times below maximum, into a dummy load. No trigger problems were observed and the waveforms were clean with no trigger breakthrough. Note that for this test, the trigger system was at full energy (0.2MW peak per ignitron).

(3) The trigger system has been run regularly and successfully for about four months.

3. A.F. HEATING SYSTEM

3.1 Introduction

Part of the bake out of the vacuum system will be by an audio frequency power supply coupled into the I_θ coil system. The resistance of the stainless steel vacuum vessel in the θ (toroidal) direction is much more constant around the machine than it is in the ϕ (poloidal) direction. Therefore, θ generated currents will achieve the more uniform heating. However, the much lower magnetising inductance in this direction means that power must be fed in at a minimum frequency of 1kHz.

3.2 Specification

A power of 25kW is required to feed into the standard 32 turn I_θ load coil system. The reflected load from the vacuum vessel is as shown in Fig 5. The power level is required to be adjustable and to be capable of being switched on and off remotely.

3.3 Circuit description

After investigation of available equipment generating continuous square waves and tests on different types of circuit, it was decided that our requirement could be met by a comparatively simple circuit that produces high energy pulses with a variable repetition rate. Each pulse of energy being at frequency of about 2.5kHz.

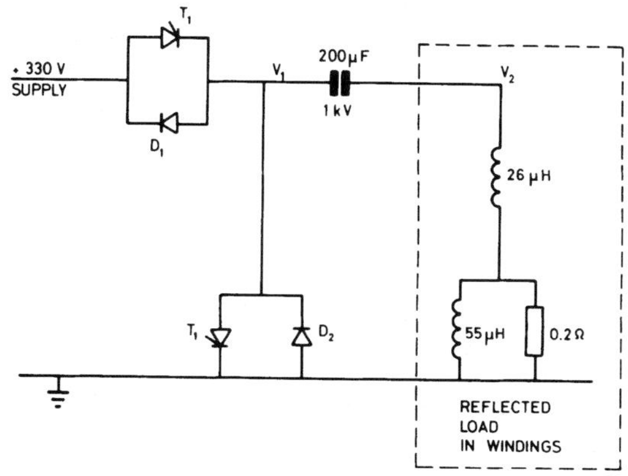

Fig. 5 A.F. Heating Circuit

Figure 5 shows the basic circuit. The fixed voltage power supply unit has a maximum rating of 30kVA at 330V but can supply repetitive peak 1kA current pulses for 1ms within its power rating. When the thyristor T_1 is triggered the voltage V_1 rises to +330V and initiates one cycle of a damped sinusoidal discharge between the 200µF capacitor and load, see Fig 6. Note that the reflected vacuum vessel impedance of 0.2Ω is the dominant impedance in the circuit. The thyristor supplies the first half cycle of this discharge and diode D_1 passes the second half cycle current. During this first cycle the capacitor charges near to +330V and this energy transfer is dissipated in the 0.2Ω reflected vacuum vessel impedance, so that no useful further transfer of energy would occur if thyristor T_1 were triggered a second time. Next T_2 is triggered repeating the above sequence as the 200µF capacitor is discharged to near zero, again dissipating energy in the vacuum vessel. This operation is repeated, T_1 and T_2 being triggered alternately to supply a train of pulse energy into the vacuum vessel.

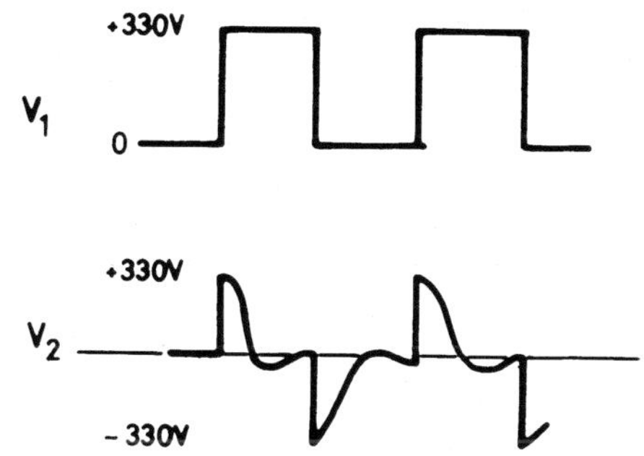

Fig. 6 A.F. Heating Waveforms

3.4 Power level control

A trigger circuit is arranged to give a train of pulses alternately to T_1 and T_2. The pulse repetition rate can be manually varied, locally to the unit, to adjust the heating of the vacuum vessel. Alternatively, a logic input to the circuit enables the pulse train to be started and stopped from the experimental control desk. Feedback from temperature sensors on the vacuum vessel will also be arranged to start and stop the pulse train, thereby enabling the temperature to be stabilised.

Each pulse delivers 10J to the load assembly. Therefore, a maximum pulse repetition rate of 2500 pulses per second produces 25kW dissipation in the vacuum vessel.

3.5 Components used

The 200µF capacitor is formed by a bank of commutation capacitors type CA12.5. The thyristors and diodes have voltage and current ratings in excess of 1.5kV and 1.5kA and are manufactured by Westinghouse.

3.6 Results achieved

A low level circuit has been built and tested satisfactorily at 100V giving 2.5kW into a dummy load. At present the full energy system is being built.

4. REFERENCES

[1] P D Wilcock: Proc 10th SOFT Conf, Padua (1978) p55.

FEEDBACK-CONTROLLED POWER SUPPLY FOR THE VERTICAL AND HORIZONTAL FIELD OF TCA

R. Keller and A. Tuszel
Centre de Recherches en Physique des Plasmas
Association Euratom - Confédération Suisse
Ecole Polytechnique Fédérale, CH-1007 Lausanne

H. Baumann and G. Irminger
Schweizerisches Institut für Nuklearforschung, CH-5234 Villigen

ABSTRACT

The transverse field system is planned for controlling the rapidly changing conditions of the initial phase and during the heating period of the TCA discharge. The power supply acts as a voltage amplifier comprising two identical parts, the sum of their currents giving the vertical field, and the difference determining the horizontal field component. It is based on the principle of a variable resistor consisting of many resistors in parallel, switched by transistors. An internal feedback loop maintains the output voltage proportional to the desired input. Preliminary tests are performed without plasma, and the response is compared with the theoretical analysis. This enables us to check the Tokamak parameters introduced in the mathematical model.

1. THE TRANSVERSE FIELD COIL SYSTEM

A description of the TCA is presented under ref. {1}. The transverse magnetic field is produced by means of eight toroidal coils of ten turns each. Fig. 1 shows how these coils are connected to the various power supplies. The system is able to produce four independent working modes: vertical, horizontal and quadrupolar fields, and an induced field for preionization. The series connected coils No 1 to 4 and No 5 to 8 are fed by two power amplifiers, the sum of their currents giving the vertical field, and the difference determining the horizontal field component.

A 16 kHz generator produces a burst of 1 msec duration preceding closely the plasma current rise. The quadrupolar field is created by a capacitor discharge of 6 msec duration starting 2 msec before the preionization. The power supplies of the vertical field are voltage amplifiers, their output impedance being negligible for frequencies less than 2 kHz. This feature allows us to use the transverse field coils independently for vertical, horizontal and quadrupolar field controls. A filter decouples the amplifiers from the coils in order to eliminate a perturbation which cannot be handled at a frequency as high as 16 kHz.

A fifth mode with the same coils is realizable: discharge cleaning. In this case the generator for preionization is plugged to 5 kHz only, and the power amplifiers have to be disconnected from the coils.

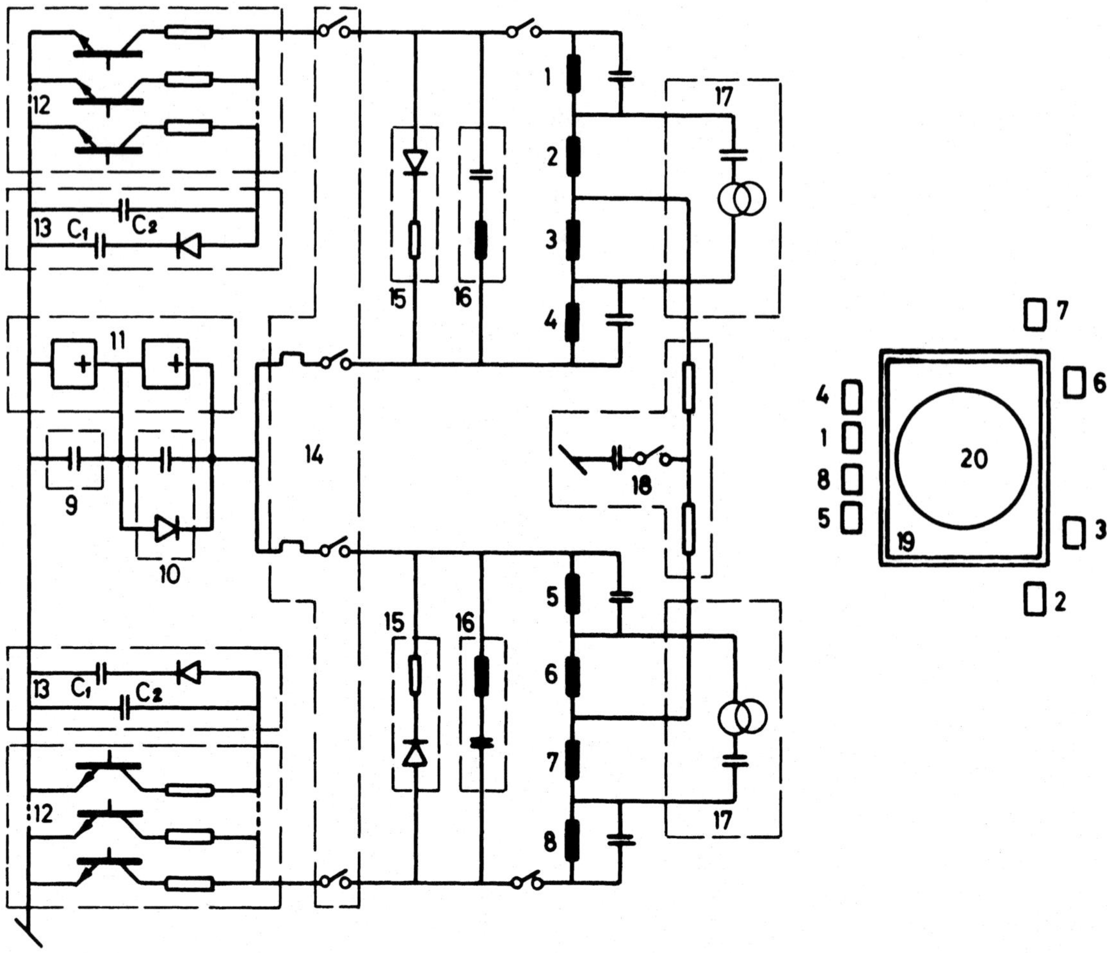

Fig. 1 Power supply of the control windings.

1 - 8	CONTROL WINDINGS
9	ELECTROLYTIC CAPACITOR BANK 4F, 400/450V
10	ELECTROLYTIC CAPACITOR BANK 40mF, 400/450V
11	CHARGING UNIT
12	VARIABLE RESISTOR
13	OVERVOLTAGE PROTECTION
14	CIRCUIT BREAKER AND OVERCURRENT PROTECTION
15	FREEWHEEL DIODE
16	16 kHz FILTER
17	16 kHz GENERATOR FOR PREIONIZATION
18	QUADRUPOLAR FIELD POWER SUPPLY
19	VACUUM VESSEL
20	PLASMA

2. FEEDBACK CONTROL OF THE VERTICAL FIELD

In order to control the horizontal motion of the plasma we use a linear combination of three variables: the time-integrated signal U_{cos} of a cosine coil, the vertical current I_v and the plasma current I_p. The method is described under ref. {2} where one can find other references about related subjects.

With an adequate choice of the coefficients of the linear combination the plasma position gets independent of $(2\,\beta_{pol} + \ell_i)$. The winding density of our cosine coil is computed in order to give zero signal when the plasma is replaced by a thin conductor centered on the vacuum vessel axis, and without vertical field. In presence of a plasma in equilibrium with the vertical field the time-integrated signal of the cosine coil contains an additional offset term. The condition of independence against $(2\,\beta_{pol} + \ell_i)$ leads to the following error signal:

$$U_{FB} = \frac{g}{A}\left[\int U_{cos}\,dt + (L_a + M_a)\,I_v - M_p I_p\right] \qquad (1)$$

A compensation of the ohmic voltage drop due to the resistivity of the vertical field coils, not mentioned here, has to be added. A = 200 is the voltage gain of the power amplifier, and $g = -500\ sec^{-1}$ stands for the overall gain of the feedback loop after integration, $L_a = 430\ \mu H$ is the inductance inside the vacuum vessel, seen from the output of the two amplifiers taken in parallel. The other coefficients are approximately $M_a = 0{,}48\ L_a$ and $M_p = 4{,}2\ \mu H$. The flat-top values of I_v and I_p are 3,9 kA and 130 kA if we assume $2\,\beta_{pol} + \ell_i = 3$.

The error voltage U_{FB} is applied at the input of the power amplifier. As the amplifier has a fixed voltage gain, it reacts like an integrator when working on an inductive load. Consequently, for a given plasma current I_p the error voltage must decay to zero, which is only possible if the vertical current I_v reaches a constant value corresponding to the equilibrium position of the plasma. But the plasma position would present a large overshoot if no driving signal were added simultaneously. The driving signal $U_1 = const.\ \dot{I}_p$ is produced by a Rogowski coil wrapped around the minor diameter of the vacuum vessel. The best calibration is obtained when the driving signal, applied to the open loop, forces the vertical current to reach the same value as in the closed loop situation.

But a preliminary test of the amplifier is recommended without plasma. In this case the cosine coil produces an integrated voltage equal to $L_a(I_c - I_v)$ where I_c is the transformed wall current of the vacuum vessel induced in the control windings. The error signal is now

$$U_{FB} = \frac{g}{A}\left[L_a I_c + M_a I_v - M_p I_p\right] \qquad (2)$$

For the driving term we find (in Volts)

$$U_1 = \frac{30}{A} \cdot 10^{-6}\dot{I}_p \qquad (3)$$

The plasma current can be simulated with a function generator. The chosen waveform is $I_p = 100\ kA\ \left[1 - \exp(-250t)\right]$. Thus the driving signal injected into the system (see ref. {2}) amounts to

$$U_1 = 3{,}7\,\exp(-250t) \text{ and } -\frac{g}{A} M_p I_p = 1{,}1\ \left[1-\exp(-250t)\right] \qquad (4)$$

The load forms a network comprising the following elements: An external inductance of 570 μH connected in series with the internal inductance of 430 μH of the vacuum vessel. And the latter is shunted with 0,28 Ω which represents the transformed resistivity of the vessel.

Fig. 2 shows the observed response of the closed loop perturbed by the above driving signal. The 1st trace is the error signal. The 2nd trace is the input voltage of the power

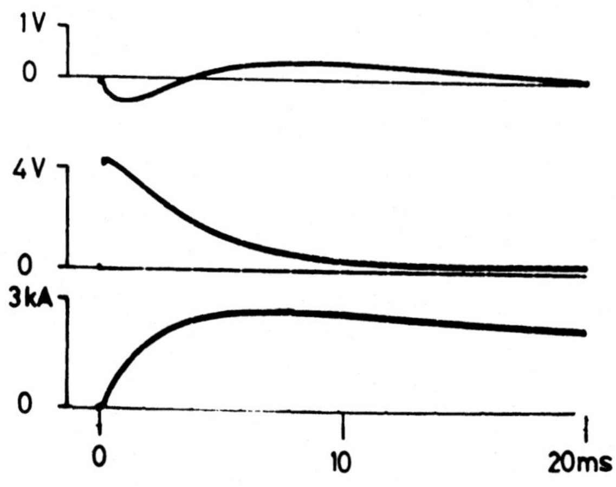

Fig.2 Observed response

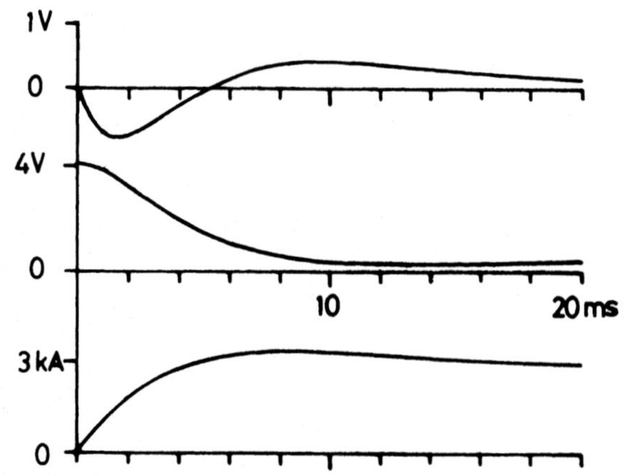

Fig.3 Calculated response

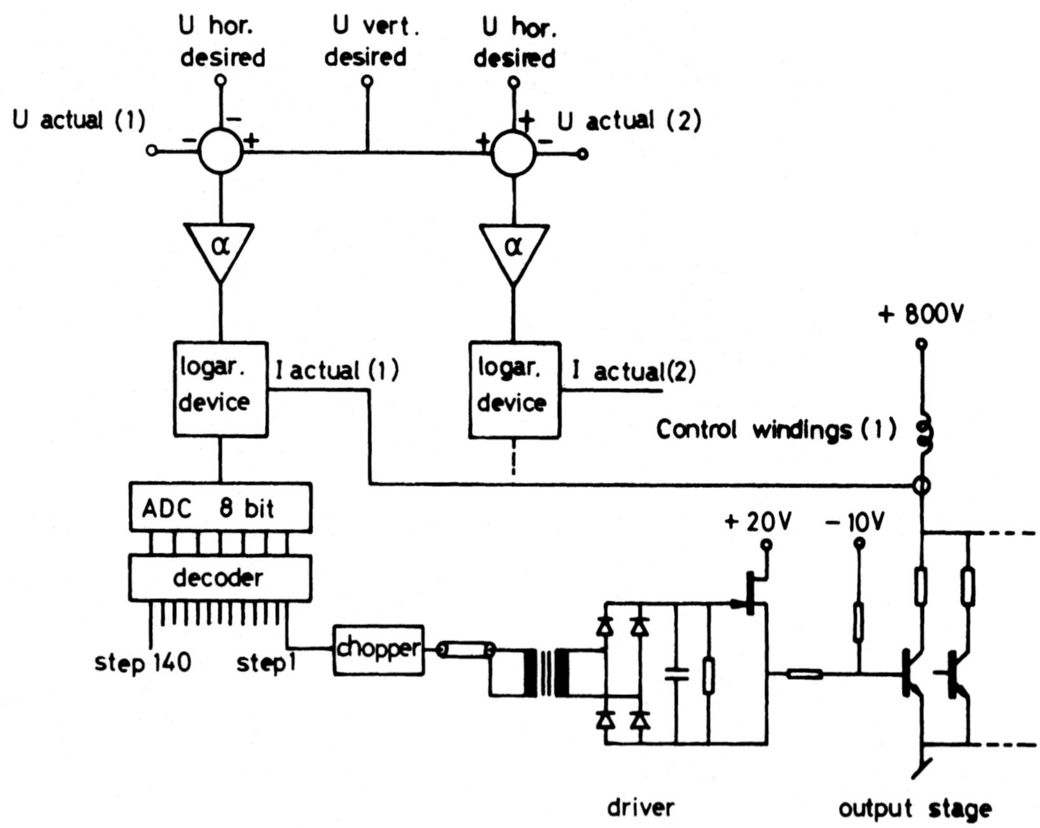

Fig. 4 Block diagram of the feedback circuitry

amplifier. The 3rd trace is the vertical field current of both amplifiers. Fig. 3 shows the calculated behaviour of the same quantities. The compensation of the ohmic voltage drop due to the coil resistivity is included.

The good agreement between the observed and the calculated response proves that the power amplifier behaves like a linear element in the feedback loop.

Similar measurements of the vertical mode (horizontal field) were performed. We observed good agreement with the theory.

3. THE POWER SUPPLY

The power supply has been constructed by the Swiss Institute for Nuclear Research (SIN). Fig. 4 shows the block diagram of the apparatus. The two amplifiers are fed by a 800V electrolytic capacitor bank consisting of two parts in series. The main bank of 4 Farad produces the current during the flat-top of the plasma current, for which 400 V is sufficient. The smaller bank of 40 mF discharges quickly but provides an additional 400 V during the current rise, when a greater inductive voltage is required to build up the current in the control windings. This small bank is bridged by a diode when it has discharged. Initially a current rise of 1600A/msec is reached, and the peak current is limited to 4000 A for both amplifiers taken in parallel. The maximum field is 700 Gauss corresponding to 130 kA plasma current. The maximum current can be sustained during 0,2 sec. The electrolytic capacitor bank is separated in many groups protected by individual fuses. See also ref. {3}.

The power supply is based on the principle of a variable resistor comprising many resistors connected in parallel by transistors which operate only in the switching mode. Each amplifier consists of 1000 transistors and resistors forming 140 groups which are switched on in a consecutive order, providing an almost continuously variable resistance. The resistance varies exponentially with the number of switched steps. The number of steps is determined by an ADC referencing an analogue control input. An internal feedback loop containing a logarithmic device assures that the output voltage is indeed proportional to the input signal, and the complete amplifier thus behaves as a high-power voltage amplifier at zero output impedance.

The linearised feedback control works as follows: The resistance R of each amplifier varies exponentially with the number of steps n

$$R = R_o \exp(-kn) \qquad 1 < n < 140 \tag{5}$$

The best fit for this expression gives $R_o = 7,3\ \Omega$ and $k = 0,0372$. Let $U_v = U_o - U_R$ be the coil voltage. U_o denotes the voltage of the capacitor bank. $U_R = IR$ is the voltage existing at the controlled resistance.

A regulator compares the actual voltage U_v with the desired value U_c and multiplies the difference by a transfer factor α having a high static gain.

A logarithmic device creates the function

$$E = \ell n \frac{R_o I}{\alpha(U_v - U_c)} \tag{6}$$

I is the load current, measured by a transformer followed by an integrator. E is the normalized driving voltage of the ADC. The linear character of the converter may be expressed by kn = E. Now we close the loop and we find

$$U_R = \alpha(U_v - U_c) \tag{7}$$

Thus the system is linearised. We see immediately that for $|\alpha| >> 1$ the amplifier performs its duty: $U_v \xrightarrow{=} U_c$.

For the vertical field control U_v is in fact the mean voltage of the two sets of control windings. In the case of horizontal field control the voltage difference of the two sets is, of course, compared with the desired signal for this mode.

Many practical requirements do not permit to reach the above ideal relations:

1. At the start I = o the logarithmic device would be saturated if a bias ΔI of about 4% of the maximum current were not added to the measured I value. This bias value was experimentally adjusted as the best compromise.

2. It is necessary to render the waiting period I = o insensitive to an unavoidable ambient noise level. So a negative bias voltage ΔE is superimposed to the driving signal of the ADC. 2% of E_{max} is sufficient. This bias is automatically maintained with the aid of a slow auxiliary control loop to stabilize the output of the regulator against drift.

3. Formula (5) gives a conductance of $1/7\ \Omega^{-1}$ at the first step n = 1. But during the waiting period called n = 0, when all transistors are switched off, the conductance equals zero. This first jump of conductance is much bigger than the next one. It causes a low frequency relaxation due to the inductive load. This trouble disappears if we flatten the beginning of the function (5) by reducing the first step to $1/100\ \Omega^{-1}$.

Furthermore, the gain of the regulator has to be relatively low in the high frequency domain, in order to avoid a switching instability. The suitable transfer function is

$$\alpha = 200 \frac{1 + 2.10^{-4} i\omega}{1 + 10^{-2} i\omega} \qquad (8)$$

Various precautions have been taken to protect the transistors. In case of a fast switch-off signal not all transistors get blocked simultaneously because the storage time is distributed at random. The mean storage time at 8A of the Darlington transistors from Motorola, type MJ10011, is 20 μsec. Fast power FETs are used for the drivers. A capacitor C_1 of 150 μF is connected to the variable resistance, which hinders the voltage from rising too fast. So the last conducting transistor never exceeds the limiting current of 16A. A dead time of 60 μsec is introduced by the capacitor.

When the overcurrent switch interrupts the load current a freewheel diode in series with 0,1 Ω allows the current to decay. But the current cannot be diverted at once from the transistors to the freewheel diode. The stored magnetic energy in the cabling must be absorbed by another capacitor C_2 as big as 400 μF. This capacitor is normally charged and a diode in series prevents the variable resistance from overvoltage, but isolates the capacitor from the variable resistance during the normal operating conditions.

In order to achieve a clean control signal for the drivers a galvanic isolation is provided between the ADC and each driver, by using a 1 MHz chopper and an isolation transformer followed by a rectifier.

4. REFERENCES

[1] CHEETHAM, A.D., et al., Lausanne Laboratory Report LRP 162/80 (1980)

[2] KELLER, R., LISTER, J.B., TUSZEL, A., Lausanne Laboratory Report LRP 166/80 (1980)

[3] SCHENK, G., PACHER, G., PACHER, H., Symposium on Fusion Technology, Padova, 4 - 9 September, 1978. Proceedings pp 1019-1024.

HIGH-SPEED THYRISTOR POWER AMPLIFIER CONTROL FOR THE JT-60 POLOIDAL FIELD

I.Takahashi A.Nabae R.Shimada
Technological Univ. of Nagaoka Japan A.E.R.I.
(Nagaoka-shi Japan) (Tokai-mura Ibaragi-ken Japan)

ABSTRACT

In undertaking basic research into a high power thyristor amplifier system for the JT-60, we have developed direct computer control employing modern control theory, and have experimented on a small model system. This paper describes how to attain the required high speed and accuracy and the hardware. To simulate the actual system, a model system composed of a flywheel driven a-c generator, a computer-controlled thyristor amplifier and three coupled magnetic field coils was used. From the experimental results, 0.2% static accuracy, 0.5% dynamic accuracy and 3 msec control delay were obtained.

1. INTRODUCTION

In order to further research into nuclear fusion in Japan, the JT-60 project is being developed, and recently a part of the system has been constructed and tested. Fig.1 shows the JT-60 Poloidal field system, excited by five thyristor power amplifiers supplied from a flywheel (1.3GJ) driven synchronous generator (500MVA). The total capacity of the thyristor amplifiers will exceed 1200MVA, making it the largest system in the world. Moreover, the amplifiers are required to control to within a 5msec time delay and 0.5% accuracy. [1]

This paper proposes the following new control schemes to accomplish the aims: (1) To regulate the coil current as above, we have applied minimum time control theory to the thyristor amplifiers for controlling plasma position; (2) The magnetic coupling between the Poloidal coils causes interference between the coil currents, and a control error can be effectively eliminated by decoupling control; (3) Because of the rapid changes in the output voltage of the thyristor converter, the voltage and frequency of the synchronous generator fluctuate ±10% and ±1~2Hz respectively. In order that the thyristor amplifier may follow the voltage source, the gate

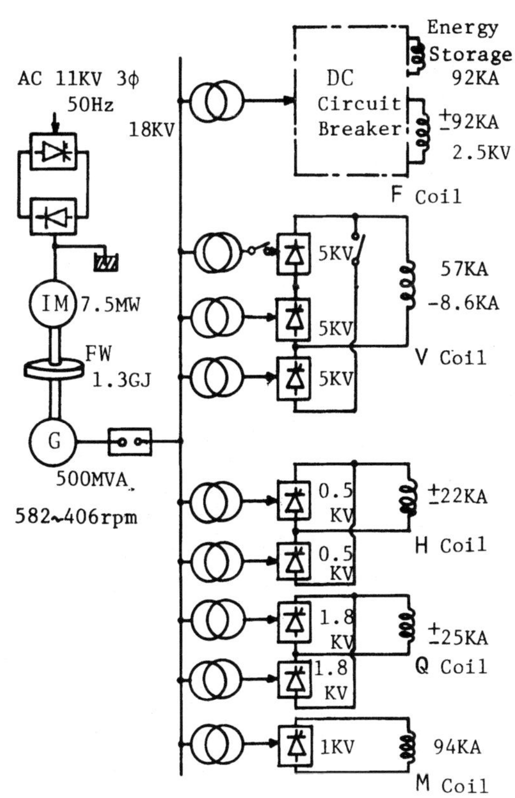

Fig.1. JT-60 Poloidal field system.

control circuit is digitalized so as to give a quick response to the variations. In this paper, we describe the solution of these problems and the experimental confirmation on a small model system using the developed theories and techniques.

2. Thyristor amplifier minimum time control

An asymmetrical thyristor converter composed of two symmetrical converters in series has better power-factor characteristics than a symmetrical converter, the converters being triggered at different gate angles α_1 and α_2 respectively. The output voltage waveform of this converter is shown in fig.2, where one of the firing angles is constant. In the case of a small rate of change of control angle, the mean voltage over a sampling interval T can be expressed approximately by the stepwise sample-hold wave shown as a dotted line in fig.2. The stepwise wave obtained resembles the sample-hold waveform in a discrete control system. Initially, we shall assume a simple R-L series circuit as a load, for which the transfer function is written as follows:

$$G(s) = 1/(1 + s\tau) \qquad (\tau = L/R) \tag{1}$$

Using modern control theory, the system is able to follow the input signal after only one sample, unless the system is saturated. [2] Fig.3 shows the proposed block diagram for the sampling response of the first order system expressed in eq.(1). The converter has a maximum allowable voltage saturation limit $\bar{e}_{max}$ and a minimum limit $\bar{e}_{min}$. We represent the normalized values of these and the output voltage $\bar{e}$ as follows:

$$I_M = \bar{e}_{max}/R, \quad I_m = \bar{e}_{min}/R, \quad m' = \bar{e}/R \tag{2}$$

If the system in fig.3 is not saturared, that is $m = m'$, the output current i is expressed by the first order differential equation

$$pi = (1/\tau)(m - i), \quad p = d/dt \tag{3}$$

and we can represent the state vector as

$$V = (i_r, m, i)^t \tag{4}$$

where i_r denotes the thyristor amplifier reference current. The state transition matrix Φ and transition matrix B at the sampling instants are expressed as follows:

$$\Phi(t) = \begin{bmatrix} 1 & 0 & 0 \\ 0 & 1 & 0 \\ 0 & 1 - e^{-t/\tau} & e^{-t/\tau} \end{bmatrix}$$

$$B = \begin{bmatrix} 1 & 0 & 0 \\ 1 + K_V & 0 & -K_V \\ 0 & 0 & 1 \end{bmatrix} \tag{5}$$

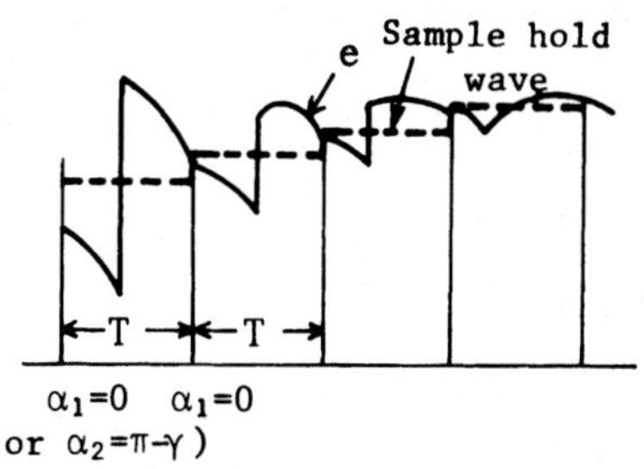

Fig.2. Output voltage and sample-hold approximation.

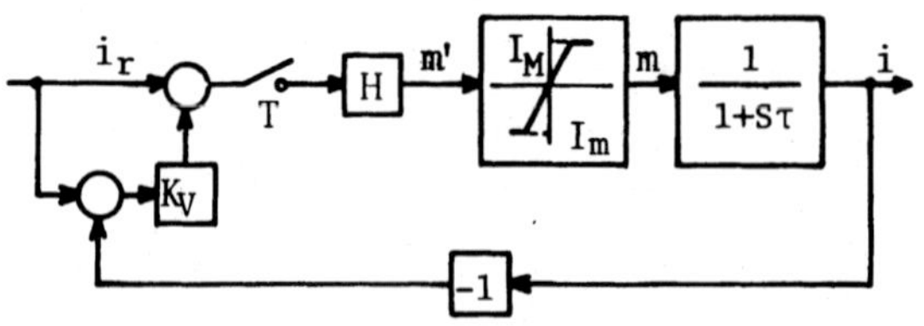

Fig.3. Block diagram for optimum control.

where K_V is the gain factor. The vector at $t = \overline{k+1}\,T$ is given by

$$V(\overline{k+1}\,T) = \Phi B V = \begin{bmatrix} i_{rk} \\ i_{rk} + K_V(i_{rk} - i_k) \\ \{i_{rk} + K_V(i_{rk} - i_k)\}(1 - e^{-T/\tau}) + i_k e^{-T/\tau} \end{bmatrix} \qquad \left(\begin{matrix} i_{rk} = i_k(kT) \\ i_k = i(kT) \end{matrix}\right) \tag{6}$$

To obtain the minimum time and deadbeat control, the relation $i_{k+1} = i_{rk}$ must be satisfied in eq.(6). The gain K_V can then be determined by solving eq.(6), which yields

$$K_V = e^{-T/\tau} / (1 - e^{-T/\tau}) \tag{7}$$

The unsaturated system can be settled in one sample period. K_V is a constant conveniently determined only by T and τ. For the saturated system, a one sample response time is obtained after it comes out of the saturated condition.

3. Decoupling control for a magnetically coupled system

Consider a system with magnetic coupling, the characteristic equations of the system being as follows:

$$\begin{bmatrix} e_1 \\ e_2 \\ \cdot \\ e_i \\ \cdot \\ \cdot \\ e_n \end{bmatrix} = \begin{bmatrix} R_1 + pL_1 & pM_{12} & \cdots & pM_{1n} \\ pM_{21} & R_2 + pL_2 & \cdots & pM_{2n} \\ \vdots & \vdots & \ddots & \vdots \\ pM_{n1} & \cdots & \cdots & R_n + pL_n \end{bmatrix} \begin{bmatrix} i_1 \\ i_2 \\ \cdot \\ i_i \\ \cdot \\ \cdot \\ i_n \end{bmatrix} \tag{8}$$

The system can be divided into n magnetically decoupled systems similar to those in the preceding section, since on n-th order system which is controlled by n signals can have n state variables set to any values in one sample period. We now examine the relations between the applied voltages of the coupled system e_i ($i=1\sim n$) and those of the decoupled system e_i' ($i=1\sim n$), as shown in fig.4. From eq.(6), the optimum voltage in the decoupled system of the i-th coil at t=kT, e_{ik}' is as follows:

$$e_{ik}' = \{i_{rik} + K_V(i_{rik} - i_{ik})\} R_i \tag{9}$$

while the characteristic equation of the coil is

$$e_{ik}' = (R_i + pL_i) i_i \tag{10}$$

Approximating the current i_i as i_{ik} during the k-th sample interval, eq.(10) can be rewritten as

$$pi_i = (1/L)(e_{ik}' - R_i i_{ik}) \tag{11}$$

Combining eqs.(8) and (11), the control voltage e_{ik}, the mean value of e_i for the k-th sample interval of the coupled

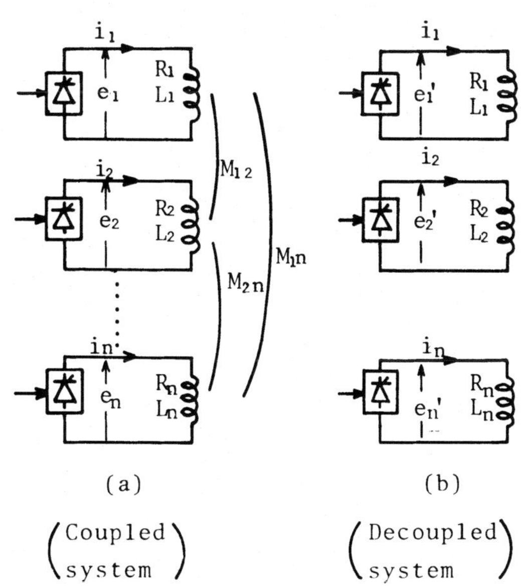

Fig.4. Magnetically coupled and decoupled systems.

system, is given as follows:

$$e_{ik} = e_{ik}' + \sum_{j=1, j\neq i}^{n} (M_{ij}/L_i)(e_{jk}' - R_j i_{jk}) \qquad (i = 1,2,\ldots,n) \tag{12}$$

We shall now consider a system that has n' shorted coils ($e_{n-n'+1} = 0, \ldots, e_n = 0$) such as the plasma or vacuum chamber shown in fig.4. Since the system has only n-n' control signals, the currents in the remaining n' coils cannot be controlled in one sample interval. Fortunately it is not necessary to control the currents of these coils in the actual system. The matrix may be easily solved for the mean voltage $e_{1k} \sim e_{(n-n')k}$, using e_{ik}' in eq.(9) as before, and we obtain

$$\begin{bmatrix} e_{1k} \\ e_{2k} \\ \vdots \\ e_{n-n',k} \end{bmatrix} = \begin{bmatrix} A_{11} & \cdots & A_{1,n-n'} \\ \vdots & \ddots & \vdots \\ A_{n-n',1} & \cdots & A_{n-n',n-n'} \end{bmatrix} \begin{bmatrix} e'_{1k} \\ \vdots \\ e'_{n-n',k} \end{bmatrix} + \begin{bmatrix} B_{11} & \cdots & B_{1n} \\ \vdots & \ddots & \vdots \\ B_{n-n',1} & \cdots & B_{n-n',n} \end{bmatrix} \begin{bmatrix} R_1 i_{1k} \\ \vdots \\ R_n i_{nk} \end{bmatrix} \tag{13}$$

The elements of the matrices are functions of the inductances L_i and M_{ij}.

4. Digital gate control using a microcomputer [3]

The optimum voltage can be calculated by a minicomputer (PDP 11/34) in 500~600 μsec, with the aid of a digitalized calculation table.

The digital gating circuits, comprising a zero voltage detecting circuit, a frequency multiplier, and a microcomputer gate controller, are shown in fig.5. A trigger pulse at control angle α is obtained by comparing the counter contents, which increase in a sawtooth manner, with the digitalized control angle α as calculated on the minicomputer, the pulse being generated at the time when both values coincide with each other, as shown in fig.6. In a three phase system ,three counters Ca,Cb,Cc are needed as shown in fig.6. The maximum value of each counter is set at 2^8(8bit), which corresponds to a control angle $\alpha = 2\pi$. To synchronize these counters to the line frequency, f, they are clocked at a frequency of $2^8 f$ and reset at the zero voltage point ($\alpha = 0$).

The zero voltage detector operates with noisy and fluctuating voltage,and so a high response low pass filter with no phase displacement has to be used. The proposed filter is composed of an active low pass filter with a minimized phase error (+1°~-2° for 78~54 Hz) and a high response compensating circuit made up of an F-V converter and nonlinear element, as shown in fig.7. Since the phase error of the filter depends on the input frequency, the error can be compensated for

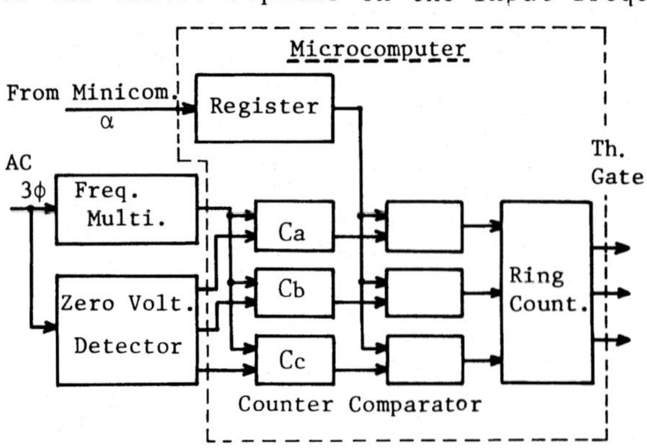

Fig.5. Digital gate controller.

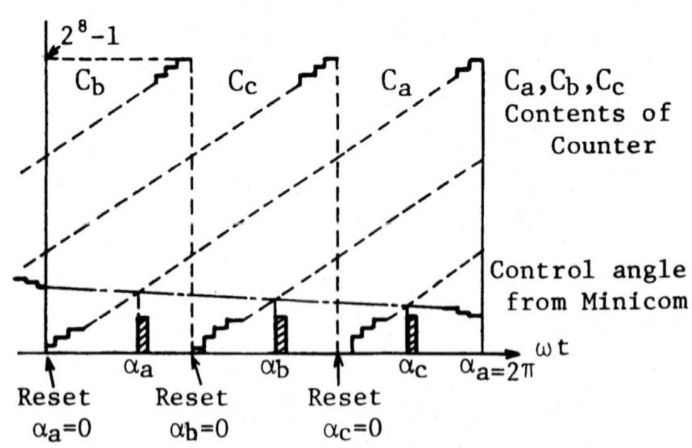

Fig.6. Principles of the digital gate circuit.

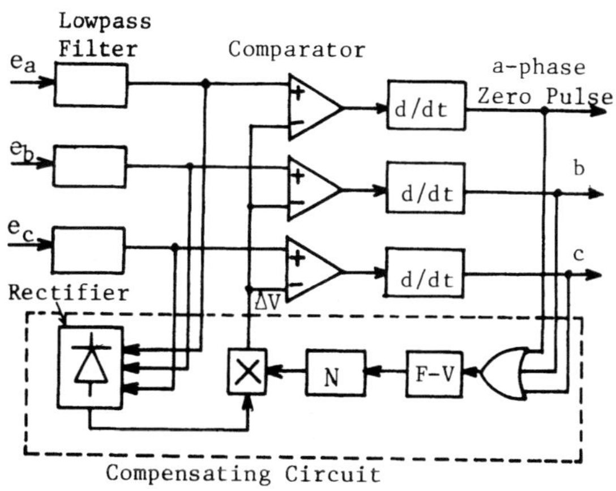

Fig.7. Zero voltage detector.

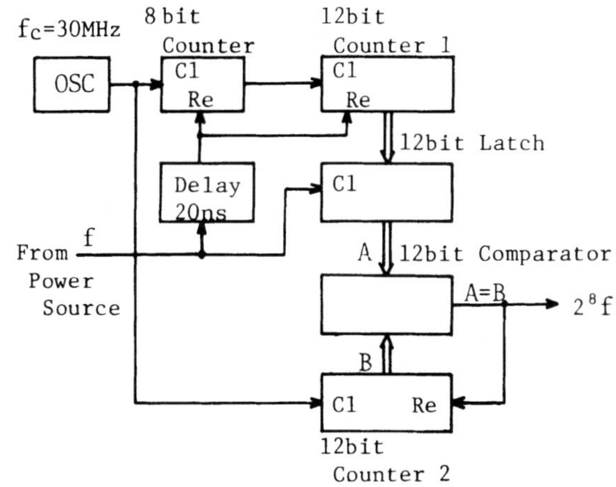

Fig.8. Frequency multiplier.

by biasing the zero voltage detecting comparator by a voltage ΔV corresponding to the phase error, using a nonlinear element N whose characteristics resemble those of the filter phase error. From experimental results, phase errors of within 0.5% are obtained for the fluctuating voltages described in the previous section. Frequency multipliers are used for the clock pulse train which decides the trigger times of the thyristors. So as to minimize the error in the trigger times, a high response of the frequency multiplier is indispensable. The principles of the digital time interval measuring circuit shown in fig.8 are applied. The line frequency signal is f, and is multiplied by 2^n to get a clock frequency fc in the range 10~30MHz. Counter 1 is clocked at $fc/2^n$ and reset by f. The maximum value, $fc/2^n f$, is stored in a latch circuit and is compared with the value of counter 2 which is clocked at fc. When both values are equal, counter 2 is reset to zero. The frequency of the reset pulse is thus $fc/(fc/2^n f) = 2^n f$, and frequency multiplication is accomplished. The microcomputer gate control circuit is composed of a Z-80 C.P.U., and three parallel I/O (PIO) and memory ICs. A six-phase asymmetrical thyristor converter can be controlled using this circuit to within 55 μsec. The three counters Ca,Cb and Cc shown in fig.6 correspond to three registers in the microcomputer, which are incremented pulses at a $2^8 f$ repetition rate at the non-maskable interrupt input.

5. Experimental system and results

Fig.9 shows the overall system used in our experiment. The system is composed of a simulated flywheel made using d-c motor speed control, a three-phase a-c generator with high speed exciter, the thyristor amplifier discussed above, and a magnetic field coil with three coils. The flywheel characteristics can be obtained by decreasing the d-c motor speed in the same way as the removal of energy from the flywheel would decrease its rotation speed. To improve the dynamic response and static accuracy of the a-c generator, we adopted a high speed self-exciting system and a voltage feedback system, as conventionally used. From our experiments, it was found that the static accuracy obtained was better than 1% and the dynamic response was improved to 0.2sec.

The thyristor amplifier was controlled by a computer hierarchy comprising one minicomputer and four microcomputers. Coil currents i_1, i_2 and i_3 were detected and isolated, and then converted to a 12 bit digital signal. Fig.10 shows the experimental results for the system. The waveforms are coincident with the simulated ones to good accuracy. Considering the following characteristics

Fig.9. Experimental system.

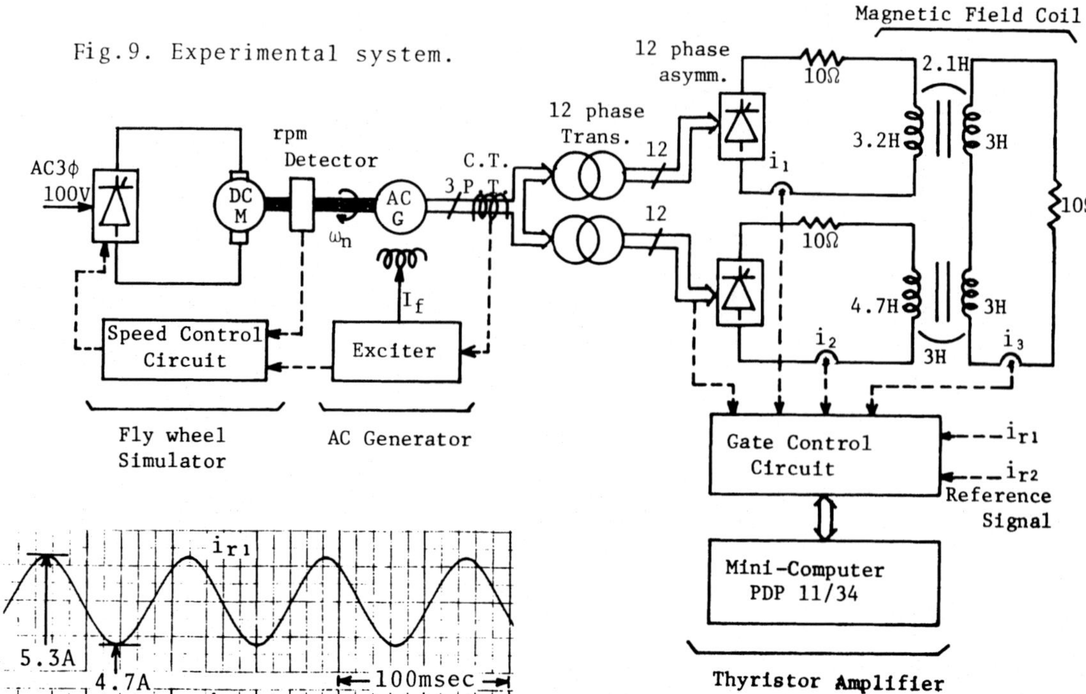

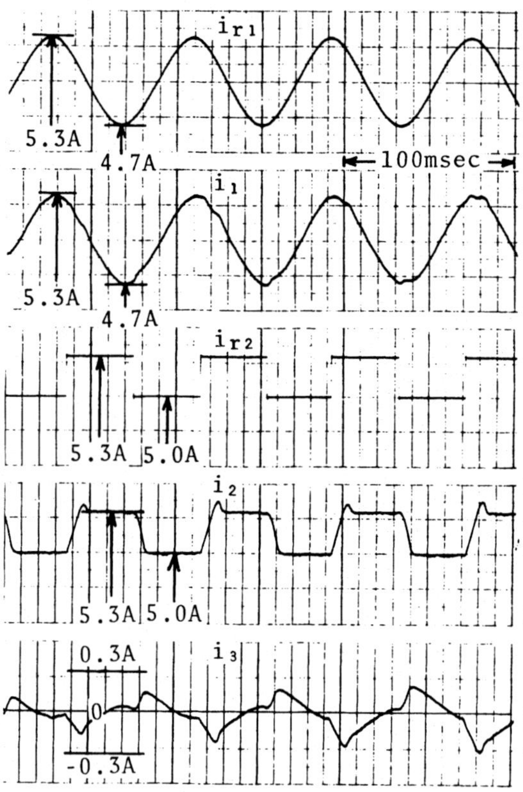

Fig.10. Responses of experimental system.

with regard to input signals i_{r1} and i_{r2}, a small coupling effect appears in i_1 at the point of the step change in the other coil. This effect is, however, less than 0.5%, and be of no account in the actual system. Static accuracy of 0.2% and control delay of 3 msec were obtained. The proposed schemes make it possible to optimize the control of the high power thyristor amplifier.

6. Acknowledgement

The authors would like to express their appreciation to Dr.M.Yoshikawa and Dr.S.Tamura, of the Division of Large Tokamak System Development at the Japan Atomic Energy Research Institute.
This work was carried out under a financial grant from the Ministry of Education.

7. References

[1] Annual Report of Fusion Research and development Center, JAERI-M8661(1980)
[2] I.Takahashi,Trans, I.E.E.Japan VOL. 54-B9 (1979)
[3] I.Takahashi, 1979 National Conv.Inst.Elec.Engrs.Japan, No. S-8-6

A THYRISTOR BREAKER OF 1.5 10^9 V.A. FOR THE POLOIDAL FIELD SYSTEM OF TORUS SUPRA

B. Bareyt, C. Leloup, E. Rijnoudt

ASSOCIATION EURATOM-CEA SUR LA FUSION
Département de Recherches sur la Fusion Contrôlée
Centre d'Etudes Nucléaires
Boîte Postale n° 6. 92260 FONTENAY-AUX-ROSES (FRANCE)

ABSTRACT

The poloidal field system of Torus Supra has an inductive storage of ~ 38 MJ, which has conducted to research the best solution for a D.C. breaker (55 kA., 27 kV.)
A solid-state breaker has been chosen.
The working principles have been tested on a small size model. The final circuit breaker will contain a large number of thyristors in series and in parallel ; the critical problem lies in the series arrangement. A test unit for full tension has been constructed. In this unit the thyristors are submitted to the maximum current as well as to the maximum voltage. The surges measured during tests are not higher than the calculated values. A synthetic circuit of the poloidal field system has been used for test under the final working conditions.

1. GENERAL DESCRIPTION AND PRINCIPLES OF OPERATION OF THE POLOIDAL FIELD SYSTEM [1]

The main components of the poloidal field system of Torus Supra are :

- a saturated iron core transformer,
- a single poloidal field coil system ; all coils located outside the T.F. coils are used simultaneously for premagnetization, ohmic heating, vertical field amplitude and index control as well as feedback against horizontal and vertical plasma displacements (see fig. 1).
- a system of thyristor power supplies ; one of them "Go" feeds all the coils in parallel and the others provide additional control voltage between different coils (fig. 2).
- a DC thyristor breaker "T" in order to force the current, reached at the end of the premagnetization, through the resistor "R_2" producing the high voltage required for the start up of the plasma discharge ($\gtrsim$ 100 V/turn).

The D.C. thyristor breaker for T.S. is rated at 55 kA, 27 kV. This breaker is put in parallel to the main power supply Go. When the current has reached the chosen value, the power supply is pushed into the converter mode, producing thus the voltage needed to transfer the current from Go to T, where T acts as a fly-wheel circuit. When the current is completely transfered on T (less than 2 ms) a fast opening switch "IM" is opened (no breaking ability is required) in order to disconnect Go during the high voltage operation.

Then the fast make switch CR_1 is operated and the precharged condenser C_1 (160 KJ, 30 kV) is discharged by ignitrons in order to drive the breaker current to zero, (with $dI/dt \approx -120\ A/\mu sec.$ when I_T reaches zero) and maintains a negative voltage accross T during $\approx 120\ \mu$sec. With this system the thyristor breaker has to conduct the full current during a short time (20 to 30 msec), given by adding to the opening time of IM the closing time of CR_1 triggering times including.

By this method, studied under contract by the Jeumont Schneider Cy. an overcurrent can be driven through the thyristors, leading thus to a serious cost reduction. Fast thyristors have been preferred because in case of using standard thyristors, the reduction of the reverse voltage during turn-off time, to avoid unwanted plasma breakdown obliges to multiply by about 10 the condenser bank value.

2. BREAKER MODELS AND TEST STAND

2.1. - A small scale model T_1 (1 thyristor) has been used to test the use of thyristors for D.C. circuit breaking. During these tests a 400 A current from a 1.8 kJ inductive load has been interrupted under 1000 V [2].

2.2. - Full voltage model (T_2).

The final circuit breaker will contain a large number of thyristors in series and in parallel ; the critical problem lies in the series arrangement. A test unit for full voltage has been constructed by the CGEE-ALSTHOM Cy. In this unit the thyristors are submitted to the maximum running current per thyristor. This thyristor valve is rated for 27 kV using 24 2-inch fast thyristors (τ_q = 40 μsec) in series, giving a safety factor of two. A current of 4500 A, 40 msec should not heat the thyristor by more than 40°C per pulse (nominal current 1000 A).

To reduce the dispersion of the over voltage due to the $\frac{dIrr}{dt}$ at 0 current the thyristors have been tested and chosen with a recovered charge value "Q_{rr}" as similar as possible ($Q_{rr} \approx 372$ μC $\pm$ 12 μC at $\frac{dI}{dt}$ = 20A/μsec). At the same time a selection control of the finger voltage was made, to ensure simultaneous triggering at low voltage. Each thyristor is equiped with an over voltage damping circuit. The thyristor valve has a self triggering protection against direct overvoltage.

2.3. Test stand

Each of the thyristor valve characteristics has been tested independently after assembly in the factory :

- direct and reverse static voltage distribution (14 kV on each half valve),
- triggering under low voltage (400 V),
- triggering under high voltage (3 kV per half valve),
- turn-off without saturable inductance protection ($\frac{dI}{dt}$ = - 27 A/μsec. recovery time = 40 μsec),
- reverse voltage distribution during the turn-off time (dispersion $\lesssim$ 20 %).

In order to test all the characteristics simultaneously, a synthetic circuit has been prepared, modelling the Torus Supra poloïdal field circuit (full voltage and 1/10e in current ; see fig. 3).

- Main generator "G_o" is a thyristor controlled rectifier 4500 A,526 V, 1 sec.
- Load : "L_p" = 22 mH, R_p = 94 mΩ.
- Resistor : "R_2" = 6÷9Ω.
- Condenser banc "C_1" : up to 48 μF, adjustable step ~ 2 μF, triggered by ignitrons.
- Selfinductance : "L_1" up to 1500 μH adjustable every 50 μH.

The electro-mechanical switches : "IM", CR_1", CR_{cc}" are already dimensioned for the final poloidal field system of Torus Supra.

- IM A.E.G. : GA 736 ; overcurrent 1 sec, 37 kA ;
  opening time 9 $\pm$ 1 msec. including triggering time ;
  voltage (open) 36 kV ;
  alternative : Driescher CY ; 75 kA, 1 sec. ; 10 msec. 45 kV.

- CR_1 and CR_{cc} : Driescher fast make switch :
50 kA - 1 sec. ; total closing time 10 msec. ; voltage (open) 40 kV.

The command and control system using a micro processor is also a model for the final control system, especially for the fault conditions : protection of G_o against over-voltage, reignition of T, current limitation, etc ... [3]

3. RESULTS OF TESTS

3.1. - Transfer of the current from G_o to T.

This phase is illustrated in fig.4A, at the scale of the figure it looks instantaneous. The duration of the transfer takes not more than 1 msec. as shown in fig. 4B.

3.2. - Adjustment of $\frac{dI}{dt}$ at turn-off and of the recovery time τ_r.

The adjustment of $\frac{dI}{dt}$ and τ_r is obtained by a proper selection of C_1 and L_1 ; the results of the computer analysis are summarized in fig. 5.

Two typical experimental results are shown on fig. 4 corresponding to interrupted current amplitudes of 2.4 kA (fig. 4A) and 3.9 kA (fig. 4B) : C_1 = 30 μF (precharge voltage 14.4 kV for fig. 4A and 23 kV for fig. 4B) L_1 = 720 μH. The results for 4A are :

$\frac{dI_o}{dt} = - 10{,}5\ \text{A}/\mu\text{sec}$ at turn off, τ_r = 127 μsec, maximum blocking voltage 1.25 kV, direct voltage of the transfert of the current to R_2 = 14.4 kV.

For 4B : $\frac{dI_o}{dt} = - 17{,}5\ \text{A}/\mu\text{sec.}$ at turn-off τ_r = 125 μsec.;blocking voltage 5 kV ; reverse voltage 23.4 kV.

The experimental points for 4A and 4B are shown on fig. 5, close to the calculated values.

3.3. - Over voltage limitation

The damping circuit against over voltage has been studied on the small scale model T_1. The maximum over voltage occurs after the first zero of the valve current, during the negative current in the thyristor, corresponding to the outflow of the electrical charges in the thyristor. The valve current and voltage are shown in fig. 6A ; experimental results with valve T_1 without damping circuit show the over voltage peak. The effect of a damping circuit RC is illustrated on fig. 6B.

4. CONCLUSION

The tests have shown that the use of a thyristor D.C. circuit breaker in parallel to the main generator to reduce the cost of the breaker by limiting to 30 msec. the conduction time, is feasible. The small conduction time permits to multiplie the nominal thyristor current four times (with a safety factor of $\gtrsim$ 2), reducing the cost of the valve to a price approching other types of breakers. The transfer operation is fast and the protection of the main generator by a fast switch is working in a satisfatory way. Presently the maximum transferable current is 4000 A due to load limitation. This value has been obtained. The first series of tests has reached values ($\frac{dI_o}{dt} = - 17.5\ \text{A}/\mu\text{sec.}$; τ_r = 120 μsec.) close to the maximum figures of the valve.

References :

[1] AYMAR, R., LELOUP, C., PARIENTE, M., Xth Symposium on Fusion Technology, 4-9 September 1978, Proceedings pp 939-944.

[2] BAREYT, B., LELOUP, C., RIJNOUDT, E., JACOUD, A., Association Euratom-C.E.A., DPh-PFC, 1198 (1979).

[3] BOURON, A., CAMBIEN, L., CHARET, M., TOSOLINI, Y., Association EURATOM-CEA, DRFC, 1209 (1980).

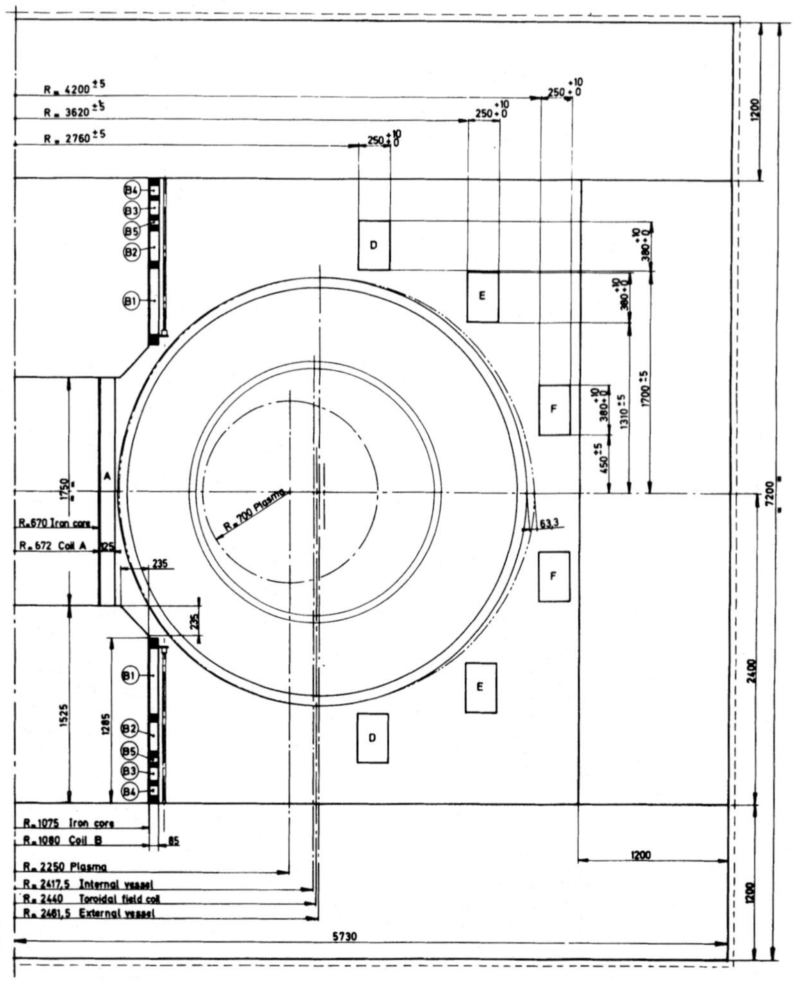

FIG : 1

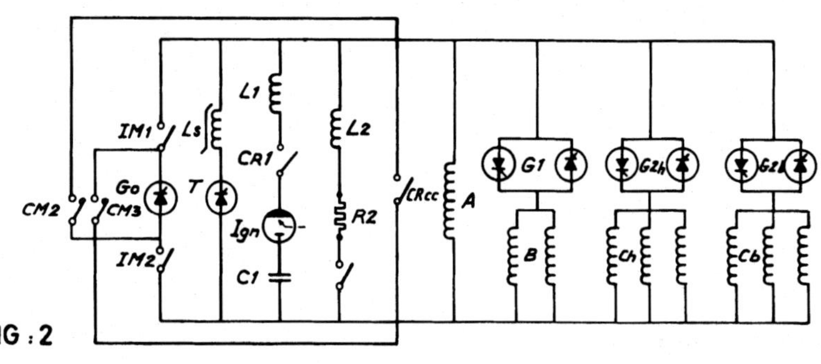

FIG : 2

SIMPLIFIED DIAGRAM OF THE POLOIDAL FIELD CIRCUIT OF TS

TEST CIRCUIT

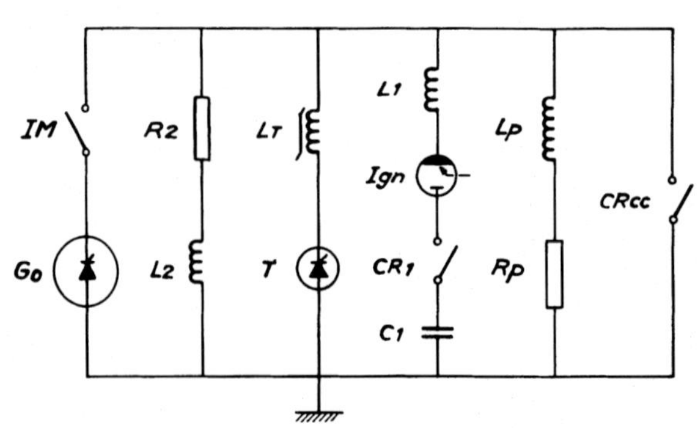

FIG : 3

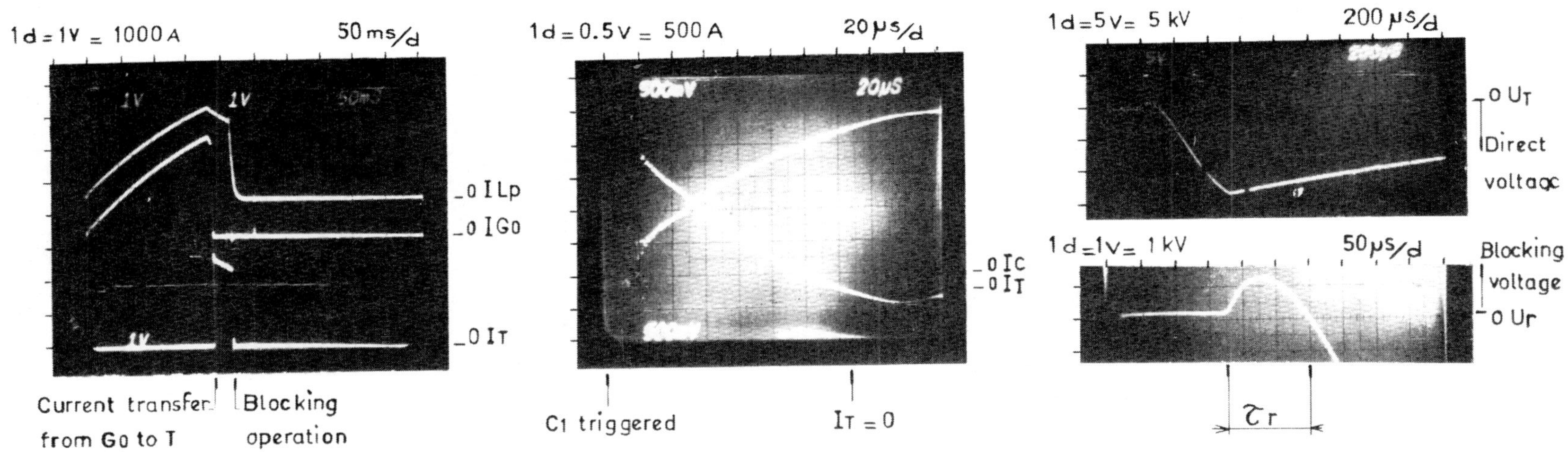

4A Experimental results for an interrupted current of 2.4 kA : C_1 = 30 μF ; 14.4 kV ; L_1 = 720 μH ; τ_r = 127 μsec ; $\frac{dI_0}{dt}$ = -10.5 A/μsec conducting time of the valve 30 m sec

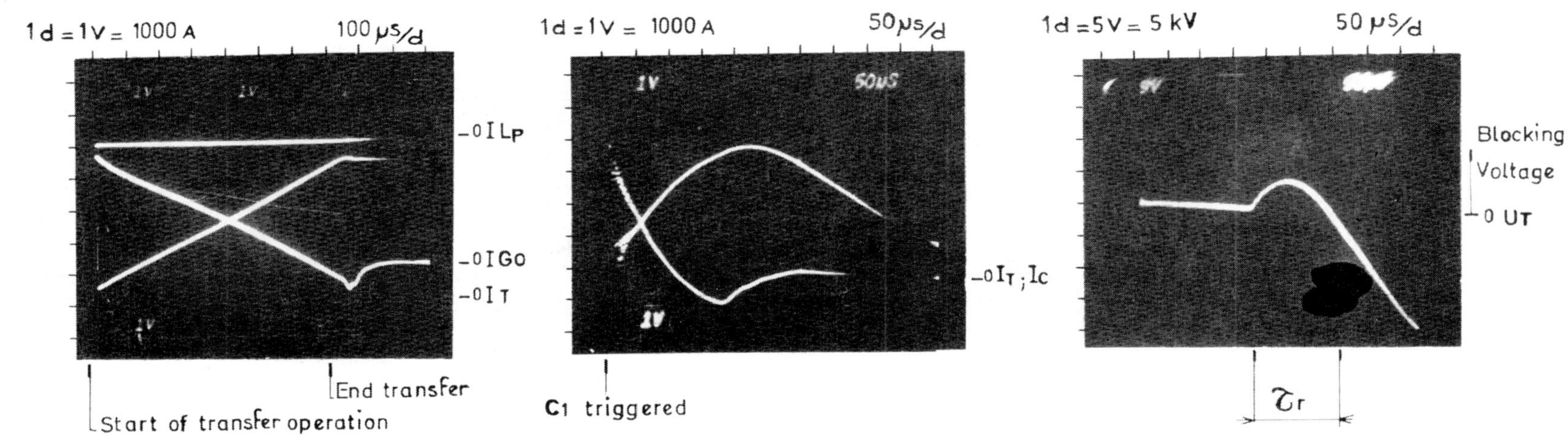

4B Experimental results for an interrupted current of 3.9 kA : C_1 = 30 μF ; 23 kV ; L_1 = 720 μH ; τ_r = 125 μsec ; $\frac{dI_0}{dt}$ = -17.5 A/μsec

$\frac{V_c}{I_o} = 6\ V/A$ – $R_1 = 15\,m\Omega$ – $R_2 = 6\,\Omega$ – $L_2 = 70\,\mu h$ – $L_T = 5\,\mu h$ – $L_p = 22\,mh$ – $R_p = 0.1\,\Omega$

L_1 and C_1 (μH and μF)

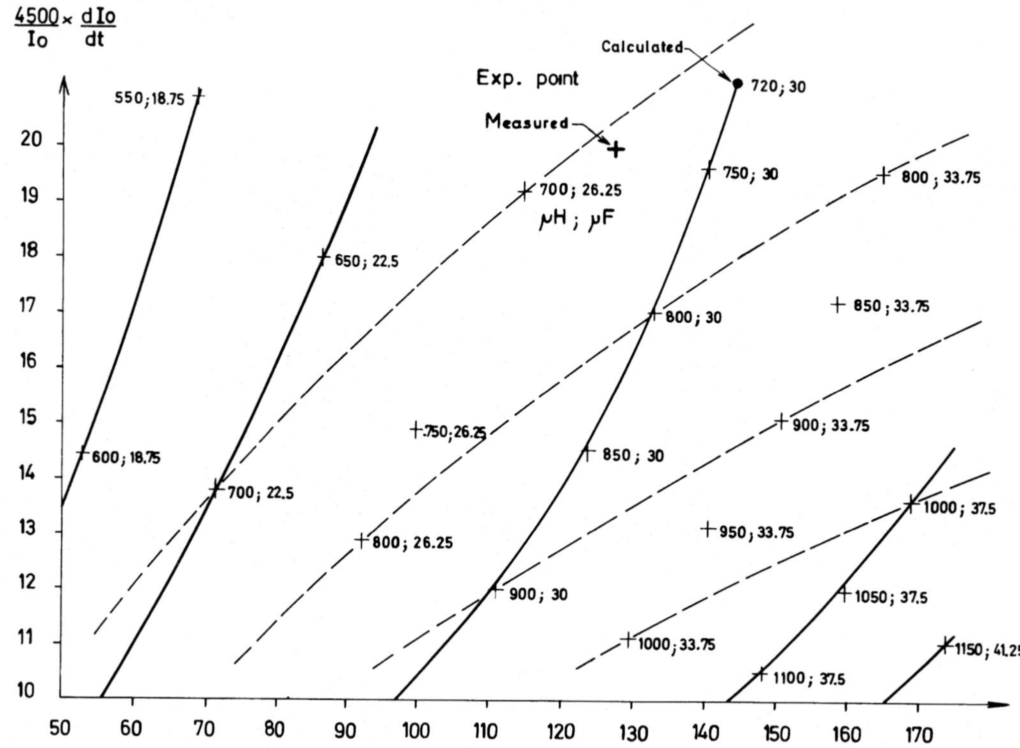

FIG_5

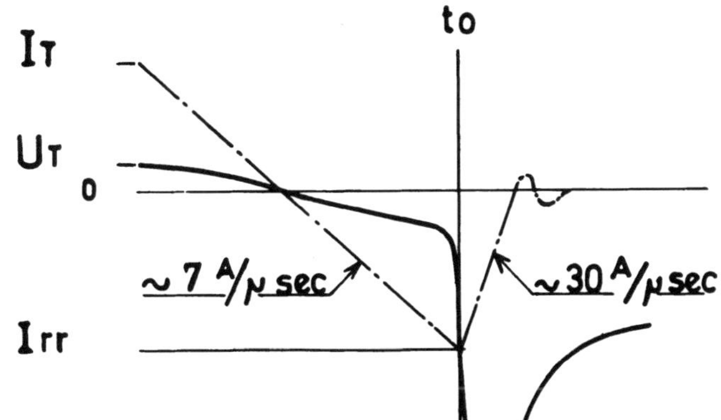

6A

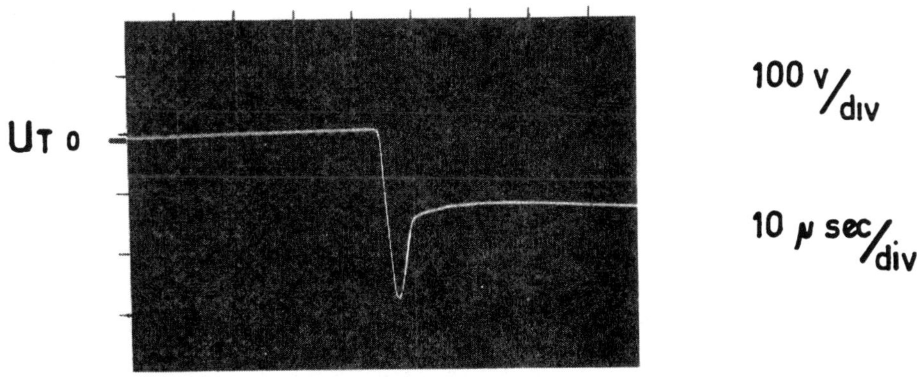

6B Without damping circuit

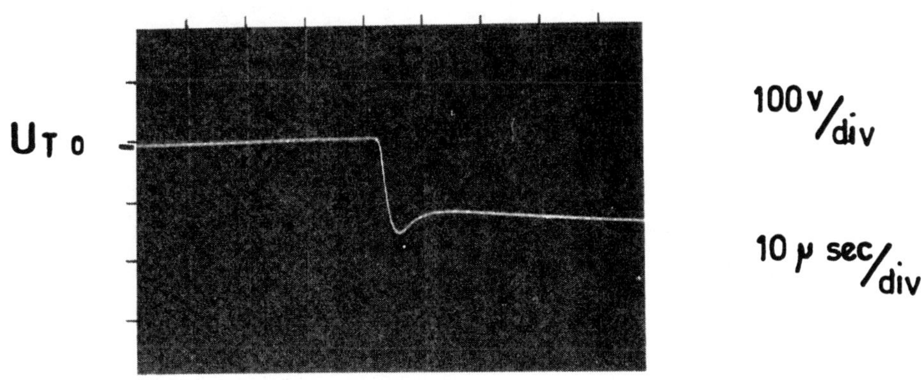

6B With damping circuit

JET TOROIDAL FIELD POWER SUPPLY:

CAPABILITIES AND LIMITATIONS

K Selin

edited by T Eriksson

JET Joint Undertaking, Abingdon, Oxfordshire, UK

ABSTRACT

Limiting factors decisive for the maximum current and flat top time of the JET Toroidal Field power supply are presented. Analyses are made showing the effects on the maximum achievable flat top time when a more rapid pulse repetition rate than normal is considered.

For operation with maximum field current available, a longer flat top time is achieved by a delayed switching-in of the Flywheel Generator Convertor in respect to the Static Unit(s).

1. INTRODUCTION

The JET Toroidal Field power supply consists of a Flywheel Generator Convertor (FGC) connected in series with one (or two) silicon controlled rectifiers (Static Units) supplied from the 400kV network. The Toroidal field load time constant is about 10 seconds.

There are basically three limitations of the flat top time of the Toroidal field current pulse:

(i) Thermal limitations, index ∂, based on the i^2t value per pulse relating to both the FGC and the Static Unit(s).

(ii) Kinetic limitation, index T, being reached when the Generator has decelerated from rated maximum speed, 225 rpm, to half speed.

(iii) Voltage limit, index τ, when the FGC voltage together with the Static Unit(s) no longer can supply a voltage necessary for maintaining the load current.

The capability of supplying a certain Toroidal field coil current with a certain flat top time has been evaluated. The following combinations of power supplies have been considered:

- The Flywheel Generator Convertor and two Static Units.
- The Flywheel Generator Convertor and one Static Unit.
- The Flywheel Generator Convertor only.
- Two Static Units.

2. ASSUMPTIONS

The voltage e_1 of the Static Unit is described by:

$$e_1 = e_o - ki \qquad (1)$$

where e_o represents the Static Unit no load voltage and k expresses the internal impedance of the Static Unit and the 400/33kV stepdown transformer.

The value of e_1 is calculated during the rise time R of the toroidal current, when no loads other than the Static Unit are connected to the 400/33kV transformer.

During the flat top other loads are increasing the voltage drop in the 400/33kV transformer changing the value of e_1.

Typical values are given during the rise time R, $e_1 = 2.0815 - 0.006695i \quad t \leq R \quad (2)$

and during the flat top $e_1 = 1.832 - 0.004343i \quad t > R \quad (3)$

The generator voltage is increased linearly as:

$$e_g = \frac{n}{225} 12.86t, \qquad 0 \leq t \leq 0.7s \tag{4}$$

$$e_g = \frac{n}{225} 9 \qquad i \leq 30kA \tag{5}$$

$$e_g = \frac{n}{225} (11.43 - 0.08108i) \qquad 30 < i \leq 67kA \tag{6}$$

$$e_g = \frac{n}{225} (21.46 - 0.2308i) \qquad i > 67kA \tag{7}$$

The load is assumed to be a resistance of 0.064Ω in series with an inductance of 0.66H.

$$e_1 + e_2 + e_g = 0.064i + 0.66 \frac{di}{dt} \tag{8}$$

The rated energy to be extracted from the Generator through a speed drop from 225 rpm to half speed is:

$$\tfrac{1}{2}J\left(225 \frac{2\pi}{60}\right)^2 (1 - \tfrac{1}{4}) = 2600 \text{ MJ} \tag{9}$$

When an energy W_g is extracted from the Generator the speed drops from 225 rpm to n.

$$\tfrac{1}{2}J\left(225 \frac{2\pi}{60}\right)^2 \left[1 - \left(\frac{n}{225}\right)^2\right] = W_g \tag{10}$$

At the end of the current rise, at t = R, the speed is n_R

$$\frac{n_R}{225} = \sqrt{1 - \frac{3}{10400} W_{gR}} \tag{11}$$

The energy extracted from the Generator during the rise of TF current.

$$W_{gR} = \int_0^R e_g \, idt \tag{12}$$

Assuming the speed has dropped to 112.5 rpm at t = R + T, the energy limited flat top time is:

$$T = \frac{2600 - W_{gR}}{e_{gf} i_f} \tag{13}$$

(e_{gf} = FGC flat top voltage)

(i_f = flat top TF current)

At t = R + τ the Generator speed n_τ is so low that the FGC voltage no longer can support the load current. The speed n_τ is derived from:

$$e_{gf} = 0.064i_f - e_1 - e_2 = e_{g\tau} \tag{14}$$

where in equations (5-7)

$$n = n_\tau$$

The voltage limited flat top time τ is:

$$\tau = \frac{1}{e_g\, i_f} \left\{ \left[1 - \frac{(n_\tau)^2}{(225)} \right] \frac{10400}{3} - W_{gR} \right\} \qquad (15)$$

The power supply is rated for 67kA during 20 seconds. The current paths during the current build-up and decay are assumed to be identical. The load time constant is 10 seconds. Hence the ∂ values are valid also for a load (TF-coils) of such rating. Therefore the flat top time, thermally limited, is:

$$\partial = \frac{1}{i_f^2} \left(67^2 \times 20 - \int_0^R i^2 dt - 5 i_f^2 \right) \qquad (16)$$

3. CALCULATED FLAT TOP TIMES AND FLAT TOP LIMITATIONS

Table I shows maximum TF current for different circuit configurations with flat top time limitations.

∂ = thermal limit

T = lack of generator energy

τ = lack of voltage

Connecting the FGC and two Static Units (Table I, 1st column) will give limitations in FGC voltage ($\tau < T < \partial$) for currents > 76.5kA.

The pulse length can be marginally increased (available voltage and energy increased) by switching in the Static Unit(s) ahead of the FGC. The 2nd column of Table I shows that a flat top time of 6.2 sec can be ascertained for i_f = 75kA by switching in the FGC with a delay of 10 seconds in respect to the two Static Units. When the FGC and one Static Unit are available, the flat top time shown in the 3rd column can also be increased using the same method.

During the current decay the Static Units may be by-passed and the Generator Convertor be free wheeling through Generator de-excitation, which will increase the values of ∂ in Tables I and II.

The underlined figures in Table I show the flat top time limit for a number of Toroidal Field currents.

TABLE I

Flat Top Current Capability for Different Configurations of the TF Power Supply System

	FGC + 2 Static Units			2 Static Units + FGC Delayed 10 sec			FGC + 1 Static Unit			FGC only			2 Static Units
i_f	∂	T	τ	∂	T	τ	∂	T	τ	∂	T	τ	∂
kA	s	s	s	s	s	s	s	s	s	s	s	s	s
20	219	∞	-	221	∞	-	219			219	97	127	218
30	94	∞	-	92	∞	-	94	355	487	94	39	51	92
40	50	∞	-	47	∞	-	50	58	81	50	19	24	46
50	30	∞	-	28	∞	-	29	22	31	28	8.7	10	-
53	26	181	268	24	218	304	25	17	24	24	6.4	6.9	-
60	18	37	56	17	46	65	17	8.7	12	-	-	-	-
67	13	15	24	12	19	28	11	2.8	3.6	-	-	-	-
70	10	10	16	9.6	14	20	-	-	-	-	-	-	-
75	6.7	4.1	5.8	6.2	7.3	9.0	-	-	-	-	-	-	-
76	5.6	2.9	3.5	5.4	6.1	6.7	-	-	-	-	-	-	-
77	3.7	0.87	0.3	4.4	4.8	4.2	-	-	-	-	-	-	-

4. PULSING THE TF POWER SUPPLY AT INCREASED REPETITION RATE

The TF power supply is designed for 67kA during 20s every 10 minutes. For example a tripling of the pulse repetition rate would allocate only 6.7s of 67kA to each pulse. Part of this current carrying capability is taken up by the intervals of current increase and decrease (unaltered current paths) with the remaining being the flat top. The maximum flat top time following a Generator acceleration time of 5 minutes and of 3 minutes is shown in Table II.

Table II. Maximum flat top time of toroidal field current (seconds)

i_f	Acceleration time A = 5 mins					Acceleration time A = 3 mins				
kA	Flat top time (s)					Flat top time (s)				
30	∂=54	∂=53	∂=54	T=23	∂=53	∂=30	∂=29	∂=30	T=12	∂=29
40	∂=25	∂=23	∂=25	T=9.9	∂=21	∂=13	∂=10	∂=12	T=3.5	∂=2.9
50	∂=13	∂=11	T=10	T=3		∂=5.3	∂=3.7	T=2		
53	∂=11	∂=9.2	T=7.1	T=1.3		∂=3.8	∂=2.4	T=0.2		
60	∂=6.3	∂=5.2	T=2				∂=0.1			
67	∂=3	∂=2.2								
70	T=0.8	∂=1.1								
Scheme	FGC +2SU	FGC +2SU FGC DELAYED	FGC+1SU	FGC	2 SU	FGC +2SU	FGC + 2SU FGC DELAYED	FGC+1SU	FGC	2 SU

With a shorter acceleration time the energy available is naturally less. But in the majority of cases an increased repetition rate is thermally limited.

5. REFERENCES

[1] Selin, K I. JET Group Note JDN/G(80)138.

[2] Selin, K I. JET Group Note JDN/G(80)141.

TEST RESULTS OF THE HELIOTRON E POWER SUPPLY AND CONTROL SYSTEMS

K.Murai, S.Koseki, S.Suzuki, Y.Ueyama
Hitachi, Ltd.
Hitachi, Ibaraki, Japan

K.Uo, A.Iiyoshi, T.Obiki, O.Motojima
Kyoto University
Uji, Kyoto, Japan

ABSTRACT

The power supply and control systems were developed for the Heliotron E, an experimental apparatus designed to investigate the possibility of a toroidal steady fusion reactor. They have the following special features and functions. (1) A 330 MVA MG set is used to feed all the power needed, in consideration of the small power receiving capacity of the system, relative to the large load.A vertical type generator with large inertia is used,which does not require a flywheel. (2) Thyristor type power supplies are used because they are convenient for control and maintenance. In particular, for the joule heating power source, a 40kV, 6kA thyristor circuit breaker is used in consideration of economics and performance. (3) A control system of the joule heating power source was developed,whereby plasma current is built up and maintained.

Coil current flow tests were finished successfully in July 1980 and various kinds of test data were obtained. Especially one-turn voltage of the joule heating coil at current interruption showed the possibility of good plasma current start up.

1. INTRODUCTION

The Heliotron is an apparatus designed to realize a high temperature plasma using the Heliotron magnetic field with a large rotational transform and shear as a magnetic vessel.

First studies with this apparatus were begun in 1958. From 1970, experiments with the Heliotron D and DM have been carried out.[1] Based on the accomplishments of those studies, construction of the Heliotron E was started in 1976 and completed in March 1980. In April 1980 site tests and experiments were started.

The Heliotron E has the following experimental objectives: 1) confinement and stability experiments on ohmically heated plasma, 2) additional heating experiments, 3) divertor experiments, 4) confinement and stability experiments on joule current-less plasma and so forth.[2]

The machine parameters are as follows.

1) Toroidal component of the helical field produced by the helical coil		2	Tesla
2) Toroidal coil field		0.6	Tesla
3) Vertical coil field		0.37	Tesla
4) Vacuum chamber	Major radius	2.2	m
	Minor radius	0.21 m x 0.40	m
5) Flat top time		0.5	sec
6) Repetition time		10	min

In this paper we describe the power supply and control systems for the Heliotron E. The Heliotron E has helical, vertical, toroidal, auxiliary vertical and joule heating coils. Energy for these coils is supplied by a motor-generator set. Thyristor converters are applied for the power supply system. Especially the inductive energy storage type using a thyristor circuit breaker is adopted in the joule heating power source for plasma current start up. Since the frequency varies during one experiment, the thyristor converters must be controlled accordingly.

2. POWER SUPPLY SYSTEM

2.1 Overall construction

Fig. 1 shows a one-line circuit diagram of the Heliotron E power supply system. Characteristics of this system are as follows. (1) Because the energy receiving capacity of 5.7 MVA is small compared with the load capacity of 330 MVA, a vertical type motor generator set with a large flywheel effect is used to supply all the power required. For the same reason, a self-exciting type is chosen for the excitor of the generator. (2) For the convenience of maintenance and improvement of control performance, the coil power sources are all of the thyristor type.

Typical coil conduction patterns are shown in Fig. 2. To minimize the generator capacity, current is raised in steps.

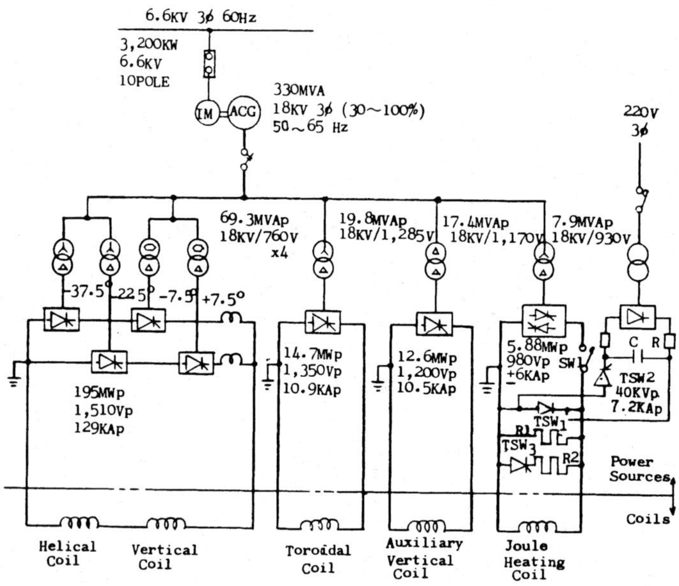

Fig. 1. One-line Circuit Diagram of the Power Supply System

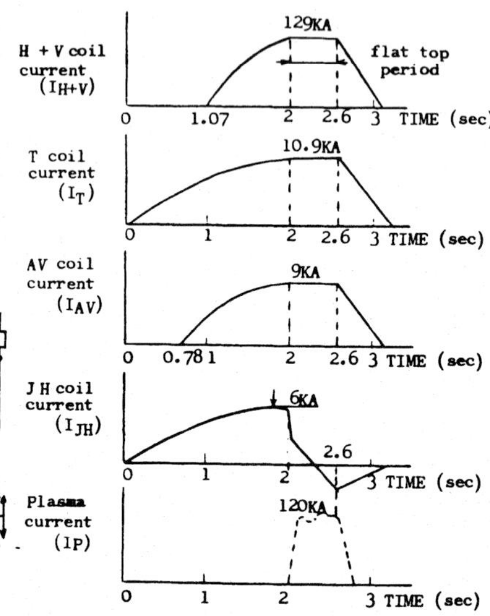

Fig. 2. Typical Current Pattern of Each Coil

2.2 Motor generator system

In order to supply all the energy of 300 MJ for the coils with minimum size, the motor generator was designed for a top speed of 650 rpm.

The generator is a vertical type with inertia of 1,400 t-m^2 and the drive motor is controlled by the liquid resistor method. Because the generator has a large inductive type load, a non-linear resistor is applied as the discharge resistor of the generator field circuit. With this resistor, the output voltage of the generator is decreased rapidly and the output energy disappears swiftly when the line short circuit occurs.

2.3 Thyristor converters and the thyristor switch circuit

6-pulse thyristor converters are used as the power supply for the toroidal coil(T coil), auxiliary vertical coil (AV coil) and joule heating coil (JH coil). The thyristor converter for the helical and vertical coil (H+V coil) has the largest capacity of all the thyristor converters. In order to reduce the direct current ripple and minimize the effect on the generator, a 24-pulse thyristor converter is used. There in this converter, four groups of 3-phase bridge circuit with 15° phase difference are connected in series and parallel.

The thyristor switch in the joule heating power source is composed of a main thyristor switch (TSW_1) of 40 KV, 6 KA and an auxiliary thyristor switch (TSW_2) of 40 KV, 7.2 KA.

When the JH coil current is interrupted, the AV power source is subjected to a voltage spike of about 5 KV and the AV coil current varies. Therefore the AV power source is constituted by 2 series thyristors of 3.5KV.

3. CONTROL SYSTEM [3]

This system controls the following three important processes: (a) acceleration and excitation of the motor generator, (b) start up and maintenance of plasma current, (c) plasma confinement. For accurate control of the above, the power sources described in the previous section must be properly controlled.

Acceleration of the motor generator is controlled by a liquid resistor method. Start up and maintenance of plasma current are controlled by means of the joule heating power source. Plasma confinement is controlled by the power sources of the helical, vertical, toroidal and auxiliary vertical coils.

3.1 Control of the motor generator

In order to minimize the excitation loss and the iron loss of the generator, the excitation system is operated only during the coil excitation phase. From a labor saving point of view, not only the motor generator, but also its auxiliary machines are operated automatically.

3.2 Control of plasma current start up and maintenance by the joule heating power source

For the control of plasma current start up and maintenance, the preprogramming control method of the JH coil current without detecting plasma current was developed. Fig. 3 shows the circuit operation of this method. The time-to-time operation of the circuit will be discussed by dividing it into the three modes, i.e. coil excitation, plasma current start up and plasma current maintenance.

First, by the joule heating power source (JH PS), current is made to flow in the JH coil. When the JH PS voltage is reversed, the main thyristor switch TSW_1 is fired. Then the JH PS current is commutated to the TSW_1 and the JH PS is cut off. (Coil excitation mode)

Next, the auxiliary thyristor switch TSW_2 is fired and TSW_1 current is interrupted, whereupon current is transferred to resistor R_1 and coil current abruptly decreases. Thus a high voltage is generated across the coil terminals and plasma current is built up. When the plasma current reaches the specified level, the time constant regulating resistor R_2 is connected to lower the generated voltage and, at the same time, suppress the rise of the plasma current. (Plasma current start up mode)

Next, before the plasma current begins to decrease, the JH PS is connected again. In this case the thyristor converter is connected in converse-parallel to the JH PS. Then the JH coil current is reversed to minus and controlled to keep the time derivative of the JH coil current constant. Thus the plasma current is maintained further. (Plasma current maintenance mode).

The waveforms of the coil voltage, coil current and plasma current in the three modes are shown in the time chart of Fig. 4.

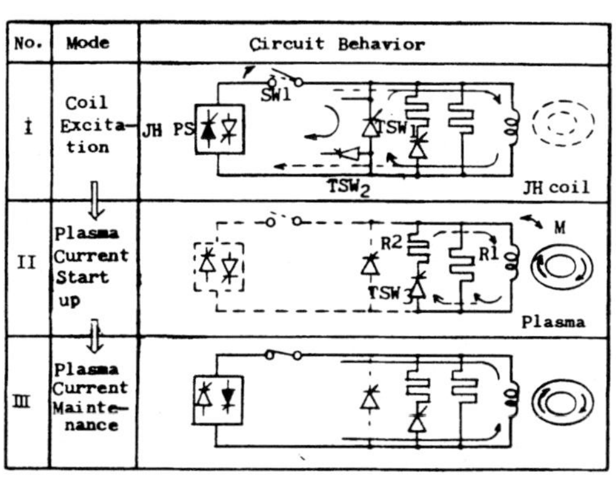

Fig.3. Circuit Behavior of JH Power Source

Fig. 3. Circuit Behavior of JH Power Source

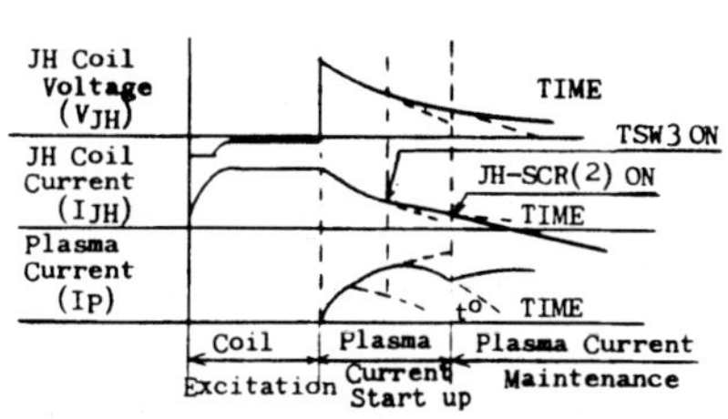

Fig.4. Time Chart of Plasma Current Control

3.3 Plasma confinement control

In order to confine plasma, the current drift in all coils must be held within a determined value in consideration of mutual interference between plasma and coils.

For the control of the confinement coil power source, the current feedback control system was adopted in order to simplify the circuit configuration as far as possible.

The control system was considered by simulating all modes from current rise through fall in all coils.

Fig. 5 shows some results of the computer simulation. Current waveforms of coils and plasma are obtained. Current of the confinement coils are kept constant by the feedback control. Plasma current is built up by the start up control, and the current is held in the range determined value by the maintenance control. In plasma production, confinement coil current fluctuates because of mutual interference with the plasma, but the maximum current drift of each coil is kept within the target of ±2.0% through the plasma flat top period.

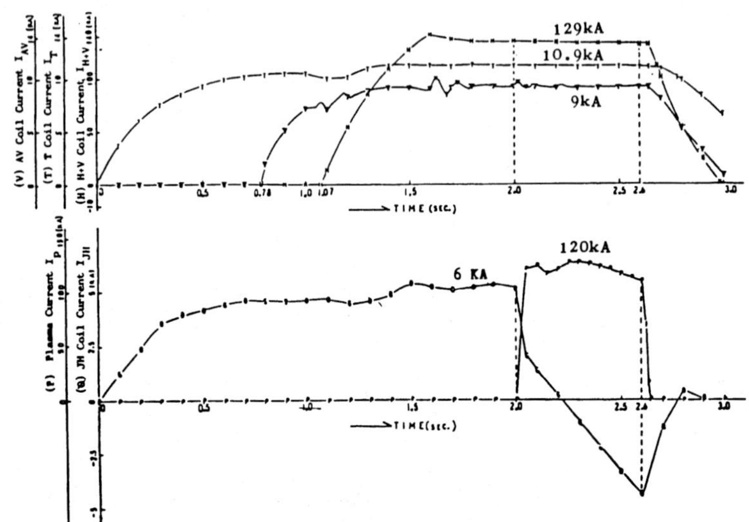

Fig.5. Simulation Result of the Power Supply and Control Systems

4. TEST RESULTS

The following tests were undertaken to confirm the characteristics of these systems.

4.1 Test results of the motor generator

In order to make sure the characteristics of the motor generator, full load operation test was done. Fig. 6 shows the test results according to the following conditions: (1) generator voltage; 18kV, (2) repetition time; 10 minutes, (3) maximum speed of generator; 636 rpm, (4) each coil current;100%. The frequency of the motor generator decreased from 63.6 to 54.6 Hz. This decrease of frequency satisfied the specification of the motor generator, i.e. from 65 to 50Hz. The voltage of the motor generator varied according to the fluctuation of the coil current, but it was kept within the target of ±10%.

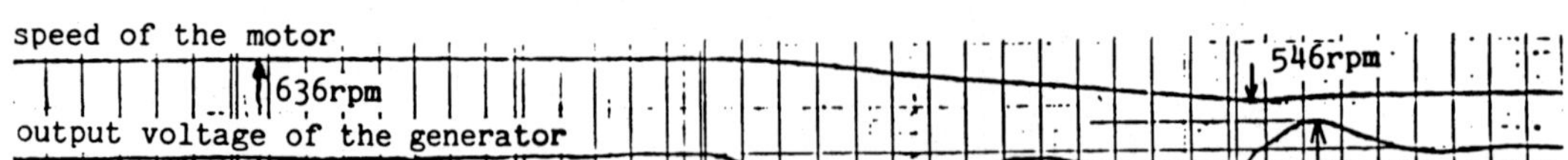

Fig. 6. Test Result of the Motor Generator

4.2 Test results of the thyristor converters and the thyristor switch circuit

In order to make clear the ability of the thyristor converters and the thyristor switch circuit, coil excitation tests were done. Fig.7 shows results of the H+V coil excitation test. In this test case, the control circuit of the H+V coil thyristor converter were adjusted to satisfy two conditions: (1) to reach the flat top level within a minimum interval of 0.93 sec, (2) to keep the current overshoot within 5%. The first condition was chosen from the view point of the coil rating capacity, while the second was chosen in order to keep the magnetic force under a permitted level.

Fig.8 shows the result of the JH coil excitation test. In this test case, the interval from TSW_1 fired to TSW_1 current interrupted is 80 msec and decrease of the coil current during the time is less than 10%. This result satisfies the designed value. The JH coil current waveform shows that the coil current abruptly decreases by TSW_1 interruption and, after a few msec, it decreases slower by TSW_3 fired. Then it is controlled to keep the time derivative of the JH coil current almost constant.

Fig. 9 shows the waveform of the JH coil voltage at the moment of interruption. The voltage is the maximum value of 36KV. The plasma one-turn voltage according to the value is about 200V, showing the possibility of good plasma current start up. The waveform of the JH coil voltage at the moment of interruption agrees well with the results of computer simulation and verifies that the thyristor switches realize high accurate control of the plasma current.

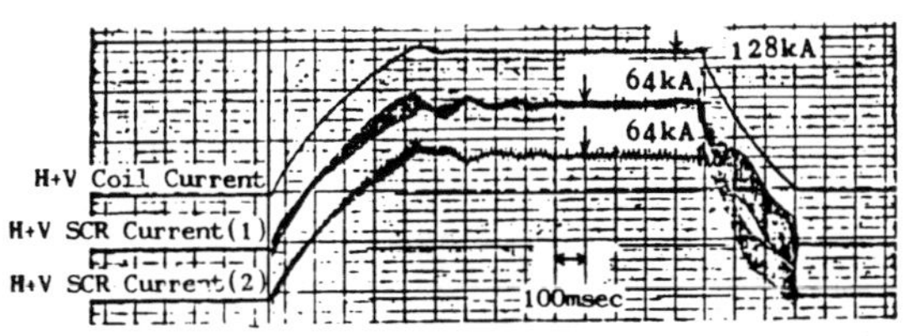

Fig.7. Result of the H+V Coil Conduction Test

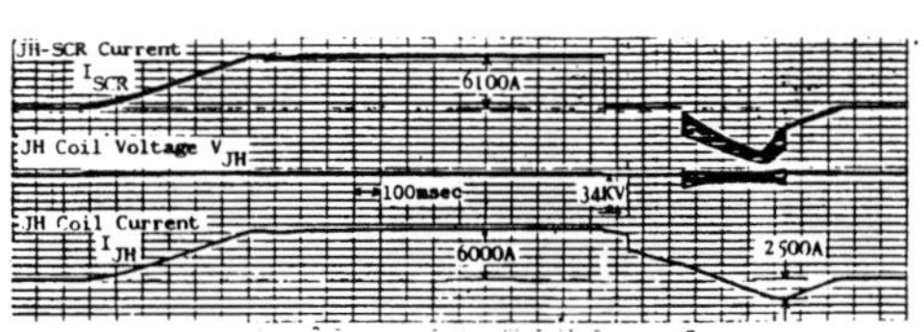

Fig.8. Result of the JH Coil Conduction Test

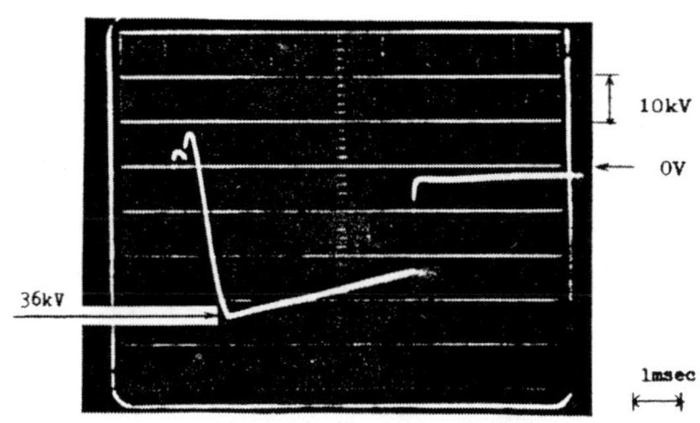

Fig.9. Waveform of the JH Coil Voltage at the TSW_1 Current Interruption

4.3 Overall test results

After the coil excitation tests, overall operating tests were done according to the coil current patterns of Fig.2 to make sure the characteristics of overall system. Fig.10 shows the results of the test according to the following conditions: (1)generator voltage;14.4KV, (2) repetition time;8 minutes, (3)maximum speed of the motor; 610 rpm, (4)each coil current;79%.

During the H+V coil current rise and the JH coil current interruption, other current fluctuate by mutual interference among coils, but the maximum current drift in the flat top period is kept within the target of ±2.0%. Good characteristics of the coil current control system are obtained.

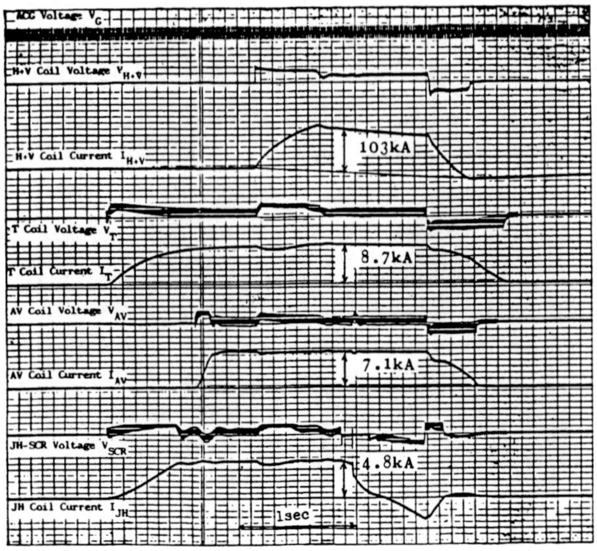

Fig.10. Result of the Current Flow Test in Every Coil

5. CONCLUSION

This paper described the construction and the test results of the Heliotron E power supply and control systems. In the joule heating power source for plasma production, the inductive energy storage type was adopted using thyristor switches. Test results agree well with designed value and show the possibility of good plasma current start up. Results of current flow test in every coil agree with computer simulation and show the possibility of good plasma confinement.

REFERENCES

[1] K. Uo, et al., in Proc. of 7th Int. Conf. on Thermo Plasma Physics and Controlled Nuclear Fusion Research, 1978, IAEA-CN-37/L-1.

[2] K. Uo, O. Motojima, A. Iiyoshi, : Heliotron E Project, 10th Symposium on Fusion Technology (Padova, Sept. 1978)

[3] K. Uo, et al.,: Development of the Heliotron E Power Supply and Control System, 10th Symposium on Fusion Technology (Padova, Sept. 1978)

A 150 KW LOW FREQUENCY OSCILLATOR FOR WALL CLEANING AND PREIONISATION OF THE TCA TOKAMAK

A. Lietti
Centre de Recherches en Physique des Plasmas
Association Euratom - Confédération Suisse
Ecole Polytechnique Fédérale de Lausanne
CH-1007 Lausanne / Switzerland

ABSTRACT

A single oscillator induces pulses of alternating loop current at differing frequencies for cleaning and preionisation.

Since an arrangement of external coils coupled to the discharge torus performs a number of operations, pneumatic switches and electrical filters have been used to separate the circuits. Preliminary measurements concerning gas breakdown and energy transfer are reported.

1. INTRODUCTION

Wall cleaning and preionisation in Tokamaks are usually carried out by separate devices. In the TCA Tokamak {1} the same apparatus performs both operations by means of low frequency oscillations. The efficiency of wall cleaning by means of low frequency oscillating discharges is a well-established technique {2}, {3}. An oscillating discharge is also useful, in Tokamak experiments, to make the gas breakdown easier, thus reducing the volt-seconds of transformer flux to build up the current peak.

In principle, when the frequency of the discharge increases, the gas breakdown becomes easier, but the electrical stress produced by the e.m.f. induced in the poloidal coils increases at the same time, for a constant amplitude of the discharge.

Since we need an alternating loop current lower for the preionisation than for the cleaning discharge, it follows that we obtain optimal performances when the frequency of the oscillating discharge is higher for preionisation than for cleaning. Note that this philosophy minimizes at the same time the jitter of the gas breakdown at the start up of the preionisation process.

2. APPARATUS

Fig. 1 shows the electrical diagram of the apparatus.

2.1 The oscillator

The oscillator is constructed using two commercial tubes for industrial heating (B.B.C. ITL 30-2) in a self-oscillating push-pull arrangement with pulsed grid control. The energy flow from the mains supply is averaged by a delay line. The frequency of the clean-

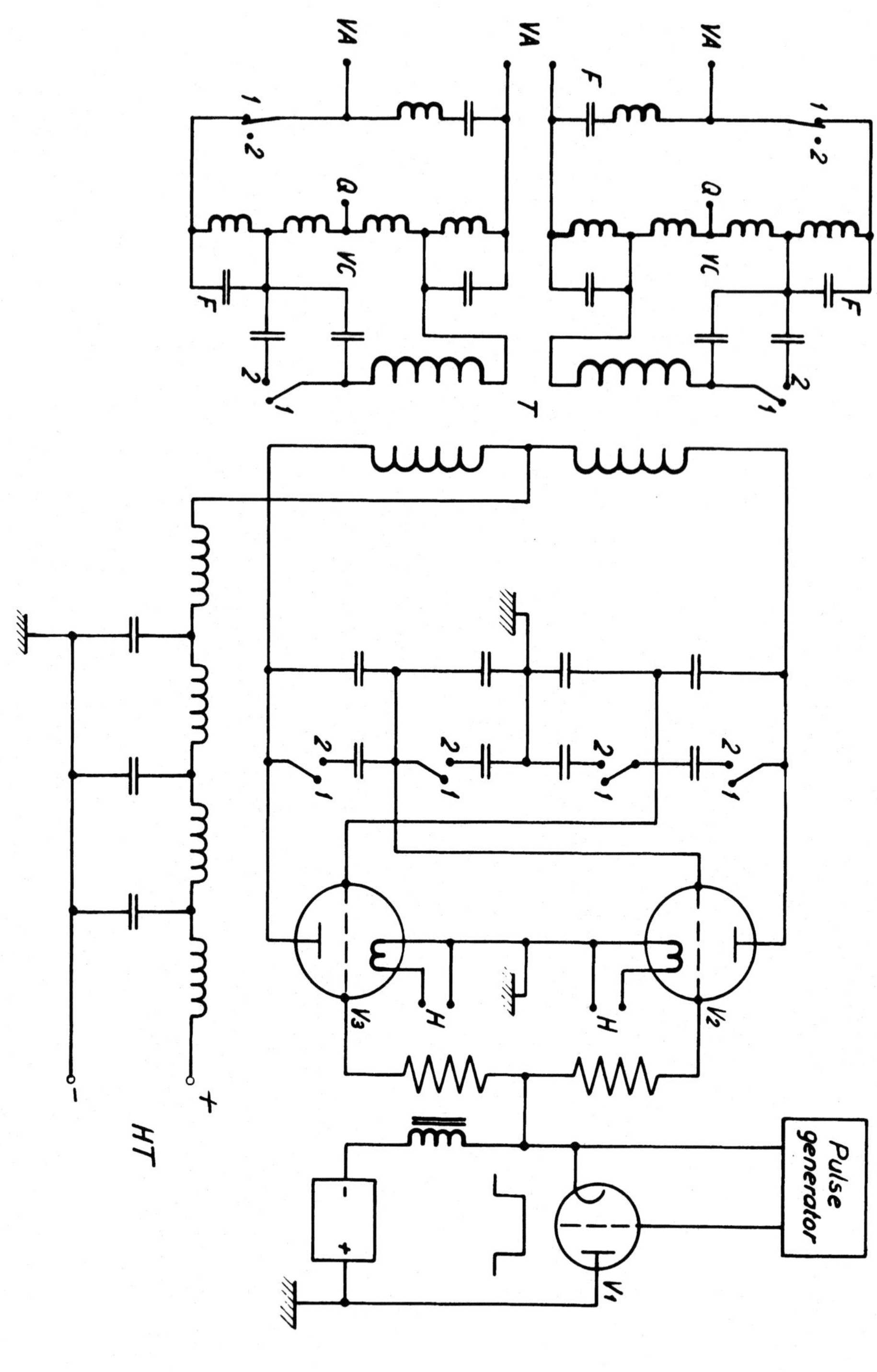

Fig. 1: 4-15 kHz, 150 kW, Pulsed Oscillator (Pos. 1 - Preionisation; Pos. 2 - Cleaning)

T	- output transformer	V_2-V_3	- ITL 30-2 triodes
VC	- vertical field coils	H	- filament heating
F	- 15 kHz filters	Q	- input of the quadrupole field current
V_1	- grid modulator	VA	- to the vertical field generator

ing pulse is 4.3 kHz, the duration can be varied from 5 to 20 ms, and the repetition rate from 1 to 10 Hz. A much shorter pulse is necessary for the preionisation (2 ms - 15 kHz).

The frequency change is carried out by pneumatic switches. An output aircore transformer avoids undesired loops and allows a correct matching of the oscillator to the Tokamak windings. The oscillator was tested on a dummy load, producing pulses of 150 kW average power.

2.2 The coupling circuits

An arrangement of external coils produces the coupling to the discharge torus. A special circuit allows the same coils arrangement to perform the following operations : wall cleaning, breakdown and initial preionisation, production of the vertical stabilizing field and of the quadrupole field, which has been provided in order to allow a stable and controlled start-up of the discharge. Fig. 1 shows the coil arrangement. Special care has been taken to reduce unwanted coupling by using pneumatic switches and electrical filter to separate the circuits. During the cleaning discharge, with a current surface density of 18 A/cm induced in the gas, the e.m.f. induced in the adjacent windings does not exceed 300 Volt per turn. More information can be found in ref. {1} and {4}.

3. EXPERIMENTAL

3.1 Preliminary test

In order to test the whole system, we made some preliminary discharges in hydrogen gas. The cleaning discharges have been performed with a filling pressure of $2 \cdot 10^{-4}$ Torr and a toroidal field of 500 gauss. The gas breakdown was optimized by applying a constant vertical field of a few gauss.

The preionisation discharges have been performed with a filling pressure of $3 \cdot 10^{-5}$ Torr and a toroidal field of 7500 gauss. No quadrupole field was applied.

3.2 Measurements

Fig. 2 shows the loop voltage pulses. The time when the discharge starts is also indicated. The breakdown occurs with negligible jitter at a peak loop voltage of about 50 $\hat{V}$. The oscillating preionisation pulse was essential to initiate the discharge in these preliminary experimental conditions. In fact the loop voltage step of 30 V due to the variation of the current in the ohmic heating coil is not sufficient to produce a breakdown without the auxiliary oscillating pulse.

The cleaning pulse induces an average current of 2 kA in the gas. This current results in a linear current density of 18 A/cm at the limiters. We estimate that the intensity of the discharge is sufficient to heat the plasma to 2 ÷ 3 eV, thus optimizing the wall cleaning, in agreement with previous experiments {2}. Measurements in order to confirm this will start soon.

3.3 The energy involved in the cleaning process

An estimation of the average power delivered to the plasma from the cleaning pulse has been made by measuring the energy flow at the output of the oscillator as follows. First we performed the measurement in normal operation. We found that the total power delivered from the oscillator was 100 kW. We made a second measurement without hydrogen filling, thus measuring the electric losses in the Tokamak circuits. From the difference we estimate that about 10 kW are transferred to the plasma, that means 10% of the available energy. We think

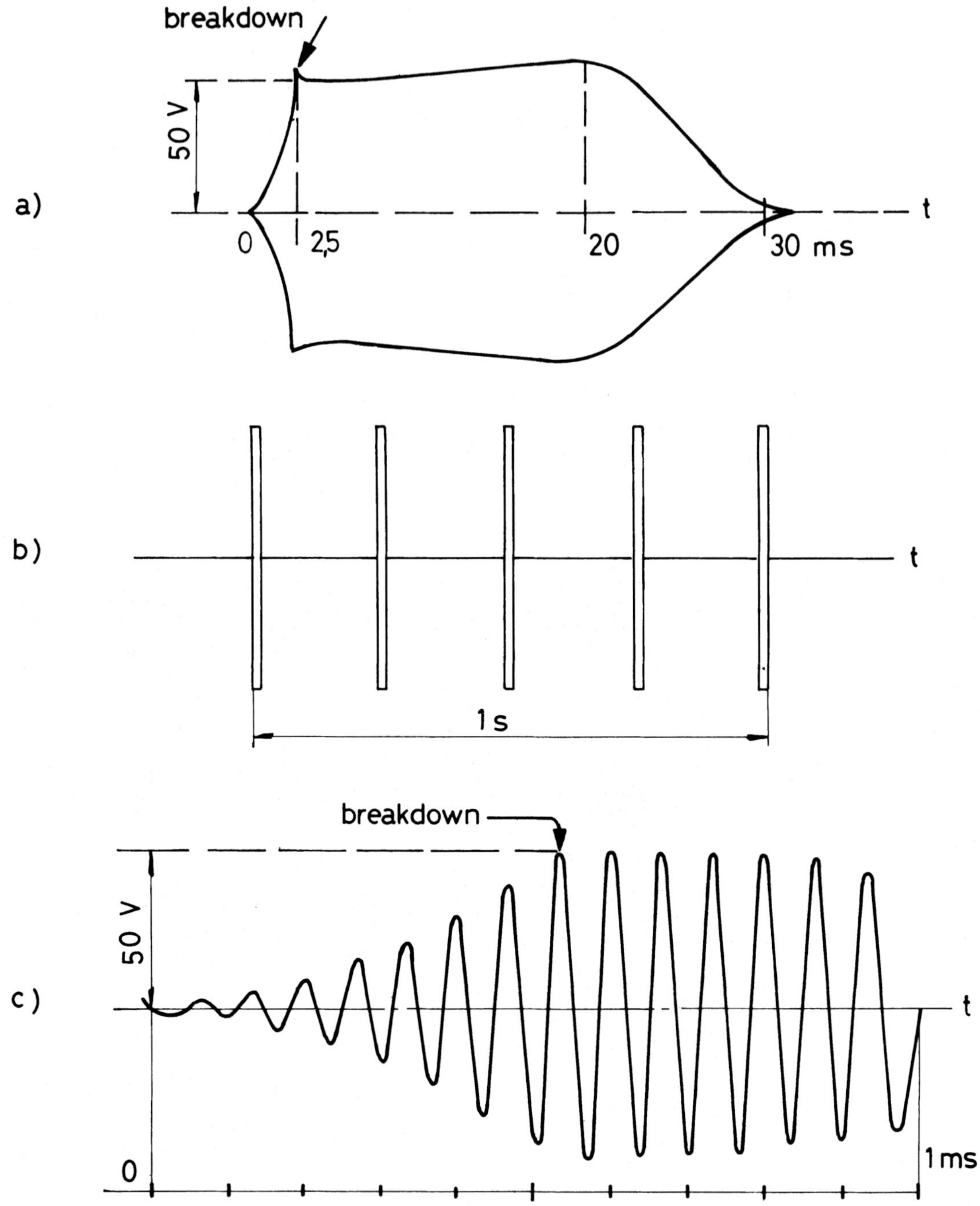

Fig. 2: Loop voltage pulses

a) Envelope of the cleaning pulse - Frequency 4.3 kHz

b) Sequence of cleaning pulses

c) Beginning of the preionisation pulse

that the most important source of losses originates from the image currents flowing on the surface of the vacuum vessel. The vessel is made of stainless steels, the joints are electrically soldered. A confirmation of this view is obtained by measuring the increase in temperature of the chamber, produced by the skin current without gas filling. We mention that in this way it is possible to heat the torus to about 100°, in order to increase water desorption. We estimate that the power transferred during the pulse per unit of surface of the vessel is 2 W/cm^2. Consider now the estimation of 10 kW transferred to the plasma. This results in an energy transfer per pulse of 200 joules. Note that the energy which is necessary to ionize and heat the plasma at 3 eV, assuming an electron density of 10^{12} cm^{-3}, is only a few joules. This implies an high energy transfer from the plasma to the wall, according to the cleaning process.

4. REFERENCES

{1} CHEETHAM, A.D. et al., 11th Symposium on Fusion Technology, Oxford, 15-19 September, 1980.

{2} OREN, L. TAYLOR, R.J., Nucl. Fus. 17, 1143-1151 (1977).

{3} DYLLA, H.F. et al., Report P.P.P.L. 1485 (1978).

{4} KELLER, R. and TUSZEL, A., 11th Symposium on Fusion Technology, Oxford, 15-19 September, 1980.

JT-60 Ohmic Heating Power Supply

N. Miya, R. Shimada, T. Aoyagi, S. Tamura
Division of Large Tokamak Development
Japan Atomic Energy Research Institute
Tokai-mura, Naka-gun, Ibaraki-ken
319-11 JAPAN

ABSTRACT

The JT-60 ohmic heating power supply system is now being manufactured. The system has an additional inductive energy storage coil other than the ohmic heating coil to make a stepped build-up of plasma current and is designed to meet a variety of operations required for the JT-60 experimental program. In this paper, the overall design and operation of the system are described. A detailed description is also given of some important components whose performances need to be tested under various conditions.

1. INTRODUCTION

One of the basic points of the JT-60 ohmic heating (OH) circuit system is its capability of making a stepped build-up of plasma current with the aid of an inductive energy storage (IES) coil other than the OH coil of the tokamak. The advantage of this scheme is to reduce the operation duty imposed on DC circuit breakers of the OH system. It is also expected to save the total inductive energy required to build up a given plasma current [1].

We have investigated the OH circuit design based on this idea and selected two stage build-up system for JT-60 taking into account of the total magnetic energy and the interrupting capability of the vacuum circuit breakers developed for this purpose. Additional functions which will be needed for the plasma current control have been designed for this system to meet a variety of experimental requirements as much as possible. These functions which will be operated during the build-up and the flat top phase of the plasma current are also the characteristic points of the system.

To complete the system, various high techniques are required. One of the most critical components to be developed is the DC circuit breakers. The other components are large thyristor convertor banks, equipment for measuring a large current and voltage, and equipment for over voltage protection. Since some of the operation performances required for these components are quite important and unexperienced, it is necessary to test their performances at the full rating using a prototype unit preceding the manufacture of them.

This paper is mainly concerned with the design of the OH circuit and a detailed description of some important components mentioned above.

2. DESIGN OF THE OH CIRCUIT

The OH circuit system of JT-60 makes the plasma current build up to 2.7 MA in 100 mS-2 S and sustains it for 5-10 S. The volt seconds required during the operation were investigated assuming various operation modes and plasma parameters. The design value of the maximum available volt seconds consumption for JT-60 is 25.8 V.S. The OH coil (L_{oh}) and the IES coil (L_s) store the magnetic energy of 35 MJ each. As shown in Fig. 1, the OH coil current is forced to swing negatively by the action of the IES coil, thus the plasma current will rise up in two steps. Figure 1 also shows a typical wave form in the operation of the system.

A circuit diagram of the OH coil power supply is shown in Fig. 2, and the rating of main components is summarized in Table 1. The resistors R1 and R2 can be varied stepwise in the

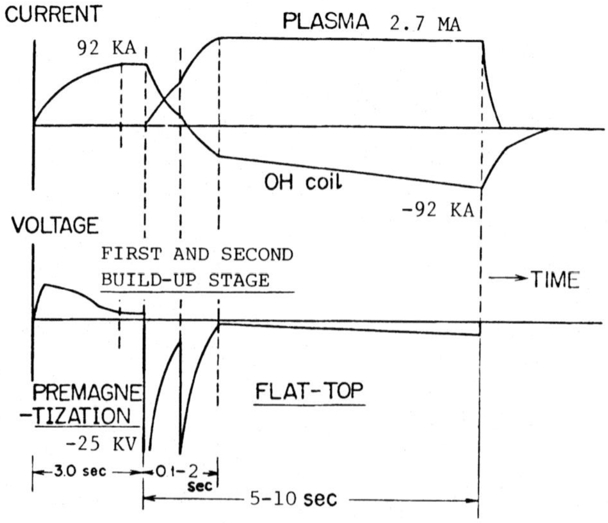

Fig.1 Four Phases in JT-60 Operation

range of 70 mΩ- 1 Ω and 70 mΩ- 0.5 Ω respectively, and it is possible to control the time constant and the break down voltages for plasma current initiation with the maximum voltage across the OH coil of 25 KV. These resistors are composed of several elements, and the design has been done carefully to minimize their residual inductances which cause over voltage in L_{oh} and also to reduce their heat loads during the operation. The value of L_s can also be varied stepwise to be used as a ripple filter to be mentioned later.

Major functions of the OH circuit are: (1) The control of voltage across the OH coil can be made even in the plasma current build-up time. In the case of slow build-up, namely the rise time of longer than ∿0.5 S is required, the power from the generator will be used to energize the OH coil directly through the thyristor convertor. This function will enhance the controll-ability of plasma current and is introduced by connecting the resistor R1 and the thyristor convertor PSF1 in series; (2) Voltage ripples from the thyristor convertor will be reduced. In the flat top phase, the voltage ripples will disturb the plasma measurements, especially the one turn loop voltage measurement. The thyristor convertors have been designed to reduce them

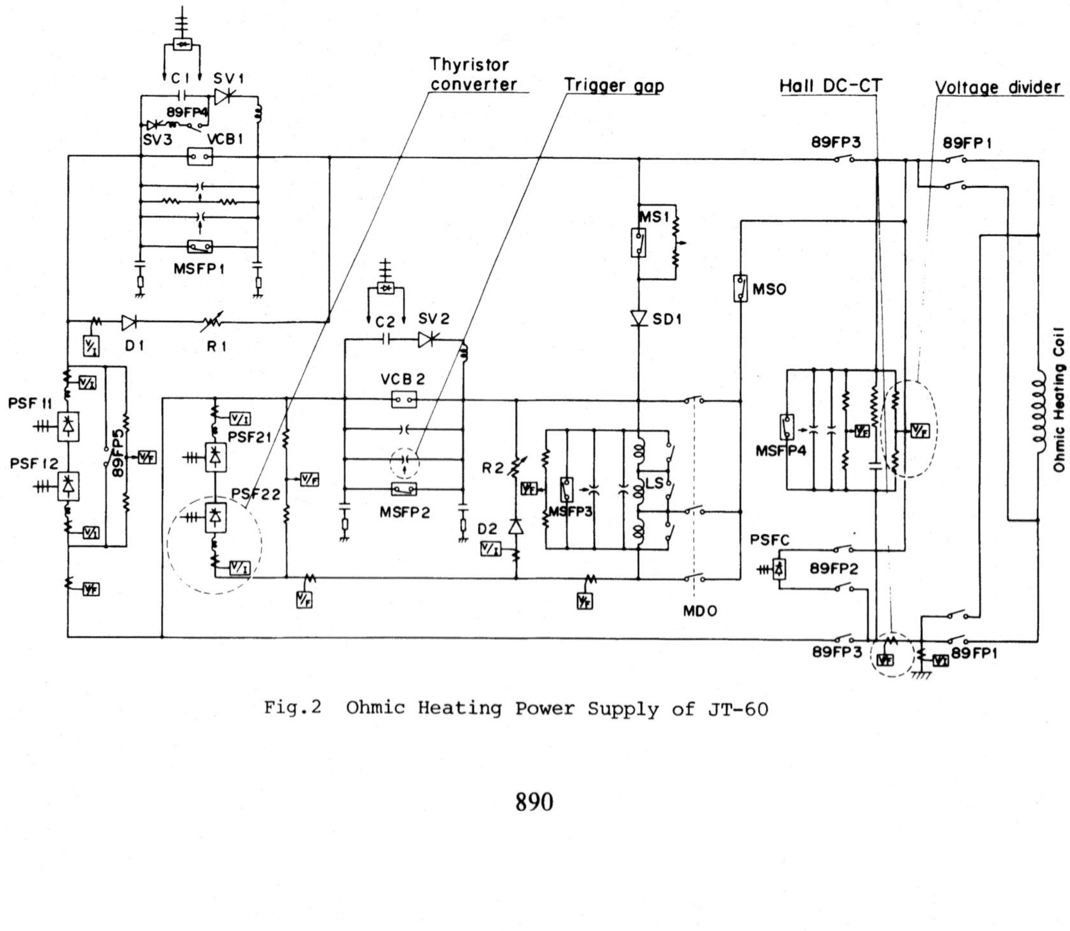

Fig.2 Ohmic Heating Power Supply of JT-60

Table 1 Specification of Main Components of Ohmic Heating Power Supply

DC circuit breaker		
VCB1	Vacuum circuit breaker	25 KV - 92 KA
	2 series 4 parallel	
VCB2		15 KV - 92 KA
Commutation circuit		
C1	Condenser Bank	1,000μF - 25 KV
C2		1,700μF - 15 KV
SV1	Thyristor switch	25 KV - 92 KA
SV2		15 KV - 92 KA
SV3	(for discharge cleaning)	25 KV - 92 KA
Switchgear		
MS0	High speed make switch	92 KA - 25 KV
MS1		92 KA - 25 KV
MSFP1		92 KA - 25 KV
MSFP2		92 KA - 15 KV
MSFP3		92 KA - 15 KV
MSFP4		92 KA - 25 KV
89FP1 ∿ 5	Connect switch	92 KA - 25 KV
Diode switch		
D1	Diode switch	92 KA - 25 KV
D2		92 KA - 15 KV
SD1		92 KA - 25 KV
Thyristor convertor		
PSF11 ∿ PSF22	24-pulse converter	1,383V - 101.2 KA
Energy storage coil		
Ls	Iron core reactor	92 KA - 1.2, 2.4, 4.4 mH
Time constant controller		
R1	Stainless steel	0.07 - 1 Ω
	Coaxial pipe	variable in 20 steps
R2		0.07 - 0.5 Ω
		variable in 7 steps
Low power discharge cleaning power supply		
PSFC	Thyristor converter six-pulse	2,500 V - 6 KA

by using the multi-pulsed operation of convertor up to 24-pulse. But this is found to be insufficient, because the ripples on the one turn loop voltage at the vacuum vessel are calculated as high as 1 V. The OH circuit in the flat top time has a low-pass filter composed of R1 and L_s, and is expected to work satisfactorily to attenuate the voltage ripples; (3) The thyristor convertors PSF1 and PSF2 can be controlled in a low-rating circulating current mode through R1. This function can reduce the heat losses in the resistor R1, and also can protect the OH coil from over voltage owing to the plasma disruptions during the flat top phase.

The circuit operation is divided into four phases as shown in Fig. 1 and a detailed description of the operation cycle and simplified circuit corresponding to these phases are summarized in APPENDIX.

3. DESCRIPTION OF SOME COMPONENTS

3.1 DC circuit breaker

The specification of DC circuit breaker for JT-60 is as follows:

Interruption current	92 KA
Hold off voltage	25 KV
Number of operation	4,000 Times/Bottle
Pulse repetition	1/600 Sec

In order to develop a complete DC circuit breaker, a research and development work for the vacuum circuit breaker (VCB) and the thyristor circuit breaker (TCB) was carried out in

1975 -1976. We have chosen VCB as the JT-60 OH circuit breaker from the viewpoints of operational performance and economical consideration. A near full scale inerruption test was performed to demonstrate the capability of VCB. This test showed the interrupting ability of VCB up to 130 KA-44 KV [2,3,4], and it is already successfully operated with the maximum capability of 125 KA-25 KV for the E-coil breaker of Doublet III.

3.2 The thyristor convertor

The outline of the thyristor convertor PSF2, which is now being manufactured, is build up by 8 blocks of six-pulse thyristor bridge convertors as shown in Fig. 3. The specifications are the same as those used in PSF1. PSF2 is divided into two sections, the lower side (PSF21) and the upper side (PSF22) section, and each section consists of four six-pulse bridge convertors and operates in 24-pulse mode individually. The PSF22 No. 4 unit is the one of the four six-pulse convertors of PSF22. The shift of firing timing among four banks in PSF22 is 15 degrees and the shift between the upper side and the lower side on the same branch, e.g. PSF22 No.4 and PSF21 No. 4 is 30 degrees. The rating of the six-pulse convertor is shown as follows:

PSF22 No. 4	Six-pulse thyristor bridge convertor
Rating Power	35 MW
Rating Voltage	1,383 V
Current	25.3 KA
Thyristor	SF3000GX21 (4,000 V -3,000 A)
Structure	2S-7P-6A

The cross current limiting reactor is connected in series to the above six-pulse convertor and its rating was carefully determind considering the parallel operation of four convertors. The PSF22 No. 4 unit consists of two sections. One is the thyristor convertor and the other is the DC reactor with DC current detector provided for protections. The thyristor convertor is composed of six modules that correspond to six arms of the six-pulse convertor.

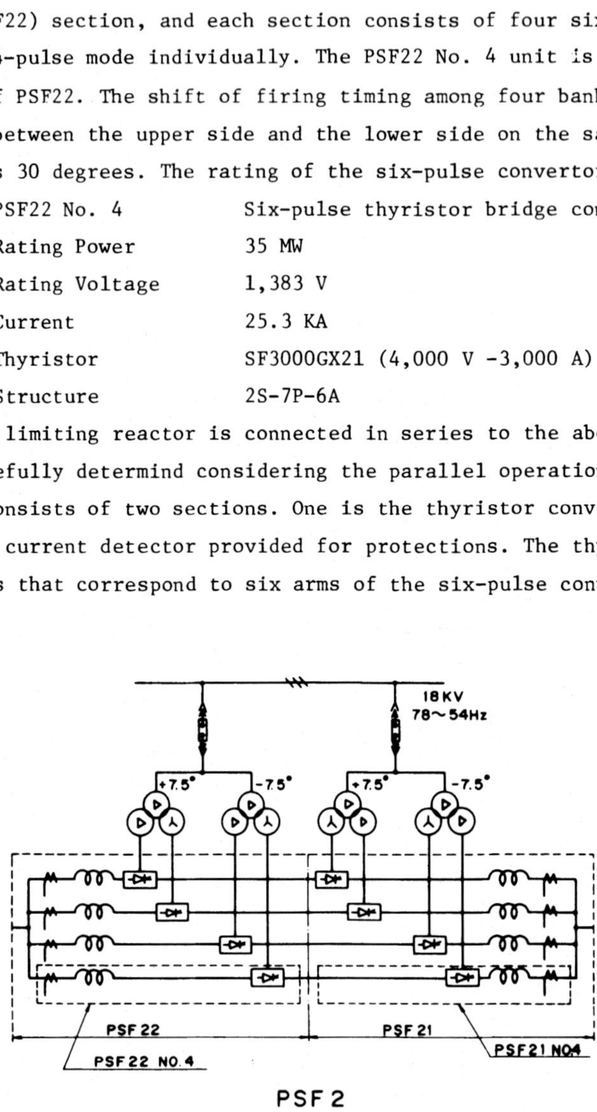

Fig.3 Circuit Diagram of PSF2

3.3 Hall DC-CT

The DC current measurement is indispensable to control the coil current and to protect the DC power supply. It is conventionally made by use of a shunt resistor or a DC-CT which would be a magnetic amplifier type, and sometimes a Logowsky coil with integrator is applied.

The poloidal field power supply requires difficult specifications to measure DC current. Concerning to JT-60, the following specification is required:

Rating current	DC ±100 KA
Accuracy (linearity)	±0.1 - 0.5 %
Response	at least 1 KHz

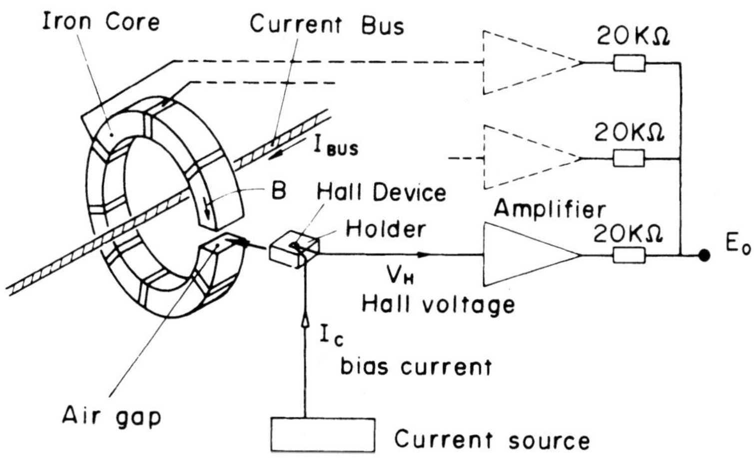

Fig.4 Schematic Diagram of Hall DC-CT

and the conventional DC current measurment equipment can not fulfill above specifications.

The Hall DC-CT solves the problem and achieves the desired performance with measuring the magnetic field produced by the DC current by means of Hall device that converts the strength of magnetic field to voltages (the Hall effect). As shown in Fig. 4, the Hall DC-CT has four components. They are as follows:

1. Current-magnetic field conversion; The iron core with eight small air gaps in series makes the magnetic circuit and surrounds the current bus. The magnetic flux density in each of air gaps can be considered approximately uniform and can be measured with the Hall devices. The core size and gap length are determined by the rating current.
2. Magnetic field - voltage conversion; The Hall device is housed in each air gap and converts the strength of magnetic field to voltages.
3. Amplifier; The amplifier composed of solid state circuit amplifies the Hall voltage (0.5- 1 V) to the values needed for control and measurment.
4. Current source; The current source is a stabilized power supply to provide a constant current in the Hall devices.

When DC current flows in the bus, the magnetic flux is induced in the iron core by the DC current. This magnetic flux is proportional to the DC current as far as the magnetic circuit remains in the linear region. Setting position of the DC-CT in the OH circuit is shown in Fig. 2. The specification is summarized in Table 2.

Table 2 Specification of Hall DC-CT

Conditions	Temperature	-10 ∿ +40°C
	Humidity	max. 90%
	Magnetic field intensity	max. 100 Gauss
Ratings	Current	±100 KA (peak)
	output voltage	±10 V (max.)
	Output impedance	1 KΩ (min.)
	Overcurrent	1 MA - 1 sec
	Insulation	AC 50 KV
Characteristics	Error	±0.5% (±10 ∿ 100 KA)
		±10% (±DC 100 ∿ 200 KA)
	Step response	0.2 ms
	Frequency response	DC - 5 KH (-3 db)
	Temperature coefficient	0.02% / C°
	Magnetic field coefficient	±0.1% at 100 gauss

3.4 Trigger gaps for protection

Nitrogen gas gap with trigger electrode is connected to the protection switchgear in parallel as shown in Fig. 2. It makes the trigger-delay time of this protection circuit to the order of micro-seconds. The rating of the trigger gap is shown as follows:

Operation region	DC 15 KV -55 KV
Current	92 KA-20 mSec
Over current	500 KA-20 mSec
Insulation	AC 25 KV-1 minute
Trigger delay time	Shorter than 10 μS
Operation times	10 Times (Filling gas N_2 is exchanged in every time and electrode in every ten times)

In case of the field distortion trigger type, the voltage region with the short delay time is in (0.9- 0.3)times of break down-voltage V_B. It is convenient to set the break-down voltage at 50 KV ±5 KV, while the operation voltage is 30 KV (0.6 v_B). The delay time from a start pulse is less than 10 μS. Electrode structure of the trigger gap is shown in Fig. 5.

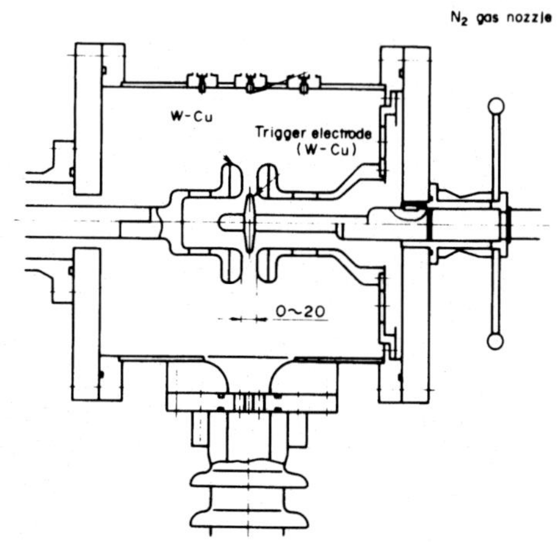

Fig.5 Electrode Structure of Trigger Gap

3.5 Voltage divider

The specification of voltage divider used in the ohmic heating power supply as shown in Fig. 2 is the rating voltage of 25 KV, accuracy of ±1 % and frequency responce of DC - 10 KHz. Requirement for the frequency responce is determined considering the pulse voltages (50 μSec rise) which may appear in the DC circuit breaker and/or thyristor convertors.

It requires the response of 10 KHz at high voltage level. As shown in Fig. 6, the voltage divider has five series resistors and one resistor with a voltage convertor. The temperature characteristics of resistor must be about 100 PPM/°C to realize the accuracy of 0.5 % within 50°C rise. The output voltage is transmitted through the fiberoptic cables from the viewpoint of the noise protection.

The reduced output voltage by the known factor is converted into two types of optical signals, analog signal line and pulse signal line:

1. Analog signal line; The light intensity modulated by the input voltage directly transmits through the fiberoptic cable. This line is expected to be useful for the high frequency part of signals.
2. Pulse signal line; The light pulse modulated by the V/F convertor transmits through another fiberoptic cable to the receiver that has an F/V convertor. The output voltage of F/V convertor is used to correct the static drift of analog line and improves the

accuracy of the voltage measurements.

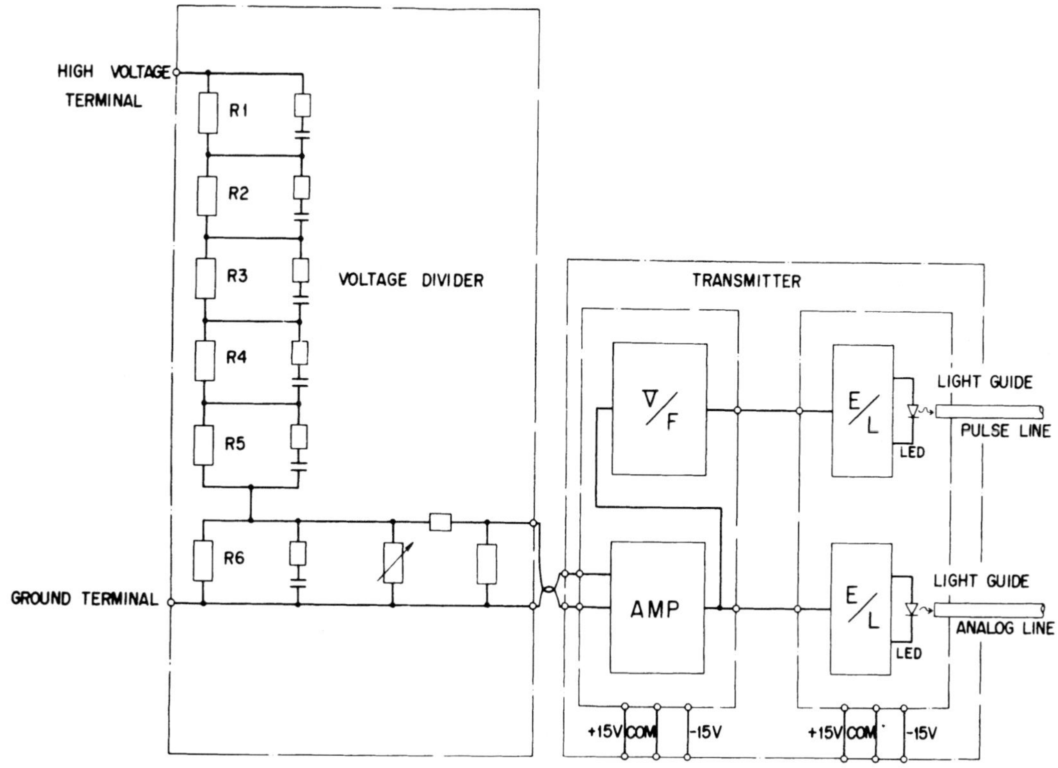

Fig.6 Voltage Divider and Transmitter

4. CONCLUSION

The design of JT-60 ohmic heating circuit system has been finalized. In this circuit, the control of voltage across the OH coil by thyristor convertor is possible even in the plasma current build-up phase and the voltage ripples which appear across the OH coil are also reduced. These operations will be used optionally to enhance operation performance depending on the various conditions and be expected to increase experimental flexibility.

A large thyristor convertor for pulsed performances and noncommercially available equipment to be used in the JT-60 ohmic heating power supply have been developed and their prototypes were manufactured. These prototype components are fully tested to obtain detailed information of their performances and accordingly to finalize the specifications for manufacture. Optimization and improvement will be made, if necessary, after examining the test results.

5. REFERENCES

[1] R. SHIMADA, K. TANI, S. TAMURA, "Excitation of Plasma Current in Tokamak Devices by Inductive Energy Storages", Trans. IEE of Japan, Vol. 99-A, No. 2 Feb. 1979.

[2] R. SHIMADA, et al., "Development of High Current DC Circuit Breaker for Large Tokamak Device", Proc. 10th Symposium on Fusion Technology, Padova, Italy 1978, Vol. 1, P. 401.

[3] S. TAMURA, R. SHIMADA, et al., "Parallel Interruption of Heavy Direct Current by Vacuum Circuit Breakers", IEE Trans. on Power Apparatus and System, Vol. PAS-99, No. 3 May/ June 1980.

[4] R. SHIMADA, et al., "Synthetic Test Methods of High-Direct-Current Circuit Breaker", Proc. IEE, Vol. 126, No. 10 Oct. 1979.

APPENDIX

The ohmic heating circuit operation of JT-60 is divided into four phases. The operational sequence and simplified circuits corresponding to these phases are described in detail.

(a) Premagnetization (See Fig.a)

Switches MS_0 and MS_1 are open. Before discharge, the PSF1 and PSF2 premagnetize L_{oh} and L_s respectively. The current in L_{oh} and L_s are controlled to be preset current value by the thyristor firing delay angle control. After the charging time of 2 Sec, there is a flat top of maximum 1 Sec, the poloidal field circuit has to wait for start until the toroidal field reaches to its prescribed value.

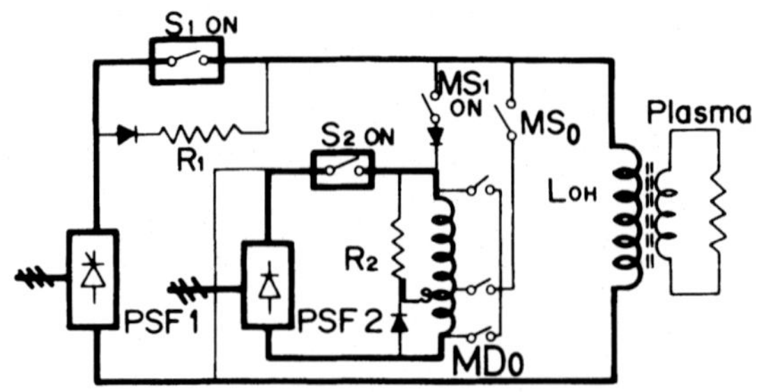

Fig.a Premagnetization

(b) The first stage (See Fig.b)

By opening the S_1 (VCB1), the coil current transfers to R1. Therefore the resistor R1 is inserted between PSF1 and L_{oh} in series. The voltage across L_{oh} is inverted up to 25 KV within 1 mS. The plasma current starts to rise due to the induced voltage. During the build-up phase, PSF1 has capability to vary the voltage across L_{oh} within its maximum voltage of ±2,500 V.

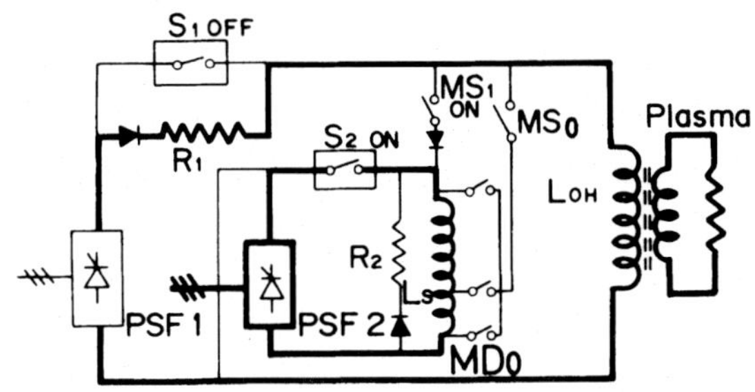

Fig.b First Stage

(c) The second stage (See Fig.c)

The switch MS_1 is closed. Sequentially S_2 (VCB2) is interrupted, and the current in L_s flows to R1 and R2. The voltage of OH coil rises again and accordingly the plasma current builds up in second stage. The current in L_{oh} consequently turns to be negative.

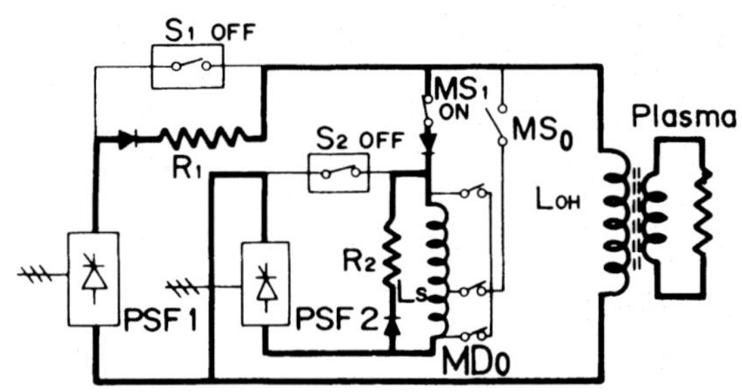

Fig.c Second Stage

(d) Reconnection (See Fig.d)

When the coil voltage becomes below the voltage of PSF2 or the plasma current reaches to a preset value, the make-switch MS_0 is closed. The plasma current is built up further, if necessary, or maintained mainly by PSF2. In the case of reconnection of the thyristor convertor, there is an application mode* depending on the selections of switch MD_0.

* Ripple filter mode; the PSF1 turns to be by-pass pair mode. A low-pass filter composed of R1 and L_s connected at the middle point reduces the voltage ripples caused by the thyristor convertor. The voltage ripples from the 24-pulse rectifier are attenuated to 3 % when L_s is 1.4 mH.

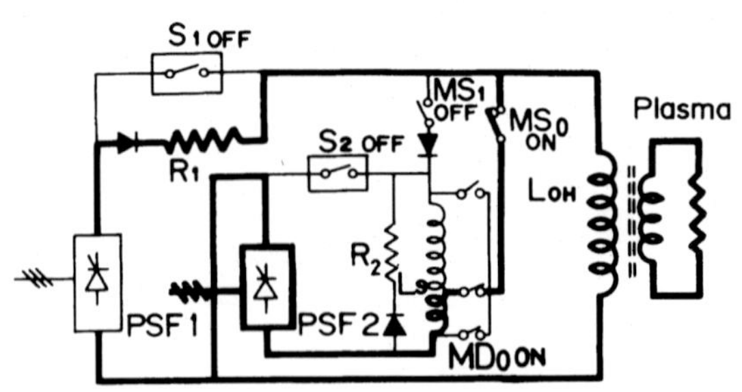

Fig.d Reconnection

OPTIMIZED OHMIC HEATING SYSTEM WITH VACUUM INTERRUPTER FOR THE W VII-STELLARATOR

D. Dorst, J. Sapper
Max-Planck-Institut für Plasmaphysik
EURATOM-Association, D-8046 Garching

ABSTRACT

The ohmic heating circuit of the W VII stellarator has been optimized to the experimental requirements using the experience gained from 23 000 shots over a period of four years. Converters in antiparallel connection with a ± 27 kA at 2.5 kV capability supply the coil system for 1.5 s. A vacuum interrupter with commutation circuit acts as a DC circuit breaker (10 kV) during the plasma current rise.

A fast mechanical switch in parallel with an HV ignitron steps down the resistance in the ohmic heating circuit automatically as a function of a preselected plasma current flat top. After that an overlaid control loop guides the current control loop of the converter for plasma current regulation. Preselected current waveforms of any desired shape can be performed. All components are equipped with protection systems against malfunction.

Some details and experiences are reported concerning the vacuum interrupter and commutation circuit, the components used, the step-down switch for the resistor in the ohmic heating circuit, the protection devices and special requirements for the power supplies.

1. INTRODUCTION

The first ohmically heated plasma in the W VII-stellarator was produced in 1976. The OH transformer is described in [1]. Electrical data are given in table I. Various converter sets for power supply are assembled in a central station. Cables, bar lines, and a distribution bus bar system transfer the energy to the experiment. There a central controller runs the pulse. The time chart of a pulse is divided into five intervals: transformer loading phase, plasma initiation and current rise, passing over to plasma current flat top, plasma current regulation and, finally, deloading the transformer from reversed current. Failures or unallowed parameters stop normal operation or de-energize the system. This paper discusses essential parameters, components and system engineering of the OH circuit of W VII (fig. 1).

TABLE I

Electrical Data of the OH Transformer of WENDELSTEIN VII-A

Number of turns	162	Rectangular cooper bar (water-cooled)	
Nominal insulation voltage	30 kV/pulse	Nominal current	25(35) kA/pulse
Inductance (no load)	14 mH	Ohmic resistance	20 mΩ
Time constant	700 ms	Flux swing (+20/-25 kA)	3.7 Vs
Pulse length max.	2.5 s	Repetition rate	1 per 180 s
Short-circuit peak current	≈60 kA		

2. REQUIREMENTS AND SYSTEM DESIGN

The design values for the OH system of W VII have been derived from calculated and estimated plasma parameters. An improvement followed with the first actual plasma discharges. Special information about the influence of ripple in the primary transformer current, defined by the number of pulses of the rectifier connection, and the size of possible overvoltages in the system could be obtained. In detail, the following parameters proved to be important for the successful running of the system:

- To initiate the plasma, the secondary loop voltage is in the range of 40 V to 60 V, i.e. ≈ 5 V/m electric field force on the torus axis (discharges with H_2, He, D_2). The corresponding primary transformer voltage is about 10 kV and dI_{OH}/dt exceeds 700 A/ms. These values are well within the design parameters.

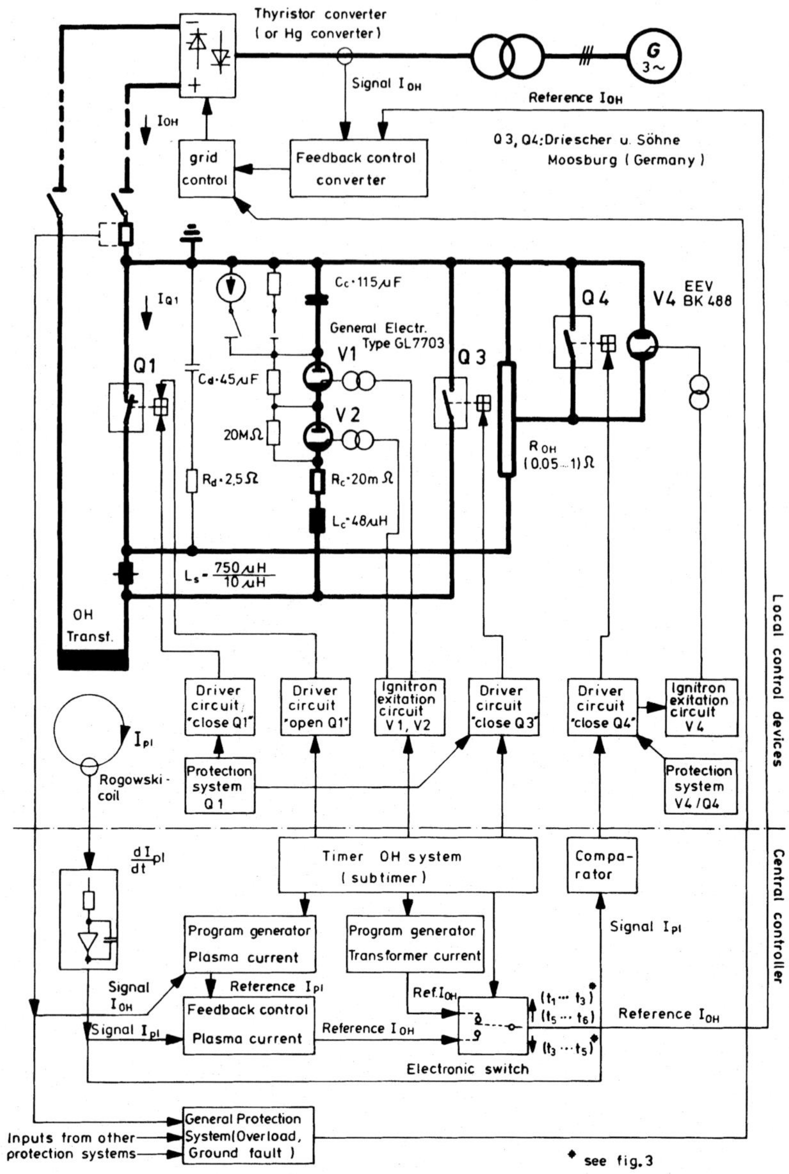

Fig.1. Control and circuit diagram ohmic heating system WENDELSTEIN VII A

- To avoid overshooting the desired flat top current, plasma current rise is stopped exactly and automatically at a preselected value by stepping down the OH resistor. This is done by a subsystem. The initial trigger is given from the central controller. When the system fails, the plasma current overshoot might cause a disruption. This has to be taken into account for the torus protection system.

- During plasma current control by the closed-loop feedback system, a secondary loop voltage of 10 V to 15 V, including a margin for dynamic acceleration of the current, is necessary. The primary transformer voltage then is about 2.5kV, the maximum dI_{OH}/dt reaches 170 A/ms. These values are well suited to the nominal data of the converter modules and also comparable with other machines [2].

- A typical plasma time constant in W VII-A is about 20 ms. A perturbation coming from the plasma can only be influenced by a current control loop if its response time is less then the given plasma time constant. This is necessary because the inductive coupling between the plasma current and the primary OH circuit in W VII is very weak (air core transformer, coupling factor k ≈ 0.25). We therefore first studied a feedback loop directly acting upon the grid control of the converters, which would eliminate all further time constants and delays in the signal transmission. A failure analysis, however, showed that it is better to use an overlay structure, working on the reference input of the internal current control loop of the converter. In case of a disturbance in the converter control or in case of a line or load short-circuit, the internal control loop limits the current to nominal values and so increases the redundancy of protection (fig. 2). If the internal loop runs near its stability limit, the system still works fast enough.

- All power supplies in the IPP converter station are bridge six-pulse bidirectional converters (mercury arc types and thyristor types). Nominal data are shown in table II. Both systems can be used for W VII. This increases the flexibility and availability, but requires the normalization of signals and mode of operation. Very exact adjustment of the grid control phase, especially at parallel or series connection of the modules, minimizes the ripple on the DC voltage and helps to overcome parasitic influences on the plasma diagnostics.

TABLE II

Electrical Data of the Converter Power Supply, OH System, WENDELSTEIN VII-A

	Mercury arc converter	Thyristor converter
Nominal voltage (module)	0.53 kV	1.25 kV
Nominal pulse current/time	15 kA/0.5 s	30 kA/0.5 s or 22.5 kA/5.0 s
AC supply (flywheel generator)	10.5 kV, 3 ~ 50 Hz	10.5 kV, 3 ~ 100 Hz
Connection of modules for W VII: Series/parallel/antiparallel, series connection with interphase transformer, twelve-pulse operation		

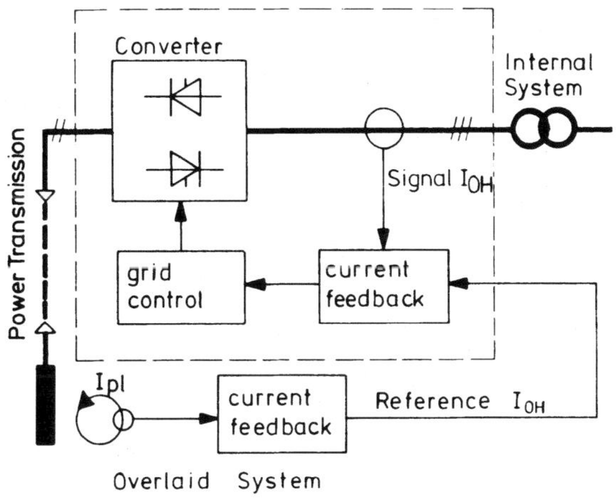

Fig.2. Overlay Structure of Plasma Current Regulation

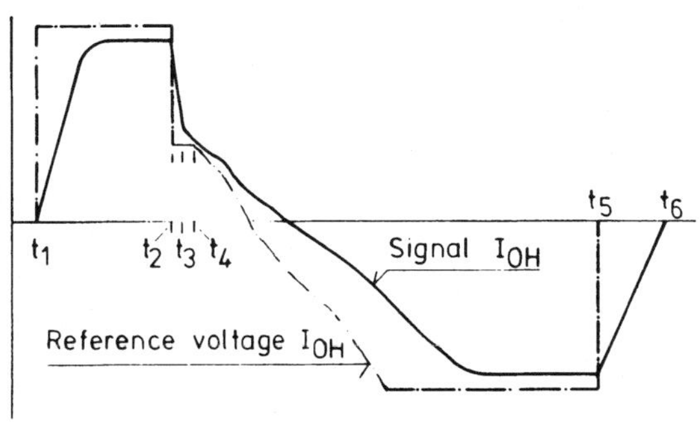

Fig.3. Time chart of an OH pulse

$t_1 \cdots t_2$ transformer loading (program)
t_2 commutation of I_{OH} into R_{OH}
$t_2 \cdots t_3$ plasma current rise
t_3 stepping down R_{OH}
$t_3 \cdots t_4$ passing over to plasma flat top.
$t_4 \cdots t_5$ plasma current regulation (feedback)
$t_5 \cdots t_6$ deloading from reversed current

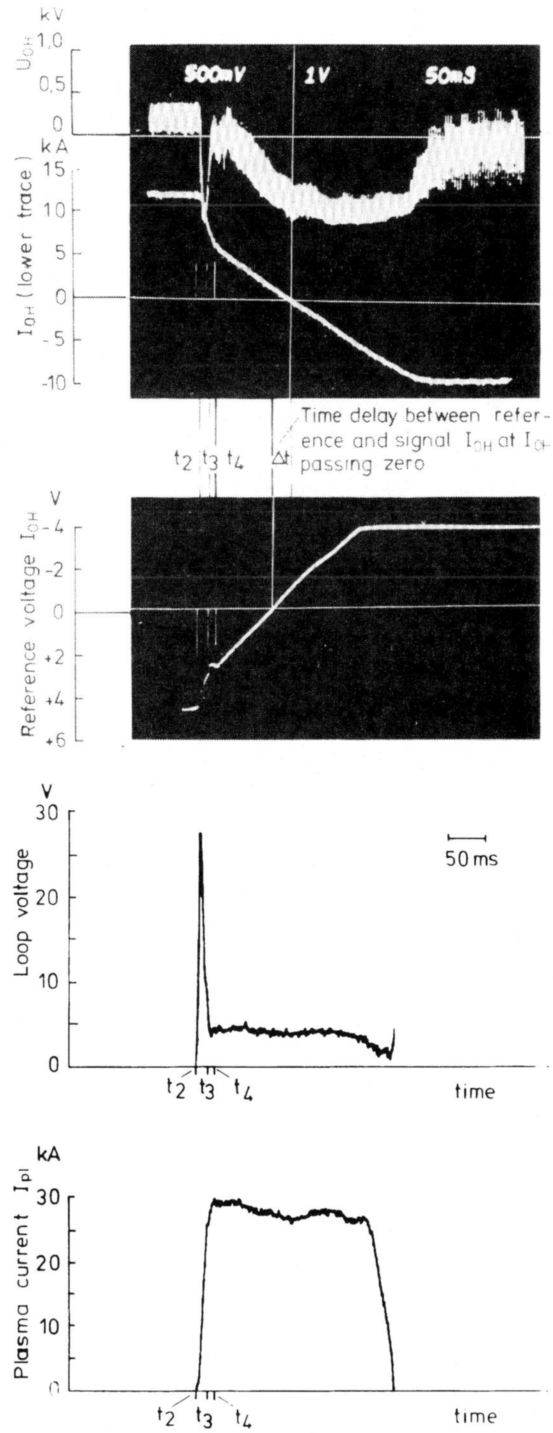

Fig.4. Oscillograms taken from one actual OH pulse
Note: The amplification of the plasma current feedback control was low during that shot

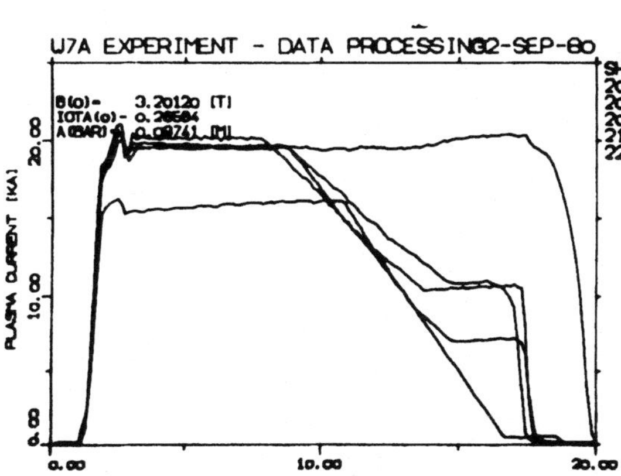

Fig.5 Various types of plasma current waveforms W VIIa

- Measurement and calculation of the stray-flux influence of the air core transformer on other equipment of the experimental device (control, diagnostics, electronic data processing) showed that it is optimal to perform the flux swing symmetrically around zero. Therefore, the converters are equipped for antiparallel operation and fitted with reactors for limited circulating-current flow. With the nominal data of the converters, a four-quadrant performance is possible without interruption at zero. The OH resistor is crowbarred near current zero.

3. SYSTEM OPERATION AND PERFORMANCE

In fig. 3 the time chart of an OH pulse is demonstrated in principle. The oscillograms in fig. 4 illustrate the essential data of an actual shot. Fig. 5 shows some plasma current waveforms of which the system is capable.

4. COMPONENTS AND THEIR PROTECTION

More than 23000 plasma discharges have been carried out with the ohmic heating system of W VII-A, 16000 with the fuse system [3], 4000 with the LMPV system [4], and, since Dec. 1979 more than 3000 using a vacuum circuit breaker [5]. According to the original planning, the LMPV circuit breaker system should serve as the standard c.b. system, however, the malfunctions of the Liquid Metal Plasma Valve (LMPV) exceeded 10 %, so it became necessary to look for another c.b. system. The reliable fuse system had too high costs in operation and maintenance. Thyristors were too expensive, an air blast circuit breaker needed additional space for installation and enlargement of the capacitor bank C_c.

The best solution we found was to install a vacuum circuit breaker (SIEMENS, type 3 AF 1756-4JA 40-0AA0-Z, 12 kV, 2500 A, three-phase) using components of the LMPV c.b. system (controls, commutation circuit) as shown in fig. 1. This vacuum circuit breaker is relatively small and cheap and has a turn off time of only 10 µs, which is similar to that of the LMPV. With one vacuum interrupter more than 3000 interruptions at currents of 12 kA to 18 kA were carried out. The total ampere-seconds (current x arcing time) was 300 000 As. Only 5 malfunctions occurred due to operator errors (wrong charging voltage for C_c). Restrikes after current interruption were not observed. The unexpectedly long contact life and the reliability of the vacuum interrupter at currents up to 18 KA was only achieved when an axial magnetic field of 0.046 T was applied to the arc in the vacuum interrupter. The influence of the axial magnetic field is discussed in [6]. It prevents the formation of an anode spot, which can be recognized in the arc voltage (fig. 6). Suppressing the anode spot reduces evaporation of anode metal, thus increasing contact life and reducing the probability of malfunctions. In fig. 7 an oscillogram taken during an interruption is shown.

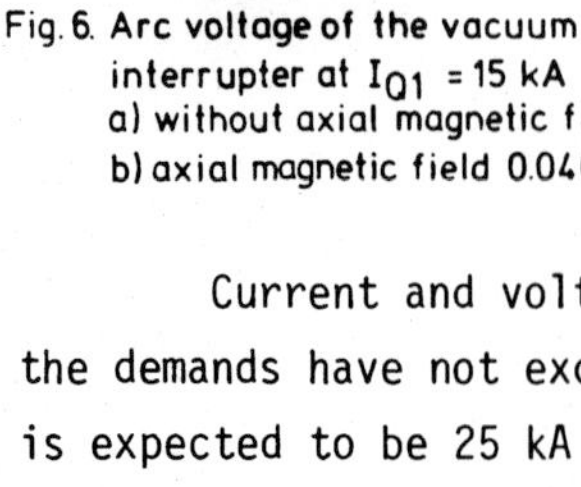

Fig. 6. Arc voltage of the vacuum interrupter at I_{Q1} = 15 kA
a) without axial magnetic field
b) axial magnetic field 0.046 T

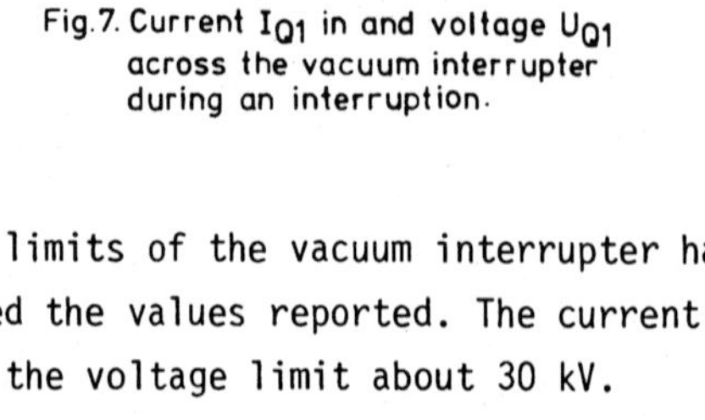

Fig. 7. Current I_{Q1} in and voltage U_{Q1} across the vacuum interrupter during an interruption.

Current and voltage limits of the vacuum interrupter have not yet been tested, because the demands have not exceeded the values reported. The current limit with axial magnetic field is expected to be 25 kA and the voltage limit about 30 kV.

For stepping down the ohmic heating resistor after starting the plasma current as mentioned in para. 2, solutions with thyristors, ignitrons,and fast mechanical switches were investigated. The most economic solution was to use an ignitron. Ignitron V4 crowbars a certain fraction of the ohmic heating resistor with a time delay of < 0.5 ms. However, the ignitron is not able to carry the current for the whole duration of an ohmic heating pulse. So a mechanical fast making switch (Q4) was connected in parallel, which closes 13 ms after triggering, taking over the current from ignitron V4. To ensure that Q4 will close, the driver circuit "close Q4" is directly triggered by the comparator. The trigger for V4 is then derived from the driver pulse which actuates Q4 (see fig. 1). This system and the capability of the ignitron is described in detail in [7].

In case of a malfunction or restrike, the arcing time of the vacuum interrupter has to be limited to 40 ms, otherwise the contact system could be destroyed. Three independent units are used for fast and reliable protection: a voltage unit, a current unit, and an energy unit. The principle of operation is shown in figs. 8, 9, and 10 respectively.

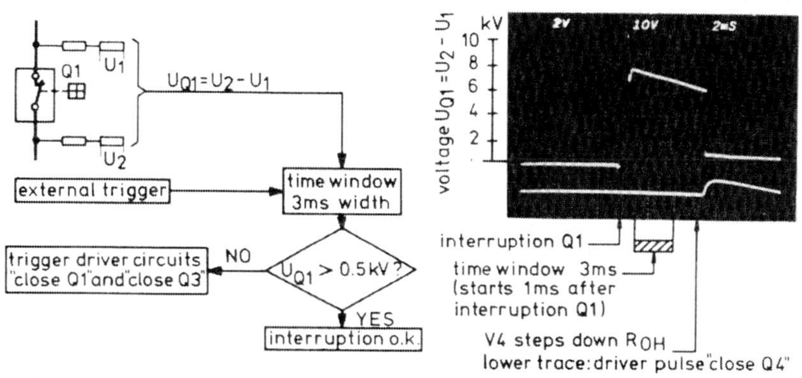

Fig. 8. Voltage unit of the protection device for Q1, principle of operation

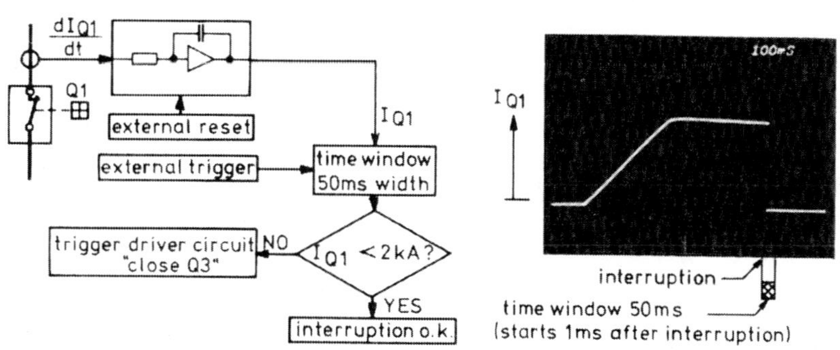

Fig. 9. Current unit of the protection device for Q1, principle of operation

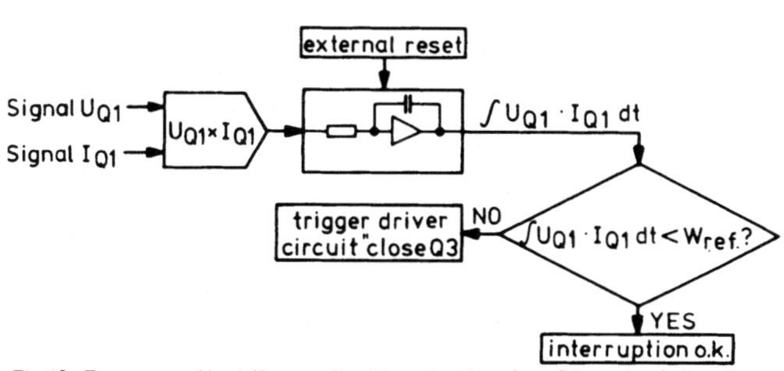

Fig. 10. Energy unit of the protection device for Q1, principle of operation

The dielectric medium for the contact gap of safety switch Q 3 is pressurized dried air (2 bar). In the open position, a breakdown between the contacts might occur when Q1 interrupts. The following arc can damage the contact system if Q3 is not closed as fast as possible. Q3 is protected against breakdowns in the contact gap by the voltage unit of the protection device for Q1.

Switch Q4 is of the same type as Q3. Protection of Q4 against breakdown in the contact gap as well as overloading ignitron V4 by an unallowed ignition is ensured by a tape wound iron core arranged at the bus bar running to V4 and Q4. When the current in this bus bar rises due to a fault-current in V4 or Q4, the iron core is rapidly driven into saturation and a voltage pulse is induced in a winding around the iron core. This voltage pulse then triggers the driver circuit "close Q4".

Surge arrestors AEG, type Hamburg, which have been developed for railway applications are used for protection against overvoltages. Four of these surge arrestors, each for 1000 V, are connected in series to protect the thyristor power supplies and switchgear. The voltage distribution across the four arrestors is additionally controlled by capacitors. If the surge arrestors are activated by an overvoltage, the power supply is shortcircuited and grounded. The tested current capability is $0.5 \cdot 10^9$ A^2s.

The general protection system comprises overload and ground fault protection of the OH system and also inputs from other protection systems (helical windings, toroidal field coils, emergency stop), which requires de-energizing the OH system. The output of the general protection system is directly connected with the grid control to stop the OH pulse.

6. CONCLUSIONS

The circuit diagram of the OH system of WENDELSTEIN VII-A, principles of control and protection have been described and some information given about the components used. The present system optimized to the experimental requirements allows an average pulse rate of about 15 shots per hour during normal operation.

7. REFERENCES

[1] JAENICKE, R., et al., 8th SOFT, Noordwijkerhout, June 17-21, 1974, Proceedings pp 119-126

[2] KARADY, G., et al., 10th SOFT, Padova, Sept. 4-9, 1978, Proceedings pp 359-364

[3] BRAUNSBERGER, U., et al., 9th SOFT, Garmisch-Partenkirchen, June 14-18, 1976, Proceedings pp 723-727

[4] ECKHARDT, W.O., et al., 7th Symposium on Engineering Problems of Fusion Research, Knoxville, October 25-28, 1977. Proceedings pp 1767-1771

[5] DORST, D., Max-Planck-Institut für Plasmaphysik, Garching, Report IPP 2/248 (1979)

[6] WARREN, R.W., Los Alamos Sc. Lab. Report LA-6909-MS (1977)

[7] MIEDANER, H., nachrichten elektronik, 3-7 (1978) 6

THE ASDEX OVERVOLTAGE PROTECTION SYSTEM

H.Rapp, Ch.Liu, H.Preis

Max-Planck-Institut für Plasmaphysik
EURATOM-Association, D-8046 Garching

ABSTRACT

The ASDEX overvoltage protection system completes the favourable properties of the ASDEX coils and rectifiers. It uses properly matched standard elements and meets the requirements for operational and fault conditions in a very economic and reliable way. Special care has been taken for the limitation of overvoltages due to major plasma disruptions without it being necessary to interrupt the experiment run.

1. INTRODUCTION

As part of the tokamak protection system which predominantly detects failures /1/, the overvoltage protection system actively prevents damage due to overvoltage. The requirements for the system are partially given by the basic properties of a magnetically confined plasma experiment:

i The coils are DC current sources with high but limited energy.

ii The coils are powered from semiconductor rectifiers which are voltage sources with possibly very high short circuit surge currents.

iii The stored magnetic energy can easily be dissipated by the coils.

iv There exists an inductive interaction between some coils and between plasma and coils. At least this is true for the individual windings even if the total coupling coefficient of the coil is negligible.

v The plasma current behaviour must be assumed arbitrary.

One can distinguish between two main reasons for an overvoltage:

1. The plasma current disruption causes induced voltages the amplitudes of which depend on the impedance distribution of the electric circuit including passive loops and power supplies. A disruption is a normal operational condition. It should not lead to a shut-down of the experiment and must not cause any damage. The protection equipment should be self-recovering for this case.

2. A technical fault occurs:

2.1 Failure of a power supply or a DC circuit element.

2.2 Flash-over from one coil to another and short-circuit of a coil or parts of it. The shut-down of the experiment and the replacement of the protection equipment is tolerated. Damage to other equipment must not occur in case 2.1. In case 2.2 damage must be limited. This kind of fault is detected by monitoring the voltage distribution of the coils /1/.

From these statements we conclude the need for nonlinear elements to limit overvoltages from the plasma current disruption and for switching elements in the case of technical faults. The last conclusion can be easily derived from the energy to be consumed and the costs for powerful nonlinear elements.

2. THE ENGINEERING CONCEPT

The overvoltage protection system of ASDEX is based on the proper combination of the following expedients:

i The power supplies, including the leads, have a very low impedance, even for transients.

ii Magnetic coupling between complete coils is negligible.

iii A metal-oxide varistor (MOV) protects against overvoltages from a plasma current disruption. The varistor rating is matched to a surge divertor.

iv The special surge divertor (SD) automatically switches if the varistor becomes overloaded. It can carry the coil currents.

v In any case of a severe technical fault the rectifier controls are triggered either for voltage inversion or simultaneous firing of the rectifier bridge branches, i.e. free-wheeling of the DC circuit.

The varistor as well as the surge divertor are low-cost, well proven elements and commercially available. The protection circuit and its characteristic curve are given in Figs. 1 and 2. It can be matched to a desired protection level by switching single units in series. The protection level of a single unit is 4oo V or 1 kV, typically. The tolerable voltage level for the ASDEX rectifiers and some of the critical coils is 2 kV and 3 kV, respectively. The data for the rectifiers and coils are given in Ref./2/.

3. THE PROTECTION ELEMENTS

The task of the MOV is threefold:

- It limits low energy recurring overvoltage pulses.
- It smoothes the voltage rise time in order to establish DC conditions for the SD breakdown. By that means the breakdown voltage of the SD has a small spread.
- It protects against the voltage caused by the protection circuit reactance X_L when the SD switches.

In cases where the operation level is not so far from the protection level, varistors are favourable due to their marked nonlinear characteristic curve (see Fig. 2). On the other hand, they can consume only little energy. The lifetime of a varistor depends largely on the pulse energy and current amplitude (Fig. 3) /3/. We therefore put two elements in parallel although this will not help decrease the varistor current. However, as soon as one varistor fails the other varistor will automatically overtake its function. By that means the lifetime of the arrangement will be extended. If a short-circuit happens during a pulse the varistor in series will become overloaded and consequently the SD will break down. The MOV chosen is a standard industrial type. There exist more powerful types but they are hard to get and expensive.

The surge divertor consists of a replaceable powerful switching element in a housing. The switching element is made of two copper discs which are insulated by polyimid foils. The foils have a hole in the centre and set the breakdown voltage which is typically 4oo V or 1 kV. The housing must withstand high mechanical and thermal stresses. The SD is an industrial product and is usually used for the protection of railway power lines. It has been tested at the Garching stellarator W VII /4/, and the housing was improved so as to match it to the tolerable high loss integral value of more than 4oo kA^2s of the SD element. By that means the SD meets the requirements of the ASDEX poloidal field coils with currents from 3o to 45 kA and time constants of some hundred milliseconds.

If the SD protects a rectifier, a series resistor limits the short circuit current from the low impedance source. This means is also useful for decreasing the SD load when it is used for coil protection. One has to decide for each particular case on the proper matching to the different requirements and effort.

In order to check the engineering concept of the overvoltage protection system a laboratory experiment has been carried out. A capacitor was discharged on a coil with a single protection unit in series. Figure 4a shows the oscillogram of the different variables and in Fig. 4b the results of a computer simulation are presented. The system worked as expected and the agreement between experimental and numerical results was good, considering the wide spread of the varistor current due to the tolerance of the characteristic curve of the MOV.

4. OVERVOLTAGE DUE TO PLASMA DISRUPTION

As mentioned above, the MOV should be capable of limiting overvoltages from a major plasma disruption. We therefore calculated the voltages at the different power supply terminals and at the terminals of groups of windings. The equivalent circuit diagram (Fig. 5) describes all poloidal field coils and the vacuum vessel /5/. The dimension of the inductance matrix is 101 x 101. It is valid for the frequency range of some kilohertz only. For higher frequencies one needs a distributed parameter model which seemed not to be necessary in our case. The validity of the simple rectifier model is permanently tested at each tokamak shot by the OH coil current breaker. As initial condition the plasma current decays from 500 kA with a time constant of 1 ms. This figure is arbitrary but reasonable considering the vacuum vessel time constant of 8 ms and the experimental results until now. The most limiting assumption is the spatial immobility of the plasma current column. However, on the time scale of interest this seems to be no problem by the same argument of the vessel time constant. The switch S was closed so as to simulate the most probable and worst condition.

Owing to the favourable configuration of the ASDEX poloidal field coils and the very low impedance of the rectifiers we did not expect overvoltages at the rectifier terminals. This has been confirmed by the numerical results. On the other hand the voltages at different windings or packages of some windings can exceed the tolerable level.In Table I the results for different cases are given. "Closed loop" means the rectifier is switched on. The influence of the large vacuum vessel is evident. One can also read the gap voltage and the very high current in the vessel /5/. The radial field and toroidal field coils are not affected, of course, due to their field configuration.

TABLE 1

Calculated Peak Voltages at the Coil Terminals and Peak Currents in the Coils during a Major Plasma Disruption with 500 kA and $\tau = 1$ ms.

	OH Coil	VF Coil	DF Coil No. 1	DF Coil No. 2	DF Coil No. 3	Vessel Gap
Open loop voltage (KV)	48	-0.3	3.3	-6.5	3.4	-0.34
Closed loop voltage (V) without vessel	160	-78	2440	-5810	3100	-
Closed loop voltage (V) with vessel	250	-22	2030	-4170	2200	151
Transient peak current(KA) with vessel	4.8	-5.85		-2.96		140

It follows that the divertor field coil No. 2 is the only endangered one, because the possible voltage level exceeds the tolerable level of 3 kV. The DF coils are inside the vacuum vessel and special care is taken to protect them from any possible damage. Neither blocking of rectifiers nor over-currents due to transients will occur.

The different DF coils are protected separately. As mentioned above the MOV should prevent the SD's breakdown if a disruption happens. Figure 6 shows the load of the MOV. An "upper limit" MOV would eventually exceed the SD breakdown voltage. For several reasons a current limiting resistor in series with the SD was avoided. We therefore studied the consequences of a SD breakdown at the DF coil No. 2 during a plasma disruption. Figure 7 shows the result. The current in the other DF coils rises for about 10 % which is already the tolerable limit. The losses in the SD reach the value of 220 kA^2s which is below the rating.

5. TECHNICAL FAULTS

Breaking of a DC current, for whatever reasons, will cause the SD to switch. Referring to the arguments given above there will arise no problems. A diode rectifier is used for powering the TF coils. No additional protection has been provided.

The simplified grounding scheme is shown in Fig. 8. All coils are grounded directly with a short and strong lead in the basement of the experiment and the building. Only the OH system is placed about 1o m off. The rectifiers are floating. The casings and supports are grounded by means of current limiting resistors. The other equipment is insulated from the casings and the vessel. Numerical calculations have been carried out for the most probable cases of a short-circuit. The results show no severe problems. Ground faults as well as any short--circuit are detected by special equipment /1/.

6. EXPERIENCES AT ASDEX

Until now about 17oo tokamak shots have been made. No malfunction of the overvoltage protection system has been observed. Disruptions on the short time scale as calculated did only occur due to technical faults. The plasma current decay time was 5 ms for 4oo kA. Owing to the very rare appearance, it was not possible to take measurements of the variables in the protection circuit. No significant signals are observed if the usual "soft" disruption happens. In the divertor mode of ASDEX disruptions can be completely avoided. Accidently the numerical results (see Table 1) concerning the voltage at the DF coil No. 2 have been confirmed by the explosion of a silicon type voltage limiter under test with a rating of 2.7 kV.

7. REFERENCES

/1/ GERNHARDT,J. et al., An Equipment Protection and Safety System for ASDEX Tokamak. 11th Symposium on Fusion Technology, Oxford, 198o.

/2/ RAPP,H., GRESSER,F., The Power Supply for the Poloidal Field Coils of ASDEX. 9th Symposium on Fusion Technology, Garmisch-Partenkirchen, 1976, p. 691-696.

/3/ Siemens Datebook, Metal-Oxide Varistors.

/4/ SAPPER,J., IPP Garching, private communication.

/5/ PREIS,H., SCHNEIDER,F., Determination of Eddy Currents in the Vacuum Vessel of an Apparatus for Investigating Controlled Nuclear Fusion, Archiv für Elektrotechnik 62, p.181-186,(198o).

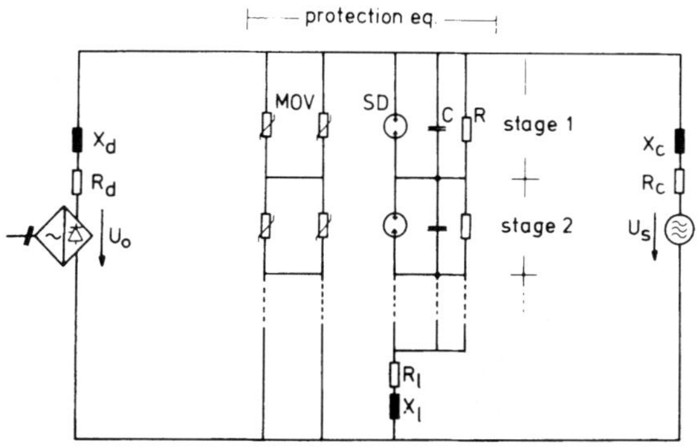

Fig. 1: Overvoltage protection equipment for a DC circuit. MOV: Metal-oxide varistor. SD: Surge divertor. C,R: Voltage divider. R_L, X_L: Current limiting resistor and reactance. U_o, R_d, X_d: Power supply voltage and impedance. X_C, R_C: Coil impedance. U_s: Surge voltage.

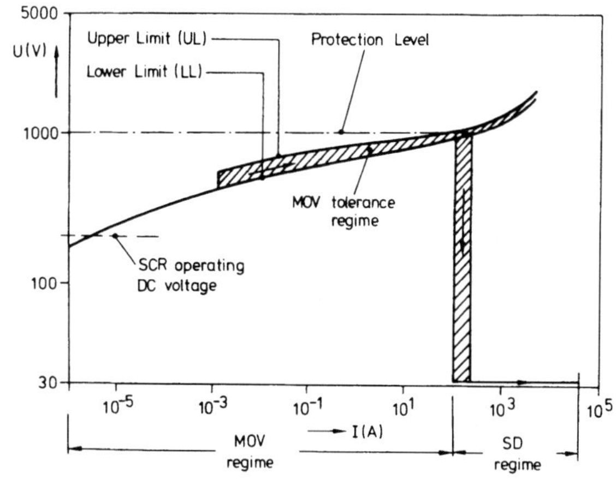

Fig. 2: Characteristic curve of a single stage overvoltage protection unit.

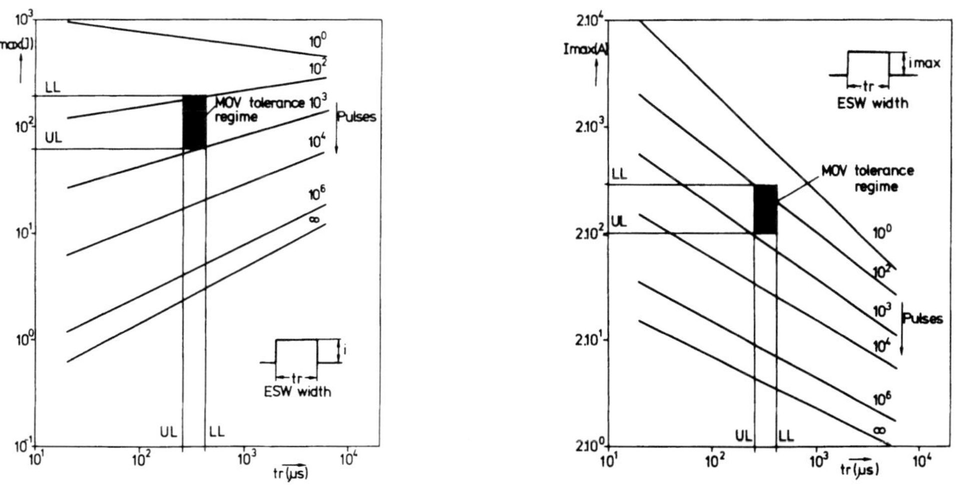

Fig. 3: Lifetime of the MOV for the ASDEX divertor field coil No.2 as estimated from Fig. 6.

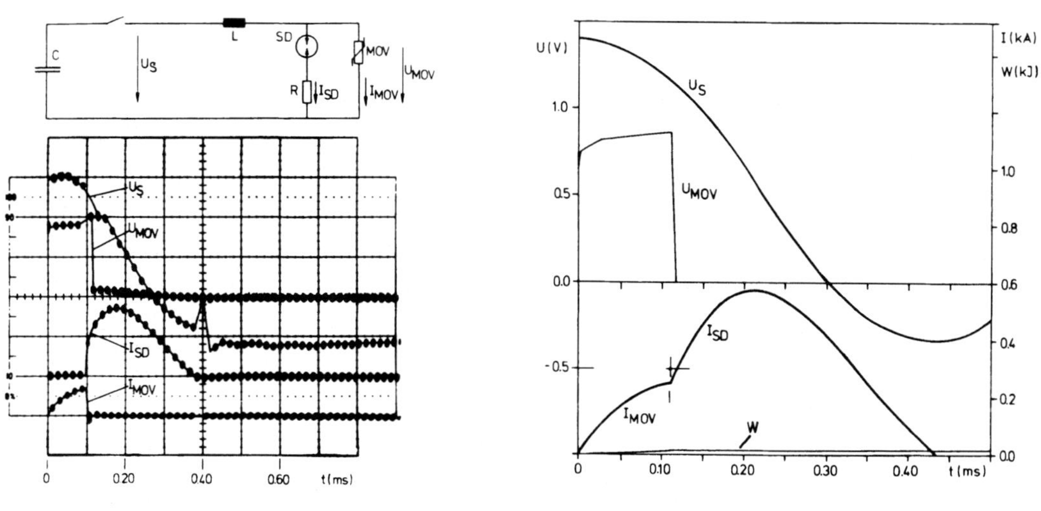

Fig. 4a,b: Laboratory test and numerical simulation of a single stage protection unit.

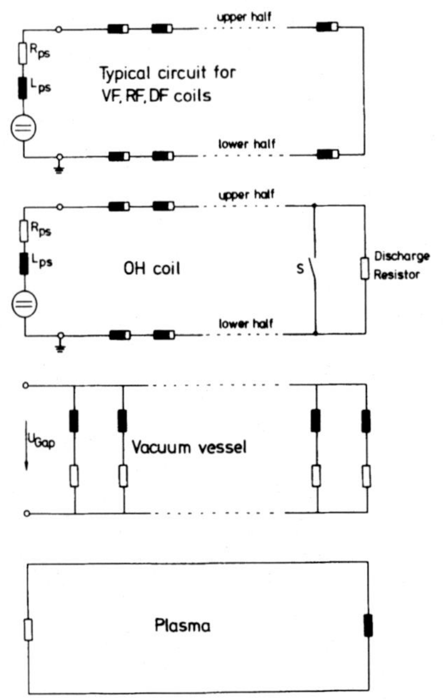

Fig.5: Simplified equivalent circuit diagram as used for the calculation of the transients caused by a major plasma current disruption.

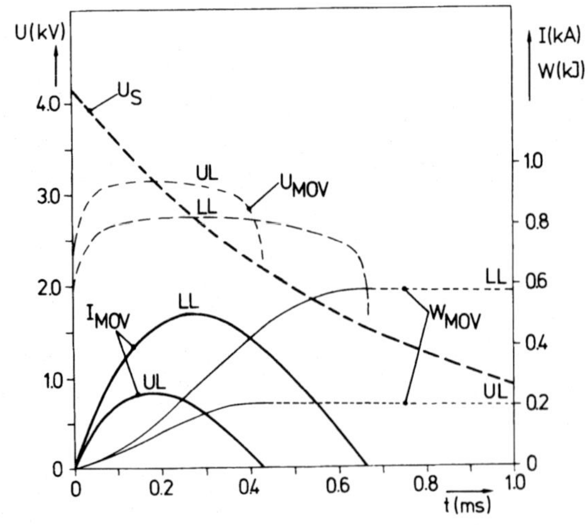

Fig.6: Calculated MOV voltage, current and energy at the ASDEX divertor field coil No.2 during a major plasma current disruption (I_p = 5oo kA, τ = 1 ms).

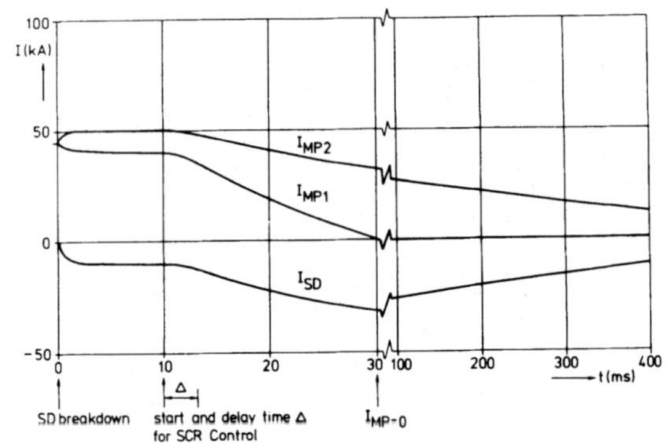

Fig.7: Calculated currents in the divertor field coils after breakdown of a surge divertor. I_{SD}: Current in the SD.

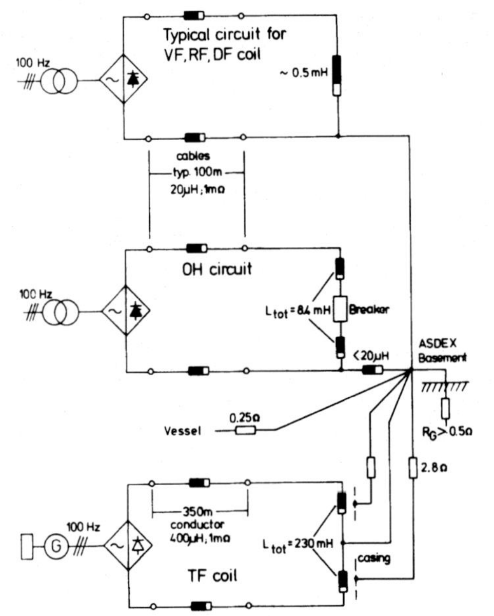

Fig.8: Simplified grounding scheme of the ASDEX coils with vacuum vessel.

A 2 MJ DRUM-TYPE HOMOPOLAR MOTOR-GENERATOR: DEVELOPMENTS AND TESTS*

S. Bolognani, F. Gnesotto, C. Monelli

Istituto di Elettrotecnica e di Elettronica dell'Università di Padova
Via Gradenigo 6/A - 35100 Padova ITALY

ABSTRACT

This paper presents the design work and the first results of a research program on kinetic energy storage and transfer in homopolar machines. A 2 MJ prototype with conventional excitation, two concentric counter-rotating aluminium alloy drum rotors, pneumostatic air bearings and solid brushes is under construction. The performances of several brush-ring materials under various operating conditions have been investigated with a suitable test apparatus and the main results are reported. A thyristor converter has been realized and tested for studying the power smoothing by the homopolar machine of a three-phase grid subject to repeated large power pulses.

1. INTRODUCTION

Homopolar machines for energy storage and transfer have been studied and developed by several laboratories. The main reasons of this interest lie in the intrinsic ruggedness of a massive rotor carrying no winding, in the absence of commutation problems and in the low armature inductance. These characteristics allow high peripheral speed (high rotor kinetic energy densities) and fast energy transfer.

Homopolar modules capable of storing some hundreds megajoules and of delivering them in times ranging between 0.1 ÷ 100 seconds could find applications both in the field of fusion research and in some industrial plants.

Fig. 1 shows the two operating principles: in fig. 1a) the store feeds directly the load (typically a large inductor for fusion research plants) after a slow charging phase; in fig. 1b) it levels the power of a grid to which a load with discontinuous power demand is connected, delivering or absorbing the power fluctuations around the mean value.

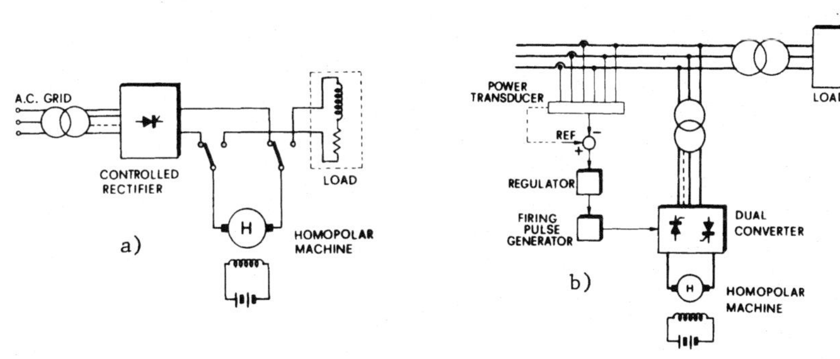

Fig. 1. Two schemes of application for homopolar machines

To verify the effectiveness of such schemes in fulfilling the requirements of load and grid and to investigate the main technological problems arising from the realization of a homopolar machine, a small-scale prototype consisting of a d.c. homopolar machine and of a four quadrant thyristor converter has been designed and is now under construction. For economical reasons a non-superconducting field winding has been adopted even if a homopolar machine with such excitation system presents voltage and efficiency values lower than those obtainable by conventional d.c. machines. This choice allows to concentrate the effort on the design and development of the rotor assembly, considered the crucial part of the system.

* Work supported by the National Research Council (CNR) as a part of its Finalized Energy Programme.

A preliminary comparison carried out between disk-type and drum-type machines did not show one solution substantially superior to the other. The drum configuration has been chosen being more convenient for our specific research programme.

The prototype has been designed with two counter-rotating vertical axis aluminium alloy drums, supported by pneumostatic air bearings, and solid brushes. Schematic views of the machine are shown in Fig. 2. The rotors are designed for a maximum peripheral speed of about 200 m/s, which corresponds to a stored energy of 2 MJ and a no-load voltage of about 125 V.

A four quadrant thyristor converter has been realized to analyze the behaviour and the effectiveness of the system shown in Fig. 1b with different control systems, load characteristics and grid requirements. In the following, the design of the machine, some experimental results about various types of brushes and sliprings and a description of the converter are reported.

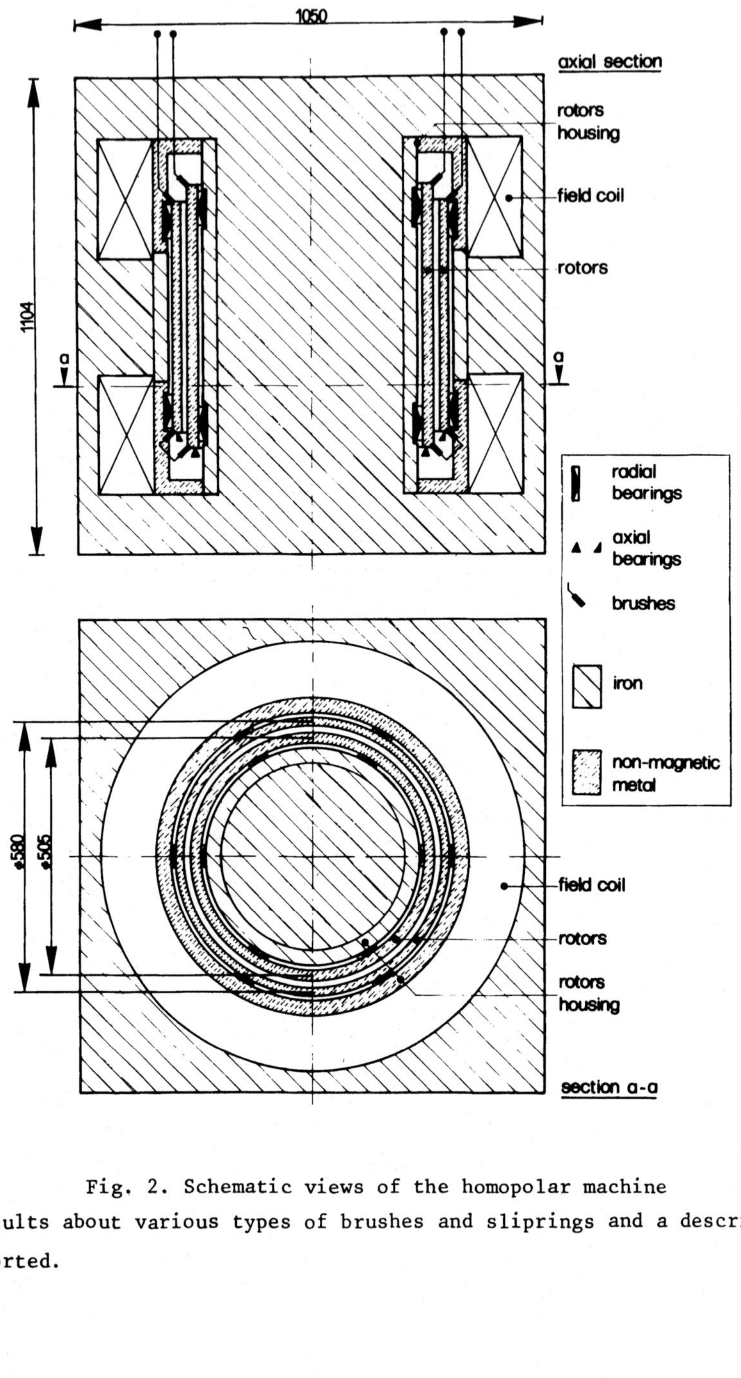

Fig. 2. Schematic views of the homopolar machine

2. THE HOMOPOLAR MACHINE

2.1 The magnet

Schematic views of the magnet are in Fig. 2. A computer-aided design of the magnetic circuit has been performed in order to maximize the e.m.f. of the machine with fixed magnet weight and exciting power. In such a study a standard code available at this Institute to analyze electrostatic and magnetostatic fields in linear non-homogeneous media has been employed. To simulate the ferromagnetic part, the magnet has been divided into several zones of different permeability: zones and permeabilities have been adjusted until the computed field enough satisfies the B-μ curve of the material.

The actual square shape of the magnet has been substituted, in order to simplify the computation, with a cylindrical equivalent, having the same iron area between the field coils and the outer perimeter. The magnetic flux path is reported in Fig. 3. The inner core is the most saturated zone, with a maximum field strength of more than 2T. Going radially towards the periphery, the iron is gradually less saturated until 0.8T. The gap-delimiting zones are not saturated (less than 1T) so that their permeability is high and most flux is driven towards the rotors. The flux cutting the rotors is 0.65 Wb - the 85% of the total produced flux - with an average density of about 0.6T and a maximum one of 0.85T. The design is such that one field line coincides with the collector surface in order to avoid field components perpendicular to the brush contact surface; and thus an e.m.f. parallel to the brush surface, that would make an undesired current circulate between the slip ring and the brush /¯1_7.

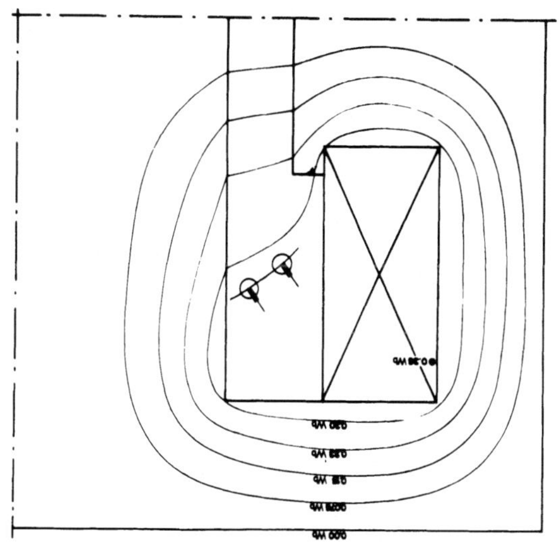

Fig. 3. Magnetic field plot

The magnet is divided in 24 horizontal sections. This solution permits the use of easy available pure iron slabs (whose square perimeter, to minimize scrap, suggested the external prismatic shape of the magnet) and makes easier machining, transportation, assembly and disassembly. The iron structure, weighing 5300 kg, constitutes itself a protection against eventual rotors failures.

Each field coil is divided in 12 independent subcoils made with water cooled hollow copper conductors. The magnetomotive force is 70.000 At/coil and the total excitation power is 35 kW.

2.2 Rotors assembly

The iron magnet acts as a rugged and stiff support structure for the rotating system of the machine (Fig. 2). To get easier maintenance and repair, bearing and brush-holders are mounted in a ferromagnetic shell which can be extracted from the cylindrical air gap of the magnet without dismantling it. The rotor material will be a high-strength aluminium alloy (Avional); although it offers lower kinetic energy densities and higher reluctance, it was preferred to the steel because it avoids magnetostatic forces due to centering errors between the magnet and the rotors. The radial force on a ferromagnetic ($\mu = \infty$) cylinder of radius R, length l and thickness t placed inside a cylindrical air gap of width g with a mean radial induction B_o and whose axis is displaced by an amount δ ($<<g$) from the magnetic axis of the system, is given by:

$$F = \frac{4\pi B_o^2 l R}{\mu_o} \frac{\delta}{(g-t)} \quad ; \qquad (1)$$

inserting the proper quantities, we obtain a radial force of about 10^5 N for a centering error of one millimeter: this figure was thought too severe for the selected radial bearings of the machine. The total mass of the two rotors can be determined from the maximum stored energy (2 MJ) and the related linear velocity (200 m/s): it is 100 kg. The mean length and mean radius of the rotors are fixed by the width of the cylindrical air gap of the magnet; their relative dimensions have been determined by the following conditions:

a) the torques on the two rotors must be equal and opposite, to avoid reaction torque on the magnet; this implies the equality of the intersected fluxes and fixes the relative length (l_{int} and l_{ext}) of the rotors;

b) the linear velocity of the rotors must be always the same: if the discharge of the machine begins with the rotors running at the same (maximum) linear speed, it is necessary that the linear accelerations be the same, or (neglecting the losses)

$$(R^2\ t\ l)_{int} = (R^2\ t\ l)_{ext} \tag{2}$$

The radial dimensions of the rotors assembly are indicated in Fig.2. The resulting maximum tensile stress on the rotors is 10.8 kg/mm^2, well below the fatigue limit of Avional. The chosen configuration requires to support the rotors on their cylindrical surfaces. The high relative speed between moving and static members prevents the use of rolling or hydrostatic oil bearings, whilst the presence of high magnetic fields in the air gap and the non-ferromagnetic nature of the rotor material don't allow the use of magnetic supports, therefore radial and axial pneumostatic air bearings have been chosen.

Axial bearings are only in the lower basis of the rotors, due to the absence of vertical forces beyond weight. The critical speed, kept low (100 rpm) by a spring mechanically in series with the axial bearings, is easily overtaken during starting, due to the strong magnetic damping of the axial vibrations. Radial bearings must be very stiff (radial electro magnetic and mechanical forces are large) and must follow the large centrifugal growth of the rotors (the radius, from null to maximum speed, increases of 0.4 mm, whereas the air bearing film is about 0.03 mm). To fulfill these requirements, each rotor is radially supported at both ends by six tilting pads (see Fig.2); their radial position is controlled by a hydrostatic system which guarantees the proper stiffness, keeping the first critical speed well over the maximum velocity (7000 rpm), and simultaneously moves in the radial direction the pads during the variations of the rotor radius. The internal rotor is supported on its internal surface, the external rotor on the external one; the air is supplied by an appropriate storage and fault protection plant.

2.3 The brush system

Liquid contacts have been rejected because of their cost, toxicity (mercury) and chemical aggressiveness (Na-K alloys) [2]; solid metal-graphite brushes have been chosen. In order to individuate the best brush-ring pair and to design a properly working brush system, able to cope with the most severe operating conditions (500 A in slow discharge mode, as in Fig. 1b, and 30.000 A in fast discharge mode as in Fig. 1a or in emergency braking) with low electromechanical losses and acceptable wear, experimental tests have been undertaken. A schematic diagram of the test system and a section of the brush holder are shown in Fig. 4. Compressed air is used to load the brush instead of metal springs; so, the critical vibrational frequency of the system is very low, the contact pressure does not depend on the brush length and the air flow through the holder contributes to cool the assembly. The slip ring material is sprayed on the periphery of a 320 mm - diameter rotating aluminium alloy disk. Different slip-ring materials and brush grades are under testing for continuous duty, at a maximum current density of about 1A/mm^2; a first set of results is reported in the following. Tests for pulse duty at higher current density will be soon carried out.

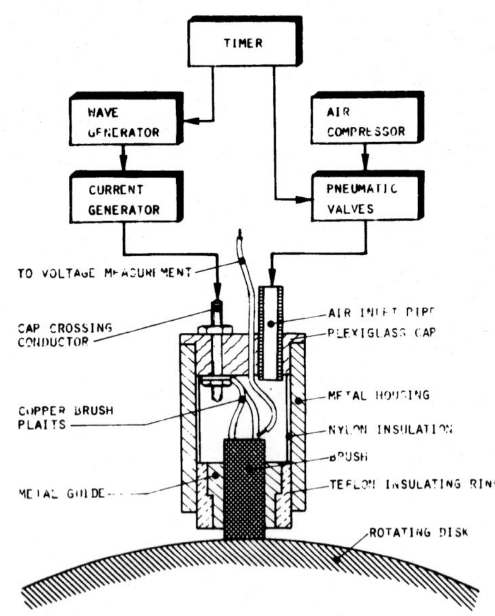

Fig.4. Brush test system and brush holder

2.4 Brush test results for continuous duty

Up to now, three metal-graphite grades have been tested: Morganite CM1S (82% metal), CM2

(82%), CM3H (75%). Brushes are 30 mm long, the contact surface is 12.5x6.3 mm. Metco Copper and Metco Monel have been already tested as ring materials. The tests have been run in air, with relative humidity varying from 30 to 50%. Surface speeds have been 100 and 200 m/s.

- Wear. The test results obtained by brush length measurements, are in table I; they are expressed by the nondimensional ratio between brush shortening and travelled distance. The measured wear results within the limits reported for similar experiments [3]. The copper ring wear, after running at 100 m/s for about 40 hours with two brushes of various types on a single track, various load pressures and currents has been 0.1 mm. Other alloys are expected to perform better.

TABLE I

Nondimensional wear rate x 10^9 (I = 75 A)

Brush grade	COPPER RING 100 m/s				MONEL RING 200 m/sec	
	0.1 MPa		0.15 MPa		0.15 MPa	
	Polarity		Polarity		Polarity	
	+	-	+	-	+	-
CM 3H	2.1	2.8	2.4	4.9	-	-
CM 1S	0.89	2.4	2.4	0.54	-	-
CM 2	0.26	0.46	0.83	1.3	0.83	0.83

- Voltage drop. The results of the voltage measurements, referred to a pair of brushes in series, are reported in table II; they are mean values, obtained by filtering the noisy voltage signal and by taking into account the scattering of the results of different tests in the same conditions.

- Friction coefficient. Its value has been obtained by power absorption measurements with brushes down and up; knowing brush pressure and speed and the power losses of the prime mover. Tests results are in table III.

TABLE II

Voltage drops [mV]

Brush Grade	0.1 MPa 25A	0.1 MPa 50A	0.1 MPa 75A	0.1 MPa 100A	0.15 MPa 25A	0.15 MPa 50A	0.15 MPa 75A	0.15 MPa 100A	Ring Material	Speed m/s
CM 3H	100	300	300	600	-	-	-	-	Copper	100
CM 1S	75	380	380	400	-	-	-	-		
CM 2	100	-	-	-	-	-	-	-		
CM2	450	900	1200	1300	250	500	-	-	Monel	100
	-	-	-	-		500	650	850		200

TABLE III

Friction coefficient

I = 0A, v = 100 m/s

Brush Grade	Copper Ring	Monel Ring
CM 3H	0.15	-
CM 1S	0.09	-
CM 2	0.12	0.10

3. THE DUAL CONVERTER

Fig. 1b shows the application of a homopolar machine as energy storage device with the purpose of smoothing discontinuous power demand of an a.c. load. The homopolar machine is connected to the a.c. grid through a dual converter apt to control the energy flow from the a.c. line to the machine and viceversa. Such a control is performed according to the result of the comparison between the istantaneous power across the measuring section of Fig. 1b and a reference signal whose value corresponds to a proper function of the same power.

In this kind of applications the quality of the power smoothing system depends strongly on the characteristics of the static converter. Moreover the latter has to operate with a d.c. load that is very similar to a capacitor and thus unusual enough for the converter applications. In order to study all these aspects a dual converter has been designed and realised. Various configurations for such converter were previously compared on the basis of economical and operational considerations and a converter with the scheme of Fig.5 has been finally chosen. It is obtained with two parallel independent dual converters, each of them built with a back-to-back connection of two thyristor bridges. The choice of a bridge (that enables a good utilization of the transformer and of the thyristors) instead of a mid-point configuration is allowed by the relatively high armature voltage of the homopolar. With the aim to limit the harmonic contents both of the a.c. line current and of the d.c. output voltage, the two dual converters are fed by 30°

shifted three phase voltages. The independent operation of the paired converters is assured by interphase reactors to which the function of limiting the amplitude of the harmonic output current is also demanded. A preliminary analysis has shown that the homopolar machine tolerates an armature current with large harmonic content and allows a mutual coupling between the reactors as illustrated in Fig. 5. This reduces the equivalent output inductance of the supply as necessary to obtain fast current variations.

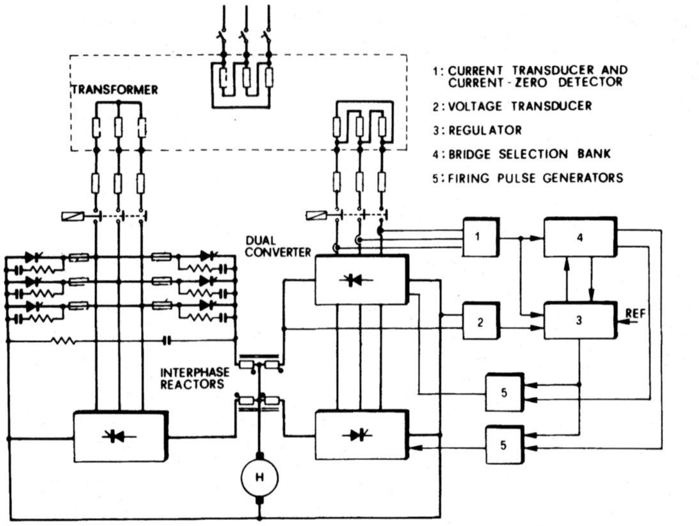

Fig.5. Dual converter scheme and control system block diagram

Different solutions for the control system can be adopted; one "without circulating current" /‾4_7 has been chosen at this stage. For testing purposes, the dual converter has been operated on several active and passive loads, also with the aim of optimizing the characteristics of its control circuit. Fig. 6 shows the time behaviour of the output current during one inversion (curve 1), measured connecting the converter with a small d.c. machine-induction motor [redacted] system, when a step variation of the reference voltage is applied to the converter control input (curve 2). Mainly because of the high value of the armature inductance of the employed d.c. machine and of the long "dead time" that is set in the "bridge selection bank", the current inversion requires some tens of milliseconds to be performed. However the low value of the inductance of the homopolar prototype and a corresponding adjustement of the control circuit would permit faster current changes and shorter inversion times. This is also confirmed by the first results of computer simulations of the whole system, performed with the aid of the SPICE program /‾5_7.

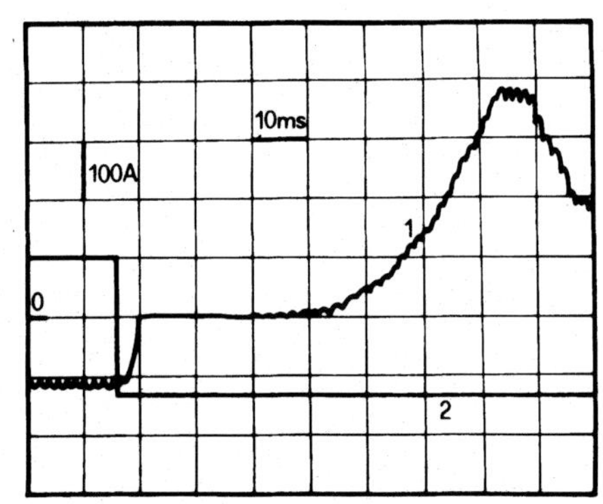

Fig. 6. Experimental results of a converter test

4. CONCLUSION

The design of a small-scale, flexible energy storage system has been described. Although the global plant is still under construction, some significant experimental results have been obtained from tests on brush-ring performances. Further investigation will pay particular attention to the operation of the bearing system and to the machine-grid interaction.

The authors wish to thank prof. G.Rostagni for his suggestions and supervision.

5. REFERENCES

/‾1_7 Electric Power Research Institute, U.S.A., EPRI ER-246 Project 469, Semiannual Report, 98-100, August, 1976.

/‾2_7 Marshall R.A., IEEE Trans., vol. PAS-85, n. 11, 1177-1188, Nov., 1966.

/‾3_7 McNab I.R., IEEE Trans., vol. CHMT-1, n. 1, 30-35, March, 1978.

/‾4_7 Pelly B.R., Thyristor Phase-Controlled Converters and Cycloconverters. Wiley, 1971, N.Y.

/‾5_7 Nagel L.W., Memorandum n. ERL-M520, 1975, Electr. Res. Lab.-Univ. of California, Berkeley.

INTERACTIONS BETWEEN THE LINE-CONNECTED STATIC CONVERTORS OF TEXTOR

B Giesen, F Petree*

Association EURATOM-KFA Jülich, 5170 Jülich, Federal Republic of Germany

*Ebasco Services, 2 World Trade Center, New York, N.Y.

ABSTRACT

This paper describes techniques used to predict and analyze the potentially harmful effects of interactions between three high-power static convertors connected to a high impedance distribution transformer, to weigh them against the benefits.

1. DESCRIPTION OF THE TEXTOR POWER SUPPLIES

The power distribution system and power convertors for the particle confinement field of the TEXTOR experiment are described in [1]. The three major loads, the toroidal and radial ("Vertical") confinement fields and the plasma current induction ("Heating") field, are fed direct current via three 12-pulse controlled self-commutating convertors and three rectifier transformers from a 20 kilovolt, three phase bus.

The overvoltage forcing factor of the power supply for the toroidal confinement field is 1.0, i.e. there is no overvoltage, and the rectifier remains phased fully forward throughout the experiment until the end, when it is inverted to commutate the load current into a "free-wheeling" diode circuit. On the other hand, the Poloidal power supplies, i.e. those for the Vertical and Heating fields, require an overvoltage forcing factor of more than 10.0, in order to raise or lower the current in .03 seconds, in circuits whose natural time constants are 0.5 - 0.8 seconds.

2. DESIGN CONSIDERATIONS

There is sufficient electrical power available in the vicinity of the TEXTOR experiment that no intermediate buffering by flywheel, batteries, or capacitor bank is necessary. Therefore the pulsed load of 75 MVA for ten seconds, every five minutes, is applied directly to the local 560 MVA grid, and the requirements of the utility company become a major factor in the design, along with the requirements of the experiment and the cost.

Distribution Transformer - A 40 MVA distribution transformer feeds power from a 140 MVA, 110 kV line to a 20 kV bus with 350 MVA short circuit capacity. The high impedance of this transformer (12.5 %) permits an economic design of the 20 kV bus, but causes the line voltage to sag 15 % during a pulse. The transformer is dedicated to the power convertors and neutral beam systems, although provision is made for connection to service loads too. Its 19 taps, which would normally compensate a variation of $\pm$ 16 % in the 110 kV source voltage, will instead be used to fine-tune the matching of the rectifiers to the net.

Rectifier Transformers - In contrast to the distribution transformer, the rectifier transformers have a low impedance (8 % of their pulsed ratings), so as to provide rapid commutation between the rectifier and bypass circuits. Six taps on each transformer, at steps of 8 - 16 %, permit tailoring each experiment to the net for maximum power factor.

Bypass - The poloidal ("Vertical and Heating") rectifiers must furnish high forcing voltages for a short time, but supply low voltages to maintain the currents during most of the pulse. The bypasses mitigate this condition of very low power factor somewhat, by providing free-wheeling paths for a portion of the load currents. They consist of thyristors connected in parallel with the outputs of the rectifier bridges. A computer simulation [3] shows that they will reduce the fundamental component of the current to the bridges by 37 %, while raising the harmonic content by 44 %.

Filter - The 2.5 MVA filter is designed to reduce the characteristic harmonic components of the current drawn from the mains by the rectifiers. It consists of three series resonant traps connected in parallel per phase, in a wye connection, tuned to the 11th, 13th, and 23rd harmonic numbers of 50 Hz. The rating of the "minimum cost" filter (by Kimbark [2]) would be about 1.1 MVA, but the assignment of reasonable "Q" values to the inductors determined the larger size. Aside from this, we make the MVA rating as small as possible, for economy but also to keep the current resonance - between its capacitance and the leakage inductance of the distribution transformer - at a frequency that is as high above the seventh harmonic as possible.

The high impedance distribution transformer would contribute excessively to the commutating impedance of the rectifiers, were it not for the filter, which reduces this contribution by inserting admittance paths in parallel. This is important for the proper functioning of the bypass circuits.

The filter also reduces the voltage distortion on the 20 kilovolt bus and 110 kilovolt mains. This is important for the proper operation of the rectifiers and other equipment connected to the 20 kilovolt bus, and for meeting the requirements of the utility in respect to voltage distortion at the mains.

Firing Control - The firing of the thyristors will be controlled by phase-locked loop circuits, synchronized to the 20 kilovolt bus voltage with a time constant of one second, more or less. These circuits accommodate to the lag in the 20 kilovolt bus in respect to the 110 kilovolt mains during the relatively slow (9 seconds) rise of toroidal current, but are assumed to be nonresponsive to the sudden leads and lags introduced by the rapid fluctuations in the poloidal currents, required by the experiment.

3. ADVANTAGES OF HIGH-IMPEDANCE SYSTEM

The high impedance of the power distribution system that will furnish current to the TEXTOR convertors provides a small amount of decoupling between the mains and the load that gives certain advantages in the design and operation of the machine.

The first advantage is that the short circuit current is reduced at the 20 kilovolt bus. This permits some economies in the component parts of this bus, but most important it is a major factor in preventing damage to the rectifier transformers in the event of a short circuit at their load side. This permits the rectifier transformers to be fabricated with minimum leakage reactance without causing extraordinary short circuit currents that requires extraordinary mechanical bracing. We feel that this enables a lower commutating impedance for the system which in turn makes the bypass circuits more effective in reducing the reactive power drawn from the mains circuits.

The next advantage profferred by the high impedance system involves the action of the poloidal convertors upon the toroidal convertor via the 20 kilovolt bus. In brief, as the poloidal currents build up they cause the bus voltage to sag, which in turn reduces the voltage applied to the toroidal field coils. This produces an "automatic flat-top" effect that, in principle, makes it unnecessary to phase back the toroidal rectifier for the flat-top. More exactly, the coupling effect permits - indeed, requires - a higher voltage connection on the toroidal fields that produces an overvoltage during the rise of the toroidal current. This overvoltage reduces the rise time of the toroidal current and the dissipation and temperature rise in the toroidal coils.

The distribution transformer is a standard unit for service in Germany, that is overloaded by some 75 % during each pulse. This may increase the probability of failure in service, and so it is important that it can be replaced quickly.

The final advantage of the high impedance system is that it reduces the MVA rating required for the filter to satisfy the requirements of the utility company for attenuation of the harmonic currents.

4. CONSEQUENCES OF HIGH IMPEDANCE SYSTEM

The three potentially harmful consequences of the high impedance system, namely, the large droop in the bus voltage, the rapid fluctuation in the phase of this voltage, and the large harmonic distortion of this voltage, need to be studied in some detail in order to take any appropriate measures to mitigate their effects. The magnitudes of these effects were estimated by simulating in models.

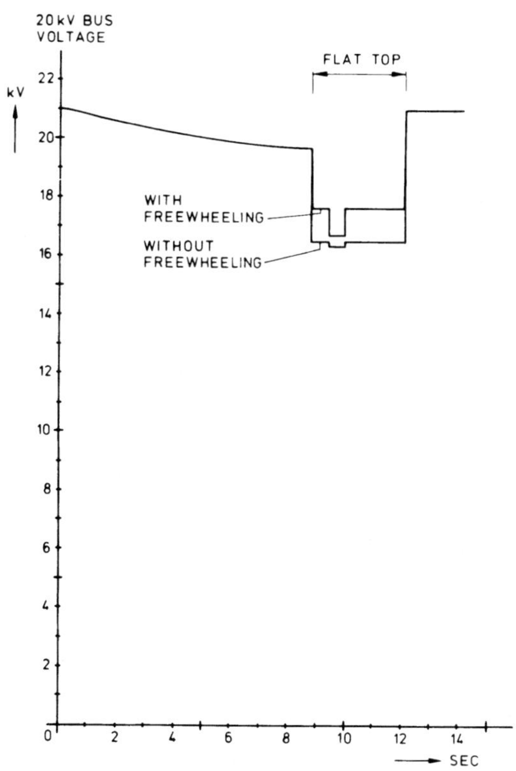

Fig. 1. Voltage fluctuation

The drop in bus voltage at the 20 kV bus is proportional to the reactive power drawn from the TEXTOR 110 kV-bus. Fig. 1 shows the 20 kV-bus voltage during an experiment for the case where no effort is made to reduce the reactive power, and when "freewheeling" diodes are used to reduce the reactive power, according to a computer model [3]. During the rise in toroidal current the reactive power rises to 10 MVAR. With no "freewheeling", when the poloidal currents come on, the reactive power increases by 48 MVAR, of which 38 MVAR is drawn by the poloidal supplies, and 10 MVAR is an increase in the toroidal reactive power due to the larger voltage drop at the 20 kV-bus. With "freewheeling", our computer simulation estimates that the reactive power will increase from 19 to 52 MVAR during the "flat top" interval. This means that the voltage from the distribution transformer drops about 16.25 %, from, say, 21 kV to 17.6 kV.

During changes in the poloidal current the "freewheeling" does not operate, so the reactive power is greater, causing the voltage at the 20 kV-bus to drop even more, to less than 17 kV. This increases the com-

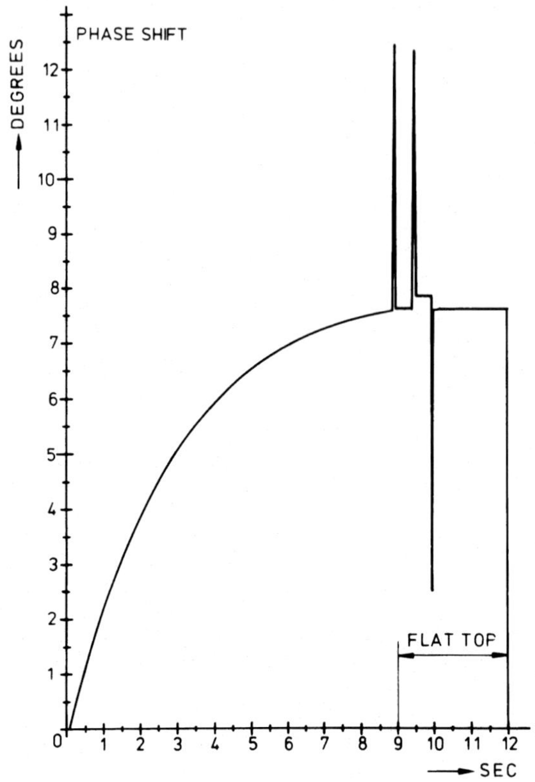

Fig. 2. Phase shift

mutating angle, and the possibility for a commutation failure, so that it becomes necessary to provide more margin against commutation failure in the inverted rectifier.

Fig. 2 shows how the voltage on the 20 kV-bus lags behind the mains voltage, by an angle proportional to the real power drawn from the mains. This real power is always positive, because the toroidal field draws some 36.5 MW, which is higher than any power returned by the poloidal supplies. If the firing generator were synchronized to the 110-kV mains, this would advance the firing angle in respect to the supply voltage seen by the rectifiers. Instead, we will synchronize to the 20 kV-bus with a slow (2 second) response, removing the effect of the slow power change in the toroidal power supply, but retaining the effects of the rapid power changes in the poloidal power supplies. The rapid fluctuation in the phase of the bus voltage affects operation at the two ends of the control range. Near the end where the control angle is zero, there is a possibility of trying to fire the thyristors while their anode voltage is too low for reliable turn-on, or is even negative. Near the opposite end, the danger of a failure to commutate is enhanced when the poloidal supplies return some net power to the mains, after an interval of operation at high β, because this is equivalent to retarding the firing angle.

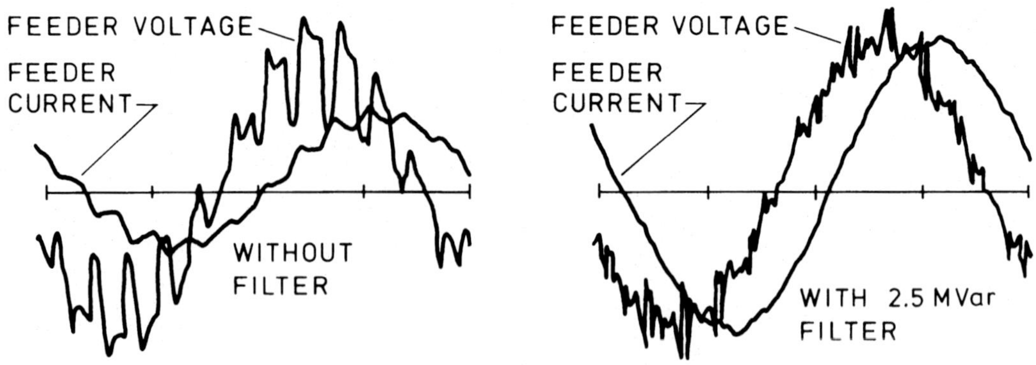

Fig. 3. Ripple voltage on 20 kV bus

Fig. 3 shows the large harmonic distortion (3 %) in the supply voltage at the 20 kV-bus [4]. It affects the synchronization of the phase-locked loops, and causes dips in the anode voltages of the thyristors that could cause some to extinguish, leaving other more tolerant thyristors overloaded.

The possibility of failure to fire or to remain conducting is eliminated by maintaining gate current in the thyristors throughout their 120 degree conduction angle. Thus, if a thyristor extinguishes owing to a dip in its anode voltage, it conducts again as soon as that voltage goes positive. The possibility that the bridge may fail to commutate properly is reduced by providing a limit stop in inversion that provides at least 20^0 extinction angle under normal operating conditions. In addition circuits will be provided that detect commutation failure and automatically fire the "freewheeling" thyristors so that they remove the load current from the bridge.

5. CONCLUSIONS

We feel that the advantages in lower cost and protection that are afforded by the high impedance power delivery system outweigh the several potential operational problems that must be accounted for due to the effects caused by the coupling that occurs at the mains. The extent that this is true depends upon the degree to which we succeed in anticipating and mitigating these effects.

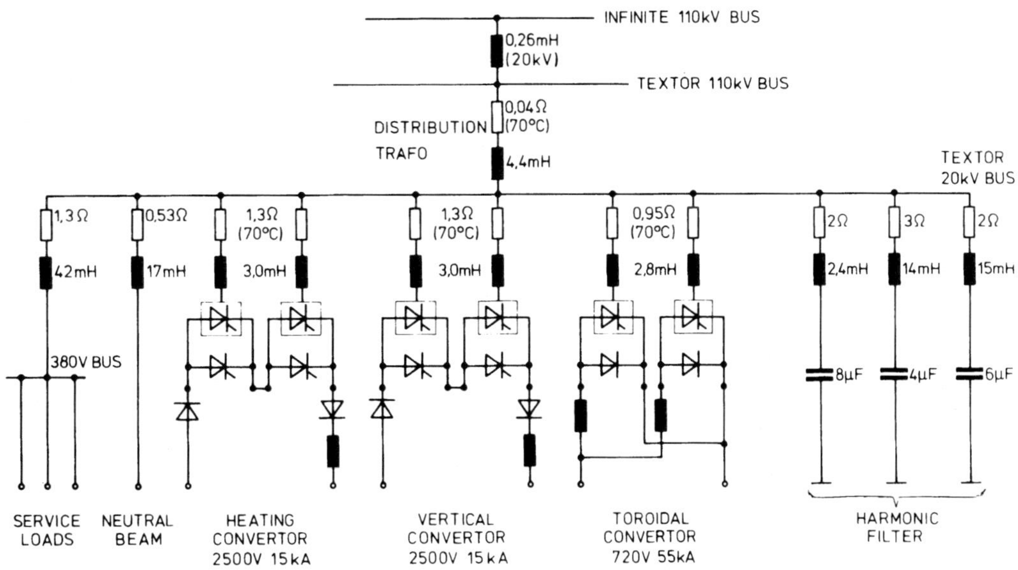

Fig. 4. Schematic diagram

6. REFERENCES

[1] GIESEN, B. and PETREE, F., The Energy Conversion and Poloidal System of TEXTOR, 8th Symposium on Engineering Problems of Fusion Research, San Francisco (1979)

[2] KIMBARK, E.W., Direct Current Transmission , Vol. 1; Chapter 8-8 (Harmonic Filters), Wiley-Interscience (1971)

[3] PETREE, F., Computer Simulation of the Effects of the Bypass Circuits in the Poloidal Power Supplies of TEXTOR, (unpublished report)

[4] BOUWKENGT, K., Harmonic Filter, Holec/Hazemeyer of Hengelo, The Netherlands (private unpublished communication)

THE GROUNDING SYSTEM OF TEXTOR

U. Schwarz C. Neumeyer

Institut für Plasmaphysik der Kernforschungs- EBASCO Services Inc., New York, USA
anlage Jülich GmbH

Association EURATOM-KFA

Abstract

The modular design of the TEXTOR grounding system is described. The resistive grounding of major components ist treated in detail. An insulation test procedure ist proposed by using the grounding system.

1. Introduction

The grounding system of the Tokamak TEXTOR meets two main requirements:

- it protects personnel working on the device against voltages > 65 V
- it keeps all voltages to ground as low as possible during operation.

The grounding system has to meet the safety standards, it must allow for easy detection of electrical faults.

2. Design of the grounding system

The grounding system has been split up into five single parts which are only connected once - at the central ground point of the device. This ground point is located near the core of the machine in the basement in order to make the connections as short as possible. It is connected solidly to the reinforcement of the concrete of the TEXTOR-pit by an open loop busbar of galvanized band steel. It is also connected to two buried steel water pipes having a length inside earth of 8 m. The ground resistance at this point is lower than 1 Ohm to remote earth.

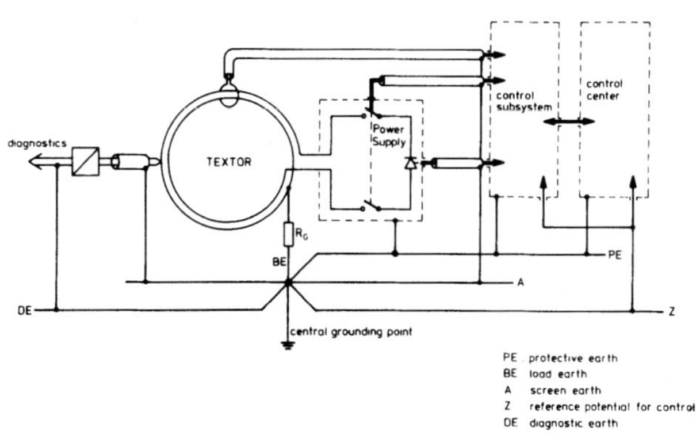

Fig.1 Grounding system of TEXTOR-schematic diagram

Fig. 1 shows schematically the five separate branches of the grounding system:

- protective earth (PE); all metal shields, casings of equipment and pipes of the water cooling system are connected to PE
- load earth (BE). All electrical power systems (mainly coil systems) of the device and structure elements inside the machine as vacuum vessel, liner, coil cases, base plate, support pillars and transformer iron core are connected to BE via suitable resistances.
- screen earth (A). It connects all cable screens - mainly those of the control cables - in such a way that no ground loops are formed
- reference earth (Z). The Z-system determines the reference potential for all control signals.

It is essential in operating the machine, that Z does not carry any transient current. It therefore has its own neutral point at the control center. Fig. 2 shows schematically that the Z-leads to peripheral control units are installed insulated and that Z is solidly grounded at the neutral point at the control center.

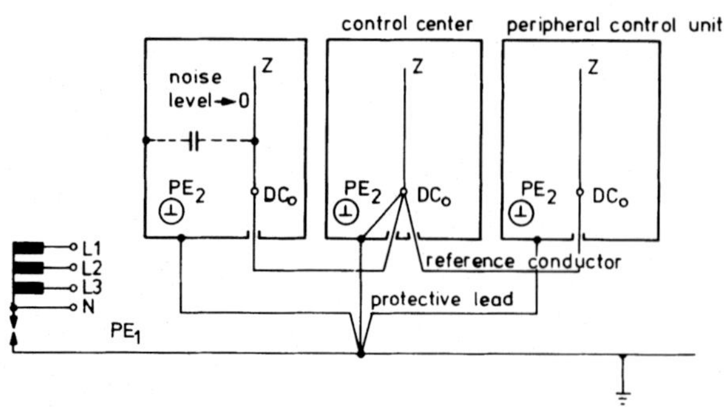

Diagnostic earth (DE). A separate grounding branch is foreseen for all diagnostic equipment. This helps reduce noise on the diagnostic leads induced by transient currents on the load earth. Also it will help in detect grounding faults .

Fig.2 Grounding of TEXTOR control system

3. Protective earth

German safety standards requirethat devices with electrical equipment be installed in such a way that personnel are protected against voltages > 65 V. For TEXTOR the protection system "Nullung" was chosen. This requires a special protective lead to which all metallic casings, pipes etc. have to be solidly connected. The ohmic resistance of this lead must be lon enough that a ground fault induces the switch off current in the defective system. This is solved by installing a copper bus-bar 4o mm x 4 mm with an ohmic resistance of about 0,1 mOhm/m separate from the electrical cables. It has only one connection to the central ground point and to the single low voltage neutrals of the three distribution transformers, which furnish the low voltage power for TEXTOR. Bore holes at every 1/2 m on the copper bus are foreseen for connecting the ground leads of the single apparatus. By the prescription that each apparatus to be grounded has to be connected directly to this bus and by the separate installed protective bus itself, ground faults and ground loops can be (hopefully) avoided.

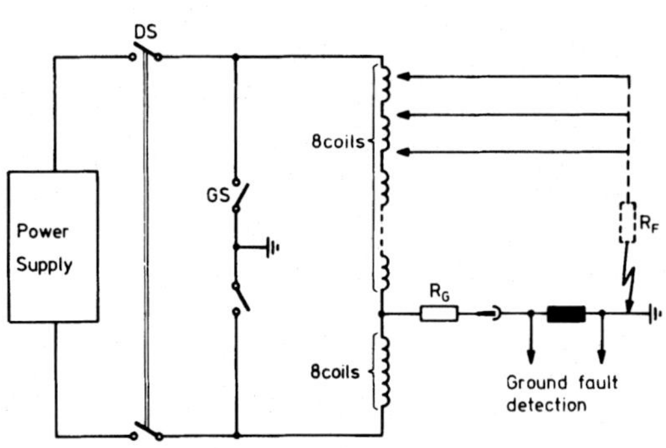

Fig.3 Grounding of Toroidal field system

Further protective devices are the grounding switches (GS) of the single coil systems. Fig. 3 shows schematically the toroidal field system of TEXTOR. A disconnect which (DS) and a grounding switch (GS) are inserted between power supply and coil. GS has to shorten the open coil terminal and to put them to ground. It can only be actuated if DS is opened. DS on the other hand can only be actuated at current zero and if GS is opened. These interlocks make sure that neither DS nor GS has to switch any current and that the power supply cannot be switched on if the coils are shorted by GS. The doors to the TEXTOR device are also interlocked with all grounding switches. They can be opened only if all grounding switches are closed. Thus it is assured that personnel working inside the machine cannot touch an open coil terminal.

4. Load earth (BE)

BE determines all electrical potentials of TEXTOR under load. It keeps all voltages to ground as low as possible. At normal load conditions BE does not carry any current other than unavoidable transient currents of short duration as for instance asymmetrical capacitive currents. It is important that all the grounded coil systems be decoupled against each other, and the ground potential should not rise higher than 65 V at a ground fault.

These requirements are met by resistive grounding of the single components of TEXTOR. The grounding resistors decouple the coil systems from each other and they limit fault currents, this limiting the ground potential rise. All grounding paths of the coil systems are equippt with a ground fault detection device which generates a signal to switch off the defective system if the current to ground exceeds a certain value. Some of the grounding paths are equipped with blocked plug connections which will be supervised by the safety system.

4.1 <u>The grounding resistance of the toroidal field system</u>
Fig. 3 shows the ground connection at the centerpoint of the load midway between the 16 TF-coils. This establishes a symmetrical distribution of potential to ground. By this method voltages against ground are kept low, but it requires floating terminals on the power supply. Further, three possible ground fault are shown. The worst case fault occurs at a power supply terminal - 8 coils are shunted by the fault at the start of a pulse.

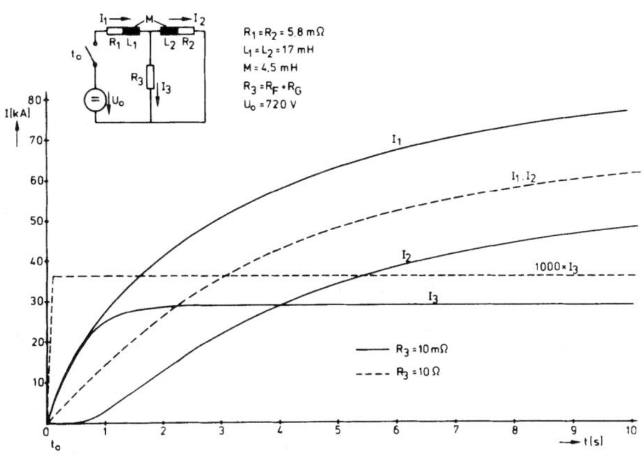

Fig.4 Toroidal Field currents - worst case ground fault

Fig. 4 shows the equivalent circuit diagram for these conditions and the current shapes at a forcing voltage of 72o V for two values of R_G assuming $R_F \ll R_G$. With R_G = 1o mOhm the fault current rises to about 3o kA within 2 s; the current I_1 exceeds 8o kA. With R_G = 1o Ohm the fault current rises much faster but is limited to 36 A. The current I_1 does not change significanlty compared with normal load conditions. The results of Fig. 4 indicate that a grounding resistance on the order of 1o Ohm ist appropriate. With the ground puls fault resistance much greater than the coil resistance (1000 : 1), the voltage distribution across the load remains linear despite the ground fault path which shunts a portion of it. The circuit may be viewed as a DC-bridge with the voltage applied to the fault branch as a function of fault location Fig. 5.

The dynamic behaviour of the fault current may be neglected. The maximum power dissipation in the fault will be 3,3 kW for this case. With a switch off time of 1oo ms for the defective system the energy deposited to the fault will be lower than 1 kWs. Fig. 5 shows further the decreasing fault current with increasing fault resist nce: a high resistance fault near the grounding point is difficult to detect.

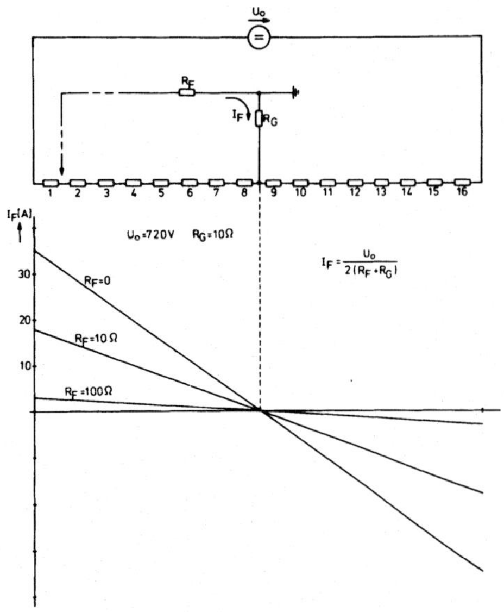

The ground fault detection system will detect a fault current down to 1oo mA; thus fault resistances ⩽ 3.6 KOhm at a forcing voltage of 72o V will be detectable. The unavoidable transient currents will not exceed a few Amperes.

Fig.5 Toroidal Field ground fault current as a function of fault location and fault resistance R_F

4.2 Grounding of the poloidal field coils

The grounding method for the poloidal field coils is the same as for the toroidal field coils, but three different voltages have to be considered:
power supply voltage, initiation voltage (2oo V/turn), disruption voltage (1.5 kV/turn). The grounding resistance of a poloidal field coil has to meet two opposite requirements. It must limit the current in the case of a ground fault driven by plasma disruption voltage, and it must allow sufficient current flow in the case of ground fault driven by power supply voltage; this is required for safe detection of the fault current.

With the chosen resistances of 5o Ohms for the magnetizing, heating and vertical field and of 1o Ohms for the correction fields, the worst case ground fault currents driven by a disruption voltage are on the order of 1oo's of Amperes. Thus the ground potential rise my exceed 65 V in this case. However, the low probability of this worst case event and the very short duration of the voltage (about 1 ms) do not warrant the use of higher ground resistances.

The self-inductances of the poloidal systems and the chosen grounding resistances give fault current time constants on the order of o.1 ms. The ratio of grounding resistance to coil resistance will be very large (1o,ooo : 1). It is there fore concluded that a fault current may reach its steady state value even within the time scale of plasma disruption and that faults do not significantly alter the linear distribution of voltage across load.

4.3 Grounding of toroidal loops

Inside the tokamak structure elements as coil cases or base plate pieces must be at ground potential. Direct grounding of these elements would create a toroidal loop which is coupled to the plasma and to the poloidal field coils, thus producing disturbing fields. Also the return yokes of the iron transformer core should not be enclosed by a ground connection.

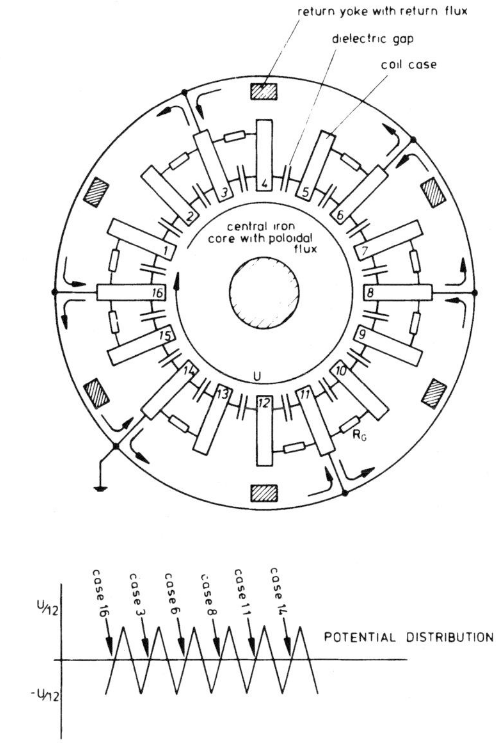

Fig.6 Grounding of Toroidal Field coil cases

Fig. 6 shows the grounding of the toroidal field coil cases which are insulated from each other by glass fiber epoxy plates. Two or three neighbouring cases are connected by a resistors of o.5 Ohm each, to create 6 isolated groups of cases. These groups are connected to ground via a ring bus outside the transformer return yokes: No return yoke is fully enclosed by a ground connection.

The voltage induced in the central "loop" is equally distributed among the 6 dielectic gaps. As each is tested for a voltage of at least 1.5 kV DC, the safety factor is comfortable, even for the case of a plasma disruption.

Under normal load conditions no significant current will flow, as all loops enclosing net flux involve a series dielectric gap. If a gap does fail the fault current is limited by the grounding resistances on a maximum value of 25o A. The energy deposited to the fault is due to the short duration of the voltage negligible. In this case the ground wire configuration shown in Fig. 6 may be rearranged to make use of one dielectric gap which was previously shorted by a case to case connection.

5. Insulation faults and tests

The design of all TEXTOR coils meets the requirement that an external flashover or breakdown should lead to a short circuit to ground.

This is important for two reasons:

- such an electrical fault will be detected and switched off by the ground fault detection systems.
- overvoltages will be led off to ground and will not propagate in an uncontrolled way.

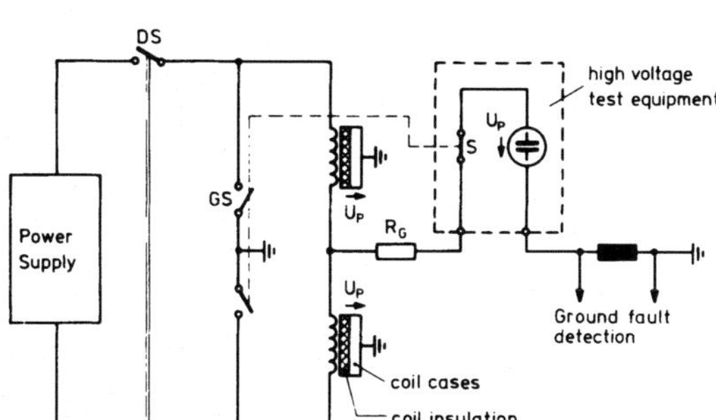

Fig.7 Insulation test arrangement for Toroidal Field coils

Therefore the toroidal field coils are surrounded by grounded cases and all poloidal field coils have on their surface a conducting paint at ground potential. This can be used for insulation tests of the single coil systems. The load earth BE of the system to be tested is disconnected by means of the plug connection. The test equipment, a high voltage source with the test voltage U_p and the switch S, ist inserted in this point. For the test switch GS must be open. The high voltage is applied across the insulation between the coil and the coil case. In case of an insulation failure the high voltage source will be switched off by the ground fault detection system. The high voltage source will have a safety system which shuts down the source if a certain current value is exceeded. As the coil systems can be tested separately, the output voltage of the test equipment can be adjusted to the different insulation levels of the single coil systems.

Conclusion

The grounding system of TEXTOR meets the different requirements by a modular design. Because of safety standards all metallic casings, pipes etc. are connected separately to a special protective ground bus. The load ground system ist realized by resistive grounding of the single systems as coils, vacuum vessel, transformator iron core, coil cases, base plate and support structure. Thus voltages under load are kept low, systems are decoupled against each other, and fault currents are limited. The grounding system can also be used for insulation tests of the mass insulation of the main components.

A STORAGE COIL ADAPTED TO A POWER PULSE SYSTEM

R. BAGET - B. THOMAZO - F. RIOUX-DAMIDAU
Laboratoire d'Electrotechnique
Université de PARIS SUD - Batiment 214 - 91405 ORSAY - France

ABSTRACT

This paper deals with a new structure of storage coil which is particulary adapted to a power pulse system. Authors present the results of a numerical study which allows to precisely calculate the better dimensions of such a coil.

1. INTRODUCTION

For some high power plasma experiments, such as imploding liners [1], a very high energy pulse has to be delivered (>> 10^6 J) in a very short time (a few μs).

In order to realize these very powerful sources, inductive storage technics [2] can be used.

These systems are easy, have high energy densities and a cost-price per Joule, lower than all other system (such as condenser banks for example).

The studied system requires three component parts :

a) an iron free asynchronous machine [3] [4] which can produce an electric power pulse in 0.1 to 1 s.

b) a coil to transiently store this energy.

c) a rapid transfer device to extract the energy out of the storage coil in a very short time (a few μs) with a high efficiency.

The purpose of this paper is to present these different parts and especially the storage coil : its structure and the results of its numerical study.

2. THE GENERATOR

That is an iron-free asynchronous machine. Its rotor is made of a solid disk-shaped metallic fly-wheel placed between two windings that have no magnetic iron structure. These windings are flat, ring-shaped and have symmetric positions with respect to the rotor.

Kinetic energy is firstly stored up in the fly-wheel ; then, the machine works as a generator and transforms that energy into electrical energy.

That system with a high magnetic induction permits

- a high energy density
- a short braking of the rotor (0.1 to 1s)
- a greater energy storage than the other systems of equal size.

The stator outputs are connected to condensers in order to balance the reactive energy of the generator and to a rectifying device which permits the transfer of the energy onto the coil.

3. THE RAPID TRANSFER DEVICE

That system can be allowed to extract the magnetic energy from the storage coil in a very short time to deliver it on to the load.

A theoretical study has shown the interest of a device compounded with two parts (fig. 1) : a mechanical switch and an absorber connected in parallel [5]. The first one carries the current during a long time (0.1 to 1 s) and next, transfers it onto the absorber which afterwards explodes, generating a high voltage. So, the energy is transferred onto the load.

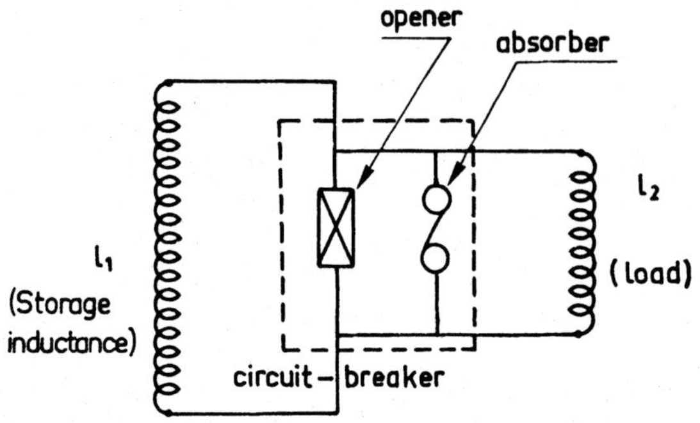

Fig. 1 : Scheme of the complete circuit

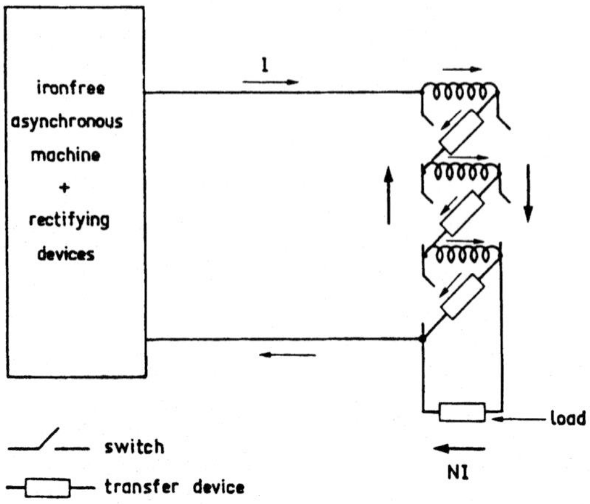

Fig.2 : Electric scheme

4. COIL

During the recovery of the magnetic energy, the current which increases in the load must be higher than the current delivered by the generator.

So, we have used a series into parallel transformation. The operation can be explained by the electric scheme of fig. 2. When the current is increasing in the coil, the turns are connected in series (switches k are opened).

At $t = \tau_1$ - the current reaches its maximum (fig.3). The switching of k connects each turn of the coil in short-circuit, keeping energy in the coil.

At $t = \tau_2$ - the transfer devices are triggered:

- the turns of the coils are connected in parallel.
- the generated voltage transfers the magnetic energy onto the load.
- by the series into parallel transformation the current increasing in the load has been multiplied by the number of turns.

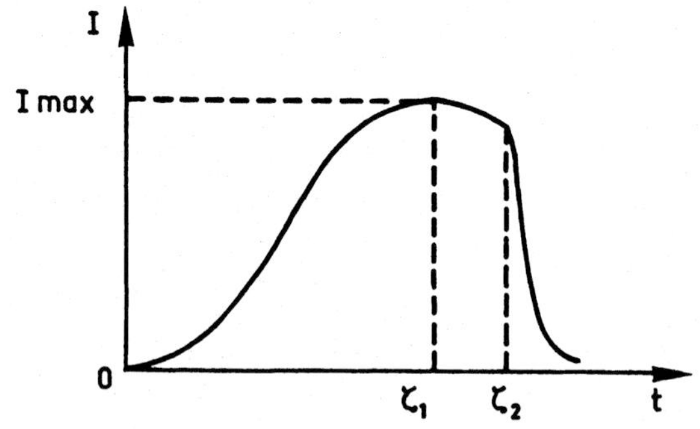

Fig. 3: Form of current in a single turn

4.1. Structure

The adopted coil form is shown on fig. 4 and 5 . Each turn presents the same spiral form. The turns are regularly wired around the central axis.

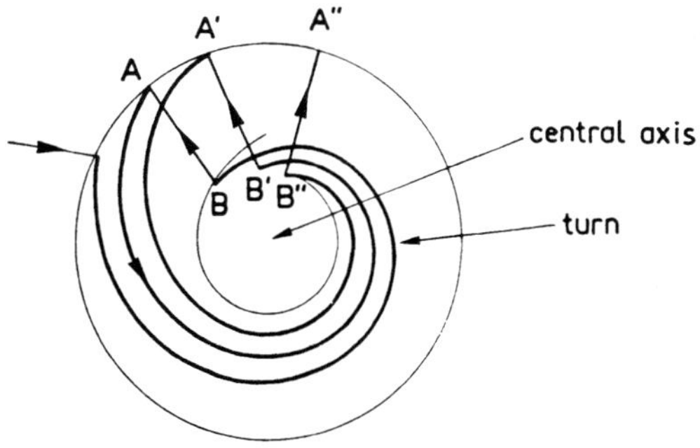

Fig.4

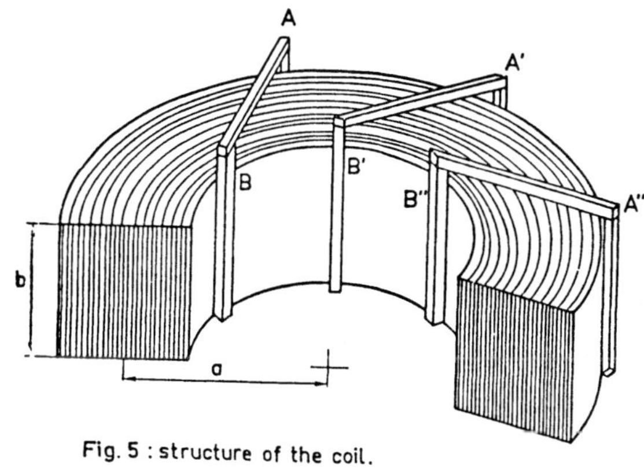

Fig. 5 : structure of the coil.

This structure permits to connect turns in series (joining points A and B, A' and B', A" and B") or in parallel (connecting points A A' A" and B B' B"). The transfer devices are placed between A and B, A' and B' etc.. and permit this transformation.

4.2. Numérical study

The storage coil must satisfy the following conditions :

- it must be adapted to the asynchronous machine.
- it must withstand an important magnetic pressure.
- it must permit a rapid recovery of the energy with a high efficiency.

Its losses may be separated as follows :

- Joule losses especially when the current is increasing in the coil (so it must have a long time constant).
- Magnetic losses during the recovery (they represent approximatively the energy of the magnetic field which is in the conductor just before the recovery).

From equations of the asynchronous iron-free machine [6] a first numerical calculation allows us to have the optimum electrical characteristics of an adapted coil. Fig. 6 shows that for one fixed time-constant there is a maximum of the energy for one inductance value.

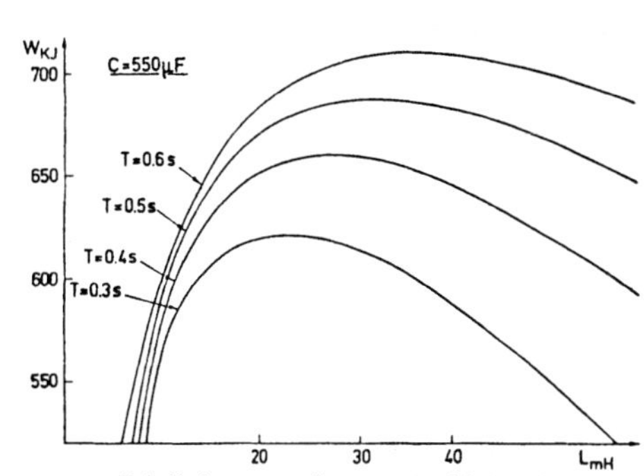

Fig6: Maximum magnetic energy stored in the coil.

To optimize the size of such a coil, the knowledge of the magnetic field chart outside and inside the coil is necessary.

The coil symmetry reduces our problem to a two dimensional resolution of the Maxwell's equations ; the partial differential equations are the following :

$$\frac{\partial^2 A}{\partial x^2} + \frac{1}{y}\frac{\partial}{\partial y}\left(y \frac{\partial A}{\partial r} \right) - \frac{A}{y^2} = - \mu_0 J$$

(the vector potential A has only one component).

Our programm is based on ADI method [7] and allows us to precisely compute A in all nodes of a rectangular mesh.

From vector potential values we can calculate :

- Inductances (total : L_T, external : L_e , internal : L_i $\quad L_T = L_e + L_i$)
- Magnetic pressures
- Efficiency of the magnetic recovery.

The results prove the interest of square cross-section coils (these coils being the more efficient for one fixed copper mass). Fig. 7 shows the variation of inductance time constant, efficiency and copper mass as a function of the two dimensions a and b of a square cross section coil.

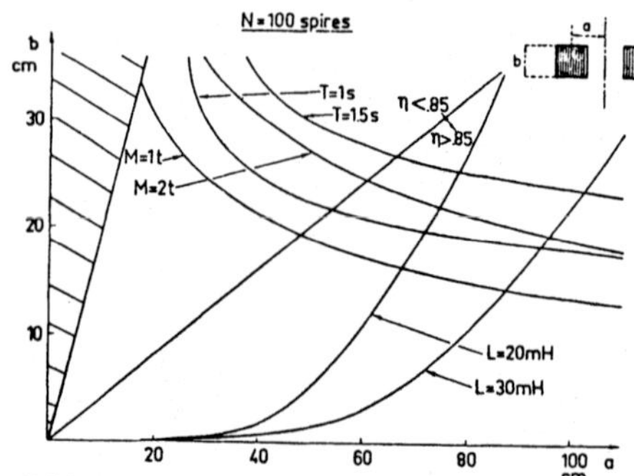

Fig7:Square cross section coil characteristics in function of its dimensions.

The fig. 6 and 7 are useful to precisely determine a coil adapted to the asynchronous iron-free machine built in our laboratory. The typical dimensions and characteristics of such a coil are :

- L_T = 30 mH T = 1s, magnetic efficiency : 80 %
- a = 0.1 m b = 0.02 m
- M = 2000 kg (copper mass)
- Number of turns = 100

5. CONCLUSION

The present study allows us to optimize the electrical characteristics, then the size, of a coil specially adapted to an asynchronous iron-free machine. These results show that the coil structure is particularly interesting for realizing high energy pulse systems.

6. REFERENCES

[1] C. RIOUX, C. JABLON, Nuclear Fusion, 15, 425 (1975)

[2] R. CARRUTHERS, Proc. of the International Conference on High Magnetic Fields MIT. Nov., 1-4, 307 (1961).

[3] C. RIOUX, F. SULTANEM, A solid rotor iron-free asynchronous machine, J.Appl. Phys. 50 (2) Feb. (1979).

[4] C. RIOUX, F. SULTANEM, Iron-free solid disk-rotor asynchronous machines, J. Appl.Phys. 50 (12) Dec. (1979).

[5] R. BAGET, F. RIOUX-DAMIDAU and R. GUILLET, A fast circuit-breaker driven by magnetic pressure, Fusion Technology (1978).

[6] F. SULTANEM, Thèse de Docteur Ingénieur, Université de Paris VI, (1977).

[7] B. THOMAZO, Thèse de 3ème cycle, Université de Paris VI (à paraître).

LARGE FLYWHEEL GENERATOR DIODE CONVERTOR FOR THE JET TOKAMAK

M Huart (EUR-EB)
JET Joint Undertaking, Abingdon, Oxfordshire, UK

E Beedham
GEC Machines Ltd, Rugby, Warwickshire, UK

ABSTRACT

Two large vertical shaft generators with integrated flywheel energy storage and output diode rectifiers will deliver DC pulses of current up to 100kA and energy up to 2600 MJ to the toroidal and poloidal field coils of the JET Tokamak.

Although a straight development from the pump storage hydro-generator, the pulsed operation of the generators (100,000 pulses), the large energy stored in the integrated flywheel (3.75 GJ) and the rapid response required from the voltage feedback loop called for careful investigation and development tests during the design phase.

Two years after the release of the Contract, a review is made here of some of the most important points of the design, now completed, as well as of the status of the manufacture. This paper is the result of a joint collaboration between the Manufacturer (GEC) and the JET team.

1. INTRODUCTION

To release the 400kV network from the high peak power, high power derivative and large reactive power associated with the supply of DC current pulses to the poloidal and toroidal field coils of the JET Tokamak, two dedicated flywheel-generator-convertors (FGC) have been adopted in the JET's power supply combined scheme [1]. To reduce the design, tool and spare costs, the poloidal and toroidal field FGC were specified identical. The energy delivered to the load during a pulse is stored kinetically in the rotor rim of the generator, acting as an integrated flywheel, during a 9 minute acceleration from idle speed (112.5 rpm) to maximum speed (225 rpm). Constant torque acceleration is obtained from a 8.8 MW, 26 poles asynchronous motor sited on top of the vertical shaft generator.

Diode rectifiers had been specified for the output convertors. They are arranged around the generator as four individual half bridges and directly connected to the generator output terminals to form two series connected full bridges.

It implied that the output voltage/current profile had to be controlled by the exciter. Typical output DC voltage and current profiles for the poloidal and toroidal field systems were specified as a contractual performance requirement.

It was also a contractual requirement that the FGC should withstand repeated number of short circuits following the closure of the crowbar switch included in both circuits to protect the coils from the effect of internal faults.

The contract was released in three stages. Stage I, released in May 1978, was concerned with the detailed design of two identical FGC. Stages II and III, released in May 1979, were concerned with the manufacture, delivery, installation and commissioning of respectively the poloidal and toroidal field FGC, and will take respectively 33 and 36 months to complete. Assembly of the first generator on site will start in November 1980.

2. MECHANICAL DESIGN & CONSTRUCTION OF THE GENERATOR

Originally a high speed, horizontal shaft machine with separate flywheels was considered but for the major size of load involved it became clear that the vertical shaft machine was the right type to use. The main considerations which led to this decision were as follows. Firstly, the high speed machine would have to run over a speed range lying between the first and second critical speeds. It would therefore have a restricted operational range and would have to run through the first critical speed during start-up and shut-down. Compared with this, the low speed machine would work entirely below the first critical speed; the speed range would therefore be determined solely by load requirements. Secondly, since the high speed machine would have separate flywheels, there would be a risk of torsional oscillations with the possibility of high shaft stresses. Compared with this, the low speed machine would have its inertia effect built into the rotor, so the risk of torsional oscillations would be minimal. Lastly, the high speed machine would use a highly stressed solid rotor in which a small material flaw could lead to disaster. The low speed machine employs a laminated rim construction using moderately stressed materials, it will thus be less at risk due to material defects.

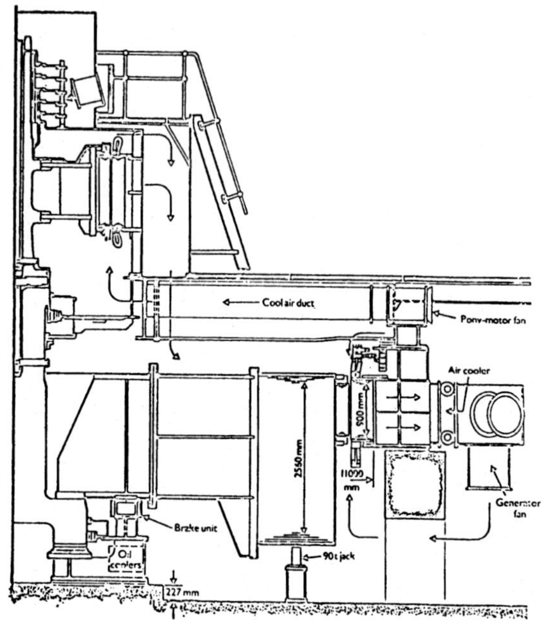

Fig 1. Cross Section of the Generator

Fig 1 shows a cross section of the vertical shaft generator and driving motor adopted. Principal dimensions of the generator, weights of major components and the total inertia are given in Table 1.

The speed ratio for a full energy pulse is 2:1. This range permits recovery of a high percentage of stored energy for pulsing (75%) but is narrow enough to be acceptable for electrical design purposes.

A design life of 100,000 pulses at the full energy rating per pulse is specified. This is the criterion against which the mechanical design of the rotor was checked.

Extensive analyses complemented by laboratory tests, were made during the design state to assess the integrity of the pole attachment and the friction rim under fatigue condition as well as that of the stator end windings under short circuit condition.

TABLE 1

Main Characteristics of the Vertical Shaft Generator

Dimensions	-	Stator core, outside diameter	-	11000 mm
		Stator core, bore	-	10300 mm
		Stator core, axial length	-	900 mm
		Rotor rim, length	-	2560 mm
Weights	-	Stator, total	-	106 t
		Rotor, comprising - wound poles	-	36 t
		rim	-	647 t
		spider	-	50 t
		shaft	-	41 t
		Rotor, total	-	774 t
		Bearing brackets, total	-	24 t

Total Moment of Inertial - 13.5×10^6 kg.m^2

The poles are secured by two T-heads located in T-slots in the rotor rim, using pairs of tapered keys to pull the pole down on to the rim. The fatigue strength of the pole fixing was determined by laboratory tests carried out on Tee-head samples of laminations and endplate material with the same geometry as that adopted in the design. Using the calculated stresses, the factors of safety based on a fatigue life of 10^5 rated cycles are estimated to be - laminations - 4.2

endplates - 4.0

Assuming the results follow a normal distribution curve, the probability of failure of either the pole endplate or the laminations was found to be well below 10^{-5}.

The rotor rim is built up of overlapping layers of medium tensile steel segments. Each segment is two pole pitches wide. The rim is clamped axially using a series of bolts through the rim itself. It is designed as a friction rim based on the axial pressure set up by the through bolts. Calculation of stresses based on this approach with the machine at 225 rpm gives factors of safety based on 0.1% proof stress of the material as follows:

Tangential stress at rim bore under poles - 3.6

Maximum stress under pole endplate - 6.6

A second method of analysis taking into account the weakening effect of joint lines and bolt holes in the rim gives a factor of safety for the maximum hoop stress of 2.6.

Fatigue tests were carried out on specimens containing a hole similar to those in the rim in order to find the stress level for a life of 10^5 cycles. These tests, performed on randomly picked rim plates, revealed a very low standard deviation, indicating a very even quality of material. This was confirmed as well by the tensile tests. Using the maximum rim stress (at rim bore) obtained by the second method of analysis, the safety factor on the mean fatigue value is 1.6. Nevertheless, assuming the results follow a normal distribution curve, the probability of failure of one rim lamination was found to be well below 10^{-5}.

A computer analysis was carried out to determine the distribution of forces on each end-winding bar and the stress and deflections which result from them. Maximum deflections were found to be small, being only 0.1 mm on a 100 kA pulse and 0.7 mm on a convertor short circuit. In both cases the maximum deflections were at the noses of the coils. Corresponding maximum stresses are 6.0 N/mm^2 and 40.1 N/mm^2. Load tests performed on similar stator bars, gave a limiting stress of 173 N/mm^2, so that the safety factors are 28.8 and 4.3 respectively.

3. ELECTRICAL DESIGN OF GENERATOR

The maximum output of the generator convertor is defined by straight lines joining the following sets of DC load ratings, with the generator running at 225 rpm.

Voltage kV	9	9	6	3	0.9	0
Current kA	0	30	67	80	85	100

This rating profile is, in effect, the saturation limit for the generator. The nominal AC rating of the generator corresponds to a DC output of 6kV, 67kA and is 4.71kV, 50.25kA, 409.8 MVA, 0.98 pf, 48 pole, 90 Hz, 225 rpm.

The generator has two separate three phase star connected stator windings each of which is connected to a three phase diode bridge rectifier. The rectifier outputs are connected in series, the link between the rectifiers being earthed.

Each stator winding has six parallel circuits evenly spaced round the machine and connected to a set of circular busrings at the top of the machine. The windings are arranged so that succeeding parallel circuits are connected into the two windings alternately.

Machine parameters, particularly inductances and time constants, affect the system's ability to cope with the required transient load changes. The important parameters have therefore been guaranteed and toleranced. Values of the major components are:

X_d	- 3.25 pu	T'_{do}	- 2.86 seconds
X'_d	- 1.14 pu	T'_d	- 0.99 seconds
X''_d	- 0.83 pu	T''_d	- 0.034 seconds

Tolerances vary, ranging from $\pm$ 7½% for X_d to $\pm$ 35% for T''_d. For performance prediction purposes a compatible set of parameters is used, taking tolerances into account, which gives the most pessimistic results. For short circuit calculations, minimum reactances and maximum time constants are used whereas for pulse load performances, minimum mutual reactances with maximum leakage reactances and time constants are used.

Two methods of computer simulation were developed to check that the generator and exciter met the transient performance requirements. In the simpler method, the convertor was assumed to behave according to the classical steady-state theory so that input AC values could be related to output DC values in terms of machine reactance and load resistance. The generator was represented by the usual two axis model. In the more detailed method, the commutation of the diodes was determined solely according to prevailing voltage and current conditions; no approximate formulae were used for AC/DC conversions. The output of the detailed program was available as instantaneous or averaged values. When applied to the P pulse profile, the results from

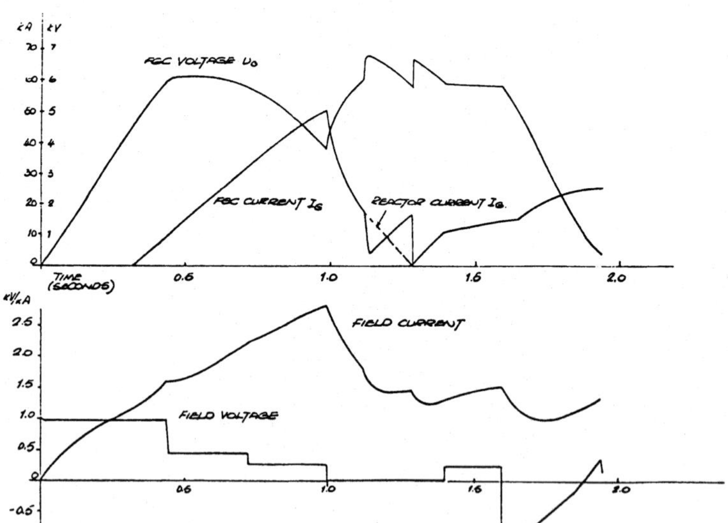

Fig 2 – P1 PROFILE – WP MACHINE STARTING AT 90Hz JET PROJECT 1114660

the two programs are in good agreement. The P profile and corresponding excitation requirements for the first two seconds, as predicted by the detailed method are shown in figure 2.

The computer programmes were also used to assess the peak fault current at various points of the load circuit and to assess the energy dissipated during various pulses in the generator windings. Some of the results obtained are:

Peak fault current:

Short circuit across convertor - 330kA

Short circuit across T system load point - 257kA (with protective inductors)

Adiabatic temperature rise:

Field winding (rectangular 67kA, 20 second pulse) - 9.7°C

Stator winding (rectangular 67kA, 20 second pulse) - 24.3°C

Damper cage (poloidal pulse) - 14.0°C

4. DIODE CONVERTOR

The two diode convertors are arranged as half bridges housed in cubicles equally spaced around the machine. Each arm consists of the parallel connections of 40 paths, each consisting of the series connections of two diodes and a fuse whose purpose is to clear a diode short circuit and allow continued operation. Safe operation is still possible with two diode paths in any one arm missing.

The convertors use a mixture of air and oil cooling. Under pulse condition, 50% of the energy is dissipated in the closed loop oil cooling system. The air flow through the convertor cubicle, which is derived from the closed loop air cooling system of the generator, is used as well to provide suitable cooling of the hole storage components.

The type of diode selected is the AEI type DS 2009. It is a capsule type device with a V_{RRM} of 4.4kV and a V_{RSM} of 4.8kV. Each capsule is clamped between the oil-cooled heat sink extrusion and individual air-cooled heat sink.

Use has been made of aluminium for both heat sinks and their heat capacities have been generously dimensioned to ensure that the diode temperature during a short circuit (no circuit breaker) is maintained at an acceptable level. During the design stage, a model of a part of a convertor stack has been tested to check the main features of the design namely the heat transfer coefficient, the heat losses to both cooling media, the thermal time constant, the heating of the hole storage components etc.

5. MONITORING SYSTEMS

A comprehensive monitoring and protection system is being provided which will link up with the overall JET systems. The only novel devices are the temperature monitors on the generator rotors and the monitoring sub-systems on the convertors.

Temperatures are to be measured by RTD on selected generator field coils and in selected motor rotor slots. The sets of signals are fed to the telemetry units which give outputs suitable for transmission by sliprings to a stator unit. Here the signals are reconverted to separate channels and fed into the monitoring system as DC voltage signals.

Failure of a diode is detected by an unbalance at the integrator on the output of a Rogowski coil placed round a convertor arm. The event is transmitted as a single diode failure. It is also stored so that subsequent failures in the same arm may be transmitted as a multiple failure indication. This signal can then be used as an urgent trip.

Other signals available to the convertor monitoring sub-system include the location of a failed diode, obtained by small Rogowski coild in each diode path, and the heat sink temperature, monitored by thermistors screwed into the top of each extrusion.

A microprocessor located in one of the cubicles is used to scan all 480 diode strings and 96 temperature monitors and to store the information. The central monitoring system is advised of any failure indications and is given a full status report on request. A complete scan of all diode strings and temperature monitors takes 12 cycles.

6. ERECTION

The FGC's will be erected in circular pits approximately 24m in diameter and 5m deep, both of which are located in the Generator Hall of the J3 building (fig 3&4). The method and sequence of erection of the generator has been greatly influenced by the lifting facilities available. In order to minimise building craneage costs the station crane capacity has been limited to 50 tonnes.

For each generator, the first operation will be to install the bottom bearing housing. This will be followed by positioning of the shaft and then the spider. The rotor rim will then be built up layer by layer, its weight being supported by a set of jacks capable of lifting the complete rotor. After insertion of the rim bolts and clamping up of the rim, the poles, complete with field coils, can be lowered into position and secured.

Fig 3. Aerial View of JET Site. (13.5.80)

For transportation purposes the stator frame will be shipped in four sections. These will be placed in position and bolted together. The complete frame will then be lifted by a set of four jacks and supported on blocks clear of the rotor. In this position the stator will be cored and wound. It will then be lowered into its final position.

The remainder of the erection programme is straightforward.

The stator jacks can subsequently be used for stator winding or rotor pole maintenance. By supporting the rotor weight on its set of jacks, the bottom bearing thrust pads can be removed. Also with the rim weight supported, the construction will allow the shaft and spider to be removed separately if it is ever required.

7. MANUFACTURING PROGRESS

At the present time, the status of manufacture of the main components is as follows:

Completed - One set of pole punchings, one set of stator punchings, one set of stator winding bars, one set of field coils, generator sole-plates, first bottom bearing housing.

In manufacture - Second bottom bearing housing, both generator shafts, both rotor spiders, rim punchings, second set of stator punchings, second set of stator and field windings. first stator frame, pony motor rotors and stators, convertor cubicles and excitation equipment.

8. REFERENCES

[1] D Ciscato, E Coccorese. "The Combined DC Power Supply for JET", 10th SOFT Conference, Padova, September 1978.

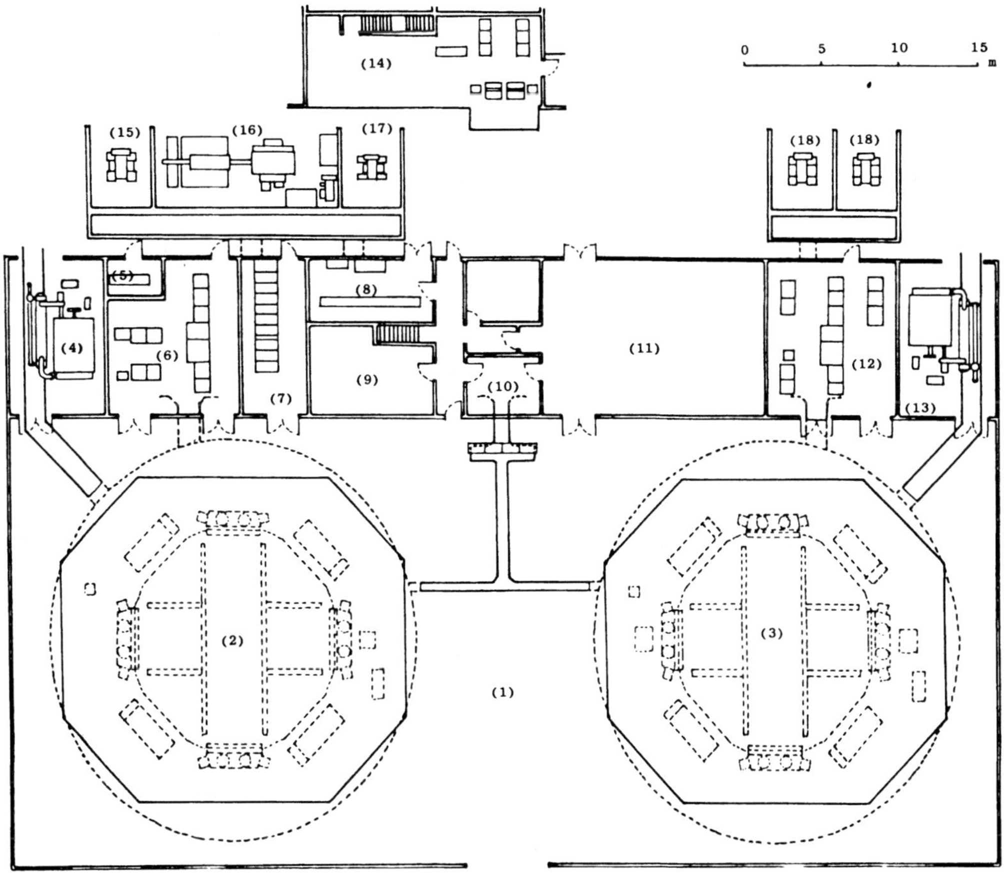

Fig 4. Station Layout

(1) Generator Hall
(2) Toroidal Field Generator Pit
(3) Poloidal Field Generator Pit
(4)(13) Liquid Controller Rooms
(5) Battery Room
(6)(12) Excitation Room
(7) 11kV Room
(8) 415V Auxiliary Room
(9)(10) Offices
(11) Workshop
(14) Control Room (1st floor)

(15)(18)(19) Excitation transformer
(16) Main 33/11.1kV Transformer
(17) 415V Auxiliary Transformer

Plasma Engineering

BEAM LINE SYSTEM OF NEUTRAL BEAM INJECTOR FOR JFT-2

K. Anno, K. Ohasa,
Y. Shimomura
Japan Atomic Energy Research Institute
Tokai, Japan

K. Masuda, O. Ohsaki, K. Yamamoto,
M. Watanabe, Y. Tanabe, M. Gotoh, S. Ioka
Toshiba Corporation
Tokyo, Japan

ABSTRACT

The neutral beam injector for JAERI Fusion Torus-2(JFT-2) was developed and is now in successful operation. This is one of the largest injection systems in Japan and was designed, fabricated and tested in about one year.

In order to obtain higher injection neutral beam power efficiency, efforts were made to design the beam line as short as possible. Since available space was limited, beam line components were designed to make their sizes compact as well as to satisfy the specifications. For design of these components, model tests and analysis were carried out.

Until now, a neutral beam power of 1.5MW was injected in a 35keV neutral hydrogen beam state for 0.05sec. It was clarified that the injection neutral beam power efficiency was about 45%. In this paper, the details of the design and test results of this beam line system are described.

1. INTRODUCTION

The development of this neutral beam injector was made from April 1979 to March 1980 in a joint cooperation of Japan Atomic Energy Research Institute (JAERI) and Toshiba Corporation. The neutral beam injector, which injects 35keV-1.5MW neutral beam power tangentially to the plasma, was developed based on scientific research and developments of ion source [1], and on technological developments for beam line system and power supply system [2] [3].

The design features of the beam line system and test results are described in the following sections.

2. OUTLINE OF BEAM LINE SYSTEM

The beam line system consists of two beam lines, a expansion chamber, a main vacuum pumping system, a hydrogen gas supply system, etc.

One beam line has two ion sources, where each source can extract max. 37 keV-28A hydrogen ion beam for 0.05 sec.

Major specifications and parameters are given in table I. Basic schematic diagram and bird's-eye view of the beam line system are shown in fig. 1 and fig. 2.

About 70% out of hydrogen ions extracted from ion sources are converted into neutral atoms by charge-exchange processes in the neutralizer cell. The residual ions deflected by the magnetic field of the ion deflection magnet hit the surface of beam dump and lose their energy.

Total power of ion and neutral beam is measured with the calorimeter inserted in the beam line chamber and injected power into the plasma with that in the JFT-2 torus, the so-called "calorimeter in JFT-2". Beam profile and power distribution are also measured by movable thermal probes attached to the drift tubes and above-mentioned calorimeter in JFT-2. Furthermore, several other major components are installed in the beam line; steering mechanism of beam direction, high speed shutter, magnetic shields, etc.

3. BEAM LINE SYSTEM DESIGN

The beam line system is designed so that the injection power efficiency and H_2 gas flow rate into the JFT-2 torus satisfy the specifications in table I.

To get optimum condition for the system, the gas balance equations in this system are solved in terms of supplied H_2 gas flow rate per ion source, and then the total H_2 gas flow rate into the JFT-2 torus and beam power losses and efficiencies are calculated as well as pressure distribution along beam path. The pressure distribution and the injection power efficiency are shown in fig. 3 and fig. 4 respectively.

It was found from the analytical results that, in the case of the beam energy of 30keV and the beam divergence of 1.3 degree, the optimum condition for the maximum injection efficiency would be realized if H_2 gas is supplied into ion source by the amount of 10 to 12 Torr·ℓ/s per ion source. It gives the following design values; neutralization efficiency of about 70%, reionization loss of about 14%, the injection power efficiency of 54% and the total H_2 gas flow rate into the JFT-2 torus of less than 4 Torr ·ℓ/s.

Similarly, the injection power efficiency is estimated to be 47% when the beam divergence is 1.6 degree.

4. BEAM LINE COMPONENTS

In order to minimize the beam divergence loss and to increase the injection power efficiency, the beam line components are designed to be as compact as possible. Thus obtained beam line length is 2.5 m.

4.1 Steering mechanism of beam direction

The beam half-widths as a function of distance along the beam axis are very important in this system. For a specified neutral beam power, this requires a compact steering mechanism capable of steering beam direction more than 2 degrees in steradian with good accuracy, while supporting an ion source of about 150kg without a little deflection due to its weight. As a R&D work a life-size model of the steering mechanism was performed and the final design was determined based on this result.

4.2 High speed shutter

To minimize the H_2 gas flow rate into the JFT-2 torus, two sets of high speed shutter are mounted, which operate in accordance with the turn on of the arc power supply and work within the time less than 0.5 sec including mechanical dead time.

The R&D for a life-size shutter model (see fig. 5) was made to develop the bearings and the welded bellows of long service life which work in a high vacuum of less than 10^{-4} Torr.

4.3 Magnetic components

(1) Magnetic shields

To prevent ion beam deflection at ion sources and neutralizer cells due to leakage magnetic field from the JFT-2 of about 100 gauss, it is required to restrict the magnetic field strength in less than 1 gauss in these areas.

Since the space surrounded by magnetic shields is about 1 m^3, comparatively large size of magnetic shields is required and its dimension is 1.5 x 1.5 x 1 m.

From the field analysis by finite element method (FEM) and test results by a small-sized model, a double rectangular shielding structure is employed, which consists of 14 mm - thick outside plates made of soft magnetic iron and 4 mm - thick inside plates made of high permeability metal.

(2) Ion deflection magnet

One deflection magnet is utilized to deflect two ion beams from two neutralizer cells, therefore its gap becomes rather wide, 540 mm.

To certify perfect deflection of ion beams to the beam dump, detailed magnetic field distribution analysis with three dimensional FEM was performed and good agreement between analysis and test results was confirmed, using a small size model.

4.4 Heat removing components

This injector consists of various heat removing components such as calorimeter, beam dump, thermal probe and calorimeter in JFT-2. These components are exposed to neutral and/or ion beam, and have functions of measuring beam power, power distribution (beam profile), and removing beam energy as heat. Thermocouples are attached to these components to measure the temperature.

(1) Calorimeter

The calorimeter is specified to withstand 10^5 beam shots for the conditioning of the ion source during all operating period. The walls exposed to the beam are composed of two oxygen free copper plates with cooling channels.

The surface of the wall is inclined with respect to the beam axis to form a V-shape so that maximum heat flux is reduced from 18 kW/cm^2 to 5 kW/cm^2 on the wall surface. The analytical result of the temperature at the heating surface and cooling surface in the conditioning mode is shown in fig.6.

(2) Thermal probe

Movable probes attached to the drift tube are used for search of the neutral beam profile along the vertical and horizontal directions. From the data of the temperature rise of the molybdenum sphere exposed to the beam, the beam profile can be estimated.

(3) Calorimeter in JFT-2

This calorimeter is used for the measurement of injected neutral beam power and power distribution in the torus.

From the temperature rises of ten molybdenum projections on a molybdenum plate exposed to beam, power distribution can be estimated.

5. VACUUM SYSTEM

This vacuum system consists of a large expansion chamber and main vacuum pumping system, whose key components are two turbo-molecular pumps (pumping speed 3500 ℓ/s x 2 sets). This system is designed under the following conditions ; in the case of H_2 gas inflow through four ion sources of maximum 60 torr ℓ/s during 0.1 sec every 1.5 sec in the conditioning mode, the pressure in the expansion chamber is kept in a range of 10^{-4} Torr and this H_2 gas can be exhausted perfectly by the start of the next shot.

Besides, auxiliary vacuum pumping system composed of a small size turbo-molecular pump is provided so that each ion source or the gas supply system can be evacuated independently.

Thirty titanium-balls (Ti-balls) are installed in the expansion chamber and in the beam line chambers to restrain light impurities.

5.1 Expansion chamber

The expansion chamber is a large cylindrical shall made of stainless steel, 2.3 m indide diameter, 5m high and 20 m^3 in volume. The volume is decided to keep the pressure below 1 x 10^{-3} Torr. The expansion and the beam line chamber are bakable up to a maximum temperature of 120°C.

5.2 Main vacuum pumping system

In order to exhaust H_2 gas inflow perfectly for 1.5 sec, net pumping speed for H_2 at room temperature is required to be more than 5000 ℓ/sec in the pressure range below 1 x 10^{-3} Torr.

Besides, the ultimate pressure in the beam line system is also required to be below 5 x 10^{-7} Torr.

6. TEST RESULTS

After the final assemble of the neutral beam injector, vacuum test, measurement of beam power and power distribution, magnetic shielding test, etc. were performed and it was confirmed that these results satisfied design specifications for the beam line system.

6.1 Vacuum test

The test of pumping characteristics, such as pumping time, net pumping speed, ultimate pressure, etc., as well as residual gas analysis by mass analyzer were performed both in factory and in site.

From this test results, it was clarified that net pumping speed for H_2 was about 5300 ℓ/sec, and that ultimate pressure in the beam line system was 2~4 x 10^{-8} Torr after 70 hours of pumping, including 46 hours of baking up to 120°C.

Further, from the gas analysis, it was also clarified no air was detected and clean vacuum was obtained.

6.2 Measurement of beam power and power distribution

Ion and/or neutral beam power and power distribution were measured with the components described in sec. 4.4, and by plasma heating experiments.

It was found that the beam divergence was about 1.5 ~ 1.6 degree, total injected neutral beam power from four ion sources was about 1.5 MW and the injection beam power efficiency eas approximately 45%.

6.3 Test for magnetic shields

The effect of magnetic shields was examined with three dimensional magnetic probe in the simulated magnetic field generated by two large circular coils in the factory and after ward in the actual magnetic field produced by ion deflection magnet, toroidal field coils and 150 kA plasma current of the JFT-2 tokamak. The test results in the site are shown in fig. 7. The field intensity in the magnetic shields was restricted to be less than 1 gauss as expected.

7. CONCLUSIONS

Design features and test results of the JFT-2 NBI beam line system have been described. From the results of tests and experiments, it was confirmed that the design for the beam line system was suitable, and that this system satisfied the design specification.

This beam line system has been operated for five months, and plasma heating experiment has been performed successfully [4] [5].

The authors are convinced that this system will make great contribution for further NBI system development.

8. REFERENCES

[1] Y. Ohara et al., 8th Symp. on Eng. Problems of Fusion Research, California, 13-16 November, 1979, Proceedings pp 198-202.

[2] A. Miyahara et al., the same as ref. [1], pp 1390-1397.

[3] S. Matsuda et al., to be published in the proceedings of 11th Symp. on Fusion Technology, Oxford, 15-19 September, 1980.

[4] N. Suzuki et al., IAEA-CN-38/T-2-3.

[5] Y. Shimomura et al., JAERI M 9065 (1980).

TABLE I

Specifications and Design Features of JFT-2 NBI Beam Line System

1. Specifications for beam line system	
o Injection power efficiency	> 50% for beam divergence of 1.3° > 44% for beam divergence of 1.6°
o Total H_2 gas flow into JFT-2 torus	< 5.5 Torr-ℓ/s for four ion sources
o Repetitive interval and total number of shots	
• Injection mode	2.5 min, 3 x 10^4 shots
• Conditioning mode	1.5 sec, 1 x 10^5 shots
2. Ion source [1]	
o Type and number of sets	Modified duoPIGatron, four sets
o Working gas	Hydrogen
o Ion beam energy per ion source	Max. 37keV–28A–0.05 sec
o Duty	1/30
o Beam divergence	1.3 degree (design value)
3. Beam line	
o Total length of beam path	2.5 m
o Steering mechanism of beam direction	> 2 degrees in steradian
o High speed shutter	
• Working time (to full open or full close)	< 0.5 sec
• Shut-up conductance	< 200 ℓ/s (for H_2)
o Magnetic shields	Shield factor : > 100
o Ion deflection magnet	> 500 gauss at the center of gap
o Calorimeter	max. heat flux : 5.1 kW/cm^2
o Movable thermal probe	max. heat flux : 12 kW/cm^2
o Calorimeter in JFT-2	max. heat flux : 5.5 kW/cm^2
4. Vacuum system	
o Expansion chamber	Type : cylindrical shell(2.3 m inside dia. x 5 m high) Volume : about 20 m^3
o Main vacuum pumping system	
• Main vacuum pump	Two turbo–molecular pumps (pumping speed 3500 ℓ/s x 2)
• Net pumping speed	> 5000 ℓ/s (for H_2)
• Ultimate pressure	< 5 x 10^{-7} Torr

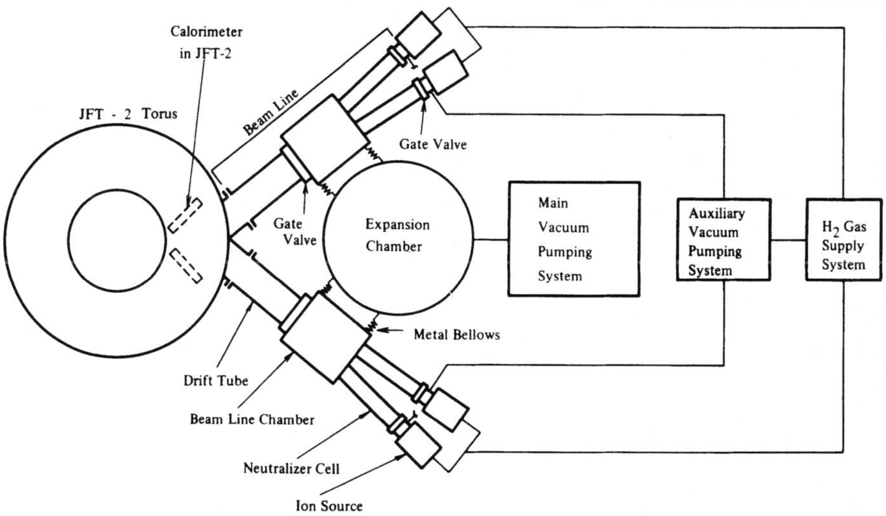

Fig. 1. Basic schematic diagram of the beam line system of JFT - 2 NBI

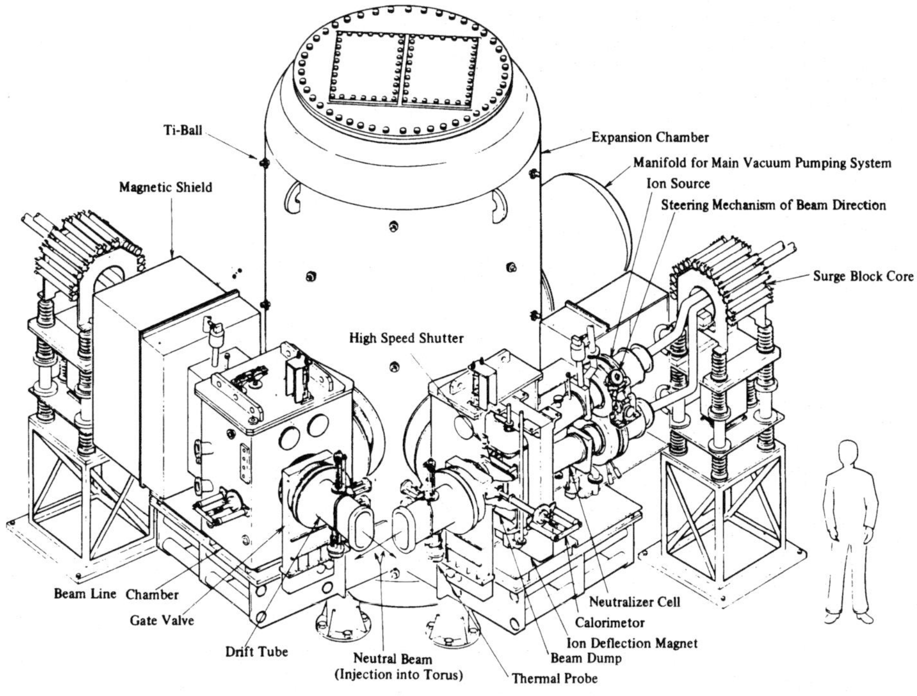

Fig. 2 Bird's-eye view of the beam line system of JFT-2 NBI.

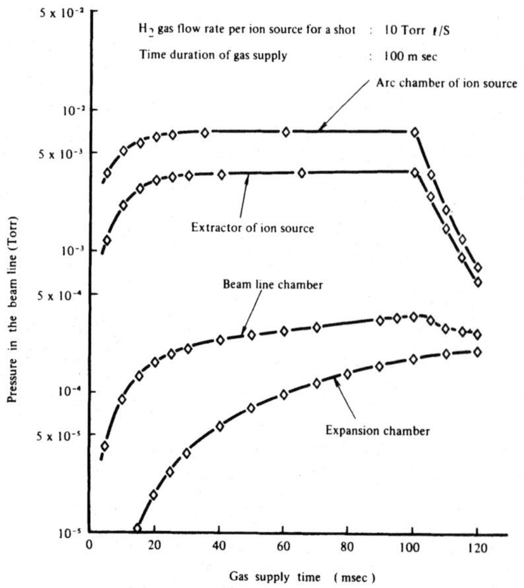

Fig. 3 Evolution of pressure distribution in the beam line.

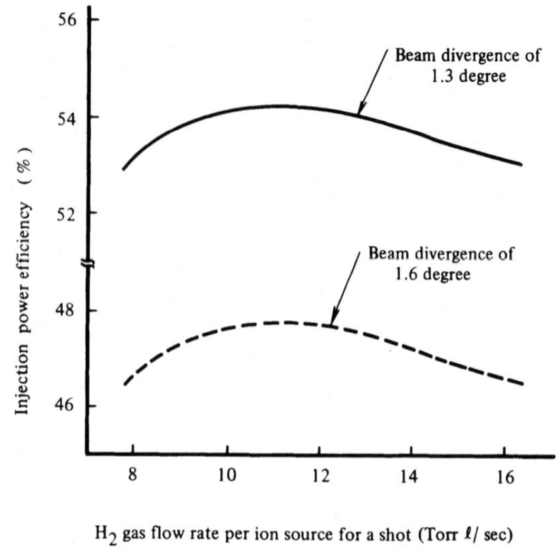

Fig. 4 Variation of injection power efficiency vs. H_2 gas flow rate depending on beam divergence.

Fig. 5 Life-size model of high speed shutter.

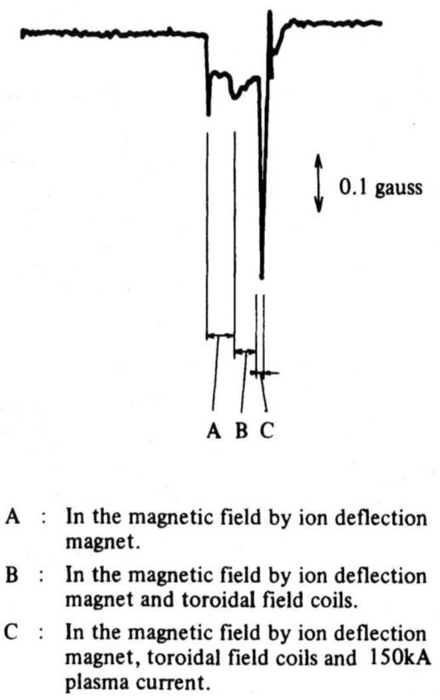

Fig. 7. Test result of the shielding effect of magnetic shields in the actual magnetic field.

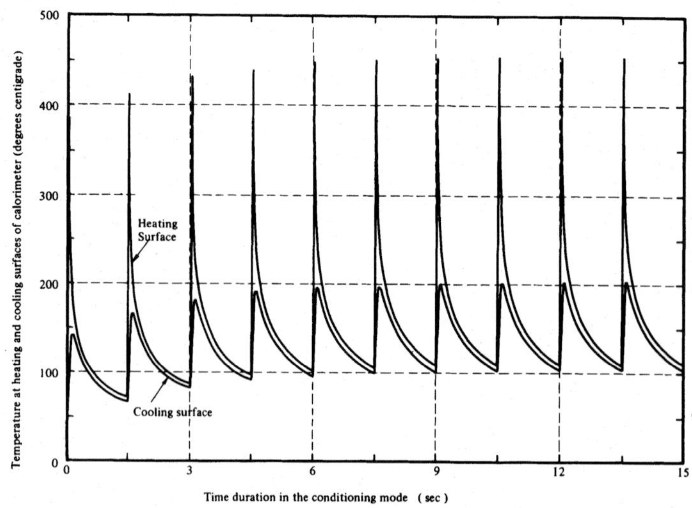

Fig. 6 Variation of temperature at heating and cooling surfaces of calorimeter in the conditioning mode.

DESIGN CONSIDERATIONS OF NEUTRAL BEAM INJECTION SYSTEM FOR AN IGNITION TOKAMAK

H Yamato, K Hashimoto, K Hayashi and T Sugawara
Toshiba Research and Development Center
Toshiba Corporation
4-1 Ukishima-cho, Kawasaki, Japan

ABSTRACT

A reference design of 20 MW - 150 keV neutral beam injector unit is presented after discussion of problems of ion source, neutralizer and residual ion beam handling from the standpoint of system integration. A direct converter of residual ion beams deflected magnetically from the neutral beamline is proposed.

1. INTRODUCTION

Heating power of 50 - 100 MW will be necessary to ignite a tokamak fusion reactor. When neutral beam injection is used, beam energy will be at least 150 keV with a pulse length of about 10 seconds. Purpose of this paper is to propose a 20 MW unit of injector on the basis of the previous design studies.[1, 2]

In the present study, it is supposed that the following are probable needs for unit injector:

(a) Injection is made at nearly perpendicular to a plasma current. Counter injection is excluded.

(b) Capability per unit injector is 20 MW D^0-beam with energy of 150 - 200 keV. Physical size of module is within the cube of 10 metre side.

In the next section, requirement for ion source is discussed from the standpoint of system integration. It is found in the previous analysis that the current density of the ion source is limited by heat removal limit. Heat load is most severe at the wall of arc chamber by the backstream electrons produced in the acceleration space. This heating is observed experimentally by Okumura.[3]

The third section deals with the direct energy converter as a residual ion beam handling device. Several schemes of inline converters have been proposed so far, for example, a "whole beam" converter by Barr,[4] a converter having a suppressor and a collector for each beamlet by Thompson,[5] a converter with immersed suppressor by Raimbault.[6] We choose a whole beam converter because it seems appropriate for high power beam application. We propose a direct converter, in which ion energy is recovered after deflected from neutral beamline by a magnet instead of an inline converter.

A reference design of injector employing this method of converter is proposed in the fourth section.

2. REQUIREMENT FOR SYSTEM INTEGRATION

When the power output is given, the physical size of a neutral injector is determined by following items; (1) current density of ion source, (2) length of neutralizer, (3) space for residual ion beam handling, (4) opening to pump the gas from neutralizer and the gas generated in ion beam collector and (5) capacity of pumping system.

In order to reduce the length of neutralizer, additional gas feed into the neutralizer has been employed in the previous design[2]. This method results in a considerable increase of gas flow from the neutralizer. We reconsider the parametric relation between heat load and operation pressure of the ion source and the length of the neutralizer assuming that the gas is fed only from the ion source. This provides the least gas feed.

The gas flow from the neutralizer Q can be derived from several conditions. We suppose that the ion source grid has a certain transparency. The length of neutralizer ℓ is approximately proportional to the inverse of the pressure in the source plasma chamber p_0, because the neutralizer should have a certain gas thickness; $\ell \propto 1/p_0$. The power flux density carried by the backstream electrons W is proportional to the pressure p_0 and to the ion beam current density in the grid aperture j_0; $j_0 \propto W/p_0$. Since the total beam current is supposed to be given, the aperture area must be proportional to the inverse of the current density j_0. Thus the scaling law on the gas flow is given by

$$Q \propto p_0/j_0 \propto p_0^2/W$$

From these relations, the choice of operating conditions of ion source is practically restricted in the design. In fact the increase of the pressure leads to considerable increase of gas flow, which requires the increase of the gas flow conductance of evacuation opening and the increase of the space for pump. On the contrary, the decrease of the pressure causes the extension of neutralizer. Figure 1 shows parametric relations for 150 and 200 keV - 20 MW injector composed of 7 injection lines. In this calculation, we assumed that the aperture of ion source grid is rectangular shaped with shorter side of 10 cm, the neutralizer has the gas thickness of 90 % equilibrium cell, the length of acceleration-deceleration space is 5 cm for 150 keV source and is 6 cm for 200 keV source, and that the grid has 50 % transparency. From the standpoint of system integration, it is not practical to employ the gas flow above 100 Torr·ℓ/sec, the ion beam current density below 0.1 A/cm^2, and the neutralizer longer than 6 m.

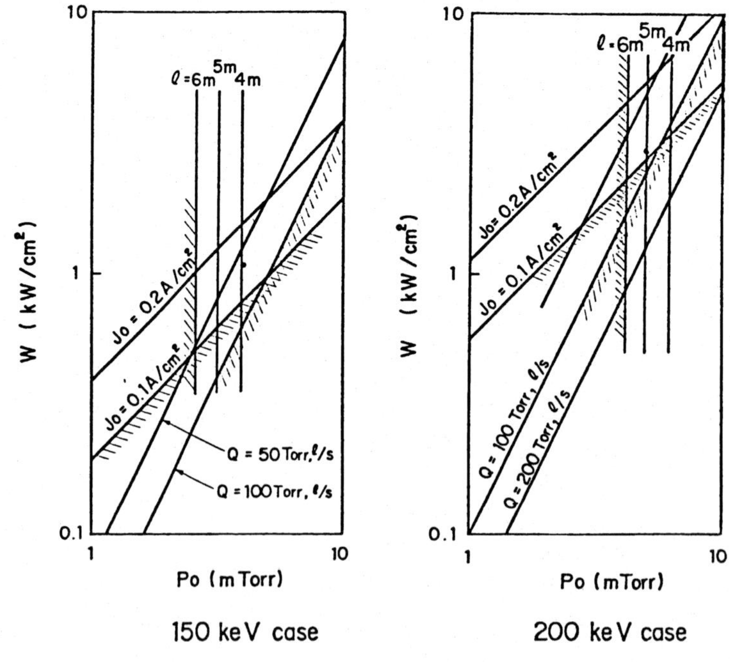

Fig. 1. Relation between ion source conditions and injector dimension for 150 keV and 200 keV beam injector with neutral output power of 20 MW. The black dot denotes the operation point employed in the reference design.

These conditions are also illustrated in the figure. It should be noted that the intense power flux carried by backstreaming electrons must be dumped on the end plate of source plasma chamber. The black dot in Fig. 1a denotes the operation point employed in the reference design, corresponding to p_o of 1 mTorr, W of 1 kW/cm^2 and Q of 62 Torr ℓ/s.

When 200 keV beam is used, 1 kW/cm^2 operation is very difficult as is shown in Fig. 1b. If 3 kW/cm^2 is permitted, operation pressure of 3 mTorr is a possible solution. Another possible solution is extremly low pressure operation of 2~3 mTorr and heat loading of 1 kW/cm^2. However, operation of the ion source in these conditions seems to be impracticable.

3. RESIDUAL ION BEAM HANDLING

In the first place we consider the secondary particle production in the deceleration space of the converter. To prevent the entrance of slow electrons from the neutralizer into the converter, the electrostatic suppressor or the magnetic suppressor are used. In the case of the converter with electrostatic suppressor, the major part of the secondary ions produced by primary ion beam and gas molecule collision hits the suppressor and yields secondary electrons on the surface. These electrons are accelerated to collide with the converter collector.

The collector heat load due to this process can be approximately estimated by using a simplified model; all the secondary ions are collected on the suppressor, and the electric field is uniform within the deceleration region. Let us write the primary ion beam energy under the suppressor and on the collector and the representative length of the deceleration space as E_s, E_c and ℓ, respectively. The collector heat load by impact of electrons from the suppressor P_e is given by

$$\frac{P_e}{I_b \cdot E_c} = n_g \, \ell \cdot \langle \gamma_i \cdot \sigma_+ \rangle \cdot \frac{E_s - E_c}{E_c}$$

where n_g is the gas density, γ_i is the electron yield by energetic ion impact on the suppressor, σ_+ is the secondary ion production cross-section for D^+ - D_2 collision, and the bracket designates the average over the energy range of $E_c \sim E_s$. The yield γ_i does depend on the electrode material and on the surface state. For simplicity, we assume that the yield is independent on the impact energy over the interested energy region and is equal to 2 ∿ 3. Thus we get

$$\frac{P_e}{I_b \cdot E_c} = (4\sim 6) \times 10^{-13} \cdot n_g$$

for E_s/E_b=1.3, E_c/E_b=0.05, the primary beam energy into the converter E=200 keV and ℓ=20cm. In the equation n_g is in $molec/cm^3$. Now suppose that the power P_e is allowed to be equal to the heat load provided by inflow of primary beam into the collector. The gas density limits allowed for this is

$$n_g = (1.7 \sim 2.5) \times 10^{12}\ \text{molec/cm}^3 = (5.1 \sim 7.6) \times 10^{-5}\ \text{Torr}$$

In the case of the converter with magnetic suppressor, the secondary ions produced in the deceleration space are accelerated to go in the direction of an ion source through the suppressor. When the energy flux of the backward ion beam is permitted up to 20 % of entrance ion beam power from the standpoint of wall heat load of neutralizer, we get the gas density limits:

$$n_g = 4.1 \times 10^{12}\ \text{molec/cm}^3$$

for E_b=200 keV, E_c/E_b=0.05 and ℓ=20 cm.

The gas generated on the converter collector is evacuated through several opening to the exterior of the converter to keep the pressure in the converter. The presence of electrode does restrict the conductance of gas flow. The pumping speed is practically restricted to a conductance corresponding to the cross-sectional area of the neutral beam. This means a condition on the ion beam current density into the converter:

$$j < \frac{1}{4} n_g \cdot v_g \cdot e = 16 \sim 25 \text{ mA/cm}^2$$

where we use $v_g/4$=31 litre/cm^2.sec. This current density is so low that the ion source area required becomes huge. Thus we find that the inline converter system is incompatible with a compact beam line formation for our intense beam injector.

We should point out other two problems involved in the inline converter system;

(1) The numerical simulation on the inline converter reveals that the uncollected ion beam passing through the converter reaches up to 10 - 20 % of primary ion beam current. This part of ion beam must be removed before the drift duct.

(2) The presence of a half and a third energy ion beam needs to make a deceleration space longer, because of setting the electrodes held at intermediate voltage. Hence the pressure restriction in the converter becomes more severe than that mentioned above.

Above discussion of the inline converter lead us to employ a beamline system with magnetic deflector. in which residual ion beam energy is recovered by converter after deflection. Since the converter in this case is not closely coupled with neutralizer, the current density of unneutralized ion beam can be decreased before the converter. The methods are the utilization of the natural beam spread along a long path, and the vertical and horizontal defocus by using the fringe field of deflection magnet. As a result, the current density restriction does not impose the extention of ion source grid area in practice. In the next section, the reference design is described for nearly perpendicular 150 keV - 20 MW/port injector.

4. REFERENCE DESIGN

The injector module shown in Fig.2 consists of seven beamlines. Major components are ion source, neutralizers, a deflection magnet, direct converters and cryopumps.

Residual ion beams are collected by a direct converter after deflection of 180° by a magnet. Deflection angle of 180° is chosen to decrease the width of the module. The beam with a half and a third energy are separated by this magnet and collected by beam dumps. Physical dimensions and duty of each component are summarized in Table 1.

Tank pressure of 5 × 10^{-5} Torr was determined from the standpoint of heat load to a collector of converter. Causes of collector heat load are the residual power of the converted ion beam, the power of the secondary electrons accelerated in the ion deceleration space, and the power of undecelerated neutral beam generated after the deflection magnet. We permitted the heat load due to the last cause up to 100 W/cm^2. Total heat load to the collector results in 184 W/cm^2 for the case of above tank pressure. In this estimation, we used the following model of converter with magnetic suppressor;

deceleration voltage	145 keV
deceleration space	20 cm long
opening between pole faces of magnet	40 cm high

The reionization loss of neutral beam is 10 % between neutralizer and drift duct.

The evacuation opening between neutralizer and magnet is chosen to have a nearly equal condactance to the pumping speed of main cryo-pump.

The chart of gas and power flow are given in Fig.4 and 5. Major part of deuterium gas from the converter is evacuated through a beam entrance aperture. Overall power efficiency of this module is expected to be 28 %.

5. CONCLUSION

A beamline system to ignite a tokamak plasma was developed from investigation of ion source heat load and performance of residual ion beam collector.

Beamlines of 150 keV and 200 keV were compared. In 150 keV system, beamline may be constructed by ion sources with operating pressure of 4 mTorr and heat load of 1 kW/cm^2. Low ion source pressure and large gas flow out of neutralizer should be permitted to achieve 1 kW/cm^2 at 200 keV.

When an inline converter is employed, the current density of ion source is limited by the conductance of neutral gas. Handling of a half and a third energy particles and residual ion passed the converter are also difficult problems in an inline converter.

A reference design of 20 MW injector module with a 180° deflection magnet and direct converter was presented. Physical size of the injector may not increase too much with application of a magnetic deflector.

REFERENCE

[1] Hashimoto, K. et al., Proc. 8th Symp. on Eng. Problems of Fusion Research, San Francisco, p. 1398 (1979)

[2] Fusion Research and Development Center, JAERI-M 8518 (1979)

[3] Okumura, Y. et al., Rev. Sci. Instrum., 51, 728 (1980)

[4] Barr, W. L. et al., Proc. 7th Symp. on Eng. Problems of Fusion Research, Knoxville, p. 308 (1977)

[5] Thompson, E. et al., Proc. 6th Symp. on Eng. Problems of Fusion Research, San Diego, p. 907 (1975)

[6] Raimbault,P. and Fumelli, M., Joint Varrena-Grenoble Int. Symp. on Heating in Toroidal Plasmas, p. 71 (1978)

TABLE I Design Parameters

Main parameters:		Deflection Magnet:	
Neutral beam power	20 MW	Larmor radius	200 cm
Beam energy	150 keV	Magnetic flux density	0.6 kG
Beam species	deuterium		
Length of beam line	880 cm	Direct converter:	
		Suppressor	magnetic
Ion source:		Recovery voltage	145 kV
Extraction current	90 A	Pressure	73 μTorr
Ion current density	130 mA/cm^2	Recovery efficiency	85 %
Operation pressure	4 mTorr		
Gas efficiency	49 %	Pumping system:	
No. of acceleration stage	3	Main cryo-pumps	
Grid transparency	50 %	Pumping speed	1.9 Mℓ/s
Extraction area	138 x 10 cm^2	Pressure	50 μTorr
		Auxiary cryo-pump	
Neutralizer:		Pumping speed	92 kℓ/s
Length	380 cm	Pressure	63 μTorr
Gas line density	0.37 Torr·cm		
Neutralizing efficiency	30 %	Power efficiency	28 %
Cross-section			
Entrance	138 x 10 cm^2		
Exit	122 x 12 cm^2		
Pressure			
Entrance	2 mTorr		
Exit	92 μTorr		
Beam divergent loss	9.8 %		

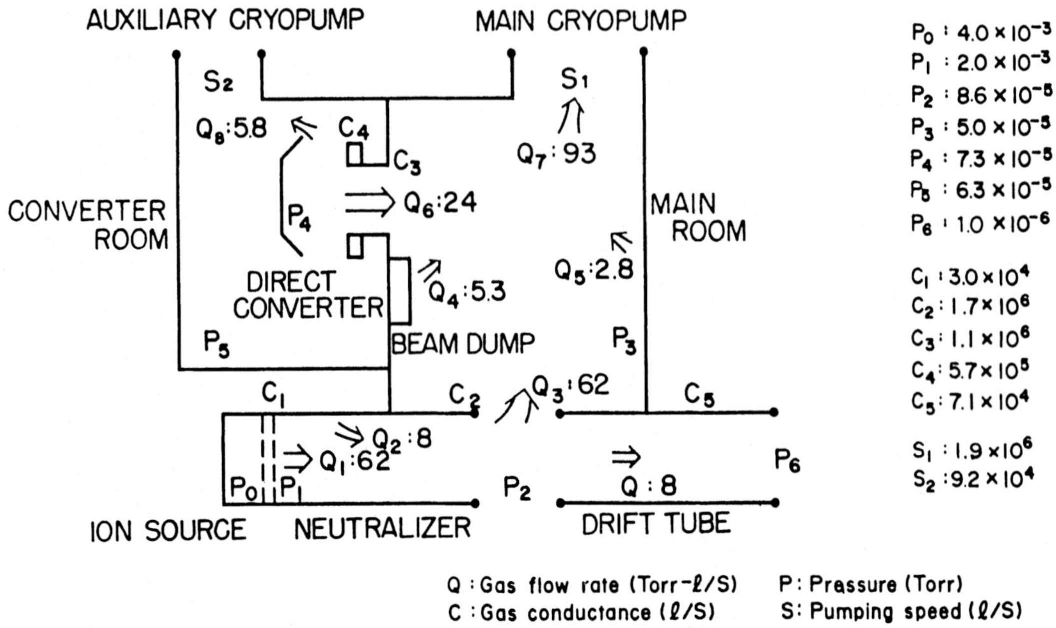

Fig. 3. Diagram of gas flow for 150 keV - 20 MW injector.

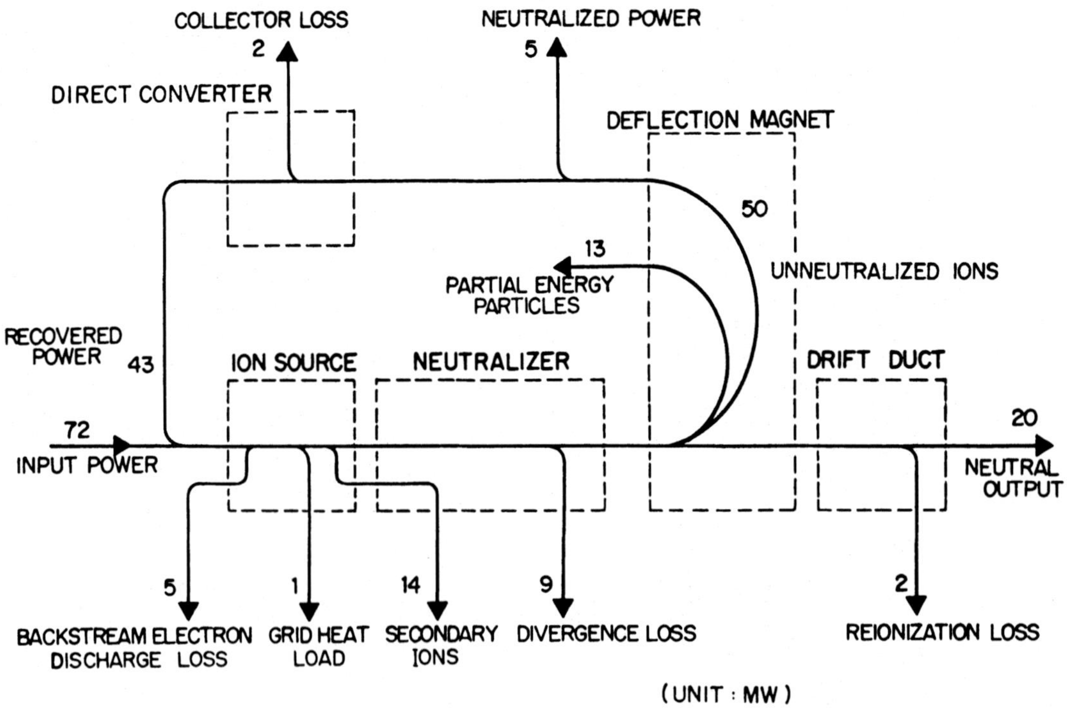

Fig. 4. Diagram of power flux for 150 keV - 20 MW injector.

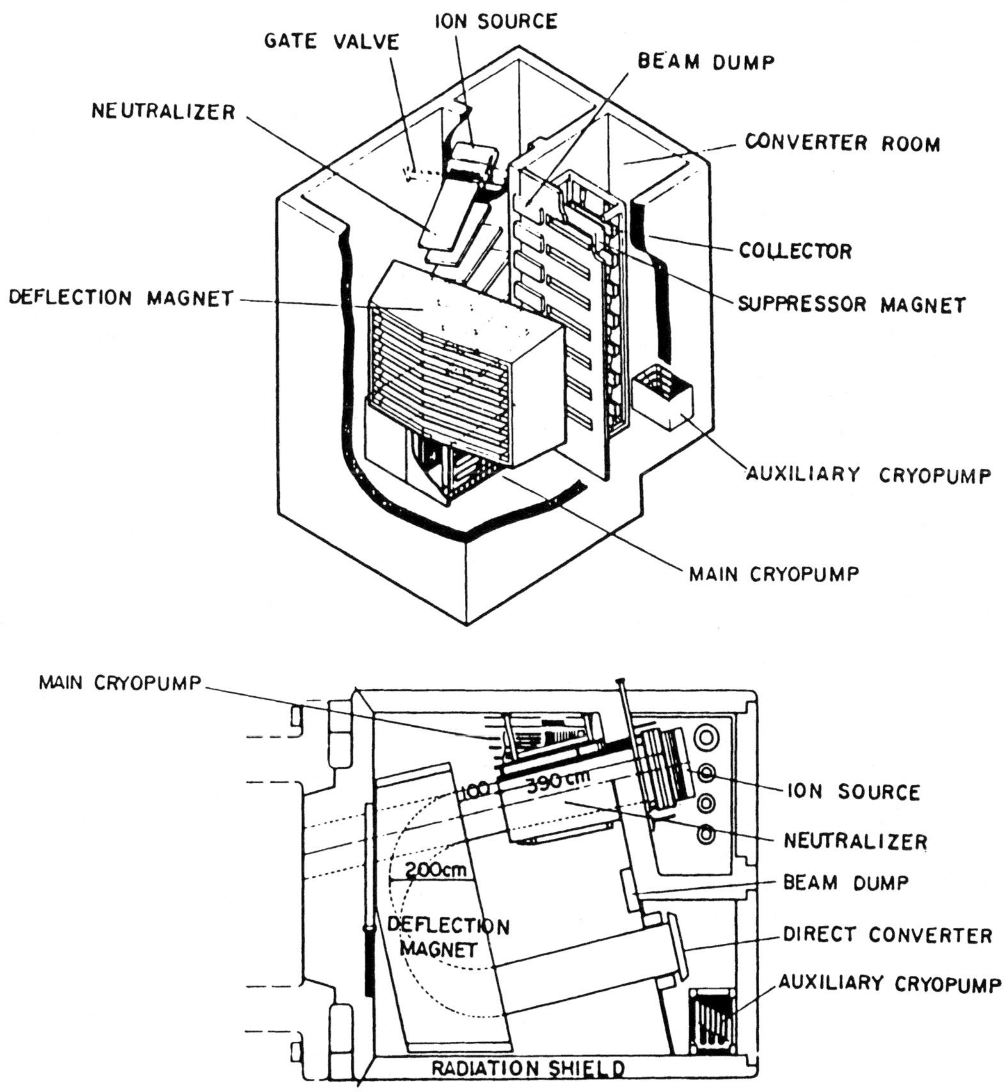

Fig.2 Whole aspect of 150 keV - 20 MW injector.

DEVELOPMENT OF EXTRACTION ELECTRODE SYSTEMS FOR 80 KV LONG PULSE OPERATION ON THE CULHAM MEGAWATT BEAM LINE

Neutral Injection Development Group
Culham Laboratory, Abingdon, Oxon, OX14 3DB, England.
(Euratom/UKAEA Fusion Association)

ABSTRACT

The Megawatt Beam Line facility at Culham Laboratory is designed for the testing of extraction systems for long pulse, 80 kV, operation to meet the requirements of neutral injection systems for the JET Experiment. The facility configuration and the range of diagnostic methods in use to characterise the beams produced are briefly described. Results of operation with multi-slot and circular aperture extraction systems are reported, and comments on future work are made.

1. INTRODUCTION

The Megawatt Beam Line neutral beam test facility at Culham Laboratory is designed to produce ~ 1 MW of ions and neutrals at an energy of 80 kV for a period of ~ 1 second. The programme is oriented towards meeting the requirements of neutral injection systems for the JET experiment. The upgrade of the test facility to the 5 MW total power, 5 second pulse length appropriate to the JET specification is described elsewhere in these proceedings [1]. The project forms part of a collaborative programme on neutral injection between Culham, Fontenay-aux-Roses and JET.

2. FACILITY DESCRIPTION

The Beam Line is shown schematically in fig. 1. It consists of a four-electrode extraction system fed by a magnetic multipole or "bucket" source. The extracted ion beam is then neutralised (the figure illustrates an early design of segmented neutraliser) and passes into a cryopumped drift region, in which most of the diagnosis of the beam characteristics is carried out. The power in the beam is absorbed on a water-cooled calorimeter. No provision is made for deflection of the un-neutralised portion of the beam, or for direct recovery.

The various components of the Beam Line have been described in detail in previous conference reports, including the overall design philosophy and the power supplies [2], the cryopumps and tests carried out on the pumps [3,4], the high voltage protection system based on a crossed-field interrupter [5], and the calorimeter system [6].

The facility has been operated in the period since the 10th SOFT Conference in 1978 to power levels in excess of 1 MW at 80 kV for pulse lengths of ~ 0.1 second (limited by the use of un-cooled extraction electrodes), and at energies in excess of 3 MJ for long pulses (up to 5 seconds) at reduced power (limited by the capability of the calorimeter). Operation with and without neutraliser gas has been investigated.

3. DIAGNOSTIC EQUIPMENT

A wide range of diagnostic equipment is installed on the Beam Line, in order to fully investigate and diagnose the beams produced. This equipment falls into several areas, as follows:

(a) Ion source and extraction system diagnostics, allowing information on all currents, voltages, gas flows, temperature monitors, etc. to be logged and displayed in the control room for system conditioning and operation. This includes the transmission of signals from high voltage to ground via optical links.

(b) Visible imaging diagnostics, based on closed circuit video cameras and recorders, allowing the beam and various internal and external Beam Line components to be displayed in the control room. Fig. 2 shows a photograph of an unneutralised beam, imaged predominantly at the H_β wavelength, recorded from a video monitor screen.

(c) Beam timing and alignment diagnostics, allowing the operators to run perveance matched beams. Here use is made of information from video signals by monitoring the light intensity along various TV lines (fig. 3), the main calorimeter (for high power operation), information from the diagnostic calorimeter (see fig. 1 and ref. 6), light profiling of the extracted beam, and the signal from a fast X-ray monitor.

(d) Spectroscopic diagnostics, allowing the beam to be characterised at the H_α (656.3 nm) wavelength. A scanning Fabry-Perot system is used for Doppler-shift analysis on the full and fractional energy components of the beam, and the unshifted background emission (fig. 4). This allows the divergence of the various components to be studied, and the species mixture of the neutralised beam to be determined. Compact Fastie-Ebert monochromators are used to study the time dependence of the emission from the various Balmer lines to investigate, e.g. gas evolution during a pulse (fig. 5).

(e) Charged particle diagnostics, using magnetic momentum analysis techniques to separate the different ion species in the beam (fig. 6), and retarding field analysis to investigate the properties of the plasma in the neutraliser.

(f) Impurity diagnostics, using the momentum analyser to monitor high mass ions, the Fabry-Perot system to monitor fractional energy hydrogen atoms produced from the breakup of water or hydrocarbons (not present in the trace in fig. 4, due to the cleanliness of the system at that time), and a fast-scanning quadrupole mass spectrometer to carry out residual gas analysis in the Beam Line vacuum tank.

4. OPERATION WITH A MULTI-SLOT EXTRACTION SYSTEM

Initial operation of the Beam Line was with a tungsten electrode nine-slot extraction system. A cross-section of the electrodes in shown in fig. 7. The system was allowed to expand in the direction parallel to the slot as they heated during the 1-second pulse lengths used during this work.

The extracted current levels achieved with the 9 slots, with open area of 72 cm^2 and a transparency of 40%, are shown as a function of voltage in fig. 8. The design current levels are also indicated. The measured hydrogenic species ratio at the higher current levels was approximately H_1^+: H_2^+: H_3^+ = 0.70: 0.15: 0.15, but there was also impurity present, at mass 17 - 18 corresponding to water vapour. A total species mixture of 0.66: 0.13: 0.13: 0.08 leads to good agreement with the design values, as indicated in fig. 8.

It is commonly observed in slot extraction systems that the divergence in the direction parallel to the slot is less than that perpendicular to the slot, and this was observed on the Beam Line system. The variation of the perpendicular divergence with perveance is shown in fig. 9, for data taken at 62 kV, using light profiling of the total beam. Scaling to 80 kV suggests minimum divergence values of 0.6^0, consistent with calorimetric measurements performed on a single slot in separate experiments. The divergences of the different energy components were also measured using Doppler-shift spectroscopy, and these suggest perpendicular divergences at 80 kV of 0.75^0, 0.9^0 and 1.1^0 for the full, half and one-third energy components respectively. Thus, we conclude that the divergence from optimised slot electrodes in the perpendicular direction is $\sim 0.6^0 - 0.7^0$.

5. OPERATIONS WITH A CIRCULAR HOLE EXTRACTION SYSTEM

A hexagonal array of 72 circular holes, based on the optimised design from beam optics studies, has been used for the bulk of the work on the Beam Line. A cross-section of the electrodes is shown in fig. 10. Consistent operation at 80 kV, above the 1 MW level, has been achieved with this system. The highest power recorded to date is 1.2 MW (76.8 kV, 15.4 A, 100 mS, 190 mA cm^{-2}), under un-matched conditions.

The extracted current levels under matched conditions are shown in fig. 11. The actual values of extracted current are very close to the designed values, when corrected for the species mix and impurity fractions shown in fig. 12. A point to note is the relatively low levels of impurity achieved in the system, also indicated by the absence of a peak characteristic of low energy hydrogen from high mass impurities in spectroscopic traces (fig. 4). The proton fractions of 70% are expected to be bettered in larger sources of the JET size [7].

Neutralised beams at perveance match have been operated on the system. Almost complete breakup of the molecular ion peaks has been achieved (fig. 6) and studies are concentrating on the determination of the gas target thickness and comparison with calculation.

The beam divergence has been measured by Doppler-shift methods. The data shows low values of divergence at low energies ($< 1^0$). The values are lower than those obtained with the slit apertures (fig. 9) and should in principle lead to higher transmission in a beam line. They do not scale as $E^{-\frac{1}{2}}$ as indicated by the measurements on single apertures (fig. 13). The variation with gas flow rate shows higher values at low flow rates as observed in single aperture studies and the absolute values are reasonably consistent. However, the minimum value obtained with increasing flow rate is only $0.8 - 0.9^0$. It is believed that this is due to gross electrode misalignment due to over-heating of the electrode assembly. The grids are essentially only radiatively cooled and the beam forming and gradient grid receive a large heat input from the plasma source. In particular, the gradient grid, which has negligible cooling, has reached temperatures of 100^0 above ambient during system conditioning. The effects of this, and of possible plasma source non-uniformity, are being investigated.

Power loading measurements have been carried out on the electrode system, and the results are compared with those from a 6-aperture beam optics experiment in Table I. In view

TABLE I

Measurements of Power Loading due to
Gas from Source as a Fraction of Beam Power

Component	Six Apertures	72 Apertures
No. 1, beam forming grid	$6 \pm 3 \times 10^{-4}$	2.3×10^{-4}
No. 2, gradient grid	2.4×10^{-4}	2.9×10^{-4}
No. 3, suppression grid	3.6×10^{-4}	2×10^{-4}

of the low power levels measured it was decided to extend the pulse length on the Beam Line, still using uncooled electrodes. This was carried out at reduced power level (up to 60 kV, 10 A) , extending the pulse length to 5 seconds (3 megajoule). The limiting factor was the temperature rise on the calorimeter/beam dump. The current pulse could be maintained reasonably flat by tuning the filament controls, and the beam divergence was constant, though its value was still worse than the design value. 1-second operation has been achieved at a level of 80 kV, 12.5 A (1 Megawatt).

6. FUTURE STUDIES

The objectives of the programme in the time remaining before the Upgrade of the Facility will be to clarify the problems of beam divergence, and to measure the power level in back streaming electrons and the total power accountability. In addition, the limits of operation of uncooled electrodes will be investigated.

7. REFERENCES

[1] Culham Laboratory Neutral Injection Development Group. A Neutral Beam-line Installation for Testing Injector Systems in the Long Pulse, Multi-megawatt Regime. These proceedings.

[2] COUPLAND, J. R., et al. 9th Symposium on Fusion Technology, Garmisch-Partenkirchen, 14-18 June 1976. Proceedings pp 903-908.

[3] COUPLAND, J. R., HAMMOND, D. P., and THOMPSON, E. Ibid, pp 79-84.

[4] COUPLAND, J. R., HAMMOND, D. P., and THOMPSON, E. 7th Symposium on Engineering Problems of Fusion Research, Knoxville 25-28 Oct. 1977. Proceedings pp 1796-1800.

[5] BERNHARDT, T., et al. 10th Symposium on Fusion Technology, Padua, 4-9 Sept. 1978. Proceedings pp 301-306.

[6] HAMMOND, D. P., and BROOKES, C. E. Ibid, pp 307-312.

[7] GOEDE, A. P. H., and GREEN, T. S. 8th Symposium on Engineering Problems of Fusion Research, San Francisco, 13-16 Nov. 1979. Proceedings pp 680-684.

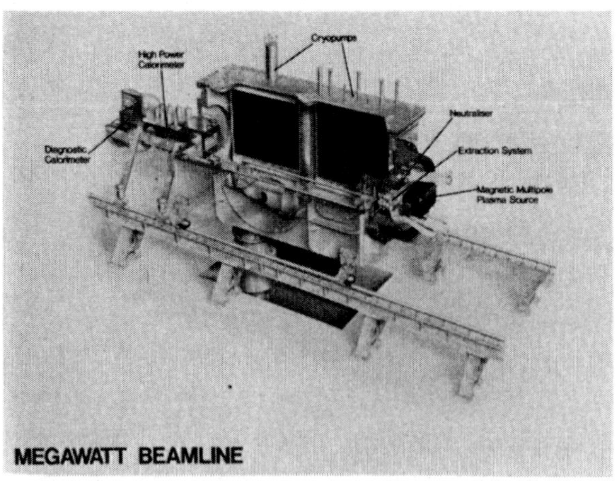

Fig. 1. Culham Laboratory Megawatt Beam Line facility schematic.

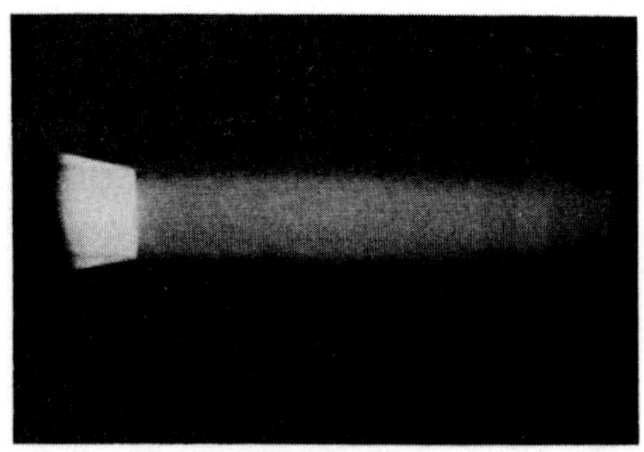

Fig. 2. Unneutralised ion beam, exiting the neutraliser at the left of the picture. The calorimeter is off-screen at the right.

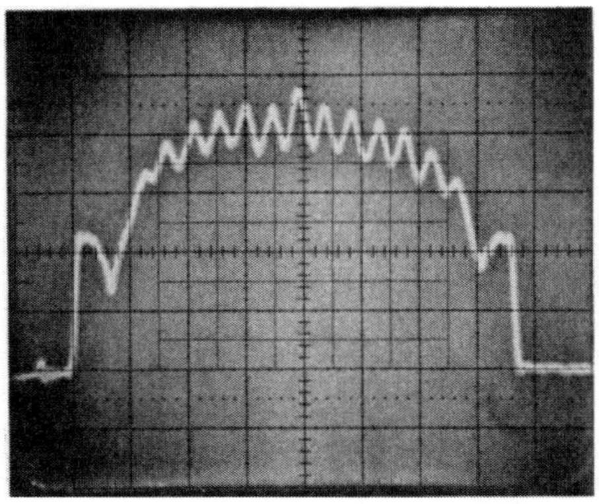

Fig. 3. Light intensity along a TV raster line of an unneutralised beam just after extraction, showing the structure produced by the extraction electrode geometry.

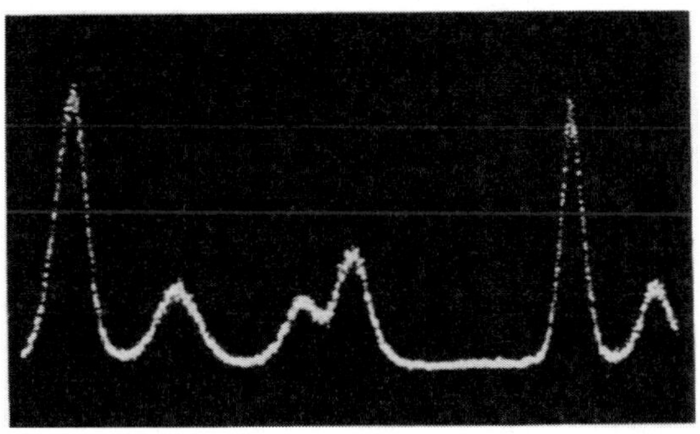

Fig. 4. H_α spectrum of a beam, showing (from left to right) the unshifted background peak, the full-, half-, and third-energy peaks, and the unshifted peak of the next order.

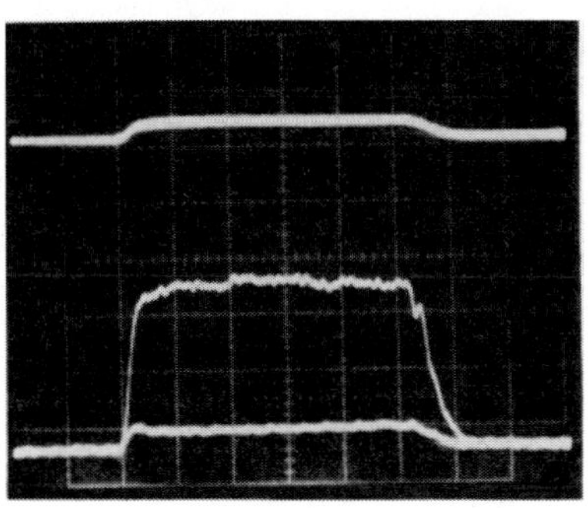

Fig. 5. H_α signal during a pulse. The top trace is the extracted ion current signal, and the bottom two traces are the light intensity for a neutralised (top) and unneutralised beam (bottom), at the same sensitivity.

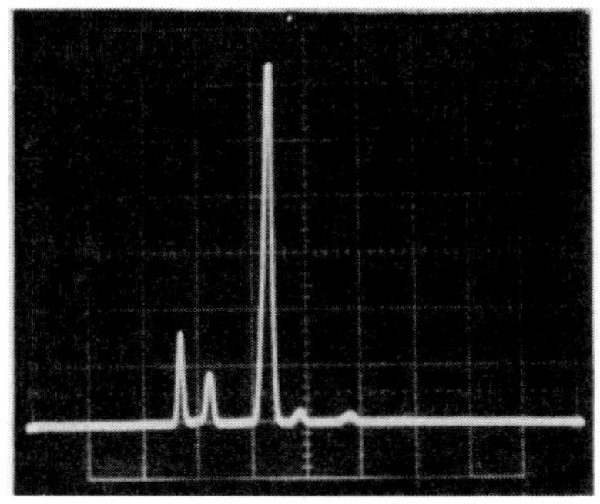

Fig. 6. Magnetic momentum analysis of the changed fraction of a neutralised beam. The three main peaks are the third, half and full energy protons.

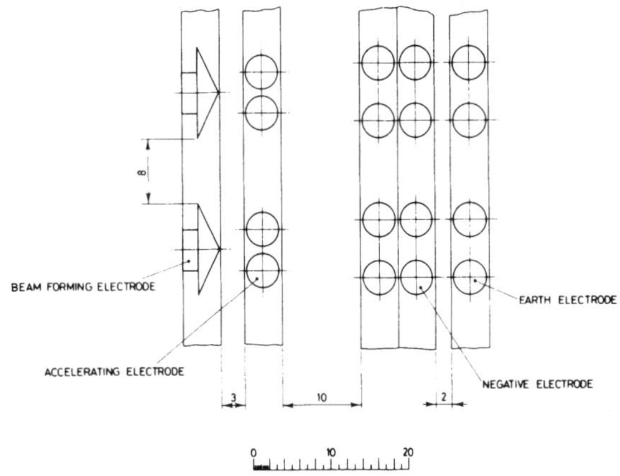

Fig. 7. Cross-section of multi-slot extraction electrodes (dimensions in millimetres).

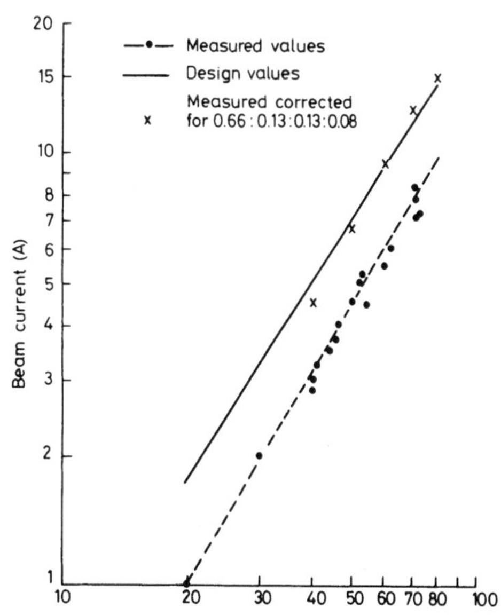

Fig. 8. Current-voltage levels achieved with the multi-slot system.

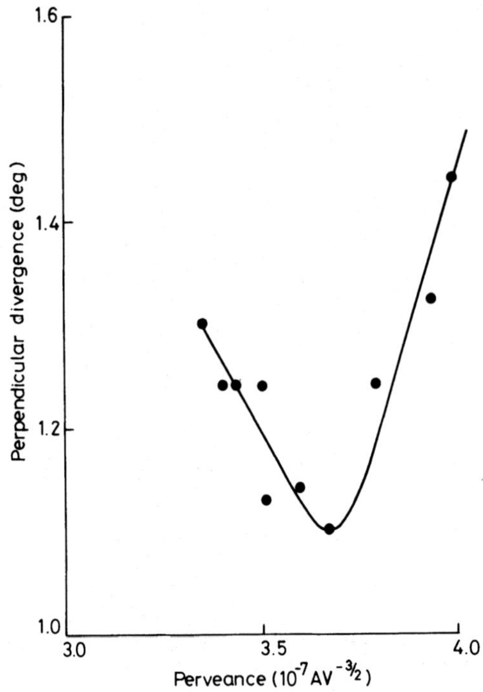

Fig. 9. Variation of perpendicular divergence with perveance for the multi-slot system.

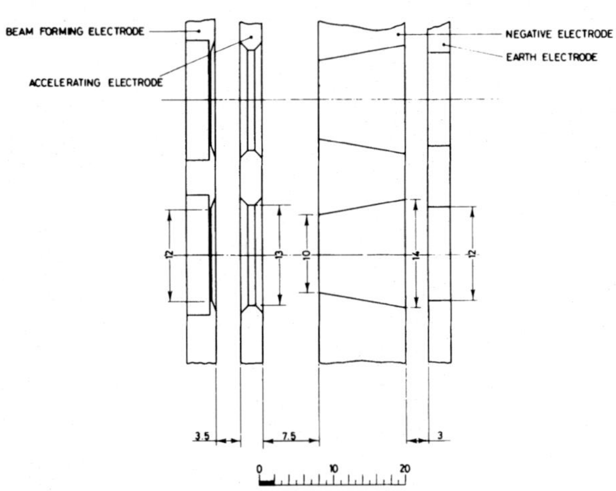

Fig. 10. Cross-section of circular aperture electrodes (dimensions in millimetres).

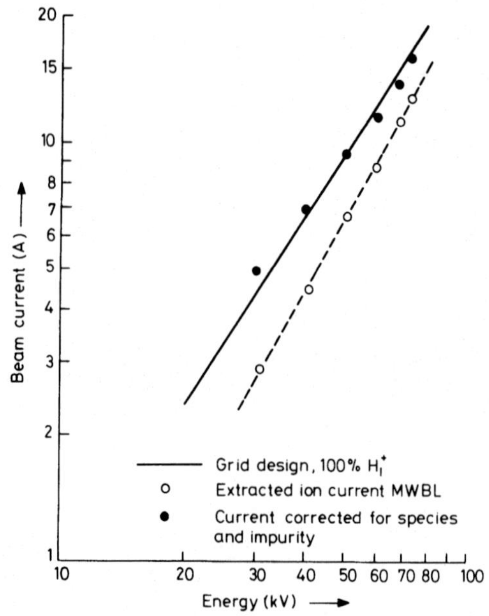

Fig. 11. Current-voltage levels achieved with the circular aperture system.

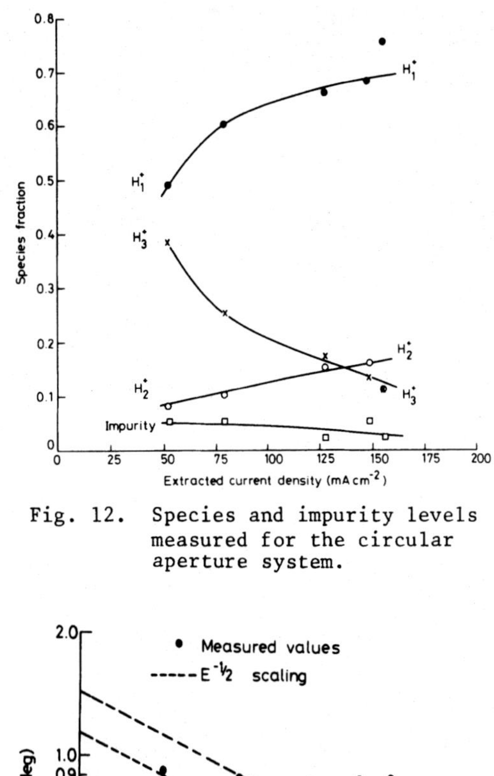

Fig. 12. Species and impurity levels measured for the circular aperture system.

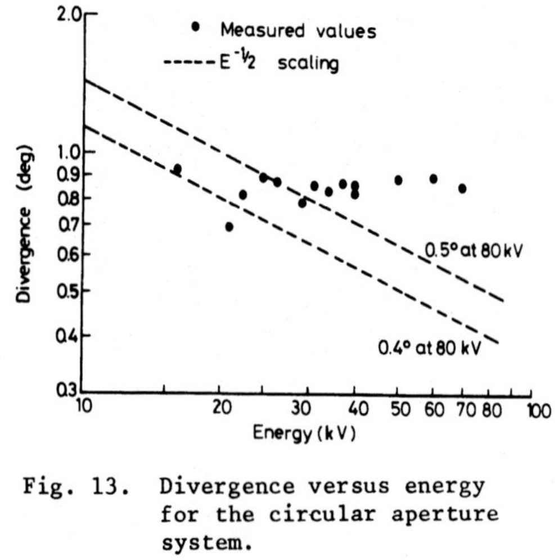

Fig. 13. Divergence versus energy for the circular aperture system.

TRANSIENT SIMULATION ANALYSIS OF GAS FEED INTO THE NBI NEUTRAL CELL

A. Maekawa
Energy Research Lab.
Hitachi, Ltd.
Hitachi, 316 JAPAN

S. Isobe
T. Uede
Y. Ono
Kokubu Works, Hitachi, Ltd.
Hitachi, 316 JAPAN

T. Obiki
A. Sasaki
A. Iiyoshi
K. Uo
Kyoto University
Uji, 611 JAPAN

ABSTRACT

The optimum feed conditions of hydrogen gas into the neutral cell of a Neutral Beam Injector (NBI) are examined using transient simulation analysis of the vacuum system. In the analysis, a neutral cell is directly connected to an ion source without an evacuation port, just as in the usual NBI.

Total efficiency is as low as gas pressure in the ion source. If a gas inlet into the neutral cell is added and its position is closed to the ion source, there is a range of gas pressures, where the decrease in total efficiency without a gas feed into the cell is compensated for and total gas consumption per pulse is lower than for other gas inlet positions in the cell. If the gas is fed into a part of the neutral cell alone, total efficiency in certain gas pressure ranges is higher than in any of the above cases, and gas efficiency is always 1.

1. INTRODUCTION

The vacuum system in the NBI must meet the following requirements:

(1) to dispose of a large quantity of gas leaked from the extraction electrodes of the ion source;
(2) to hold moderate gas pressures in the neutral cell ; and
(3) to keep a high vacuum on the beam line from the output of the neutral cell to the plasma experimental device.

Until a few years ago, the working gas pressure in an ion source was around one Pa. Thus, the NBI in Tokamak devices of recent design is that in which the cell is directly connected to an ion source without an evacuation port between them.

However, there has been remarkable progress in ion source structure [1], in which a magnetically confined plasma has been put into practical use. Such a new ion source has the merits of extracting a large ion beam current due to its large arc efficiency, and of working in the vicinity of 0.3 Pa. The behavior of the ion source at such low gas pressures is expected to reduce the arcing in extraction electrodes.

When the new ion source is operated at low gas pressure in the NBI, the neutralization efficiency η_{10} is reduced as the gas pressure in the neutral cell is reduced. (All symbols are defined in Section 5, Appendix.) The total efficiency η_T is also reduced, because η_T is the product of η_{10}, $(1 - \eta_{01})$ and η_G. Therefore, the gas feed method into a neutral cell has been examined in this paper in order to increase total efficiency.

An analysis of feeding gas into a neutral cell has been given by Feist [2]. In his analysis, gas is fed into the center of the neutral cell by considering symmetry, because the cell is put into a vacuum box and both cell ends are evacuated. However, when the particle inlet of a neutral cell is directly connected to an ion source without evacuation, as in this paper, an optimizing analysis of gas feed conditions is important.

2. CALCULATION METHOD

2.1 Comparison of caluclation methods

Analysis by Feist [2] uses steady state vacuum calculations, which requires iterative calculations for non-linear terms. In this paper, transient vacuum calculations are used, because the iterative calculations for non-linear terms are performed at the same time as the iterative calculations to solve differential equations.

Reports by Lee, et al. [3], Sredniawski [4, 5] and Maekawa, et al. [6] have described transient calculations for analyzing a complicated vacuum system. These include a treatment of the neutral cell as a single vacuum tubing [3, 4, 5]. Analysis of the gas inlet position in the neutral cell is also conveniently done using the code in which the neutral cell is divided into five parts and each part is connected to a reservoir with variable vacuum conductances and a puffing valve [6]. It is this approach that is used for calculations in this paper.

2.2 Details of calculation method [6]

Fig. 1 shows the network and symbols used in the calculation. The following assumptions were made.

(1) All the hydrogen ions are protons.

(2) The beaming effect [2] at the beam outlet of the neutral cell is not considered.

(3) The gas pressure P_T in the plasma experimental device and the timing chart for various valves and the ion beam are the same as in reference [6].

(4) The values of vacuum conductances are calculated for molcular flow [7].

(5) The values of cross-sections for 30 keV hydrogen are as follows [8]:

$$\sigma_{10} = 4 \times 10^{-20}\ (m^2) \text{ and } \sigma_{01} = 1.5 \times 10^{-20}\ (m^2).$$

The equations relating gas flow rates and gas pressures, and calculated conditions, are almost the same as in reference [6], except for the gas feed conditions shown in table I. The following equations relating to efficiencies, gas flows and ΔP_J are used in the code.

$$\eta_{gas} = \frac{Q_{JH}}{Pos(Q_{RJ}) + Pos(-Q_{J1})} \quad \cdots\cdots (1)$$

$$\eta_{10} = \frac{\sigma_{10}}{\sigma_{10}+\sigma_{01}}[1-\exp(-(\sigma_{10}+\sigma_{01})R_N)] \quad \cdots\cdots (2)$$

$$R_N = 2.65\times10^{20}\int_{\text{Neutral cell}} P\,dl \quad \cdots\cdots (3)$$

$$\eta_{01} = 1-\exp(\sigma_{01} R_{BL}) \quad \cdots\cdots (4)$$

$$R_{BL} = 2.65\times10^{20}\int_{\text{Beam line}} P\,dl \quad \cdots\cdots (5)$$

$$(1-\eta_{01}) = (1-\eta_{01A})(1-\eta_{01B})(1-\eta_{01D}) \quad \cdots\cdots (6)$$

$$\eta_T = \eta_{10}(1-\eta_{01})\,\eta_G \quad \cdots\cdots (7)$$

$$U_T = \int_0^{2.3} Pos(Q_{BD})\,dt \quad \cdots\cdots (8)$$

$$\Delta P_J = 1 - \frac{Min(P_J(t=0.1-0.3))}{Max(P_J(t=0.1-0.3))} \quad \cdots\cdots (9)$$

$$U_R = \int_{t_g}^{0.3}(Q_{RJ}+Q_{R1}+Q_{R2}+Q_{R3}+Q_{R4}+Q_{R5})\,dt \quad \cdots\cdots (10)$$

The focusing characteristic of the ion beam would be varied during the pulse if the value of ΔP_J is large. Therefore, the minimum ΔP_J is always chosen by adjusting t_g during all calculations.

3. RESULTS OF CALCULATIONS

Results of calculations made with the conditions in table I are shown in figs. 2-5.

Fig. 2 shows total efficiency $\eta_T(t=0.3)$. In case (A), in which the hydrogen gas is fed into the ion source alone, total efficiency η_T is reduced as the gas pressure in the ion source is reduced. The decrease of η_T depends on the decrease of neutralization efficiency by the decrease of gas pressure in the neutral cell. In case (B), hydrogen gas is simultaneously fed into the ion source and part 1 of the neutral cell. In certain gas pressure ranges in the ion source, the total efficiency η_T in case (B) is larger than in (A).

The sum U_R of gas flows into ion source and neutral cell during one pulse, calculated using eq (10), is shown in fig. 3. When the gas flow into the cryopump is increased, heat loading to liquid helium in the cryopump is also increased, so that frequent regeneration of the cryopump is necessary. Case (B) has the drawback of a larger gas flow than in case (A). In case (C), hydrogen gas is simultaneously fed into the ion source and part 3 of the neutral cell. Case (C) is worse than case (B), because while the former shows a small improvement in total efficiency and in neutralization efficiency, it has a very large gas flow from the reservoir.

Cases (D) and (E) are shown in figs. 2 and 3. In case (D), hydrogen gas is fed to part 1 of neutral cell alone. In case (E), the gas is fed to part 3 alone. Large total efficiencies in both cases are remarkable.

Gas efficiency η_{gas} of the ion source, defined by eq (1), is shown in fig. 4. Gas efficiencies are nearly constant and the same in cases (B) and (C). This means that gas flow rate into the ion source is nearly constant and the same, even if the gas flow rate into the neutral cell is varied. Gas efficiencies η_{gas} in cases (D) and (E) are 1.

Fig. 5 shows the time dependencies of gas pressures in NBI, where $C_{R1} = 2.5 \times 10^{-5} m^3 s^{-1}$ in case (B). Gas pressure P_D in the beam duct is increased during t=0.1 and 0.3 s, because high energy ions and neutrals hit the wall in the beam duct.

4. CONCLUSIONS

In the usual NBI, which has a neutral cell directly connected to an ion source without evacuation, transient vacuum analysis was performed for cases of gas feed into a specific part of the neutral cell.

The analysis provided information about gas pressures, efficiencies and gas flows.

The following conclusions were derived from the analysis.

(1) When the gas is fed into the ion source alone, total efficiency is as low as the gas pressure in the ion source.

(2) When the gas is fed into a part of the neutral cell and the ion source simultaneously, there is a range of gas pressures where the total efficiency and gas efficiency are higher than in (1) above.

(3) Total gas flow from the gas reservoir is low when the gas inlet position in the neutral cell in (2) above is close to the ion source.

(4) When the gas is fed into a part of the neutral cell alone, there is a range of gas pressures where the total efficiency is higher than in (1) and (2) above, and gas efficiency is slways 1.

5. APPENDIX: SYMBOLS

A : relating to vacuum box A

B : relating to vacuum box B

C : Vacuum conductance ($m^3 s^{-1}$)

$\bar{C}$: Vacuum conductance during some period ($m^3 s^{-1}$)

D : relating to beam duct

E : relating to the position between the extraction electrode system of ion source and neutral cell

H : relating to ion beam or neutral beam

i (i = 1, ..., 5) : relating to No. i part of neutral cell

J : relating to ion source

j : Hydrogen ion current (A)

l : Length (m)

Max (P_J(t=0.1-0.3)) : Maximum value of P_J during ion beam extraction period (t=0.1 -0.3 (s))

Min (P_J(t=0.1-0.3)) : Minimum value of P_J during ion beam extraction period (t=0.1-0.3 (s))

P : Gas pressure (Pa)

Pos(x) : Pos(x) = x (if $x \geqq 0$); Pos(x) = 0 (if $x \leqq 0$)

Q : Gas flow rate (Pa $m^3 s^{-1}$)

Q_{JH} : Converted hydrogen (H_2) gas flow rate

corresponding to hydrogen ion (H^+) current of j (A) ; ($Pa\ m^3\ s^{-1}$)

R : relating to reservoir tank of hydrogen gas

R_{BL} : Line integral of hydrogen gas density in beam line

R_N : Line integral of hydrogen gas density in neutral cell

S : Evacuation speed ($m^3\ s^{-1}$)

T : relating to plasma experimental device

t : Time (s)

t_g : Starting time of gas puffing from reservoir, in order to give minimum value of ΔP_J (s)

U_R : Hydrogen gas flow per pulse from reservoir to NBI ($Pa\ m^3$)

U_T : Hydrogen gas flow per pulse from NBI to plasma experimental device ($Pa\ m^3$)

V : Volume (m^3)

ΔP_J : Variance of gas pressure P_J in ion source during ion beam extraction period (t=0.1-0.3 (s)); (Pa)

σ_{01} : Cross-section of ionization (m^2)

σ_{10} : Cross-section of neutralization (m^2)

η_G : Geometrical injection efficiency of neutral beam

η_{gas} : Gas (availability) efficiency of ion source (given by eq (1))

η_T : Total efficiency (given by eq (7))

η_{01} : Re-ionization efficiency (given by eq (4))

η_{10} : Neutralization efficiency (given by eq (2))

6. REFERENCES

[1] Green, T.S., 10th Symposium on Fusion Technology, Padova, 4-9 September, 1978. Proceedings pp 873-936.

[2] Feist, J.-H., 7th Symposium on Engineering Problems of Fusion Research, Knoxville, 25-28 October, 1977. Proceedings pp 1554-1556.

[3] Lee, A.L., et al., 6th Symposium on Engineering Problems of Fusion Research, San Diego, 18-21 November, 1975. Proceedings pp 833-846.

[4] Sredniawski, J., 7th Symposium on Engineering Problems of Fusion Research, Knoxville, 25-28 October, 1977. Proceedings pp 1806-1810.

[5] Sredniawski, J., Princeton Plasma Physics Laboratory TFTR Report EP-090 (1977)

[6] Maekawa, A., et al., 8th Symposium on Engineering Problems of Fusion Research, San Francisco, 13-16 November, 1979. Proceedings pp 518-523.

[7] Roth, A., Vacuum Technology. North-Holland Publishing Co., Amsterdam (1976)

[8] Barnett, C.F., et al., Oak Ridge National Laboratory Report ORNL-5206 (1977)

TABLE I Calculated Conditions for Gas Feeds

Cases	Gas feeding positions from hydrogen reservoir	Vacuum conductances [$m^3\ s^{-1}$; $t_g \leq t \leq 0.3$ (s)]					
		C_{RJ}	C_{R1}	C_{R2}	C_{R3}	C_{R4}	C_{R5}
(A)	Ion source	varied	0	0	0	0	0
(B)	Ion source and part 1 of neutral cell	2×10^{-5}	varied	0	0	0	0
(C)	Ion source and part 3 of neutral cell	2×10^{-5}	0	0	varied	0	0
(D)	Part 1 of neutral cell	0	varied	0	0	0	0
(E)	Part 3 of neutral cell	0	0	0	varied	0	0

All values of vacuum conductances in table I are 0, when $t < t_g$ or $t > 0.3$(s).

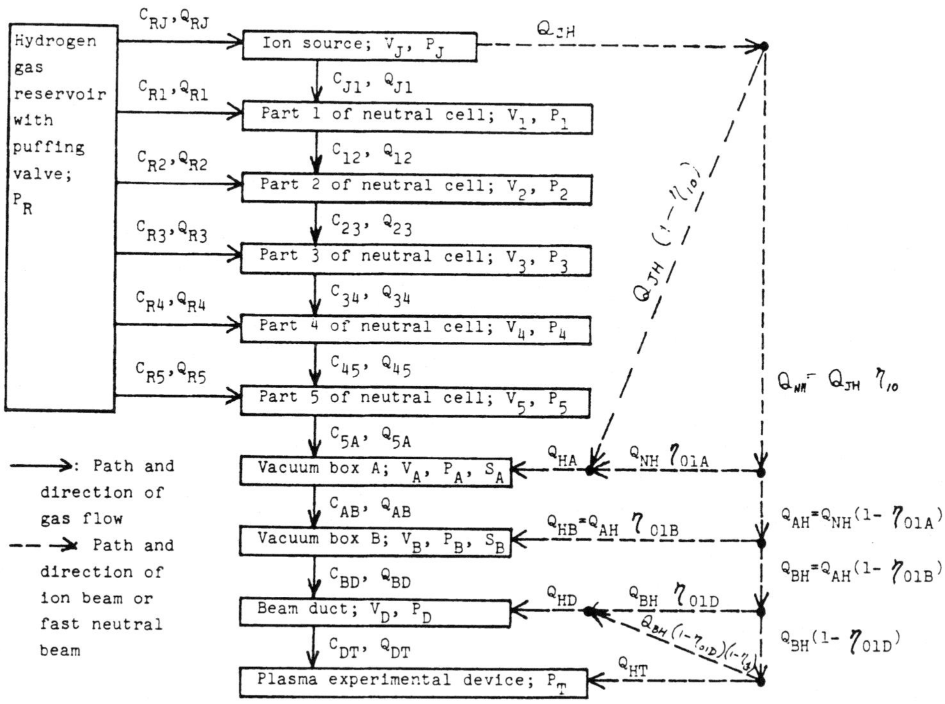

Fig. 1 Network and symbols used for the transient calculations of gas pressures in the NBI

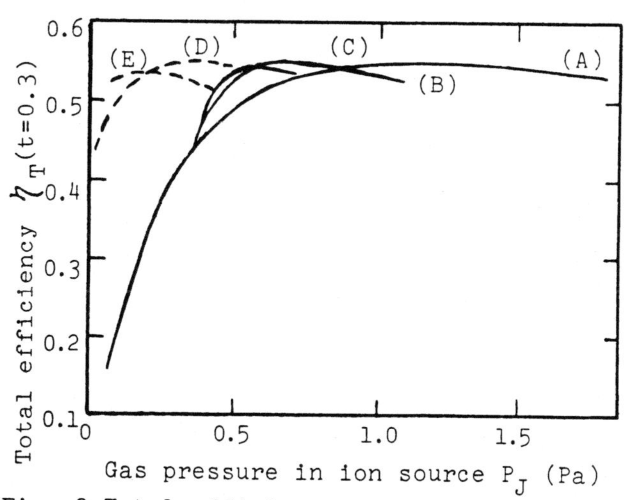

Fig. 2 Total efficiency versus gas pressure

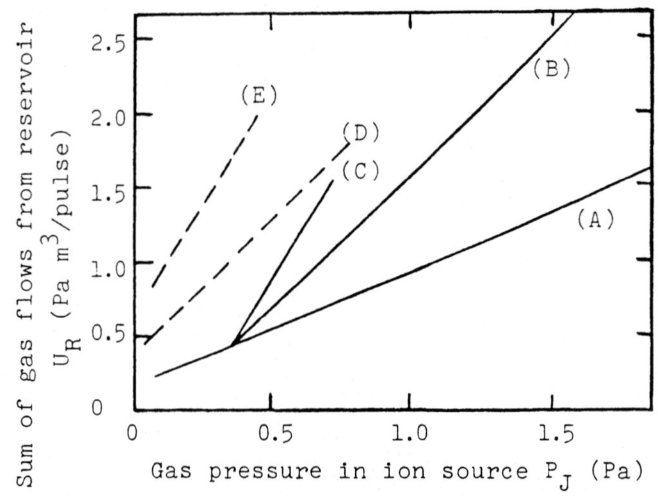

Fig.3 Sum of gas flows from reservoir versus gas pressure

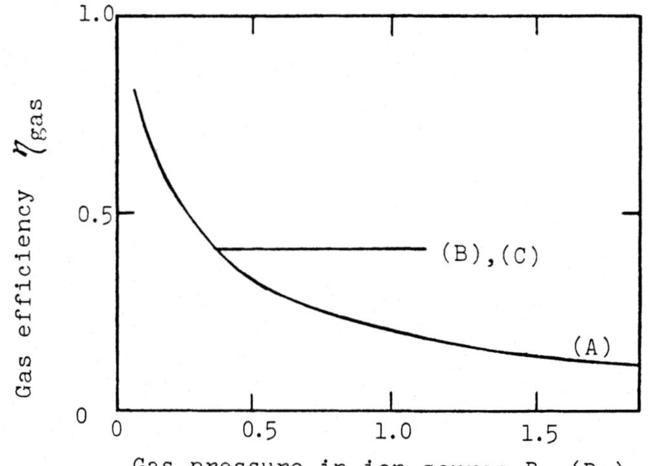

Fig. 4 Gas efficiency versus gas pressure

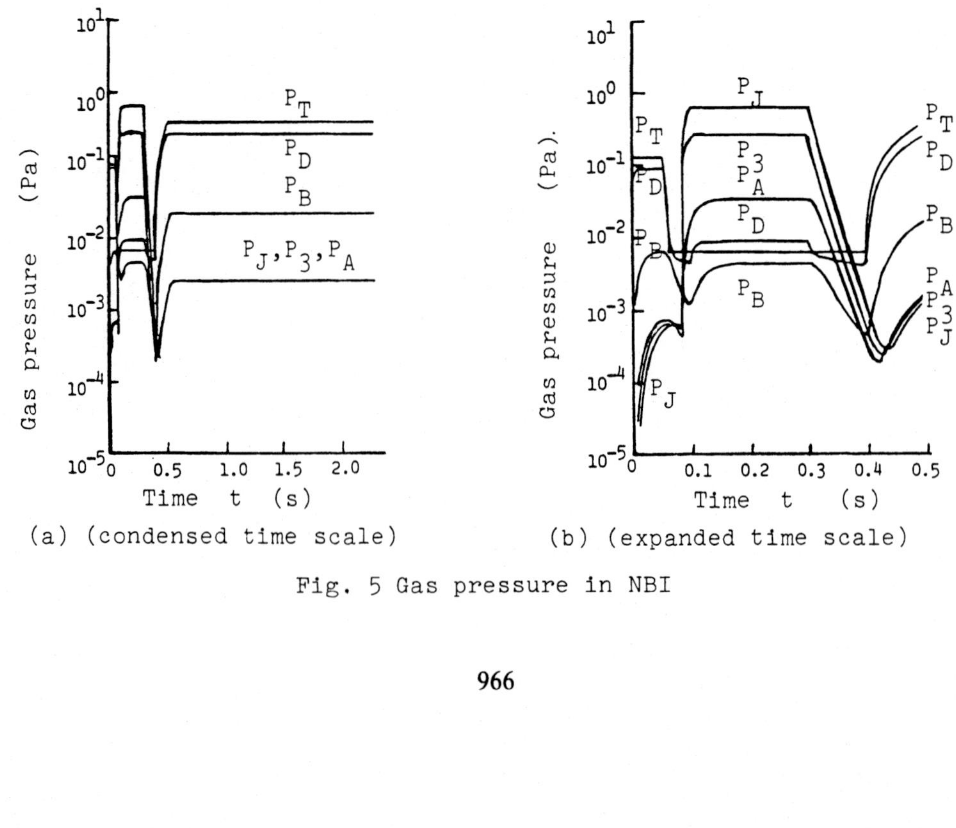

Fig. 5 Gas pressure in NBI

NEUTRAL BEAM INJECTOR PERFORMANCE ON THE PLT AND PDX TOKAMAKS

G. Schilling, D. L. Ashcroft, H. P. Eubank, L. R. Grisham, T. A. Kozub,
H. W. Kugel, J. Rossmassler, M. D. Williams
Plasma Physics Laboratory, Princeton University
Princeton, New Jersey 08544 USA

OVERVIEW

Beams of fast neutral atoms injected into initially cold target plasmas have aided in the study of the properties of hot confined plasmas approaching reactor-like conditions. Neutral beam injection has been used as a plasma heating tool on the PLT tokamak for about three years now, resulting in a plasma ion temperature of 7 keV, above the minimum reactor ignition requirement, and driving the bulk plasma deep into the collisionless reactor-like regime [1]. Injection heating has been applied to plasmas in the PDX tokamak, starting in March 1980. The parallel conditioning and shakedown procedures of the first two injectors and the tokamak discharges are now yielding initial information [2].

1. INJECTION SYSTEM DESCRIPTION

The mechanical and electrical systems to a large extent have been described conceptually and in initial tests by the beam developers and electrical engineers. Our emphasis here, therefore, will be to point out those system features unique to multiple injector operation on the tokamaks, general injection problems including component failures and improvements, and system achievements to date.

The aim of the PLT beam heating experimental program has been primarily the achievement of high plasma temperatures with the corresponding reduction in plasma collisionality characteristic of the expected reactor regime. This favors the classical tangential injection geometry, where adequate beam absorption is retained even at low plasma densities of the order of 10^{13} cm^{-3}. Lack of knowledge about any effects of plasma rotation led to the tangential injection in opposed pairs, as is shown in fig. 1. The PLT injector design goal was initially set at 4 MW H^o total power into the torus. This was reduced to 3 MW H^o during the injection line design phase when the line was lengthened by 75 cm to add a bending magnet to deflect the unneutralized beam fraction. The desired power per injector of 0.75 MW H^o was expected to be reached by a 22 cm diameter straight bore multiaperture accelerating structure rated at 40 kV/60 A/0.3 sec [3]; this power was achieved on the ORNL test stand, which geometrically simulated PLT but could not simulate the magnetic and electrical tokamak interactions, at the operating parameters of 40 kV/70 A/0.1 sec [4].

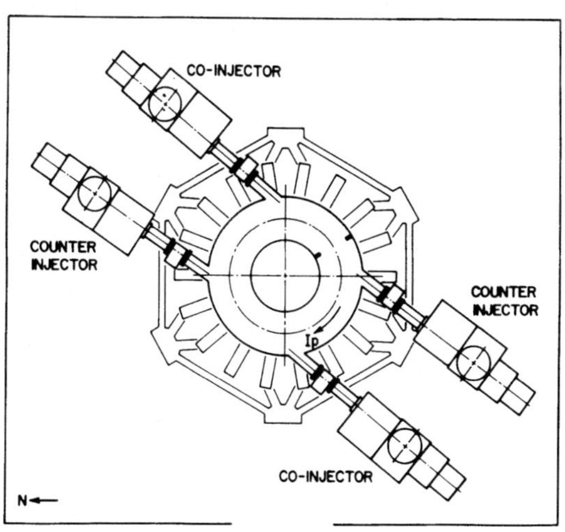

Fig. 1. PLT schematic indicating direction of plasma current and position of neutral beam injectors.

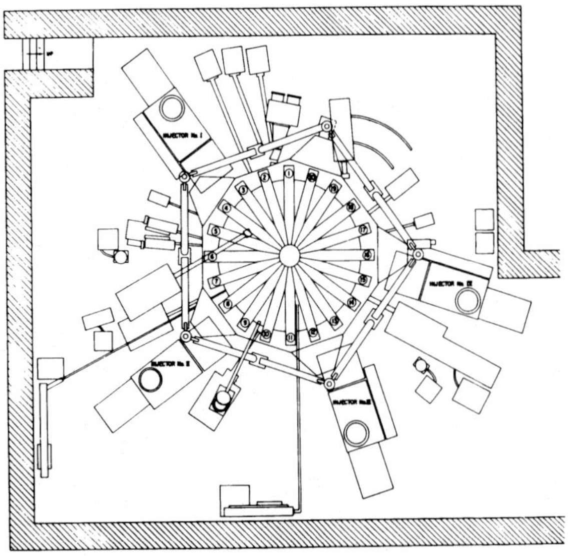

Fig. 2. PDX schematic indicating position of the four neutral beam injectors.

The PDX beam heating experimental program is aimed toward the achievement of the highest possible plasma pressure, proportional to the product of plasma density and temperature. The problem now is one of adequate beam penetration to heat the plasma core at the higher plasma densities of the order of 10^{14} cm^{-3}. This favors the radial injection geometry shown in fig. 2, with a slight tangential component, 9 degrees, to avoid fast ion trapping in local toroidal field mirrors. Since the tangential component is small, resultant plasma rotation is also small, and all injectors may be aimed in the co-direction, avoiding the large fast ion orbit loss cone of radial counter-injection. The PDX injector design goal was set at 6 MW H^o total power into the torus [5]. The corresponding power per injector of 1.5 MW H^o was expected to be reached by an ion optically improved tapered bore multiaperture accelerating structure, 30 cm diameter, initially rated at 50 kV/100 A/0.5 sec [6]. The power level achieved on the ORNL test stand, this time simulating the PDX geometry but again not simulating the magnetic and electrical tokamak interactions, was 1.5 MW H^o at the operating parameters of 50 kV/90 A/0.3 sec and 2.0 MW D^o at 53 kV/85 A/0.1 sec [7].

2. PLT INJECTORS

The PLT neutral beam injectors were developed and fabricated by the Fusion Energy Division of the Oak Ridge National Laboratory. The injection line is that logical geometric sequence of hardware reflecting the physics requirements of the simplest possible positive ion based injector; in order: low pressure diffuse arc plasma generator, electrostatic multiaperture ion beam accelerator, neutral gas charge exchange cell, residual ion deflecting magnet and ion dumps, and fast neutrals drift duct to the torus with beam conditioning and diagnostic calorimeters [8]. The injection line is shown schematically in fig. 3.

The modified duoPIGatron plasma generator has illuminated the accelerator structure in an adequate fashion. The 85% H^+ atomic ion yield has been verified, which is beneficial for beam penetration, but the injectors have not yet reached the 0.75 MW H^o power goal, with possible, but so far unmeasured, plasma density variations being at least one reasonable mechanism.

The stability of the plasma generator arc discharge has become our greatest obstacle to reliable high power source operation once the accelerator grids are conditioned. Direct arc faulting, primarily in the form of almost total arc current flow to anode one, terminates the injection pulse. An unstable or noisy arc discharge modulates the plasma density at the accelerating grids, resulting in corresponding fluctuations in beam divergence. The resulting spray particles within the accelerating structure liberate occluded gas, heat the grid surface, and generate secondary particles, all of which increase the probability of high voltage breakdown. The plasma generators were subsequently modified in the spring of 1980 [9], with the results that arc conditioning time has been shortened considerably, arc faults have practically vanished, the discharge appears to operate more stably, and our duty cycle at the highest power levels has increased considerably.

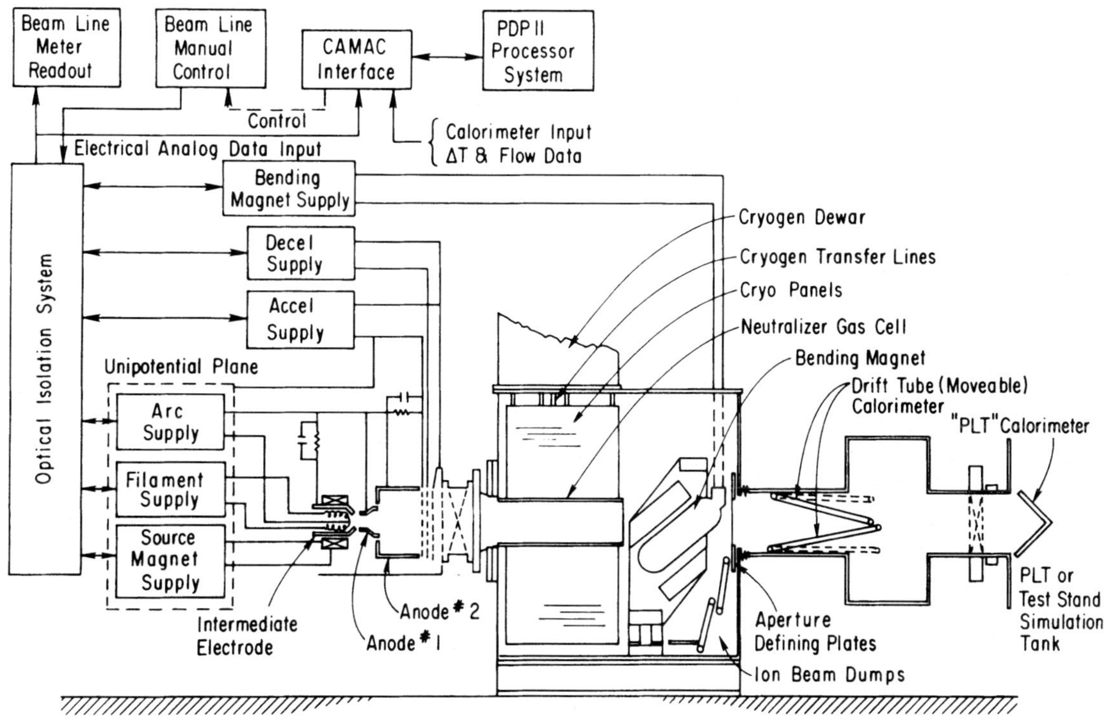

Fig. 3. PLT Neutral Beam Injector

The ion accelerating structure was altered in the late fall of 1978 through the replacement of the original straight-bore aperture grids by the newer tapered-aperture grids developed for ISX [10]. The Pierce-like geometry results in reduced beam divergence, although at a slightly reduced current resulting from a reduction in aperture area. The result is approximately the same power into the PLT torus, but with considerable reduction in spray power onto the drift duct walls. Operation at the new perveance optimum of 40 kV/53 A for our usual 0.15 sec pulse length results in approximately 0.65 MW H^o power into the PLT torus. Operation with deuterium gas results in approximately 20% lower current at higher neutralization efficiency; most of the injectors have operated at 44 kV/51 A/0.15 sec individually, resulting in 1.0 MW D^o power into PLT. In general, we have observed little systematic change in power transmission with increasing beam energy, i.e., I·V·t.

Neutral beam power transmission to the torus is measured as the ratio of neutral power measured calorimetrically on the PLT calorimeter, which is sensitive to all magnetic and electrical tokamak interactions, to the total neutral power generated, $0.95 \cdot F^o_\infty \cdot I \cdot V$. The average fraction for the four injectors is approximately 55%, H^o and D^o, with preliminary measurements of slightly higher values for the first injector using the modified plasma generator. A Gaussian model for the total beam shape suggests that the full-width-half-maximum power points equal the PLT torus aperture geometrical acceptance, ±1.7°, which is up to three times larger than the single-aperture divergence measured for these tapered apertures [11]. Although several possible mechanisms come to mind, we presently have no measurements supporting or refuting any of them. This problem must continue to receive careful investigation, especially in view of the increasing costs and complexity of the higher energy beams for future experiments.

The increase in beam duty cycle at the 40-45 kV power level has resulted in a corresponding decrease in the lifetime of downstream components. As an example, within the six months following the North injector plasma generator modification and its first achievement of 0.9 MW D^o power into the torus early in 1980, two drift duct calorimeter halves became damaged to the point of releasing cooling water into the vacuum system, and swirl tubes in the full energy ion dump were melted through with similar results. Turnaround time for repair was two to four weeks in each case. At the lower earlier power levels inadequately shielded water headers and swirl tube braze joints had already become damaged to the point of leaking on several injectors. One instance of swirl tube fatigue was observed on several tubes in the East injector drift duct calorimeter, where vibration from the highly turbulent water loosened the internally swaged ribbon, causing it to chisel through the copper tube into the vacuum.

Initial overheating of the drift duct walls, uncooled stainless steel, by spray beams caused surface and volume gas release as well as vacuum leaks resulting from seal warping. This was brought under control through the reduction in aperture at the main tank exit, addition of water-cooled copper baffles to the exit of the drift duct pump box, addition of external water cooling lines, and improvement in beam optics with the tapered aperture grids. Residual beam interaction with the drift duct now causes observed reionization losses of less than ten percent when the beams are injected into the PLT plasma.

Injection line vacuum pumping of the excess hydrogen or deuterium gas is performed by a liquid helium cooled cryocondensing pump straddling the neutralizer cell. The measured pumping speed in the injection line is 280,000 ℓ/s for H_2. With hydrogen beams LHe consumption is 5 ℓ/h, while with deuterium beams the consumption drops to the thermal radiation level of 3 ℓ/h. The integral dewar is refilled on a batch transfer basis from a central 4000 ℓ dewar with a transfer efficiency of 50 percent. We have found this technique to be financially advantageous over a closed-loop system and to be simple and reliable. We continue to be plagued by small cryogenic vacuum leaks which close at room temperature; presently, one injector is base pressure limited to $5 \cdot 10^{-7}$ torr by a LN_2 leak while another is limited to $3 \cdot 10^{-6}$ torr by a LHe leak. Since typical tank pressure rises to $1\text{-}2 \cdot 10^{-4}$ torr during a beam pulse, the leaks contribute to LHe consumption but have not yet interfered with the injection experiments.

3. PDX INJECTORS

The PDX injection line is shown schematically in fig. 4. The basic similarity to that of PLT is readily apparent. The most obvious changes are those related to the containment of 5 MW of extracted beam power for up to 0.5 sec, as compared to 2.4 MW for 0.3 sec on PLT [12]. The full-energy ion dump is now designed as a slanted "vee" semi-inertial calorimeter fabricated from 1.9 cm thick OFHC copper plate, a result of the PLT swirl tube braze joint difficulties. The drift duct calorimeter is now also of semi-inertial design, fabricated from 1.9 cm thick OFHC copper, but now is an optically tight straight "vee" which is moved out of the beam path as a unit when injection into PDX is performed. This latter change is the result of occasional inadequate overlapping of the two corresponding calorimeter halves on PLT, with beam heating of the torus valve sufficient to cause warping and vacuum leaks.

Although the radial injection geometry of PDX allows the drift duct to be shortened and increased in diameter as compared to that on PLT, doubling of the beam power for almost twice the pulse length would increase the beam energy onto the thermally integrating drift

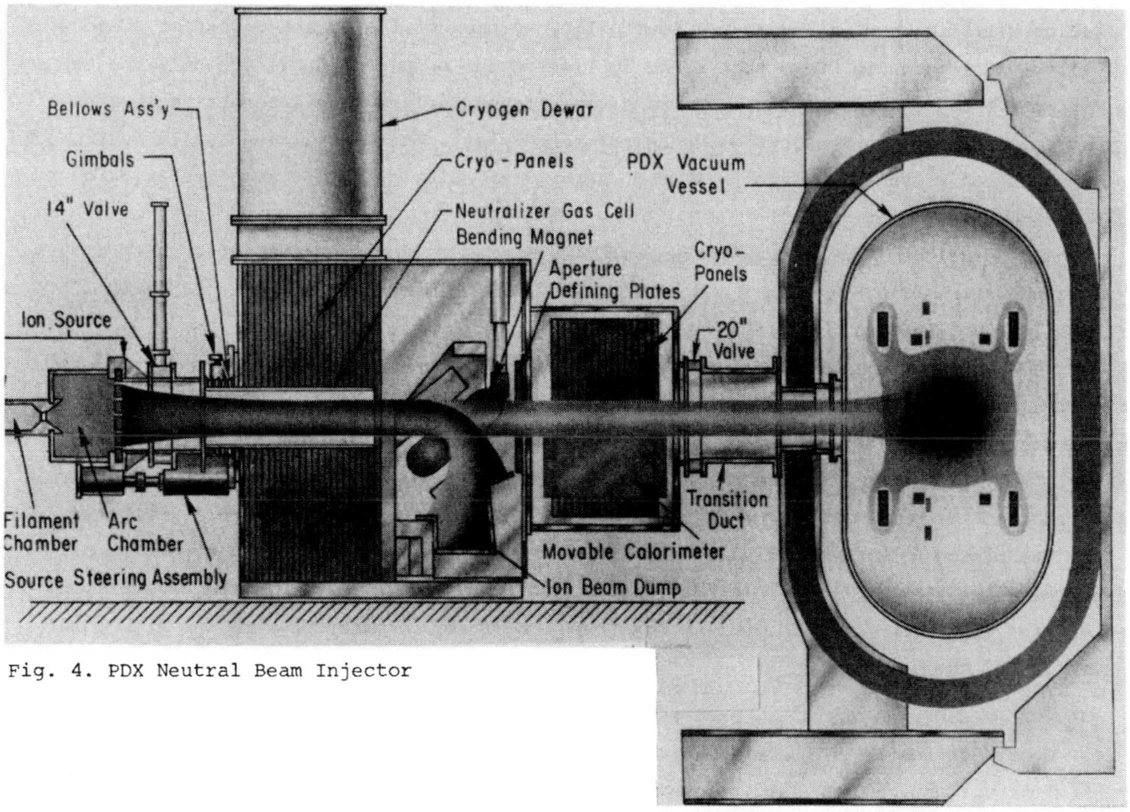

Fig. 4. PDX Neutral Beam Injector

duct walls by almost a factor of four over PLT, assuming equal spray beam interception. This increase and the resulting gas evolution were estimated to cause unacceptable reionization losses for pulse lengths greater than 0.5 seconds in the model of Stewart [13]. The drift duct was therefore enlarged considerably and a second cryopump was installed, fed from the dewars of the main pump through cryogen supply and return lines internal to the vacuum tank. The internal demountable cryogenic vacuum joints proved to be unreliable; in addition, several welds cracked in the LN_2 feed lines at the cryopanels. The no-gas LHe consumption increased from 5 ℓ/h to 8 ℓ/h. The existing two injectors on PDX were therefore installed without drift duct active pumping. This approach allows the plumbing problems to be studied off-line, with retrofitting at a later time. The remaining two injectors are scheduled to be installed on PDX this coming winter. Preliminary results on drift duct outgassing and reionization losses will be presented in the Bull. Am. Phys. Soc.

Since a potential problem with radial beam injection is "shine-through" onto the torus inner wall, the desire for local wall armor was combined with power measuring capability in the form of a fixed inner torus wall calorimeter [14]. Beam shine-through may be measured for each injected pulse, but in contrast to the movable PLT torus calorimeter, injected beam power may only be measured without a plasma present. This latter condition may mask some injector-tokamak magnetic or electrical interactions.

4. ELECTRICAL SYSTEM

The electrical system for the two injector arrays is almost identical [15], with each array having separate decel and auxiliary supplies and time sharing the accel high voltage supplies. This latter approach is practical since the PLT and PDX experiments operate on an alternating weekly schedule.

The "accel" high voltage circuit is unique in that no primary AC voltage regulation is used [16]. Output high voltage regulation is accomplished by a phase-controlled ignitron rectifier bridge. A series hard-tube modulator accomplished fine-voltage regulation as well as fast turn-off in case of ion source fault [17]. Proper timing of the rectifier phase control allows a flat-topped DC voltage to be applied to the modulator tube, avoiding the turn-on and turn-off transients resulting from transformer impedance and line interaction with the tokamak power supplies. The initial gated reference phase control will eventually be replaced by a closed loop voltage regulator. Stored energy available in case of accelerator grid fault is about 20 joules.

The "decel" voltage supplies are essentially straightforward commercial units with a fixed output transformer-rectifier and series tetrode modulator for all voltage regulation, current limiting, and pulse timing.

The plasma generator arc supplies consist of a twelve-pulse transformer-rectifier with L-C-L output filtering. The SCR rectifier is gated on and off to provide the arc pulse; arc voltage variation is provided by a motor-driven autotransformer on the arc transformer primary. Arc faults cause the rectifier SCR's to be gated off and a crowbar SCR to safely discharge the energy stored in the filter. Arc current stabilization is purely resistive.

The plasma generator auxiliary supplies for filaments and source magnet are simple transformer-rectifier supplies with motor-driven autotransformer primary voltage variation. They are fed from a separate isolation transformer to avoid interaction with the high power consumption of the arc load.

5. INSTRUMENTATION AND CONTROLS

Communication to and from the plasma generator and its supplies floating at the accel potential is via optically linked telemetry [18]. Analog signals are sampled by a 120 kHz bandwidth voltage-to-frequency converter. The resulting pulse train drives a light-emitting diode, whose output is transmitted via a light pipe to a photo-diode and then reconverted. Digital signals bypass the conversion but are transmitted identically. This system has performed well in the sense of adequate bandwidth and general reliability, but high voltage faults at higher power levels still occasionally damage the low-level integrated circuits in those sensing amplifiers which float at high voltages.

Beam power is determined calorimetrically. Calorimeter water flow, as measured by the output frequency of a turbine flowmeter, is multiplied by the integral of the water temperature pulse, performed by counting the output of a voltage-to-frequency converter. The resulting total energy is then displayed in the control room. This system [19] has behaved well in general, exceptions being the frequent need for manual zero-setting resulting from water pressure and temperature fluctuations and sensitivity to turbine signal jitter.

Plasma generator and ion accelerator waveforms as well as power depositions are now routinely sampled by an automated data acquisition system [20] based on a CAMAC interface to a PDP 11/34 processor. Waveforms and samples are collected for each injector and then archived with the corresponding tokamak shot. Initial automated source conditioning experiments have been carried out, but work on active participation by the computer must still be extended.

6. SUMMARY

The injection of fast neutral beams has proved to be a valuable tool for heating confined plasmas. Injection of up to 2.5 MW H^o and up to 3.4 MW D^o into the PLT tokamak has resulted in the achievement of an ion temperature of 7 keV, with the bulk of the plasma in the region of low collisionality characteristic of expected reactor operation. Initial injection of up to 2 MW D^o with the first two PDX injectors has allowed the investigation of heated divertor plasmas to commence.

Many difficulties have appeared in the coupling of arrays of high-power injectors to the tokamaks; none have proved insurmountable. The problems of "large" neutral beam divergence as represented by the presently achieved fraction of power transmitted to the plasma as well as the containment of higher power levels for longer pulse lengths appear especially worthy of increased investigation. The results achieved so far have easily justified these efforts.

ACKNOWLEDGEMENTS

We gratefully acknowledge the long hours our technical staff has invested in maintaining, repairing, and improving the injection hardware. The continued support and encouragement of Drs. M. B. Gottlieb, H. P. Furth, D. M. Meade, and W. Stodiek are appreciated.

This work has been supported by U. S. Department of Energy Contract No. DE-AC02-76-CH03073.

REFERENCES

[1] STODIEK, W., et al., in Plasma Physics and Controlled Nuclear Fusion Research (Proceedings 8th International Conference, Brussels, Belgium, 1980) paper IAEA-CN-38/A-1.

[2] MEADE, D., et al., in Plasma Physics and Controlled Nuclear Fusion Research (Proceedings 8th International Conference, Brussels, Belgium, 1980) paper IAEA-CN-38/X-1.

[3] TSAI, C. C., et al., Seventh Symposium on Engineering Problems of Fusion Research, Knoxville, 25-28 October, 1977. Proceedings pp 278-283.

[4] DAGENHART, W. K., et al., Seventh Symposium on Engineering Problems of Fusion Research, Knoxville, 25-28 October, 1977. Proceedings pp 533-538.

[5] SIMON, E. E., Eighth Symposium on Engineering Problems of Fusion Research, San Francisco, 13-16 November, 1979. Proceedings pp 1370-1374.

[6] MENON, M. M., et al., Eighth Symposium on Engineering Problems of Fusion Research, San Francisco, 13-16 November, 1979. Proceedings pp 656-660.

[7] GARDNER, W. L., ORNL, private communication.

[8] JOHNSON, R. L., et al., Seventh Symposium on Engineering Problems of Fusion Research, Knoxville, 25-28 October, 1977. Proceedings pp 530-532.

[9] SCHILLING, G., to be published.

[10] FUSION ENERGY DIVISION Annual Progress Report, Oak Ridge National Laboratory Report ORNL-5549 (1978). p. 112.

[11] GRISHAM, L. R., et al., Rev. Sci. Instrum. 48, 1037 (1977).

[12] GARDNER, W. L., et al., Eighth Symposium on Engineering Problems of Fusion Research, San Francisco, 13-16 November, 1979. Proceedings pp 972-975. JOHNSON, R. L., et al., Ibid. pp 1375-1378.

[13] STEWART, L. D., et al., Eighth Symposium on Engineering Problems of Fusion Research, San Francisco, 13-16 November 1979. Proceedings pp 845-848.

[14] ULRICKSON, M., KUGEL, H. W., Bull. Am. Phys. Soc. 24, 998 (1979).

[15] KLOIBER, F. W., Eighth Symposium on Engineering Problems of Fusion Research, San Francisco, 13-16 November, 1979. Proceedings pp 1256-1259.

[16] ASHCROFT, D. L., ZUVERS, H., Seventh Symposium on Engineering Problems of Fusion Research, Knoxville, 25-28 October, 1977. Proceedings pp 1147-1150.

[17] PEARSON, G. G., Eighth Symposium on Engineering Problems of Fusion Research, San Francisco, 13-16 November, 1979. Proceedings pp 2017-2018.

[18] WRIGHT, R., E., Seventh Symposium on Engineering Problems of Fusion Research, Knoxville TN, 25-28 October , 1977. Proceedings pp 1579-1581. SILVERMAN, K. B., Eighth Symposium on Engineering Problems of Fusion Research, San Francisco, 13-16 November, 1979. Proceedings pp 1917-1921.

[19] SILVERMAN, K. B., Seventh Symposium on Engineering Problems of Fusion Research, Knoxville, 25-28 October, 1977. Proceedings pp 418-420.

[20] STOOKSBERRY, R. W., et al., Topical Conference on Computerized Data-Acquisition Systems in Particle and Nuclear Physics, Santa Fe, 1979. IEEE Trans. on Nucl. Sci., Vol. NS-26, 4,August 1979, pp 4623-4632.

CONSIDERATIONS AND CALCULATIONS FOR THE NEUTRAL-INJECTION SYSTEM IN Z E P H Y R

J.H. Feist*, W. Herrmann, E. Speth
Max-Planck-Institut fuer Plasmaphysik
Euratom-Association, D-8046 Garching

* on leave of absence in LBL, Berkeley,CA.

ABSTRACT

The heating of the ZEPHYR plasma to ignition temperatures shall be accomplished by neutral injection and subsequent compression of the plasma. 25 MW of neutral power are required. A parametrical study of beam transmission was performed to find limitations on the type and arrangement of sources imposed by the small width of the porthole to the torus and the small height of the plasma target. A reference beam line has been designed and the power transmission calculated as a function of gas flow through the sources. The limitations mentioned make it necessary to design the beam lines as compact as possible to avoid intolerable losses in the duct or power outside the plasma target. Depending on the reliability of a single source, the number of sources that are available in at least 75 % of the discharges has been calculated for different numbers of installed sources.

1. INTRODUCTION

To obtain ignition in the ZEPHYR experiment, its ohmically preheated plasma has to be heated from external sources to the ignition temperature. As external sources mainly HF methods and neutral injection are considered. In case of neutral injection, heating is only effective if the injected particles can penetrate to the centre of the plasma. For the dimensions and densities of an ignition experiment, deuteron energies of a few hundred keV would be necessary if the plasma should be directly heated by the neutrals. As large powers at these energies are not yet available the heating scheme with neutral injection has therefore been modified. After ohmic heating and neutral injection, the final ignited state when α-particle heating is dominant is achieved by compression of the plasma. The compression factor for the neutral injection-heated ZEPHYR is chosen such that the plasma before compression can be sufficiently penetrated by 160 keV deuterons. Large powers of neutrals at this energy are expected to be available when ZEPHYR goes into operation.

25 MW of deuterons are required to heat ZEPHYR [1], of which at least 60 % have to be of full energy. The plasma target at which this power should be aimed, is rather small. Vertically it contains the inner 1/3 diameter of the plasma, i.e. the target has a height of ± 21 cm. Horizontally the tangential radii shall be between 28 cm and 86 cm. At smaller tangential radii first orbit losses are too large, at larger radii the particles do not penetrate to the plasma centre to a sufficient degree.

This paper contains a description of a reference beam line design in chapter 2 and the particle transmission calculation through the duct and to the target for different beam line lengths and divergences in chapter 3. In chapter 4 the actual neutral power through the duct is calculated for the reference beam line described, taking into account gas flows, pumping speeds, reionization losses etc.

Although the reference beam line shows an arrangement of three sources in the horizontal plane, this does not mean that others, or even a vertical arrangement of sources are excluded or less attractive.

2. REFERENCE BEAM LINE

The design of a reference beam line for the following calculations is shown in Fig. 1 in a top view. Three sources (1), (2) in the ground potential boxes (3) can be steered by bellows. The sources are connected via remote flanges (4) to valves (5). The beam line box contains the neutralizer (6), diaphragms (7), (9), the reflecting magnet (8) and the calorimeter (11). A diaphragm (13) shapes the beam before it enters the duct. The duct contains a fast shutter (14), an absolute valve (15) and a bellows (16), that connects the beam line to the porthole. Cryopumps (19) pump the three separated vacuum regions. In this horizontal arrangement of the sources the beams cross in the port but are focussed at the plasma, 1.5 m beyond the port. In principle, an arrangement with three sources in the vertical plane could look very similar with slightly increased problems in the duct area, as the three beams in this case should cross at the plasma target.

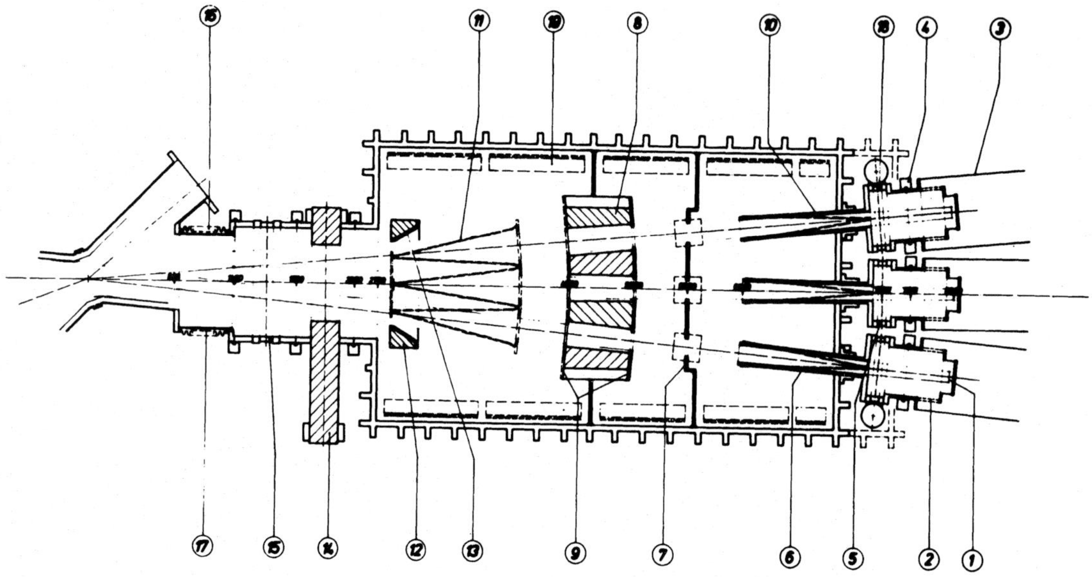

fig. 1

3. TRANSMISSION CALCULATIONS

For the calculations it is assumed that the sources are at a distance of 75 cm from each other, the beams cross in the porthole at a distance L from the sources and the individual beams are focussed at the plasma 1.5 m beyond the porthole; the horizontal $^1/_e$ divergence for a Gaussian profile is α and the vertical divergence is β, x the vertical dimension and y the horizontal dimension perpendicular to the axis of the middle beam. If not stated otherwise, the following values are taken: L = 8 m; $\alpha = 0.5^0$, $\beta = 1.5^0$. These numbers might be representative for a slotted grid system. Fig. 2 gives the relative power distribution P(y)/p along y (integrated over x) for different distances L between the sources and the porthole. Fig. 3 gives the same distribution for L = 8 m for varying α. The porthole plane is perpendicular to the axis of the middle beam. For a porthole width of $\pm$ 12.5 cm as it might be available for ZEPHYR, values of L $\leq$ 8 m and $\alpha \leq 0.6^0$ seem necessary to hold the losses in the port below a few percent. A smaller divergence, however, could compensate for larger distances and vice versa.

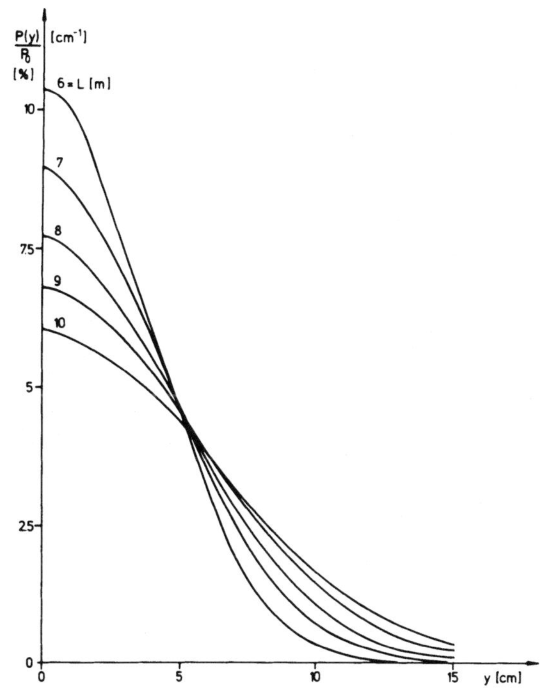

fig 2

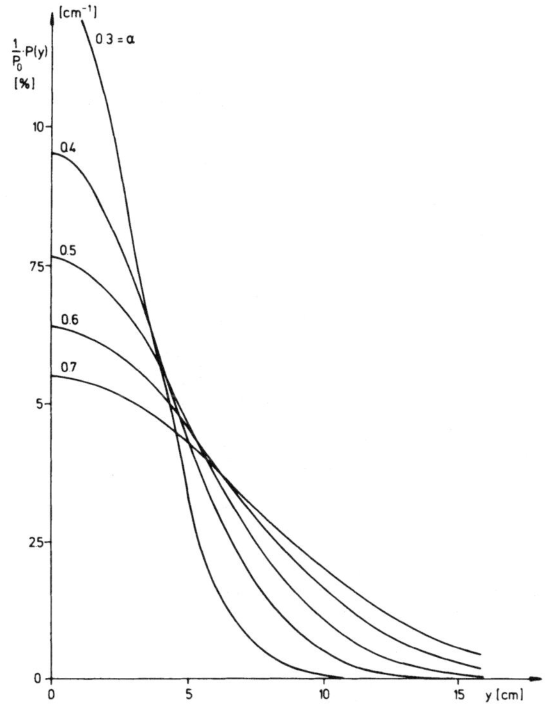

fig 3

As the port has a height of $\pm$ 35 cm, it does not very much restrict vertical divergences or beam line lengths. The main limitations on the vertical distribution are posed by the small vertical size of the plasma target. Fig. 4 gives the percentage of the neutral current I/I_0 through the plasma target with a height of $\pm$ 23 cm and a width of $\pm$ 50 cm and the percentage of the losses I_L/I_0 in the port, that is 30 cm wide and 70 cm high as a function of the distance L. Even for L = 8 m a substantial fraction of the power is outside the target. The losses in the port are mainly caused by the small width of the port, not by the height. For the calculation of the current to the target, the restrictions by the port are not taken into account. Fig. 5 gives the current to the target and the losses in the hole as a function of the vertical divergences. For distances of 8 m the vertical divergences should be below 1.2^{0} to get at least 90 % of the power to the target.

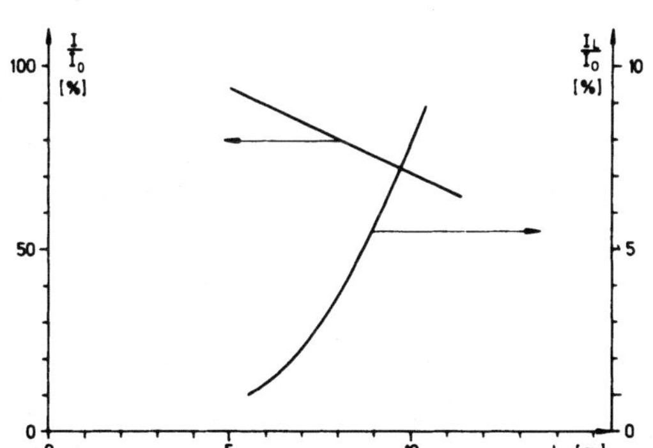

fig 4

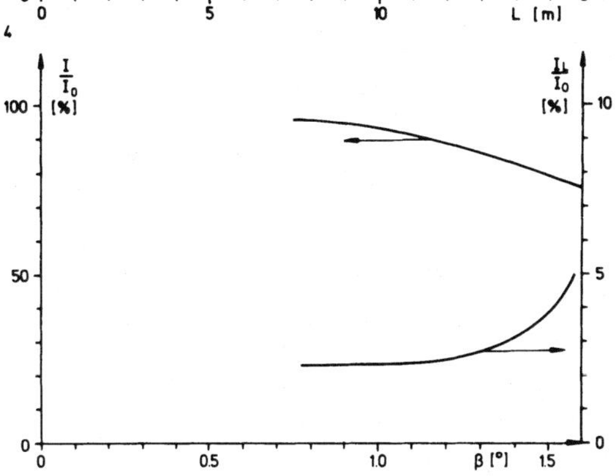

fig.5

For a vertical arrangement of sources with equal divergences in horizontal and vertical direction ($\alpha = \beta$), which might be favourable for an extraction system with holes instead of slots, the relative currents to the target and the port losses are given as a function of ß and L in Figs. 6 + 7. Here the small plasma target does not give problems up to $\alpha \sim 1^{0}$. The width of the porthole, however, requires a divergence of less than 0.6^{0} to hold the losses in the port below a tolerable limit (for L = 8 m).

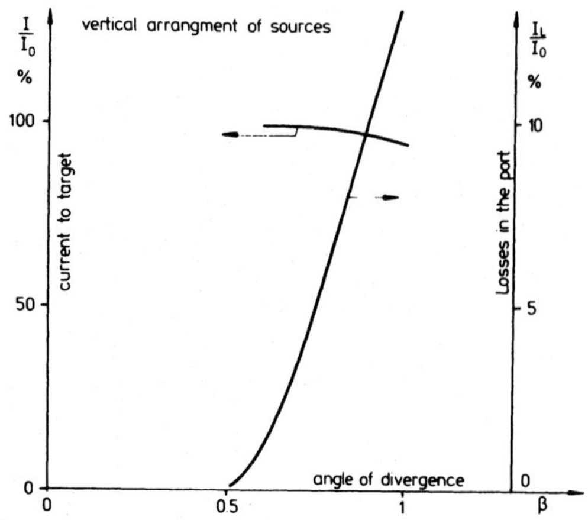

fig 6

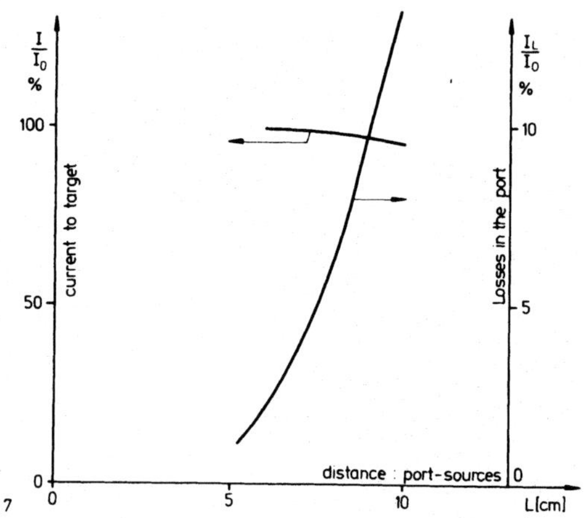

fig 7

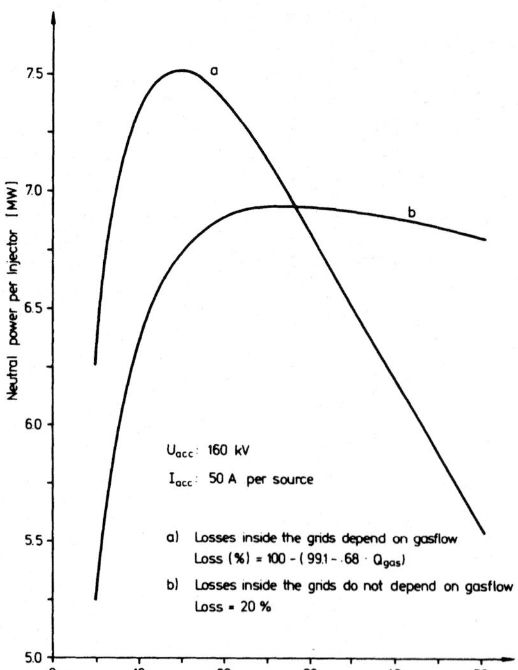

fig 8

4. POWER TRANSMISSION

The neutral power transmitted to the plasma is not only determined by divergences and diaphragms, but also by effects like neutralization rate, grid losses and reionization that depend on the gas dynamics. In a computer program described in more detail in [2], the power to the port has been calculated as a function of the gas flow through the sources for an arrangement as shown in Fig. 1. The result is given in Fig. 8. Here it is assumed that each source delivers a current of 50 A. Two cases are calculated: In case a) the grid losses depend on the gas flow as indicated in the figure, and in case b) the grid losses are constant with 20 % of the current. The maximum in curve a) results from the fact that the neutralization rate improves and saturates with the increasing gas flow, the grid losses then, however, become dominant. The not so pronounced maximum for case b) is due to the increasing reionization losses. These results are obtained for cryopumps, installed as shown in Fig. 1, with a pump height of ∿ 2.5 m. At the maximum of the curves the power lost in reionized particles is less than 200 kW for the three beams. The geometrical losses for this run were assumed to be 2.8 % in the port. The total power transmitted reaches values of more than 7 MW. These values seem to be rather optimistic. It has been assumed that the total power has a Gaussian distribution. As, however, part of the power might be carried in the wings of the distribution and as even for Gaussian distribution a larger fraction of the power is outside the plasma target (see Fig. 4 + 5), it would be safer to expect not more than 5 MW of neutral power per beam line to the plasma target.

Under the assumption that with three sources per beam line 5 MW of neutral power can be transportet to the plasma, the installation of 5 beam lines would be necessary. Whether this number is sufficient, however, depends on the reliability of the sources.

As the number of discharges in ZEPHYR is limited, it is required that the full power be available in at least 75 % of the discharges. Fig. 9 gives the number of sources $N_{0.75}$, that are available with at least 75 % probability as a function of the reliability of a single source for different numbers of sources installed. From here on one can see that with a single source reliability of $\underset{-}{\sim}$ 86 % 15 sources are available with the required probability, if 18 sources are installed.

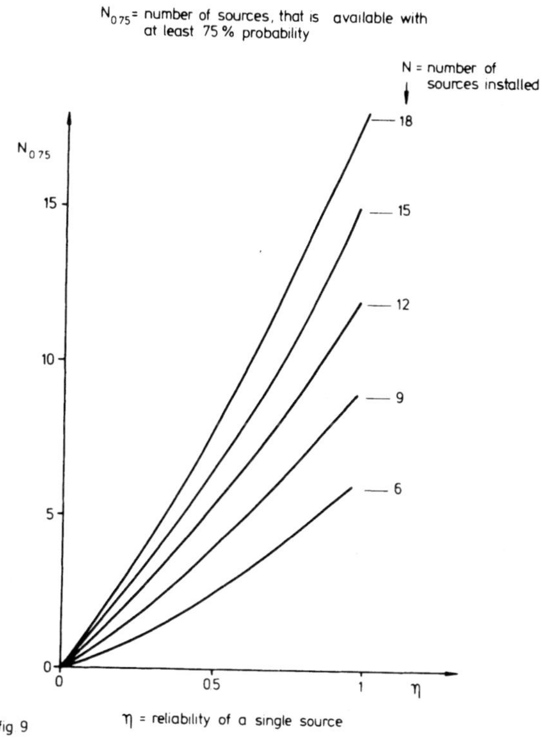

fig 9

5. CONCLUSION

Limiting factors for the choice of sources and their arrangement on beam lines are the small width of the porthole and the height of the plasma target. These make it necessary to look for a design of the beam line as compact as possible. It is shown that for a beam line length L $\leq$ 8 m and divergences of $\leq 0.6^0$ in the horizontal and $\leq 1.2^0$ in the vertical direction transmission losses can be kept at a tolerable level for the chosen arrangement of three sources on one beam line. Taking into account grid losses and reionization losses and making conservative assumption about the non-Gaussian part of the beams results in about 5 MW of neutral power transmitted to the target for 50 A of drain current per source. Reliability considerations show that probably six rather than five beam lines have to be installed if the single-source-reliability is around 85 %.

REFERENCES

[1] O. Gruber, K. Lackner, G. Lister, Internal ZEPHYR Report no. 16, Garching, August 1980

[2] LBL Workshop, Internal ZEPHYR Report no. 15, Garching, February 1980

This work was performed under the terms of the agreement of association between the Max-Planck-Institut fuer Plasmaphysik and EURATOM.

PHYSICS DESIGN CALCULATIONS FOR THE JET NEUTRAL INJECTORS

H. Altmann, A. Boschi, P. Brenan, H. D. Falter, A. P. H. Goede, R. Haange, R. S. Hemsworth, W. Obert, V. Simone, D. Stork, E. Thompson.

JET Joint Undertaking,
Abingdon, Oxfordshire, OX14 3EA, England.

ABSTRACT

The first results of physics design calculations for the JET neutral injectors are presented with the reference to beam transmission and geometric losses, vacuum calculations and reionization losses and beam blocking. The pulse length and the beam transmission required to achieve the specified performance considerably exceed that of present generation beam lines. An installed pumping speed of $\gtrsim 4 \times 10^6$ ℓ s^{-1} will be required to keep total reionization losses to $\sim$ 10%. Preliminary calculations indicate that beam blocking should not occur.

1. INTRODUCTION

The specification of the JET neutral injection system initially calls for 5 MW of full energy 80 KeV H^o atoms delivered to the plasma through a single port and a pulse duration in excess of 5s. The basic configuration is shown in figs. 1 and 2. The large injector box 7m high, 3.6m long and approximately 3m wide, contains eight beams which are grouped in pairs via four charged beam dump magnets. The magnets. which can be independantly energised, deflect the full energy ions through 90°. The 1/2 and 1/3 energy ions resulting from the break up of molecular ions are deflected through 270° onto a second beam dump which also collects the negative ions which undergo 90° reflection.

The horizontal axis of the injector box will be offset from the axis of the radial injection port on the torus by 19.5^o to give quasi-tangential injection for all eight beams. A large fraction of the most tangential column of beams misses the inner wall of the torus and will make two passes through the plasma. The more normal beams make only a single pass through the plasma before striking the inner wall of the torus. In order to simplify as much as possible the assembly, repair and maintenance of the system, most of the major components inside the vacuum tanks - the magnets, beam dumps and and some of the beam scrapers will be mounted as a single integral structure from the removable lid of the vacuum tank. The primary ion beams will be generated by cooled accelerator structures under development at Culham and

Fontenay-aux-Roses (F.A.R.) as part of the Joint Collaborative Programme on research and development for JET neutral injection. Each accelerator is rated at 60 amp extracted hydrogen ion current at 80 KeV, but the basic design is compatible with operation in deuterium at up to 160 KeV [1].

2. PHYSIC DESIGN CALCULATIONS

We have analysed the performance to be expected from this design and its sensitivity to various assumptions in the analysis. These on-going studies have concentrated on calculations of the beam transmission through the defined geometry and on the neutral gas pressure distribution through the beam line. These two factors determine, to first order, the over-all power transmission through the system and govern the engineering design of the various components.

2.1 Beam Transmission

Computer codes have been developed which enable us to determine the profile of beam power-density at arbitrary positions along the beam line for assumed values of source position, angular aiming direction, degree of overall focussing and beamlet divergence (inherent divergence of the ion beam).

Owing to relatively restrictive input geometry on the tokamak of $\pm 0.9^{\circ}$ in the horizontal plane, which is imposed by the size of the input duct and the requirement of quasi-tangential injection, there is a high premium on achieving a very low divergence in the primary ion beam. As shown in fig. 3, $\pm 0.7^{\circ}$ beamlet divergence (as achieved on the MW beam lines at Culham and F.A.R.) could lead to $\sim$ 4% power loss in the input duct of the tokamak. This rises to $\sim$ 15% if the divergence is degraded to $\pm 1^{\circ}$. The above values are quoted for the most tangential and hence most restrictive access, they are almost a factor two lower for the more normal injection angle.

By careful optimization of beam scrapers positioned at the exit of the injector box, the total power lost in the duct can be reduced to 1.2%, with only a small additional reduction in overall beam transmission, assuming $\pm 0.7^{\circ}$ beamlet divergence.

In addition to minimising the total power deposition in the duct region which is governed by the restricted geometry in the horizontal plane, the position and angular alignment of the beams in the vertical plane has also been optimised. This results in acceptable power-densities at both the inner wall of the torus (for the tangential beams) and on the walls of the duct.

Power-density contours have been computed at several points along the beam line and the profile of peak beam power-density is shown in fig. 4 for a fully optimised system. Note that the peak power-density, occurs outside the injectors and beyond the region of beam scrapers etc. It should also be noted that this computation includes a more realistic model of the neutral beam, in that the power and divergence of the 1/2 and 1/3 energy components are also included.

Using these results, the calorimeter profile has been designed to the approximate "S" shape shown in fig. 2 to reduce the peak power-density of $\sim$ 5.8 KW cm^{-2} at normal incidence to a design limit of $\lesssim$ 1 KW cm^{-2}.

2.2 Pressure and Gas-flow Calculations

The profile of neutral gas pressure throughout the injector is required to calculate the loss in beam transmission due to re-ionization of the fast neutrals by collisions on the background gas. Extensive calculations have been performed to predict the gas flow required through the neutralizers to convert 22% of the full energy ions to fast atoms (this is $\sim$ 80% of the theoretical maximum value). For an optimised pressure profile in the neutralizer, which results in $\sim$ 2 mtorr pressure at the accelerator, the total input required for the eight neutralizers is 230 torr $\ell\ s^{-1}$ [2].

A model has also been developed to predict the streaming component of the gas leaving the neutralizer [2]. The effect of streaming could have a deleterious effect on the injector performance in that it effectively enhances the gas conductance of baffles in the beam line thus reducing the effectivness of differential pumping stages. The results of these calculations can be used as input data for a computer code which calculates the pressure profile in the injector box. This calculation, based on the assumption of molecular flow, includes the effect of finite conductance between the various sources of gas and the cryopumps which cover the side walls of the box.

In addition to the gas introduced by the neutralizers, the gas load due to charge beam disposal is also included by assuming that each ion (H^+, H_2^+ or H_3^+) impacting on a surface, results in one molecule of hydrogen being released. Results of the study to date show that:-

(1) total re-ionization losses are not very sensitive to the accuracy of the predicted degree of streaming.

(2) of the several variations in beam line design considered, (using different configurations of the beam bending magnet) the one outlined above gives marginally lower re-ionization losses.

(3) in order to keep total re-ionization losses to $\lesssim$ 11% and those after the deflection magnet (which are deposited in the input duct) to $\lesssim$ 6%, will be necessary to have a total installed pumping speed of $\gtrsim$ 4 x 10 $\ell\ s^{-1}$ in each injector box (see fig.5). This speed cannot be achieved using conventional chevron shielded cryo-pumps on the side walls of the box (cf TFTR) but will necessitate the use of an advanced design of "open structure" pump.

A major uncertainty in the present calculations is that they are based upon the assumption that the background gas is at room temperature. If, for example, the gas accommodates to a temperature mid-way between room temperature and 77°K due to collisions with the extensive liquid nitrogen cooled surfaces inside the tank, then the gas density and subsequently the re-ionization losses, will be increased to $\sim$ 14%.

2.3 Input Duct and Beam Blocking

Using computations of the stray magnet field of JET in conjuction with various configurations of magnetic shielding of the injector box, we have calculated the trajectories of re-ionized fast atoms and subsequent power deposition profile in the duct/rotary valve region. An example of the computed trajectories is shown in fig. 6. These computation have demonstrated the necessity of cooling the inside of the whole of this region and give a first indication of the required geometry of magnetic shielding.

Analytic estimates have been made of the so-called beam blocking effect for the JET geometry. Using an extension of a previous calculation [3] to include the effect of direct interception of the neutral beam, we find that there should not be a catastrophic attenuation of the beam provided that the gas re-emission coefficient from the duct walls can be kept to a value of the order of a few molecules per incident fast particle. The gas re-emission coefficient in this context includes all possible sources of gas, e.g. thermal desorption from the bulk material in additon to gas weakly bound on the surface of the duct. Nevertheless, some care must be taken in the choice and pre-treatment of the surfaces in this region in order to minimize the out-gassing under particle bombardment (cf conditioning of the wall of the Tokamak itself) and also in provision for handling the power which is deposited in this region.

3. SUMMARY

The JET neutral injection system is based on the following assumed parameters:-

Beam Energy	80 KeV
Extracted Current	60 amp per source
Pulse Length	> 5 sec.
Species	$\geq 80\%\ H^+ \leq 15\%\ H_2^+ \leq 5\%\ H_3^+$
Total Installed Pumping	$\gtrsim 4 \times 10^6\ \ell\ s^{-1}$
Beamlet Divergence	$\pm\ 0.7^o$ for full 80 KeV energy H^o $\pm\ 1.0^o$ for 1/2 and 1/3 energy H^o
Neutralizer Thickness	300 mT. cms.
Peak Neutralizer Pressure	2 mT

Using this data, the predicted performance of a single injector box is as follows:-

Total Extracted Power	38.4 MW
Ionization and Charge-Exchange in Accelerator	$\sim$ 7%
Total Re-ionization	$\sim$ 10% from thermal gas loads $\sim$ 3% from gas generated in the duct
Geometric Transmission	$\sim$ 89% (assuming 5% loss from abberations)
Total Losses	$\sim$ 28%
Total Neutral Power to Torus	8.4 MW
Total Power at Full Energy to Torus	5.3 MW

4. REFERENCES

[1] Neutral Injection Development Group (Culham Laboratory), Poster Session CP1 - Plasma Engineering I. This Conference.

[2] Boschi, A., Poster Session AP1 - Vacuum System. This Conference.

[3] Hemsworth, R. S., Culham Laboratory Report CLM-R 162 (1977).

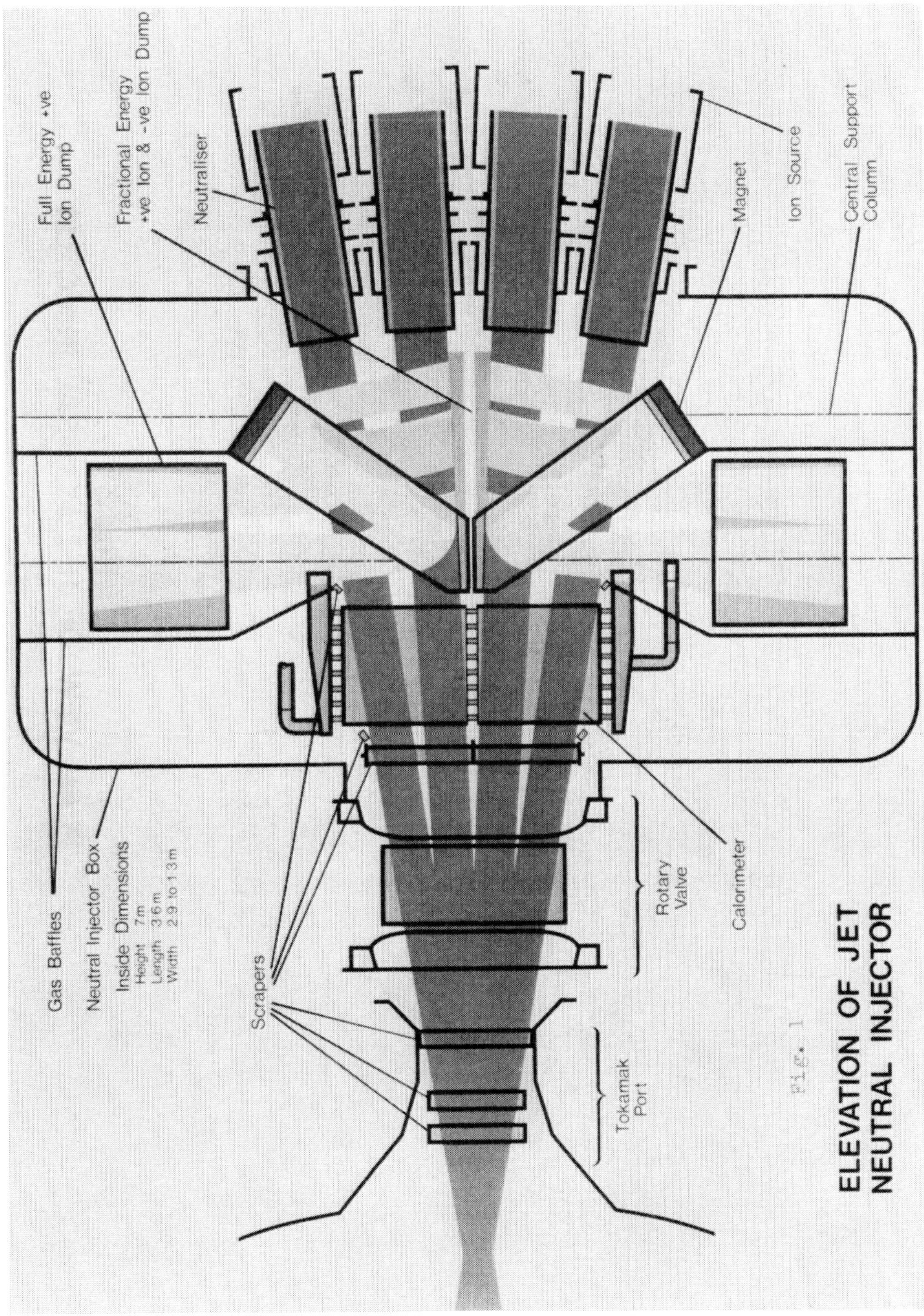

Fig. 1

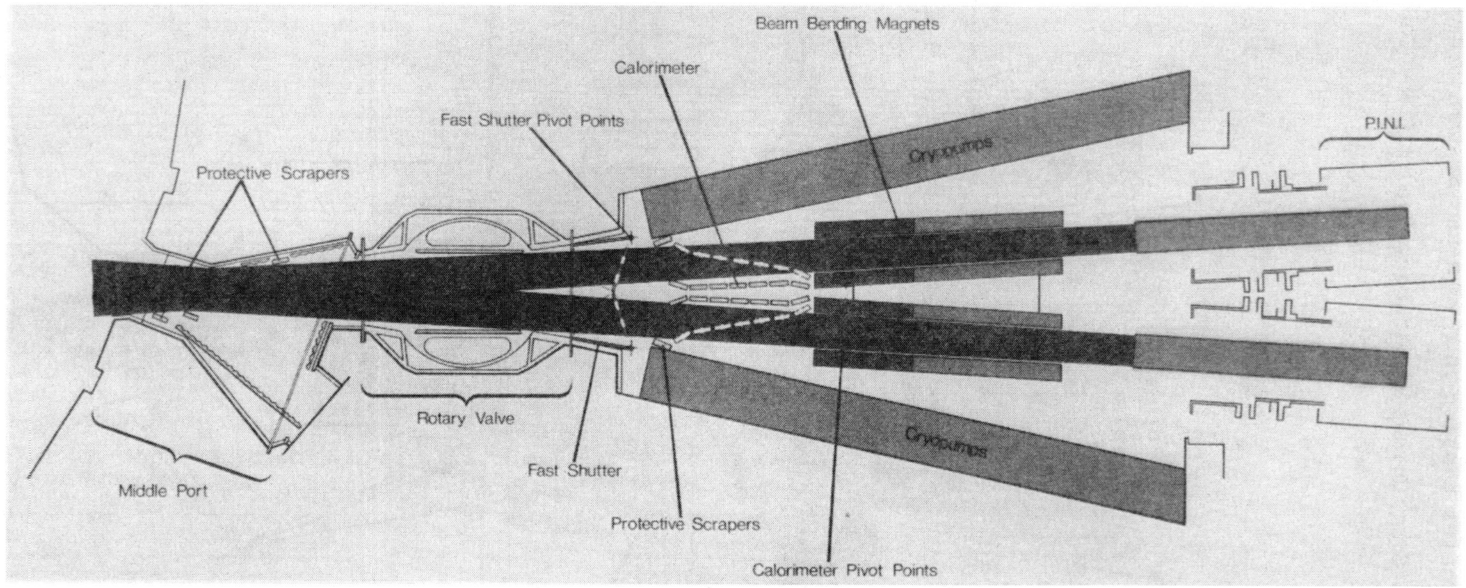

Fig. 2
PLAN VIEW OF JET NEUTRAL INJECTOR

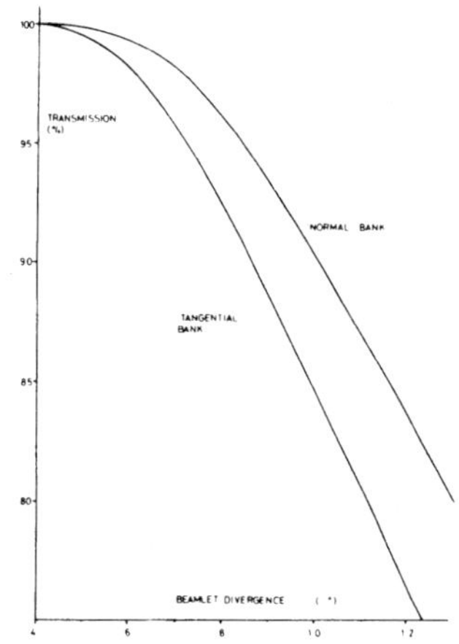

FIG. 3 BEAM TRANSMISSION V'S BEAMLET DIVERGENCE

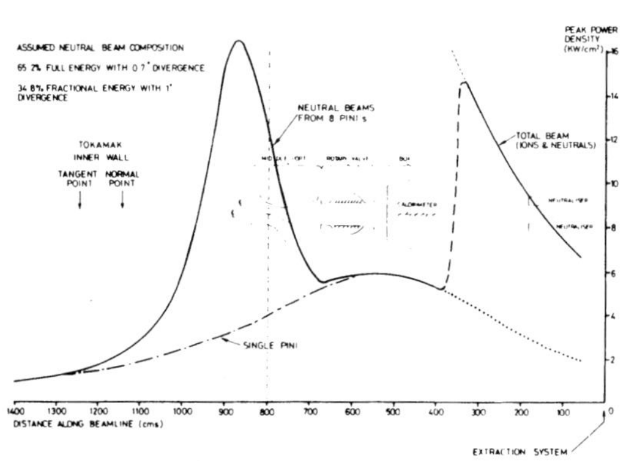

FIG. 4 PEAK POWER DENSITY V'S DISTANCE FROM THE ION SOURCE

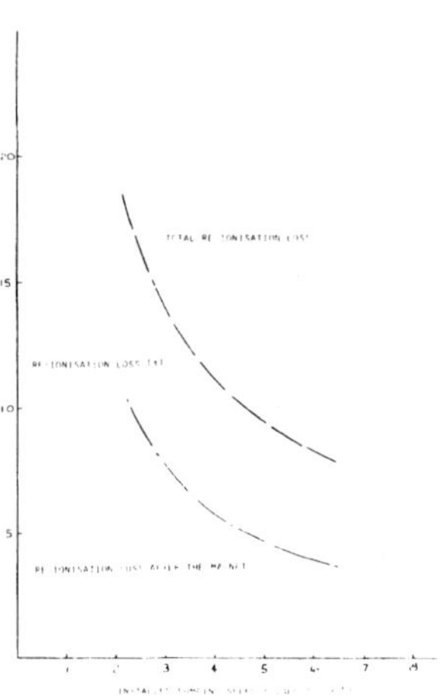

FIG. 5 RE-IONISATION LOSS V'S INSTALLED PUMPING SPEED

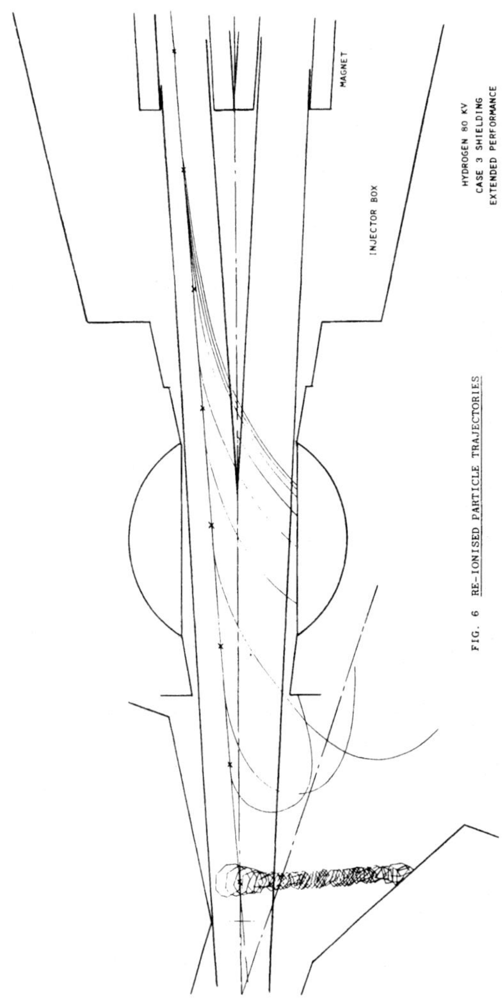

FIG. 6 RE-IONISED PARTICLE TRAJECTORIES

HIGH POWER-DENSITY STEADY-STATE HEAT SINKS

E.B. Deksnis[+], H. Fränkle[++], J.L. Hemmerich,
P.H. Kupschus, C.N. Meixner
Kernforschungsanlage Jülich GmbH - IPP/ZBB, Association
EURATOM-KFA, P.O. Box 1913, D-5170 Jülich, F R G
[+] visiting scientist from Canatom, Montreal, Canada
[++] Messerschmitt-Bölkow- Blohm, München

ABSTRACT

Increasing pulse times and power of plasma heating by neutral-injection pose requirements that reach or exceed capabilities of stationary inertial targets. Inertially-cooled and actively cooled rotating devices are viable alternatives. Model experiments and calculations show that a compact actively-cooled target 350 mm Ø, 700 mm long rotating at 3000 RPM will handle heat fluxes up to 120 MW/m^2 under steady-state conditions. Further improvements seem possible by applying a metal-hydride coating; estimated power handling is 400 MW/m^2 for a beam spot 0.1m x 0.1m.

1. INTRODUCTION

Present and future fusion devices although designed for moderate power densities, typically 1 Mw/m^2, contain areas where much greater heat-fluxes are expected, especially in auxilliary heating equipment. Neutral-injectors contain several components, eg. ion-dumps and diagnostic calorimeters for which peak fluxes can be several orders of magnitude greater than average values.

Inertial targets that are cooled between successive shots are used at present for relatively short beam-pulses, eg. /1/. Such targets typically have to be tilted at nearly grazing angles of incidence in order to handle large power-densities. It will be shown that a rotating inertial target is capable of taking comparable fluxes at normal angles of incidence.

A rotating continuously-cooled target is to be used in a neutral-injection test-stand at the IPP Jülich. As additional heating of the TEXTOR plasma /2/ by injection of neutral hydrogen beams (80 A, 50 keV ion beam, with 1.3-1.5 Mw neutral power per ion source delivered to the plasma) is intended for pulse times of 3-10 seconds ,ion-dumps and calorimeters have to withstand surface heat loads of 60-120 MW/m^2 under quasi-steady conditions.

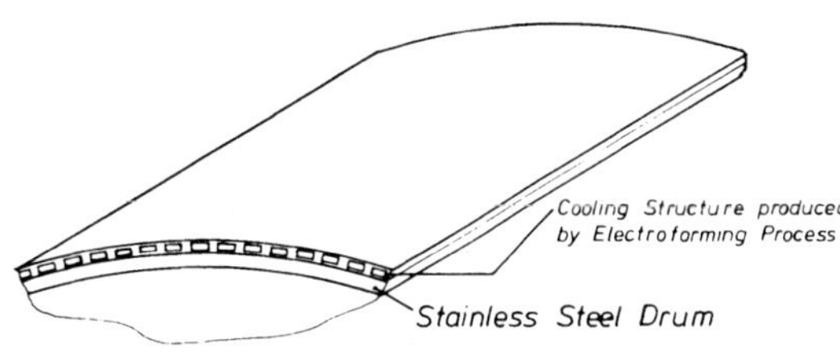

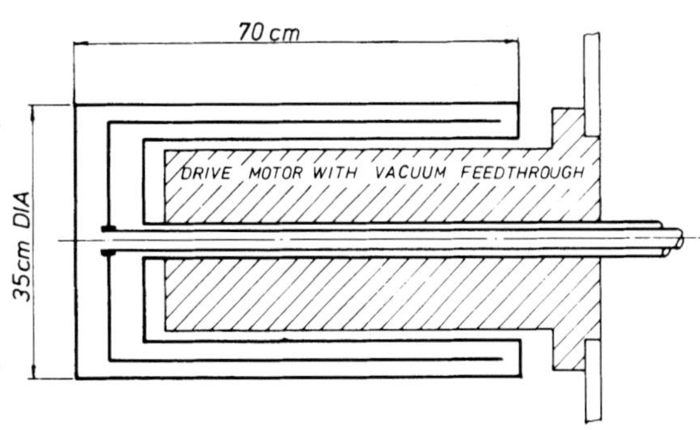

Fig. 1. Continuously-cooled Rotating Target

The first targets are at present being manufactured using a galvanic deposition process /3/ to build up a copper layer with small (2,4 mm square) coolant passages over the outside of a stainless-steel drum, cf. fig.1. A self-contained motor will spin the target at

50 Hz; other features include a rotating vacuum and coolant feed through which has been shown to work under high-vacuum conditions /4/. The target is 350 mm Ø, 700 mm long.

2. Rotating Inertial Targets

If the target shown in fig.1 has a thick outer layer and coolant is circulated only in-between shots, the response to surface heat deposition is completely inertial. The well-known result for surface temperature rise ΔT_s due to a constant heat flux Q incident on a semi-infinite slab with thermal diffusivity α, thermal conductivity k, ie /5/,

$$\Delta T_s = \frac{2Q}{k} \left(\frac{\alpha t}{\pi}\right)^{\frac{1}{2}} \quad (1)$$

may be used to give ΔT_s for a target rotating steadily through a beam spot of fixed size and constant flux density. Fig.2 below shows one

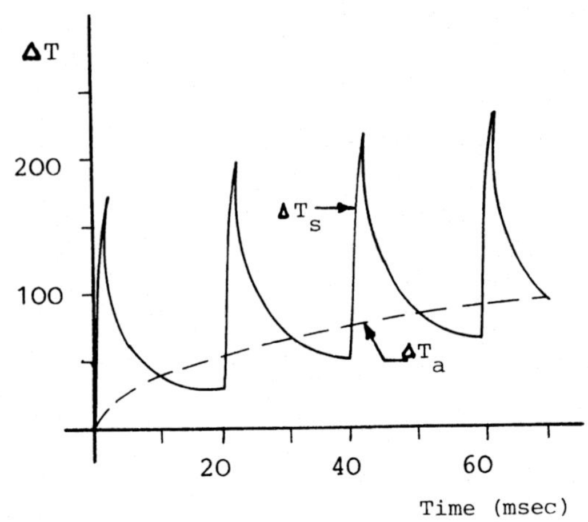

Fig.2. Surface Temperature Time-history, OFHC copper, 120 MW/m^2, 10% duty cycle, 50 Hz

temperature time-history. The quantity

$$\Delta T_a(t) = \frac{2Q}{k} \left(\frac{t_i}{t_s}\right) \left(\frac{\alpha t}{\pi}\right)^{\frac{1}{2}} \quad (2)$$

where t_i/t_s is the duty cycle is also plotted in fig.2.

One criterion for estimating power-handling capabilities is the maximum flux that produces surface melting of the bulk material. Another criterion is based on the thermal stress estimated by

$$\sigma = \frac{\varepsilon E \, \Delta T_s}{(1-\nu)}, \quad (3)$$

where ε is the thermal expansion coefficient, E Young's Modulus, ν Poisson's ratio. Tab.1 is based on σ reaching ultimate stress values.

Table 1

Materials for Rotating Inertial Targets, 50 Hz, 10% duty cycle, 3 seconds irradiation

Material	Melting (Flux, MW/m^2)	Ultimate Stress (Flux, MW/m^2)
Carbon	175	33.
OFHC Copper	165	20.
Molybdenum	190	18.5
Titanium	65	15

It is evident that stress limitations appear to be more severe than the condition of local melting. It is necessary, therefore, to provide some stress relief by structuring the surface of a target.

3 Continuously-cooled Rotating Targets

The limitations given above for inertial targets prohibit their use over pulse-times intended for TEXTOR injectors. A water-cooled rotating target, cf. fig. 1, will accomodate larger heat fluxes.

In our evaluation of power-handling capabilities we first write down some analytical results for a one-dimensional model where convective heat-transfer occurs with coefficient h. The maximum heat-flux Qm that can be absorbed without boiling is

$$Qm = h\,(T_s - T_b), \quad (4)$$

where T_b is the bulk temperature of coolant. With cooling channels a larger value of Qm can be achieved. Fig. 3 gives the results from a numerical evaluation of two-dimensional heat-transport in the geometry shown in fig. 1.

Targets for the Jülich test-stand have an estimated value of h = 0.040 MW/m^2 ^{o}C. The numerical results plotted in fig. 3 indicate that some local boiling heat-transfer is required in order to absorb 120 MW/m^2 at 50 Hz rotation and 10 % duty cycle.

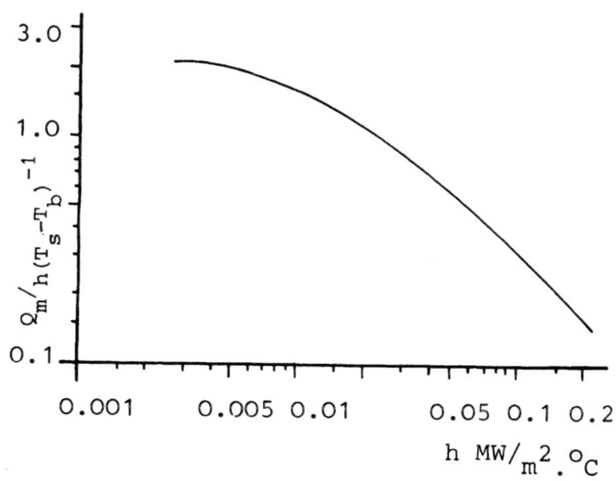

Fig.3. Efficiency of coolant passages, non-boiling heat- transfer, passages 2.4 mm square, 1.2 mm ribs, copper.

No correlations appear to exist that would allow accurate estimates of h where heat-transfer occurs over only part of the wetted perimeter. Experiments, one set of data of which are shown in fig.4, were done where only 50% of the circumference of a round tube was ohmically heated. Pipe-flow correlations give (at most) h = 0.025 MW/m^2 ^oC as compared with a measured value of 0.06 MW/m^2 ^oC.

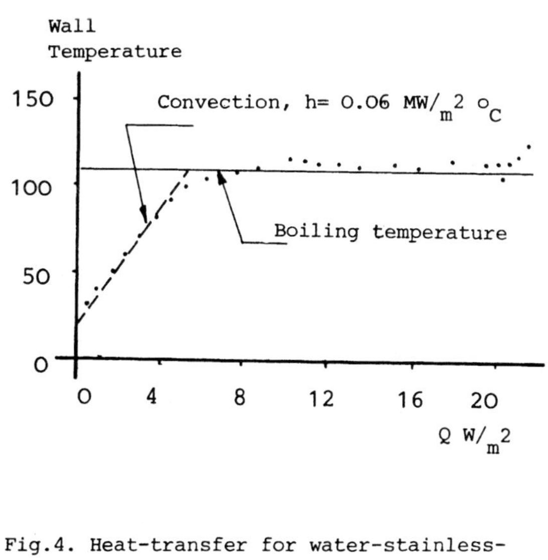

Fig.4. Heat-transfer for water-stainless-steel; tube 8.0 mm OD, 7.8 mm ID, ohmically heated over 50% of circumference, heated length 50 mm at 0.21 m from inlet.

Rotating targets undergo many stress cycles per shot, consequently, fatigue will be a problem. Fig.5 shows fatigue results for a beam 400 MW/m^2, 10% duty cycle, 50 Hz rotation, spot 100 mm wide. Stress relief due to curvature of channel lids greatly enhances useful lifetimes.

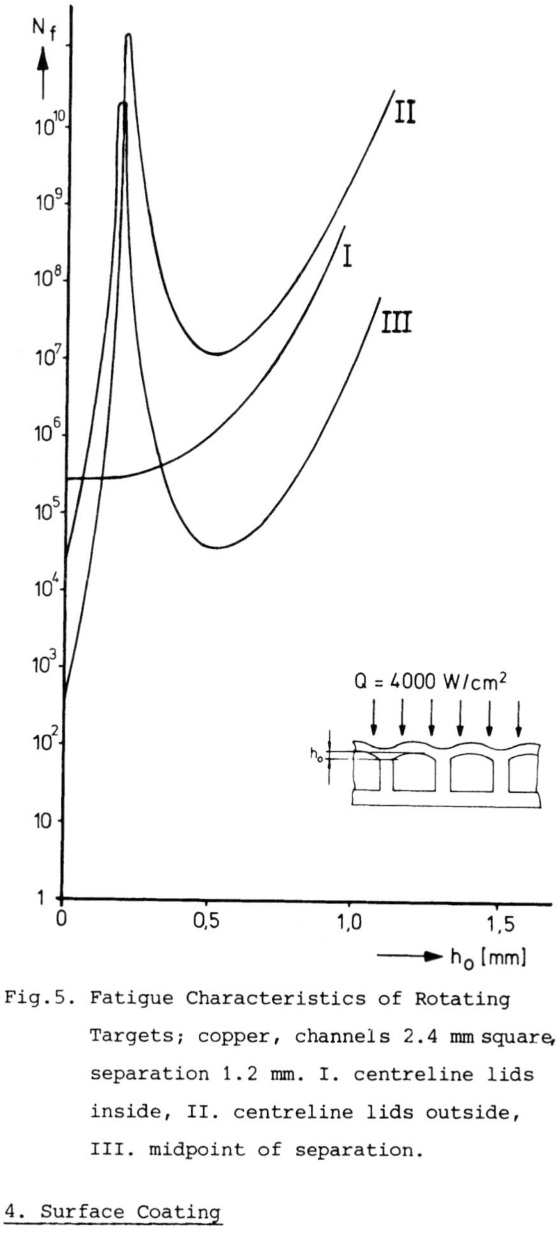

Fig.5. Fatigue Characteristics of Rotating Targets; copper, channels 2.4 mm square, separation 1.2 mm. I. centreline lids inside, II. centreline lids outside, III. midpoint of separation.

4. Surface Coating

The outer surface temperature of a rotating target in steady-state consists of a mean constant temperature upon which fluctuations are super-imposed at the rotation frequency. For our target passing through a beam spot with Q = 120 MW/m^2, duty cycle 10%, 50 Hz rotation speed, the mean temperature is estimated at 200^oC, the fluctations at $+160^oC$, -15^oC. A surface layer of metal hydride would reduce this cycling by taking advantage of the large enthalpies of formation. If our target is coated with 30 μm of yttrium hydride ($YH_{2÷3}$) which has dissociation temperatures around 200°C, then fluctuations of temperature at the copper-hydride interface are reduced to + 25 °C -10°C. An estimate for titanium hydride ($TiH_{\sim 1.5}$) gives similar results, ie a 50 μm thick layer should be able to handle flux densities of up to 400 MW/m^2 with fluctuation of ± 50°C.

These estimates assume that dissociation under dynamic conditions is reversible with sufficiently short time-constants. This would have to be confirmed experimentally. It is supported by measurements of depth profiles in tritiated targets at the Rotating Target Neutron Source /7/ which can not be explained by diffusion theory based on the heat capacity of undissociated hydride.

5. Conclusions

Rotating targets offer several advantages over stationary targets: compactness, spreading out hot spots by continuous movement of target surface, improved sputtering propperties when the beam is taken at locally normal angles of incidence.

An important design consideration for rotating targets is optimization of fatigue lifetime. Our analyses are investigating further

(1) convective heat-transfer characteristics

(2) boiling heat transfer (critical heat flux)

(3) surface hydride layers

6. References

1. MEIXNER, Ch., KUPSCHUS,P., Workshop on Neutral Injection into TEXTOR, Jülich, May 24, 1978.
2. ULRICKSON,M., KUGEL,H., APS Plasma Physics Division Meeting, Boston, November 1979.
3. HEMMERICH,J., KUPSCHUS,P., DEKSNIS,E., FRÄNKLE, H., Jahrestagung Kerntechnik '80, Tagesbericht, Berlin, March 1980
4. HAUBOLD, G., KÜSSEL,E., JOSWIG,G., KFA Technische Information No. 20.
5. CARSLAW,H.S., JAEGER, J.C., Conduction of Heat in Solids, Oxford University Press, 1959.
6. ECKARTT,D., VENUS,G., JET Technical Note (74) 9, 1974.
7. BOOTH,R.,DAVIS,J.C., etal., Nuclear Instruments and Methods, 145, 25-39, 1977.

INVESTIGATION OF THE RF-ION SOURCE RIG 10 FOR NEUTRAL INJECTORS

J. Freisinger, J. Lensing, H.W. Loeb, S. Reineck, and R. Thueringer
Ist Institute of Physics
Heinrich-Buff-Ring 16, D-6300 Giessen, FRG

ABSTRACT: In comparison to dc-ion sources our rf-source RIG 10 has some advantages (no discharge electrodes immersed in the discharge plasma; homogeneous beam profiles; simple assembly etc). - The profiles of the current density and the radial distribution of the electron temperature have been measured by floating double probes in dependence on a superposed external magnetic field. The radial current density profiles depend on the chamber length, the discharge pressure, and the discharge power. The ion current density increases linear with the discharge power. - By superposing the special inhomogeneous axial magnetic field (some 10^{-4} Tesla) we reached at 6 cm vessel length very plane radial density profiles (better than 5%) over a cross-section of about 7 cm with current densities of 280 mA/cm^2 (at 4,5 kW rf-discharge-power). - A simple theoretical model has been developed explaining the discovered dependencies. - Preliminary results of an ion extraction system (similarity laws) are given.

INTRODUCTION

Additional heating of Tokamak plasmas can be achieved by injection of intense and high energetic beams of neutral particles. For this purpose ion sources are needed, which yield a high deuteron (proton) fraction in the beam.

Neutral injection is a new field of application of rf-ion sources (1)(2). In comparison with dc-discharge sources there are some advantages (1) of our rf-ion source called RIG 10 (Radio-Frequency Ion Generator of 10 cm in diameter). This type of ion source has no discharge electrodes immersed in the discharge. The discharge chamber is made of quartz. Therefore, the recombination rate of atomic hydrogen ($2H \rightarrow H_2$) at the quartz walls is much less than at metallic ones (3). The mode of working of the rf-source is described in (1) and (4).

1. PART I: EXPERIMENTAL SET UP AND SOME BASIC PERFORMANCE DATA

1.1 Desgin of RIG 10 and RIG 10 R

Concerning the mechanical structure of the assembly, the rf-ion source RIG 10 is very simple and rigid (1)(2). In connection with the installation of a more powerful rf-generator (maximum rf-power 20 kW) the design of RIG 10 has been modified (5). Fig. 1 shows the modified ion source RIG 10 R for diagnostic investigations at higher discharge powers. The extraction system of RIG 10 (extraction anode, extraction cathode, and decel-electrode) has been replaced (Fig. 1) by a plasma holder and radially movable double probes for determining plasma density, the electron temperature, and the ion beam profile (current density) over the cross section of the beam extraction area. Also the feed system has been changed. In contrast to RIG 10 (1)(5) the hydrogen gas no more enters the discharge vessel via the distributor at the upper lid of the quartz tube. It is now fed into the discharge region in the up-stream direction at the circumference of the plasma holder (Fig. 1). By this, the design of the discharge chamber, made of quartz, has been simplified importantly.

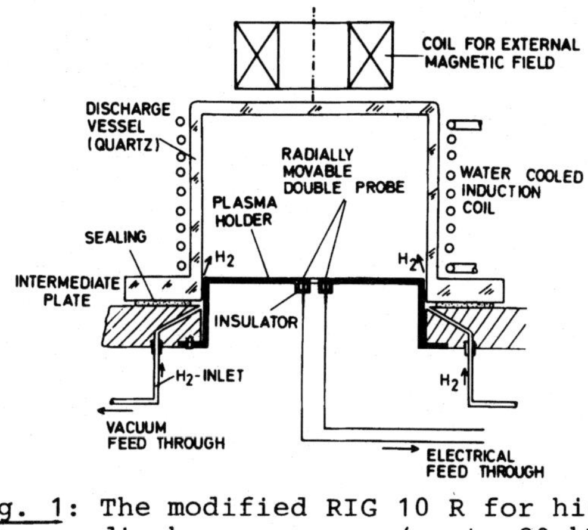

Fig. 1: The modified RIG 10 R for high discharge powers (up to 20 kW)

1.2 Superposed External Magetic Field

The effect of a weak superposed stationary magnetic field on the ion current density is very distinctive. The ion current density increases at the discharge tube axis by 25% as depicted in Fig. 2, when the field intensity of

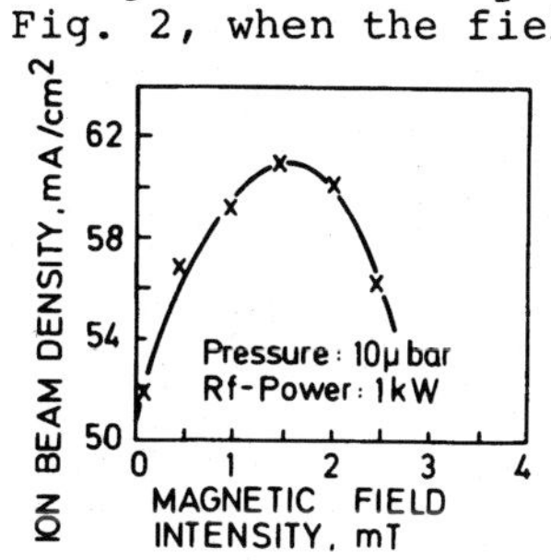

Fig. 2: Influence of a small external magnetic field on the ion current density at the discharge chamber axis

1.5 mT is applied. A possible explanation of this relationship is given in (6).

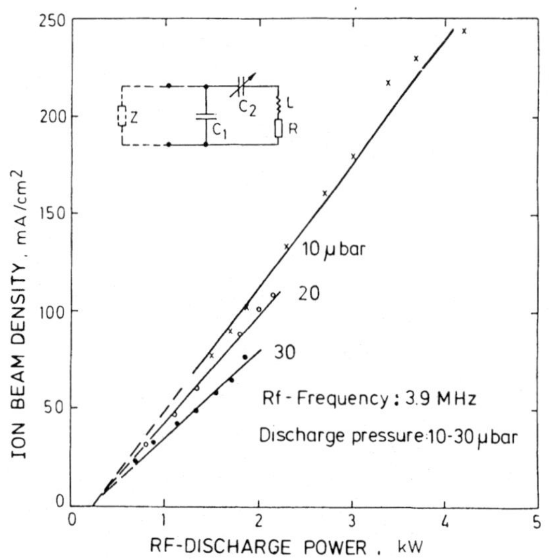

Fig. 3: Matching of the 20 kW-rf-generator to the discharge. In the accomodation connecting (upper left) the symbols mean:

- Z: characteristic output impedance of the 20 kW-rf-generator (50 Ω)
- L: induction coil (10 μH)
- R: resistance of the rf-discharge
- C_1: fixed condensor (1000 pF)
- C_2: variable condensor (25...10,000 pF)

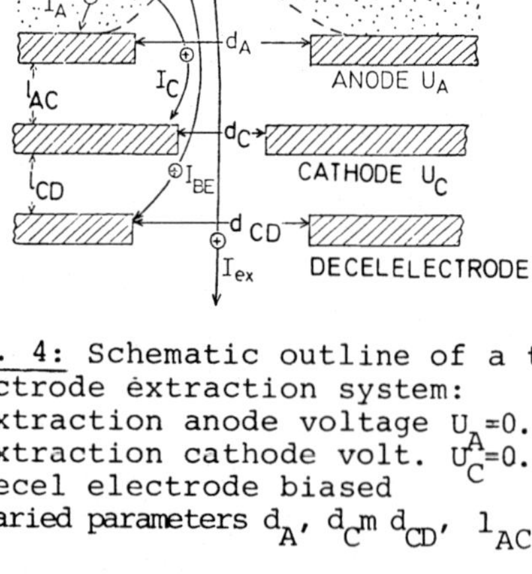

Fig. 4: Schematic outline of a three electrode extraction system:
extraction anode voltage U_A=0...10 kV
extraction cathode volt. U_C=0..-2.5 kV
decel electrode biased
varied parameters d_A, d_Cm d_{CD}, l_{AC}, l_{DC}

1.3 Power Dependency

The procedure of matching the 20 kW-rf-generator with the discharge plasma has been very lengthy. At last, the accomodation connecting - shown in Fig. 3 (upper left) - turned out to be the best one. Z denotes the characteristic output impedance of the rf-generator, L the induction coil, R the resistance of the rf-discharge, C_1 a fixed condenser, and C_2 a variable condensor. - The ion current density increases linear with the discharge power as plotted in Fig. 3. Parameter is the discharge pressure, which varied between 10 and 30 μbar. The slope of the ion current density curve is steepest at a discharge pressure of 10 μbar. For only 4 kW a current density of 250 mA/cm^2 (Fig. 3, see also Fig. 10) is reached. This very important result is much better than we expected by extrapolation of previous measurements (1)(4).

1.4 Extraction Measurements

In order to find an extraction system, which is best adapted to the RIG 10, we built a variable three electrode system for investigation. Fig.4 gives a schematic outline and a list of the varied parameters. It is essential that the distance between extraction anode and extraction cathode l_{AC}, the most important geometric parameters, can be varied continually from the outside. Thus, it becomes possible to vary the extraction geometry while the discharge is running undisturbed under constant conditions.

In Fig. 3 (and Fig. 10) the ion current density - determined by double probe measurements - is a linear function of the rf-discharge power. The main aim of the following investigations is to provide a beam current of high density. It is, therefore, very important to determine, what ratio of the probe-measured current density can be extracted. - From all these measurements it could be concluded, that the relation between the extracted and the probe-measured current density is always a linear one, dependend only on the geometric parameters of the extraction system (similarity law). An example of this relationship is given in Fig. 5 for three similar extraction geometries (d_A/d_C = 2; l_{CD} = d_C). Another preliminary result can be seen here: The anode hole diameters D_A were varied between 4 mm and 10 mm. In this range the extracted current density (for a constant probe-measured current density) increases when d_A is extended.

The most important geometric parameter is the distance between extraction anode and extraction cathode l_{AC}. As can be seen in Fig. 6, the extracted current density j_{ex} increases noticable with decreasing l_{AC}. But the obtainable gain for j_{ex} by decreasing l_{AC} is limited, because the resulting increase of the beam divergence causes a considerable rise of the leakage current to the extraction cathode and decel-electrode. The extractable ion current density is 60% to 75% of the probe-measured current density. But as the results up to 1 kW rf-power show, that the ratio of extractable current density per probe-measured current density becomes higher with increasing rf-power, we expect even better results when the experiments are continued at a higher power level.

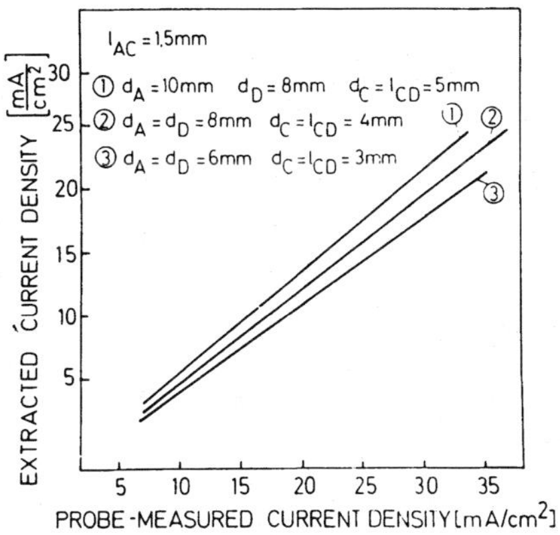

Fig. 5: The extracted current density in relation to the probe-measured current density for three similar extraction systems

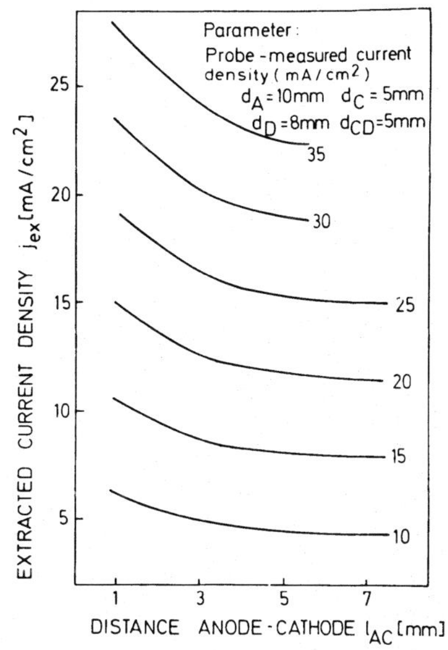

Fig. 6: Extracted current density in dependence of the anode-cathode distance

2. PART II: INFLUENCE OF THE DISCHARGE CHAMBER LENGTH ON THE ION-CURRENT-DISTRIBUTION AND THE POWER CONSUMPTION OF THE SOURCE

2.1 General Remarks

Assuming ionizing electron-neutral collisions to be the only carrier production process in the discharge and wall recombination (besides excitation) to be the only loss mechanism (which is allowed in the considered pressure range of about 10 µbar), the discharge behaviour is dominately influenced by the ratio of plasma volume to vessel surface.

Searching for an optimum (plane) radial current density distribution means an optimization of the chamber length (at a fixed vessel diameter), respecting given boundary conditions like rf-power, discharge pressure, and a superposed weak magnetic field.

2.2 Experimental Results

We measured the radial distribution of the ion saturation current to the double probe (y-axis) directly by coupling the radial probe position with the x-axis of an xy-plotter. The variation of the discharge vessel length was realized by stepwise moving the bottom (molybdenium). At constant rf-power and fixed discharge pressure, we registered an extreme dependency of the radial current density distribution on the vessel length (Fig. 7).

With shorter chamber lengths the total ion current density as well as the radial profile inhomogeneity increases. With regard to the extraction of an (intense) homogeneous ion beam, the crosses indicate in Fig. 7 where the variations exceed the maximum tolerable 5%-limit.

As at short vessel lengths the current density reduction at the vessel axis overcomes 5% (relative to the maximum), extraction is only reasonable for the greater chamber lengths (80/90 mm). Being interested in both, high current-densities at very homogeneous distributions across the extraction area, we have to smooth the profiles at shorter vessel lenghts. By former investigations (1) we know, that an inhomogeneous axial magnetic field raises the current density at the axis while decreasing it near the cylinder wall. Fig. 8 shows the change of the radial profiles with increasing magnetic field strength for three different vessel lengths (40/70/90 mm). 1 Ampere coil current corresponds to $6 \cdot 10^{-4}$ Tesla in the center of a 70 mm-vessel, 3.5 cm from the top (see sketch in Fig. 8).

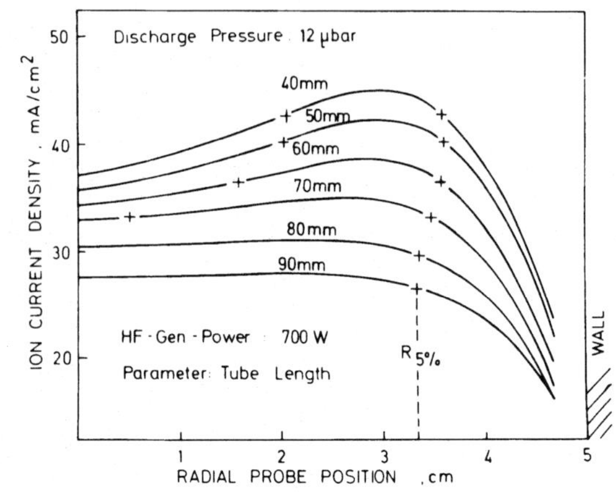

Fig. 7: Radial ion current density profiles across the discharge bottom at various vessel-lengths (by double-probe measurements).

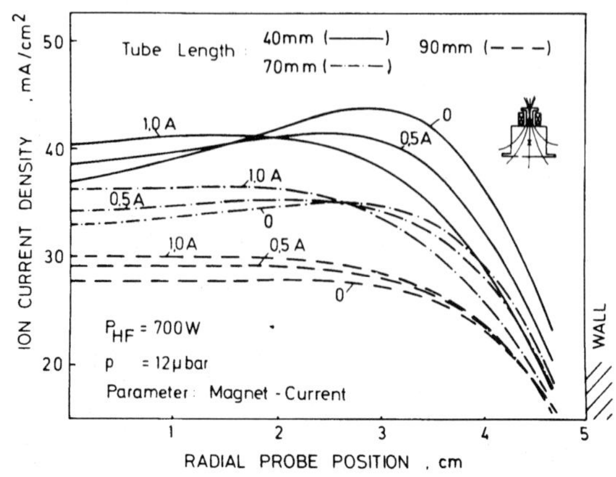

Fig. 8: Radial current density profiles under the influence of an inhomogeneous axial magnetic field (sketch) for different vessel lengths (40/70/90 mm).

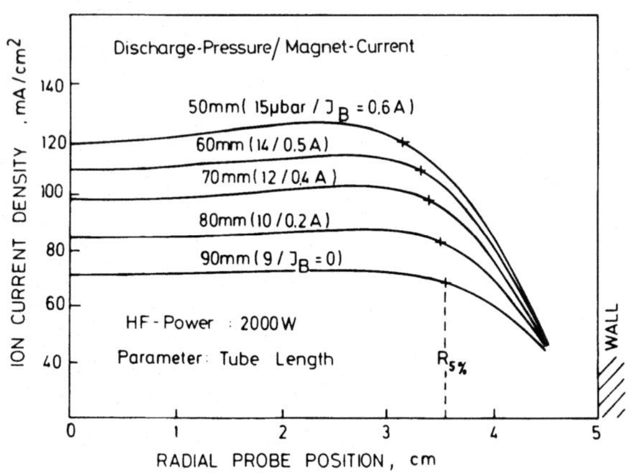

Fig. 9: Radial current-density distributions for various discharge chamber lengths with optimized pressure and magnetic field strength, respectively. (The crosses indicate the limitation of the useful extraction area, where the profile variation exceeds 5%.).

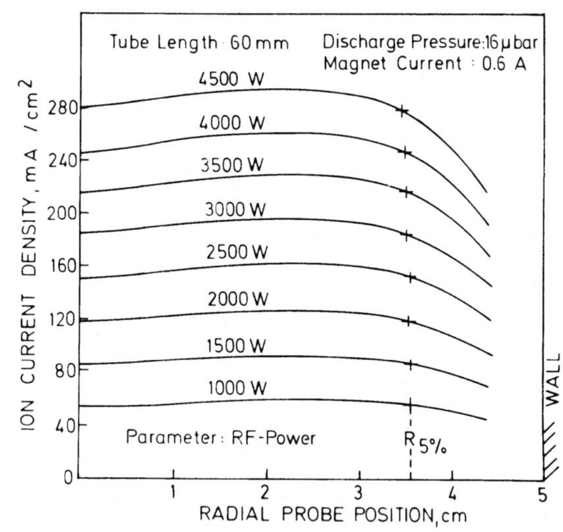

Fig. 10: Current density profiles at a constant vessel length and a fixed magnetic field strength for different rf-powers. (The magnetic coil current is equivalent an auxiliary magnetic field strength of 0.15 mT)

The special magnetic field smoothes the radial distributions sufficiently, but too high field strengths lead to an earlier current decrease at greater radial positions (near the wall). Hereby, the available extraction area becomes reduced extremely. While we need higher magnetic coil currents for profile planing at shorter vessel lengths, they are disadvantageous in the case of long vessels (reduction of the extraction area). Depending on the chamber length there exist optimum magnetic field strengths, reducing the respective profile variation just below the 5%-limitation. We found optimum discharge conditions for each vessel length, delivering plane current density distributions with variations less than 5% over maximum cross sections (Fig. 9). Whereas the total current density raises with shorter vessel lengths, the useful extraction area will be reduced continuously (proportional to the square of the 5%-radius). By integrating over the appropriate cross section, the chamber length of about 60 mm turns out to be the best one (with regard to a maximum extractable ion current).

Being interested in very high current densities of about 200 mA/cm^2 for neutral injection in fusion devices, we have to increase the applied rf-power. Fig. 10 shows the optimized current density profiles up to 4,5 kW rf-power for a vessel 6 cm in length. (The current of 0.6 A of the auxiliary field coil corresponds to a field induction strength of $1.5 \cdot 10^{-4}$ Tesla at the bottom of the vessel.) Under stationary conditions we gained current densities of 280 mA/cm^2 at 4,5 kW rf-discharge power (Fig. 10), measured in the ion entrance of the suggested extraction system. Hereby the quite plane density profile (variation ≤ 5%) enables an extraction area of 7 cm in diamter (i.e. 70% (!) of the vessel diameter).

2.3 Theoretical Model

The experimental results show firstly an overall increase of the attainable current density with shorter chamber lengths and secondly an increase of the radial profile inhomogeneity (Fig. 7). - Both effects can be explained by a simple theoretical model based on the arc-theory of LANGMUIR and TONKS (7).

Here we will restrict ourselves to the first effect, considering that a discussion of the second topic (radial inhomogeneity) has been published already (6). Calculations about the absolute size of the ion current density were only possible for mean values.

The total power consumption of the discharge results from the losses of excitation, dissoziation, ionization, and the ion current to the negative charged wall.

Assuming no volume recombination of the ionized charge-carriers in the pressure range considered (about 10 μbar), the losses by ionization and ion wall currents can be combined to a single wall loss term. Accelerated by the negative wall potential (floating potential) $U_W = -1/2\ kT_e/e\ (\ln 0.4\ m_i/m_e)$, the ions transport the ionization energy and additionally the collected energy from the potential gradient to the wall

$$P_{wall} = j_i/e\ (W_I + \varkappa kT_e)\ A_{vessel} \qquad (1)$$

(j_i = ion current density, e = elementary charge, W_I = ionization energy, $\varkappa \sim 3.75$ for hydrogen, A_{vessel} = vessel surface)

The power consumption by excitation and dissoziation can be written in the following way:

$$P_E = \int_V \frac{\delta^2 N_E}{\delta V \delta t} W_E \ dV \cong n_o n_e < \sigma_E v > V_{vessel} \qquad (2)$$

$$P_D = \int_V \frac{\delta^2 N_D}{\delta V \delta t} W_D \ dV \cong n_2 n_e < \sigma_D v > V_{vessel} \qquad (3)$$

(N = absolute particle number, n_o/n_2 = neutral/H_2-particle density, n_e = plasma density, v = electron velocity, $< \sigma_E v > \equiv$ Maxwellian impact integral; E and D denote excitation and dissoziation).

Thereby you have to take appropriate mean values for the excitation energy W_E (resp. dissoziation energy W_D) as well as for the plasma density n_e. As the Maxwellian impact integral is depending on the electron velocity, we have to determine the corresponding electron temperature. Combining the ion-balance-equation $\tau_i^{-1} = n_o <\sigma_I \mathbf{v}>$ (τ_i = mean ion lifetime, n_o = neutral-particle density, $<\sigma_I \mathbf{v}>$ = ionization reaction rate) with an expression from LANGMUIR and TONKS (7) for the mean ion lifetime, valid in the medium discharge-pressure range (ion's free path $\lambda_i \leq$ discharge tube radius r) we get an estimation for the mean electron temperature T_e as a function of the vessel length ℓ:

$$(0.8\ T_e - 4.0) 10^3/s \cong 1.9 \frac{\sqrt{\lambda_i}}{n_o} (r^{-3/2} + \ell^{-3/2}) \sqrt{\frac{kT_e}{m_i}} \qquad (4)$$

(T_e in 10^4 K, n_o in 10^{14} cm^3, k = Boltzmann-constant.) The T_e-dependent impact integral $<\sigma_I \mathbf{v}>$ was approximated linear, which is valid for both, H-atoms and H_2-molecules in the electron-energy range of about 10 eV (typical experimental value).

Knowing the dependency $T_e(\ell)$, which is in rather good agreement with the experiments (8), we are able to calculate the specific power consumption of the discharge as function of the vessel length. We get:

$$\frac{P_{rf}}{j_i} = 2\pi r^2 [(U_I + 3.75 \frac{kT_e(\ell)}{e})(1 + \frac{\ell}{r}) + + n_o \frac{\varepsilon}{\sqrt{T_e}} U_E <\sigma_E \mathbf{v}>(T_e)\cdot\frac{\ell}{2} + n_2 \frac{\varepsilon}{\sqrt{T_e}}\cdot U_D <\sigma_D \mathbf{v}>(T_e)\frac{\ell}{2}] \quad (5)$$

(P_{rf} = rf-power, j_i = ion current density, $U_{I,E,D}$ = voltages, corresponding to resp. process energy, $\varepsilon = 1.8\cdot 10^{-2}\ K^{1/2}\cdot s/m$, n_2 = H_2-molecule density).

Fig. 11 shows the calculated specific discharge power consumption per current density unit depending on the chamber length. In good agreement with the experiments, the relation describes the increase of ion current density with shorter vessel lengths at constant rf-power. The specific power consumption varies from 21 Watt per mA/cm^2 at 40 mm tube-length to 30 W per mA/cm^2 current density at 90 mm discharge length. Beyond that we learn, that the wall-losses (incl. ionization) are the dominant factor (72 - 65%) followed by excitation (24 - 29%) and dissoziation (4 - 6%), corresponding to chamber-lengths from 40 to 90 mm.

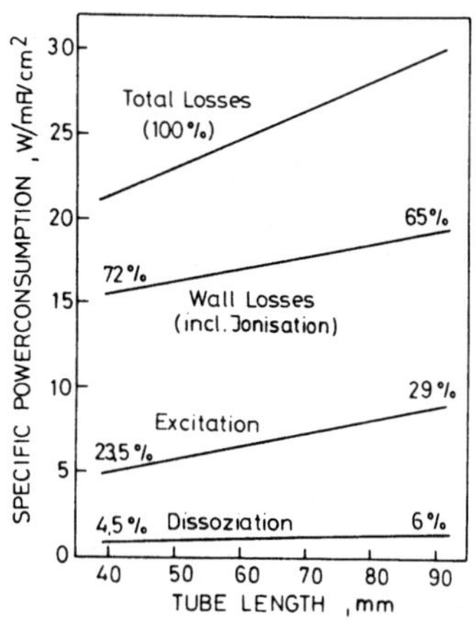

Fig. 11: Calculated specific power consumption of the discharge as function of the tube length

3. ACKNOWLEDGEMENT

This work is supported by the Deutsche Forschungsgemeinschaft (German Research Agency). The investigations described in chapter 1.4 are part of the Doctorate Thesis of the second author, and PART II of the paper is an extraction of the Doctorate Thesis of the last author (8).

The paper is dedicated to Professor Dr. Dr. E.h. W. HANLE on occasion of this 80th birthday.

4. REFERENCES

1. J. FREISINGER, S. REINECK, H.W. LOEB: "An Rf-Ion Source for Neutral Injection" Proc. 10th Symp.Fusion Tech., Padova (Italy), Pergamon Press, Oxford, p. 251 (1979)
2. J. FREISINGER, S. REINECK, H.W. LOEB: "The Rf-Ion Source RIG 10 for Intense Hydrogen Ion Beams" Journ. de Phys. 40 (suppl.no. 7), p. C7 - 477 (1979)
3. L. VALYI "Atom and Ion Sources" John Wiley & Son, London (1977)
4. J. FREISINGER, S. REINECK, H.W. LOEB: "Diagnostics in the RF-Ion Source RIG 10 Proposed for Neutral Injection" Proc. 4th Internat. Symp. Plasma Chem., edited by S. Veprek and J. Herz, University Press, Zuerich, p. 635 (1979)
5. J. FREISINGER, J. LENSING, S. REINECK, and H.W. LOEB: "Diagnostics in the Rf-Ion Source RIG 10 for Intense Ion Beams" Proc. 6th Internat. Conf. Gas Discharges and their Applications, Edinburgh , 8 - 11 September 1980
6. R. THUERINGER, J. FREISINGER, AND H.W. LOEB: "Optimization of the Discharge Chamber Length of the Rf-Ion Source RIG 10" Proc. 6th Internat. Conf. Gas Discharges and their Applications, Edinburgh, 8 - 11 September 1980
7. TONKS, L., LANGMUIR, J.: "A General Theory of the Plasma of an Arc" Phys. Rev. 34, 876 (1929)
8. THUERINGER, R. Doctorate Thesis Giessen (June 1980)

A NEW PROCEDURE FOR THE DESIGN OF THE EQUILIBRIUM-FIELD-CURRENTS FOR NON-CIRCULAR PLASMAS

S. Bobbio°, E. Coccorese°, R. Martone°°

° Univ. di Napoli, Ist. di Elettrotecnica; via Claudio 21, Naples (Italy)

°°Univ. della Calabria, Dip. Elettrico; Rende,Cosenza (Italy)

ABSTRACT

A numerical procedure for designing the external currents required to maintain in equilibrium a plasma of given cross-section (even non-circular) in an axisimmetric device is presented. Having approximately assigned the current density distribution on the plasma cross-section, the proposed procedure determines firstly an exact equilibrium configuration having a current-density "profile" within the plasma very similar to the assigned one. Then, the external currents necessary to produce the required Maintaining-Equilibrium-Field are calculated in such a way as to minimize a suitable functional of the currents itselves (e.g. the total Ampere-turns).

1. INTRODUCTION

The design procedure of the external currents required to maintain in equilibrium the plasma ring in an axisimmetric device depends on the kind of problem with which one is faced when studying the plasma equilibrium configuration.

In fact, if a Free-Boundary-Problem (FeBP) has been formulated to analyze the equilibrium, the external currents capable of producing the required Maintaining-Equilibrium-Field (MEF) have been "a-priori" assigned. Then, having chosen in an arbitrary way (at least in the framework of an ideal MHD theory) the two functions $A(\psi)$ and $B(\psi)$, one looks for the plasma equilibrium configuration which is compatible with both the assigned external currents and the specified functions $A(\psi)$ and $B(\psi)$. In such cases the plasmas cross-section is not a-priori assigned: it is an output of the adopted computing procedure. To avoid the risk of converging to the trivial solution, the values of some suitable functionals of the solution are usually assigned (e.g. the total plasma current, the β values, etc.).

When the equilibrium is studied on the basis of a Fixed-Boundary-Problem (FxBP) [1], [2], [3], the plasma cross-section is "a-priori" assigned. Then, having chosen in some way the two functions $A(\psi)$ and $B(\psi)$, one has to solve the equilibrium equation

$$\frac{\partial^2 \psi}{\partial r^2} - \frac{1}{r}\frac{\partial \psi}{\partial r} + \frac{\partial^2 \psi}{\partial z^2} = -2\pi\mu_o \cdot \left[A(\psi)\, r^2 + B(\psi)\right]$$

with the boundary condition

$$\psi = \text{const}$$

on the given plasma boundary. Having found (in an analytical or numerical way) the solution $\psi(r,z)$, it is possible to calculate all the physically interesting quantities within the plasma(i.e., the current-density, the pressure, the magnetic field) by using the following relations:

$$J = A(\psi)\cdot r + B(\psi)/r$$
$$A(\psi) = 2\pi \, dP/d\psi$$
$$B(\psi) = \mu_0/4\pi \, dI^2/d\psi$$

and the equations of Magnetostatics. In particular, one can calculate the toroidal current-density J within the plasma and its contribution to the Poloidal-Magnetic-Field. Then, MEF can be evaluated by subtrating this contribution to the total Poloidal-Field associated with the Flux-Function $\psi(r,z)$. Finally, the design of the external currents required to produce the calculated MEF can be carried out. When calculating these currents, the procedure can lead to a non-well posed problem; however, it has been shown [4] that Tikhonov's Regularization Method can be used to calculate the particular external current distribution which produces the required MEF and minimizes a suitable functional of the currents itselves.

Both the described procedures have some advantagesanddisadvantages: the first one (based on a FeBP) has the advantage of freely assigning the position and the values of the external currents, but has the drawback of a considerable computational complexity since the equilibrium analisis is generally non-linear. The second procedure requires less heavy computations when analyzing the equilibrium, but implies an additional numerical work to determine the external currents. Moreover, both procedures require of a-priori assigning the two functions $A(\psi)$ and $B(\psi)$, what heavily conditions the kind of resulting equilibrium configuration. On the other hand, since these two functions have a little physical meaning, it is generally very difficult to choose them in such a way as to arrive to equilibria showing technically interesting profiles of pressure, current density and magnetic field. This difficulty is expecially evident when designing devices with strongly-non-circular plasmas(e.g. "doublet" devices [5]) or when analyzing equilibria which, even if circular, present unusual profiles (e.g., Flux-Conserving-Tokamaks).

In this work, a new numerical procedure is presented which is based on a FxBP and permits of calculating the external currents distribution capable of producing equilibrium configurations showing previously specified profiles. The procedure does not require of firstly assigning the two functions $A(\psi)$ and $B(\psi)$: therefore, it permits of a-priori specifying (at least approximately) the current-density profile within the plasma.

2. DESCRIPTION OF THE PROCEDURE

The proposed procedure initiates by assigning the plasma cross-section and the toroidal-current-density profile one wishes to obtain within the plasma.

By choosing, as usual, a cylindrical coordinate system r,z,φ with the z-axis coinciding with the symmetry axis, let $J_o(r,z)$ be the function (expressed analytically or numerically) describing the desired toroidal current-density distribution. It is well known that J_o can not generally verify the equilibrium condition; that means that it is impossible to find two functions $A(\psi)$ and $B(\psi)$ such that the condition

$$J_o = A(\psi)\cdot r + B(\psi)/r$$

is verified everywhere within the plasmas. Nevertheless, it is always possible to determine a suitable external currents distribution capable of making the plasma boundary coinciding with a vectorial line of the resulting magnetic field (RMF) produced by both the $J_o(r,z)$ toroidal-current density and the previously mentioned external currents. Indeed, it is well known that there is an infinity of external currents distributions capable of doing that: they differ for a distribution producing zero-field within the plasma.

Let $\psi_o(r,z)$ be the flux-function describing RMF and γ_k the generic vectorial line

$$\psi_o(r,z) = \psi_k$$

developing within the plasma. Then, let $J_o(s_k)$ the function giving the values $J_o(r,z)$ takes on the points of γ_k (having indicated with s_k a curvilinear abscissa on γ_k). Now, two numbers A_{1k} and B_{1k} can be determined in such a way as the function

$$J_{1k} = A_{1k}\cdot r + B_{1k}/r$$

has a "distance" from $J_o(s_k)$ as small as possible. Of course, this distance can be chosen in several ways; in the following, the norm of the maximum for the difference was assumed:

$$\left\| J_{1k} - J_o(s_k) \right\| = \max \left| J_{1k} - J_o(s_k) \right| .$$

The same procedure can be carried out for each vectorial line within the plasma, arriving to determine two functions

$$A_1(\psi) \quad \text{and} \quad B_1(\psi)$$

defined by associating to each value ψ_k the corresponding values A_{1k} and B_{1k} in the previously desribed way. Now, let $J_1(r,z)$ be the function

$$J_1(r,z) = A_1(\psi)\cdot r + B_1(\psi)/r.$$

This new current density distribution can be substituted to the $J_o(r,z)$ and the procedure can be iterated. Convergence is achieved when the distance between J_{k+1} and J_k vanisches for $k\to\infty$. A discussion about the convergence of this procedure can not be reported in this paper, but is presented in [6].

Finally, the part of the procedure devoted to calculating at each step the external currents capable of making the plasma boundary coinciding with a vectorial line of RMF, remains to be described. In this respect, the procedure, taking into account the geometrical constraints due to the presence of the various components of the machine, automatically defines the region where the equilibrium coils can be allocated. Such a region is then subdivided in N_e elementary coils that numerically represent the possible conductors of the equilibrium winding. The flux equilibrium condition is imposed on N_b points of the plasma boundary, with $N_b \ll N_e$. A linear programming method is then applied to find out a configuration optimal with respect to a certain "quality" index (such as amount of conducting material, total Ampere-turns, etc.). It should be noticed that technical constraints (e.g. max current density allowed) are considered by the procedure.

3. NUMERICAL RESULTS

Several sample cases were presented to demonstrate the possibilities of utilizing the proposed procedure and to check its correctness and accuracy.

In a first case, the procedure was applied to JET; to this purpose, a "D" plasma cross-section was considered whose dimensions correspond to those of JET. Then, an initial "flat" current-density distribution of the type

$$J_o = A_o \cdot r + B_o / r$$

with $A_o = 8.42 \cdot 10^3$ A/m^3 and $B_o = 7.65 \cdot 10^5$ A/m, was assumed to initiate the procedure. Just one iteration was sufficient to reach very good convergence (within 0.1%) and many following iterations did not modify the results. Moreover, the functions $A(\psi)$ and $B(\psi)$ automatically chosen by numerical procedure have the constant values $A = 8.46 \cdot 10^3$ A/m^3, $B = 7.65 \cdot 10^5$ A/m, in very good agreement whith the previous values.

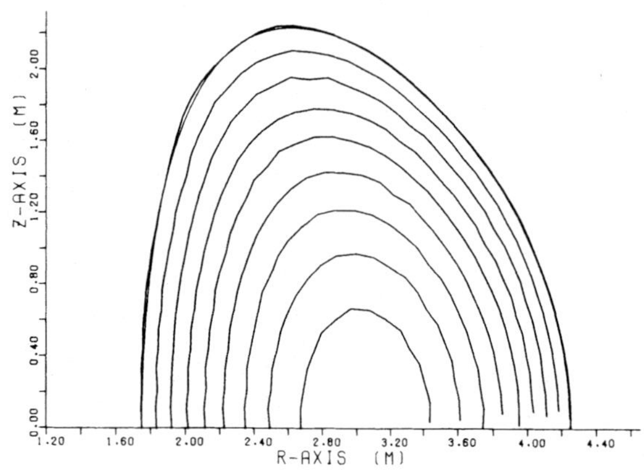

Fig.1 - Equil. Field Map of JET; I_{pl} = [illegible] MA

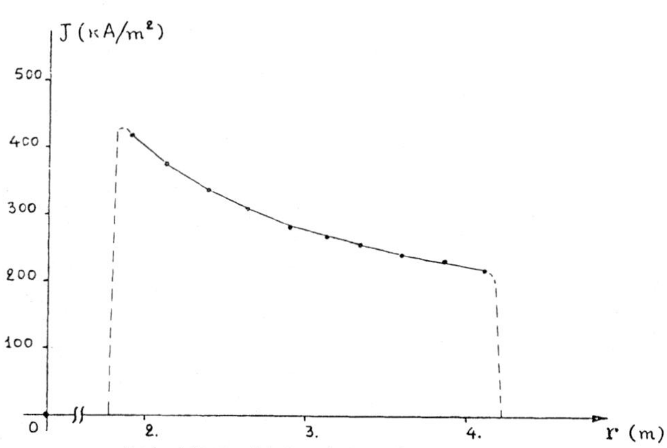

Fig.2 - Behaviour of J(r) on the Equat. Plane in JET

In fig. 1, the field map for the JET plasma is reported, while in fig. 2 the current density distribution is shown on the equatorial plane.

A second sample case refers to a "doublet" plasma cross-section: an uniform initial current distribution was assumed on the whole section. Also in this case, in spite of the existance of a "separatrix" in the magnetic field, the convergence was very good: 4 iterations were sufficient to reach the convergence (wi-

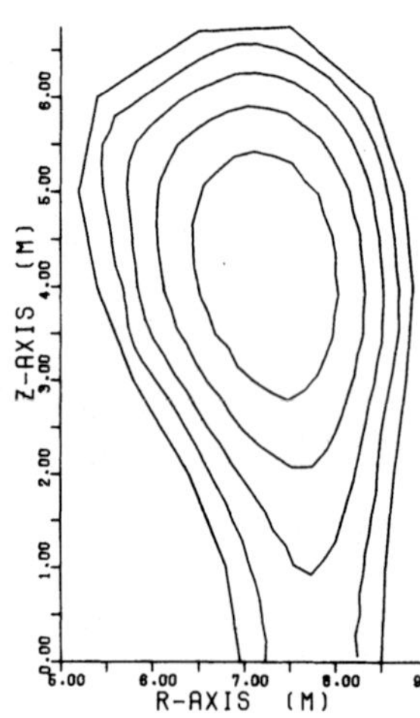

Fig. [illegible] - Equil. Field Map of a "Doublet" Plasma; I_{pl}=1A

thin 0.5%). In fig.3, the field map is reported, while in fig. 4 the computed functions A(ψ) and B(ψ) are shown for this case.

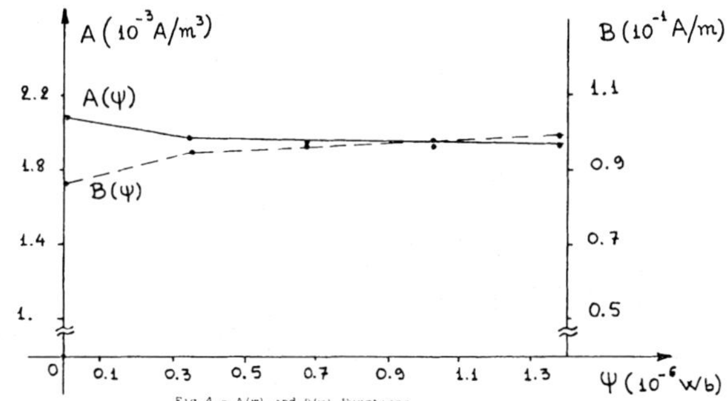

Fig.4 - A(ψ) and B(ψ) Functions for "Doublet"

Another sample case refers to the NET/INTOR geometry, with an uniform assigned profile. In fig.5 it is shown the equilibrium field map while in fig. 6 the computed A(ψ) and B(ψ) are plotted. Finally fig. 7 shows the external currents required for the equilibrium (evaluated taking into account the geometrical constraints of the design) and the Vertical-Field map.

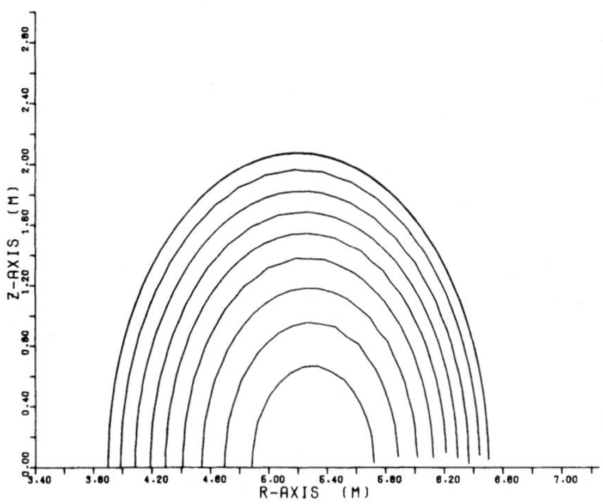

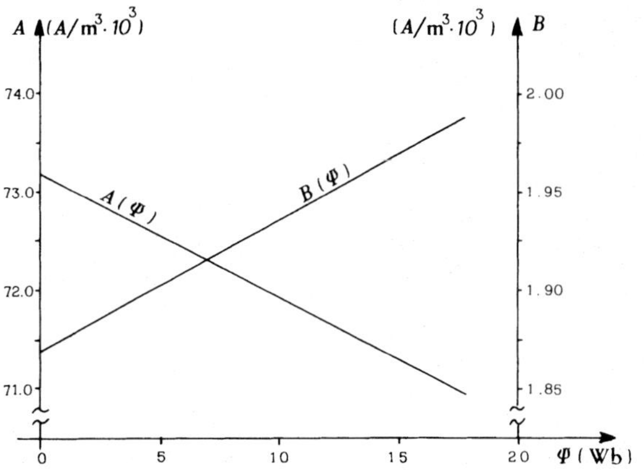

Fig.6 - A(ψ) and B(ψ) functions for NET/INTOR

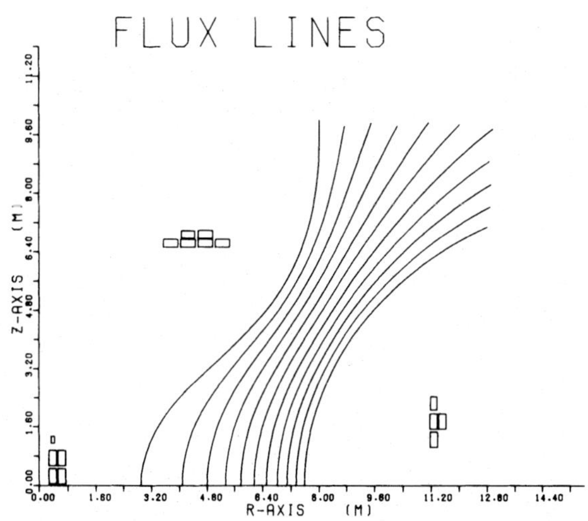

4. CONCLUSIONS

The proposed procedure appears particulary useful to calculate the external currents required to produce the Maintaining - Equilibrium-Field since it permits to optimize the currents distribution taking into account the technical constraints required by the design.

As far as the equilibrium analysis is concerned, the procedure:

- reachs convergence in few iterations (3÷6) with good accuracy ($\leqslant$ 1%);

- allows to consider any current-density distribution within the plasma (at least, if non "pathological") being confident that an exact equilibrium configuration exists very near to the assigned one.

Finally, it seems intersting to point out that the proposed procedure does not require an "a priori" assigment af the β_p value: in fact, β_p is an output for each run. Therefore, this procedure seems to be interesting also for the analysis of flux-conserving equilibria.

REFERENCES

[1] SHAFRANOV V.D.:"Equilibrium of a plasma toroid in a magnetic field", Sov. Phis., JETP, 37 (1959)

[2] MASCHKE E.K.: "Exact solution of the MHD equilibrium equations for a toroidal plasma", Plasma Physics, 15 (1973)

[3] GREENE J.M., JOHNSON J.L., WEIMER K.E., "Tokamak equilibrium", The Physics of Fluids, 14 (1971)

[4] ZACHAROV L.E.: "Numerical methods for solving some problems of the theory of plasma equilibrium in toroidal configurations", Nuclear Fusion, 13 (1973)

[5] HELTON F.J., WANG T.S.: "MHD Equilibrium in non-circular Tokamaks with field shaping coil systems", Nuclear Fusion, 18 (1978)

[6] BOBBIO S., COCCORESE E., MARTONE R.: "A new procedure for the numerical computation of axisymmetric MHD equilibria in non-circular plasmas", to be published.

ANALYTICAL EVALUATION OF THE INDUCTANCES AND $\frac{\delta L}{\delta R}$ MATRICES IN JET *

S. Bobbio
Istituto di Elettrotecnica dell'Università di Napoli
Via Claudio 21, Naples (Italy)

* Work performed under JET Contract G-IY-451.

ABSTRACT

An analytical method is presented to evaluate the inductances-matrix and its derivative with respect to a "rigid" horizontal displacement of the plasma in JET. The results of the calculations of the self-and-mutual inductances between the coils, the vessel-and-structural elements and the plasma are reported. Finally, a comparison is presented between the obtained analytical results and the corresponding ones obtained in a numerical way (via EQUILI-TOPE-POTENT codes). The good agreement between these results allows to use with confidence the analytical calculations even in those cases for which numerical results are not available.

1. INTRODUCTION

An analytical method is presented to evaluate the inductances and $\frac{\delta L}{\delta R}$ matrices in JET. The following assumptions are made:

- toroidal axisymmetric geometry,
- $\mu \to \infty$ within the Iron, except in the central core,
- saturated central Iron-core, with a permeability $\mu_1 (\cong 1.5\,\mu_0)$
- geometrical configuration as in Fig.1.

No particular assumption is required as regards:

- the shape of the plasma,
- the current-density distribution,
- the shape of the vacuum-vessel and that of the mechanical structure as in Fig.1.

2. FORMULATION OF THE PROBLEM

Consider a situation like that sketched in Fig.1. To perform the calculations in toroidal geometry, an extension of the Roth method [1] (valid for cylindrical geometry) is proposed. To this purpose, at a first stage of the analysis, a completely saturated central core is assumed ($\mu_1 = \mu_0$); then, the obtained results are corrected to take into account the permeability ($>\mu_0$) of the Iron-core. Assuming a cylindrical coordinate-system as shown in Fig.1, consider the vector-potential $\bar{A}$ (0,0,A) associated with such an axisymmetric configuration. The magnetostatic equations take the form:

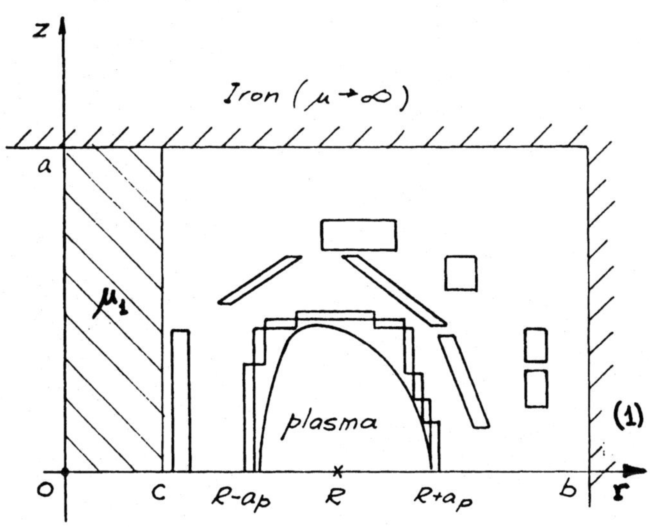

Fig 1

$$\frac{\partial^2 A}{\partial r^2} + \frac{1}{r}\frac{\partial A}{\partial r} + \frac{\partial^2 A}{\partial z^2} - \frac{A}{r^2} = -\mu_0 G(r,z) \qquad (1)$$

having indicated with G(r,z) the current-density distribution in the plasma, in the coils and in the vessel/structure elements.

The boundary conditions deriving from the above-mentioned assumptions (with a completely saturated Iron-core) are:

$$\left.\frac{\partial A}{\partial z}\right|_{r=0} = 0 \quad (2), \qquad \left.\frac{\partial A}{\partial z}\right|_{z=0} = 0 \quad (3), \qquad \left.\frac{\partial A}{\partial z}\right|_{z=a} = 0 \quad (4),$$

$$\left[\frac{A}{r} + \frac{\partial A}{\partial r}\right]_{r=b} = 0. \qquad (5)$$

The A potential can be expressed in the form:

$$A = \sum_{m=1}^{\infty} \sum_{n=1}^{\infty} B_{mn} \cos(mz) \cdot J_1(nr) \tag{6}$$

where J_1 is the first-order Bessel's function. It is easy to verify that such an expression of the potential satisfies conditions (2) and (3) for any values of m and n. To satisfy condition (4), it results:

$$\left.\frac{\partial A}{\partial z}\right|_{z=a} = -\sum_m \sum_n m B_{mn} \sin(ma) \cdot J_1(nr) = 0$$

and then:

$$m_h = (h-1)\frac{\pi}{a}, \quad h = 1, 2, \ldots\ldots \tag{7}$$

In a similar way, to satisfy condition (5), it results:

$$\left[\frac{A}{r} + \frac{\partial A}{\partial r}\right]_{r=b} = 0 \quad \Rightarrow \quad \frac{1}{b} J_1(nb) + n J_1'(nb) = 0. \tag{8}$$

It is easy to verify that condition (8) is equivalent to the following equation:

$$J_0(nb) = 0 \tag{8'}$$

Indicating with α_k $(k=1,2,\ldots)$ the zeros of this equation and posing:

$$n_k = \frac{\alpha_k}{b}, \quad k = 1, 2, \ldots \tag{9}$$

one can conclude that the double series:

$$A = \sum_{h=1}^{\infty} \sum_{k=1}^{\infty} B_{hk} \cos(m_h z) \cdot J_1(n_k r) \tag{10}$$

with

$$m_h = (h-1)\frac{\pi}{a}, \quad n_k = \frac{\alpha_k}{b}, \quad h, k = 1, 2, \ldots\ldots$$

(the α_k being the zeros of eqs. (8) and (8')), satisfies all the boundary conditions (2) $\div$ (5).

Now, taking into account the following orthogonality property:

$$\int_0^1 t\, J_1(\alpha_m t) \cdot J_1(\alpha_n t)\, dt = \begin{cases} 0, & \text{for } m \neq n \\ \frac{1}{2}\left[J_1(\alpha_n)\right]^2, & \text{for } m = n, \end{cases}$$

(where $\alpha_1, \alpha_2, \ldots$ are the zeros of eq.(8)), it can be shown [2] that the double series (10) is a Fourier-Dini's series which, under assumptions quite similar to those of the Fourier-Bessel's series, can be used to represent any given function F(r,z). To verify the magnetostatic equation, one can substitute eq.(10) in eq. (1), arriving to the following equation:

$$\sum_{h=1}^{\infty} \sum_{k=1}^{\infty} B_{hk} \cos(m_h z)\left[n_k^2 J_1''(n_k r) + \frac{n_k}{r} J_1'(n_k r) - \frac{1}{r^2} J_1(n_k r) - m_h^2 J_1(n_k r)\right] = -\mu_0 G$$

and then:

$$\sum_{h=1}^{\infty} \sum_{k=1}^{\infty} B_{hk} (m_h^2 + n_k^2) \cos(m_h z) \cdot J_1(n_k r) = \mu_0 G(r,z). \tag{11}$$

In conclusion, to satisfy this equation, it is sufficient for the coefficients B_{hk} to be those of the Fourier-Dini's expansion of the function $\mu_0 G(r,z)$. To calculate these coefficients, one can multiply both sides of eq.(11) by the term

$$r \cos(m_h z) \cdot J_1(n_k r)$$

and integrate over the whole domain [0,a;0,b]. It results:

$$\text{(12)} \begin{cases} (m_h^2 + n_k^2) B_{hk} \dfrac{ab^2}{4}\left[J_1(\alpha_k)\right]^2 = \mu_0 \displaystyle\int_0^a \int_0^b G(r,z) \cos(m_h z) \cdot r \cdot J_1(n_k r)\, dr\, dz, & \text{for } h \neq 1 \\ n_k^2 B_{1k} \dfrac{ab^2}{2}\left[J_1(\alpha_k)\right]^2 = \mu_0 \displaystyle\int_0^a \int_0^b G(r,z)\, r\, J_1(n_k r) \cos(m_h z)\, dr\, dz, & \text{for } h = 1 \end{cases}$$

In order to evaluate the right hand side integrals, three different cases have been considered:

a) the coils and the vessel/structure elements with a rectangular cross section,

b) the plasma (with arbitrary cross section),

c) the structure-elements with non-rectangular cross section.

As far as the case a) is concerned, considering the ith conductor and assuming an uniform current-density in it, with reference to the symbols used in Fig.2, it results:

for $h \neq 1$:

$$\text{(13)} \quad B_{hk} = \frac{4\mu_0}{ab^2} G_i \frac{b_i + b_i'}{m_h \cdot n_k} \cdot \frac{1}{m_h^2 + n_k^2} \cdot \frac{\left[J_0(n_k b_i) - J_0(n_k b_i')\right] \cos\left[\frac{m_h}{2}(a_i + a_i')\right] \sin\left[\frac{m_h}{2}(a_i' - a_i)\right]}{\left[J_1(n_k b)\right]^2}$$

and, for h = 1 :

$$\text{(14)} \quad B_{1k} = \frac{\mu_0}{ab^2} G_i \frac{b_i + b_i'}{n_k} \frac{1}{n_k^2} (a_i' - a_i) \frac{J_0(n_k b_i) - J_0(n_k b_i')}{\left[J_1(n_k b)\right]^2},$$

having indicated with G_i the current-density in the ith conductor. As far as the case b) is concerned, a "flat" current density distribution is assumed; therefore, the current density

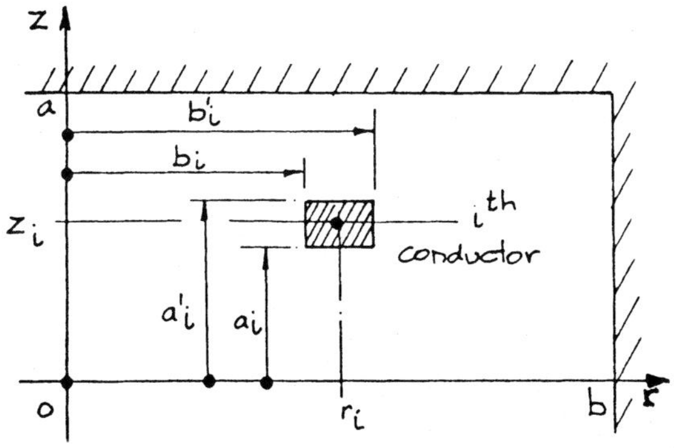

Fig 2

within the plasma is assumed to be:

$$G_p(r,z) = Ar + B/r$$

with A and B constants of given values. Substituting this expression in eqs. (12) and indicating with $Z_p(r)$ the function which describes the plasma boundary, it results:

for $h \neq 1$

$$B_{hk} = \frac{4\mu_0}{a b^2}\,\frac{1}{m_h^2 + n_k^2}\,\frac{1}{[J_1(\alpha_k)]^2}\,\frac{1}{m_h}\,S_1 \quad (15)$$

with

$$S_1 = \int_{R-a_p}^{R+a_p} \sin\left[m_h z_p(r)\right](Ar^2+B)\,J_1(n_k r)\,dr$$

and, for $h = 1$:

$$B_{1k} = \frac{2\mu_0}{a b^2}\,\frac{1}{n_k^2}\,\frac{1}{[J_1(\alpha_k)]^2}\,S_2 , \quad (16)$$

with

$$S_2 = \int_{R-a_p}^{R+a_p} (Ar^2+B)\,J_1(n_k r)\,z_p(r)\,dr .$$

Finally, as far as the case c) is concerned, assuming an uniform current density distribution within each ith conductor and indicating with $z_{i1}(r)$ and $z_{i2}(r)$, respectively, the "upper" boundary and the "lower" one of the conductor (see Fig.3), it results:

for $h \neq 1$

$$B_{hk} = \frac{4\mu_0}{a b^2}\,\frac{1}{m_h^2 + n_k^2}\,\frac{G_i}{[J_1(\alpha_k)]^2}\int_{r_{min}}^{r_{max}} \left\{\sin(m_h z_{i1}) - \sin(m_h z_{i2})\right\} r\,J_1(n_k r)\,dr \quad (17)$$

and, for $h = 1$:

$$B_{1k} = \frac{2\mu_0\,G_i}{a b^2 n_k^2 [J_1(\alpha_k)]^2}\int_{r_{min}}^{r_{max}} (z_{i1} - z_{i2})\,r\,J_1(n_k r)\,dr \quad (18)$$

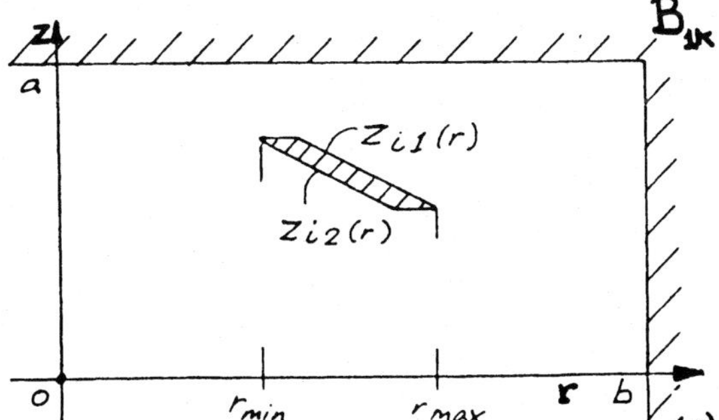

Fig.3

where r_{min} and r_{max} indicate the extremal values of r for the ith conductor of the c) type).

Having calculated the B_{ji} coefficients, it is easy to evaluate the mutual inductances between the currents. As far as the case a) is concerned, the mutual inductance M_{ji} between the "j" and "i" currents is given by:

$$(19)\quad M_{ji} = \frac{\pi}{G_i(a'_i - a_i)(b'_i - b_i)}\sum_h\sum_k B_{hk}\cos(m_h z_j)\,r_j\,J_1(n_k r_j),$$

where the B_{hk} coefficients are expressed by eqs.(13) and (14). In order to correct eq.(19) (which holds for a completely saturated central core), to take into account the permeability of the Iron, it is sufficient to notice that the geometry of the problem assures that the field is almost "vertical" in the central core region. Therefore, it is possible to express the M_{ji} inductance in the following way:

$$M_{ji} = \frac{\pi}{G_i(a'_i - a_i)(b'_i - b_i)}\sum_h\sum_k B_{hk}\cos(m_h z_j)\,r_j\,J_1(n_k r_j) + \left(\frac{\mu_1}{\mu_0} - 1\right)\sum_h\sum_k B_{hk}\,c\cdot J_1(n_k c) \quad (19b)$$

where c is the radius of the central core. This relation gives also, as a particular case, the self-inductance of the ith couple of conductors (one in the "upper" semiplane and the other in the "lower" one).

As far as the case b) is concerned, the mutual inductances between the plasma and the other conductors can be evaluated by means of the following relation:

$$M_{pi} = \frac{2\pi}{I_p}\sum_h\sum_k B_{hk}\cos(m_h z_i)\,r_i\,J_1(n_k r_i) , \quad (20)$$

where the B_{hk} coefficients are expressed by eqs. (15) and (16).

In order to evaluate the plasma self-inductance L_p , it is required to take into account the different links between the field-lines within the plasma and the plasma current distribution; to this purpose, for sake of simplicity, the plasma can be considered subdivided in N_p conductors (with rectangular cross-section) whose centers have coordinates r_i, z_i ,each carrying a current I_i. On the basis of these assumptions, L_p can be expressed as:

$$L_p = \frac{4\pi}{I_p^2}\sum_{i=1}^{N_p/2} r_i I_i \sum_h\sum_k B_{hk}\cos(m_h z_i)\,J_1(n_k r_i) \quad (21)$$

where the B_{hk} coefficients are expressed by eqs.(15) and (16).

As far as eqs. (20) and (21) are concerned, it is possible to take into account the permeability of the central core in a way which is similar to that used for eq. (19 bis).

As far as the case c) is concerned, the mutual inductances between the structural elements and the other conductors can be evaluated as follows:

$$M_{ji} = \frac{\pi}{I_i} \sum_h \sum_k B_{hk} \cos(m_h z_j) r_j J_1(n_k r_j) \qquad (22)$$

where the B_{hk} coefficients are given by eqs.(17) and (18) and the current I_i is given by

$$I_i = S_i G_i \qquad (23)$$

where S_i is the cross-section area of the ith conductor. As a particular case, eq.(22) allows to calculate also the self-inductances of the structural elements of c) type.

Finally, the derivatives of the inductances with respect to a radial displacement of the plasma have to be evaluated. Assuming a "rigid" displacement model, these derivatives can be expressed as:

$$\frac{\delta M_{\ell i}}{\delta R} = \frac{2\pi}{I_\ell} \sum_h \sum_k \frac{\delta B_{hk}}{\delta R} \cos(m_h z_i) r_i J_1(n_k r_i) \qquad (24)$$

where the terms $\frac{\delta B}{\delta R}$ are given (see App.) by:

for $h \neq 1$

$$\frac{\delta B_{hk}}{\delta R} = \frac{4\mu_o}{a b^2} \frac{1}{m_h^2 + n_k^2} \frac{1}{[J_1(\alpha_k)]^2} \frac{n_k}{m_h} \int_{R-a_\ell}^{R+a_\ell} \sin[m_h z_\ell(r)](Ar^2+B) J_o(n_k r) dr \qquad (25)$$

and, for $h = 1$:

$$\frac{\delta B_{1k}}{\delta R} = \frac{2\mu_o}{a b^2 n_k [J_1(\alpha_k)]^2} \int_{R-a_\ell}^{R+a_\ell} (Ar^2+B) z_\ell(r) J_o(n_k r) dr . \qquad (26)$$

To calculate $\frac{\delta L_\ell}{\delta R}$, it should be possible to follow the same method, but, for sake of simplicity, it is preferable to use another method, although less rigorous, but much simpler.

For this purpose, the plasma self-inductance L can be expressed as:

$$L_\ell = L_e + L_i = L_e + 2\pi R \cdot \ell_i^p \qquad (27)$$

where L_e is the "external" inductance, and ℓ_i is the "internal" inductance of the plasma per-unit-length. Due to the "rigid" displacement model assumed for the plasma, it seems reasonable to assume:

$$\frac{\delta \ell_i}{\delta R} \cong 0 \qquad (28)$$

and then:

$$\frac{\delta L_\ell}{\delta R} \cong \frac{\delta L_e}{\delta R} + 2\pi \ell_i \cdot \qquad (29)$$

Neglecting the terms in $(\Delta R)^2$, it results:

$$\frac{\delta L_e}{\delta R} \cong \frac{1}{R-a_\ell} L_e \qquad (30)$$

and L_e is given by:

$$L_e = \frac{2\pi(R-a_\ell)}{I_\ell} \sum_h \sum_k B_{hk} J_1[n_k(R-a_\ell)]. \qquad (31)$$

Finally, by substitution of eqs. (30) and (31) in eq. (29), it results:

$$\frac{\delta L_\ell}{\delta R} \cong \frac{1}{R-a_\ell} L_e + 2\pi \ell_i = \frac{2\pi a_\ell}{R I_\ell} \sum_h \sum_k B_{hk} J_1[n_k(R-a_\ell)] + \frac{1}{R} L_\ell . (32)$$

4. NUMERICAL RESULTS

To calculate the values of the B_{hk} coefficients given by eqs. (13) ÷ (18) and those of the inductances (together with their derivatives), a very simple numerical code was set up (INDUCT). In this code, for sakc of simplicity, all series have been calculated summing up their first 400 terms (20 values for the h index and 20 for k): the estimated "truncation-error" is lower than 1%. The integrals contained in eqs.(15), (16), (24) and (25) have been calculated numerically in the usual way.

In Tab.1 the complete inductance matrix resulting from the analytical calculations is reported for a circular plasma, with reference to the following parameters:

a = 3.5 m; c = .72 m; $\mu_1 = 1.5\mu_o$

which appeared to be the ones giving the best agreement with the numerical results reported in Tab. 1' [3]. The symbols refer to Fig.4.

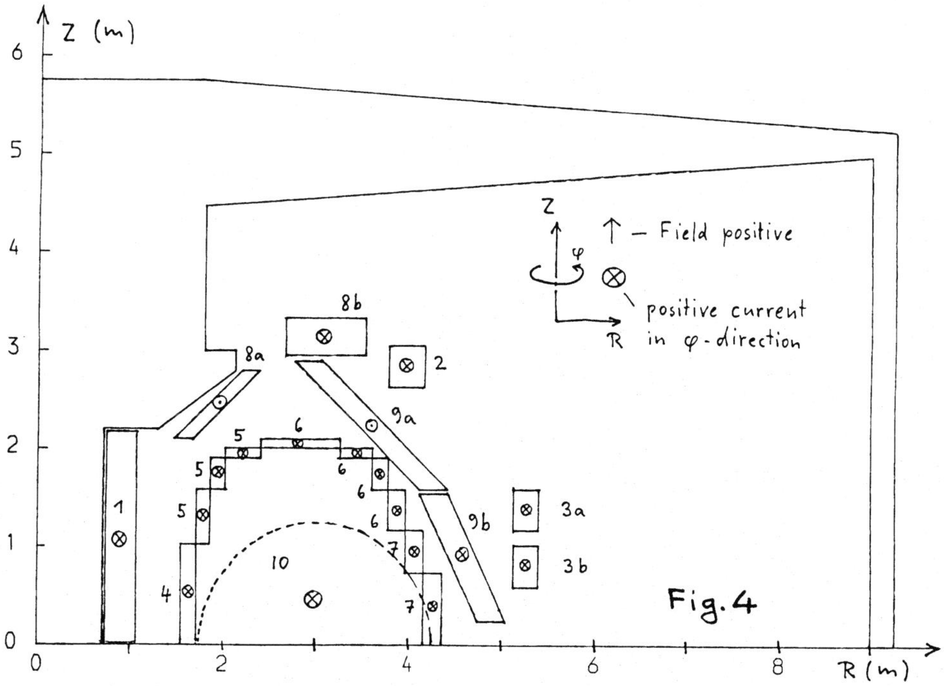

Fig.4

Single - Turn - Inductance - matrix for the circular plasma: Aij (in μH).

Tab.1 (Analytic results)

i \ J	1	2	3	4	5	6	7	8	9	10
1	0.82	0.66	0.69	0.84	0.72	0.69	0.71	0.02	0.038	0.72
2	0.56	13.55	8.20	1.42	2.34	6.40	6.52	4.57	-0.26	4.56
3	0.63	8.25	18.63	1.70	2.38	7.04	10.83	2.97	5.90	6.50
4	0.90	1.58	1.83	3.17	1.59	1.72	1.91	0.30	0.27	2.17
5	0.65	2.35	2.37	1.40	3.04	2.48	2.36	0.24	0.002	2.31
6	0.60	6.39	7.06	1.57	2.52	7.49	6.61	2.47	0.39	4.99
7	0.64	6.57	10.85	1.82	2.38	6.84	13.03	2.35	5.05	7.64
8	0.06	4.67	2.99	0.33	0.28	2.47	2.36	6.88	-1.27	1.51
9	0.073	-0.27	5.91	0.30	0.025	0.42	5.10	-1.30	7.64	2.25
10	0.72	4.56	6.50	2.17	2.31	4.99	7.64	1.51	2.25	7.37

Tab.1' (Numeric results)

i \ J	1	2	3	4	5	6	7	8	9	10
1	0.72	0.65	0.66	0.77	0.75	0.69	0.67	0.03	-0.05	0.75
2	0.63	13.06	7.98	1.56	2.84	6.73	6.40	5.25	-1.40	4.18
3	0.64	7.97	17.28	1.69	2.72	7.07	9.98	3.40	3.73	5.31
4	0.80	1.65	1.80	3.01	1.75	1.81	1.91	0.36	-0.01	2.43
5	0.74	2.84	2.73	1.67	3.51	3.00	2.67	0.29	-0.41	2.58
6	0.68	6.78	7.15	1.72	3.01	7.97	6.95	2.76	-0.52	4.71
7	0.68	6.64	10.41	1.86	2.77	7.12	13.03	2.69	3.61	6.38
8	-0.06	4.29	2.72	0.15	-0.10	2.09	2.02	7.04	-1.09	1.03
9	0.00	-0.84	4.95	0.15	-0.16	0.07	4.56	-1.00	6.81	1.36
10	0.75	4.18	5.31	2.43	2.58	4.71	6.38	1.03	1.36	5.97

The comparison between the two reported Tables show that the agreement is satisfactory, except for the structural elements **"9"**. However, it has to be noted that the numerical results present an inconsistency between the A_{ij} and the A_{ji} values for the structural elements (while this inconsistency disappears for the other conductors). For these reasons, the analytical results (which present a good agreement between the A_{ij} and the A_{ji} values) seem to be probably more realistic than the numerical ones, as far as the structural elements are concerned. Finally, in Tabs.2 and 3 , the values of the plasma inductances (self and mutual) and those of their derivatives are reported, with reference to the following parameters of a " D "-shaped plasma:

$a = 3.5$ m; $c = .72$ m; $\mu_L = 1.5\,\mu_0$; "flat" current profile;
plasma major radius: 2.96 m; plasma r-semiaxis: 1.25 m; plasma z-semiaxis: 2.1 m; I_p=3.7 MA
$\beta_p = 1.4$.

4. CONCLUSIONS

The inductances and $\frac{\delta L}{\delta R}$ matrices have been analytically evaluated for JET. For this purpose, some simplifying assumptions have been made as far as the geometry of the system is concerned. The most important of them refers to the shape of the saturated central Iron-core, which was assumed cylindrical (i.e.,without the "centre-piece").

The agreement between the analytical results and the available numerical ones seems very satisfactory also as far as the derivatives are concerned.

Tab.2 (μH)

$A_{10,1}$	=	0.69;
$A_{10,2}$	=	4.65;
$A_{10,3}$	=	6.21;
$A_{10,4}$	=	2.00;
$A_{10,5}$	=	2.42;
$A_{10,6}$	=	5.20;
$A_{10,7}$	=	7.06;
$A_{10,8}$	=	1.57;
$A_{10,9}$	=	1.77;
$A_{10,10}$	=	5.54.

Tab.3 (μH/m)

$\frac{\partial A_{10,1}}{\partial R}$	=	- 0.027;
$\frac{\partial A_{10,2}}{\partial R}$	=	1.53;
$\frac{\partial A_{10,3}}{\partial R}$	=	3.07;
$\frac{\partial A_{10,4}}{\partial R}$	=	-0.20;
$\frac{\partial A_{10,5}}{\partial R}$	=	0.0;
$\frac{\partial A_{10,6}}{\partial R}$	=	1.51;
$\frac{\partial A_{10,7}}{\partial R}$	=	3.77;
$\frac{\partial A_{10,8}}{\partial R}$	=	0.57;
$\frac{\partial A_{10,9}}{\partial R}$	=	1.79;
$\frac{\partial A_{10,10}}{\partial R}$	=	3.80.

APPENDIX

To evaluate the derivative:

$$\frac{\delta}{\delta R}\int_{R-a_p}^{R+a_p}\sin[m_h z_p(r)](Ar+\frac{B}{r})\,r\,J_1(n_k r)\,dr$$

it shall be born in mind that, when a "rigid" displacement ΔR is imposed to the plasma, the functions $z_p(r)$(which describes the plasma shape) and $Ar+B/r$ (which describes the current-density distribution) "translate" together with the plasma, while the function $r\,J_1(n_k r)$ remains "fixed". Moreover, the limits of integration become $R - a_p$ and $R + a_p$ and then it results:

$$\frac{\delta}{\delta R}\int_{R-a_p}^{R+a_p}\sin[m_h z_p(r)](Ar+\frac{B}{r})\,r\,J_1(n_k r)\,dr = \int_{R-a_p}^{R+a_p}\sin[m_h z_p(r)](Ar+\frac{B}{r})\frac{d}{dr}[r\,J_1(n_k r)]\,dr.$$

AKNOWLEDGEMENTS

The author wishes to aknowledge drs. E.Bertolini, D.Ciscato and P.Noll for many fruitful discussions.

REFERENCES

[1] K J Binns and P J Lawrenson : "Analysis and computation of Electrical and Magnetic Field Problems" Pergamon Press, 1963, pag.109.

[2] G N Watson : "A treatise on the Theory of Bessel's Functions"; Cambridge University Press, 1958.

[3] P Noll : Private communication.

PLASMA EQUILIBRIUM CONTROL IN DOUBLET III

R. Stambaugh, S. Adcock, R. Callis, J. deGrassie,
J. Luxon, P. Rock, and J. Wesley
General Atomic Company

S. Seki, M. Shimada and H. Yokomizo
Japan Atomic Energy Research Institute

ABSTRACT

The Doublet III tokamak (R = 1.43 m, a = 0.45 m, b/a = 3.1) is operated by General Atomic Company (GA) for studies of vertically elongated plasma cross-sections by research groups from GA and the Japan Atomic Energy Research Institute (JAERI). Stable ohmically heated discharges with a duration of up to 0.8 seconds have been obtained with a variety of plasma shapes. The JAERI group has produced diverted and non-diverted dee-shaped plasmas, the latter with elongations of up to 1.6. The GA group has produced droplet plasmas (two separated teardrops) with elongations of up to 1.5, doublet plasmas with an internal separatrix and elongations of up to 3.2, and an expanded boundary configuration. Record values of plasma current have been achieved: 1.0 MA in the dee and 2.2 MA in the doublet at a toroidal field of 2.4T. Dees with q(a) = 2.6 and diverted dees with q(a) $\approx$2 have been obtained.

Important features of the machine which contribute to effective shape control are: 1) a parallel-connected set of field shaping coils (F coils) which fully and closely surround the vacuum vessel, 2) a continuous resistive vacuum vessel with a 1 msec flux penetration time, 3) SCR and fast switching chopper power supplies in series with the F-coils for active shape control, and 4) feedback regulation of the position of the outermost plasma flux surface. The latter feature, although not required for stable operation, improves discharge reproducibility and greatly simplifies tuning for the desired plasma shape.

The control signals used for the flux surface position regulation are derived from measurements of the poloidal field and flux obtained with an array of sensors located immediately outside the vacuum vessel. The close proximity of the sensors to the plasma surface ($\lesssim$ 0.2a) allows the position of the plasma surface to be accurately computed with simple analog circuitry. The short time constant of the resistive vessel allows for stable high-gain closed loop control without elaborate pole-zero compensation. The small-signal response time is typically less than 10 msec. Control of the plasma position to ±0.5 cm is routine.

1. INTRODUCTION

The plasma equilibrium control system or field shaping system on Doublet III has been able to control many diverse equilibrium configurations with plasma current up to 2 MA and plasma volumes up to 17 m^3. Principal configurations produced have been: doublet plasmas [1], elongated dee plasmas [2], and both [3] a single null divertor and an expanded boundary configuration with the divertor coils <u>outside</u> the vacuum vessel. All of these configurations produce good plasma confinement and can be run in a steady state with the shot duration (0.8 sec) limited only by the pulse length capability of the Ohmic and field shaping power supplies. There are two key elements to the flexibility of the equilibrium control system. The first is the use of a sufficient number (24) of field shaping coils close enough to the plasma to control in detail the shape of the plasma boundary flux surface. The second feature is the extensive use of active shaping power supplies for dynamic control of the plasma shape; e.g., droplet plasmas (two disconnected current channels) can be merged into a doublet, and the divertor can be turned on and off in the same shot.

2. F-COIL SYSTEM

The essential concept of the Doublet III field shaping system is direct and local control of the boundary poloidal flux distribution. In contrast to the circular plasma situation [4], no analytic theory is available to specify the control fields for more general plasma shapes. By use of an MHD code [5], however, the plasma equilibrium flux surfaces are sufficiently determined if the poloidal distribution of magnetic flux on the plasma boundary and one or two parameters of the plasma current profile are known. From MHD code studies of a desired configuration, one can find the boundary flux values that need to be established at the flux measuring loops in Fig. 1. These flux values can then be produced in a shot by control of the 24 field shaping coils. Since the field shaping coils are only about 0.2 a ≃10 cm from the plasma edge, an F-coil may be considered to individually control the local plasma boundary flux as measured on the one turn flux loop attached to its front face. The close proxmity of the F-coils also minimizes the required Ampere-turns in the coils and the forces on them. The 1.0 MA dee (Fig. 1) requires only 0.6 MA-turns ($\lesssim$ 10 kA shaping supply current) in the field shaping system.

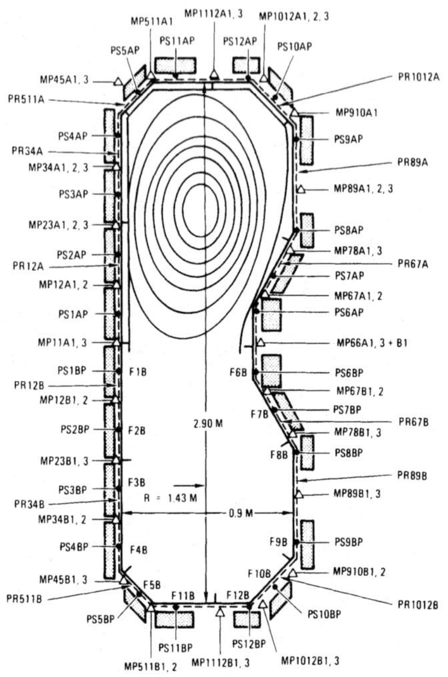

Fig. 1. Doublet III F-coil and magnetic diagnostic locations. F denotes a field shaping coil; PS denotes a flux loop, MP denotes a magnetic probe at toroidal azimuth (1,2,3 = 67°, 247°, 307°); PR denotes one of 12 segments of a full Rogowski loop. The plasma configuration is the 1.0 MA dee with elongation of 1.6. The F coil labeling is symmetric across the midplane.

A typical electrical connection of the poloidal field system is shown in Fig. 2. Each F-coil is wound in two 29 turn halves with the center at a virtual voltage zero. Active power supplies are connected at this virtual voltage zero to minimize the terminal potential, although the supplies can float to 30 kV. All 96 coil leads are brought out to a patch panel where copper bus is used to vary the connection scheme. About two hours are needed to make a major configuration change. The F-coil system is inductively decoupled

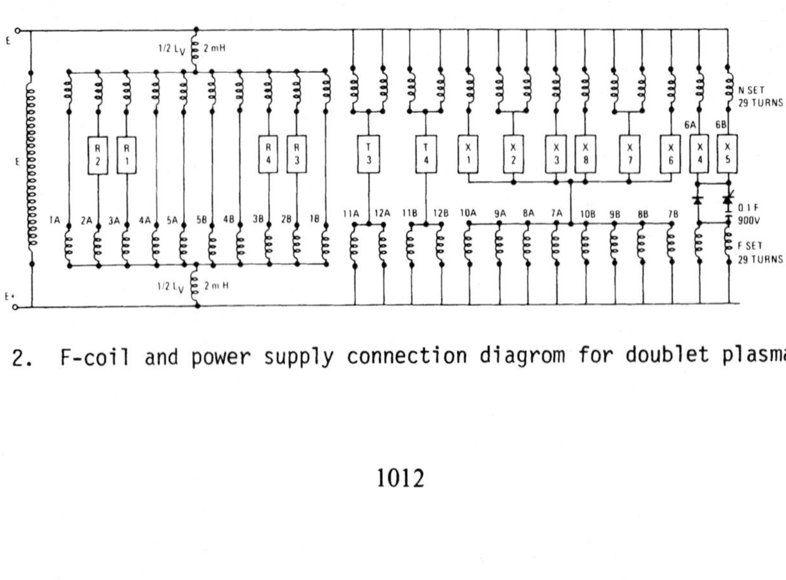

Fig. 2. F-coil and power supply connection diagrom for doublet plasmas

[6] from the E-coil by virtue of the spatial distribution of the E-coil windings the E/F turns ratio (60.5/58) and the connection of the F-coils to the E-coil.

3. RESISTIVE VACUUM VESSEL

The Doublet III vacuum vessel is made of two layers and one strengthening corrugation of 1.5 mm Inconel with no insulating breaks. The resistance of the vessel is 0.25 mΩ. On the discharge flattop the vessel current is about 8 kA and has negligible effect. At breakdown, (with 130 V per turn) the vessel current is 520 kA and has an appreciable shaping effect in the early phase of doublet shots (Sec. 5). The flux penetration time of the vessel is 1 msec. Since no feedback stabilization is required in any of the four major configurations and since 100 Hz is a maximal necessary rate for dynamic configuration changes, the vessel does not restrict the useful bandwidth of the control system.

4. CLOSED LOOP FEEDBACK CONTROL

In the early phase of the shot, the F-coil currents are established mainly as inductive images of the plasma current, in close analogy with the action of a copper shell. The centering vertical field is inductively established by the series choke L_v on the inner coils. However, since the L/R time of the F-coils is about 0.24 sec and the shot duration is 0.8 sec, the F-coils must also be actively driven. Any two parallel connected coils will develop a flux difference due to resistive flux decay equal to $2\pi N(\psi_1 - \psi_2) = \int_0^t (I_1R_1 - I_2R_2)\,dt$ unless power supplies make up the IR drop ($\lesssim$ 270 V) in each coil. The supplies R_1 and R_2 in Figs. 3,5,6 basically serve this resistive compensation function. Their voltage programs are derived from direct feedback regulation of fluxes (e.g., $\psi_2 = \psi_3 = \psi_4$). Actively driven F-coils are also needed for precise positioning of the plasma inside the limiter aperture (e.g., radial position control), for aspects of the boundary flux control that depend strongly on the time evolving plasma current profile (e.g., ψ_6 in the doublet), and unnatural features of the configuration (e.g., divertors).

4.1. Precision Control Signals

For precise position control, the small distance between the flux loops (Fig. 1) and the limiter make possible accurate local linear extrapolations of the boundary flux from the flux loops to the limiter using the flux gradient measured by the local magnetic probes or partial Rogowski coils. Various analog circuits have been constructed based on this technique. A signal RPA (RPB) computes the distance between the outer plasma surface and the outer limiter near coil 8A (8B) by dividing the difference in extrapolated limiter fluxes ($\psi_8^{LIM} - \psi_3^{LIM}$) by the flux gradient near coil 8. RPA is used to feedback regulate the radial position of dees and diverted plasmas and the upper lobe of doublets; RPB is used on the lower lobe of doublets. A signal HA (HB) computes the distance from the top (bottom) of the plasma to the limiter by dividing the difference in extrapolated limiter fluxes ($\psi_{TOP\ (BOT)}^{LIM} - \psi_3^{LIM}$) by the flux gradient on the top (bottom). The radial positions and vertical positions of the plasma are regulated to ± 0.5 cm.

Another class of signals can be derived from computer experiments. The MHD code was used to simulate singlet equilibria in the upper half of the vessel with various vertical distances Z_p from the midplane. From the computed magnetic probe signals the expression

$$Z_p = 30\ \text{cm}\left(\frac{\text{MP 34A} - \text{MP 12A}}{\text{MP 34A}}\right) + 70\ \text{cm}$$

was derived, and constructed in an analog circuit. For doublet equilibria, an expression (P_s) for the fraction of the poloidal flux inside the doublet separatrix was found to be

$$P_s = 100\% \left(2.0 - 1.15 \ \frac{PR\ 12B}{PR\ 34B}\right)$$

This simple relationship is accurate to ±10%.

4.2. Active Power Supplies and Feedback Electronics

The capabilities of the shaping power supplies are summarized in Table 1. The line-commutated rectifiers are commercial units. The chopper power supplies are built at our laboratory and have been described elsewhere [7]. They consist basically of a thyristor switching network that can apply positive voltage by switching a charged 0.25 F electrolytic capacitor bank in series with the F-coil or negative voltage by switching a varistor in series with the F-coil. The average supply voltage is determined by controlling the duty cycle between positive and negative voltage in response to a ±10 V analog signal. The X supply connection in Fig. 3, with the arrow directed into the F-coil, denotes that positive supply voltage increases F-coil current flow anti-parallel to the plasma current, thereby pushing the plasma away from that F-coil, and driving the corresponding ψ value more negative. Conversely, negative voltage pulls the plasma closer to the F-coil and drives the ψ value more positive. The T-choppers, an early design, use resistors instead of varistors; the pulling voltage in that case is just the IR drop on the resistor and can be unacceptably weak if the supply must regulate a low F-coil current.

TABLE 1

Active Shaping Supplies

Number	Type	V, I, Time	Full Power Bandwidth
2	T-choppers	± 300 V, + 3 kA, 1 sec	1 kHz
8	X-choppers	± 600 V, + 3 kA, 1 sec	1 kHz
4	U-choppers	± 300 V, ± 2.5 kA, 1 sec	1 kHz
1	D-LCR*	± 600 V, + 10 kA, 1 sec	60 Hz
1	V-LCR*	± 600 V, + 15 kA, 1 sec	60 Hz
6	R-LCR*	+ 150 V, + 3 kA, dc	60 Hz

*LCR-Line Commutated Rectifier.

The feedback control electronics consists of NIM packaged differential amplifier modules, analog divider modules, summing amplifier modules, and waveform generators (W.G.). The bandwidth of all the control electronics exceeds 10 kHz. The waveform generators provide a D/A conversion from a 128 word memory, generally clocked at a 10 msec/address rate. The memories are loaded from the control computer via interactive graphic software on a Tektronix storage display terminal. In this way, arbitrary waveforms can be specified for a boundary flux, radial position, or any control variable. Waveform generators also control the plasma current waveform and the gas puffing system throughput. The timing system for triggering the power supplies on or off, triggering the waveform generators, or relay closures is built on two levels. The highest level distributes ASCII encoded event triggers around the facility. A second level timing station is housed in a CAMAC crate and consists of a timing receiver to decode the master event triggers and one or more JORWAY 220-A delay generators. All timing settings are done through the control computer.

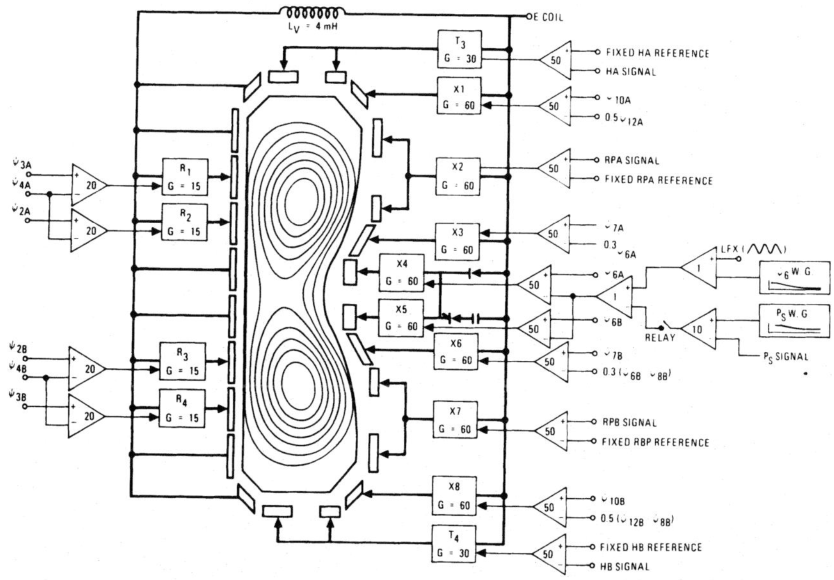

Fig. 3. Control schematic for doublet. Voltage gains are indicated on the power supplies and differential amplifiers. Heavy lines are power connections. Thin lines are control signal paths.

5. FULL CONTROL SYSTEM DESCRIPTIONS

A complete schematic for doublet plasmas together with the doublet equilibrium configuration is shown in Fig. 3 and Fig. 4 shows the waveforms this system produces. The R supplies on the inner wall serve the resistive compensation function previously discussed. Figure 5g shows the close tracking of the flux values on the inner wall.

The value P_S is controlled by X_4 and X_5 on the coils 6A and 6B. Indirect control of P_S is provided by regulating ψ_6 to follow a waveform generator program. The program was constructed by trial and error to deliver approximately the trajectory of P_S in Fig. 4c. Fine control of P_S is provided by closing a higher gain feedback loop regulating P_S at 250 msec into the shot. The P_S reference is also derived from a waveform generator. The degree to which ψ_6 follows the programmed ψ_6 is displayed in Fig. 4d. After the P_S feedback loop closes it dominates the final ψ_6 reference. Dynamic shape control is exhibited by the gradual lowering of P_S from an initial droplet plasma (two disconnected current channels) to a steady doublet with an exact P_S of 75% as derived from MHD code fits.

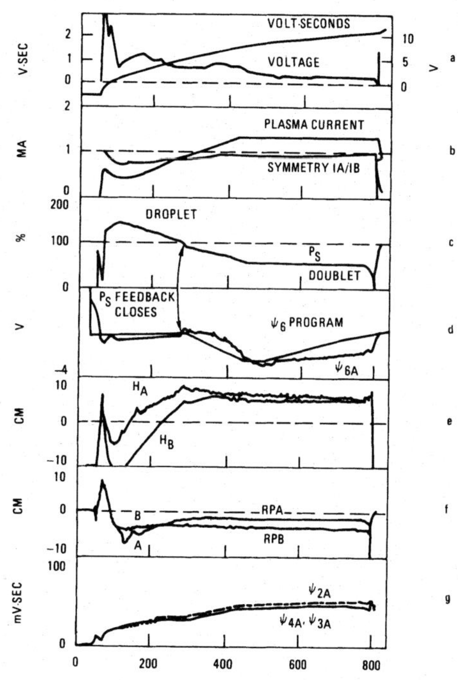

Fig. 4. Time evolution of a doublet shot

This merging technique has proven easier than controlling $P_S \lesssim 100\%$ from shot initiation for two reasons. The droplet phase is used to adjust the gas puffing waveforms to settle the two current channels into well controlled, low impurity level plasmas before merging them. Secondly, at breakdown, there is a strong tendency for the two current channels to collapse into a short vertically unstable ellipse ($P_s = 0$) due to the attraction of the large outside midplane vacuum vessel current. In order to ensure an early droplet configuration, a strong repelling field in the midplane, negative ψ_6 in Fig. 4d, is established before breakdown and at breakdown ψ_6 is quickly driven further negative by discharge of the 900 V 0.1 F electrolytic bank C_D.

The T-choppers on the top and bottom coils regulate HA and HB. Figure 4e shows that due to the strong initial repelling field present at the midplane the values HA and HB are poorly controlled until the ψ_6 programming alows P_s to drop below 100%. Before 200 msec the plasmas are limiting on the top and bottom limiters; eventually, regulation is achieved with about a 7.0 cm gap between the plasma and the top and bottom limiters.

The choppers X_2 and X_7 regulate RPA and RPB. Initially, the top lobe swings too far inside due to the top plasma current being lower than the bottom plasma current by about 25% (see Fig. 4b). On the flattop, both RPA and RPB are regulated at different values which were found empirically to optimize up/down current symmetry (Fig. 4b). The choppers X_1, X_3, X_6, and X_8 serve to interpolate the boundary flux between coils 12A, 8A, 6, 8B, and 12B.

The control schematic and the plasma equilibrium configuration for the single null divertor are shown in Fig. 5. Most of the lower coils are left open. The outer plasma field line is pulled into a divertor configuration around F-coils 1A and 1B by the reverse connected supplies R_3 and R_4. These supplies drive divertor Ampere-turns equal to 58% of the plasma current. The quantities regulated are $\psi_{1A} = \psi_{4A/0.6}$ and $\psi_{1B} = \psi_{1A}$. Some downward force on the plasma is also produced with coils 6A and 6B by T_4 which is commanded to full negative voltage, (i.e., a 0.25 Ω resistor is on the F-coil 80% of the time). Initially T_4 is programmed to push the plasma upward to ensure vertical stability of the plasma early in the shot. Because the plasma is pulled downward strongly by the midplane coils, the D supply with high (-600 V) inversion or pulling voltage has to be employed on the upper coils 10 and 12 to regulate vertical position (Zp). Radial position regulation is obtained using the V supply

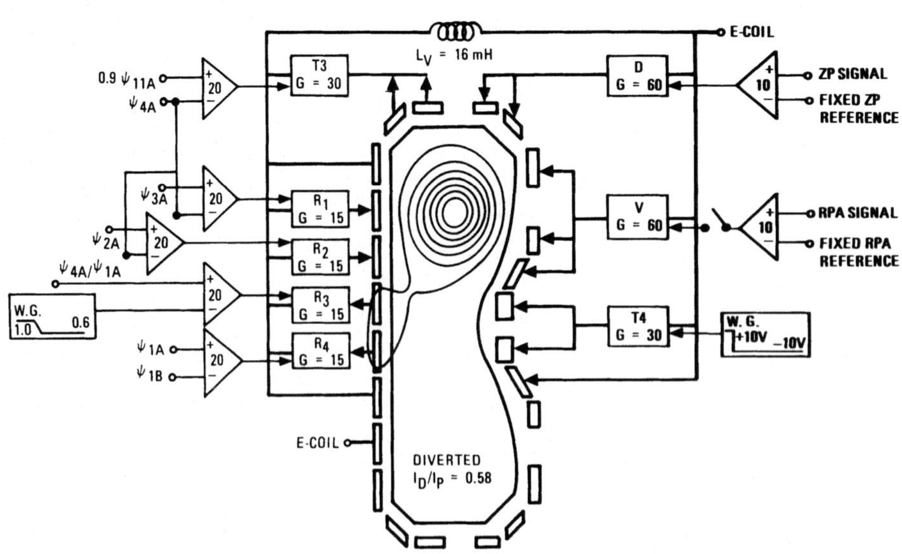

Fig. 5. Control schematic for the single null divertor and elongated dee configurations

driving F7, 8, and 9 in parallel. This same control arrangement can produce the elongated dee shown in Fig. 1 by merely setting $\psi_{1A} = \psi_{4A/0.7}$.

The control schematic and plasma equilibrium for the expanded boundary plasma are shown in Fig. 6. In this configuration the outer plasma field lines are separated from the upper chamber limiters and diverted into the bottom of the chamber where they essentially become lost in a weak poloidal field, highly radiative region. The plasma is pulled downward on both the inside (coils 1B, 2B) and the outside (coils 6B, 7B). The fluxes ψ_{1B} and ψ_{2B} follow the master control waveform ψ_{6B}. The outermost plasma field lines are steered through the mid-plane region by control of ψ_{1A} and ψ_{6A} and expanded in the bottom of the vacuum chamber with the remainder of the lower coils. As with the single null divertor, a supply capable of -300 V, a U-chopper was needed on the top coils to regulate HA. Early in the shot, upward force on the plasma is exerted by all the lower coils to ensure vertical stability until the configuration is fully developed.

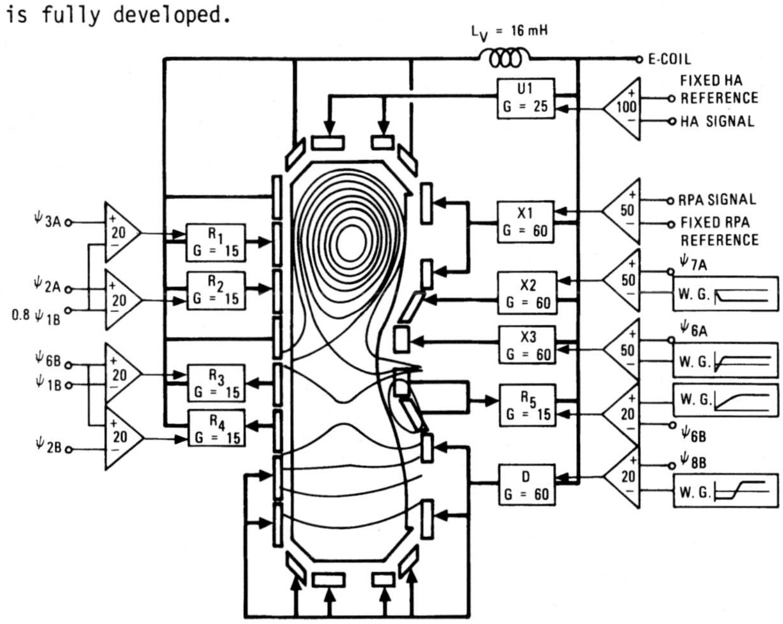

Fig. 6. Control schematic for the expanded boundary configuration

6. CONCLUSION

The Doublet III field shaping system provides precise, dynamic control of many diverse plasma equilibria. The flexibility of the control system and F-coil arrangement allows the physics research teams from General Atomic (GA) and the Japan Atomic Energy Research Institute (JAERI) to alternate running weeks on the machine, changing plasma configurations on a daily basis to support a varied program of tokamak research on dee and doublet configurations and research on two kinds of simplified divertor configuration. Record plasma currents of 1.0 MA in a dee and 2.2 MA in a doublet have been produced.

7. ACKNOWLEDGEMENT

Research carried out under the Fusion Research Cooperation Agreement on Doublet III between the United States Department of Energy and the Japan Atomic Energy Research Institute, Contract No. DE-AT03-76ET51011.

8. REFERENCES

[1] WESLEY, J.C., et al., "Shaping and Characteristics of Ohmically Heated Noncircular Plasmas in Doublet III," IAEA-CN-A-3, July 1980.

[2] KITSUNEZAKI, A., et al., "First Results of Dee Experiments in Doublet III," IAEA-CN-38/N-3, July 1980.

[3] NAGAMI, M., MAHDAVI, M., et al., "Simplified Poloidal Divertor Experiments in Doublet III," IAEA-CN-38-0-2, July, 1980.

[4] MUKHOVATOV, V.S., and SHAFRANOV, V.P., "Plasma Equilibrium in a Tokamak," Nuclear Fusion 11, 605-633 (1971).

[5] McCLAIN, F.W. and BROWN, B.B., "GAQ, A Computer Program to Find and Analyze Axisymmetric MHD Plasma Equilibria," General Atomic Company Report GA-A14490 (1970).

[6] CALLIS, R.W., "Doublet III Baseline Design Description," General Atomic Company Report GA-A13996, June 1976.

[7] ROCK, P.J., and WESLEY, J.C., "1.5 Megawatt D. C. Chopper Power Supplies for Plasma Shape Control on Doublet III," Proceedings of the 8th Symposium on Engineering Problems of Fusion Research, Nov. 1979.

EXPERIMENTAL RESEARCH ON POWER SUPPLY AND PLASMA CONTROL USING SIMULATOR

K. Yabuno, A. Ueda, R. Saito, I. Nagata,
K. Murai, T. Iba, S. Hatakeyama

Hitachi Ltd. Hitachi-shi,
Ibaraki-ken, Japan

Summary

A simulator for the power supply system of a nuclear fusion device, equipped with a simulated plasma circuit, vacuum vessel circuits and load coils is constructed and various power supply control experiments are carried out. The control of plasma position and the shape of a plasma cross section, and the control of a power supply system connected to many loads, is attempted. As a result, it is found that plasma can be maintained in a stable condition by controlling power supplies, even when a disturbance field occurs in the plasma.

Introduction

This paper describes the experimental results obtained with a power supply simulator for a nuclear fusion experimental device. Power supply and plasma control in experimental nuclear fusion devices have been studied with computer-based calculation, but there are many difficult problems that cannot be resolved in digital simulations, because fusion experimental devices are very large.

The best way to resolve these problems is to perform experimental research using a simulator. However, simulating plasma behavior is difficult with an experiment device.

In our simulator, plasma current is simulated with a coil. Plasma behavior and disturbance fields in plasma are simulated with a HITACHI HIDIC-80 control computer. The calculation of plasma position and plasma shape is also performed by the control computer. The algorithms used in the calculations are based on theoretical expressions. Power supplies are controlled with a controlled variable that is obtained with the results of the calculations. As a result, plasma position, plasma shape, and power supply control could be effectively carried out.

The classical PI control method is effective for plasma control in the main control loop and the addition of a D control method to the PI control method effectively reduces plasma displacement. A minor ACR loop is also provided in the control system, in addition to the main control loop.

As a nuclear fusion device generally consists of many loads and as they are connected to convertor in a power supply, the interaction between power supplies when a disturbance field occurs is studied with a simulator. Although disturbance fields occurred in the plasma, five convertors could be satisfactorily controlled.

Simulator Configuration

The simulator consists of a power supply system, loads and a computer control system. Fig. 1 shows the exterior of the simulator. The simulator control panel is shown on the left side of the photograph and the HIDIC-80 computer system on the right side. Fig. 2 shows the load coils used in the simulator. Convertors and loads are placed behind the control panel and the computer system. A 20 kVA generator is placed in another room.

Fig. 3 is the one-line diagram of a power supply system, including the load coils. The AC generator is connected to every load coil through thyristor convertors. The ohmic

Fig.1. Power Supply Simulator for Plasma Control

Fig. 2. Load Coils of Simulator

heating coil (OHC) system is provided with thyristor circuit breakers. Loads include the plasma, vacuum vessel, ohmic heating coil (OHC), vertical field coil (VFC), horizontal field coil (HFC), quadruple field coil (QFC) and divertor coil (DVC). The loads are composed of coils with iron cores, as shown in fig. 3, and are electromagnetically coupled to each other. A plasma current is simulated with a coil with an iron core. As a vacuum vessel current consists of uniform, poloidal, quadruple and saddle components, these four current components are also simulated with four coils with iron cores.

Fig. 4 is a block diagram of the computer control system. This system can be devided roughly into the HIDIC-80 computer and PI/O. The calculation of plasma position and plasma shape are performed with this computer system. The coil current data in the computer system

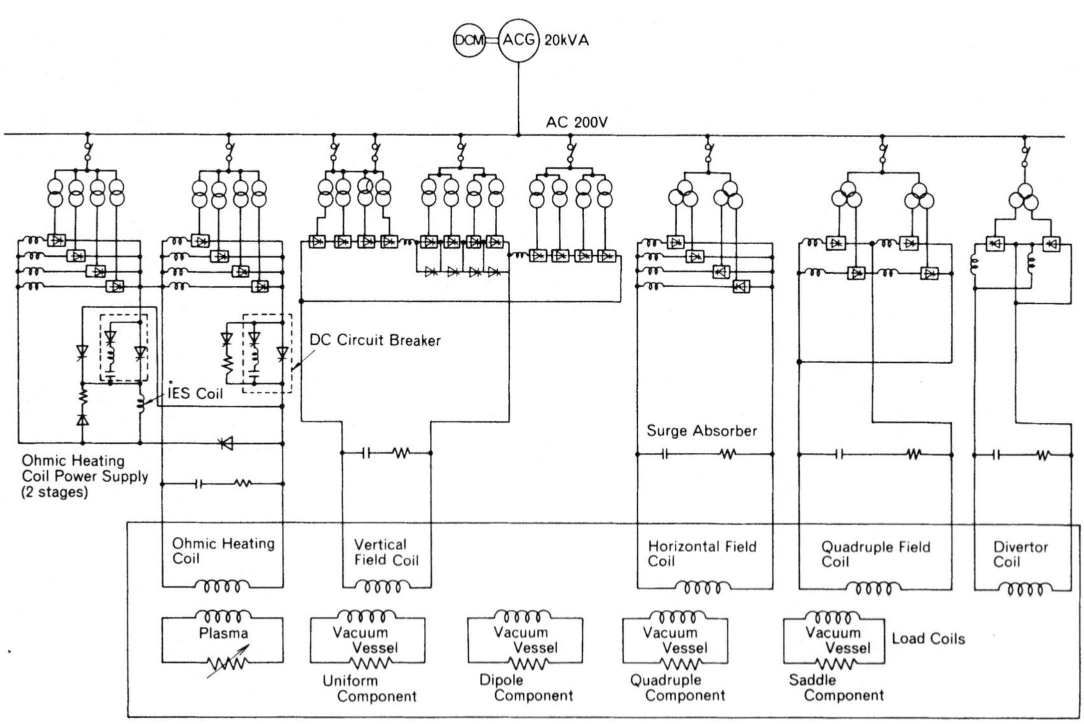

* IES Coil : Inductive Energy Storage Coil

Fig. 3. One-line Diagram of Power Supply System

are determined by sampling coil currents. Plasma position and plasma shape are calculated after AD conversion of the input coil currents and, then the controlled variable used to drive the thyristor convertors is calculated from the degree of deviation observed in plasma position and shape.

In addition to the main control loop provided within the control computer, a minor ACR loop is also constructed in the power supply controller. Horizontal and vertical plasma position control and the shape control of plasma cross section are obtained with the separate control loop shown in fig. 4. However, as plasma behavior effects plasma position and shape, all convertors in a power supply should always be controlled through the main control loop during a disturbance. Every convertor in a power supply is controlled by a voltage pattern at the plasma current buildup stage and by a change from a voltage pattern to a feddback loop at the beginning of the flat-top stage. The minor ACR loop is operated throughout all operational periods.

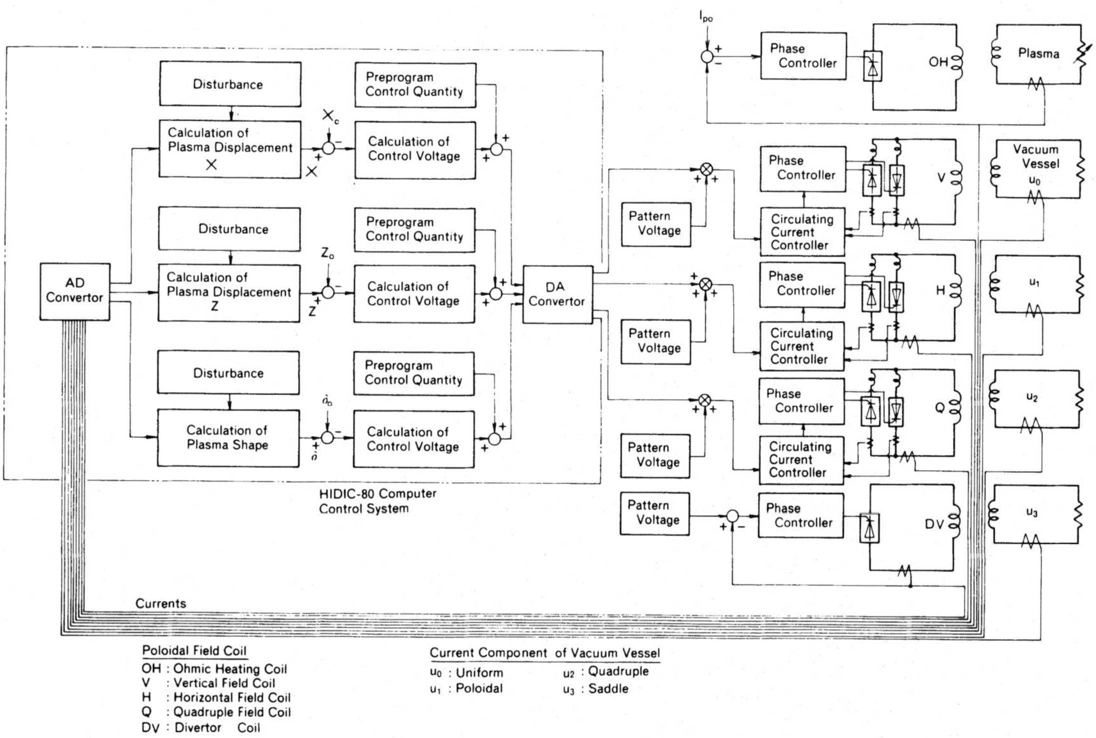

Fig. 4. Block Diagram of Control Computer and Power Supply Control System

Disturbance fields, which have an effect on plasma position and shape are determined by HIDIC-80 computer system software. A preprogram controlled variable is determined by the computer system and added to the controlled variable determined through the main feedback control loop.

Circuit Operation

The power supply of the poloidal field coils which consist of ohmic heating coil (OHC), horizontal field coil (HFC), quadruple field coil (QFC) and divertor coil (DVC), in the simulator are driven from the beginning of the coil current buildup stage to the end of the plasma shutdown stage.

Every convertor in a power supply can increase the coil current linearly from zero to a reference current value in about 100 ms. The coil current is controlled in a constant current mode for about one second. Plasma current is increased to a reference current value during a 150ms period after OHC current commutate from a thyristor circuit breaker to a load resistor and plasma initiation begins. Furthermore, in order to control plasma current under steady state conditions, convertors in a power supply are designed to control the current of their respective coils in the constant current mode for about two seconds. After that, the system current is reduced to zero.

Fig. 5 shows the experimental results obtained in the case where a VFC current flows in a negative direction during the coil current buildup period. The VFC power supply drives the coil current in the reverse direction. OHC, HFC, QFC and DVC currents are increased to the reference current value in a positive direction in order to store energy in coils.

The VFC, HFC and QFC power supplies are operated in a circulating current control mode. The control block diagram for this mode is shown in fig. 6. This control method allows even small circulating currents to be stably controlled. A circulating control current can be effectively reduced by increasing the coil current. Consequently, the sum of the current I_f of convertor-1, which is carried out by a rectifier operation, and the current I_r of convertor-2, which is carried out by a invertor operation, is controlled to a constant value. When the current I_f exceeds a set value, the circulating current is constant.

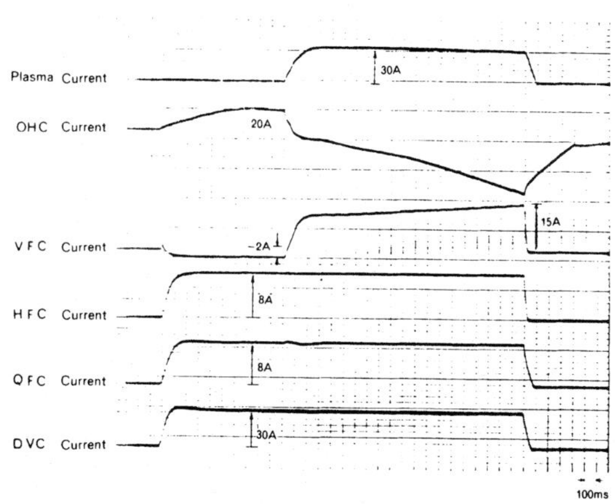

Fig.5. Experimental Results of Load Coil Current Flow Obtained with Simulator

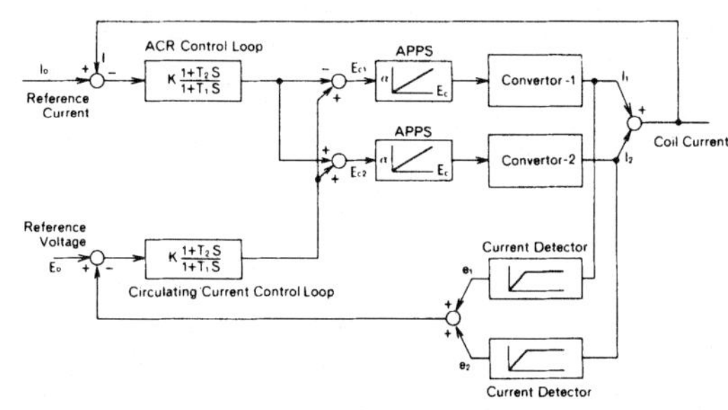

Fig. 6. Circulating Current Control Block Diagram

Therefore, a VFC current can be smoothly switched from negative to positive by the VFC power supply shown in fig. 5.

A plasma current is increased linearly to 30A during plasma buildup periods. During the flat-top stage, plasma current is controlled by operating the OHC power supply in a constant current mode. Other coil currents are controlled by operating the power supply in a constant current mode during the driving period.

Plasma Control

Control of Horizontal Plasma Displacement

Horizontal plasma displacement can be controlled by the vertical field coil power supply. The horizontal position of plasma is generally determined by the Mukhovatov and Shafranov expression [1], as a function of plasma current and the currents from every coil in a digital simulation:

$$\frac{\mu_0 I_p}{2}\left(\ell n \frac{8X}{r_p} + \Lambda_1 - 0.5\right) + 2\pi X B_z = 0 \qquad \cdots\cdots(1)$$

where I_p is plasma current; r_p is the minor radius of the plasma; Λ_1 is a function of poloidal beta β_p and plasma internal inductance ℓ_i per unit length.

However, it is desirable to introduce an approximate expression derived from eq(1), in order to shorten the calculation time in the experiments:

$$X \doteqdot X_o \frac{\Sigma a_k I_k}{\Sigma b_k I_k} \quad \cdots\cdots (2)$$

where a_k and b_k are constants determined by device structure and the data of JT-60 was referred to.

The vertical magnetic field, which exerts a force on plasma is given by a linear function of all the coil currents

$$B_z = \Sigma \nu_k I_k + \Delta B_{dv} \quad \cdots\cdots (3)$$

where ν_k is field gain; I_k is coil current; ΔB_{dv} is a disturbance field.

The disturbance field ΔB_{dv} is artificially produced with software using the control computer or with hardware using field changes in the load coil. Experiments are conducted with software. Plasma current, vacuum vessel currents and coil current data are input to the control computer of the simulator. Horizontal plasma displacement is calculated by sampling control computer input data with current detector. The computer can scan current data in 1.67ms. An actual system, such as the JT-60, will have a sampling time of 1ms.

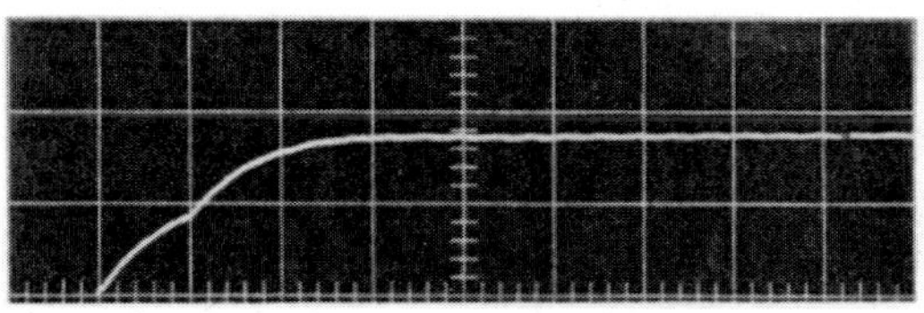

(a) I_p: Plasma Current
(10A/div, 50ms/div)

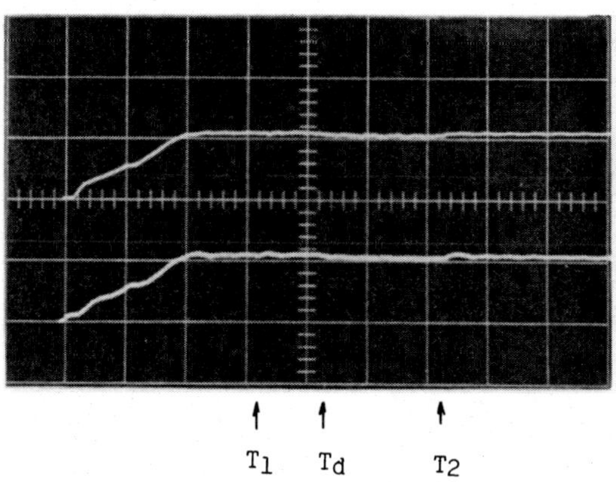

(b) E_{co} : Control Voltage (1V/div, 50ms/div)

I_v : VFC Current (5A/div, 50ms/div)

T_1-T_2: Feedback Control Period

T_d : Disturbance Field Applied

Fig. 7. Plasma Current, Control Voltage for VFC Power Supply and VFC Current

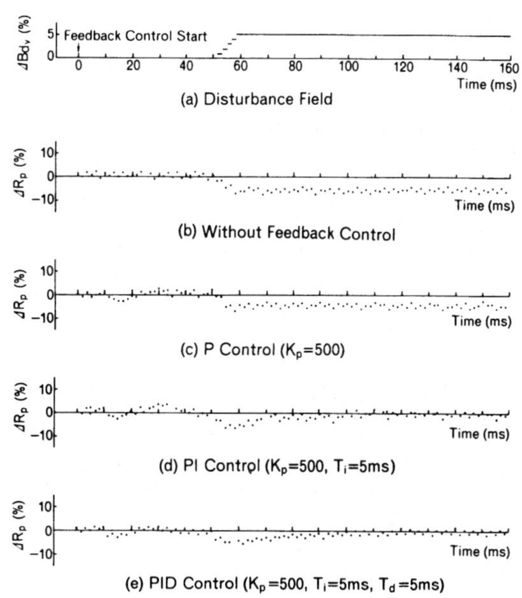

ΔR_p : Deviation of Horizontal Plasma Position to Reference Position
K_p : Proportional Gain
T_i : Reset Time
T_d : Rate Time

Fig. 8. Results of Horizontal Plasma Position Control Experiments Obtained with Simulator

Because the horizontal plasma position is calculated with eq(2), the control voltage P_m for the power supply is calculated with the following expression [3] :

$$P_m = K_p \Delta X_m + K_i \sum_{k=o}^{m} \Delta X_k + K_d (\Delta X_m - \Delta X_{m-1}) \quad \cdots\cdots(4)$$

$$\Delta X_m = X_o - X \quad \cdots\cdots(5)$$

where K_p, K_i and K_d are proportional, integral and derivative coefficients respectively; X_m is the deviation; and Xo is the plasma reference position.

Plasma current, VFC current and the control voltage for the VFC power supply are shown in fig.7. Feedback control is carried out from T_1 to T_2.

Fig. 8 shows the results obtained in horizontal position control experiments. The OHC and VFC power supplies are operated during the experiments.

Fig. 8 (a) shows a disturbance field wave form that is assumed to be 5% of B_z. The disturbance field is applied in T_d from feedback control start. Fig. 8(b) shows the plasma deviation when the main control loop is not used. Plasma deviation is 7%. ΔR_p are expressed as a percentage of a horizontal reference position X_o. Fig.8(c) shows plasma deviation from the reference position when the main feedback loop with P control is used. This indicates that plasma position control is not complete. But the PI control method is effective, as shown in fig. 8(d). The main control loop with PID control is more effective in the reduction of the amplitude of the plasma deviation ΔR_p, as shown in fig. 8(d).

Control of Plasma Vertical Displacement

Vertical plasma displacement can be controlled by the horizontal field coil power supply. Vertical plasma position is, for example, determined by the following expression [4]:

$$Z = \frac{X_o}{n B_{vo}} (B_h + B_u + \Delta B_{dh}) \quad \cdots\cdots(6)$$

where B_h, B_u and ΔB_{dh} are horizontal field, the field produced by vacuum vessel current and disturbance field respectively; X_o is a plasma reference position; B_{vo} is a vertical reference field.

Calculation of vertical plasma displacement is performed with eq(2). The decay index n is determined with the following expression:

$$n = - \frac{X}{B_z} \cdot \frac{\partial B_z}{\partial X} \doteqdot X \frac{\Sigma \alpha_k I_k}{\Sigma \beta_k I_k} \quad \cdots\cdots(7)$$

where α_k and β_k are constants, and coil current I_k represents the VFC currents in the experiment.

Fig. 9 shows the experimental results obtained. The initial position of plasma before the start of feedback control is about 28mm below the X-axis. However, the vertical position is controlled to zero by the feedback control of the HFC power supply.

Furthermore, when the horizontal disturbance field ΔB_{dh} is added to the horizontal field B_h at T_d, the vertical position changes from zero to 7mm. But the vertical position is returned to zero by the PID feedback control, as shown in fig.9(b). This proves that vertical position control can be performed by the HFC power supply.

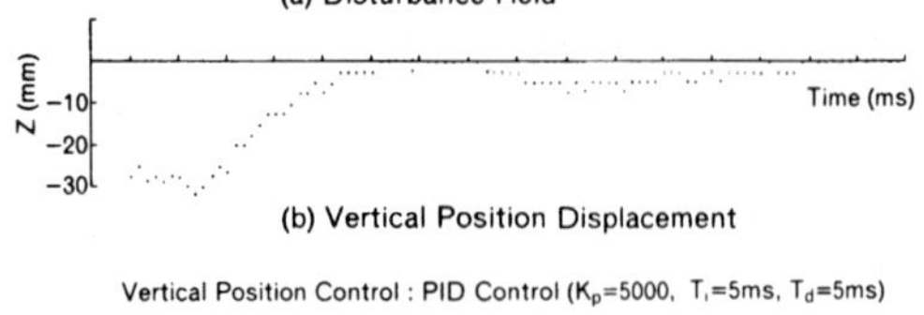

Fig.9. Results of Vertical Plasma Position Control Experiments Obtained with Simulator

Control of Shape of Plasma Cross Section

The shape of a plasma cross section can be controlled by a quadruple field coil power supply. A change δ in a plasma cross section is affected by the decay index n, horizontal plasma position X, poloidal beta β_p, plasma internal inductance l_i, and plasma minor radius r_p, as shown in the following expression [2]:

$$\delta = \delta\,(\,n,\ X,\ \beta_p,\ \ell_i,\ r_p\,) \qquad \cdots\cdots\cdots\cdots(8)$$

As the theoretical eq(8) is very complex, we used an approximate expression in the experiments. Consequently, δ can be represented as a function of the decay index n and the horizontal plasma position X:

$$\delta \fallingdotseq (\,n + K_1\,)(\,K_2 + K_3\,\Delta R_p\,) \qquad \cdots\cdots\cdots\cdots(9)$$

where K_1, K_2 and K_3 are constants determined by β_p, l_i and r_p, which we assumed were invariant.

As the decay index n is affected not only by the QFC current but also by the VFC current as represented by eq (7), we can make the statement that plasma shape is affected by a vertical field component. Fig. 10 shows the results obtained in the shape control experiments of plasma cross section when the vertical disturbance field ΔB_{dv} occurred. The initial change in shape is about 13mm before the start of feedback control, as shown in fig. 10. We assumed that the reference value of δ was 20mm in the experiment. Consequently when the disturbance field B_{dv} is added to the vertical field B_z at T_d, the horizontal plasma position changes and plasma shape changes according to the decay index n. However, plasma shape can be controlled by the QFC power supply and approaches the plasma shape reference value of 20mm when PID feedback control is used.

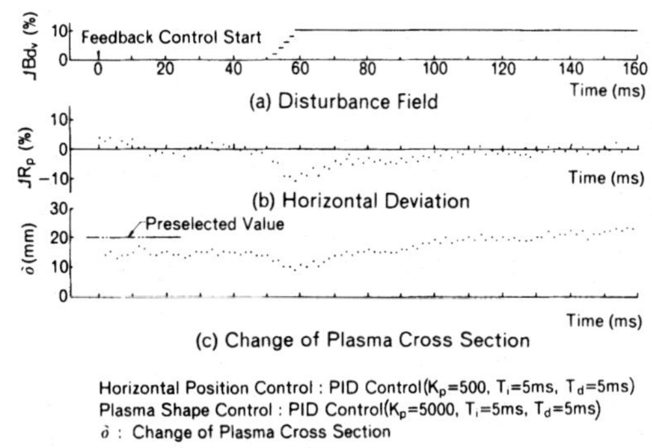

Fig. 10. Results of Plasma Cross Section Shape Control Experiments Obtained With Simulator

Power Supply Control

An experimental nuclear fusion device generally consists of many coils connected to power supplies. When these load coils are fed power from a generator, it is possible that sudden changes in load current or voltage will affect power supply control.

Fig. 11 shows the plasma and coil current wave forms obtained in experiments with the simulator when the VFC current suddenly changed 10% at T_o during the flat-top stage. Fig. 12 shows the

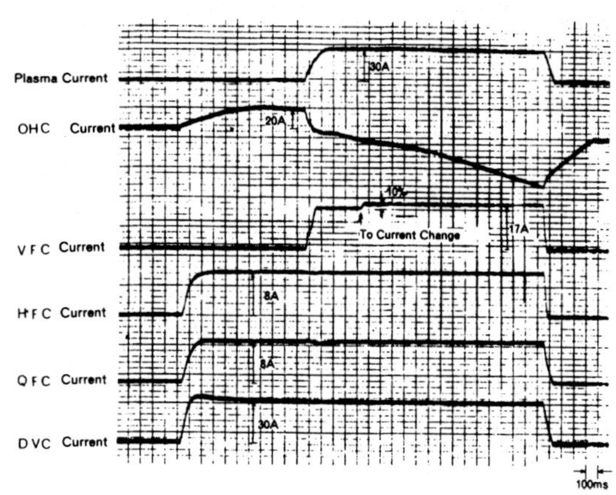

Fig.11. Influence of Sudden VFC Current Change on the Control of Other Coil Currents

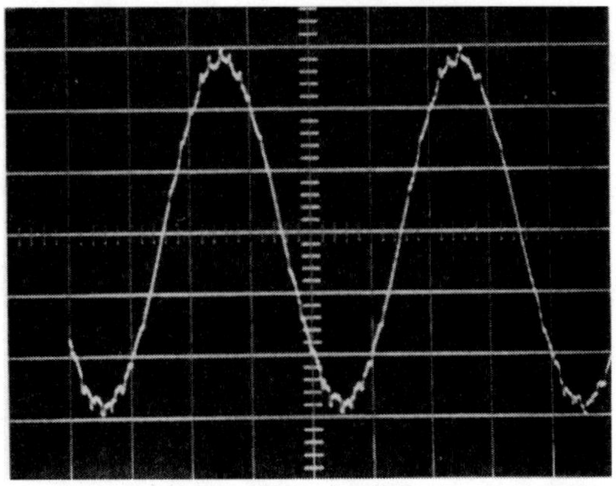

Fig.12. Output Voltage of AC Generator (100V/div, 5ms/div)

output voltage wave form for the AC generator.

As shown in fig. 11, even if the VFC current changes suddenly, other coil currents retain unaffected, which proves that the output voltage of a generator is influenced very little sudden changes in load currents.

Conclusion

A power supply control simulator, which is a valuable piece of equipment in the design of nuclear fusion device power supply systems, has been discussed. Power supply control of plasma position and shape is proved possible even when disturbance fields occur in plasma.

Acknowledgement

We would like to express our gratitude to Mr. K. Onishi, Mr. H. Amano, and Mr. Y. Kazawa of Hitachi Ltd. for the advice they extended to us in this study.

Reference

(1) Mukhovatov, V.S., Shafranov, V.D.; Nuclear Fusion 11, 665 (1975)

(2) Suzuki, Y., Ninomiya, H., Ogata, A., etc; Japanese Journal of Applied Physics, ;6, No.12, Dec., 2237-2244 (1977)

(3) Matsuzaki, Y., Fujisawa, N., etc; Proceedings of the 8th Symposium on Engineering Problems of Fusion Research, 1899-1907 (1979)

(4) Suzuki, Y., Ninomiya, H., etc; JAERI-M 6050 (1975)

OPERATION OF THE ASDEX FEEDBACK SYSTEM

F. Schneider

Max-Planck-Institut für Plasmaphysik
EURATOM-Association, D-8o46 Garching

ABSTRACT

The divertor tokamak ASDEX became operational at the beginning of 198o. With its plasma current feedback, position feedback as well as density feedback, it is presently possible to preprogram discharges with durations of greater than 3 seconds. The position of the plasma is controlled to ± 0.3 cm (with R_p = 165 cm) and the density of up to about 3.5 x $1o^{13}$ cm^{-3} within ± 4 %. The reproducibility of the plasma current (3oo kA) is about ± 1 %.

The capability of the feedback system is such to sufficiently reduce the effect of major disruptions so that rapid decays in plasma current can be avoided.

1. INTRODUCTION

As ASDEX /1/,/2/ became operational, attempts were made to control the position of the plasma with a widely variable, computer supported open loop control circuit. Because of the constantly varying plasma parameters, it was impossible to obtain reproducible discharges with a duration of greater than 1oo msec. However, when the position feedback was introduced (Fig.3) it was immediately possible to maintain the plasma current for longer than 1 sec.

A considerable improvement in the reproducibility was then achieved through the introduction of the magnetic limiter (divertor). The current in the multipole windings is maintained proportional to the plasma current.

Further improvements were achieved through the plasma current regulator (Fig.4) and the density regulator (Fig.5). The desired values of the plasma current, the density and the position of the plasma can be varied in time through preprograming by the computer via a specially designed CAMAC module /3/.

Presently under development are regulating circuits which will attempt to control MHD instabilities before they develop into disruptions.

2. DISCUSSION OF THE REGULATION CIRCUITS

2.1 Vertical field feedback (Fig.3 upper part)

This circuit is essentially composed of the following elements:

a) A set of flux monitoring loops and field strength probes(Ref./3/)

b) Specially developed, integrating amplifier with automatic offset control and optical coupling (T_I = 0.8 msec, drift ± 0.1 %/sec)

c) Analog computer for determination of: Plasma position, regulation error, plasma current and internal plasma parameters ($\beta + li/2$)

d) Proportional and derivative action controller (Ref./4/)

e) Computer controlled vertical field program(Ref. /3/)

f) Thyristor amplifier (12 pulse x 1oo Hz, 4o kA, 1 kV(Ref./4/,/5/)

g) Vertical field coils (0.43 mH, 2.2 mΩ)

h) Vacuum vessel between vertical field coils and plasma (time constant T_v = 1o msec)

i) Plasma (R_p = 165 cm, a = 4o cm, I_p = 5oo kA)

The calculated regulation (Ref./4/) could be experimentally verified. The regulation settings are currently:

$$F = K_p \cdot \frac{1 + j\omega T_r}{1 + j\omega T_1} \qquad T_r = 3 \text{ msec}, \quad T_1 = 0.3 \text{ msec}, \quad \text{loop gain } K = 9$$

As expected, the recovery time could not be decreased substantially below the vessel time constant of 1o msec (see Fig.8). On the other hand, because of the choice of voltage control by the thyristor amplifier and by the vacuum vessel, the stabilization is so good that even disruptions can be controlled without considerable loss of the plasma current (see Fig.7).

The new method for determining the plasma position /3/ was thereby proved very useful. As representative values, the magnetic flux difference and the field difference are given in Fig.6. From the comparison of these two values, the influence of the internal plasma parameters is clearly seen /6/.

2.2 Radial Field Feedback (Fig.3 lower part)

The radial field control loop is not significantly different from the vertical field control loop. In particular, instead of the thyristor-amplifier, a linear transistor-amplifier (4 kA, 6o V) is used.

During operation we were surprised by an unexpected effect. Although the radial field penetrates the vacuum chamber in 2o msec, the experimentally measured plasma shift response time was almost 1 second.

An explanation for this result is found in estimating the passive feedback (Fig.O): A relatively small radial field results in a large vertical displacement of the plasma, so that even a small vertical displacement induces eddy currents in the vessel, which result in a large opposing radial field. This high loop gain effect is the basis for the multiplication of the time constant.

Fig.O Ref./4/,/11/

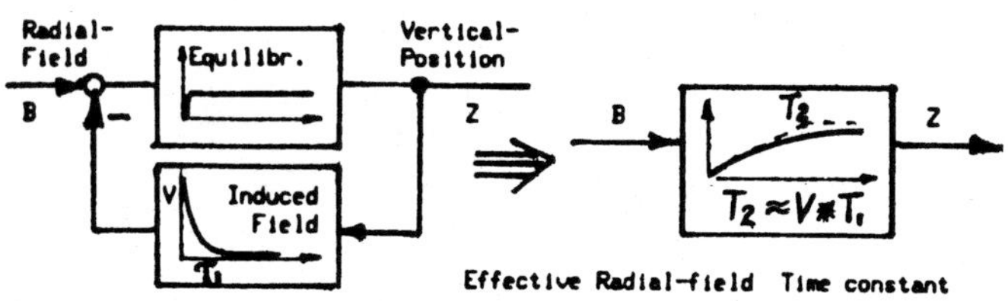

2.3 Plasma Current Feedback (Fig.4)

Even though a thyristor is also used for the OH-coils, as for the vertical field coils, the two feedback circuits are basically very different. A computer controlled program is used for coarse control of the vertical field, while fine adjustments are made to maintain the flux difference, calculated by the analog computer, at zero. The plasma current controller on the other hand compares the measured plasma current directly with the programmed curve. The regulation settings have been chosen as follows:

$$F = \left(1 + \frac{1}{j\omega T_n}\right) \cdot \frac{(1 + j\omega T_r)}{(1 + j\omega T_1)} \qquad T_n = 5o \text{ msec}, \quad T_r = 7 \text{ msec}, \quad T_1 = 1 \text{ msec}$$

It is not stability that causes the main problem in this circuit loop, but the limited capacity of the thyristor amplifier. Not only the unforeseen changes of inductivity and resistance of the plasma, but also the switching of resistors in the OH-circuit induce high voltage jumps which can only be compensated to a small extent by the thyristor amplifier.

In case the desired plasma current program (for example during ignition phase) does not closely coincide with the resulting plasma current, the regulator then will be over controlled and afterwards some time is required to recover. Figure 1 shows the situation when the current program was too steep, while it was too flat in Fig. 2. The interference at about 0.8 sec is caused by closing the OH-breaker at a time when the OH current was not exactly at zero.

2.4 MHD Mode Feedback

Shortly before plasma disruptions occur, an increased activity of MHD modes is observed. We therefore planned to generate some magnetic effects on the modes - being the cause for the disruptions - by means of some kind of helical windings. Yet, calculations have shown that this idea requires very fast amplifiers with unrealistically high power.

A much simpler method had then to be realized(Fig. 4 dotted line): The plasma current is changed when MHD modes occur, thus the current profile and q also changes and the modes will hopefully be smeared out. Unfortunately, there was not sufficient experimental time to show results.

2.5 Density Feedback (Fig.5)

This circuit loop is the most simple one, yet it shows a very complicated behaviour. First the Piezo-valve is non-linear. Second,the penetration time as well as the streaming of the particles from the plasma is very complex. They depend for example indirectly on the capactiy of the divertors. The problem of non-linearity of the Piezo-valve during the first experimental phase was solved by means of frequency modulated impulses.

Fortunately, however, it became possible to use a simple feedback regulator with a high loop gain (K > 35) which permits all linear and non-linear effects to be regulated satisfactorily. Figure 7 demonstrates, however, that much time is required (7o msec) to recover the density loss after a disruption. Figure 1 shows a typical density curve programmed with a platform and two successive ramps.

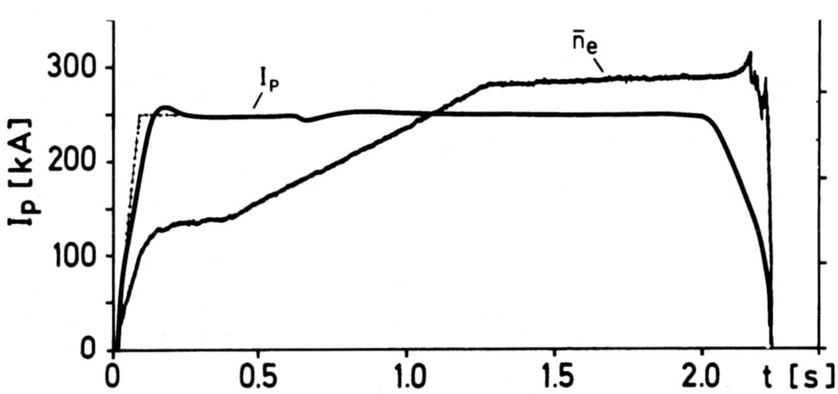

Fig.1: Typical shot for tests at the density limit.

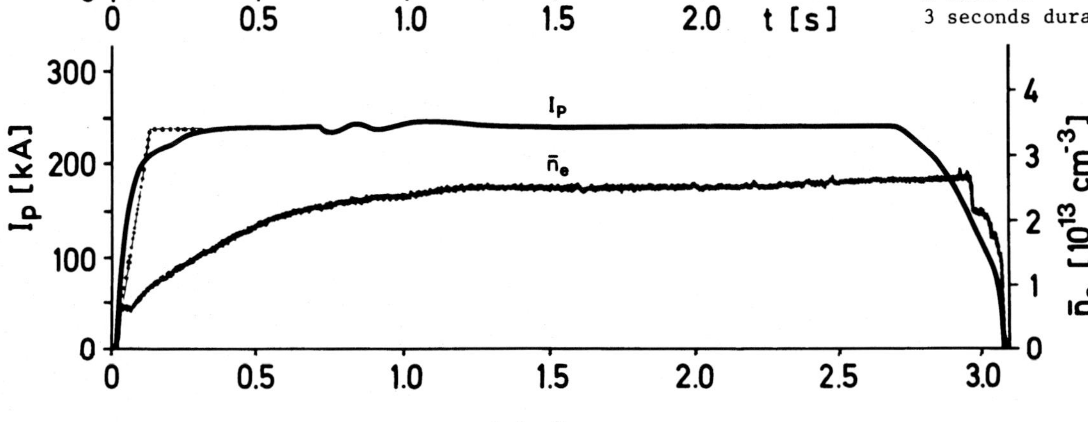

Fig.2: A shot of 3 seconds duration

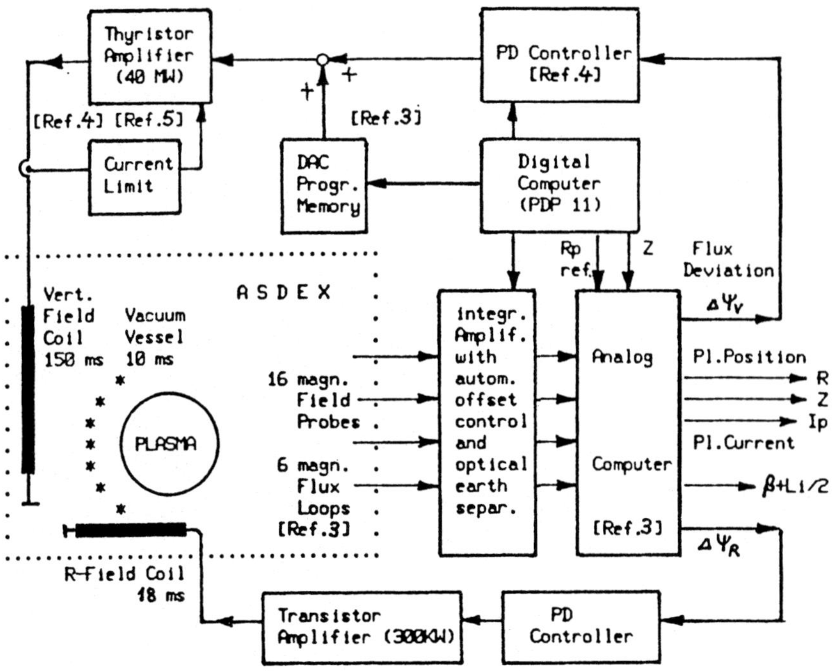

Fig.3 Plasma POSITION Feedback

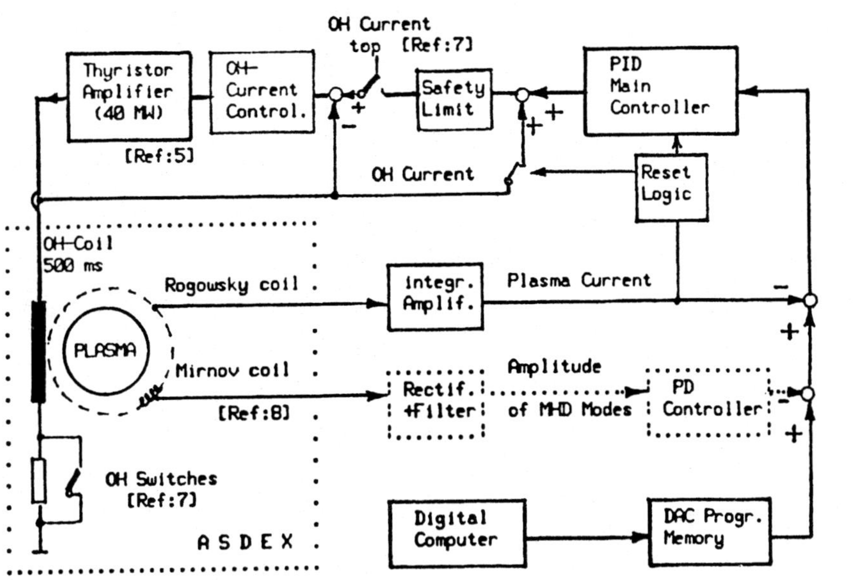

Fig.4 Plasma CURRENT Feedback

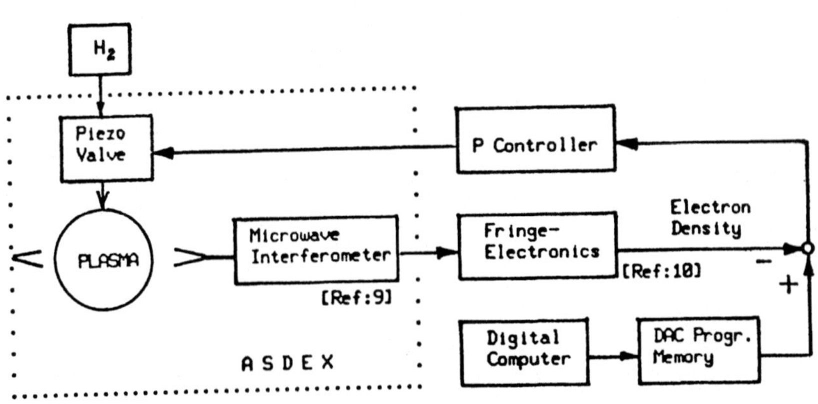

Fig.5 Plasma DENSITY Feedback

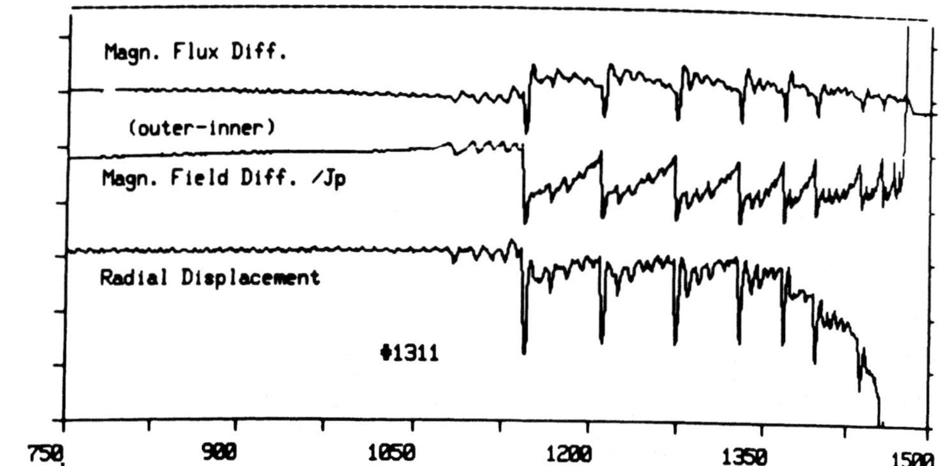

Fig.6

Correlation of Feedback Controlled Parameters of an ASDEX Shot with "Disruptions"

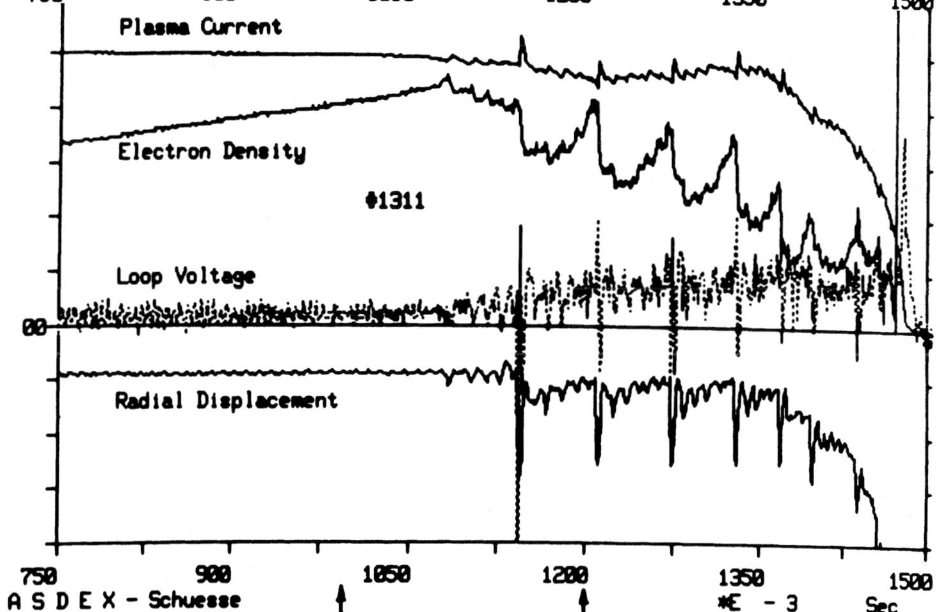

Fig.7

Detail = Fig.8

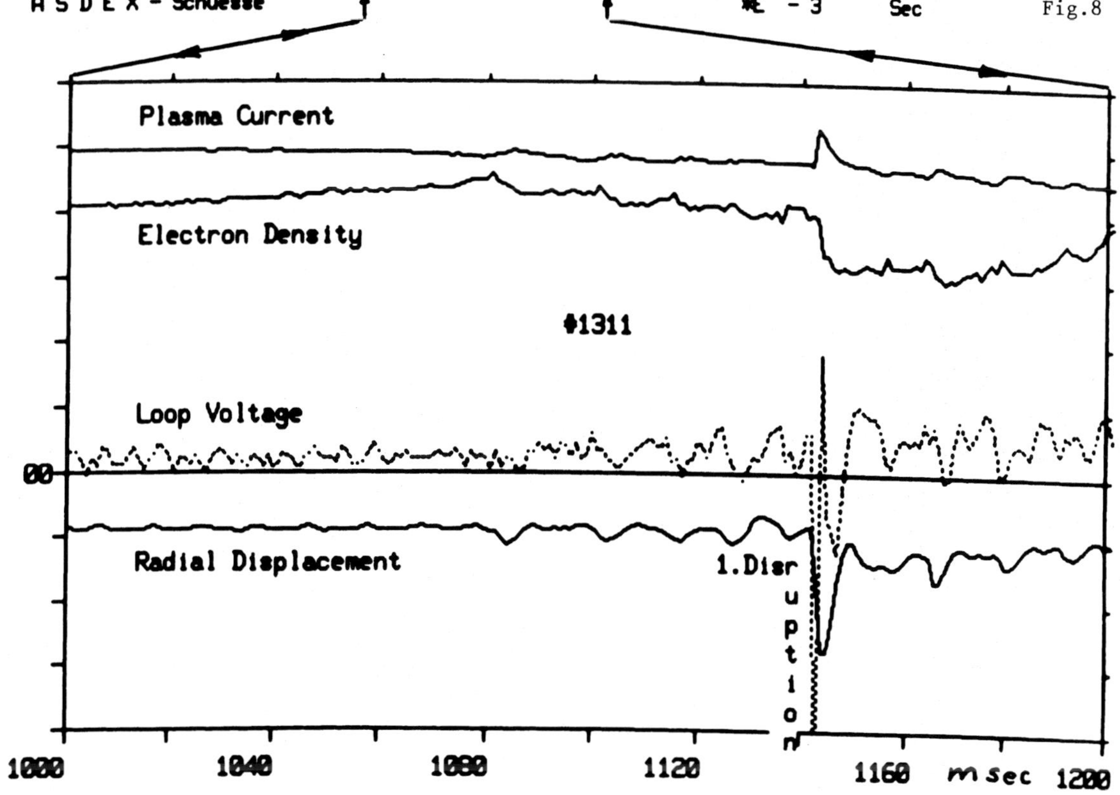

Fig. 9: View of feedback electronics in the ASDEX main control room

3. CONCLUSION

The ASDEX feedback system has proved to be capable of satisfactory control of plasma position, density and current with a large degree of flexibility.
The regulation of the current and density profiles of the plasma will be a future task.

4. REFERENCES

/1/ KEILHACKER, M. et al., "Impurity Control Experiments in the ASDEX Divertor Tokamak", 8th Int.Conf.on Plasma Physics and Controlled Nuclear Fusion Research, Oxford, 1-1o July, 198o. Proceedings, paper IAEA-CN-38/O-1.

/2/ WESNER, F. et al., "Completion and Start-up-Phase of the ASDEX Tokamak", 11th Symposium on Fusion Technology, Oxford, 15-19 September, 198o. Proceedings

/3/ SCHNEIDER, F., "Novel Method of Determining the Plasma Position and its Application to the ASDEX Feedback System", 1oth Symposium on Fusion Technology, Padua, 4-9 September,1978 Proceedings pp 1o13

/4/ SCHNEIDER, F., GRESSER, F., "Investigations on the Feedback System of ASDEX", 9th Symp. on Fusion Technology, Garmisch-Partenkirchen, 14-18 June, 198o. Proceedings p.783.

/5/ RAPP, H., GRESSER, F., "The Power Supply for the Poloidal Field Coils of ASDEX",9th Symp. on Fusion Technology, Garmisch-Partenkirchen, 14-18 June, 198o. Proceedings p.691

/6/ ALBERT,D.B., "Analysis of the Position Control System of ASDEX Tokamak" IPP Garching Report IPP III/61 (198o)

/7/ WEDLER, H., "First Experimental Results with the ASDEX Ohmic Heating System",11th Symp. on Fusion Technology, Oxford, 15-19 September, 198o. Proceedings

/8/ GERNHARDT,J. et al., "Design of Magnetic Probes for MHD Measurements in ASDEX Tokamak", IPP Garching Report IPP III/59 (198o)

/9/ LISITANO,G., "Schlieren Measurements of Plasma Cinematics during Current Disruptions in Tokamaks", 9th Europ.Conf.on Controlled Fusion and Plasma Physics, Oxford, 17-21 September, 1979. Proceedings paper EP7 p 148.

/1o/ GROENING,D.E. et al., "Unmodulated Interferometer", IPP Garching Report IPP III/41 (1978).

/11/ PREIS,H., SCHNEIDER,F., "Bestimmung von Wirbelströmen im Vakuumgefäß einer Apparatur zur Erforschung der kontrollierten Kernfusion", Archiv für Elektrotechnik 62 (198o)

PLASMA FEEDBACK CONTROL IN JET

D Ciscato, L de Kock, P Noll
JET Joint Undertaking, Abingdon, Oxon, UK

ABSTRACT

The paper outlines the present status of design for plasma current and plasma position feedback control in JET.

The magnetic measurements used for feedback control are described. A mathematical model of the system, based upon rigid displacement of the plasma, has been established. The radial position feedback loop is briefly discussed.

1. INTRODUCTION

In the JET experiment the plasma is governed by currents in a number of poloidal field coils outside the vessel/toroidal field coil assembly (fig. 1). The coils are grouped into 3 sets and connected with 3 controllable power supplies (fig. 2). Coil functions are:

Magnetising coil (M)	premagnetisation current build up current sustainment
Vertical field coil (V)	radial position current built up plasma shape
Radial field coil (R)	vertical position

The coils M and V are electrically connected and magnetically coupled. The major part of the vertical equilibrium field is provided from the energy in the flywheel generator FCG of the ohmic heating circuit, whereas the vertical poloidal field amplifier (PVFA) provides only corrections to the radial plasma position.

The P3 section of M coil allows adjustment of the stray field to values favourable for plasma breakdown.

The plasma shape and elongation depends on turn ratio between sections P2, P3, P4 of the V-coil. Various ratios can be chosen prior to each experiment.

The Table 1 shows the main parameters of the power supplies and the range of selectable turns in the poloidal field coils.

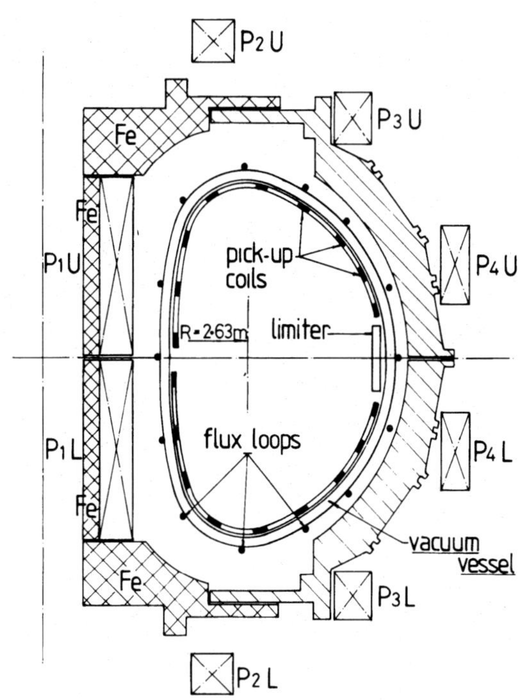

Fig. 1. Schematic section of the JET apparatus

TABLE 1

POWER SUPPLIES		COIL TURNS		
FGC	2.6 MJ 80 kA at 3 kV V_0 = 9 kV (I=0) I = 100 kA (V=0)	M	P1u P3u	284 1-8
PVFA1 +PVFA2	25 kA/25 s 35 kA Peak +2.3 kV/-1.8 kV 24 Pulse	V	P2u P3u P4u	3-20 1-60 33-62
PRFA	±3 kA, ±1 kV 12 Pulse	R	P2u P3u	8 or 16 20

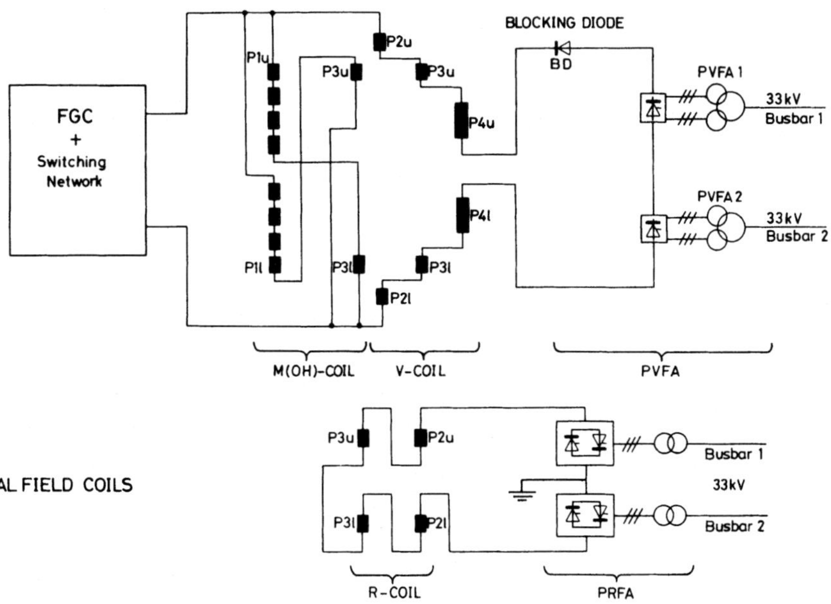

Fig.2.
POLOIDAL FIELD COILS

The poloidal vertical field amplifier (PVFA) has a restricted control range at the start of the discharge where up to 40 kV are applied at the M-coil, whereas PVFA range is ±2 kV.

As far as the R-coils are concerned this circuit is decoupled and independent of the other circuits as long as vertical plasma displacements are small.

The fig. 3 shows typical waveforms, voltages and currents during a JET pulse.

2. FEEDBACK CONTROL: MAGNETIC MEASUREMENTS

The following 3 major feedback loops are foreseen for the plasma control in the JET experiment:

- control of plasma current
- control of radial plasma position
- stabilisation of vertical position

The basic feedback signals for position control are derived from poloidal flux measurements in the vicinity of the plasma using flux loops and discrete poloidal field pick up coils. This method has been applied on ASDEX [1]. This method is particularly suited for non-circular plasma shape as foreseen in JET.

The magnetic measurements comprise (fig. 1):

- Flux loops at 14 poloidal positions outside of vessel. Where full loops are not possible, differential loops will be used on each vessel octant giving flux differences $(\psi_i - \psi_k)/8$ between any two positions i and k.
- Poloidal field pick up coils (discrete) at 18 poloidal positions inside vessel, one set per octant. A pick up coil is illustrated in the fig. 4.
- Rogowski coils around the vessel, mounted along the toroidal field coils. They give signals of the sum of the plasma and the vessel current.

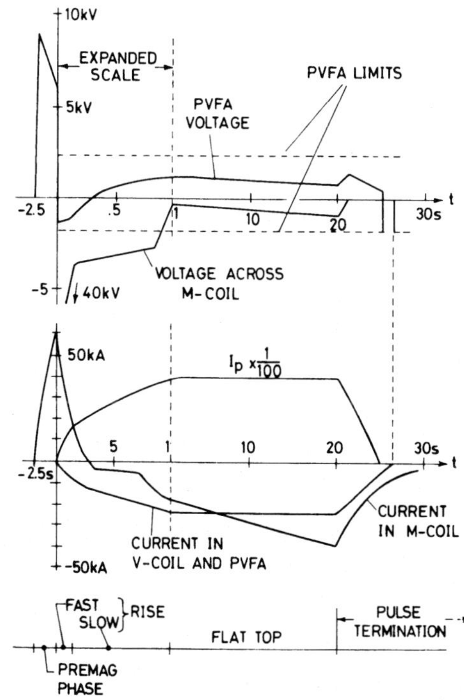

FIG. 3. VOLTAGE AND CURRENT WAVEFORMS (TYPICAL)

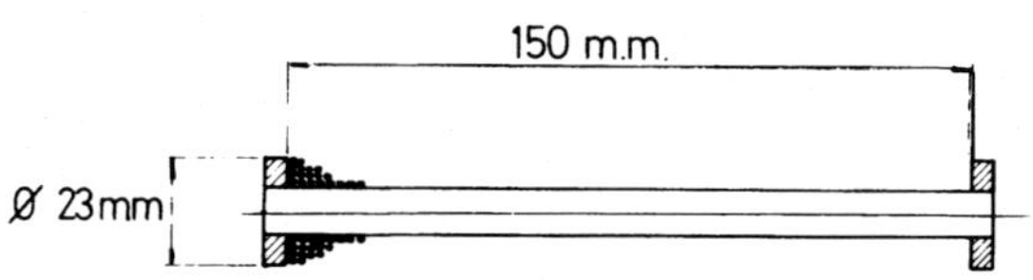

Fig. 4. Discrete Magnetic Pick-up Coil

Number of turns: 4 layers of 100 turns
Effective area: $0.08\ m^2$
Wire: mineral insulated coaxial cable
O.D. 1.5 mm $^{+0}_{-0.02}$
Core diameter 0.5 mm
Coil carrier: stainless steel 316L

As regards to radial position control the fig. 5 shows the measurement principle foreseen in JET: The last closed magnetic surface touching the limiter at R_2 ($\approx$ plasma boundary) shall be controlled to have an inner intersection with Z=0 at distance D_1^* from a reference radius R_1. The flux difference $\delta\psi$ between R_2 and $R_1^* = R_1 + D_1^*$ is measured using loop signals ψ_{L1}, ψ_{L2} and B_{POL} signals B_1, B_2 around Z=0.

$$\delta\psi = \psi(R_2) - \psi(R_1^*) \approx (\psi_{L2} - \psi_{L1}) + 2\pi(R_2 d_2 B_2 - R_1(d_1 + d_1^*)B_1) + \tau(\dot{\psi}_{L1} - \dot{\psi}_{L2}) \quad (1)$$

The first term represents the difference of loop signals, the second term is the first order correction derived from pick up coil signals, the last term is a dynamic correction due to vessel currents. Higher order terms due to field gradients are ignored.

The relation of $\delta\psi$ to the displacement variable:

$$I_p \delta R_p = I_p (R_{p1} - R_1^*)/2$$

is given by

$$\delta\psi = 4\pi R_1 B_1 \delta R_p \approx \text{const}\ I_p \delta R_p \quad (2)$$

The error signal $\delta\psi$ will be used as feedback signal for the radial position control.

The inductances L_{ik} and their variations $\partial L_{ib}/\partial R_p$ have been computed numerically and analytically [4]. The permeability distribution in the iron is for 18 V_S fore flux (from full equilibrium computations); it is assumed = constant during a radial displacement.

For small displacements $\delta R_p = R_p - R_o$ from an equilibrium position R_o of the plasma, the force balance equation can be written in the form (assuming plasma mass = 0):

$$\delta R_p \left(\frac{\partial B_v}{\partial R} + \frac{\partial B_v}{\partial R_p} - \frac{\partial B_{eq}}{\partial R_p} \right)_{R_o} = B_{eq}(I_p;R_o) - B_v(I;R_o) \quad (4)$$

where B_v = vertical field due to all external sources including iron "image current" fields of the plasma.

B_{eq} = required vertical equilibrium field

$\frac{\partial}{\partial R}$ = gradient

$\frac{\partial}{\partial R_p}$ = variation with plasma position

It is worth noting that the term $\frac{\partial B_v}{\partial R_p}$ is due to the presence of iron.

With linearisation of the currents $I_j = I_{jo} + \tilde{I}_j$ and using the fact that $\frac{\partial B_v}{\partial R} = \frac{\partial B_{eq}}{\partial R} \propto I_{po}$ at $R_p = R_o$ one obtains

$$\sum_{j=1}^{10} F_j \tilde{I}_j + F_p^{`} I_{po} \delta R_p = 0 \quad (5)$$

F_j are normalised forces related to $\partial L_{jp}/\partial R_p$, $F_p^{`}$ is the normalised differential self force of the plasma. $F_j$ and $F_p^{`}$ have been deduced from magnetic field computations.

If the circuit equations (3) are combined with the force balance (4), the displacement δR_p appears always in the form $I_p \delta R_p$ which is proportional to $\delta\psi$, the output of the magnetic measurement handling. Division of $\delta\psi$ by I_p is only required for the display of $\delta R_p(t)$.

The feedback radial position control during the current rise has been studied [5] with PD control from displacement signals short lasting ($\sim$ 10 ms) excursions $\delta\hat{R}_p \approx$ 10 cm and $\hat{V}_A$ up to 30 kV were found.

The stability of the radial position has also been investigated in the limiting case where eddy currents in vessel and shell were neglected. Stability was found both for the rigid displacement and the "peel off" plasma models.

5. STABILISATION OF THE VERTICAL PLASMA POSITION

For the vertical position control the same flux loop/pick up cool set will be used but with different signal combinations.

The vertical position is basically unstable due to the positive gradient $\partial B_{eq}/\partial R$ of the equilibrium field for the elongated JET plasma and due to the iron circuit. The open loop instability growth time is of the order of the vessel time constant ($\sim$ 5 ms).

Preliminary studies have been carried out on the basis of a similar model as applied for radial position control. A full size D shaped plasma has been assumed.

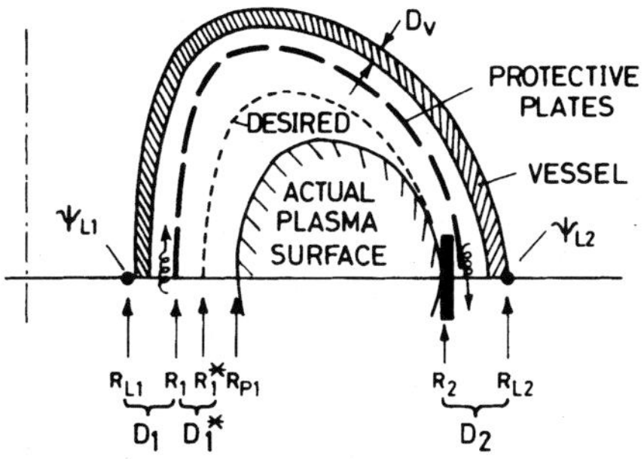

Fig. 5. Measurement principle for radial position feedback control

Other applications of magnetic measurements

- derivation of plasma boundary shape
- derivation of plasma current centre and higher moments
- combination with MHD codes to derive consistent plasma properties
- measurement of magnetic field fluctuations

3. FEEDBACK CONTROL: BLOCK DIAGRAM

The block diagram for the radial position control is shown in the fig. 6.

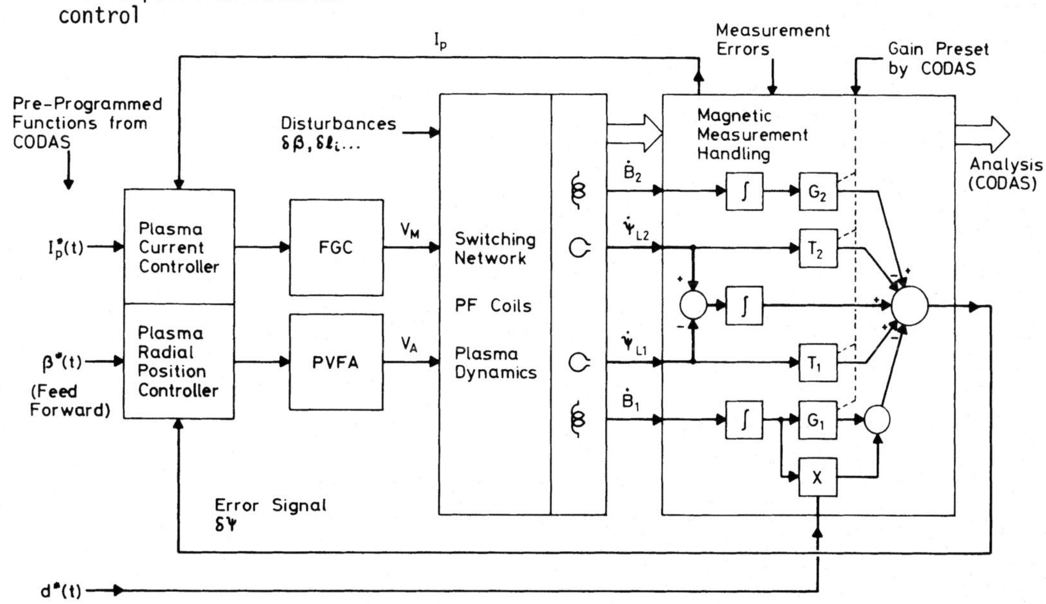

Fig.6 PLASMA CURRENT AND RADIAL POSITION CONTROL
$\beta^*(t)$ = Expected β from Power Injection
$d^*(t)$ = Desired Plasma/Wall Separation $R_1^* - R_1$

The magnetic measurement handling system, indicated in the fig. 6, illustrates how the feedback signal $\delta\psi$, eq (1), is obtained from flux loop and pick up coil signals. $D^*(t)$ is the desired plasma separation from an inner reference radius R_1, shown in the fig. 5. This parameter, as well as the plasma current reference signal $I_p^*(t)$, are given as pre-programmed reference functions from the JET control and data acquisition system (CODAS) and converted to analog form during the pulse under control of the CODAS timing system. In the simplest case D^* is taken = const, which corresponds to constant plasma/wall distance. For the current build up phase it may be advantageous to pre-program $D^*(t)$ in such a way that the plasma minor diameter increases monotonically with the current.

The plasma current feedback and the radial position feedback loops are expected to interact with each other because of the poloidal field coil connections and the presence of the plasma. The controllers are shown in one unit to simplify the inclusion of decoupling elements if required (multivariable controller).

In order to reduce the transient displacement of the plasma when additional power is injected into the plasma a feedforward signal $\beta^*(t)$ can be applied. $\beta^*(t)$ represents the anticipated increase of the plasma poloidal beta.

Important factors for design and specification are:

- destablising effect of the iron circuit
- eddy currents in the vessel and the mechanical structure
- changing plasma parameters, in particular in the current build up phase.

These factors impose conditions on the response time and dynamic range of power supplies and on the design of the controllers.

The FGC and PVFA systems are provided with their own automatic Voltage/Current regulator. Their dynamics will be taken into account in the design of the overall feedback loops.

Analog and digital techniques will be used for the plasma current and position controller and for magnetic measurement handling so that all parameters can be preset and monitored by CODAS.

4. MODEL FOR RADIAL POSITION AND CURRENT FEEDBACK CONTROL

As an aid for the design of feedback control loops a simplified model has been set up of the plasma and poloidal magnetic field system.

For the plasma various models have been proposed, such as a "peel off model" [2], where a constraint $R_p+a \approx$ const is imposed due to the plasma/limiter interaction (a = minor plasma radius), or adiabatic model [3] where adiabatic expansion is assumed during an outward displacement. For JET a rigid displacement model has been considered where the plasma shape and current profile do not vary during a radial displacement. In some respects this model lies between the above mentioned models.

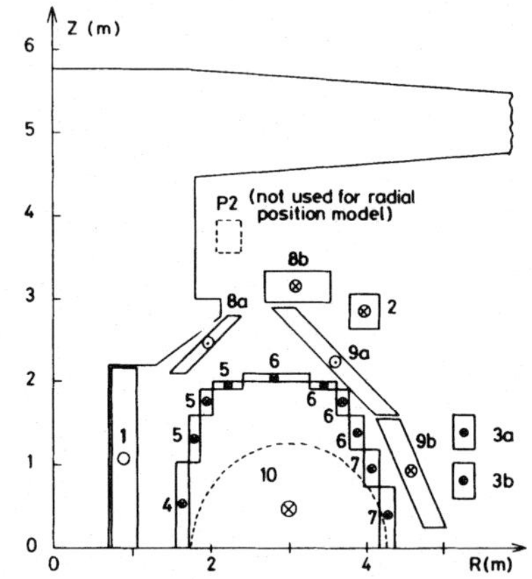

Fig. 7. Model of iron, PF coils, vessel and mechanical structure

Fig. 7 shows the current carrying elements considered in the model. They are simulated in axi-symmetric elements:

PF coils P1, P3, P4:	i = 1,2,3
vessel:	i = 4,5,6,7
inner mechanical structure:	i = 8a,8b
outer mechanical structure:	i = 9a,9b
plasma:	i = 10

The current densities in all elements except in the plasma are assumed uniform.

Structure elements have poloidal gaps, hence $I_{8a}+I_{8b} = 0$ and $I_{9a}+9_{9b} = 0$ is imposed.

The circuit equations are

$$L\dot{I} + \dot{R}_p \frac{\partial L}{\partial R_p} I + \eta I = V \qquad (3)$$

where L = inductance matrix

$\frac{\partial L}{\partial R_p}$ = matrix of inductance variation with plasma position R_p

η = resistance matrix (diagonal)

$I = \|I_1 \ldots I_{10}\|^T$ = vector of currents

$V = \|V_1 \ldots V_{10}\|^T$ = vector of coil voltages, related to applied voltages at M, V coils: $V_1 = V_M$, $V_2 + V_3 = V_M - V_A$

In an investigation of various feedback schemes [6] including the simulation with a detailed converter model, perturbations leading to displacements up to δZ = 15 cm were found to be easily stabilised with a 50 Hz 12 pulse bidirectional thyristor AC/DC converter.

The open loop instability has also been considered for cases where the plasma is built up from small size to full size D shape. It was found that the growth time for small size is not smaller than $\sim$6 ms, whereas the growth time for a full size plasma is $\sim$13 ms (eddy current effects of the mechanical structure have been ignored). The stabilisation by feedback appears, therefore, feasible for this particular operating mode which may be of interest for JET.

6. CONCLUDING REMARK

The FGC power supply is being manufactured and the PVFA has been ordered. Future work will concern the final specification of the PRFA, the design of feedback controllers using the described models and extensions thereof, and the electronic design of signal handling.

7. REFERENCES

[1] Schneider, F., Tenth Symposium on Fusion Technology, Padova, 1978. Proceedings pp 1013-1018.

[2] Blum, J., Dei Cas, R., Morera, J.P., and Plinate, P., Tenth Symposium on Fusion Technology, Padova, 1978. Proceedings pp 999-1006.

[3] Gran, R., Rossi, M.I., and Sobierajski, F., Seventh Symposium on Engineering Problems of Fusion Research, Knoxville, 1977. Proceedings pp 104-111.

[4] Bobbio, S., JET Contract G-IY-451, this Symposium, Oxford 1980. Paper CP.1-20.

[5] Green, B.J., Perrone, M.R., Noll, P., Dokopoulos, P., Ninth European Conference on Controlled Fusion and Plasma Physics, Oxford, 1979. Contributed Papers p 160.

[6] Schruttke, W., Report on work performed under JET Study Contract F45-GY-254, TU Braunschweig, October 1978.

PARTICLE AND ENERGY UNLOADING FROM A TOKAMAK USING SCRAPE-OFF LIMITERS COMBINED WITH HELICAL ISLANDS

G Fuchs, H G Goetze, D Reiter
Institut für Plasmaphysik der Kernforschungsanlage Jülich GmbH,
Association EURATOM-KFA, 5170 Jülich, Federal Republic of Germany

ABSTRACT

Following proposals for the scrape-off limiter, the helical divertor and their combination, this paper describes the construction of the vacuum and limiter part of a magnetically shielded scrape-off limiter.

The dimensions in the particular construction given are chosen to fit a q = 4 magnetic island suitable for a helical divertor at the TEXTOR boundary.

Based on recent experiments and computer simulations the energy flux density striking the limiter is estimated.

From this we derive the temperature distribution within the limiter and the permissible power loading. The particle unloading capability is calculated using a Monte Carlo simulation.

1. INTRODUCTION

Unloading from Tokamak discharges can be accomplished by use of divertors as was demonstrated experimentally for the bundle divertor on DITE and the poloidal field divertor on ASDEX, DIVA and PDX. Drawbacks of these divertor schemes are that they either use large currents in the divertor coils leading to large forces onto the conductors or require a topologically complicated arrangement of the coils and the blanket as far as a reactor is concerned.

A more recent divertor scheme is the helical divertor [1]. The idea here is to disturb the flux function near a rational q-surface at the plasma boundary by using a resonant helical magnetic field (multipole order 2q). This resonant field can be produced by helical currents which follow a magnetic field line on the rational q surface (to be disturbed) in parallel and are of the order of 1 % of the plasma current,on a reactor the conductors can be located outside the blanket.

The superpositioning of the helical multipole field and the tokamak field results in a helical island structure (see fig. 1), which can be understood as a sub-torus wound q-times around the main torus. The plasma diffuses across the separatrix, is then guided (following field lines) along the edge of the island and can be collected at the rear island side as seen from the plasma core. A scrape-off limiter [2] (cf. fig. 1) is employed to collect the particles, whereby the limiter area where the plasma recombines is optically shielded from the plasma core and the neutrals emerge from the collecting surface preferentially towards the pump holes. At the same time the island structure protects the limiter edge from an excessive power load. [3]

The two major parts of a helical divertor are: the helical coils together with their power supply, and the scrape-off limiter(s) together with the pumping system.

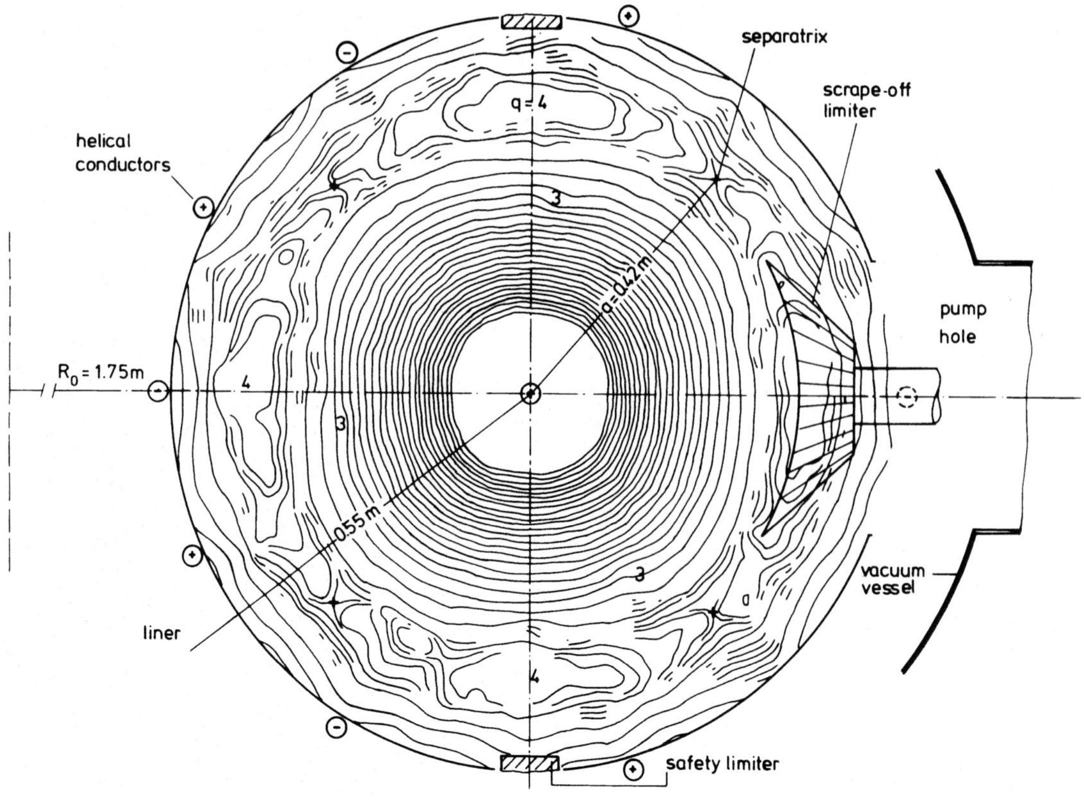

Fig. 1. Flux surfaces with helical island structure at the q = 4 surface as calculated according [4]. The poloidal field is that of a wire on axis plus the equilibrium vertical field. The helical windings follow a field line at the undisturbed q = 4 surface at the same poloidal angle $I_{hel} = 10^{-2}\ I_p$.

This paper deals with the limiter part suitable for the q = 4 surface at the TEXTOR (R = 1,75 m, a = 0,5 m [5]) boundary. Because the concept of the helical divertor is not yet tested experimentally and the physics of the boundary is only partially understood the design should give a large degree of experimental freedom. This is achieved by a plug on unit.

2. DIMENSIONING OF THE LIMITER

The plasma diffuses from the plasma core across the separatrix and then into two directions; namely, towrads the island center and towards the liner. At the same time it streams along the field lines until it hits a limiter. Both processes together lead to a characteristic plasma decay length δr in the scrape-off layer.

The island has a minor diameter which follows from distrubance theory to be approximately [6]

$$\Delta r = \sqrt{\frac{2B_h R}{mB_\varphi}\ \frac{dr}{d\iota}} \tag{1}$$

where B_h is the radial component of the helical field, m the poloidal mode number and ι the rotational transform. One can also give an approximate formula for the rotational transform within the island strucutre (subtorus)

$$\iota_i = \frac{m\,R\,B_h}{B_\varphi\,\Delta r} \tag{2}$$

The number of toroidal turns a field line takes to encircle the island is

$$N = (\iota_i \cdot \iota)^{-1} = q_i \cdot q \tag{3}$$

$N \cong 28$ for $m = 4$. At the plasma boundary both Δr and N are insensitive to the plasma current density profile and the helical current location [7].

Knowing N, δr can be estimated using the arguments as given in [8]. For TEXTOR we expect $\delta r \cong 4$ cm for $N \cong 28$. Δr can be made slightly larger than $2\delta r$ before eq (1) fails and the island becomes ergodized. The scrape-off limiter should have a radial extension $h \geqq 2\delta r$ and it should be positioned such that starting about at the island center it extends to the outside towards the liner at least δr beyond the separatrix (cf. fig. 1). The poloidal extension b has to be made large enough so that every field line which encircles the island center once strikes the limiter at least once. The toroidal extension has to be long enough that the impinging energy is handled. For a reactor this might require following the whole subtorus, because otherwise large power densities could occur at the ends (field line topology, angle of incidence).

3. SPECIAL LIMITER AND PUMP DESIGN FOR TEXTOR

For TEXTOR we designed the limiter and the pumping system in such a way that it can be plugged onto a porthole 850 mm x 650 mm (cf. fig. 2). The limiter head (cf. fig. 3) is shaped like a chantearella mushroom in order to enhance the scattering of the recombined plasma into the pump hole. It consists of a trunk which can be cooled, and has several grooves into which the pieces forming the limiter can be pushed. The pieces are fixed by a plate screwed on top. After a break of the vacuum the pieces can be replaced from inside the torus which allows testing different materials. In case graphite is used we would make the head from one piece.

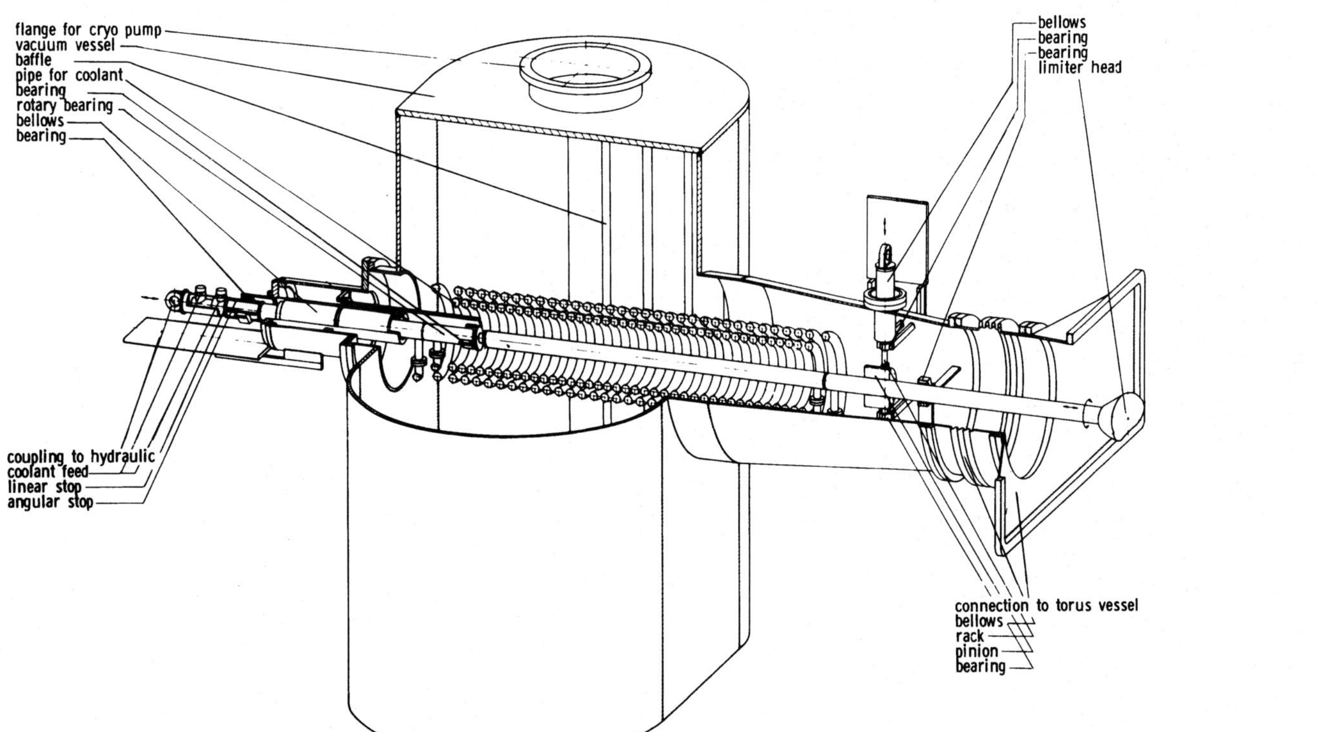

Fig. 2. Vacuum system together with the scrape-off limiter, its lateral and angular motion drive and the cooling

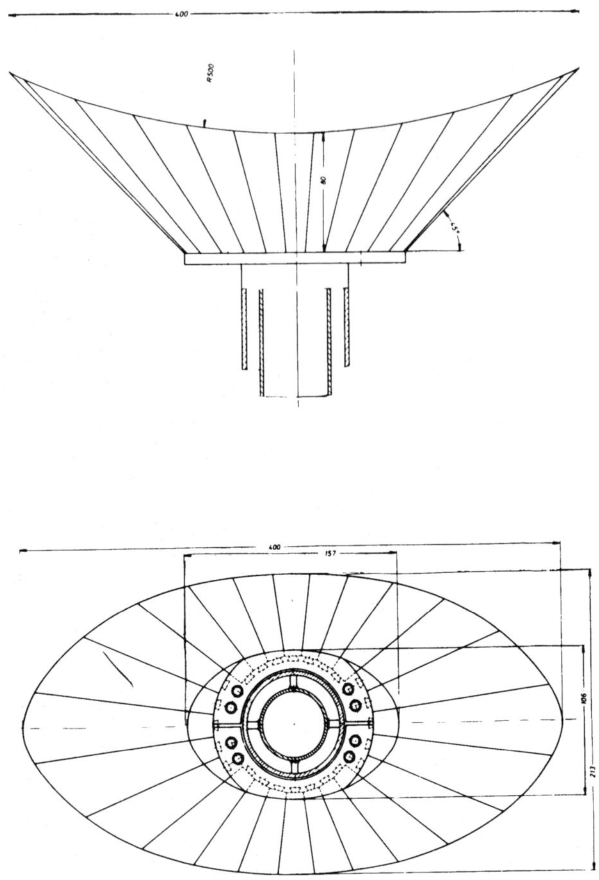

a

b

Fig. 3. Scrape-off limiter head

To provide some experimental freedom the radial position and the poloidal width of the limiter can be altered between subsequent shots. The latter is possible because the limiter can be turned 90° and has an elliptical cross section (cf. fig. 3b). The adaption to the plasma cross section is best if the large diameter of the elipse is in the plane of a poloidal cut. The concept requires that a lateral motion, an angular motion and a coolant are brought through the vacuum wall under UHV conditions to the same place. To solve this problem the limiter is mounted onto a rod which is guided in three bearings. The lateral motion feed through is made with a bellows similar to the TEXTOR main limiters. To allow for rotation the rod is made of two pieces joined with a bearing. The part containing the limiter has a pinion which can be rotated with a rack, reaching via a bellows through the vacuum wall. The two motions are driven hydraulically. All bearings within the vacuum are coated with titanium carbide to avoid cold welding.

The coolant (gas or water) flows in the limiter-rods. The break is bridged by tubes wound in spirals. The length of these springs is determined by the torque force and the required twist angles of 45° in both directions.

The vacuum system consists of a tank housing 2 cryopumps (10^5 l/s each [9]) and is prepumped with a turbo pump ($> 1{,}5 \cdot 10^3$ l/s). The cryopumps are optically shielded by baffles from the hot plasma. The vacuum tank is connected to the torus vessel via a tube containing a bellows to compensate for thermal expansion. Two ports allow one to observe the limiter.

4. ESTIMATE OF THE PERMISSABLE ENERGY LOADING

If the helical island is in the proper position around the scrape-off limiter the power load maximum will be behind the limiter edge and the power density falls off towards the edge and the liner. The power distribution is expected to vary with experimental conditions. In the absence of measurements or a reliable theory we took an estimate which we assume is a good approximation for a variety of experimental conditions. We used

$$Q = Q_o \, e^{-(r-a)^2/\delta r^2} \quad (4)$$

for the case where the island is properly posisioned; if not (e.g. during start up), we use

$$Q = Q_0 \, e^{-(r-a)/\delta r} \tag{5}$$

whereby Q_0 is determined from the total power load $P = \int Q \, dr$.
The temperature distribution was calculated for two representative limiter pieces using a two-dimensional finite element code incorporating radiative cooling. A special example is shown in fig. 4. If we assume that a Mo limiter can sustain about 2500 °C at the surface it can absorb about 2.6 MW over 4 s if it is positioned broadside on; if it collects mainly at the bow (90° rotation), it can absorb only 1.1 MW.
If the island is not in position (eq (5)), the permissable power is reduced by a factor 5.

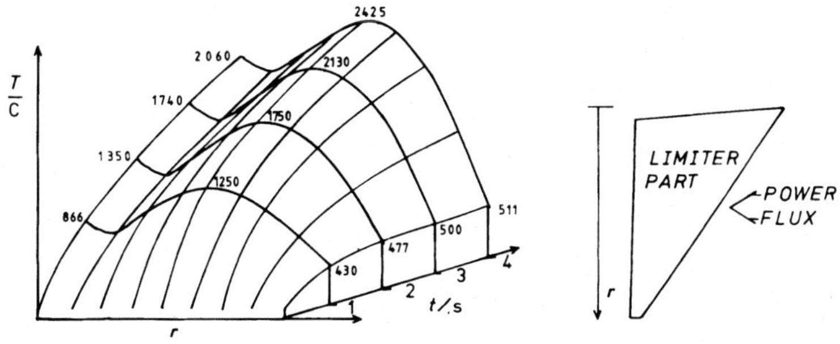

Fig. 4. Limiter surface temperature versus the distance from the edge and the time. Power flux according eq (4). P = 2.6 kW, a = 4 cm, limiter (Mo) exposing its broadside

5. ESTIMATE OF THE UNLOAD EFFICIENCY

To estimate the unload efficiency of the helical divertor we used a Monte-Carlo code based on [10] but extended it to 3 dimensions. For the plasma we use a cylindrical model and assume that the plasma parameters n, T_e, T_i fall off parabolically from the center to the separatrix (a). The values at the center are chosen based on measurements of comparable tokamaks; measured values at the boundary (a) are either not existent or are controversal between different authors. We based our assumptions on [8] . In the scrape off layer we assume a Gaussian distribution centered at a (like eq (4)) but used different decay lengths δr for n, T_0, T_i. The flux onto the limiter surface can then be calculated. The resulting neutral flux is distributed partly according the Behrisch matrix [11] the rest is reemitted with surface temperature. The particles are then traced and a record is made whether they are ionized by electrons or charge exchange,or hit a wall and are reflected (according [11]), or enter the pumpholes. The same is done for the different generations of charge exchange neutrals which undergo a random walk from their birthplace. The unload efficiency to the porthole is then calculated as the ratio of the neutral flow into the port to the flow onto the limiter.

We find that, depending on the assumed plasma parameters, 30 ... 40 % of the particles enter the pumphole whereby the contribution from reflected particles is larger by about a factor 2 as compared to re-emitted ones. In fact re-emitted particles lead to charge exchange and the high energy neutrals have a sizeable probability of entering the porthole. The unload efficiency is reduced due to backstreaming from the pumping chamber. With the two 10^5 l/s cryopumps we expect a net unload efficiency of 10 %.

6. CONCLUSIONS

A design for the limiter - and vacuum part of a helical divertor for TEXTOR tokamak was presented. We expect that the scrape-off-limiter can take about 2,6 MW over 4 s and can unload 10 % of the ions striking its surface. For TEXTOR this is sufficient to unload the energy - and particle - flux from two neutral injection ports 1,3 MW at 50 kV each (provided the density in the scrape-of-layer exceeds 10^{12} cm^{-3}). Because the helical conductors can be mounted outside the blanket, the scheme is also suitable for a reactor.

7. ACKNOWLEDGEMENT

We thank A. Sievers for the finite element calculations.

8. REFERENCES

[1] KARGER, F. and LACKNER, K., Phys. Let 61A, 385 (1977)

[2] BIEGER, W., DIPPEL, K.H; FUCHS, G., WOLF, G.H. Int. Symp. Plasma-Wall-Interaction, Jülich, 18-22 October, 1976. Proceedings P. 609. Pergamon

[3] WOLF, G.H; Plasma Phys. and Controlled Nucl. Fusion Research, Innsbruck 22-30 Aug. (1978)Proc,V.I p. 826

[4] FUCHS, G., STEFFEN, B.; J. Nucl. Mat.92, 93 (1980)

[5] CONRADS, H., Int. Symp. Plasma-Wall-Interaction, Jülich, 18-22 October, 1976. Proceedings p. 711.

[6] RECHESTER, A.B. and STIX, T.H., Phys. Rev. Let. 36, 587 (1976)

[7] BELITZ, J.H. and QUIEDE, J., Internal Report IPP-KFA, Jülich (1980), unpublished

[8] FUCHS, G., NICOLAI, A. (to appear) Nucl. Fus. 20 (1980)

[9] HEMMERICH, J.L. et al., Proceedings 8th Symp. on Engineering Problems of Fusion, San Francisco (1979), IEEE 79-CH-1441-5 NPS

[10] HUGHES, M.H., POST. D.E., J. Computational Phys., 28 43 (1978)

[11] BEHRISCH, R., Tokamak Reactors for Breakeven, Erice 21. Sept. - 10. Oct. 1976 e.d. H. Knoepfel Pergamon Press (1976)

HEAT REMOVAL IN INTOR VIA A TOROIDAL LIMITER

P. Mioduszewski

Kernforschungsanlage Jülich GmbH, Association EURATOM-KFA, D-5170 Jülich, FRG

ABSTRACT

In the present paper the potential of removing about 100 MW of thermal plasma power via a toroidal limiter in INTOR is studied. The heat flux distributions on various limiter configurations are calculated and the thermal response of a graphite tile limiter is estimated on the base of a one-dimensional heat conduction approach. The evaporation rates which have to be expected for the given energy flux densities and radiation cooled graphite tiles are evaluated. According to the present understanding it should be possible to remove 100 MW power from the INTOR plasma via a radiation cooled toroidal limiter.

1. INTRODUCTION

Although some form of magnetic divertor is probably required in INTOR [1] to provide for helium exhaust, the potential of material limiters is worthwhile to be discussed by several reasons. If non-divertor solutions for ash removal could be found, limiters seem to be most attractive for heat removal and wall protection because of their simplicity as compared to magnetic divertors. In addition, one or the other type of a wall protection device might be necessary anyway to handle the excessive heat loads during disruptions, to absorb the neutral beam shine-through or simply to act as a back-up for the divertor. In INTOR the fusion power during burn is expected to be in the range of 600 MW with a corresponding thermal wall load of about 120 MW. Presumed that the limiter has to be designed to take up to 80 % of the total thermal plasma power, it has to accomodate about 100 MW of thermal power. This power cannot be removed with poloidal limiters like in present day machines, and methods have to be found to spread the heat over a surface area as large as possible. For this purpose the potential of toroidal limiters has been studied with respect to an application in TFTR [2] and other tokamaks of the next generation. In the present paper we study the applicability of this concept to INTOR with special emphasis on steady state operation, on the heat load distribution due to non-circular plasma cross-sections, and on impurity generation.

2. PLASMA-LIMITER INTERACTION MODEL

The calculations of the heat flux distribution on the limiter are based on the following assumptions: The plasma energy flows - via heat conduction and convection - radially out of the plasma core into the limiter scrape-off layer. Within the scrape-off layer the energy flows along the magnetic field lines to the limiter surface. The e-folding length δ of the scrape-off layer for typical INTOR plasmas has been studied in [3] to yield between about 3cm and 5cm. For our purpose we choose a pessimistic value of δ = 3 cm.

For steady state heat removal the energy flux density should not exceed 5 - 7 MW/m^2. To remove 100 MW in INTOR, a limiter surface of 15 - 20 m^2 is required. This can only be achieved if the power flux in the scrape-off layer hits the limiter surface - in the poloidal projection - by oblique incidence, as illustrated in fig. 1.

Fig. 1 shows a poloidal projection of the plasma with minor radius a and two extreme cases of limiter configurations. In the most simple case the poloidal projection of the limiter is a straight tangential line and in the other case the limiter surface is curved with radius R_L and wrapped around half of the plasma column. We study the surface heat load for these two extreme examples and some other cases in between. To avoid excessive heat loads to the first wall at the limiter edge the limiter blade is extended over 3 e-folding lengths of the scrape-off layer to absorb 95 % of the total power. The location of the limiter is preferably at the inboard side of the torus. This way several tasks can be accomplished by the limiter: normal limiter action, protection of the first wall from hard disruptions [4], absorption of the injected neutral beam shine-through.

3. HEAT LOAD DISTRIBUTION ON THE LIMITER

Under the assumption that the energy flux density $\dot{q}(\varrho)$ in the scrape-off layer falls off exponentially with the distance into the scrape-off layer, we get

$$\dot{q}(\varrho) = \dot{q}_o \exp\left(-\frac{\varrho - a}{\delta}\right) \tag{1}$$

where ϱ is the polar radius of the limiter and q_o is the maximum power density at ϱ = a. The corresponding poloidal projections are:

a) for the straight tangential limiter:

$$\dot{q}(Y) = \dot{q}(\varrho)\cdot(d\varrho/dY) \tag{2}$$

b) for the curved tangential limiter:

$$\dot{q}(S) = \dot{q}(\varrho)\cdot(d\varrho/dS) \tag{3}$$

The arc length on the curved limiter given by $S=R_L \cdot \alpha$ is referred to as perimeter S. We insert $\varrho=\sqrt{Y^2+a^2}$ for the straight limiter and $\varrho=\sqrt{R_L^2+\varrho_o^2-2R_L\varrho_o\cos(S/R_L)}$ with $\varrho_o=R_L-a$ for the curved limiter into eqs. (2) and (3) respectively. Integration of both equations over the poloidal and toroidal directions yields the scale factor q_o and consequently the energy flux distribution as a function of the limiter perimeter S or Y respectively:

$$\dot{q}(Y) = \frac{P}{4\pi(R_o-a)\cdot 0{,}95\delta}\,\frac{Y}{\sqrt{Y^2+a^2}}\,\exp\left(-\frac{\sqrt{Y^2+a^2}-a}{\delta}\right) \tag{4}$$

$$\dot{q}(S) = \frac{P}{4\pi(R_o-a)\cdot 0{,}95\delta}\,\frac{\varrho_o \sin(S/R_L)}{\sqrt{R_L^2+\varrho_o^2-2R_L\varrho_o\cos(S/R_L)}}\,\exp\left(-\frac{\sqrt{R_L^2+\varrho_o^2-2R_L\varrho_o\cos(S/R_L)}-a}{\delta}\right) \tag{5}$$

where P is the thermal power to the limiter and the factor 0.95 results from integration over 3 e-folding lenghts δ. For the assumed INTOR parameters P = 100 MW, a = 1,64 m and δ = 3 cm we get the energy flux distribution as a function of the perimeter S or Y respectively as shown in fig. 2. Since the energy deposition is symmetrical to S = 0 (Y = 0) we have plotted only one half of the power deposition profiles. The profiles no. 1 through 4 correspond to the curved tangential limiter with various limiter radii R_L and profile no. 5 corresponds to the straight tangential limiter.

Fig. 2 gives estimates for the heat load on the limiter when the plasma cross-section is circular. Actually the plasma cross-section in INTOR will be non-cicular and - as a consequence - the contact area between plasma and limiter will get larger with increasing non-circularity to help spread out the power flux. Although the shape of the INTOR plasma is presumably not exactly elliptical, we demonstrate the effect of noncircularity by applying an ellipse with minor axis b and major axis a and a/b = 1.6. For symplicity the ellipse can be approximated by the corresponding osculating circle with radius a^2/b yielding a deviation from the ellipse smaller than 6 % for $\varphi \leq 30^\circ$ (cf. fig.1). This approach cannot be applied for the case indicated by curve no. 1 in fig. 2 where $\varphi = 90^\circ$.

For INTOR the minor axis of the elliptical plasma cross-section is assumed to be b = 1.3 m and the elongation $\varepsilon = 1.6$ which gives $a^2/b = 3.328$ m. Replacing (R_o - a) in eqs. (4) and (5) by (R_o - b) and a by a^2/b elsewhere we get some estimates for the limiter heat load due to elliptical plasma cross-sections. The result is plotted in fig. 3. As can be seen the maximum energy flux density is lowered by about a factor of 1.6 as compared to the circular case. Since the actual cross - section of the INTOR plasma is probably more D-shaped rather than elliptical with the flat side of the D in contact with the limiter, the non-circularity effect on the heat load is likely to be more pronounced than demonstrated here.

4. THERMAL RESPONSE OF THE LIMITER

For a proper calculation of the thermal response of the limiter we need to employ a two-dimensional heat conduction formalism corresponding to the heat flux distribution given by figs. 2 and 3. At this stage and for first estimates, however, the following procedure seems to be good enough: We assume that the whole limiter is made of graphite tiles - as proposed for TFTR - 10 x 10 cm^2 and 2 cm thick. It is further assumed that each graphite tile receives a mean energy flux density P_o and that heat conduction between different tiles is negligible. As two special examples we consider the cases represented by curves no. 4 and 5 in fig. 3. The tiles are numbered 1 through 11 along the perimeter as shown in fig. 3. For our computation we pick the corresponding tiles with the highest heat loads. For reasons of maintenance and repair the graphite tiles are preferably cooled by radiation cooling, whereas some kind of backing plates have probably to be cooled actively in order to avoid intolerable heat loads on the first wall.

First we check on the temperature oscillations due to the pulsed heat load with 100 s pulse time and 40 s dwell time. An isolated graphite tile of the dimensions given above is irradiated with the energy flux density P_o. With rising temperature it also looses energy by radiation to the surrounding medium at T_o, according to $P_R = 2\,\varepsilon\sigma(T^4 - T_o^4)$, where the emissivity for graphite is taken to be $\varepsilon = 0.8$, the Stefan-Boltzmann constant $\sigma = 5.67 \cdot 10^8$ W/m^2K^4 and the factor 2 is due to radiation from both sides. For the lumped thermal capacity approach the heating up of a tile with mass m, specific heat c and surface area A is given by

$$mc\,\frac{dT}{dt} = A\left\{P_o - 2\varepsilon\sigma(T^4 - T_o^4)\right\}. \tag{6}$$

Integrating this differential equation yields the heating time t_H as a function of temperature T. Substituting $(P_o/2\varepsilon\sigma + T_o^4) = x$ we get:

$$t_H = \frac{mc}{2A\varepsilon\sigma}\left\{\frac{1}{4x^{3/4}}\ln\frac{x^{1/4}+T}{x^{1/4}-T} + \frac{1}{2x^{3/4}}\operatorname{arctg}\frac{T}{x^{1/4}}\right\} + C \tag{7}$$

where C has to be determined from the initial condition $T = T_i$ at $t = 0$. Assuming that the initial limiter temperature is equal to the first wall temperature we choose $T_i = T_o = 800$ K.

The cooling down between burn times can be calculated by means of eq. (6) with $P_o = 0$. If cooling down by radiation begins at $t = 0$ with initial temperature T_i, integration of eq. (6) with $P_o = 0$ yields

$$t_{cool} = \frac{mc}{2A\varepsilon\sigma}\left\{\frac{1}{4T_o^3}\ln\frac{T_o+T}{T_o-T} + \frac{1}{2T_o^3}\operatorname{arctg}\frac{T}{T_o}\right\} + C \tag{8}$$

where, again, C has to be determined from $T = T_i$ at $t = 0$. Heating up and cooling down of the graphite tiles for $P_o = 4$ MW/m^2 and 2 MW/m^2 is shown in fig. 4. During the first cycle steady state temperature is reached fully at 4 MW/m^2 and up to 95 % at 2 MW/m^2. Therefore, the steady state temperature is taken to estimate the erosion of the tiles due to evaporation.

For a one-dimensional approach we start with the assumption of simple, separated tiles and the heat load distributions given in fig. 3. By attributing average power densities P_i to the individual tiles-according to fig. 3 - the corresponding steady state temperatures are computed. For a given temperature of the graphite tile the vapour pressure and the corresponding evaporation rate can be calculated [5] for each individual tile. In fig. 5 the mean tile temperatures along with the corresponding evaporation rates are plotted for the limiter configurations 4 and 5. For these plots it can be seen that radiation cooling is not trivial for all limiter configurations considered here. At T = 2500 K the vapour pressure of graphite is several mTorr, representing an impurity source for the plasma. To estimate the total amount of generated impurity atoms the evaporation rate of each tile is integrated over the toroidal circumference. Assuming a particle confinement time of $\tau_p = 1$s and the total number of plasma particles in INTOR of $N = 3 \cdot 10^{22}$ the resulting impurity levels are 27 % for limiter configuration no. 5 and 1 % for limiter no. 4. Whereas 27 % of C-atoms is clearly not acceptable for a burning fusion plasma the impurity level of 1 % - as produced by limiter no. 4 - seems to be tolerable. One has to keep in mind that this is the impurity production by evaporation only, other processes like sputtering and arcing are not considered.

5. CONCLUSIONS

The present estimates show that, on the basis of the present model, high thermal power levels of the order of 100 MW can be removed via toroidal limiters in INTOR. If the limiter shape is chosen appropriately, the local power flux density can be kept at a tolerable level. For the concept of radiation cooled graphite limiters fairly high evaporation rates at elevated graphite temperatures have to be taken into account. If the maximum local power deposition on the limiter is below 3 MW/m^2, the total evaporation rate is low enough to keep the carbon content in the plasma below 1 %. For calculations more accurate than the estimates presented here the understanding of plasma-limiter interaction must be improved by conducting appropriate experiments. After the limiter action and corresponding heat fluxes are evident, a more accurate analysis including two-dimensional heat conduction codes can be carried out.

6. REFERENCES

[1] "International Tokamak Reactor", Zero Physe, Vienna, 1980

[2] SCHMIDT, J.A., TFTR Physics Report no.12, PPPL, 1979
BUDNY, R., J. Vac. Sci. Tech. 17 (1980) 298

[3] HARRISON, M.F.A., HARBOUR, P.J., HOTSTON,E.S., MORGAN, H.G.
4 th Int. Conf. on Plasma Surface Interactions in Controlled Fusion Devices, Garmisch-Partenkirchen, April 21 - 25, 1980

[4] KLÜBER, O., CORTI, S., Gernhardt, J., KARGER, F., LISITANO, G., MEISEL, D., SESNIC, S., 5 th Conf. Plasma Physics and Controlled Nuclear Fusion, Tokyo, 11 - 15 Nov. 1974.

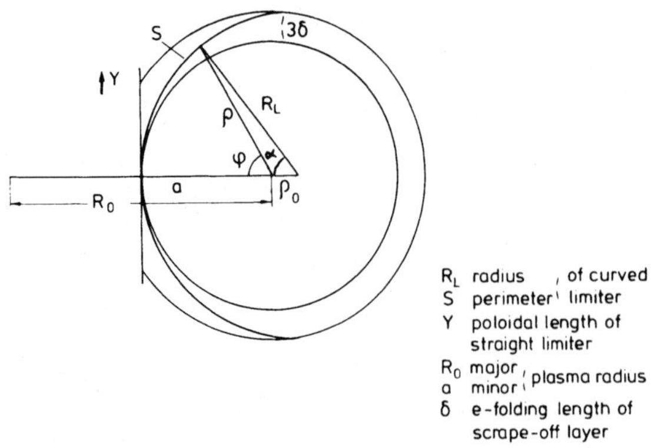

Fig. 1: Geometry of plasma column and limiter configurations

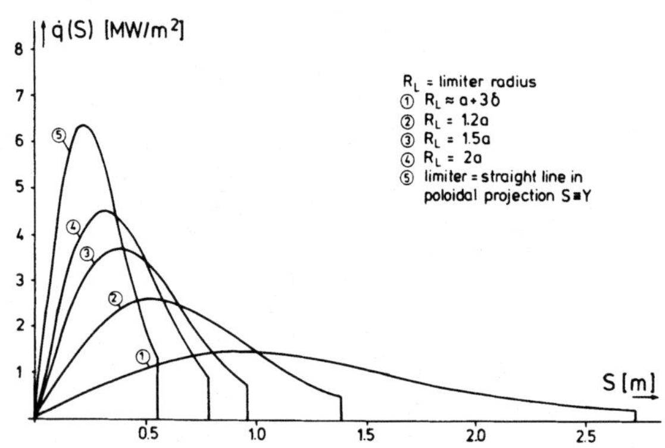

Fig. 2: Heat load to various limiter configurations for circular plasma cross section

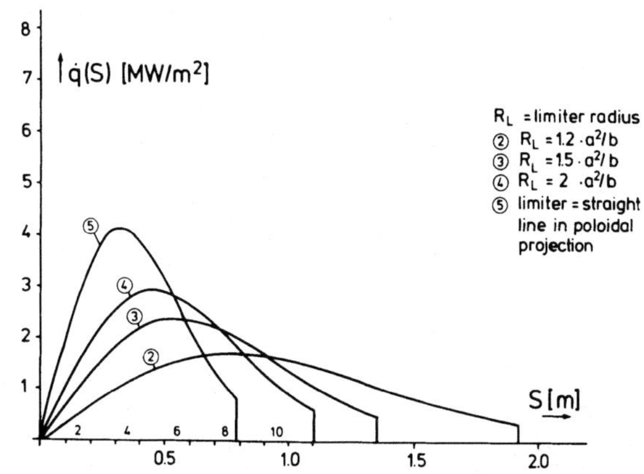

Fig. 3: Heat load to various limiter configurations for elliptical plasma cross-section

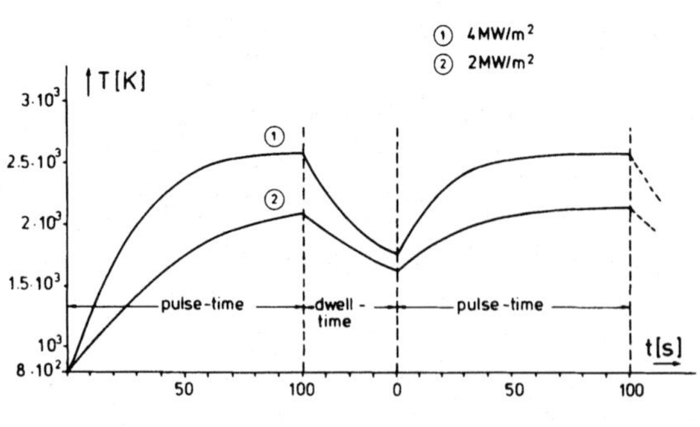

Fig. 4: Heating and cooling of isolated graphite tile

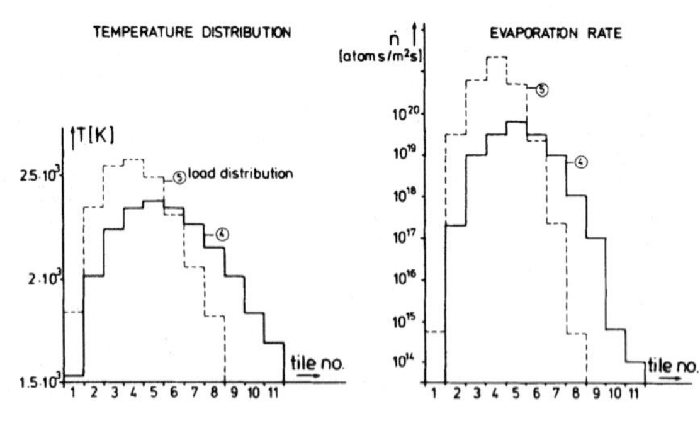

Fig. 5: Temperature distribution and evaporation rates on graphite tiles

DESIGN CONSIDERATIONS AND LAYOUT OF THE JET LIMITERS

K. J. Dietz
The JET Project, Abingdon, OX14 3EA, U.K.

ABSTRACT

The design of the limiters for the basic operation of JET is presented. It is laid out for a thermal load of 4 MWm^{-2} of 20 s duration. The limiters are actively cooled, their surfaces are of Inconel 600. The further development into limiters for extended performance is discussed.

1. INTRODUCTION

To achieve in JET [1] near reactor conditions, additional heating of tens of megawatts is required [2]. In accordance with results of existing tokamaks [3], [4], [5] it can be assumed that up to 50% of this power will be deposited on the limiters resulting in loads up to 5 MWm^{-2}. In case of additional heating by neutral beams the load on the limiters is nearly doubled [2] due to the simultaneous injection of particles with unwanted energies. The power load is mainly carried by particles, resulting in flux densities of 10^{19} - 10^{21} $cm^{-2}s^{-1}$. Severe surface erosion can be expected.

During the build-up of the plasma current runaway electrons can be produced. They hit the limiters locally with energies up to 10^8 Ws. Surface cracking and melting will occur [4]. During disruptions [6] of the plasma current the decay of the poloidal field in the limiters produces mechanical forces on the structure. The design of the limiters has to take into account all these effects.

2. DESIGN CRITERIA

Many data (e.g. thickness of the scrape-off layer, energy distribution of the particles hitting the limiter, time of current disruption) which are required for the layout of the limiters are only known approximately. Thus, as a first step, limiters for the basic performance [2] of JET (2.7 MW ohmic heating, 10 MW additional heating, 18 MW total injected power) are designed. The layout for the extended performance (4 MW ohmic heating, 25 MW additional heating, 45 MW total injected power) can be based on the experience gained with the first limiters.

Eight limiters (400 mm x 800 mm each) located toroidally in the mid-plane of the torus at the outer wall will be built into JET for the ohmic heating phase (Plasma power $\sim$ 2.7 MW). They will be increased to sixteen for the basic performance.

A schematic view of the limiters is presented in figs. 1 (limiter face) and 2 (limiter support). Each limiter face consists of eight limiter elements.

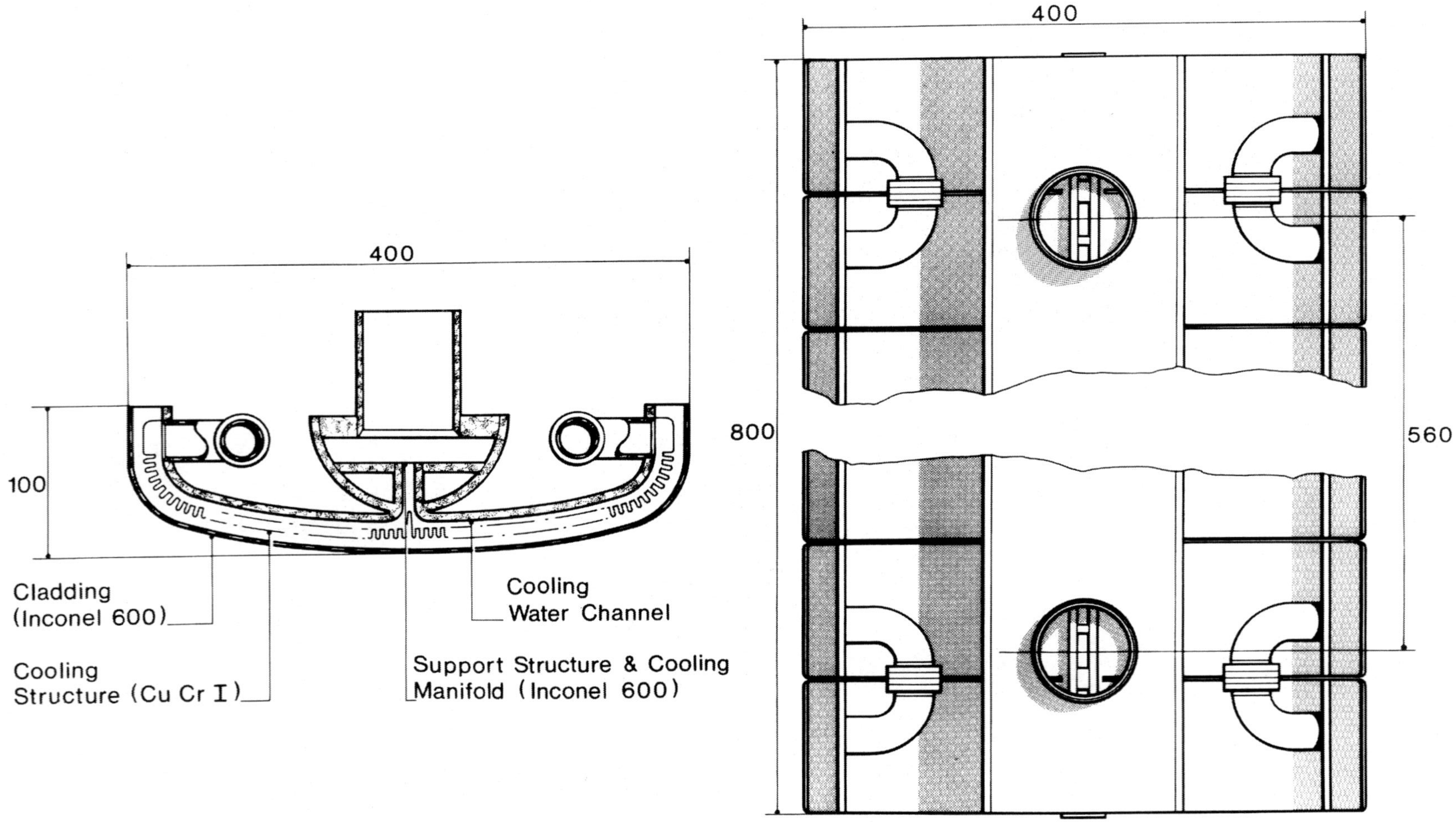

Fig. 1. Limiter Face

The cooling structure is the main part of each element. A sheet of Inconel 600 (1 mm thickness) is explosively bonded to the surface, a 5 mm thick plate of Inconel 600 forms the back. The back plates are welded to a beam which serves as cooling water manifold. The beam is welded to the limiter support. According to fig. 2 the support tubes are situated inside the protection and guide tubes.

To allow for 40 mm of axial movement between the limiter face and the vacuum vessel, the support tubes are connected via bellows to the protection tubes, which are joined by a sealing weld to the torus structure.

2.1 Thermal load

Heat loads on the limiters of 4 MWm^{-2} (basic performance) and later 10 MWm^{-2} (extended performance) for a time up to 20 s inhibit the use of inertia cooling. The application of small cooling pipes [7] proved to be impractical. The system chosen is of the Hypervapotron type (Thomson-CSF) which handles loads up to 20 MWm^{-2}. The cooling structure consists of a 9 mm thick front plate with fins protruding into the cooling water channel (fig. 1). The water flows perpendicular to the fins. Heat transfer occurs mainly by nucleate boiling and pool boiling. Due to the required high thermal conductivity and mechanical strength at elevated temperatures the material for the cooling structure will be precipitation hardened CuCrI.

Finite element analysis of heat transfer and mechanical stresses in the elements of the limiter face was carried out, the results support the choice of this cooling system.

During each discharge steady state conditions are reached after 2 s. The heat load is limited to 4 MW/m^2 by thermal stress in the copper-inconel composite due to the low thermal conductivity of Inconel. Additionally low cycle fatigue has to be taken into account. Stress analysis including all forces to assess this problem is in preparation.

2.2 Shape of the surface

The power distribution on the limiter was calculated following the Schmidt's model [8]: the particle density in the scrape-off layer decreases with 1/e (T_e= const). The particles are flowing parallel to the field lines. Finite Larmor radii effects [4] are not considered..

The peak power density, for the limiter shape chosen, ranges from 3×10^6 Wm^{-2} (thickness Δ of scrape-off layer 20 mm) to 4×10^6 Wm^{-2} (Δ = 10 mm) for additional heating of 10 MW. (18 MW total injected power [2]). This corresponds to an average load on the limiters of 2×10^6 Wm^{-2} (50% of total injected power).

2.3 Surface cladding and runaway electrons

As a copper surface is not very well suited to face the plasma, the cooling structure is explosive bonded with a sheet of Inconel 600 of 1 mm.

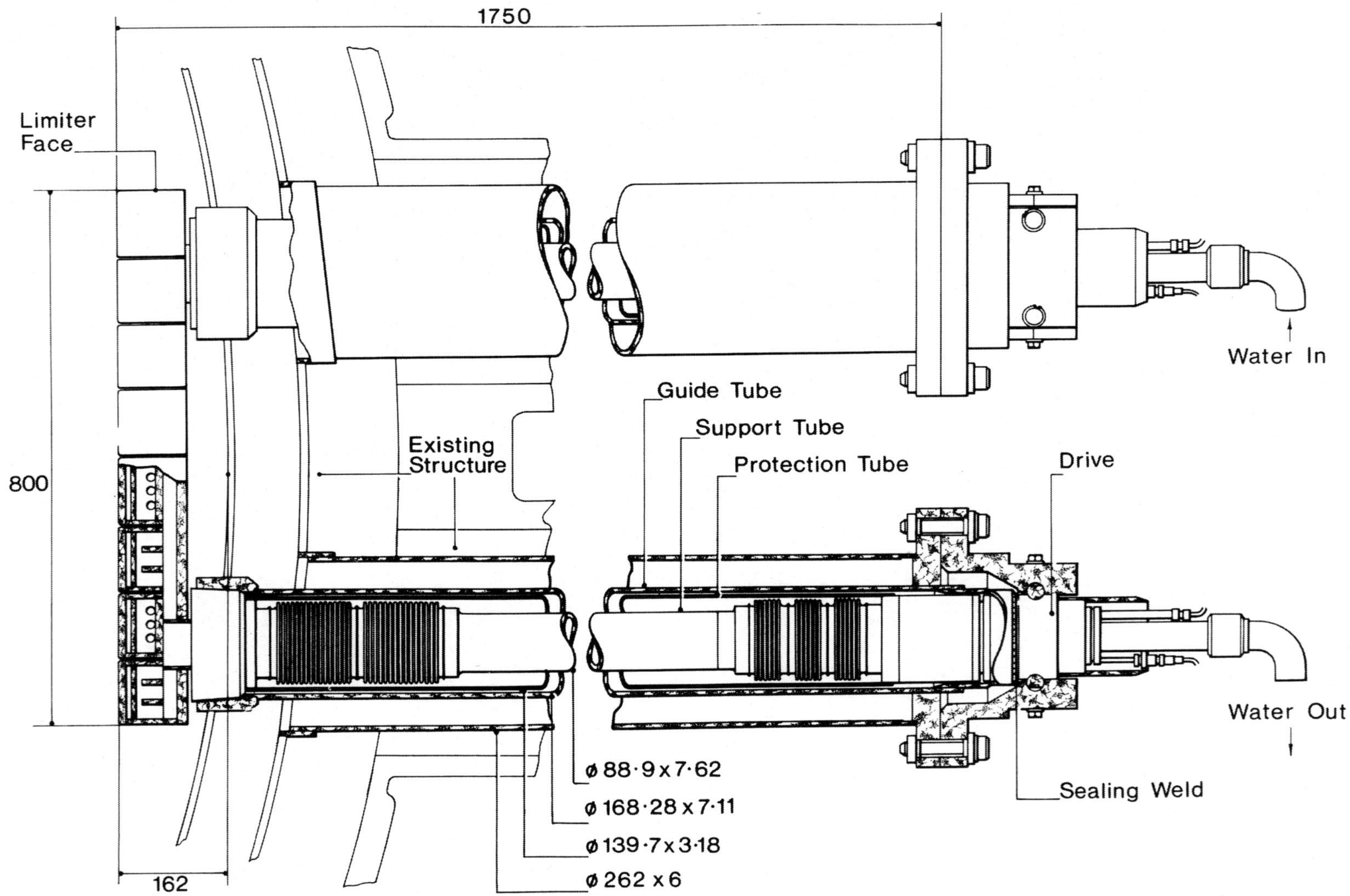

Fig. 2. Limiter Support

thickness. The cladding serves also as a sacrificial layer for ablation cooling during runaway discharges. In this case Inconel is evaporated on Inconel. The lifetime of the cladding is calculated to be 200-500 runaway discharges. Assuming that one runaway discharge occurs every 50 discharges, the lifetime of the cladding with respect to runaways will be between one and two years.

The thickness of the cladding is a compromise between the requirements for runaway discharges and the thermal stresses in the cooling structure.

2.4 Low-Z coatings

The thickness of oxide or carbide coatings is limited to 20-50 μm due to thermal shocks. Such thin coatings are easily eroded by sputtering. For example a TiC coating of 50 μm thickness would last less than 750 discharges in JET at extended performance (10 MWm^{-2}, energy of particles hitting the limiter 500 eV).

A thickness of 1 mm is required for one year's lifetime of the coating but then the heat load is limited to 2×10^6 Wm^{-2}. As a consequence in situ deposition of coatings in tokamaks had to be developed before the scheme of thin coatings will be useful for JET.

As long as these methods are unavailable only claddings with at least 1 mm thickness and thermal conductivities ≥ 50 Wm^{-1} K^{-1} can be used. Candidates for such applications are Be, C, V, Ni and Mo.

Carbon, however, must be protected against methane formation, Beryllium has its environmental probelms, Nickel is magnetic, so that in the immediate future V and Mo seem to be favoured, neither of them is a low-Z material. On a long term Beryllium seems to be a suitable material, at least at the time when tritium operation forbids any access to the vacuum systems.

2.5 Disruptive discharges

The following model was used to calculate the forces on the limiters during a disruptive discharge: The plasma current disappears immediately. The decaying magnetic field in the limiter structure produces eddy currents at the surfaces. Their magnetic moments together with the toroidal field generate torques on the limiter elements.

Currents up to 10^4 Amps and torques up to 160 Nm result for some surfaces of the limiter elements. They are cancelled to a high degree by torques at the opposite side,thus inducing stresses up to 10^8 Nm^{-2} in the material. The net torque on a limiter structure of eight elements is about 300 Nm. This can be taken by the support.

2.6 Instrumentation and remote maintenance

Each limiter is movable by 40 mm in radial direction and is equipped with six thermocouples in the limiter face, one in the cooling water inlet and one in the outlet. Additionally one of the limiters will be observed by an infra-red camera. By measuring the energy deposition on the limiters they can be positioned to take equal loads.

The lifetime of the limiters is limited, thus they must be remotely exchangeable. This can be done by breaking one weld as indicated in fig. 2, pulling the limiter faces including support and protection tubes into the vessel and taking them out subsequently through one of the major ports.

3. LIMITERS FOR EXTENDED PERFORMANCE

The limiters can only be tested in JET. The limiters for extended performance will be designed based on the results (thickness of scrape-off layer, energy distribution on the limiters, resistance to disruptions and runaways): The surface cladding will no longer be Inconel but one of the materials mentioned in 2.4. The thermal problems, which are the major ones in the whole limiter design, will then be similar to those for the first limiters. Reliable information on the scrape-off layer would allow us to improve the shape of the surface thus decreasing the peak load. The problems related to runaway and disruptive discharges will be nearly the same as for the basic performance.

4. CONCLUSIONS

The limiters designed for the basic performance of JET will meet all the requirements pointed out in this paper. For their design the work related to the limiters of the other large tokamaks could not be directly incorporated because of the different pulse durations and limiter positions.

For JET, the principle of inertia cooling cannot be used and because of the high dose of particles hitting the limiter thin coatings seem to be impractical at least as long as their in situ deposition is only in an experimental status.

Thus for JET the development leads to claddings with thicknesses of 1 mm or larger. These claddings will be low Z materials with a high thermal conductivity. Hydride formation is not expected because of surface temperatures of about 1000 K during discharges at extended performance.

Independent of possible developments in additional heating for JET [9] limiters of the described kind will be built into JET. They are designed to be a stepwise approach to the final limiters.

5. REFERENCES

[1] The JET Project — Report EUR 5516e, 1976

[2] BICKERTON, R.J. — Proc. 9th Europ.Conf.Contr.Fusion and Plasma Phys. Invited papers, (1979), 401

[3] DESCHAMPS, P. — Private communication

[4] COHEN, S.A. et.al. — Report PPPL-1671, 1980

[5] SABADO, M.M. et.al. — Report GA-A 15633, 1979

[6] CALLEN, J.D. et.al. — Plasma Phys. Contr. Fusion Res. Vol.I (1978), 415

[7] TRACHSEL, C.A. et.al. — J.Nucl.Mat. 85 + 86 (1979) 203

[8] SCHMIDT, J. — Bull. Am. Phys. Soc. 28 (1979) 970

[9] REBUT, P.H. — This conference

TRAIL--A TOKAMAK RAIL GUN LIMITER*†

W-S. Yu, J. R. Powell, and J. L. Usher
Department of Nuclear Energy
Brookhaven National Laboratory
Upton, NY 11973

ABSTRACT

An attractive new limiter concept is investigated. The TRAIL (Tokamak Rail Gun Limiter) system impacts a stream of moderate velocity pellets (100 to 200 m/sec through the plasma edge region to absorb energy and define the plasma boundary. The pellets are recycled after cooling, to the injector of an E-M mass accelerator. Heat fluxes of $\sim$30,000 W/cm^2 can be readily accommodated by the pellets, with very low recirculating power requirements ($\sim$ 0.1%) for the accelerator. The mass accelerator velocity requirements are well within the present state of the art (several Km/sec). Accelerators injecting pellets at $\sim$ 1 Km/sec can be used to control local plasma temperature and current profiles and to act as energy absorbers to shut down the plasma without damage to the first wall if a plasma disruption occurs.

1. INTRODUCTION

Present tokamak experiments generally control the plasma boundary through the use of uncooled solid limiters. These are not acceptable for future fusion devices since operation for more than a few seconds will destroy them. Numerous theoretical and experimental studies have been made of magnetic divertors (poloidal and toroidal) which, in principle, could both control the plasma boundary and reduce impurity injections from the first wall. However, it is not certain that divertors will work; even if they are feasible, they greatly complicate reactor design and increase cost.

Recognizing the problems with divertors, the STARFIRE design has proposed an actively cooled limiter using thin wall ($\sim$ 1 mm) tantalum structures cooled by high velocity water at relatively low temperatures. This concept, while it can probably accommodate the calculated heat loading of $\sim$ 400 w/cm^2, has very little margin to accommodate higher local heating rates and is vulnerable to erosion from charge exchange neutrals and plasma dumps.

In this paper, we propose a new limiter concept, termed TRAIL (Tokamak Rail Gun Limiter). An E-M mass accelerator continuously launches a stream of small pellets at moderate velocities (e.g., in the range of 100 to 300 meters/sec) through the plasma edge region. The moving pellets cool the edge plasma and define its boundary much as conventional material limiters do. TRAIL has several important advantages over an actively cooled limiter, however:

1. It can handle much higher heat fluxes (approximately two orders of magnitude higher),
2. There is no possibility of a coolant leak to the plasma chamber, and
3. Since pellets are reformed before recycling to the E-M mass accelerator, material damage (due to radiation, thermal cycling, etc) is not of concern.

The TRAIL concept can have important benefits beyond providing a more attractive limiter, however. When necessary, pellets could be injected at appropriate locations, locally cooling the plasma to control its temperature and density profile. This would help to prevent plasma disruptions. Finally, high velocity pellets ($\sim$ 1 Km/sec or greater) could

*Work performed under the auspices of the U. S. Department of Energy, Washington, DC.
†By acceptance of this article, the publisher and/or recipient acknowledge the U.S. Government's right to retain a nonexclusive, royalty-free license in and to any copyright covering this paper.

[1] PETERSON, D.R., Impact Fusion Workshop, Los Alamos, 10-12 July 1979. Proceedings pp 234-44.

[2] RASHLEIGH, S.C., Workshop on Gun and Launch Applications of Electromagnetic Pulse Power, Annapolis, 5-6 December 1978. Proceedings pp 149-51.

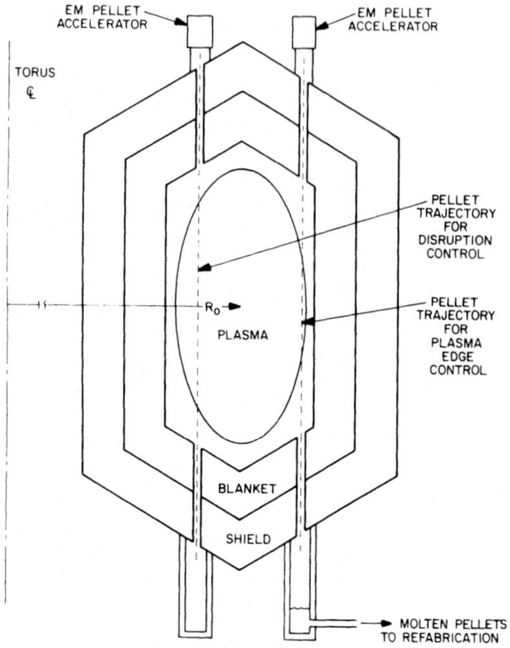

Fig. 1. Trail system for plasma control.

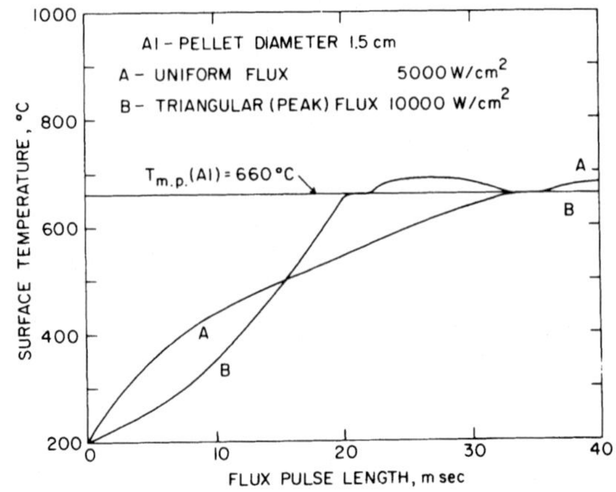

Fig. 2. Surface temperature of Al pellet.

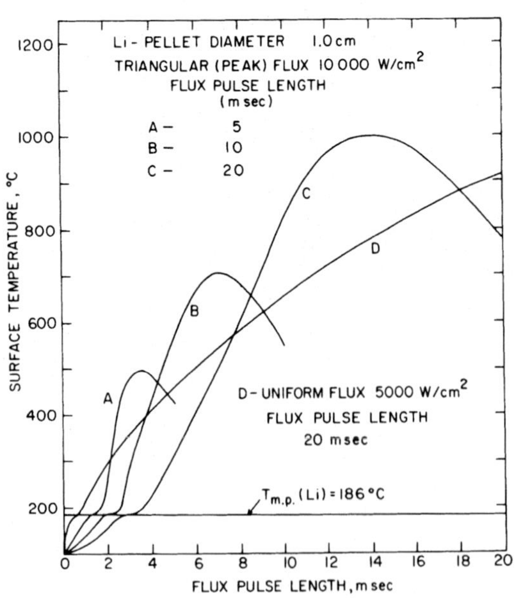

Fig. 3. Surface temperature of Li pellet.

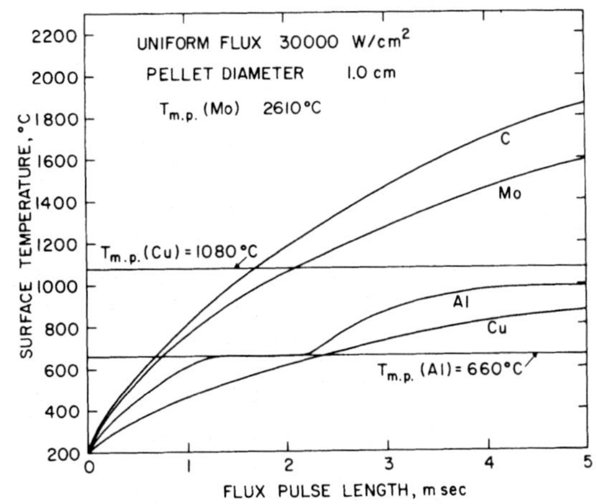

Fig. 4. Surface temperature of Al, Cu, Mo, and C pellets.

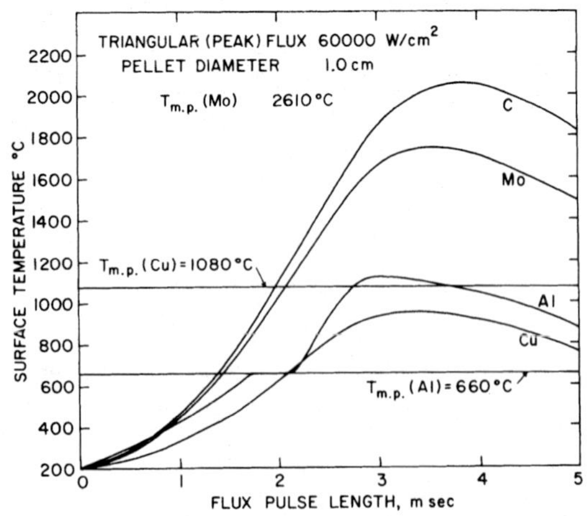

Fig. 5. Surface temperature of Al, Cu, Mo, and C pellets.

be injected when a disruption began; this would rapidly remove plasma energy and prevent it from damaging or destroying portions of the first wall. In this application, TRAIL would function in a similar manner to the SCRAM system in a fission reactor.

E-M mass accelerators are under development for a variety of defense applications. Generally, these involve acceleration of relatively massive projectiles (e.g., in the range of 100 to 1000 g) to velocities of several Km per second. Experiments with rail guns have demonstrated acceleration of projectiles of several grams to velocities approaching 10 Km per second. In general, the mass accelerators required for TRAIL applications are much less demanding than those being developed for defense applications and could be developed relatively quickly for future fusion experiments.

2. DESCRIPTION OF THE TRAIL CONCEPT

Fig. 1 shows an idealized version of the TRAIL system applied to Tokamak reactors. For boundary layer control, the pellets are fired vertically downward past the outer edge of the plasma along a constant toroidal field trajectory (i.e., a constant radius trajectory). Constant field trajectories are probably necessary for electrical conducting pellets (e.g., Al or other metals), but non-conducting pellets (e.g., graphite, which has much lower conductivity) could be fired on non-constant magnetic field trajectories. Firing conducting pellets on non-constant field trajectories would cause the trajectory to curve, due to field gradient forces. In principle, this could be compensated for, but it appears difficult and pellet conductivity and size would have to be closely controlled.

As will be shown in a later section, pellet heat fluxes of up to $\sim$30,000 W/cm^2 averaged over the trajectory at the plasma edge can be accommodated so that pellet position need not be critically controlled in its flight through the plasma edge. In contrast, actively cooled material limiters would have to be carefully positioned and controlled so that local hot spots did not occur which could cause burnout at a few hundred watts/cm.

The required pellet areal flow rate through the plasma edge is given by:

$$\frac{dA_p}{dt} = \left(\frac{3.5}{Q}\right) \frac{P_o(1 - f_R)}{q_p} \tag{1}$$

where

P_o = total reactor power, watts,

f_R = fraction of alpha particle energy radiated to wall by Bremsstrahlung,

Q = total energy released per fusion event, MeV,

q_p = integrated energy absorbed by pellet during its passage, Joules/cm^2, and

$\frac{dA_p}{dt}$ = total areal flow rate of pellets, cm^2/sec with suitable impurity seeding, $f_R \sim 0.85$; for the other parameters typically, $Q \sim 21$, $q_p \sim 400$ J/cm^2, and $P_o \sim 4 \times 10^9$ watts (commercial reactor size). The required areal flow rate is then 2.5×10^5 cm^2/sec.

The corresponding number and mass flow rates of pellets are:

$$\frac{dN_p}{dt} = \left(\frac{1}{4\pi r_p^2}\right)\left(\frac{dA_p}{dt}\right) \text{ pellets per second} \tag{2}$$

$$\frac{dM_p}{dt} = \frac{r_p}{3} \rho_p \left(\frac{dA_p}{dt}\right) 10^{-3} \text{ kg per second} \tag{3}$$

where

r_p = pellet radius, cm

ρ_p = pellet density, g/cm^3.

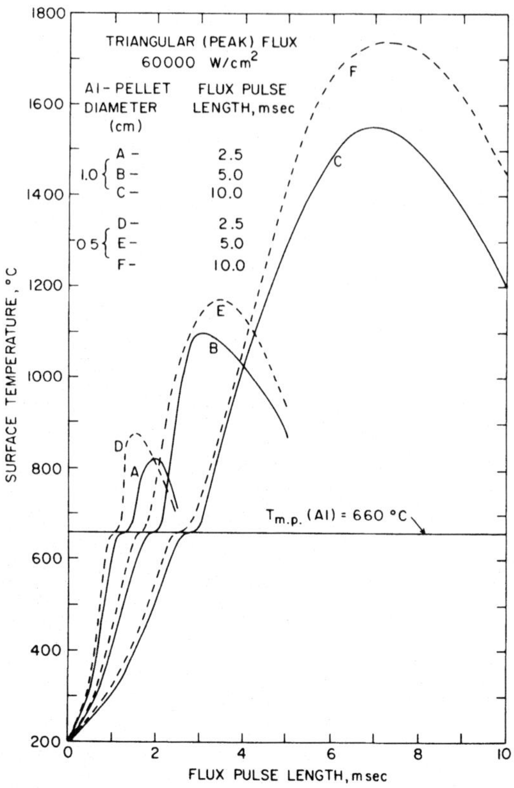

Fig. 6. Surface temperature of Al pellet.

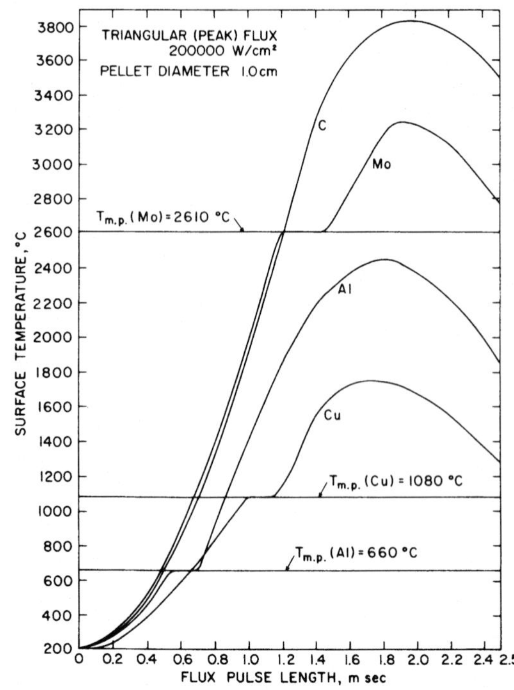

Fig. 8. Surface temperature of Al, Cu, Mo, and C pellets.

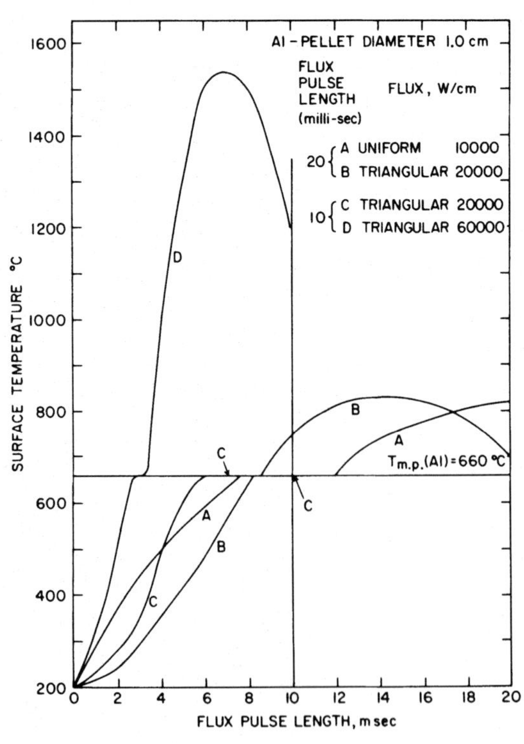

Fig. 7. Surface temperature of Al pellet.

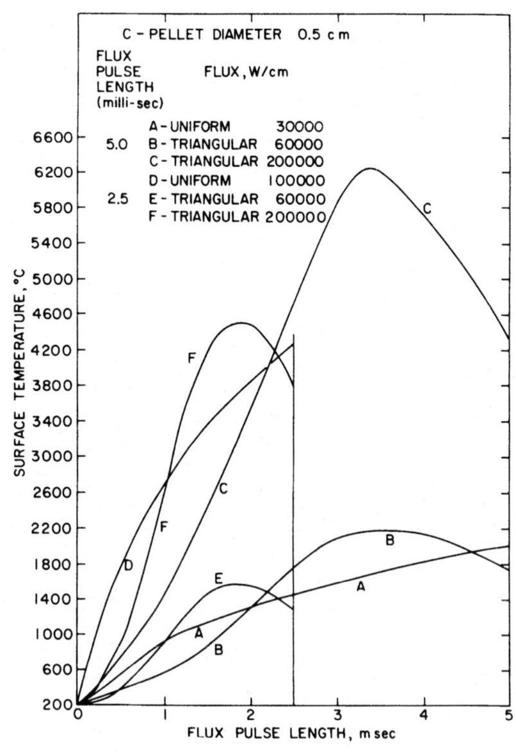

Fig. 9. Surface temperature of C pellet.

For representative 1 cm diameter aluminum pellets (ρ = 2.8 g/cm^3), 80,000 pellets per second are required, with a mass flow rate of 110 kg/sec. These can be accommodated with practical pellet accelerators as discussed in a later section.

An important parameter is the dimensionless heat absorption efficiency, E_H, which is the ratio of the heat absorbed during its passage through the edge region of the plasma to its kinetic energy. E_H is given by

$$E_H = \frac{6q\Delta x}{\rho_p r_p v_p^3} \qquad (4)$$

where

q = average heat flux to pellet during passage through the plasma, in w/cm^2,

Δx = path length through plasma, cm, and

v_p = pellet velocity, cm/sec.

Practical TRAIL systems result in E_H values on the order of 100. Further, the E-M accelerators will have efficiencies in excess of 90% (electrical to kinetic energy), and the TRAIL heat system only absorbs approximately 2% of the reactor thermal power. The recirculating power necessary to operate the TRAIL system is thus very small, on the order of 0.1% of the generated electric power.

3. HEAT TRANSFER PERFORMANCE OF TRAIL SYSTEM

Transient heat analyses were carried out for a number of pellet materials (Al, Cu, Li, C, and Mo) over a range of heat fluxes (5000 w/cm^2 to 100,000 w/cm^2), heat pulse length (2.5 to 40 milliseconds), and pellet diameters (0.5 to 1.5 cm).

The analyses assumed that only conduction occurred in the pellet, no radiation from the pellet surface (negligible compared to the applied heat flux), spherical pellets with uniform heat flux (in the time scale of plasma ion and electron thermal velocities, the pellet is essentially stationary).

Heating-5, a finite difference conduction code, is used with temperature dependent changes of state incorporated where necessary. For any given case, surface and internal temperature distributions are determined as a function of time after the heat flux is applied, along with the fraction of the pellet that melts. The corresponding amount of material evaporated is determined by the time dependent surface temperature and the associated vapor pressure. Free molecular flow of the evaporated material is assumed. The evaporation rate is taken as nv/4, where n and v correspond to values for the equilibrium vapor. This approximation fixes an upper bound on evaporation rate, since kinetic or flow impedance effects may act to lower the time evaporation rate.

Heat transfer response is evaluated for both uniform applied heat flux and a triangular-shaped heat flux with the same average value to approximate the non-constant conditions as the pellet moves through the edge region. In general, the two approximations yield similar maximum surface temperatures and melting fractions, indicating that pellet performance primarily depends on the total energy applied during the pulse, rather than on its shape.

Results for different pellet materials are shown in figs. 2 through 12 for average heat fluxes of 50,000 w/cm^2. An Al pellet barely melts on the surface (fig. 2) even for relatively long transit times (i.e., 40 milliseconds). The surface melts at $\sim$ 20 msec with a triangular-shaped pulse (peak of 10,000 w/cm^2) but quickly refreezes as heat flux drops. With a uniform heat flux of 5000 w/cm^2, melting only starts to occur after $\sim$40 msec.

Under the same heat flux conditions, pellets show much higher surface temperatures, reaching levels at which evaporation becomes a concern. This is a result of the considerably lower thermal conductivity for lithium. Accordingly, lithium pellets are

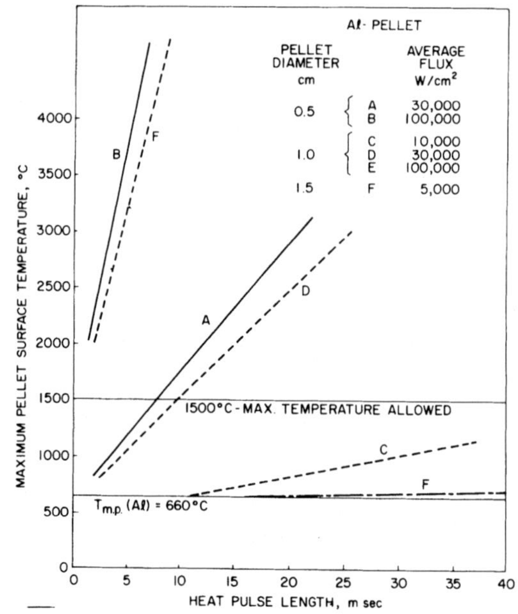

Fig. 10. Maximum surface temperature of Al pellet.

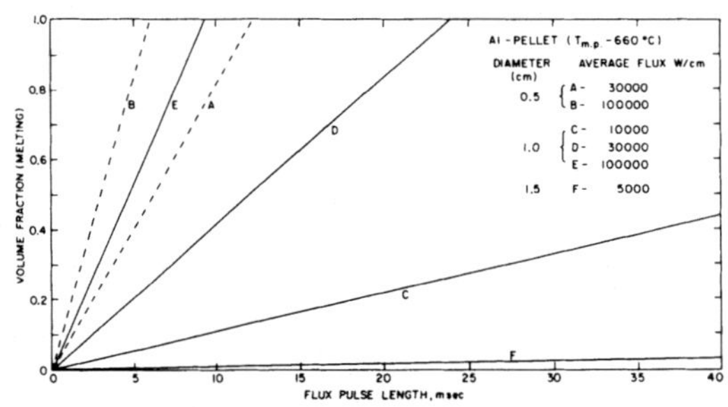

Fig. 11. Volume fraction (melting) of Al pellet.

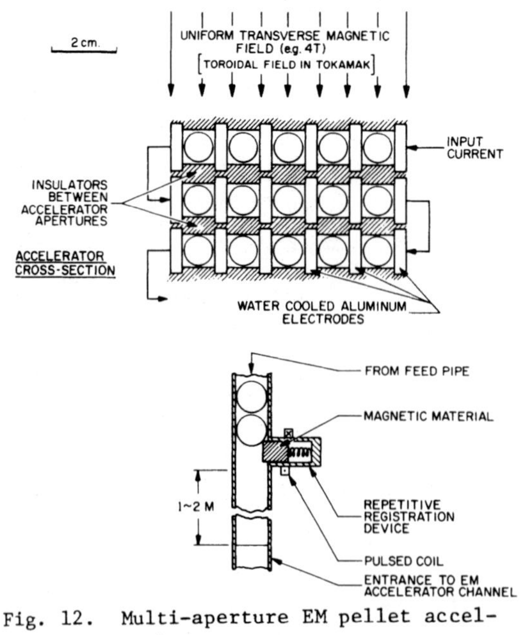

Fig. 12. Multi-aperture EM pellet accelerator.

probably only suitable for relatively low heat fluxes and/or short transit times.

Figs. 4 and 5 compare the surface temperatures of Cu, Al, Mo, and C pellets for an average heat flux of 30,000 w/cm^2 [uniform and triangular-shaped pulse of 5 msec length are compared].

Copper does not melt, aluminum does, but its surface temperature is still well below the limit (∿ 1500°C) at which vaporization becomes excessive. Both Mo and C pellets would also be suitable with Mo well below its melting point of 2610°C. Although Mo and C have much higher surface temperatures, vaporization is still very small.

To further minimize impurity affects accompanying vaporization, one could coat pellets with thin layers of low Z, high melting point materials, e.g., Be, B_4C, B, etc. Copper pellets with a few mils of Be, for example, should make a very attractive pellet. Triangular heat pulses show somewhat higher peak surface temperatures than uniform heat pulses, but only by a relatively small amount.

Fig. 6 compares surface temperatures for aluminum pellets as a function of heat pulse length (2.5, 5, and 10 msec) and pellet diameter (0.5 and 1.0 cm) for a triangular heat pulse with maximum heat flux of 60,000 w/cm^2. The limiting heat pulse length under these conditions appears to be about 10 msec. Assuming a 2 meter path length past the edge of the plasma, this corresponds to a velocity of 200 meters/sec, which would be relatively easy to achieve. For lower heat fluxes velocities of 50 to 100 meters/second appear ample, as illustrated in fig. 7. Here, it is clear that average heat fluxes of 10,000 w/cm^2 could be accommodated for periods much longer than the 20 msec limit chosen for the analyses.

Average heat fluxes substantially greater than 30,000 w/cm^2 appear difficult to accommodate, however. Fig. 8 compares the performance of C, Mo, Al, and Cu pellets to a triangular-shaped heat flux of 200,000 w/cm^2 peak and 2.5 msec duration. All exhibit an excessive temperature rise; copper and carbon pellets probably could accommodate this heat flux level if the pulse length were somewhat shorter, however. Fig. 9 shows the performance of C pellets as a function of heat flux magnitude and pulse length. Only extremely high heat fluxes of ∿ 100,000 w/cm^2 average pose problems and these can probably be avoided by using somewhat shorter heat pulse lengths.

Fig. 10 summarizes the heat transfer performance of aluminum pellets in terms of peak surface temperature during the pulse. Aluminum is a very attractive candidate for average heat fluxes up to ∿ 30,000 w/cm^2, but is not suitable for very high average heat fluxes (∿ 100,000 w/cm^2). Depending on heat flux, integrated energy absorption values of ≥ 300 J/cm^2 can be accommodated.

Fig. 11 shows the fraction of the Al pellet that would be melted as a function of heat flux and pulse length. For the higher heat flux range of 10,000 to 30,000 w/cm^2, approximately half of a 1.0 cm diameter pellet could be melted. A larger fraction, e.g., up to ∿80%, could be melted if somewhat smaller pellets were used. At low heat fluxes, a large (1.5 cm) pellet was used so that only a small fraction melted (∿ 5% at 40 msec). With longer heat pulse lengths and smaller pellets, it should be possible to melt virtually all of the pellet at an average heat flux of ∿ 5000 w/cm^2.

4. E-M MASS ACCELERATOR DESIGN

The simplest type of E-M mass accelerator is the simple rail gun, consisting of a pair of electrically conducting rails with a projectile between the rails. When a voltage and current are applied to the breech of the rail gun, a JxB force develops on the projectile causing it to accelerate along between the rails. Either electrically conducting or non-conducting projectiles can be used in a rail gun. In the former case, current flows through the projectile creating a direct JxB body force, while in the latter case, a plasma arc drives the projectile ahead of it.

In numerous experiments, Marshall [1] accelerated gram size projectiles, both conducting and non-conducting, to several km/sec using a simple rail gun. The highest velocity achieved was with a Lexan projectile at 5.9 km/sec. Tests are now being carried out at LASL to produce even higher velocities [2] using explosively driven-flux compression rail guns.

For control of the plasma edge, aluminum pellets injected at ∿ 100 meters per second appear satisfactory. A 1 cm pellet accelerated in a 0.2 meter long augmented rail gun (B_T = 3T) would require an applied current of only 1100 amp during the acceleration period of 0.004 second and would experience an acceleration of 2.5 x 10^4 m/sec^2 (2500 g).

The I^2R loss in the pellet would be trivally small compared to the kinetic energy gain (∿ 0.01 J vs ∿ 10 J). Frictional and electrical losses at the boundary between the sliding pellet and the rails will be considerably larger than the I^2R loss. These will have to be measured in an actual accelerator, but it appears likely that they will still be small compared to the kinetic energy gain.

Achieving the desired velocity and efficiency is relatively easy and straightforward in terms of present knowledge about E-M mass accelerators. The only development needed appears to be in the area of accelerators to handle high pellet flow rates.

Fig. 12 shows a design approach for a multiple aperture rail gun with high pellet flow rate. Pellets are synchronously injected at low velocity (∿ 1 m/sec) into a set of channels. A pulsed current (controlled by an SCR, for example) is applied to the accelerator when the falling pellets reach the appropriate point. The multiple channels are electrically connected in series to ensure uniform current (and acceleration) in each pellet.

5. SUMMARY AND CONCLUSIONS

The feasibility of using an E-M mass accelerator to inject streams of small pellets to act as a renewable material limiter for shaping the plasma boundary has been examined. Velocity requirements are well within current capabilities and the only development needs are for high feed rate devices.

Very high heat fluxes, up to ∿ 30,000 w/cm^2, can be handled by the pellet stream at moderate velocities, up to ∿ 200 meters/sec, without surface vaporization causing objectionable impurity effects in the plasma. The heat absorption capability of pellets is several hundred Joules/cm^2 of surface area. The heat absorption's efficiency--the ratio of heat absorbed in the pellet to kinetic energy of the pellet--is high, on the order of 100. The pellets absorb on the order of 2% of total reactor power, and recirculating power requirements to the E-M accelerator are small, on the order of 0.1% of output electric power.

Aluminum pellets appear to be very promising since they have high thermal conductivity and can be collected as a liquid after passage through the plasma edge. The liquid would be frozen into pellets and recycled to the accelerator. Copper, Mo, and C pellets were examined and also appear promising.

The TRAIL concept also appears very attractive for controlling local plasma current and temperature profiles to help prevent disruptions. If a disruption occurs, a TRAIL could inject high velocity pellets (∿ 1 km/sec) to absorb plasma energy and enhance plasma radiation. Such a safety system would prevent the disruption from causing damage to the first wall.

More detailed investigation of the TRAIL concept is recommended, including analyses of heat flux profiles for various trajectories through the plasma. If more detailed studies show TRAIL to be as attractive as it appears now, high flow E-M mass accelerators should be tested, and experiments on a small tokamak should be carried out.

LIMITERS IN THE FRASCATI TOKAMAK DEVICE

C. Alessandrini, C. Ferro, M. Samuelli

Associazione EURATOM-CNEN sulla Fusione, Centro di Frascati,
C.P. 65 - 00044 Frascati, Rome, Italy

ABSTRACT

The Frascati Tokamak is provided by a transfer system which allows the extraction of the main limiter without any relevant interruption of the experiments. During the runs the limiter was extracted several times. It showed evaporation, melting and deformations due to thermal shocks.

As a consequence, the design of the limiter was modified and three types of limiters have been used. The extraction mechanism as well as the types of limiters used are described. In addition an analysis of the observed damage is presented.

1. INTRODUCTION

In the Frascati Tokamak (FT) it is possible to study the plasma characteristics as a function of the plasma radius and of the limiter material.

The main parameters of the machine have been already reported [1].

The FT machine operates at cryogenic temperature and one cannot break the vacuum unless the vacuum chamber is heated to room temperature. To remove the main limiter without any relevant interruption of the experiment an extraction mechanism has been designed. Up to now about 3,500 powerful discharges have been obtained. During the runs the limiter was extracted several times. It showed evaporation, melting and deformations due to thermal shocks. As consequence, the design of the limiter was modified and three different limiters were used.

In this paper there are described the design of the extraction mechanism as well as the different types of limiters designed to optimize their behaviour respect to thermal deformations. Some discussion of the behaviour of the plasma and plasma-limiter interaction is also reported.

2. EXTRACTION MECHANISM

The limiter supporting structure and the transfer system are shown in Fig. 1. A screwed flange (A) separates, during operations, the torus (B) from the auxiliary chamber (C) which is normally in air. A cantilever (D) bolted on the flange (A) supports the limiter (E) by means of two insulated clamps (F). Two wires (G) and two welded feedstroughs provide electrical connections. On the side of the auxiliary chamber the flange is fastened to a carriage (H) which runs on a pair of rails (I), and is driven manually by means of a vacuum tight sprocket wheel (L).

The auxiliary chamber consists of a stainless steel tube (0.6 m I.D.) with several ports for the inspection and removal of the limiter. The auxiliary

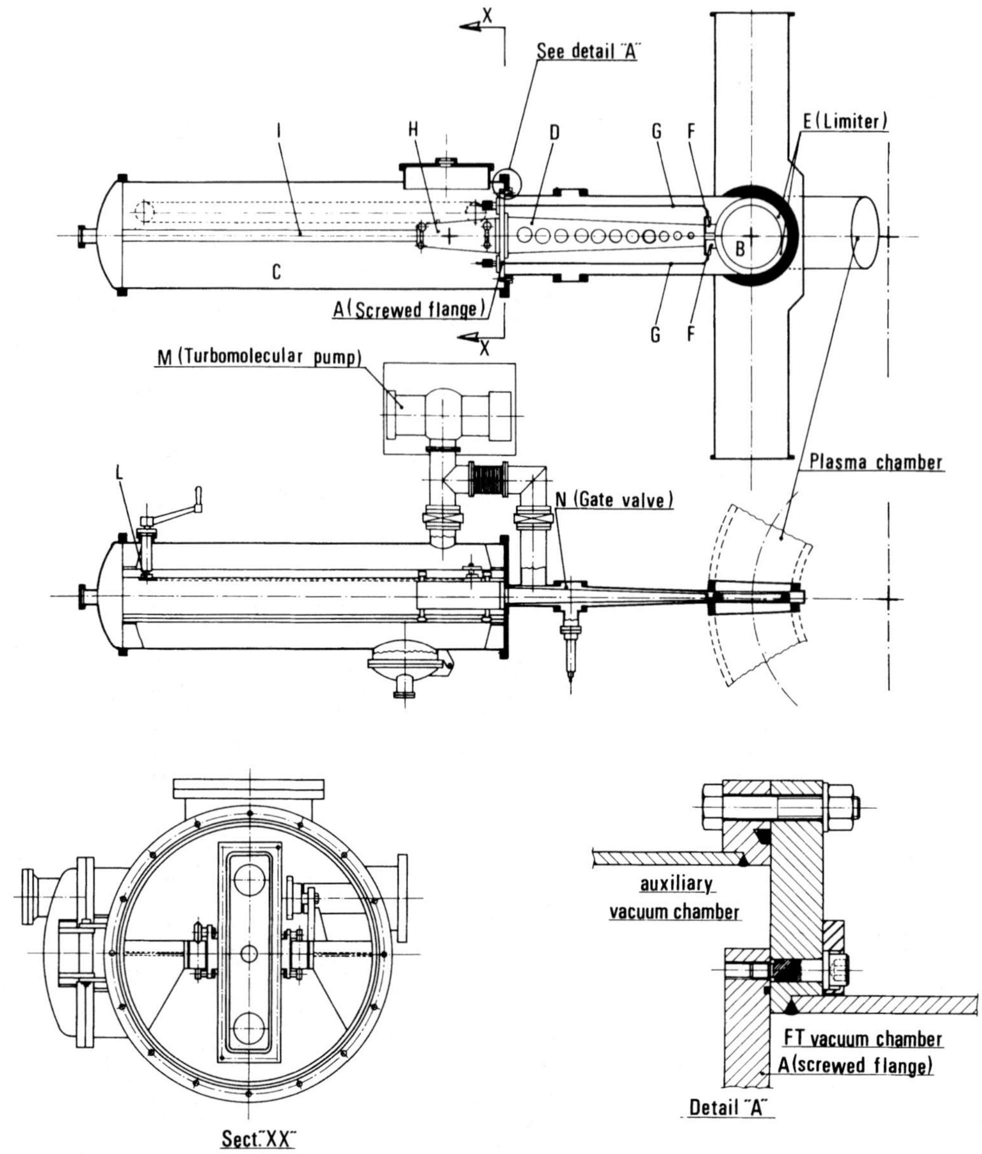

Fig. 1 Extraction mechanism

chamber may be evacuated to 10^{-6} Torr in few minutes by means of a turbomolecular pump (M).

An electro-pneumatic valve (N) (0.46 m x 0.09 m) separates the torus from the auxiliary chamber when the limiter is extracted. Special care have to be payed on the thermalization of this valve interfacing the cryogenic torus vessel to the room temperature extraction vacuum chamber. The thermal contraction of the FT machine between cryogenic and room temperature is of the order of 1 cm. The whole extraction structure is free of following the motions of the torus during the thermal cycles.

3. LIMITER

The FT poloidal limiter is obtained from two half rings which are electrically insulated with respect to the vacuum vessel and to each other. Insulation was obtained using plasma-sprayed ceramic break [2].

Beyond to determine the plasma radius and to prevent the damage of the relatively thin walled vacuum vessel, the limiter is also used for preionizing the discharge by means of a low energy spark produced between the upper and lower halves by applying a R.F. source (1.5 kV; 1 MHz) for about 30 msec before the beginning of the discharge.

Three limiters have been used up to now and the main characteristics of the discharges obtained are reported in Table I.

TABLE I

Characteristics of the discharges

Limiter N° Material	Limiter radius [m]	Number of discharges	Total Energy [MJ]	Peak Power [kW]	Average[1] Power [kW]	Discharge[1] Duration [sec]	$\overline{I_p}$[1] [kA]	B_{Tor} [kG]	$\bar{n}$ [cm^{-3}]	$T_{o\,e,i}$ [keV]
1 Mo+Inc 600	0.183	1911	275	500	250	0.3	200	40	10^{14}	0.8
2 AISI 316 L	0.183	941	280	1000	600	0.9	350	60	10^{14}	0.8
3 AISI 316 L	0.200	374	146	800	600	0.9	350	60	10^{14}	0.8

[1] The averages were calculated taking also into account anomalous discharges.

A critical parameter in the limiter life has been found to be the thermal deformations which can induce a short circuit between the two half ring and/or between them and the vacuum vessel.

3.1. Mechanical design of the limiters

Both radius and cross sections of the limiter were modified in order to optimize thermal behaviour and manufacturing facility. The limiters used are shown in Fig. 2a-b-c.

Limiter N. 1 (Fig. 2a): Two half rings (Inconel 600) support 24 pieces of molybdenum fastened by means of rivets. The molybdenum surface were curved (elliptic form) to try to maintain constant the heat flux [3]. The two half rings were not retained each other in the inner part.

Limiter N. 2 (Fig. 2-b): It consists of two stainless steel (AISI 316 L) massive half rings. No particular care was payed on the geometry of the surface facing the plasma. The upper and lower limiter segments were retained each other in the inner part.

Electrical insulation was obtained by means of alumina pieces hidden from the plasma.

Limiter N. 3 (Fig. 2-c): The upper and lower segments (stainless steel AISI 316 L) are retained each other and insulated by means of alumina pieces.

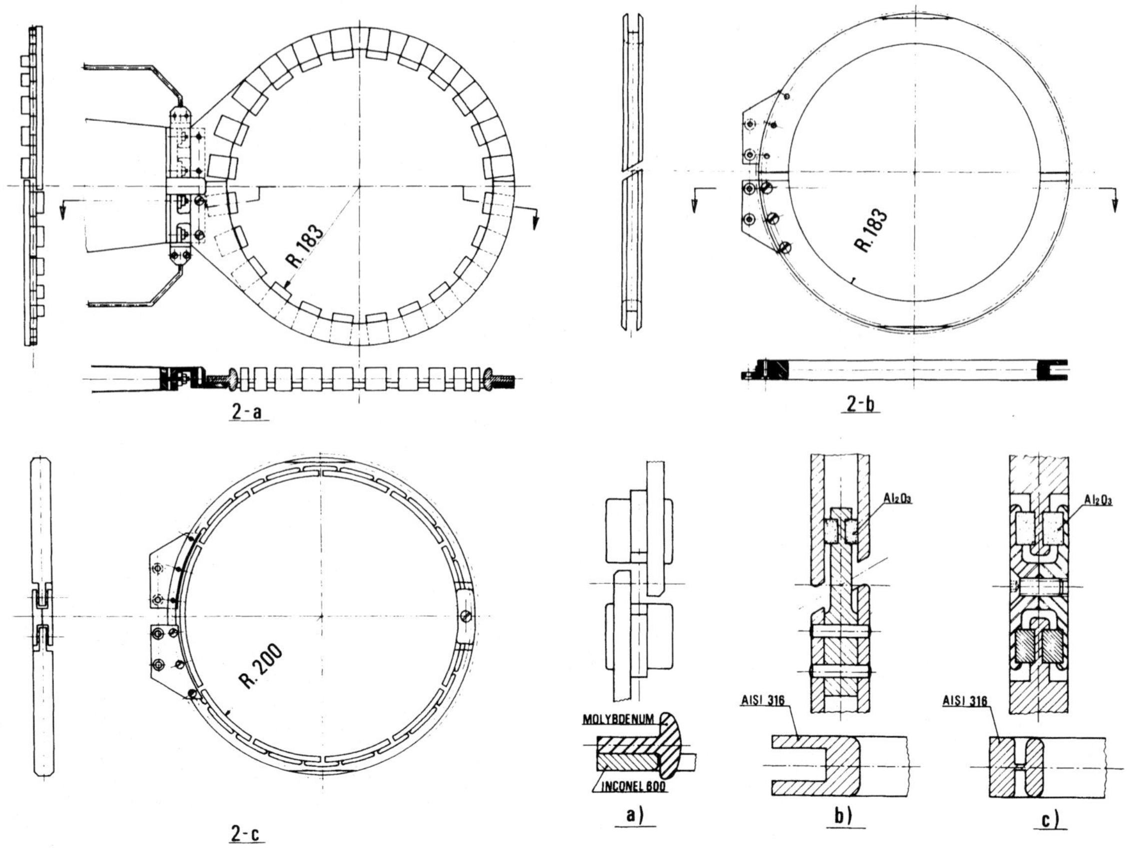

Fig.2. Mechanical design of the FT Limiters

The limiter facing the plasma is connected by means of thin bars to the massive support rings and can distort without inducing large thermal stresses in the whole structure.

4. ANALYSIS OF THE RESULTS

4.1. Conditioning of the limiters

Before being inserted into the machine the limiters were rinsed with a basic solution and then outgassed in vacuo at 800 °C. The effect of a new limiter on the discharges parameters has been analysed.

The most relevant parameter seems to be $\overline{I \cdot t}$ (the average current times the duraction of the discharge). In the early discharges plasma disruptions are mainly due to an influx of a large amounts of impurities from the limiters. The release of these impurities is strictly related to the temperature of the limiter itself which is a function of the current and duration of the discharge. The two diagrams reported in Fig. 3 show that a good conditioning of the limiter is reached in about the same number of shots. Moreover in Fig. 3-a it is visible the effect of a ten hours Taylor type discharge cleaning at 1 Hz repetition rate. After an immediate increase in the performance of the discharge the diagram shows the same trend of Fig. 3-a.

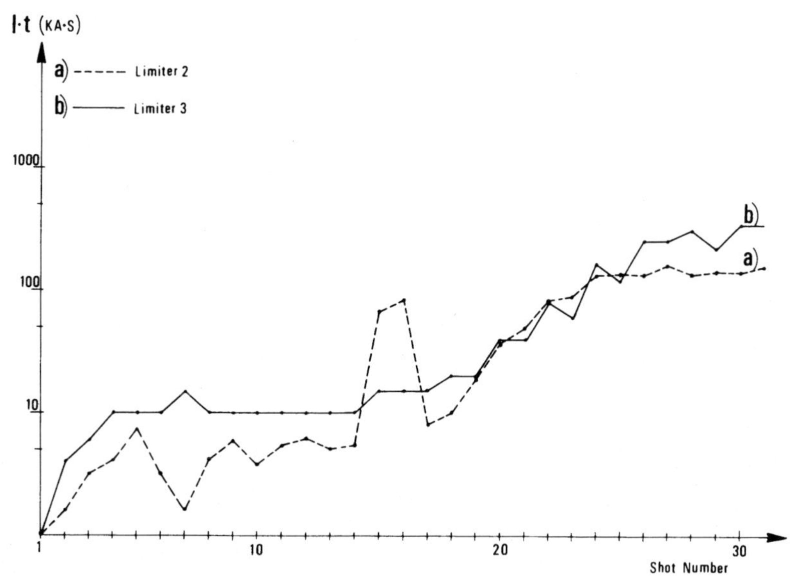

Fig. 3. Conditioning of a new limiter.The arrow indicate the ten hours, 1 Hz, Taylor type discharge cleaning.

4.2. Damage suffered by the limiters

All limiters present macroscopic damage due to heat loads during and at the end of the discharge.

The observation of topographic modification suggests a temptative division of the damages in two phases: 1) during the discharge, 2) end of the discharge.

1) During the discharge. The limiters present modifications in the whole circumference both in the electron and ion side. These modifications start from a minimum (no modification) at the line of tangency of the plasma with the limiter until a depth into the scrape-off region going from 0.4 cm in the upper external region to 1.2 cm in the electron side; in the ion side the minimum is rotated of about 180°.
The molten zones are formed by craters and microscopic droplets are observed for away from the molten zone. This fact suggests that during the discharge the energy is deposited in form of pulses of ionized particles of relatively low energy. In this case the highest temperature occurs just below the surface and crater formation process can be compared with an explosion with ejection of material in form of droplets. The absence of run away of high energy is due to the characteristics of the FT (a high density machine), where only few run away discharges were produced (less than 10%) and was confirmed both by the hard X rays monitor and by the lack of any relevant activation. Only on the limiter N.2 a small activation has been detected (0.2 mR/h). The molybdenum pieces in the limiter N. 1 present even several microcracks and were partly flaked off. These effects, not observed on the stainless steel limiter, can be attributed to the different thermomechanical properties.
No evidence of arcing could be found in the FT machine.

2) End of the discharge. In the FT the control of the plasma position is pro

vided by two coils connected in parallel with the transformer circuit at the start of the discharge. A resistor connected in series provides a rough regulation of the vertical field current. Generally at the end of the discharge the vertical field cannot provide the equilibrium and the plasma hits the inner or the outer part of the limiter. No feedback stabilization is performed. This phase of the discharge interests, from the point of view of damage, a zone situated in the meridian plane and up to an angle of about 30° both in the inner and outer part. In this zone the whole surface facing the plasma is strongly damaged. Large areas show lack of material untill a depth of 0.7 mm. Large drops on both sides of the limiter in the toroidal direction suggest that the removal of material can be attributed to flushing of molten material.
The extension of this region is related to the control of the discharge. The limiter N. 3 shows a stronger damagement in the upper part respect to the other limiters. This is probably due to a shift of the discharge in the vertical direction in the last 50 shots as a consequence of a break down in the transformer circuit.

In our opinion stronger damagements have been prevented by the effect of the massive cryogenic copper shell surrounding the vacuum vessel.

5. CONCLUSION

Some general remarks may be drawn from the analysis of the limiters:

a) Discharge cleaning techniques, which interest surface process, are not effective on the limiter where, due to large temperature gradients are involved bulk properties of the material.
b) The damage is mainly concentrated in the meridian plane. Two limiters in this region, having a depth of 3-4 cm seem to be sufficient to protect the vacuum chamber.
c) A ring limiter having a depth of 1.5 cm is sufficient in the other regions.
d) Composite limiters where the pieces facing the plasma are free to distort seems to be the best mechanical solution.
e) Curvature of the plates of the limiter facing the plasma is useful to try to know the e-folding distance of the energy. The curvature will play an importan role in cooled limiter [4].

REFERENCES

[1] The FT Device, Report 73.14, Associazione EURATOM-CNEN sulla Fusione, Centro di Frascati, Italy (1973)
[2] BOSCHI, A., FERRO, C., ORSINI, A., Tenth Symposium on Fusion Technology, Padova, 4-9 Sept. 1978, pp 855-860
[3] GINOT, P., Private Communication (1970)
[4] CONN, R.W., University of Wisconsin, Report UWFDM-339 (1979)

THERMAL CONSEQUENCES OF PLASMA DISRUPTIONS IN TFTR AND ETF

R. Budny and C. Ludescher
Plasma Physics Laboratory, Princeton University
Princeton, New Jersey 08544

ABSTRACT

We studied thermal responses of first walls for TFTR and ETF during plasma disruptions. To model the flux, we assumed the entire kinetic energy is deposited by axisymmetric horizontal displacement of the plasma. The deposition time is a free parameter. In TFTR, the minimum deposition time which does not cause the toroidal limiter to melt is 7 or 14 ms depending on whether or not the limiter is actively cooled. In ETF, the minimum time which does not cause surface melting of the cooling tubes is 80 ms.

1. INTRODUCTION

Plasma disruptions are phenomena where the plasma rapidly destabilizes or disappears. They can harm tokamaks through several effects. Rapid loss of plasma current generates large image currents that can cause large J x B forces. Rapid transfer of energy heats the first walls and can cause melting, thermal fatigue, and loss of strength.

Many of the phenomena of plasma disruptions are not well understood, and have not been measured quantitatively. For instance, we do not know whether most of the stored particle energy is radiated or convected to the first wall. If it is radiated, the flux on the first wall will be relatively small. Also we do not know much about the distribution of time scales for plasma disruptions. There appears to be a relatively long time scale, on the order 1-10 ms in PDX, for the plasma current to decay. A loop voltage spike appears with a shorter time scale. Also the soft x-ray intensity decreases on a relatively short time scale, on the order 100-300 μs in PLT and in Alcator C. This could indicate either a rapid loss of energy or a readjustment of internal profiles.

We studied the thermal responses of first walls in TFTR and ETF during plasma disruptions to determine their vulnerability. For lack of better models, we used a simple model for the heat flux deposited on the walls. We assume the entire kinetic energy stored in the plasma is deposited uniformly toroidally, as if the plasma were transported horizontally inward or outward. The time for this to occur determines the severity of the consequences. If it occurs rapidly, the deposited energy would not have time to diffuse into the first wall during the deposition, so the surface would heat up. We treated this deposition time as a free parameter.

We can quickly estimate how hot the first wall would get by assuming it is infinitely thick, that it has uniform temperature before the deposition, and that the thermal properties do not vary with temperature. The first assumption is very good for the deposition times of interest, but the other two are not. The prediction is that the temperature rise of the front surface is

$$\Delta T = \frac{2u}{\sqrt{\pi k \rho C \tau_D}} \qquad (1)$$

where u is the deposited energy per unit area, k is the heat conductivity, ρC is the heat capacity per unit volume, and τ_D is the deposition time. This equation indicates how our results scale in u and τ_D.

We used a two-dimensional, finite element heat transfer code [1] to calculate the temperature response and a two-dimensional, finite element stress code [2] to calculate the corresponding stress response. We varied the material properties with temperature. Also, we included the cooling effects of thermal radiation.

2. TFTR

A toroidal limiter will be used in TFTR to absorb heat flux, to protect the vacuum vessel, and to shape the plasma. It will cover the entire inner rim of the vacuum vessel. In one of the conceptual designs, protective tiles will be mounted on stainless steel or Inconel backing plates. Candidates for the tiles are graphite coated with titanium carbide, and copper covered with vanadium. The reasons for using graphite or copper tiles are that they have large heat capacities. The coatings reduce impurities in the plasma.

Plasma discharges will heat the limiter periodically. During the 300 s dwell time between discharges, it will cool. If a fluid is used to actively cool the limiter, it will be relatively cool or at ambient temperature at the start of each discharge. If not, thermal radiation and a small amount of conduction will cool the limiter. Its temperature will ratchet up until the radiation becomes effective. In this case, the hottest part of the limiter will be about 520°C at the start of the discharge.[3]

The temporal distribution of the heat flux on the first wall during non-disruptive operations will be approximately

$$F = \begin{cases} F_o\left(1 - e^{-t/\tau_E}\right) & t < \tau_{NB} \\ F_o\, e^{-\left(t - \tau_{NB}\right)/\tau_E} & t > \tau_{NB} \end{cases} \qquad (2)$$

with τ_{NB} the neutral beam duration (1.5 s), and τ_E the energy confinement time. The highest anticipated flux is $F_o = 1\ \mathrm{kW/cm^2}$. The particle energy stored in the plasma is proportional to τ_E. We chose a relatively high value, 0.3 s to be pessimistic about the amount of energy stored in the plasma, and thus deposited during disruptions. The flux is plotted in Fig. 1.

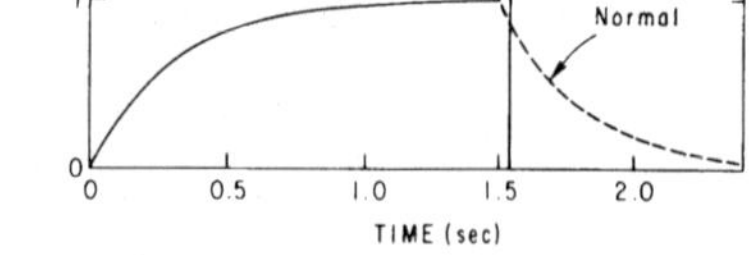

Fig. 1. Schematic of first-wall fluxes resulting from normal and disruptive plasmas.

The worst time for the disruptions to occur would be at the end of a neutral beam pulse, which would have heated up the limiter. We assume the temporal distribution is constant, as shown in Fig. 1., and that the disruption deposits $u = 150\ \mathrm{J/cm^2}$ on the hottest tiles. This would correspond to a stored energy U = 16 MJ being deposited uniformly on an area $A = 11\ \mathrm{m^2}$. U is approximately the maximum kinetic energy of the plasma. A is the area of the limiter which would be intercepted if a plasma with minor radius 85 cm collapsed horizontally inward towards the vacuum vessel centerline.

We calculated the minimum τ_D which does not melt the protective tiles. The melting temperatures are copper: 1083°C, vanadium: 1919, and titanium carbide: 3150. The graphite sublimes at higher temperatures. If the limiter is actively cooled sufficiently so that it starts the pulse at 50°C, then either design can withstand τ_D as low as 7 ms. This would correspond to fluxes of u/τ_D = 21 kW/cm^2. With lower τ_D the vanadium or titanium carbide would reach their melting temperatures. Temperature histories are shown in Figs. 2 and 3.

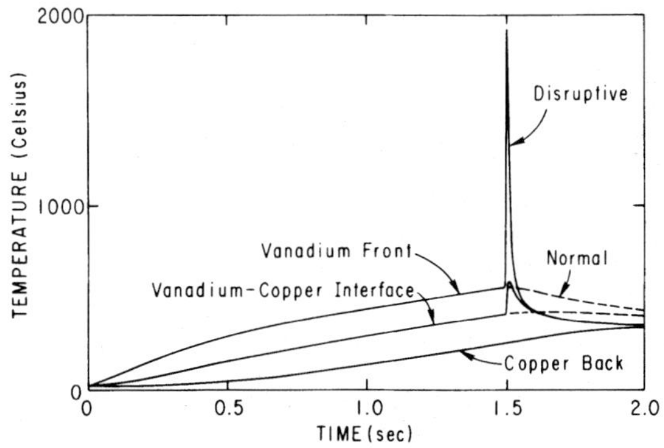

Fig. 2. Temperature histories for copper-vanadium tiles exposed to fluxes from normal and disruptive plasmas with τ_D = 7 ms.

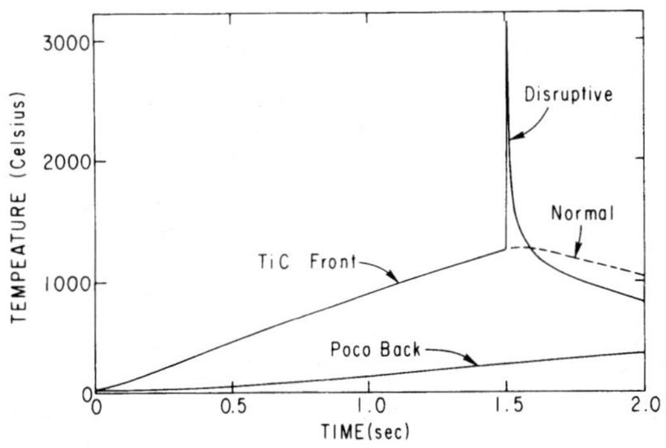

Fig. 3. Temperature histories for Poco graphite tiles exposed to fluxes from normal and disruptive plasmas with τ_D = 7 ms.

If only thermal radiation cools the limiter, it will heat up from shot to shot. In this case, there would be less reserve heat capacity, so τ_D must be higher to avoid melting. The minimum values are 12 ms for copper-vanadium and 14 ms for graphite-titanium carbide tiles. Results are in Figs. 4 and 5.

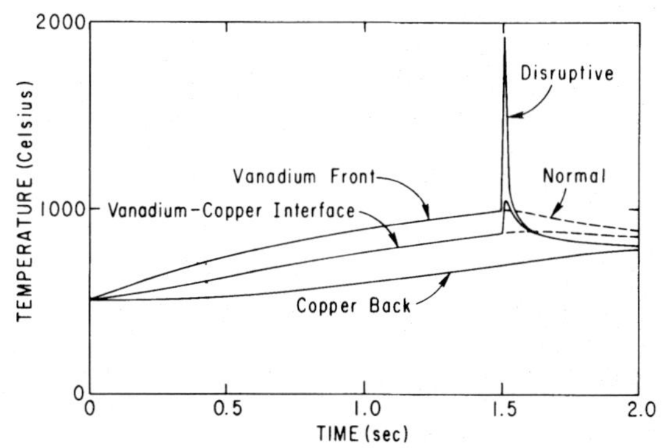

Fig. 4. Temperature histories for hot copper-vandium tiles with τ_D = 12 ms.

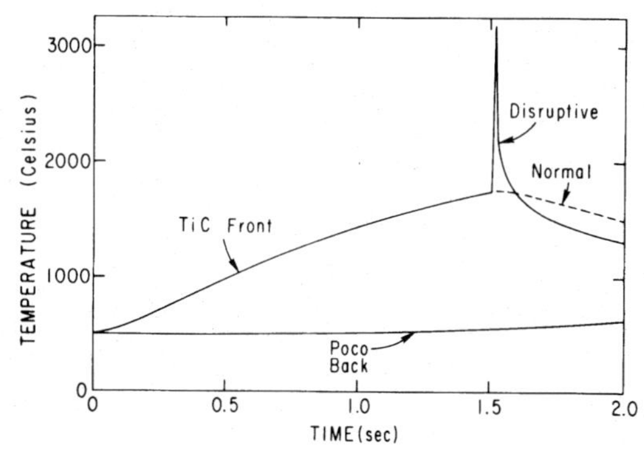

Fig. 5. Temperature histories for hot Poco graphite tiles with τ_D = 14 ms.

We also calculated stresses in graphite tiles assuming the titanium carbide coat is sufficiently thin to not effect the stresses. Figure 6 shows the nodal model for the top half of a tile. Figure 7 shows the stresses when the tile is unconstrained and is cool at the start of the discharge. The stress normal to the front surface reaches - 90 MPa during the disruption. The stress midway through reaches 18 MPa. The magnitudes of these are less than the ultimate compressive and tensile strengths respectively of Poco graphite. The stress in the back face is given also.

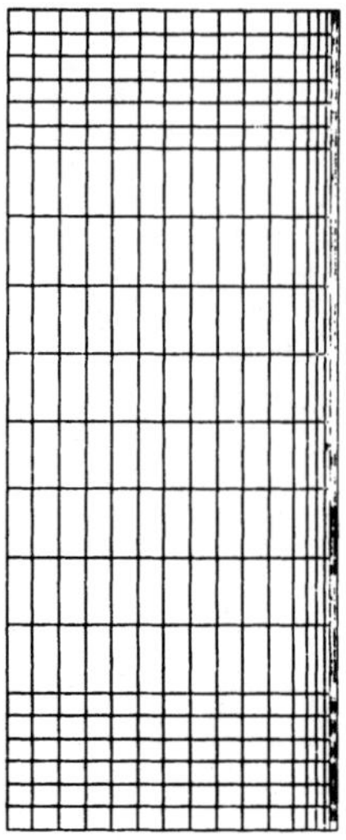

Fig. 6. Mesh used to model half of a Poco graphite tile. The flux is incident on the right.

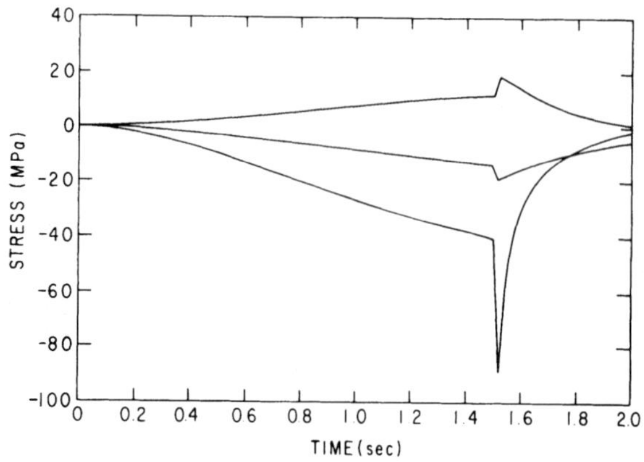

Fig. 7. Stress histories for Poco graphite tiles corresponding to the temperatures in Fig. 3.

3. <u>ETF</u>

In one of the designs for ETF, part of the first wall would consist of stainless steel cooling tubes with radius 1.27 cm and thickness 0.3 cm. The anticipated flux from normal plasmas is 100 W/cm^2. The pulse length will be long, on the order of 100 s, so the temperatures of the tubes will be at their equilibrium values for most of the pulse.

We assumed that the coolant in the tubes removes heat at the rate

$$F = h\ (T - T_c) \qquad (3)$$

with the heat transfer coefficient, h, being 1 $W/cm^2/^oC$, and the temperature of the coolant, T_c, being 50^oC. This implies that the peak temperature of the tubes will be 415^oC.

We assumed the disruption deposits $u = 300\ J/cm^2$. This corresponds to a stored energy U = 200 MJ in a plasma of minor radius 130 cm being deposited by horizontal outward transport onto the cooling tubes at a major radius of 413 cm. We used the nodal model shown in Fig. 8 to calculate the thermal and stress response. The minimum τ_D which does not melt the surface of the tube is 80 ms. Figure 9 gives thermal results at the hottest front surface, at a depth of 0.2 mm below, and at the inside surface facing the coolant.

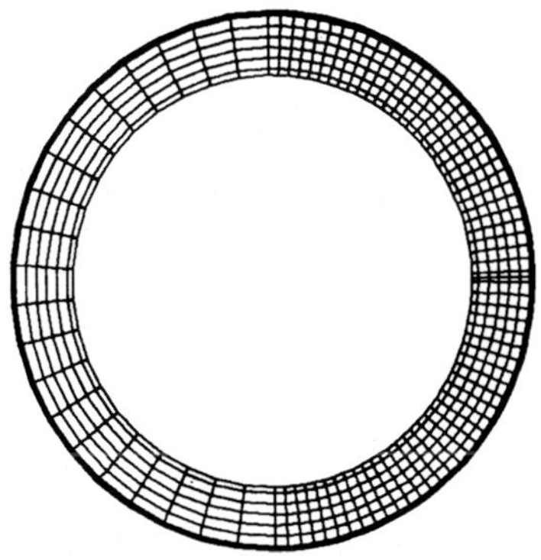

Fig. 8. Mesh used to model an ETF cooling tube. The flux is incident on the right.

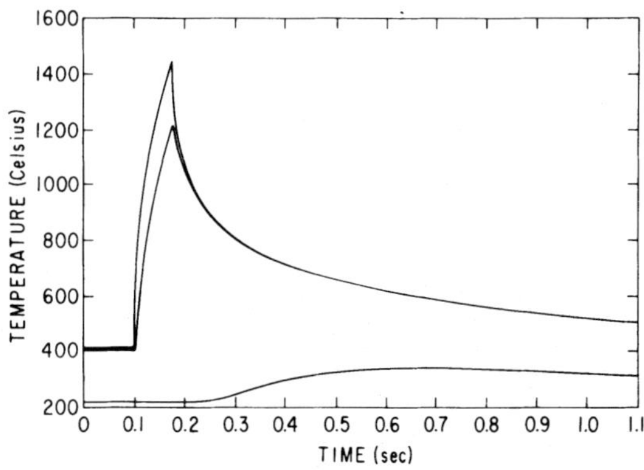

Fig. 9. Temperature histories for ETF cooling tubes exposed to disruptive plasmas with τ_D = 80 ms.

4. ACKNOWLEDGMENTS

We wish to thank J. Cecchi, J. Schmidt, M. Ulrickson, and S. Yoshikawa for helpful discussions. This work was supported by the U.S. Department of Energy Contract No. DE-AC02-76-CHO-3073.

5. REFERENCES

[1] MASON, W.E., Lawrence Livermore Laboratory Report UCID-17980, Rev. 1 (1980).

[2] HALLQUIST, J.D., Lawrence Livermore Laboratory UCRL-52678 (1979).

[3] BUDNY, R., J. Nucl. Mater. 93-94 (1980).

ESTIMATES ON A HELICAL MAGNETIC LIMITER APPLIED TO INTOR

W. Feneberg[+], G. Fuchs and G.H. Wolf

Institut für Plasmaphysik der Kernforschungsanlage Jülich GmbH
Association EURATOM-KFA, D-5170 Jülich, Fed. Rep. of Germany
+ Max-Planck-Institut für Plasmaphysik, Ass. EURATOM/IPP
D-8046 Garching b.München, Federal Republic of Germany

ABSTRACT

A helical magnetic limiter is employed to support a dense and cool plasma boundary layer which surrounds the outermost closed magnetic surface and which, via convection cooling, transmits the energy losses from the plasma to the first wall such as to improve (i) the equidistribution of the heat load, (ii) the reduction of impurities and (iii) the helium exhaust. Characteristic features of the concept of such a limiter applied to INTOR are described and some typical properties of the expected boundary layer are given. From these result estimates on the maximum heat load of the first wall and of the particles fluxes to the wall and the required pumping area.

1. INTRODUCTION

For a save decoupling of a fraction of the fusion energy to the first wall, for H_e - ash removal and for maintaining a clean plasma it is desirable
1. to have a low plasma temperature in boundary layer (low sputtering)
2. to destribute the power equally so as to avoid lokal overheating of limiters, scrape-off-limiters [1], or divertor end-plates.

Methods that have been proposed in order to achieve these goals are the cold gas blanket, radiation cooling and field line ergodization in the boundary region [2]. The latter is considered in this paper, whereby the desired magnetic field line pattern is created by a resonant helical multipole field (Fig. 1). Due to the small field perturbation the field lines have a component normal to the wall (angle $\varepsilon \ll 1$) which leads to a convective outflow with sound speed v_s. This allows to remove the energy at low temperature and high density.

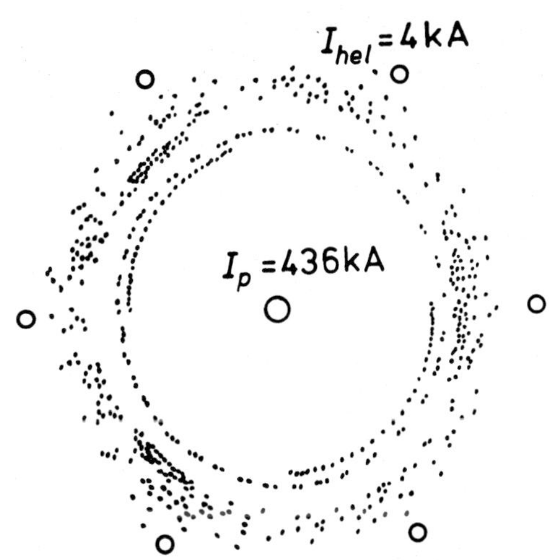

Fig.1: Example for field line ergodization produced by helical windings wound with constant pitch angle for the mode number m = 6, n = 2, onto a tokamak with R_o = 2,1 m r_L = 0,4 m, q_L = 3,5. The thickness of the ergodized layer inside r_L is $\Delta r \approx$ 7 cm. For different tokamak sizes the figure is similar for the same q and similar currents.

2. PARTICLE AND ENERGY EXHAUST CONSIDERATIONS

A fraction (20 %) of the fusion energy E_f is given to α-particles E_α and a fraction f_α of the α-energy is transported by charged particles to a wall area A which contacts the plasma. The power balance equation at this area is

$$\dot{N}_\alpha \; f_\alpha \; E_\alpha = \gamma \; j_W \; T_W \; A \tag{1}$$

where $\dot{N}_\alpha$, γ, j_W, T_W are the α-particle production rate, an energy coupling factor [3] depending e.g. on the sheath potential, the particle flux normal to A and the plasma temperature of the boundary layer respectively.

The α- particle balance at the wall is

$$\dot{N}_\alpha = j_W \, c_\alpha \, f_p \, A \tag{2}$$

where c_α is the α-particle concentration within the boundary layer and f_p is the fraction of the neutral He which can be pumped.

Due to the magnetic field disturbance the normal flux j_W is related to the convective flux along field lines.

$$j_W = n_W \, \varepsilon \, v_s \tag{3}$$

where n_W is the plasma density at the wall, its value has to be related to the core density n_C. To find this relation we have to look for the particle continuity near the wall in detail. For the ions this gives

$$\varepsilon \, v_i \frac{dn_i}{dx} = n_e n_o \langle \sigma_i v_e \rangle \tag{4}$$

and for neutrals

$$v_o \frac{dn_o}{dx} = -n_e n_o \langle \sigma_i v_e \rangle \tag{5}$$

We assume full recycling, $n_e = n_i$, and $\langle \sigma_i v_e \rangle$ the ionization rate coefficient (x is the distance from the wall). A detailed investigation had to include momentum transfer. We make the simplifying assumption $v_o = v_i$ = const., which is a pessimistic estimate.
Solving equ. (4) and (5) leads to

$$n_W = 0.5 \, n_C \tag{6}$$

Equ. (1) to (3) can be solved for

$$kT_W = \left(\frac{\dot{N}_\alpha f_\alpha E_\alpha}{A \gamma \varepsilon n_W}\right)^{2/3} \left(\frac{M}{2}\right)^{1/3} \tag{7}$$

M = Ion mass

and

$$f_p = \frac{1}{c_\alpha} \left(\frac{\dot{N}_\alpha}{\varepsilon n_W A}\right)^{2/3} \left(\frac{M \gamma}{2 f_\alpha E_\alpha}\right)^{1/3} \tag{8}$$

These functions are shown in figures 2 and 3.

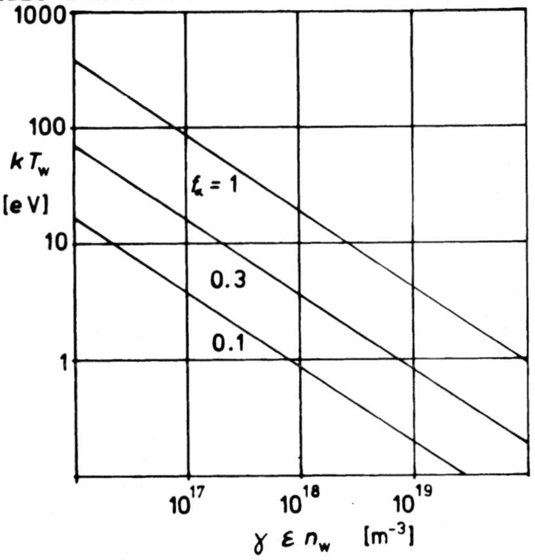

Fig. 2: Dependence of the plasma edge temperature on the edge density according equ. (7) for a given average heat flux of $q_w = .31$ W/m^2 corresponding to $\dot{N}_\alpha = 2 \cdot 10^{22}\ s^{-1}$ in INTOR

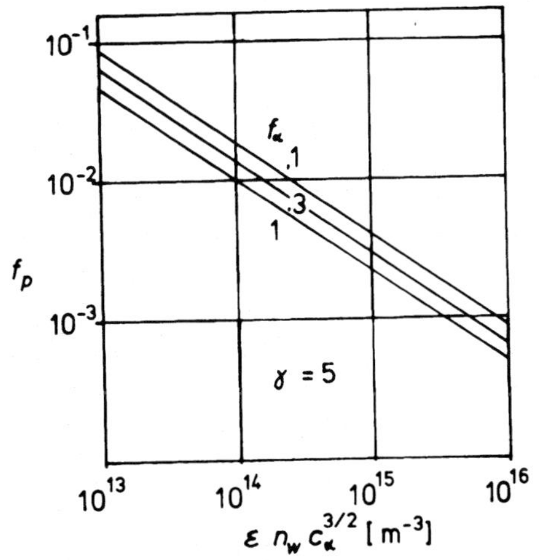

Fig. 3: Required pumping fraction according equ. (8).

3. DISCUSSION

To avoid self-sputtering the impinging impurity ion energy should be (depending on the atomic number) below 50 eV [4]. If the sheath voltage can be kept low [3], an edge temperature of 30 eV or below is desirable. This corresbonds to $\gamma \varepsilon n_w = 6 \cdot 10^{17}\ m^{-3}$ for the worst case $f_\alpha = 1$. By tracing a field line within the ergodized region (cf. fig. 1) we find $\varepsilon \approx 1{,}5 \cdot 10^{-3}$. Using $\gamma = 5$ a scrape off density of $8 \cdot 10^{19}\ m^{-3}$ is required. The value $\varepsilon n_w c_\alpha^{3/2}$ is $1{,}3 \cdot 10^{-15}\ m^{-3}$ for $c_\alpha = 5 \cdot 10^{-2}$ and the required pumping fraction $f_p = 3 \cdot 10^{-3}$ (cf. fig. 3). The value of f_p is already considerably reduced due to backstreaming from the pumpholes (factor 10 [5]), c_α is limited because otherwise the fusion reaction is extinguished due to fuel dilution, from this we conclude that a fraction of at least 3 % of the INTOR surface has to be devoted for pumping.

4. REFERENCES

1 Bieger, W.; Dippel, K.H.; Fuchs, G.; Wolf, G.H. Int. Symp. Plasma-Wall-Interaction, Jülich, 18-22 October, 1976. Proceedings P. 609. Pergamon

2 Feneberg, W. VIII European Conference on Controlled Fusion and Plasma Physics, Prague 1977.

3 Harbour, P.J.; Harrison, M.F.A.; Sanderson, A.D.; Stott, P.E.; Plasma Physics and Controlled Nuclear Fusion, Res. III, S. 431, Vienna (1979)

4 Sengoku, S.; Azumi, M.; Matsumoto, Y.; Maeda, H.; Shimomura, Y.; JAERI - memo 8131.

5 Fuchs, G.; Goetze, H.G.; Reiter, D.; This issue

HIGH POWER RF SYSTEMS OF RFC-XX

T. Aoki, K. Adati, Y. Ohkubo, T. Watari,
A. Miyahara and RFC group

Institute of Plasma Physics, Nagoya University,
Chikusa-ku, Nagoya 464, Japan

ABSTRACT

The rf system of RFC-XX is presented: Six rf systems of 1 MW level and of 7-85 MHz in frequency are built and operated simultaneously, overcoming parasitic oscillations, interferences and breakdown problems. A pair of rotating type III coils and the power feeding system for them are described.

INTRODUCTION

RFC-XX is an open ended machine which has two spindle cusps and a central cell between them. The plasma is heated by ICRH in the central cell and is plugged by rf plugging both at the two line cusps and two point cusps [1]. The magnetic field strengths are 21 kG at the line cusp and 39 kG at the point cusp and 10 kG in the central cell. Hydrogen plasma is used for the experiment. Both ion and electron temepratures are a few tens of eV before the application of rf field and the plasma density ranges 10^{11} - 10^{13} cm^{-3} at the zero magnetic field region in the cusp.

Six rf oscillators are built (fig. 1), the specifications of which are summarized in table 1. In the previous paper we reported on the oscillator and the matching ciruits for the line cusp [2]. In this paper, are discussed suppression of parasitic oscillations, a simultaneous operation of the six high power oscillators (§ 1), troubles of breakdown to which we often faced at various places in the high power circuits and the methods we found to care them (§ 2). In designing the rf launcher, we considered two things: First, by selecting the mode rotating in ion sense, we can get high heating efficiency [3]. Second, taking into account that the radial wave number is determined by the axial wave number through the dispersion relation, we can select favourable radial modes by choosing the phase relation of axially different two points. A pair of rotating type III coils and power feeding circuits for them are designed and tested to fulfil these requirements.

TABLE I

System	frequency range	output power	purpose
#1	20 - 45 MHz	1300 kW	line cusp plugging
#2	"	"	"
#3	40 - 85	700	point cusp plugging
#4	"	"	"
#5	7 - 15	400	ICRH
#6	"	"	"

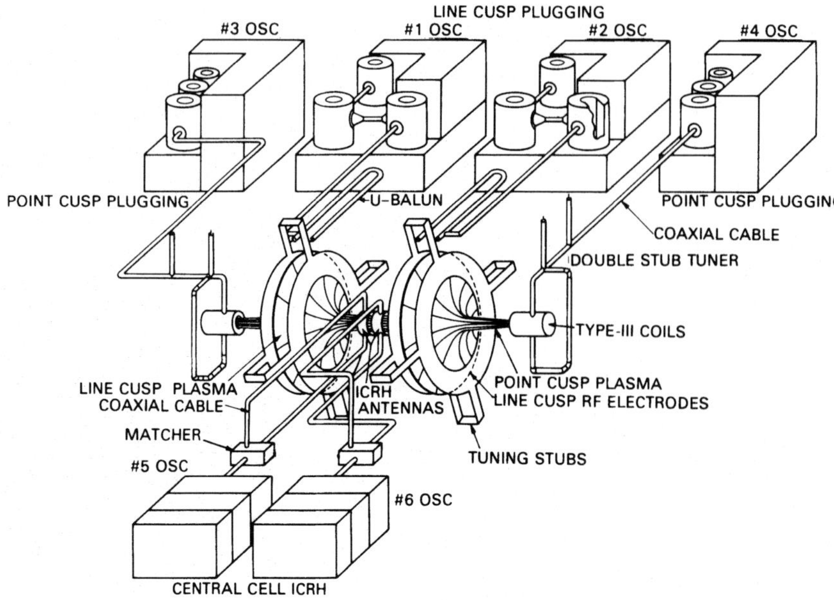

Fig.1. Whole rf system of RFC-XX

1. CONSTRUCTION OF FREQUENCY CHANGEABLE MW RF OSCILLATORS

In a frequency changeable high power rf oscillator, parasitic oscillations are apt to be caused by stray capacitances and stray inductances of tuning circuits. Grid grounded circuits are effective to prevent the parasitic oscillations. A tuning circuit composed of a short-circuited coaxial line and variable capacitors is also effective to prevent the parasitic oscillations in the plate resonant circuit.

In rf systems for point cusps (fig. 2), the control grid of TH116, a triode in the main amplifier, that of 4CW50,000E, a tetrode in the driver amplifier, and of 8F67R, tetrode in the exciter amplifier, are all grounded directly. This system requires the grid bias power supplies to have lower impedance compared to the usual one, since in this system the plate dc circuit current flows through the bias circuit. The screen grid of each tetrode is grounded through a capacitor which is formed around the socket of the vacuum tube with a metal disk and a teflon disk piled on the ground plane. With the use of above mentioned grounding method, we could build the oscillators for point cusp, their frequency range being 40 MHz-85 MHz, output power 700 kW and pluse width 100 ms.

In our system a signal of mW level from the original oscillator is amplified through multistage amplifiers to MW level, which allows easy handling of wavefrom and phase. But we lead to experience unexpected self oscillations and blocking oscillations, which may be caused by feedbacks of leakage rf field from the higher power level circuits to the lower level circuits. The original oscillators and amplifiers of lower than 10 W level are placed in a shielded room separated far apart from the high power

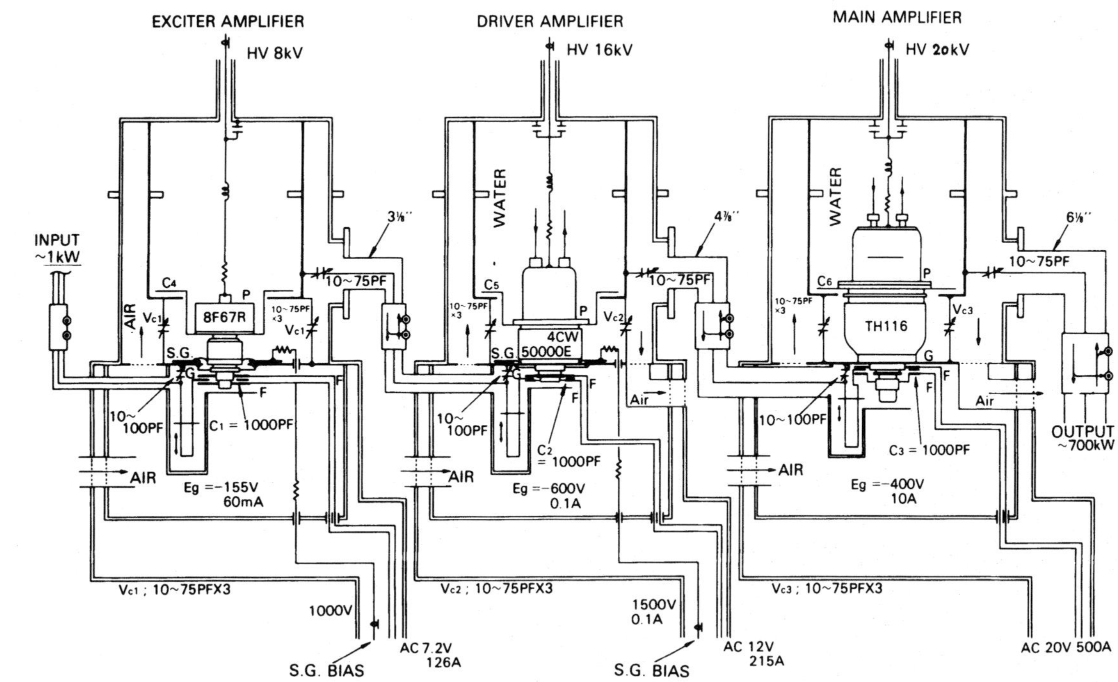

Fig.2. Schematic diagram of the point cusp power amplifier

section. The power of 10 W level is sent to the high power section through a semirigid coaxial cable which has very low transfer impedance and suffers little induction disturbances. Interferences among six systems are also overcome by this system.

2. PROTECTION AGAINST RF BREAKDOWN

The applied voltage to the plate-voltage blocking capacitor of the main amplifier in rf oscillator for point cusp (C_6 in fig. 2) is a superposition of dc 20 kV and rf 34 kV peak to peak in rated operation. At the first set up, breakdowns happened at the launcher coil or at the transmission line in operation of 400 kW level. Once a breakdown happens, the output power is reflected to give much higher voltage across the capacitor and breakdown it. This can also happens if the plasma changes to cause mismatching the circuit.

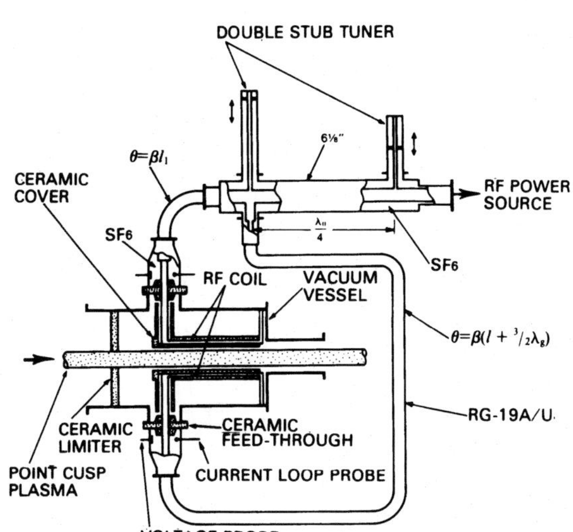

Fig.3. Matching circuit of point cusp

The launcher coil and the feed-through are encased in ceramic sheathers to protect from direct contact with dense plasma. For the matching circuit of point cusp rf

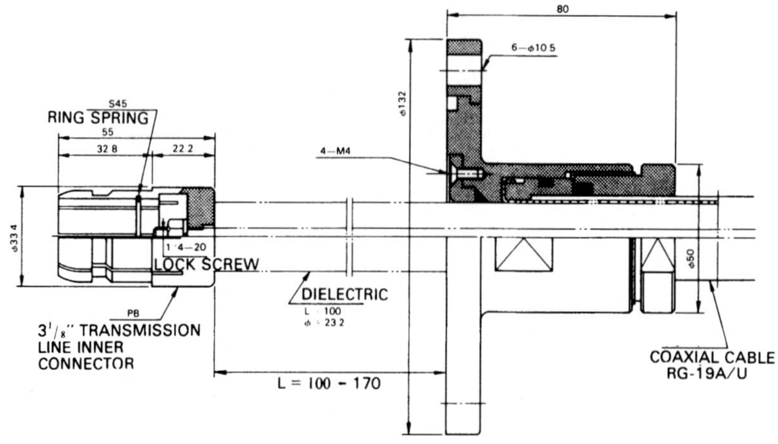

Fig.4. Structure of RG-19A/U cable connector

system, standard double stub tuners composed of $6\frac{1}{8}$" transmission line and RG-19A/U coaxial cable are used (fig. 3). The transmission lines and feed-through sections are filled with SF_6 gas of 4 kg/cm^2 in pressure. A special connector for RG-19A/U coaxial cable (fig. 4) are designed. The outer conductor is removed so as to get a distance of 10 - 17 cm between the connector for the outer conductor and that of inner conductor along the surface of dielectric (polystyrene). The dielectric of the cable is steeped in glycerine at the point where it contacts with connector of the outer conductor. The high dielectric permiability of glycerine decreases the concentration of electric lines of force. In order to increase the surface flash over voltage, the plate-voltage blocking capacitor is covered by RTV silicone rubber (room temperature valcanizing).

By these improvement, we could apply a peak rf voltage of 22 kV across each element coil of the launcher measured by the voltage probe shown in fig. 3. The plate-voltage blocking capacitor is expected to endure a peak voltage of 60 kV.

3. ROTATING TYPE III COILS AND THEIR POWER FEEDING SYSTEM

Two type III coils are combined to form a rotating type III coil by setting up at the same axial position with an azymuthal separation of 90° (fig. 5). A type III coil generates $|m| = 1$ linearly polarized wave, while the rotating type III coil can generate $m = +1$ or $m = -1$ circulary polarized wave if the amplitude and the phase of currents to each coil element are properly adjusted. A two stage power divider is prepared for this purpose (fig. 6). By the first stage, the output from the oscillator is divided into two equiamplitude, equiphase signals. One of them is divided again, after passing through a coaxial cable of length l, into two equiamplitude, dephase signals by the second stage, and the other also, after passing through a coaxial cable of length $l + \frac{1}{4}\lambda_g$, where λ_g is a wave length in the coaxial cable.

Two rotating type III coils are placed with an axial separation of 63 cm. By adjusting the phase difference δ between the input currents to the coils, we can expect to have an optimum efficiency of wave excitation. (Similar experiment has successfully been done with two type III coils [1].) The adjustment is performed with a use of a voltage controlled phase shifter inserted just after the original oscillator (fig. 6).

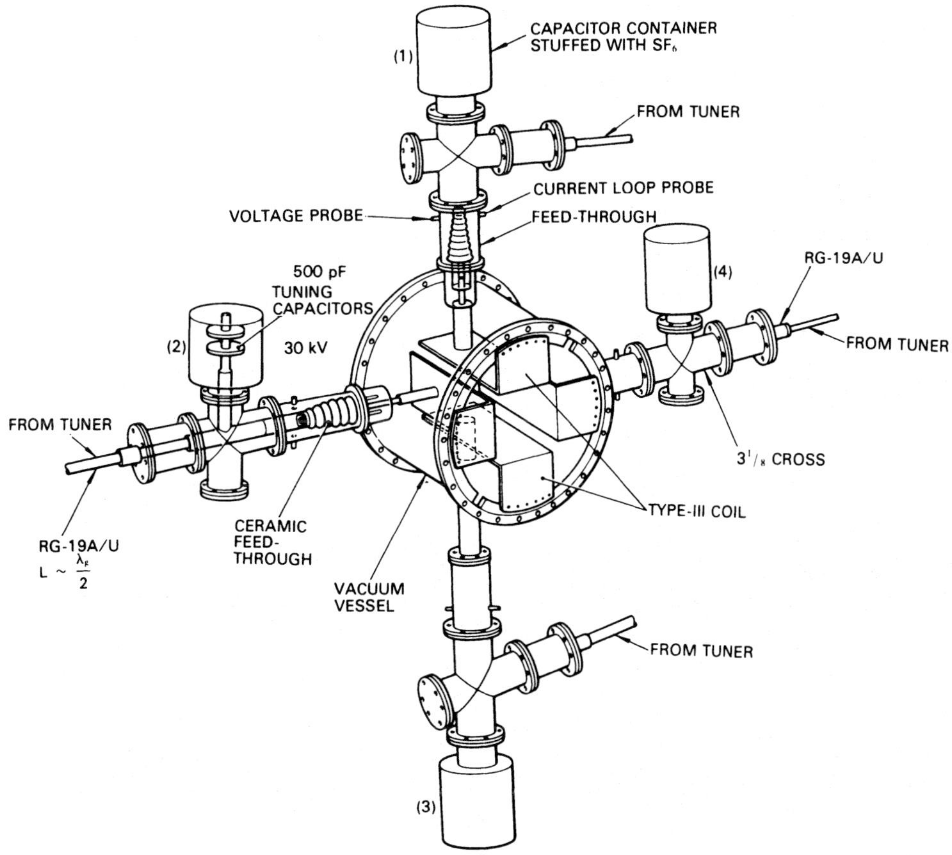

Fig.5. Structure of rotating type III coil

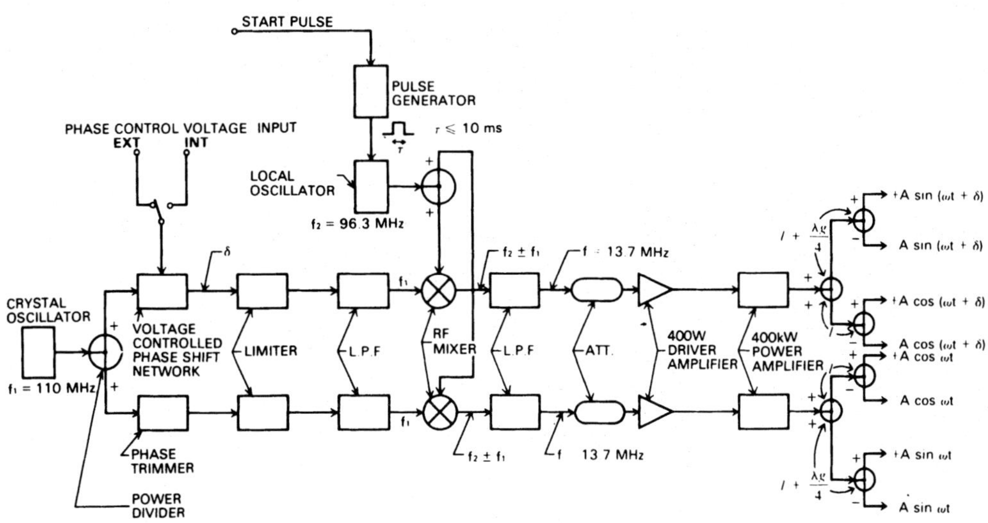

Fig.6. Schematic diagram of the ICRH power source
(for the rotating type III coils)

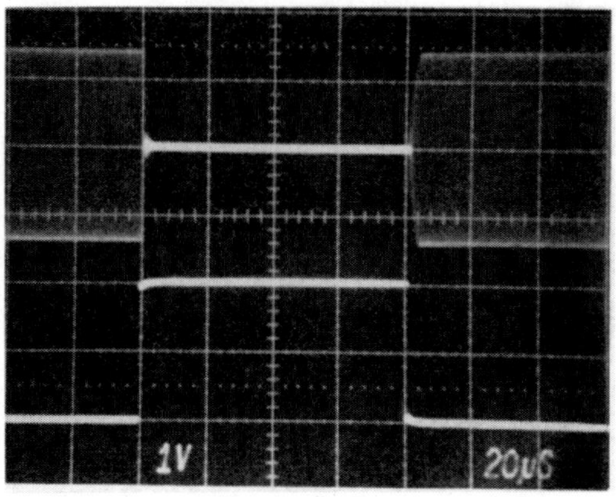

Fig.7. Time response of voltage controlled phase shifter

Typical time response of the voltage controlled phase shifter is 100 degree/μs and voltage sensitivity is 45 degree/volt. A time response is demonstrated in fig. 7. The lower trace shows the control voltage wave form applied to the phase shifter. Zero volt gives 0° phase shift and 4 volts gives 180° phase shift. The upper trace shows a supperposition of input signals to the two 400 W driver amplifier (fig. 6), where one signal has passed through the voltage controlled phase shifter and the other has not. In order to get fast response and wide range of phase adjustment, the voltage controlled phase shifter is composed of six element phase shifters each of which includes a parallel-resonant circuit loaded with a varactor diode. The phase shifter network operates in fixed frequency of 110 MHz. Output of 5-25 MHz in frequency is derived by a heterodyne method, where the 110 MHz signal is mixed with a signal of 105-85 MHz generated by a local oscillator.

As for the power measurements, directional couplers are used. We can estimate the power sent from the launcher into the plasma, P, a part of which may be used to heat the plasma and the rest may be used somewhere, in the wall of the vacuum vessel for example. A diamagnetic loop is placed in the central cell. Assuming that the diamagnetic signal, D, is proportional to the total thermal energy of the confined plasma, we obtain an apparent energy confinement time $\eta\tau_E$ from D/P. By using a pair of rotating type III coils as a launcher, we obtained $\eta\tau_E$ 1.5 times larger compared to the value obtained by a pair of type III coils in a power level of P $\sim$ 100 kW.

ACKNOWLEDGEMENT

The authors wish to acknowledge their indebtedness to Mr. Y. Kubota for his assistance in designing short-circuited resonant coaxial lines of the high power amplifiers for point cusp.

REFERENCES

[1] Adati, K., Watari, T., Aoki, T. et al., 8th International Conference on Plasma Physics and Controlled Nuclear Fusion Research, Brussels, 1980 (IAEA-CN-38/F-5).

[2] Aoki, T. et al. Proc. of 10th Symposium of Fusion Technology, 1978, p.277.

[3] Yasaka, Y. and Itatani, R., Proc. of Joint Varenna-Grenoble International Symposium, Grenoble, 1978, p.139.

TECHNICAL ASPECTS OF A GRILL COUPLING SYSTEM FOR LOWER HYBRID HEATING IN THE WEGA TOKAMAK*

M. LIPA, R. MAGNE, T.K. NGUYEN, G. TONON

ASSOCIATION EURATOM-CEA
Département de Recherches sur la Fusion Contrôlée
Service IGn - Centre d'Etudes Nucléaires
85 X - 38041 GRENOBLE CEDEX (France)

Abstract

The object of the lower hybrid heating experiment in WEGA is to heat deuterium plasmas of peak density $6.10^{13}cm^{-3}$ at a field of 2.5 T, corresponding to a frequency of 800 MHz, at powers largely exceeding the ohmic heating power. The grill and the emitter are therefore designed to couple 800 KW for 40 ms. In order to optimize the power spectrum transferred to the ions, a 4-waveguide grill, each 35 mm wide, was chosen. This necessitated construction of a new modular stainless steel chamber and a special racetrack coil, which increases the magnetic field ripple to 9 % at the limiter radius. In order to minimize reflection by the plasma, the entire grill is movable to permit adjustment of the plasma parameters at the grill mouth. Using the calculated field reflection coefficient 0.5 for this condition, the electric field in this section is 3,5 KV/cm. To avoid breakdown, the grill must be pumped down from the filling pressure (10^{-4} torr) in the 50 ms preceding application of the HF, which is shown not to be possible over the 3 m length of the grill by pumping of the plasma alone. Therefore, a system of double windows and differential pumping was chosen. The line is of modular stainless steel construction and includes a wave transformer to go from the reduced to standard dimensions.

The generator consists of 4 high-power klystron amplifiers with individual floating deck modulation. Regulation of the beam current using this system should permit regulation of the output power to about 5 %. The generator is powered by a 60 KV, 60 A, 10 s power supply. The elements of the HF system are being tested.

1. Introduction

To achieve ignition in a fusion reactor device additionned heating will be necessary. To use RF at the lower hybrid frequency seems to be one of the possibilities with the advantage that the part of the coupling structure near the plasma is simple, and contains no materials other than those of the torus. The object of the LH heating experiment in WEGA III [1] -a joint experiment of Max-Planck-Institut für Plasmaphysik Garching and the CEN-Grenoble- is to heat deuterium plasmas of peak density $6.10^{13}cm^{-3}$ at a field of 2.5 T corresponding to a frequency of 800 MHz at powers largely exceeding the ohmic heating power.

In preceding experiments, WEGA I + II [2] [3] as a Tokamak and stellarator [4] device, the RF power was coupled to the plasma by two loop antennas fed with opposing currents.

WEGA III is a Tokamak using a four waveguide grill coupler system.

2. Machine

To achieve the desired wave spectrum, the RF ports in the torus have to be situated sufficiently close to each other. This necessitated construction of a new modular stainless

* a joint experiment of Max-Planck-Institut für Plasmaphysik-Garching and CEA-Grenoble.

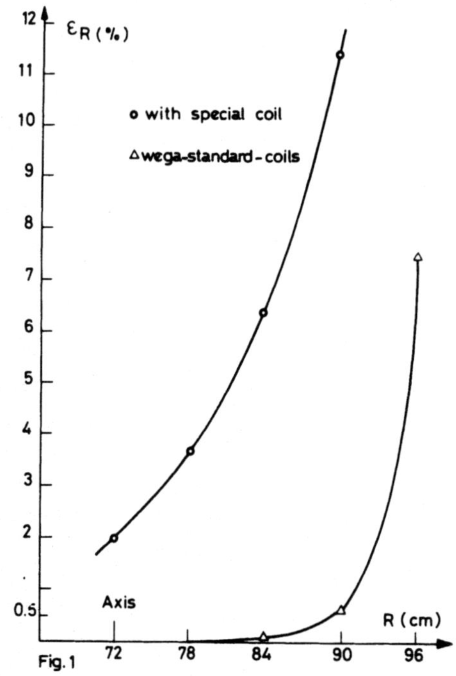

Fig. 1
Ripple of main magnetic field as a function of machine radius with WEGA-standard and special coils.

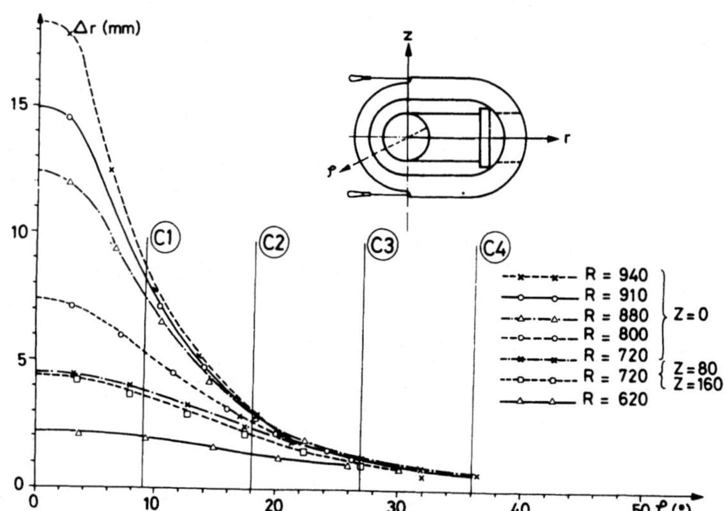

Fig. 2 Variation of magnetic lines near the special field coil.

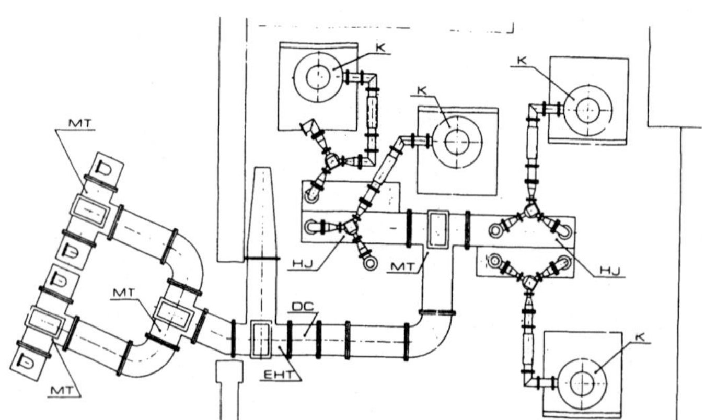

FIG.3 The 4 klystron emitter and the power transport line Implantation scheme

steel chamber.
To permit a max. aperture of the ports the angle (15°) between the ports and the shape of a new magnetic field coil were optimized together. This special race-track coil increases the "local" magnetic ripple to 9 % at the limiter radius [FIG.1]. The field lines near the special coil are calculated [5] and show a radial deviation of 13 mm [FIG.2] at the limiter magnetic surface, thereby necessitating the possibility of grill adjustment, necessary in any case to vary the distance grill-plasma.

3. Emitter

The grill and the emitter are designed to couple 800 KW for 40 msec. The generator consists of four high power klystron amplifiers, type VA 948P, powered by a 60 KV, 60 A, 10 sec power supply. The high voltage is applied via four series ballast tubes (Tetrode EIMAC Y 546) permitting simultaneous switching and regulation of the individual beam currents. The perveance of the klystrons is about 2.10^{-6} $A.V^{-3/2}$. For this perveance the efficiency is in the order of 40 percent and each of them generate 200 KW RF power at 36 KV and 14 A beam current. When tuned to an 8 MHz bandwidth, the klystrons have a gain close to 50 dB which necessites an input power of 2 watts. All of the tubes are installed vertically with the collector on the top. The electrodes (voltage between anode and cathode up to 37 KV) are immersed in oil. The body and the collector of the klystrons are water cooled by means of a heat exchanger. The output of each klystron is protected against a high voltage standing wave ratio by a circulator (insertion loss = - 0.46 dB).

4. Line

For the power transport line, standard waveguides, type WR 1150 are used [FIG.3]. The klystrons are connected to groups of two using two hybrid junctions followed by a magic-T, thereby collecting the power in a single waveguide (two bidirectional couplers allow the measurement of incident and reflected powers). This is done to permit matching of the load (plasma) to the emitter, using a single E-H tuner. Three further magic-T's re-divide the RF power into four equivalent channels with the same phase which are then connected to the standard waveguides feeding the grill by 3 1/8" coaxial cables which are pressurized with SF6. The transition to coax at this junction emitter-grill was chosen for easy assembly.

5. Coupling system

To achieve coupling to the lower hybrid wave in the plasma, a slow wave with $\vec{E}$ parallel to $\vec{B}$ must be excited. This determines the width, 35 mm, of the waveguides at the grill mouth. The waveguide spacing is 49 mm and they are driven out of phase, giving the desired spectrum [6] in parallel wavenumber [FIG.4]. The transition from the grill coupler, situated in the RF torus port, to the standard waveguide consists of the following elements [FIG.5].

- I Grill coupler
- II Intermediate waveguide
- III Wave transformer
- IV Spacer

The entire grill system is a double waveguide in pre burnished stainless steel welded construction. The grill mouth is geometrically adapted to the plasma-radius.
The intermediate waveguide is equipped for RF measurements by the provision of five regularly spaced holes of 15 mm diameter.
The wave transformer permits the transition in three steps from the grill-coupler dimensions (35 × 244 mm) to the standard waveguide (146 × 292 mm). The number, spacing, and dimensions of

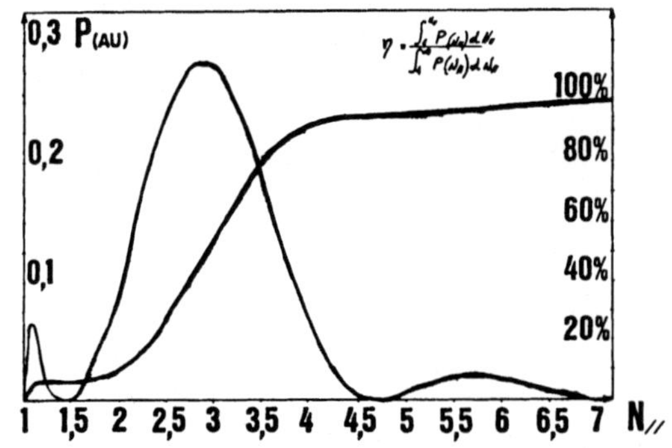

FIG. 4 **Transmitted power as a function of the parallel wave number**

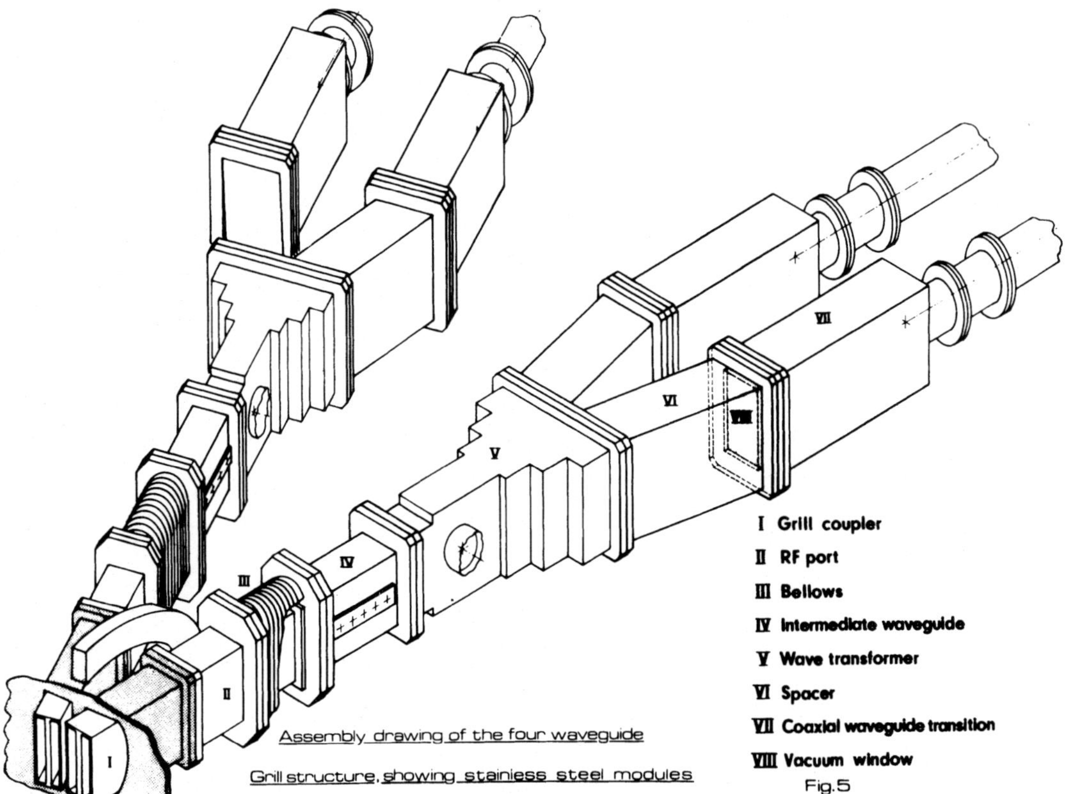

Fig.5

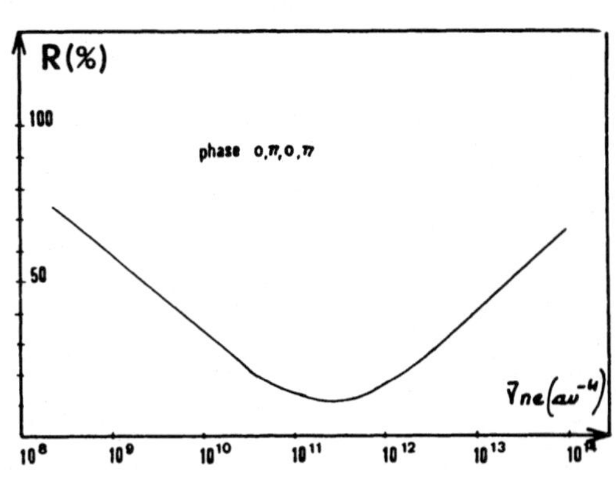

FIG.6 **Reflection coefficient as a function of the density gradient**

these steps were calculated to match the transformer at 800 MHz and minimize the local electric field. Pumping ports are provided on each side of the waveguides.
The spacer waveguide in standard size of 146 × 292 mm is closed by a resonant teflon window, which is calculated to conserve the characteristic impedance of the waveguides. This interchangeable vacuum window (thickness 30 mm) can be replaced by windows using other dielectrics. All the elements mentioned above, (except the teflon window) are metal sealed and bakeable up to 120°C. To achieve optimal RF coupling [FIG.6] the grill is supported on a bearing system and movable in the radial direction (75 mm) ; stainless steel bellows between the torus port and the intermediate waveguide are used.
The grill must be pumped down from the torus filling pressure (10^{-4} Torr) in the 50 msec preceding application of the RF to avoid breakdown due to an intense RF field [FIG.7]. On the other hand at this low pressure multipactor effect can occur if the electric field equals 9.41 KV/cm / 2n - 1, where n is the order of the multipactor effect. [FIG.8] shows that plasma pumping [7] alone over a grill length of 3 m is not sufficient. Therefore, differential pumping was chosen using a not completely vacuum tight teflon plug (length = 156 mm $\hat{=} \lambda_G/2$) near the grill mouth and a turbomolecular pump of 100 l/sec connected directly on the wave transformer. The neutral gas pressure obtained in the grill system was up to 10^{-7} torr. In addition the teflon plug was protected from bombardment by charged particles and neutrals emanating from the plasma by a 1.5 mm glass sheet slid into the waveguide.

6. Results and discussion

The experiments carried out up to now have shown that the space (1 cm) between the teflon plug and the protection window is not pumped rapidly enough to evacuate the pressure increase caused by the RF. Consequently breakdowns occured in this area, as observed on disassembly. Chemical analysis of the teflon plug's front side show metallisation from stainless steel components. Present experiments are carried out using an open grill version without the teflon plug and the grill pumping system. This is possible because of a modified gas flow program for the experiment. Low level RF measurements show that the entire transport line up to the grill is well matched,(reflexion coefficient is less than - 20 dB for each waveguide) and allow to determine the scattering matrix of the grill.
Conditioning of this grill is starting.

Acknowledgements

The authors wish to thank the WEGA team for their assistance, especially Mr. J.J. Capitain for his contributions in installing and operating the RF system and Mr. P. Pras and M. Provost for the design and assembly of the grill-coupler.

References

[1] TONON, G.F., et al., "Investigation of Lower Hybrid Heating in the WEGA Tokamak at 25 KGauss using a 4-waveguide Grill", IAEA-CN-38/T-2-2, Bruxelles (1980).

[2] PACHER, H.D., et al., "Lower Hybrid Heating in the WEGA Tokamak", IAEA-CN-37/A-43 Innsbruck (1978).

[3] SÖLDNER, F., et al., "Lower Hybrid Heating in the WEGA Stellarator", 9th Eur. Conf. on Contr. Fusion and Plasma Phys., Oxford (sept. 1979).

[4] FRITSCH, R., LIPA, M., et al.,"Technological aspects of the WEGA Stellarator", 9th Symposium on Fusion Technology, Garmisch-Partenkirchen (1976).

[5] GOURDON, C., "Programme optimisé de calcul numérique dans les config. magnét. toroïdales", DPh-PFC-EUR-CEA-FC-449 (Janv. 1968).

[6] PACHER, H.D., "Optimization of Power Spectra of Multiwaveguide Grills", Report IPP 2/247 (1979).

[7] MELIN, G., "Grill pumping and other vacuum realated problems" (Private Communication).

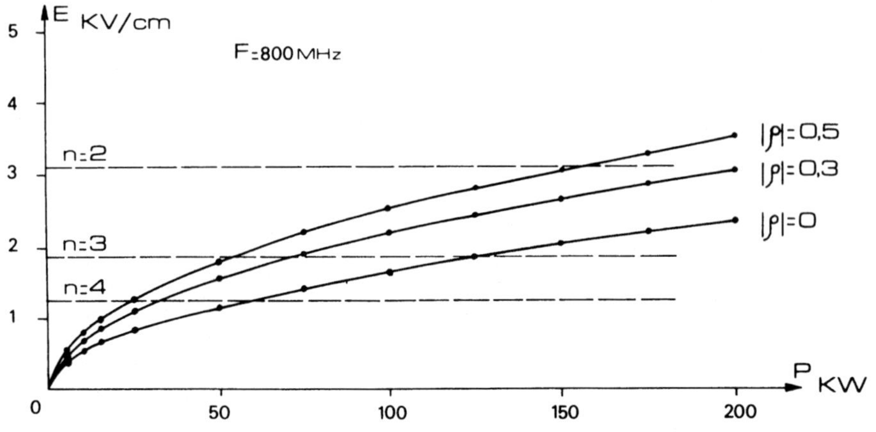

FIG. 7 Electric field at the narrowest part (λg=58,5 cm, b=3,5cm) of the grill as a function of the RF incident power per guide for different reflection coefficients compared with resonant electric field for multipactor effect of different orders.

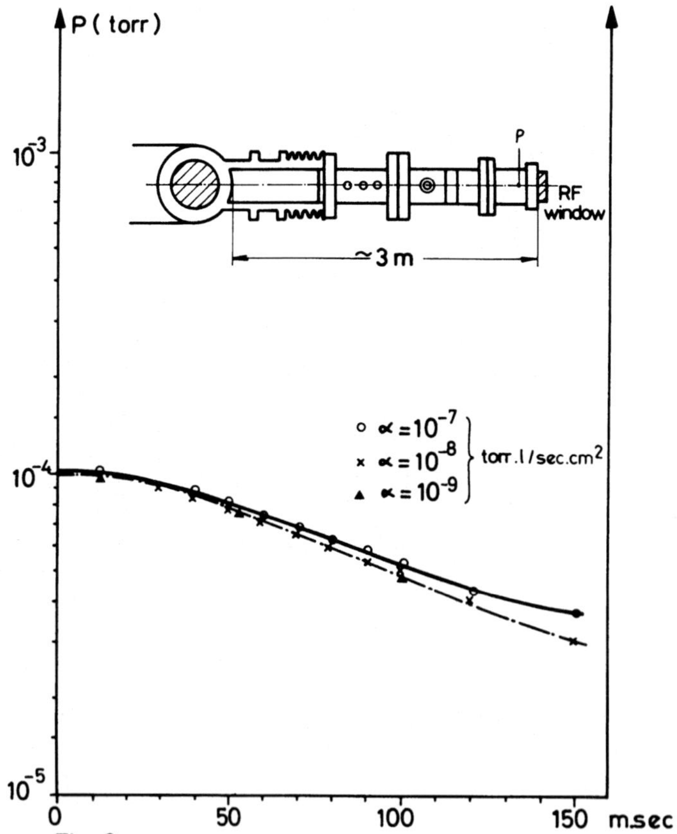

Fig. 8

Effect of plasma pumping in an „Open Grill" version calculated for several degassing rates – ∝ .

APPLICATION OF TORSIONAL TTMP HEATING IN PETULA TOKOMAK - DEVELOPMENT NECESSARY FOR LARGE MACHINES

J.P. PERIN - G. REY

ASSOCIATION EURATOM-CEA
Département de Recherches sur la Fusion Contrôlée
Service IGn - Centre d'Etudes Nucléaires
85 X - 38041 GRENOBLE CEDEX (France)

ABSTRACT

This paper describes a computer program which determines the RF current distribution and the corresponding voltage and dissipated powers in the conductors for the application of torsional TTMP heating -the results fit well with a Petula-scale experiment-. Two prototype coils have been studied and tested for Petula.

The application for JET points out the main technical problems which need further developments to ensure the technical feasibility of this method.

1. INTRODUCTION

The good agreement between theory and experiment shown for compressional TTMP heating [1], [2] in the Petula experiment leads one to expect good results in torsional mode [3]. This mode is excited by RF conductors located in the top and bottom of the vacuum vessel. These do not take up space in the median plane as for the compressional case and moreover such an arrangement is more suited to the elliptic or D-shape of the vacuum vessel (cf. Fig.1 and 2).

2. COMPUTER PROGRAM

2.1. Description : The program computes in cylindrical geometry, the RF current distributions which are flowing in the conductors : driving currents in the coil itself and induced currents in the metallic vessel which encircles the coils.

The program is written for elliptical vessels and circular cross-section coils, the conductor centres of which are located on a circle-a- (cf Fig.3). A serie of wires in parallel is used to simulate the vessel and the conductors, if the frequency is high enough and the resistance of the conductors is not too high, the field lines do not penetrate inside the conductors. In each k point (cf Fig.3), we can write that the whole of the i_j current creates no radial field :

$$\sum_{j=1}^{n} c_k^j \, i_j = 0 \qquad (1)$$

So we obtain a linear system of n equations with n unknowns, the solution of which allows us to determine the current distribution values. The trivial solution i = 0 throughout, is eliminated by taking into account the voltage source which excites the coils.

From the knowledge of the current distribution, the magnetic field, the voltage and the dissipated power can easily be deduced.

2.2. Comparison of computer program and experimental results

A Petula size scale model has been built to verify the computer program. Copper conductors of several diameters and at different positions are located in a 462 mm diameter long copper cylinder which is also used as return path for the current. The main measurements are : the magnetic field along the cylinder axis, the voltage and the current flowing in the coils when a RF generator excites the coils tuned by capacitors.

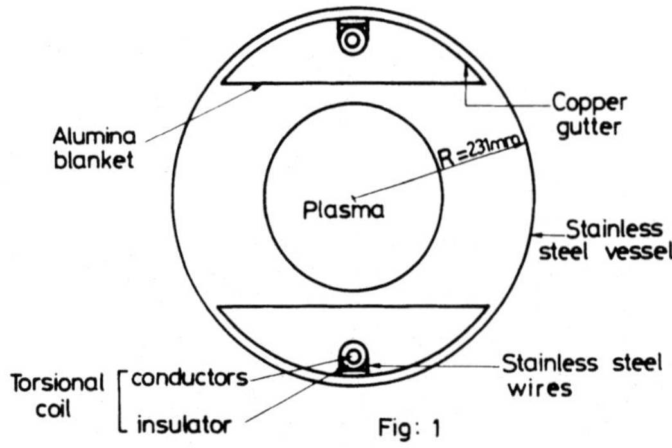

Fig: 1

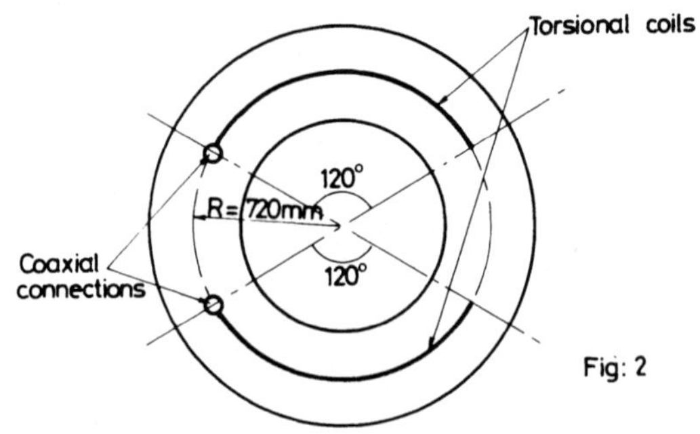

Fig: 2

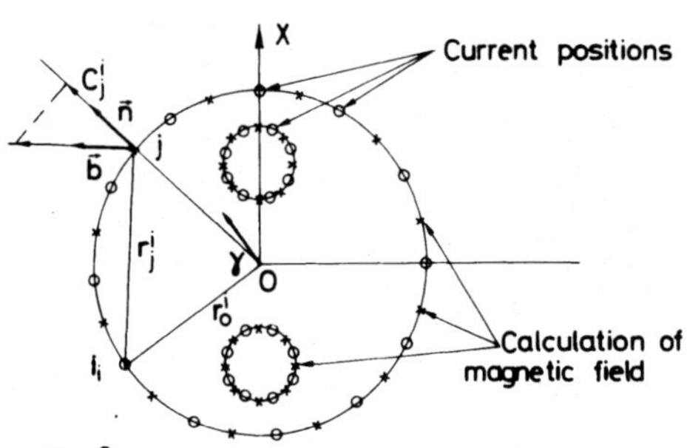

Fig: 3

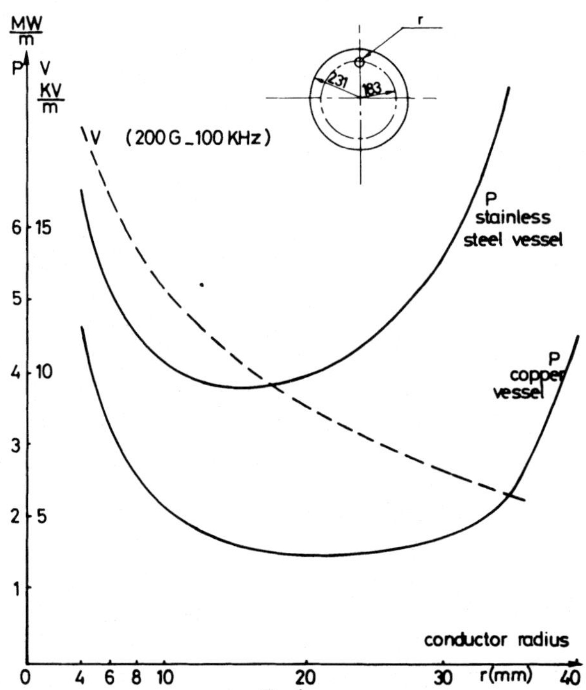

Fig: 4

The main results are summarized in the table I : the dissipated power is given for 200 G at 100 kHz.

TABLE I

Conductor diameters (mm)	12	12	12	12 3 cond //	25	36
Distance of the conductor centers from the axis (mm)	198	183	162	183	183	183
Inductance (μH) measured	1.64	1.82	1.95	1.16	1.35	1.10
Inductance (μH) calculated	1.43	1.58	1.74	0.95	1.12	0.93
$B/I\ 10^{-7}$ T/A measured	5.9	8.8	11.8	8.7	9.1	8.7
$B/I\ 10^{-7}$ T/A calculated	5.8	8.5	11.6	8.3	8.2	8.1
Coil dissipated power (MW) (calculated)	9.8	4.3	2.3	2	2.1	2
Connection and capacitor losses (MW) (calculated)	3	1.5	0.8	1.2	1.3	1.2
Sum of the calculated losses (MW)	12.8	5.8	3.1	3.2	3.4	3.2
Measured losses (MW)	16.7	7.3	4.2	4.1	5.1	4.9

We can conclude a good agreement between the computer and experimental results. The discrepancy which occurs, can be explained by the connection inductances and by the capacitors losses which are badly evaluated.

The scale model has also allowed confirmation that a sinusoidal magnetic distribution along the axis, can be given by coils which are 1.5 m long for Petula -i.e. in the top view, the azimuthal extension of the coils is 120°.

3. APPLICATION TO PETULA

3.1. Application of the computer program

The results of the program show that an optimisation of the dimensions and positions of the conductors can be reached. The following curves are plotted to get a 200 G peak -100 kHz- in Petula experiments.

On Fig.4 are plotted the voltage and losses as a function of the small radius of the conductor, for copper and stainless steel vessels.

On Fig.5, the losses and voltage are plotted as a function of the angle between the conductors for two conductors in parallel.

Clearly shown on these curves is the power gain obtained by working with a copper shell, especially if the conductors are near the metallic vessel. On the other hand, if several parallel conductors are used, the distance between conductors must be large enough to get an uniform current distribution.

Moreover the induced currents in the vessel are located mainly in the vicinity of the conductors, which allows us to use a copper gutter only at the top and the bottom of the vessel.

3.2. Prototype coils

Two prototypes has been built by the following process :

A copper gutter which ensures the return current is fixed on the vessel ; a ϕ_e = 12 mm, hollow copper conductor is encased in alumina or glass pipe. The insulator is

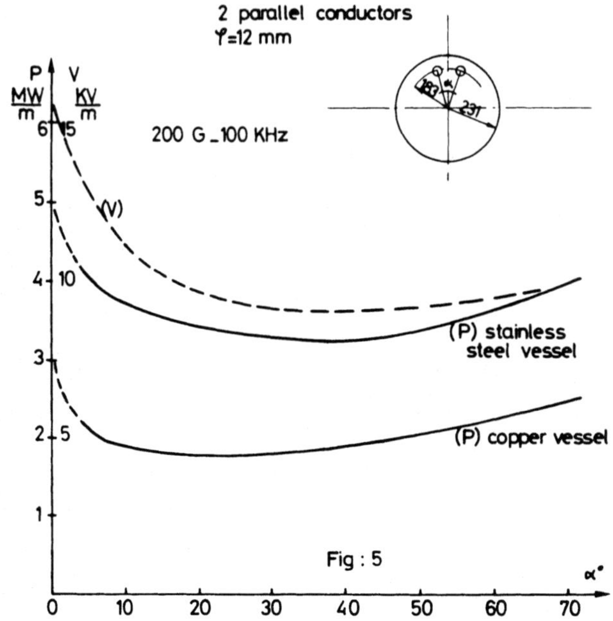

Fig : 5

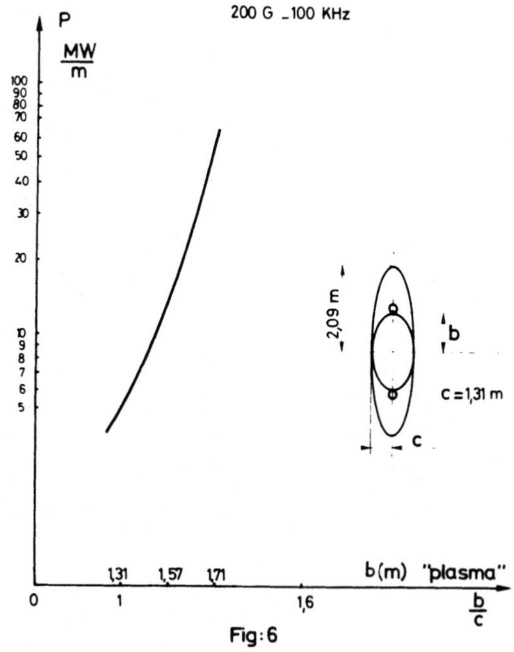

Fig: 6

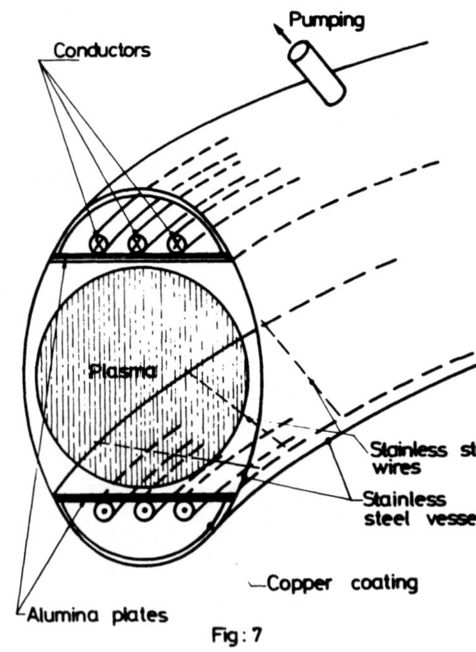

Fig: 7

fixed to the copper gutter by stainless steel wires and can be used to support the electrostatic shield ; alumina plates closes the whole to protect the coil from plasma interaction (cf Fig.4). To increase the electrical insulation, nitrogen at a pressure of 6 bars fills the dielectric pipe.

The glass prototype has been built from a hot-bent glass pipe which is a cheap but brittle design. The alumina prototype is made of eight alumina-stainless steel pipes welded together.

The stainless steel-alumina junctions allow easier welding between the elements but it seems to be limited to the Petula size coils (5 cm diameter) ; for large size Titanium has to be used.

3.3. Specifications and tests of the coils

To get 200 G peak in Petula at 100 kHz, the total losses are expected to be 10-12 MW and the corresponding voltage 12 kV.

The two prototypes have satisfied pressure and electrical tests. The power tests were cancelled following postponement of the TTMP heating in Petula.

4. APPLICATION TO JET

For the same frequency and RF magnetic field (100 kHz - 200 G), Fig.6 shows an example of a possible optimisation : the point b = 1.71 m is obtained for coils made of three, 16 cm radius, 7°5 spaced, parallel conductors, the point b = 1.51 m, for three, 20 cm radius, 10° spaced parallel conductors and the point b = 1.31 = for three, 20 cm radius, 15° spaced parallel conductors, in a copper vessel.

For the special case : 6 m long coils made by two, 20 cm radius, 30° spaced conductors the centre of which is a = 1.41 m, in an elliptical vessel (1.31 × 2.09 m). We get the table II for the arbitrary conditions : 200 G - 100 kHz.

Let us note that the scaling of the power is given by $P \sim B^2\sqrt{f}$, where B is the magnetic field and f the working frequency.

TABLE II

B/I 10^{-7} T/A	I_{peak} kA	V_{peak} by conductors kV	Coils losses MW	Connections losses MW	Capacitors losses MW	Total losses MW	Network power (MW)
1.19	170	192	32.4	16.7	65.3	∿ 140	200

The power values in the table indicate the main technical problems which need further developments :

The capacitor losses are two times higher than coil losses if $tg\delta = 10^{-3}$: industrial development is necessary to lower this value to 5.10^{-4} or less.

The connections losses are evaluated for two groups of 2 m long seven coaxials in parallel to lower their losses and inductances.

The construction of the coils has to solve the insulation, cooling and first wall problems. A conceptual design is shown in Fig.7 the vacuum conditions for the electrical insulation is ensured by additional or plasma pumping. The copper parts of the coils are protected by alumina plates. Horizontal stainless steel wires are used as electrostatic shield. Alumina blanket could be suppressed, the stainless steel wires thus acting as protection.

The use of a vessel made of low resistivity material clearly appears. Copper-stainless steel sandwich vessels could be studied for solving the RF and first wall problems .

In conclusion a global optimisation has to be made to take into account the RF parameters, the manufacturing capabilities and the plasma performance to ensure the feasibility of TTMP heating [3].

The whole of the study is discribed in the EUR-CEA Report [4].

Acknowledgements

The writers are grateful to J. Brossaud and R. Lanier for their technical assistance.

References

[1] E. CANOBBIO, Nucl. Fusion 12 (1972) 561.

[2] BARDET and al., 7th Int. Conf. on Plasma Physics and Controlled Nuclear Fusion Research, Innsbrück (1978) paper CN37Q1.

[3] M. BERNARD, Status Report on TTMP, Technical part, included in Status Report on RF Heating, Advisory Group RF Heating, EUR-FU-BRU-XII/487/80 (1980).

[4] J. BROSSAUD, R. LANIER, J.P. PERIN, G. REY, Réalisation du chauffage TTMP torsionnel. Application à Pétula. Développement pour les grandes machines. EUR-CEA Report, to be published.

AN ALL METAL ARRAY OF ANTENNAE FOR RF HEATING OF TOKAMAKS IN THE ION CYCLOTRON RANGE OF FREQUENCY

J. Jacquinot, H. Lebot, J. Adam, H. Kuus

ASSOCIATION EURATOM-CEA SUR LA FUSION
Département de Recherches sur la Fusion Contrôlée
Centre d'Etudes Nucléaires
Boite Postale n°6. 92260 FONTENAY-AUX-ROSES (FRANCE)

ABSTRACT

500 KW, the maximum available RF power, at a frequency of 60 MHz and in 50 to 100 ms pulses, has been launched in TFR plasmas using an array of 4 half turn antennae. The array has a potential power capability of 1 MW through a single port. The electrical coupling efficiency is about 90 %.

1. INTRODUCTION

Radio frequency heating in the ion cyclotron range of frequency has made, in the recent years, considerable progress in the fields of both physics and technology [1]. Consequently multi-megawatt programmes have been initiated on the large Tokamaks of several countries. It is widely recognized that a major problem is to design an internal coupling structure with the conflicting requirements of efficiency, smallness, cleanness and high degree of resistance against mechanical stresses and thermal load.

We consider that this entire antenna problem has only matured its early infant stage. In the first part of this paper we briefly relate the trial and error process that led to the first maturation of the principal conceptual element of an antenna. The importance of the Faraday shield and insulating materials are especially discussed. The following 3 parts deal with the detail antenna design, the conditionning procedure and the plasma operation of the most recent 4 elementsantenna array of TFR. On line, at the power level of 500 KW, for several months on TFR, its performance suggests new directions which will, hopefully, mature this intricate problem into an adult phase.

2. CONCEPTUAL ELEMENTS

We consider only antennae structures made of a low impedance coil which should be more properly viewed, at our working frequencies - 60 MHz - as a strip line. The antenna plasma coupling theory [2, 3] indicates 3 fundamental geometrical properties :

-i- the antenna should be as close as possible to the plasma ;

-ii- the distance between the 2 conductors of the strip line is not an essential parameter provided it does not become smaller than the separation to the plasma ;

-iii- a Faraday shield can prevent the T.M. mode to radiate and couple, eventually, to peripheral electrostatic plasma modes.

On the experimental side, several antenna designs have been tested on TFR since the early low power approach [4]. They are characterized by different screens between the RF conductors and the plasma (fig. 1). The principal conclusions are the following :

- The lateral parts of the screen are absolutely required. Design of fig. 1a is unacceptable leading to short circuit of the RF current by the plasma.
- An alumina protection between the central conductor and the screen is useless. The main purpose was to prevent the plasma particles to reach the central conductor. A special overlapping Faraday shield was shown [5] to produce the same effect. In addition it allows a lesser distance to the plasma and a faster and more thorough conditionning.
- Design of fig 1b with no Faraday shield in front of the plasma was found unsatisfactory in certain heating conditions [1] presumably in connection with peripheral electrostatic coupling. In the same design, an additional internal distributed capacitance made of simple metallic strips (8 mm wide) helped to decrease the characteristic impedance of the strip line and to improve matching between the antenna and the feeding transmission line.
- Using design of fig. 1c with a full shield, very satisfactory heating effects and limited contamination were obtained [1]. The array of antennae constructed with these principles will now be discussed in details.

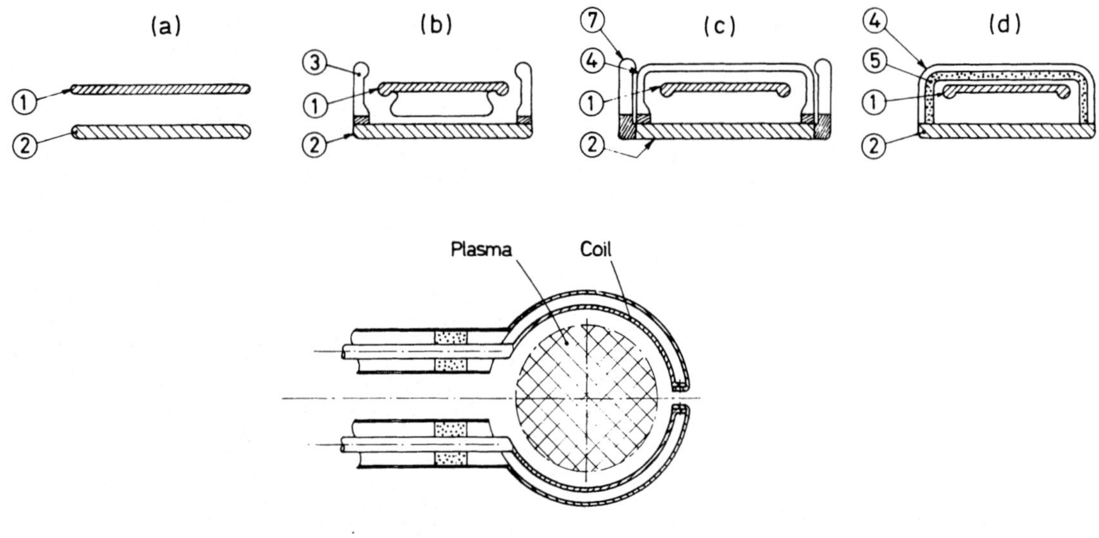

Fig. 1. - Cross section and schematics of 4 types of antennae used in TFR.
(a) bare antenna ; (b) antenna protected with slotted side shields ; (c) full shield (overlapping double screen) ; (d) full shield and ceramic casing.
(1) central conductor ; (2) return conductor ; (3) side shield with 45 degrees slots ; (4) overlapping double screen ; (5) ceramic casing ; (6) slotted internal capacitance ; (7) thick super protection with 45 degrees slots.

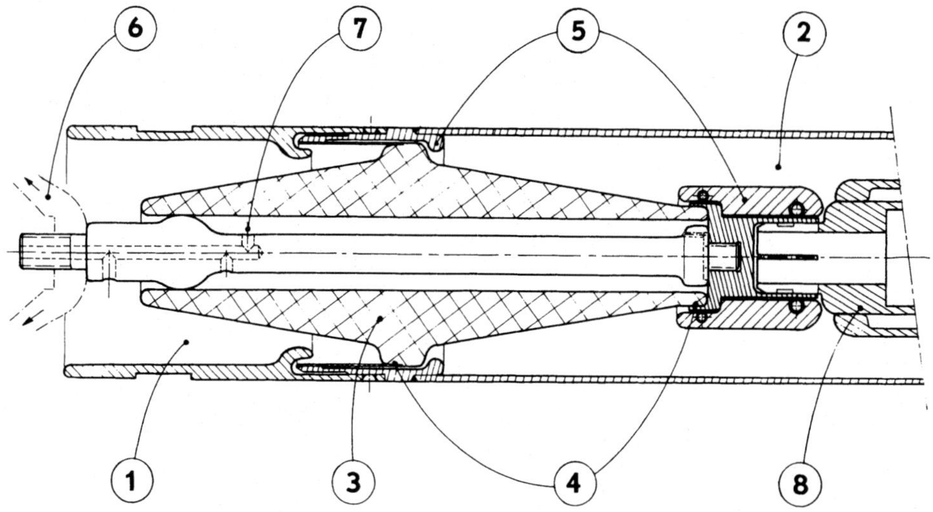

1 . VACUUM SIDE
2 . PRESSURIZED SIDE (+2 ATM.)
3 . ALUMINA
4 . $Ti\text{-}Al_2O_3$ BRAZED JOINTS
5 . CORONNA SUPPRESSORS
6 . PARALLEL INTERNAL CONNECTION OF 2 ANTENNAE
7 . PUMPING HOLES
8 . COAXIAL LINE

Fig. 3. - Coaxial vacuum - tight feed through.

3. 500 KW ANTENNA DESIGN (FIG. 2)

The antenna array is composed of 4, nearly half turn elements. The four antennae are located on each side of a single port of TFR. Each element is introduced separately in the vacuum chamber. Its width has the maximum size compatible with the introduction of the element through a standard port of TFR (10 cm wide, 52 cm high). The elements are then assembled together to form a stiff closed structure which is repelled against the port sides with braces. This arrangement leaves free the central part of the port access for diagnostic use.

The central conductors are then connected internally in parallel to form a top and bottom pair. Each pair is fed through the vacuum chamber by a 50 ohm coaxial line (8 cm O.D.) An alumina vacuum tight feed through section shown on fig. 3 isolates the vacuum chamber from the coaxial section pressurized at 2 atmospheres of Nitrogen. Care has been taken to cover the brazed joints by shaping electrodes preventing coronna. For the same anti-coronna purpose, 3 mm radius rods are placed on the edge of the central conductor. All antenna elements are made with Inconnel 600, a material requested for all inner parts of TFR. A 50 µm silver plating covers the surfaces carrying RF current in the antenna with the exception of the Faraday screen. All parts are also submitted to an extended electro-polishing treatment. The principal dimensions of one antenna element are given in table 1 (w stands for width, L for length and e for thickness).

TABLE 1

- Central conductor edge	w = 62 mm ; L = 540 mm ; e = 4 mm r = 3 mm
- Return conductor	w = 85 mm ; L = 600 mm ; e (averaged) = 13 mm
- Faraday shield 128 pieces per ant.element	w = 5.3 mm ; e = 1 mm
- Separation	- central cond. to return cond. = 20 mm - central cond. to Faraday shield = 5 mm (minimum) - between 2 layers of the Faraday shield = 2 mm

The center conductor is separated from the plasma by 3 metallic layers highly divised in the poloidal direction in order to be transparent to the T.E. electromagnetic waves : A double overlapping screen (separation 2 mm) plays the role of the shield and also contributes to the capacitance of the strip line. A thick (7 mm) lateral super-protection with 45 degrees slots every 7 mm was found necessary to absorb the plasma thermal load which appears in case of a major disruption. The inner edge of the super-protection is 2 cm in the shadow of a limiter located in an adjacent port 40 cm away from the antenna.

4. CONDITIONNING

After installation in the torus, the antenna array is baked at 200°C during 2 hours by running simultaneously the standard baking of the bellows of the vacuum chamber (350°C) and joule heating of the antenna by the RF current at the level of 1 to 2 KW CW.

It is then very usefull to remove carbon and oxygen from the antenna and its vicinity by a discharge produced by the RF itself. Stable and reproducible RF discharges correspond to the conditions :

- Continuous toroidal field : 4 KG.
- Deuterium base pressure : 0.5 to 1.5 m Torr.
- Peak RF power 100 kw ; duration : 20 ms every one or two seconds.
- Antenna and chamber temperature : 200°C.

This procedure is operated during a few hours until the production of D_2O and CD_4 stops to decrease. It seems necessary to do the cleaning only after a period of exposure to air of the torus.

A conditionning against RF arcing in the antenna is then required. It is necessary to perform this operation with short pulses (1 to 3 ms) or with a default interrupter to prevent crater formation in the antenna. With a new antenna, the first arcs occur between 7 to 10 kV. After roughly a 100 arcs the antenna becomes capable of holding 20 kV. No deterioration is observed when the pulse length is increased to 50 ms. The entire process takes between one to two hours and its benefits persist during the life of the antenna.

Even with a fully conditionned antenna, a weak multipactor effect always manifests itself by a small gas release during an RF pulse in vacuum. The gas release is maximum at RF voltages around 5 kV and decrease sharply above 10 kV. In normal operation the base pressure of TFR does not increase by more than 10^{-6} to 10^{-5} Torr and this effect cannot account

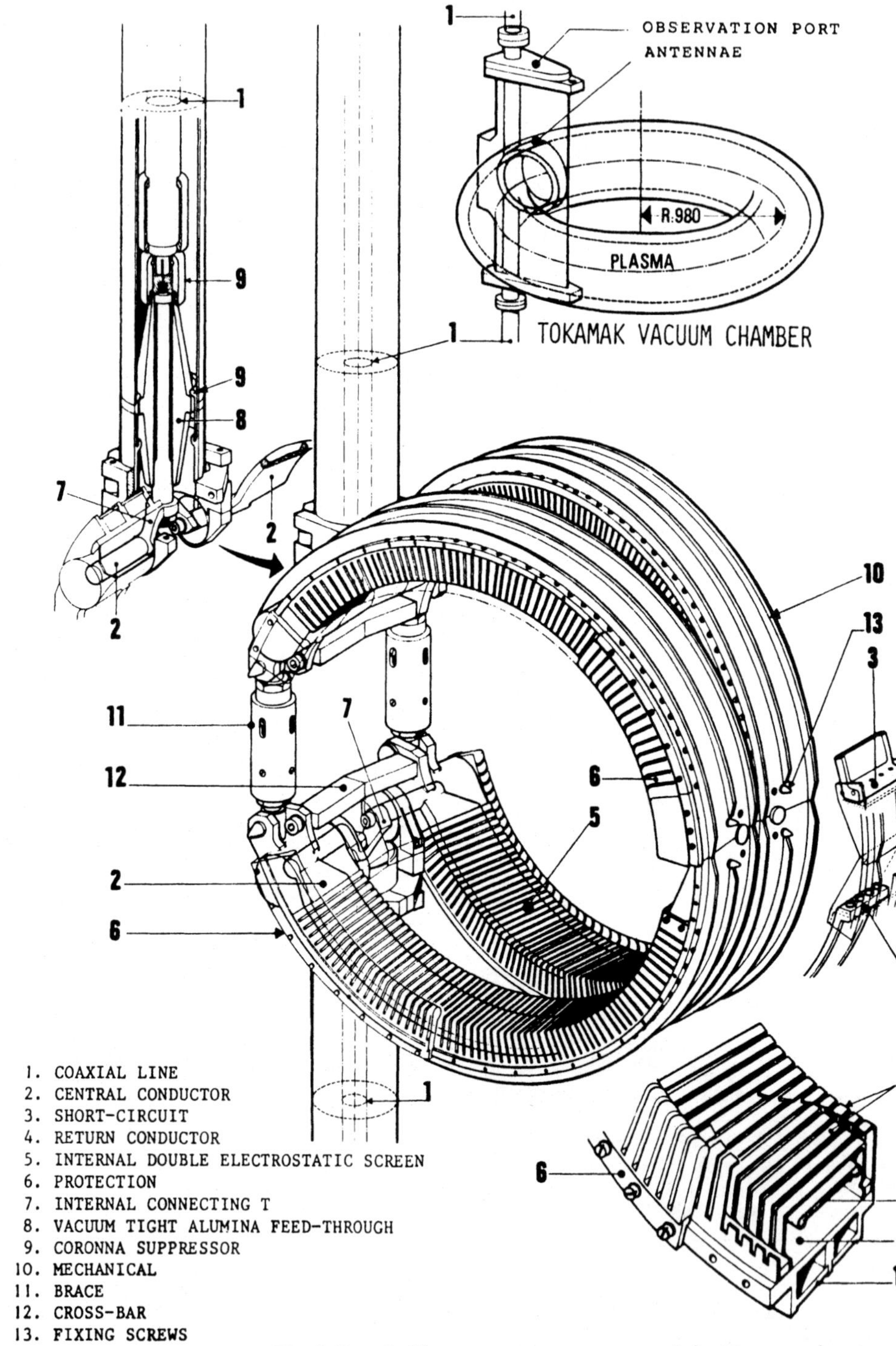

1. COAXIAL LINE
2. CENTRAL CONDUCTOR
3. SHORT-CIRCUIT
4. RETURN CONDUCTOR
5. INTERNAL DOUBLE ELECTROSTATIC SCREEN
6. PROTECTION
7. INTERNAL CONNECTING T
8. VACUUM TIGHT ALUMINA FEED-THROUGH
9. CORONNA SUPPRESSOR
10. MECHANICAL
11. BRACE
12. CROSS-BAR
13. FIXING SCREWS

Fig.2 Four half turn antenna array used in the experiments. All inconnel construction with a full Faraday shield realized with a double screen.

for the impurity release observed in RF heating experiments.

5. ELECTRICAL PROPERTIES

The electrical characteristics without plasma of a top or bottom pair, measured at the point where the central conductors are connected in parallel, are listed in table II. For these measurements the antenna is isolated from the coaxial line :

TABLE II

- Self inductance	$8\ 10^{-8}$ Henry
- Quarter wave-length resonating frequency	78 MHz
- Characteristic impedance	$Z_o = 25\ \Omega$
- Maximum impedance at 78 MHz	$Z = 2000\ \Omega$
- Phase velocity	$V\phi = 1.25\ 10^8$ m/S

The small distance between the Faraday shield and the central conductor is responsible for the slow phase velocity of this structure. As discussed in ref. [4], this property is beneficial for power coupling provided either the characteristic impedances of the coaxial feeder and the antenna are matched or the feeding point coïncides with a maximum standing voltage. The second condition is only approximately realized here. We are now studying an improved version of the design where the realization of the two conditions will be improved by using a 25 Ω coaxial feeder and by decreasing further the antenna phase velocity with an internal capacitor as discussed in section 2.

With plasma, we have not noticed significant changes in either the voltage stand off capabilities or in the reactive properties of the antennae. The major effect is an order of magnitude increase of the nominal radiation resistance defined as $R = 2\ P\ Z_0^2/U_0^2$ where P is the coupled power, U_0 the peak voltage on the feeder and Z_o its characteristic impedance. Supplying only a single top or bottom antenna pair, R increases from $R_v = 0.5\ \Omega$ without plasma to R_p=4 to 7 Ω with plasma. The plasma loading appeared to be sensitive to detailed operating conditions, in particular to plasma position and to peripheral plasma density. With 2 antenna pairs connected externally in parallel at a point of maximum voltage(fig.5), the overall loading resistance approximately doubles (8 to 20 Ω). The sensitivity to plasma conditions does not allow to determine an eventual positive mutual influence between the 2 pairs.

The coupling efficiency of the antenna defined as $\eta = \frac{R_p - R_v}{R_p}$ appears to be in excess of 90 %.

Each pair has been separately tested with plasma at the level of 400 KW without reaching a limit due to arcing. The maximum voltage on the line was 20 kV. Such a voltage applied to the entire array corresponds to a potential power capability of about 1 MW.

A series of experiments has been conducted during five months with this array. Efficient plasma heating effects and limited contamination were obtained [1]. Fig. 4 a and b represents the peak RF voltage of the waves emitted by the generator and reflected to it. The power level (500 KW) and the time evolution are quite typical of TFR operation. Arcing was a very rare event in the experimental series : about once every 50 to 100 runs ; the traces of fig. 4 illustrates the operation of the default interrupter when arcing does occur. If the reflected power exceeds an adjustable level, the interrupter opens, in 100 μS, the link to the first amplifier of the power generator during 1.5 ms. This short interruption is usually sufficient to quench the arc and allows to resume normal operation.

The coupling properties of the array did not degrade during the five months period although the super protection needed replacement in view of the melting produced by frequent major plasma disruptions.

Aknowledgements : It is a pleasure to acknoledge the efficient work of MM. Caplan, Devaux, Patris, Quet, Speziale and Vlemincx, who adjusted and installed the antenna in TFR.

References :

[1] Equipe TFR, EUR-CEA-FC-1046 (1980)

[2] Adam, J., EUR-CEA-FC-1004 (1980)

[3] Weynants,R., et al. to be published in the proceedings of the 2nd Joint Grenoble-Varenna Inter. Symp. on Heating in Toroidal plasmas, Como, September (1980)

[4] Adam, J., Darbandi, A., Jacquinot, J., and Kuus, H., proc. tenth Symp. of Fusion Technology, Pergamon Press (1978) p. 263

[5] TFR Group, Nucl. Fusion, 19, 1538 (1979).

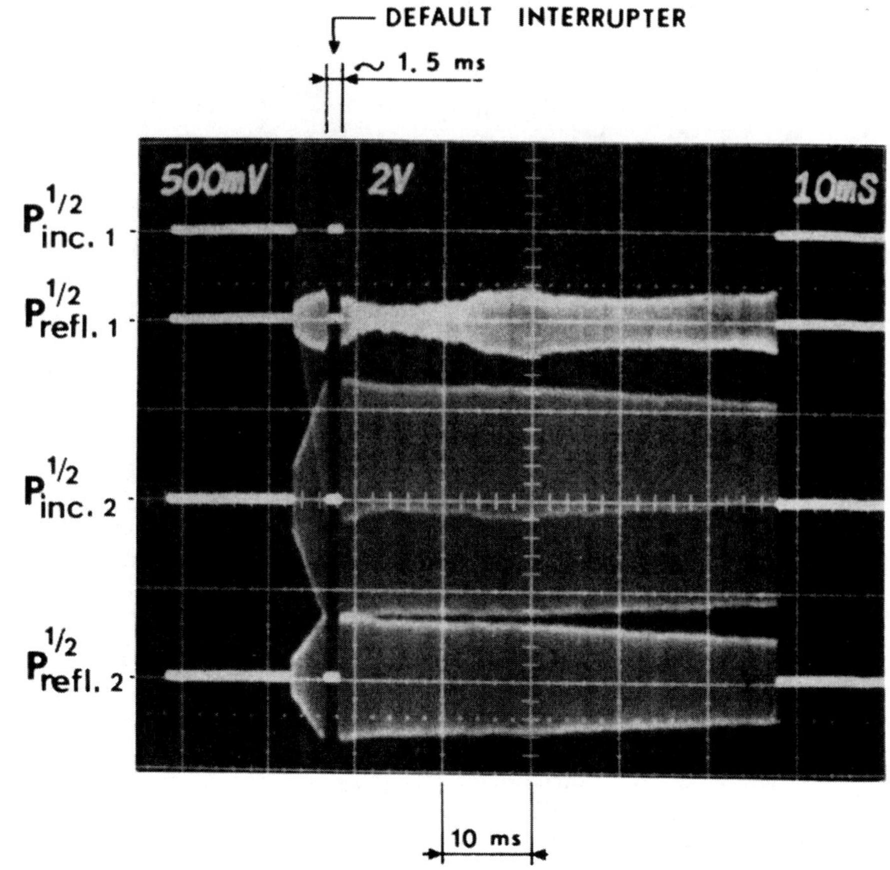

Fig. 4. - Oscilloscope traces of directionnal couplers located on the coaxial feeders (fig. 5). $P_{inc.1}$ and $P_{refl.1}$ are the powers emitted by the generator and reflected to it. $P_{inc.2}$ and $P_{refl.2}$ represents the maximum standing voltage of the incident wave and the reflected wave on the feeder.

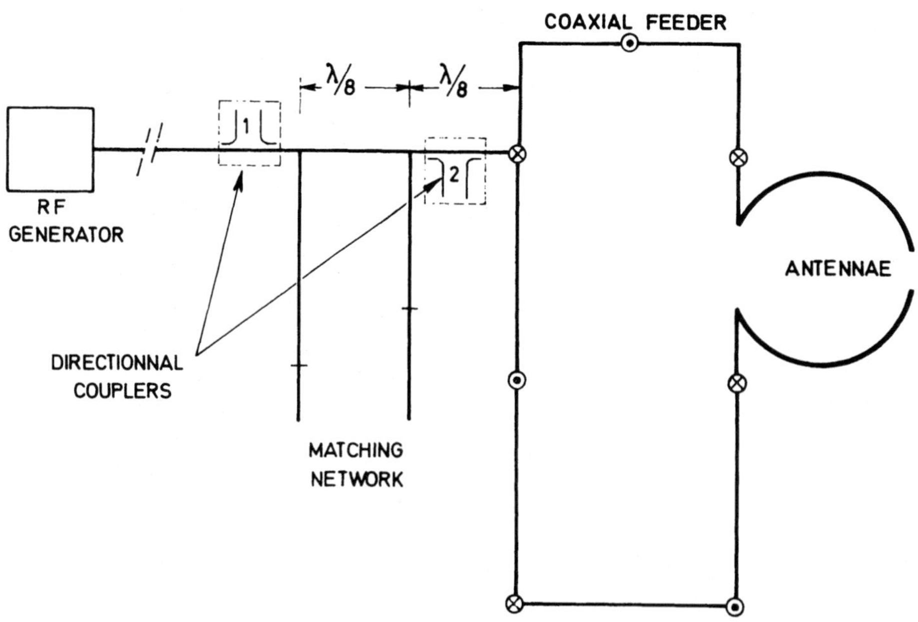

Fig. 5. - Electrical connections of the RF system. The crosses and the dots represent, respectively the locations of the maximum and minimum standing voltage of the feeder at 60 MHz.

START-UP AND HEATING OF RFP REACTORS AND LARGE EXPERIMENTS

M. Bagatin, S. Ortolani

Centro di Studio sui Gas Ionizzati (Associazione EURATOM-CNR) and

Centro 'G.Levi Cases' - Università di Padova (Italy)

ABSTRACT

Ohmic heating to ignition in an RFP is a very attractive feature from a technological point of view. Heating and energy loss mechanisms are discussed and the heating and ignition phases of an RFP reactor are studied with a time dependent O-D model. The performance of a large device, RFX, is studied on the basis of the specified design figures. The study of the current rise time and flat top phases indicates that short rise time are not necessary but the possibility to sustain the current flat top for 120 ms with the available volt-seconds is marginal when realistic loss processes are considered.

1. INTRODUCTION

The RFP is an axysimmetric magnetic confinement system [1,2,3] in which ohmic heating by large toroidal currents can be used without necessarily requiring high toroidal magnetic field. This characteristics is attractive from a technological point of view since, in comparison to a tokamak, lower fields at the wall and higher beta could be obtained in a device with no auxiliary heating systems. High current experiments (current density 2-4 MA/m^2) have had so far small dimensions ($a \lesssim 20$ cm); the only large experiment was ZETA [4] ($a = 50$ cm) but the current was rather low (typically 500 kA, $J \approx .5 MA/m^2$) so that all the RFP plasmas studied experimentally have been limited to low temperatures (100-200eV). However an RFP reactor scheme has been proposed based on physics assumptions [5] which represent large extrapolations because of the lack of experiments. Recently a large experiment, RFX (a=60cm, I=2MA), has been proposed and a set of specifications have been prescribed for the device design [6]. Also recently the ETA-BETA II experiment [7,8] has proved good confinement of an RFP plasma when a regime of reduced losses, which can be characterized by the ratio, I/N, of the current over the number of particles per unit lenght, is obtained [9,10]. SI units are used except for the temperature which is in eV.

2. STEADY-STATE CALCULATIONS

For the purpose of this section we consider ohmic and α-particle heating (P_Ω, P_α) and bremsstrahlung and radiation losses from impurities (P_{brem}, P_{imp}). A plasma with equal concentrations of deuterium and tritium is considered, with electrons and ions having the same temperature.

The radiation loss due to an impurity species is written as:

$$P_{imp} = n_e n_s L_s(T) \tag{1}$$

where L_s is a polynomial fit of a function of T obtained by an average ion model [11]. The total radiation loss P_{rad} as a function of both the temperature and the fractional impurity content, $f = n_s/n_f$, can be written:

$$P_{rad} = P_{brem} + P_{imp} = n_e^2 \left[f_b(T) + f L_s(T) \right] / \left[1 + f \bar{Z}_s(T) \right] \tag{2}$$

where $\bar{Z}_s$ is the average charge of the impurity ions of the s-species, and f_b is the function of the temperature describing the bremsstrahlung of the Z=1 filling gas. The solution of the equation:

$$P_\alpha + P_\Omega = P_{rad} \tag{3}$$

gives the equilibrium temperature as a function of I/N in terms of f. There are, in general, more than one equilibrium temperature for each given couple of values of f and I/N, the lower of which represents the maximum stable equilibrium temperature obtainable by heating a cold plasma. Furthermore each value of f determines also a critical value of I/N above which the temperature is no longer superiorly limited. The temperature corresponding to the critical I/N represents the ignition temperature for the heating and loss mechanisms considered. It is straightforward to get an approximate analytical expression for the dependence of the critical I/N on f. Because of the narrowness of the temperature interval around ignition, $\bar{Z}_s(T)$ and $\bar{Z}_s^2(T)$ can be considered with good approximation constant for each impurity species so that $Z_{eff}=(n_f+n_s\bar{Z}_s^2)/(n_f+n_s\bar{Z}_s)$ becomes a function of f only. In the same sense we can also take approximate expressions for both P_α and P_{rad} /12/. With the approximation of considering the ignition temperature as the one at which $P_\Omega = P_\alpha$, the following ignition criterion for the ratio I/N is then obtained:

$$\frac{I}{N} > \left(\frac{I}{N}\right)_O (1+fC_s)^{11/14} Z_{eff}^{-1/2} \qquad (4)$$

where C_s depends on the impurity species /13/ and the constant value $(I/N)_O=1.33\cdot 10^{-14}$A·m represents the ignition condition without impurities (f=0). Condition (4) is plotted in fig.1 as a function of both the fractional impurity content f and the effective charge Z_{eff}. When the line radiation of the impurity is neglected and only the contribution of the impurity ions to bremsstrahlung radiation is considered, it follows that $1+fC_s=Z_{eff}$, so that the condition (4) would be simplified to that given by Lawson in the RFP reactor study /5/:

$$\frac{I}{N} > \left(\frac{I}{N}\right)_O Z_{eff}^{2/7} \qquad (5)$$

For comparison this condition is also reported in fig.1 (dashed line).

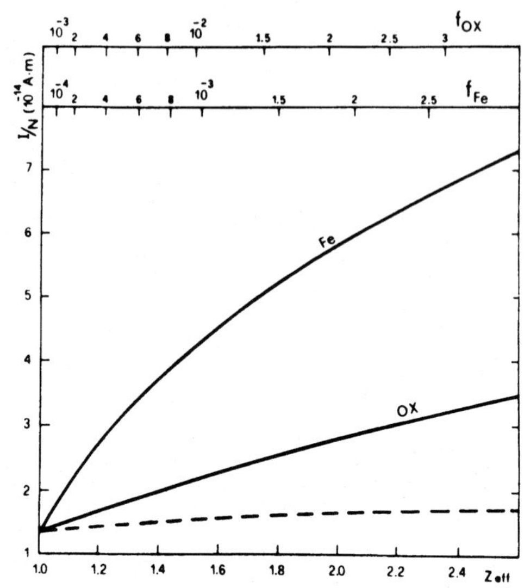

Fig.1

Another way to elucidate the importance of the I/N parameter is to consider the nτ variation as a function of the temperature at various I/N. If τ is the energy confinement time relative to all the energy loss processes with the exception of bremsstrahlung and impurity radiation we can write with $n=n_e=n_i$:

$$(3nT)/\tau + P_{brem} + P_{imp} = P_\alpha + P_\Omega \qquad (6)$$

which can be written as:

$$n\tau = F(T,\ I/N,\ f) \qquad (7)$$

The presence of impurities, as seen before, results simply in increasing the minimum I/N necessary for ignition whereas the effect of varying I/N is shown in fig.2 where nτ as a function of the temperature is drawn for various I/N with f=0. As we see when $I/N < 1.5\cdot 10^{-14}$A·m ignition is not possible at all. Above this threshold for I/N ignition (which is represented by the maximum of the curves) becomes possible at progressively higher temperatures and requires, of course, lower

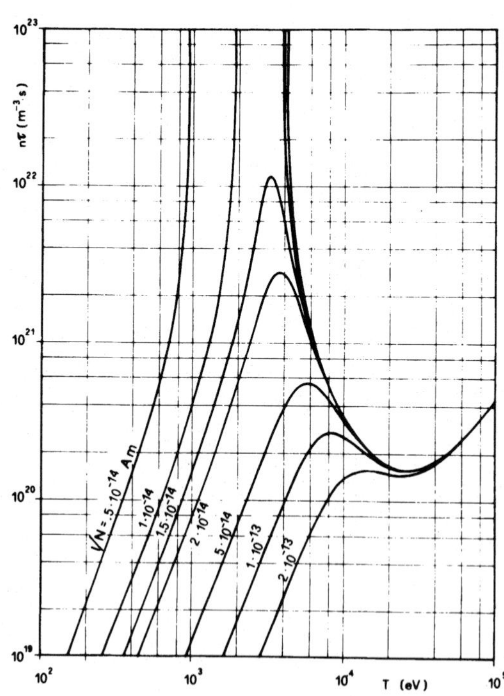

Fig.2

nτ during the ignition phase. If we think of a device with a given current, high density operation results in the difficult task of climbing the nτ slope as the ignition temperature is approached. However low density operation may result in unacceptable transport losses to obtain the required energy confinement times. In practice a range of values between $1\text{-}2\cdot 10^{-14}$A·m seems the one which corresponds to best experimental performances [9,10]. Impurity contents must then be rather small (see fig.1).

3. RFP REACTOR STUDIES

The start-up and heating phases of a proposed RFP reactor (RFPRS) [3] have been tested by a zero-dimensional, time-dependent plasma model [12]. The model accounts for the energy loss processes of thermal ionization, excitation and recombination of both the main filling gas and of the impurities, for bremsstrahlung and synchrotron radiation and for heat diffusion. A Z=1 nuclear charge gas is considered as the main filling gas. The impurities can be treated by time dependent and/or steady-state corona model equations. A 50% D-T mixture has been considered as the main filling gas (with a density of $2.5 \cdot 10^{20} m^{-3}$) and for the remaining parameters we refer to [3].

The amount of energy consumed in the setting-up phase affects significantly the reactor balance, representing at the same time one of the major uncertainties in the reactor studies. The required performance of the RFPRS is to reach 19 MA of plasma current with a K_e (defined by the ratio of the ohmic dissipation energy to the total magnetic energy in the plasma) of about 50%. These conditions have been matched with a fractional concentration of oxygen $\bar{f}_{ox} = 1.6 \cdot 10^{-3}$ which does not seem too unrealistic to invalidate the assumed energy consumption ratio. The value of f_{Fe} leading to $K_e = 0.5$ when $\hat{I}_p = 19MA$ is $\bar{f}_{Fe} = 2.8 \cdot 10^{-4}$ and is obviously much lower than $\bar{f}_{ox}$. Furthermore we notice that K_e varies rapidly around the $K_e = 50\%$ value and then even small increases in the concentration of the considered impurities would prevent to achieve both the required parameters and ignition [13].

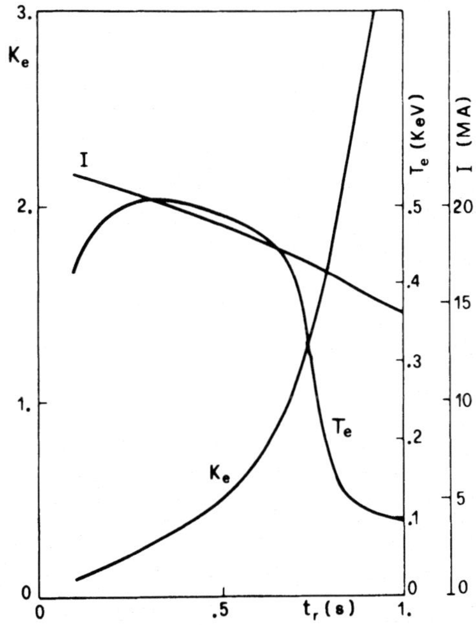

Fig.3

The variation of the current rise time has been studied in the range $t_R = 0.1$-1 s keeping constant the energy supplied and $f_{Fe} = \bar{f}_{Fe}$. Fig. 3 shows both a net increase of K_e and a drop of I_p beyond $t_R = 0.5$s, while the maximum value of T_e is reached at $t_R \simeq 0.3$-0.4s.

An example of the time dependence of the ion temperature and plasma current during the heating and ignition phases is drawn in fig.4. Ignition occurs in about 10s and the time to reach the working point at 10 keV is about 22s. If somewhat lower densities are considered ($n = 2.08 \cdot 10^{20} m^{-3}$) ignition times of ~ 6.5s are obtained.

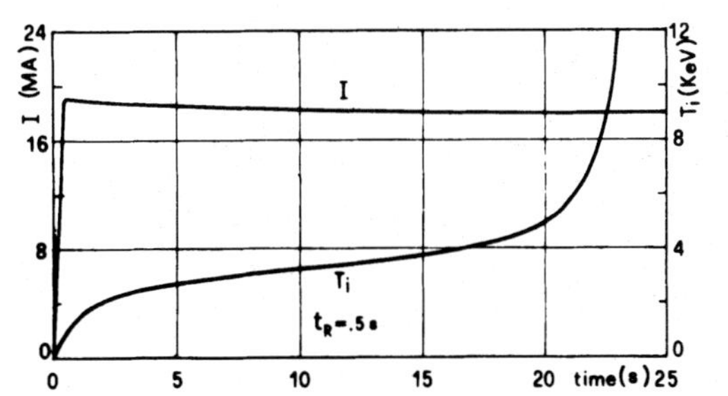

Fig.4

4. RFX PERFORMANCE STUDIES

The design specifications for the RFX device [6] give the total volt-seconds for the toroidal current circuit $\Delta\phi_{Tot} = 12.5 V \cdot s$ to produce a current of 2 MA with rise times in the range 10-50 ms and with a current flat top time of 120 ms. More specifically for the current rise phase we have [6]:

$$\Delta\phi_R = kL\ I \tag{8}$$

which with I = 2MA, L = 2.818 μH and k = 1.806 (assumed to account for the resistive volt seconds consumed during the rise) gives $\Delta\phi_R = 10.18$ Vs. The remaining volt seconds ($\Delta\phi_{FT} = 2.32 V \cdot s$) are available for the current flat top at 2MA and are consistent with the specified resistance:

$$\begin{cases} R = R_o + R_1 \cos(2\pi t/t_d) & \text{for } t_R < t < t_R + t_d/2 \\ R = \text{const} = 6 \ \mu\Omega & \text{for } t > t_R + t_d/2 \end{cases} \quad (9)$$

where $R_o = 28\mu\Omega$, $R_1 = 22\mu\Omega$, $t_d = 40$ ms.

We have investigated the obtainable plasma performance by using a O-D time dependent model /̄12̲7. In particular oxygen line radiation is computed by solving the corona model time dependent ionization rate equations, the iron radiation is evaluated in the steady state corona model limit and the energy losses associated with magnetic field fluctuations are written as /̄12,14̲7:

$$P(\Delta B/B) = 4.15 \cdot 10^{-8} (\Delta B/B)^{*} n^{1/2} j T S^{-1/2} \quad (10)$$

where $(\Delta B/B)^*$ represents the level of fluctuations at $S = 10^3$ (S being the magnetic Reynolds number). In the analysis we consider all the design specifications /̄6̲7 except for the plasma resistance which is obtained self-consistently as a result of the computer model in which classical resistivity is assumed. We assume for simplicity an exponential voltage drop during the current rise and a constant voltage during the flat top phase.

4.1 Current rise phase

The current rise time has been varied in the range 10-50ms. In fig.5 is shown the oxygen fractional concentration f_{ox} which is compatible with obtaining 2MA at the end of the current rise with the given flux swing ($\sim$10.2V·s). The deuterium filling density is $1.75 \cdot 10^{20} m^{-3}$ and full ionization is reached in $\lesssim$1ms. As we see from the figure almost 1% of oxygen is tolerable with current rise times of 50ms. Of course, when shorter rise times are used, larger oxygen contents become compatible with obtaining 2MA. In particular at $t_R \approx 10$ms concentrations $\gtrsim 3$% are possible but this seems really out of any pessimistic prevision. We then conclude that even by operating at relatively high densities and assuming complete ionization (i.e. having $I/N \approx 10^{-14}$A·m) a rise time of 40--50 ms should be sufficient to overcome light elements radiation losses.

Operating with t_R=52.5 ms and f_{ox}=0.7% we have then varied the iron content f_{Fe} and $(\Delta B/B)^*$. The current rise phase is very little affected by these two variations (see $I(t_R)$) in fig.7 and 9, and $\Delta\phi_R$ in fig.6 and 8), since both mechanisms tend to become important at relatively higher temperatures (>50 eV). Therefore about 2 MA (within 10%) are reached even if large fluctuations ($(\Delta B/B)^* \sim 5$%) or relatively high iron concentrations (1%) are present. It appears therefore that there is no predictable

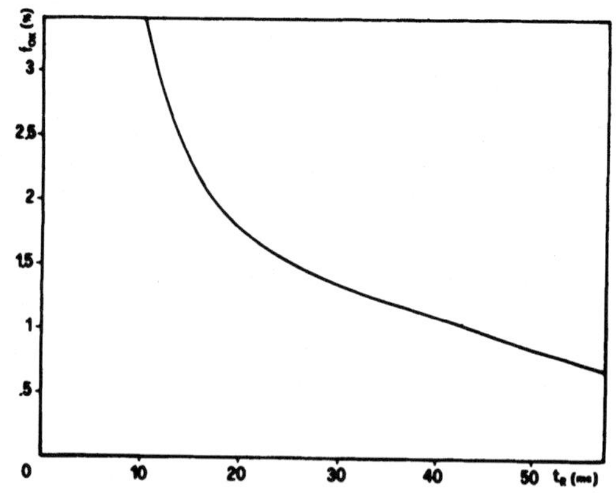

Fig.5

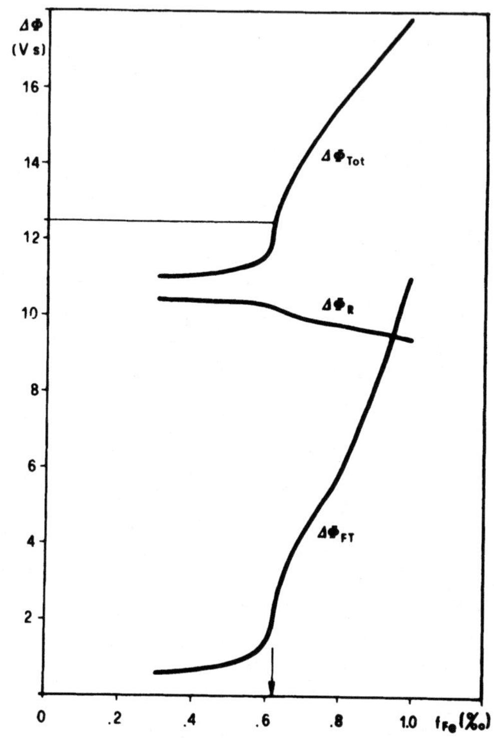

Fig.6

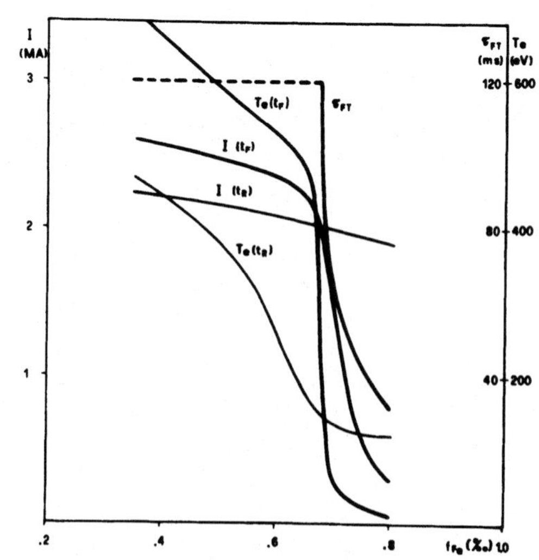

Fig.7

reason to operate in RFX with risetimes much shorter than 50 ms.

4.2 Current flat top phase

Two types of studies have been done to evaluate the performance of RFX during the current flat top phase. In a first set of runs we keep the total volt seconds constant and equal to the specified value of 12.5 We keep constant f_{ox} so to have about 2 MA at the end of the current rise and we vary f_{Fe} and $(\Delta B/B)^*$. In fig.7 we see that by increasing the iron fractional content the current flat top time τ_{FT} (defined here as the time for the current to stay constant within 10%) drops rapidly when f_{Fe} approaches 0.7‰. The temperature reached at the end of the current rise $(T_e(t_R))$ decreases as f_{Fe} increases but the ohmic heating can produce subsequent heating giving higher final temperatures $(T_e(t_F)) \gtrsim 500$eV if $F_e \lesssim 0.65$‰. Also drawn in the figure are the current amplitude at the end of the rise (which is independent of the iron content within 10%) and the final current amplitude which actually is somewhat larger than the $I(t_R)$ when the losses are small (since the applied loop voltage is constant), and which drops rapidly when excessive energy losses prevent to match the requirements for the flat top duration.

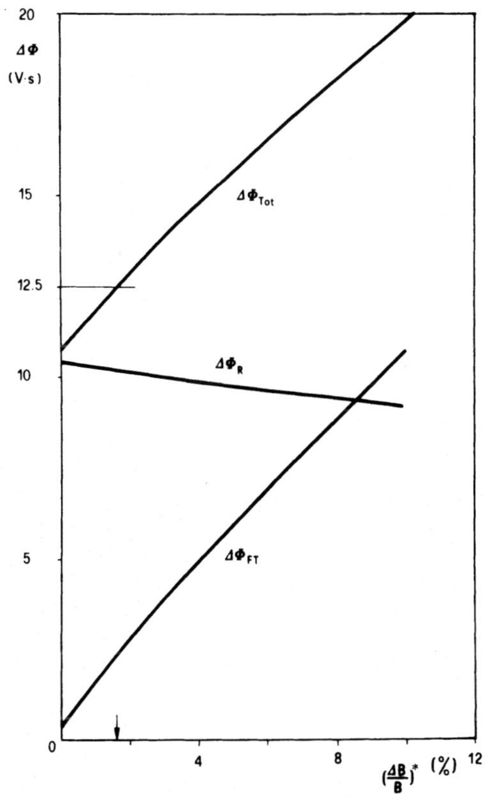

Fig.8

A similar behaviour is observed in fig.9 where the same quantities are plotted against $(\Delta B/B)^*$. Fluctuation levels higher than 2-3% appears incompatible with satisfying the current flat top requirements.

In a second series of runs we set the voltage constant at the time of peak current and deduce the required volt-second to maintain constant (within 10%) for 120 ms the current at the peak value of 2 MA. The results are reported in fig.6 and 8 where the volt-seconds consumed during the current rise, during the flat top phase and the total one are reported versus increasing iron contents (fig.6) and level of fluctuations (fig.8). From the graphs we see again as figures of $f_{Fe} \simeq 0.65$‰ and $(\Delta B/B)^* \simeq 2$% are marginal to achieve the specified performance with 12.5 V·s. Increasing f_{Fe} or $(\Delta B/B)^*$ the required $\Delta\phi_{FT}$ increase rapidly and a figure of ≈ 5V·s available for the current flat top would probably be safer.

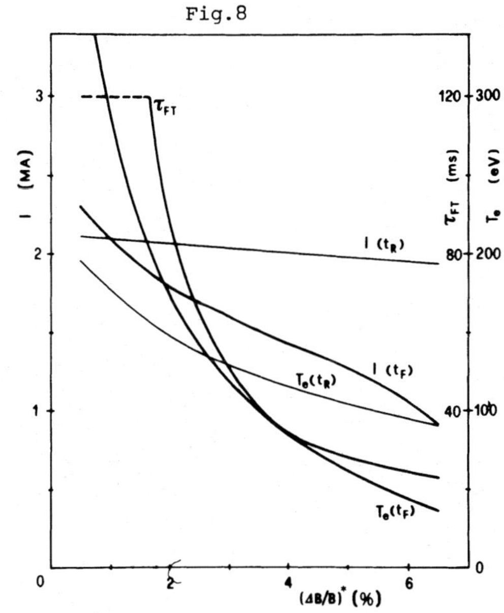

Fig.9

In fig.10 we report a comparison between the time dependence of the plasma resistance as assumed in the specifications (eq.(9)) [6] and the ones which correspond to some runs of the 0-D computer model. We must remember that in the design specifications the resistance of 50 μΩ for the rise time does not play any role since the volt-second consumption for the current rise phase is in any way specified by eq.(8). However the resistance specification (eq.(9)) for the current flat top phase is crucial. We note that in the simulations reported in fig.10 the resistance drops after about 10 ms to values of the order of 20 μΩ, but subsequently it does not decrease any more if more than 0.6‰ of iron is present or if 0.45‰ of iron and a fluctuation level of

1% are considered at the same time.

Finally in fig.11 the time evolution of the plasma current and temperature along with the supplied volt-seconds are reported, as an example, for the case with 0,7% oxygen, 0.45‰ iron and $(\Delta B/B)^{*}=1\%$, which does not satisfy the prescribed specification.

5. SUMMARY

The ohmic heating up to ignition in a Reverse Field Pinch is possible but severe limitations on the energy losses must be satisfied. Examples have been given on the critical I/N values necessary to achieve ignition if impurities are present and the $n\tau$ dependence on I/N has been discussed.

The proposed reactor scheme is capable of achieving ignition in times of the order of 10 s. In the proposed large experiment, RFX, the envisaged flux consumption during the current rise seems appropriate and there should be no need to operate with rise times much shorter than 50 ms. However the sustainement of the current flat top for times up to 120 ms seems critical with the available volt-seconds if realistic energy loss processes are considered.

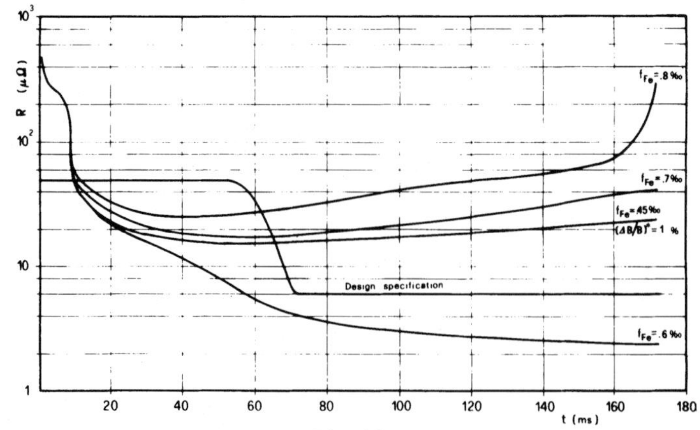

Fig.10

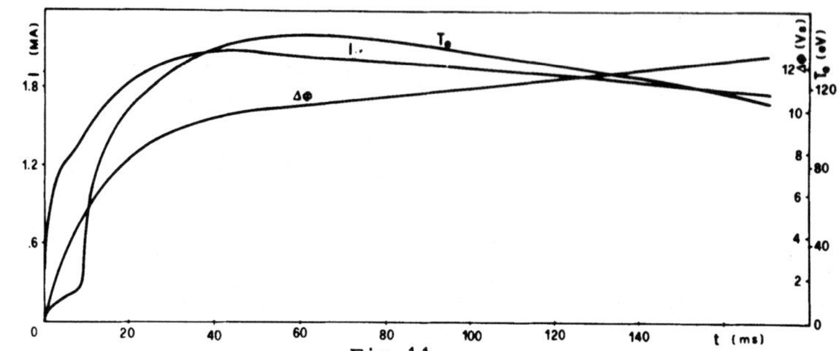

Fig.11

6. REFERENCES

[1] Proc. of the Workshop on RFP, Padova University Report UPee 78/08 Vol.I and II (1978).

[2] ORTOLANI, S., Report on the Padova RFP Workshop, Nuclear Fusion, 19, 535-544 (1979).

[3] HANCOX, R., et al., IAEA Meeting on Fusion Reactor Design, Madison, 10-21 October, 1977. Proceedings pp. 319-336.

[4] ROBINSON, D.C., KING, R.E., 3rd Int.Conf. on Plasma Physics and Contr. Nucl. Fusion Research, Novosibirsk, 1968. Proceedings, Vol.I, pp. 263-276.

[5] LAWSON, J.D., Culham Laboratory Report CLM-R171 (1977).

[6] 2 MA RFX Design Specifications agreed at the Int. Workshop held at Padova, Feb. 1980.

[7] FELLIN, L., et al., 10th SOFT, Padova, 1978. Proceedings pp. 61-68.

[8] BUFFA, A. et al., 9th Eur.Conf. on Contr. Fusion and Plasma Physics, Oxford, 1979. Proceedings Vol. 2 p. 544.

[9] BUFFA, A., et al., 8th Int. Conf. on Plasma Physics and Contr.Nucl.Fusion Research, Bruxelles 1-10 July 1980. Paper IAEA-CN-38/L-1.

[10] ORTOLANI S., ROSTAGNI G., Padova University Report UPee 80/03 (1980).

[11] POST, D.E., et al. Princeton Plasma Physics Laboratory Report PPPL-1352 (1977).

[12] BAGATIN, M., et al., Padova University Report UPee 80/05 (1980).

[13] BAGATIN, M., et al., 7th Int. Conf. on Plasma Physics and Contr.Nucl.Fusion Research, Innsbruck, 1978. Proceedings pp. 37-52.

[14] ROBINSON, D.C., 3rd Top. Conf. On Pulsed High Beta Plasmas, Culham, 1975. Proceedings pp. 273-281.

ACCELERATION AND INJECTION OF D_2-PELLETS

V. Andersen, M. Gadeberg, P.B. Jensen, P. Nielsen and H.Sørensen
Association Euratom-Risø National Laboratory, DK-4000 Roskilde, Denmark.

ABSTRACT

In the studies of pellet-plasma interactions made at Risø National Laboratory, 0.13-mg pellets of solid D_2 are accelerated to velocities up to 200 m/s and injected into a small tokamak, named Dante. The pellets are accelerated in a light-gas gun by means of a propeller gas pressure created by pulse evaporation of solid H_2. The pellets are injected into the tokamak through a 2.8-m long curved guide tube with 4-mm bore. At the moderate velocities used, injection through a 20-m guide tube proved possible without a serious loss of pellet material. A pellet gun can thus be placed in nearly any position with respect to a tokamak. Another gun is being tested where 0.01-mg pellets are accelerated to velocities above 300 m/s.

1. INTRODUCTION

One way of fuelling present tokamaks and future fusion reactors is to inject solid pellets of fuel material. According to present theories of pellet-plasma interactions such pellets should be given velocities up to 10^4 m/s in fusion reactors, whereas lower velocities suffice for present tokamaks [1]. It is important to know the pellet velocities needed for fusion reactors, and studies of pellet-plasma interactions at various plasma temperatures and densities are thus of significance.

At Risø National Laboratory these interactions are studied in a small tokamak, named Dante (Table I). In this report we shall discuss the pellet gun

R = 50 cm	Plasma duration = 15 ms	Table I.
a = 12.5 cm	T_e (X-rays) = 200 eV	Data for Dante.
B_T = 0.8 T	$\bar{n}_e = 3 \times 10^{13}$ cm^{-3}	
I_p = 20 kA		

and the problems met during pellet injection.

Deuterium is solid only at very low temperatures and the pellet should therefore be produced, handled, and accelerated at these temperatures.

2. THE PELLET GUN

One possible method for accelerating a pellet is the pneumatic one, i.e., the pellet is loaded into a gun barrel and it is then blown out by means of a propeller gas pressure applied behind it. Pneumatic guns for accelerating small pellets of H_2 and D_2 have now been built at three laboratories, Oak Ridge [2], Garching [3], and Risø [4]. At the first two locations, H_2 gas, pressurised at room temperature, is used as propellant, and velocities above 1000 m/s have been reached. At Risø the propellant gas pressure is created by pulse evaporation of a fixed amount of H_2. The velocities attained so far are 200 to 300 m/s.

The low temperature part of the pellet gun is placed below the bottom of a liquid helium cryostat (Oxford Instruments MD4A) and is protected against thermal radiation by two radiation shields, one at liquid nitrogen temperatures and the other at liquid helium temperatures. The various connections from outside, wires for electrical heaters and thermometers, capillary tubes for H_2 and D_2 gas, and rods for mechanical movements, are led in from the side and top of the cryostat by traditional methods.

The working principle for the gun may be demonstrated by means of the schematic drawing in fig. 1. The solid D_2 from which a pellet is formed is contained in a D_2 chamber (10). The solid D_2 can be extruded through a nozzle (9), a tube (8), and a punch chamber (7) to appear as a filament of solid D_2 1 mm in diameter (6) in the vacuum space below the gun set-up. The D_2 is extruded when a force is applied to the piston (11) by means of a rod coming from the top of the cryostat. During extrusion the solid D_2 should be heated to a temperature of 10 to 12 K while nozzle, tube, and punch chamber should be heated to a slightly higher temperature. If the different parts are not heated the extrusion cannot be accomplished and if they are heated too much the extrusion becomes unstable. A pressure of 100 bar is applied to the solid D_2 during extrusion.

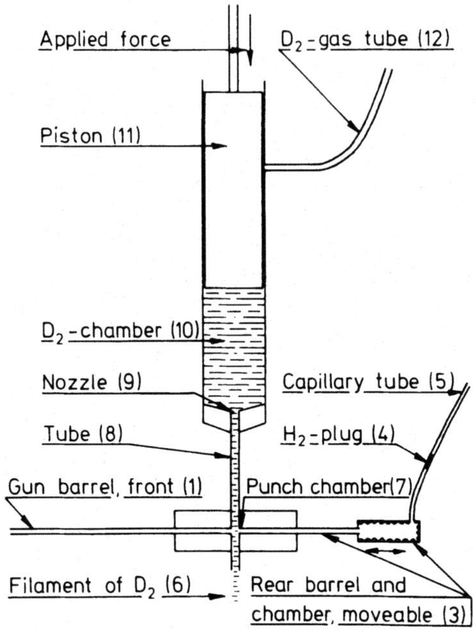

Fig. 1. Schematic drawing illustrating the fabrication of the pellet, the loading of the gun, and the acceleration of the pellet for the first gun.

As mentioned earlier the D_2 filament passes through a punch chamber. Here the filament crosses a channel in which the gun barrel is placed. The barrel is divided into two parts, front (1) and rear (3), the latter of which is moveable. During extrusion the rear barrel is drawn back. After extrusion it is moved forward and pushed through the D_2 filament so that a pellet is punched out of the filament. The pellet is pushed forward further, and when the two barrel parts meet, the pellet is deformed and squeezed into the gunbarrel. The filament is extruded through a tube with 1.2-mm i.d. The barrel has an outer diameter of 1.1 mm and an inner diameter of 0.75 mm. The pellet is thus reduced to 0.75-mm diameter when squeezed into the barrel and the length increased to 2.5 mm.

The pellet can now be blown out of the barrel when a pressure pulse is applied from the rear. The pressure is created in the chamber (3) that terminates the rear barrel. A small amount of solid H_2 is deposited inside this chamber, and it is forced to evaporation very quickly by heating the chamber with an energy pulse. We heat the chamber by means of an electrical heater wound around it. The H_2 propeller gas is led into the chamber through a capillary tube (5) coming from a reservoir at room temperature. The capillary tube should be closed near the chamber so that the propeller pressure is not lost backwards through this tube. This closing is accomplished by means of a plug of solid H_2. The temperature of the capillary tube is adjustable at the plug position. When the tube is cold at this position the propeller gas led down from the reservoir will freeze here to form the plug. The plug is removed again by the local heating of the tube and the H_2 gas will then be cryopumped by the inside wall of the chamber, after which it is ready for use as propeller gas. The capillary tube may be closed again as soon as the tube is cold. Closing of the capillary tube could also take place by means of a small mechanical valve.

The operation of the gun is accomplished in three steps each of which is run automatically. In the first step a measured amount of propeller gas is placed in the moveable chamber and the capillary tube is thereafter closed with a plug of solid H_2. The next step is the extrusion, which is observed with a TV-camera; several heaters are used here before a filament of D_2 appears. In the third step, the pellet is punched out of the filament and loaded into the gun, after which the gun is ready for firing. Before firing, the gun barrel and the punch chamber should be heated to 10-12 K. Without this heating the pellet cannot be blown out of the barrel.

The time to go through the different steps before firing totals 140 seconds. After firing, a pause of 100 seconds is required during which the chamber (3) cools down. The propeller gas is also pumped away during this interval, but this pumping takes place rather quickly since the cryostat acts as a cryopump with a large pumping area.

The performance of the gun was tested several times, and data exist for the gun's properties. The data are shown in Table II, while a shadow photograph of a pellet is shown in fig. 2.

The advantage of the acceleration method described here is that the amount of propeller gas used is fairly small, which limits the extent of differential pumping needed when injecting into a tokamak. The method apparently has two disadvantages: One is that the velocities obtained are rather moderate. The other that there is a fairly large delay between the firing pulse and the appearance of the pellet. It is possible that these two properties can be improved by using a different method for heating the propeller gas.

D_2 pellet
$2.3\ mm \times 0.6\ mm^{\varnothing}$ - $0.6\ mm^3$
0.125 mg - 2×10^{19} molec.
Velocity - 200 m/s
Spread in direction below $\pm 0.2^{\circ}$

Gun
Barrel length - 60 mm
Firing pulse - 4 to 5 Joule
Delay after pulse - 200 to 300 ms
4 to 5 minutes between shots
H_2 propeller gas - 5 bar•cm^3,
$\sim 12 \times 10^{19}$ molecules

Table II. Data for the first pellet gun.

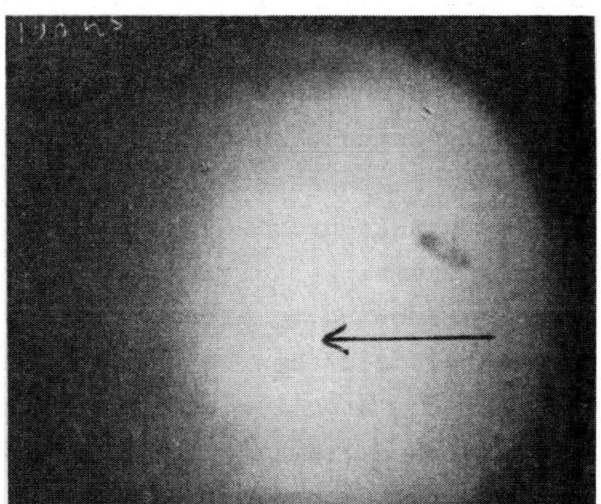

Fig. 2. A photograph of a moving pellet made with an exposure time of 100 ns. The direction of the movement is shown with an arrow.

3. PELLET INJECTION

The pellet gun should be used for firing pellets into the tokamak and with the angular spread in direction of less than 0.2° there was no difficulty in hitting a port in the tokamak with the gun placed 1- to 2-m away. To limit the flow of propeller gas from the gun a 1-mm aperture was placed a few centimeters in front of the gun's muzzle. The aperture was made by shooting several pellets through a 16-μm Al foil placed in front of the gun.

To interphase the tokamak and the pellet gun we need to know the time of appearance of the pellet and its velocity. To this end the pellet passes through an observation chamber placed immediately after the above-mentioned aperture. Two light sources are placed on one side of this chamber and opposite these there are two vertical slits, 1-mm wide and 10-cm apart. Photodiodes are placed behind these slits. When a pellet passes a slit the light intensity to a photodiode is reduced for a short period. The dip in intensity is transformed into an electrical pulse. The velocity is then inferred from the time difference between the two pulses. The signals are fed to a computer which fires the tokamak at a preset time before the arrival of the pellet.

Another problem concerns the conditions in the tokamak following the injection of the pellet. The pellet is rather large; the quantity of matter in a pellet is actually 20 times that in the tokamak gas. Only a part of the pellet material was ablated by the tokamak plasma; the rest of the pellet hit the inner wall and was scattered back into the plasma as a large number of small fragments. The pellet was thus suddenly transformed into an equal mass of smaller pellets with a much larger total surface. This, of course, is disastrous for the measurements of the interaction between the pellet and the plasma. It would improve the experiment if one could shoot a pellet through the tokamak and then collect it following this so that the remaining pellet material would not disturb the plasma. One way of doing this would be to shoot the pellet in through a port in the top and out through a port in the bottom of the tokamak. Such a vertical trajectory is also better suited for observation of the pellet-plasma interaction than is the horizontal one used originally. The pellet gun sends the pellets out horizontally. It cannot be mounted to shoot through the tokamak directly. One way of solving the problem would be to use a curved gun barrel to provide a vertical trajectory. Another way is to shoot the pellet through a curved tube of diameter larger than the pellet. The pellet would then slide on the tube wall on a layer of evaporated pellet material, hopefully without too much loss of material, and the pellet could then be guided into any trajectory. It was shown that this injection method worked well with a brass tube having a radius of curvature of 0.7 m. The tube had an inner bore of 4 mm with a smooth looking surface. The pellets are now injected through a 2.8-m long guide tube and they may be collected after they have passed through the tokamak. Another advantage of the guide tube is that it provides excellent separation between the gun and tokamak vacuum systems.

It is tempting to attempt this injection method with a longer guide tube. The method could then be used at larger tokamaks where it may be difficult to find room for a pellet gun close to the tokamak. The gun could be placed at a large distance from the tokamak without requiring a straight sight tube. The gun could be mounted outside possible radiation shields with good access for maintenance and making modifications. Furthermore, the use of guide tubes will provide great flexibility in the choice of the pellet entrance. It may even permit successive firing of several pellet guns through a single port. Finally, a long guide tube will provide good separation between the vacuums of a pellet gun and a tokamak.

Pellets were therefore shot through a 20-m long guide tube. A brass tube of an inner bore of 4 mm with a smooth-looking surface and a radius of curvature of 1.5 m was used. The pellet was photographed after having passed the tube, and its size was measured by determining the pressure increase in a known volume after evaporation. For these measurements the pellet velocity was somewhat below 200 m/s. Averaged over several shots we found an input-to-output velocity ratio of 1.4±0.1 and an average material loss of 25% at an entrance velocity of 137±22 m/s. The shadow photographs showed that very little had happened to the shape of the pellet.

It thus seems possible to inject a pellet through a long guide tube without serious loss of material, at least at moderate velocities. What happens at higher velocities we do not know, but it might be worthwhile to investigate the method at those velocities. Pellet parameters such as velocity, shape, temperature, and size could be important, while guide-tube parameters such as material, inner smoothness, radius of curvature, and tube temperature likewise would be important.

4. A NEW PELLET GUN

It was mentioned earlier that the pellets were too large for our tokamak, the amount of material in a pellet being 20 times the amount of material in the tokamak plasma. We are therefore building a new pellet gun fabricating pellets with a nominal size of 0.4 mm in length and diameter. This pellet gun is being tested and only a limited amount of data exists for it.

The gun is again built up under the bottom of a liquid helium cryostat (Oxford Instruments MD4A) and is again protected against thermal radiation by means of radiation shields. Connections from outside are led in as for the first gun.

The working principle may be demonstrated by means of the schematic drawing in fig. 3. The pellet is again made from a filament of solid D_2 (6) that is extruded from a cylinder with solid D_2 (10). However, it is now extruded through the side of the D_2 chamber (10) near the bottom. The filament (6) first passes through the nozzle (9), then through a hole in a thin moveable plate (the pellet transporter (7)),and finally through a hole in a second plate (the knife plate (8)) so that it now becomes visible outside the knife plate.

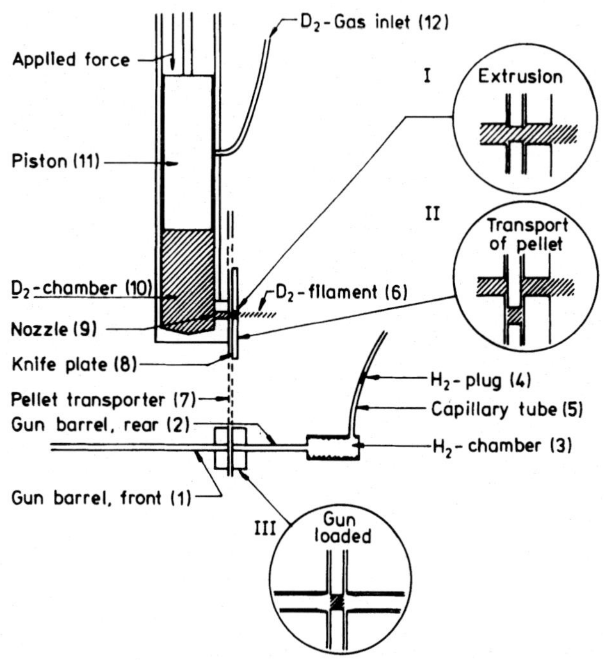

Fig. 3. Schematic drawing illustrating the fabrication and transport of the pellet, the loading of the gun, and the acceleration method for the second gun.

The gun barrel is seen in the lower part of the drawing. It is again divided into two parts, (1) and (2). The pellet transporter (7) may be moved in the space between the two barrel parts in such a manner that the hole in the transporter becomes aligned with the two barrel parts. The gun is loaded by moving the pellet from the extrusion position to the firing position by means of the transporter. This may be illustrated by means of the three small drawings at the right side of fig. 3. In the upper drawing the transporter is seen in the extrusion position. The hole in the transporter is a bit smaller than those in the nozzle and knife plate, making the positioning of the transporter less critical. In the next drawing, the transporter plate has been moved downwards and the filament has been cut by the sharp edges of the holes in the nozzle and the knife plate. A pellet now rests in the transporter. The length is equal to the plate thickness, 0.4 mm, and the diameter is equal to the hole diameter, 0.4 mm. In the third drawing the transporter has been moved to the position where the hole, and thus also the pellet, is positioned in the gun barrel. The gun is now loaded. This loading method is similar to that used at IPP Garching.

This gun is also fired by pulse evaporation of solid H_2.

The operation of the gun has so far been accomplished only manually and thus only a limited amount of data have been determined. These data are listed in Table III.

D_2 pellet - 0.4 mm × 0.4 mm$^{\emptyset}$
0.05 mm^3, 0.01 mg, 1.6 × 10^{18} molec.
Velocity - above 300 m/s
Barrel length - 57 mm
H_2 propellant gas - 5 bar•cm^3
∿ 12 × 10^{19} molecules

Table III. Data for the second pellet gun, preliminary.

5. CONCLUSION

In order to study the pellet-plasma interaction we built a tokamak, called Dante, as well as pellet guns. In this report we discussed one gun which has been in use for some time and a second which is nearly completed, together with the problems met during the injection of pellets. The results obtained to date on the pellet-plasma interaction have been published elsewhere [5,6].

6. REFERENCES

[1] CHANG, C.T., JØRGENSEN, L.W., NIELSEN, P., LENGYEL, L.L., Nuclear Fusion 20, 859-893 (1980).

[2] MILORA, S.L. and FOSTER, C.A., Rev. Sci. Instr. 50, 482-487 (1979).

[3] IPP Garching, Annual Report 1978.

[4] NORDSKOV, A., SKOVGÅRD, H., SØRENSEN, H., and WEISBERG, K.V., Risø M-2245 (1980).

[5] ANDERSEN, V., GADEBERG, M., JENSEN, P.B., and NIELSEN, P., Bull. Am. Phys. Soc. 24, 1028 (1979).

[6] ANDERSEN, V., GADEBERG, M., JENSEN, P.B., and NIELSEN, P., 1980 IEEE International Conference on Plasma Science, University of Wisconsin, May 19-21, 1980.
IEEE Catalogue No. 80CH1544-6 NPS.

CONCEPTUAL DESIGN OF CLUSTER ION ACCELERATOR WITH MOVING POTENTIAL WELLS

Hiroshi Enjoji
Department of Electrical Engineering,
Faculty of Engineering, Mie University
1515 Kamihama-cho, Tsu 514, Japan

ABSTRACT

The deuterium cluster ions of specific sizes less than 1000 atoms/unit charge(a positive elementary charge) in a beam of an equivalent current of 1000 A_{eq} are accelerated with the potential wells(troughs) in accelerated traveling waves on a delay line from 3.10 x 10^7 cm/s(1 keV/atom) to 3.10 x 10^8 cm/s(100 keV/atom) at an acceleration of 5 x 10^{12}cm/s^2. The potential depth of the wells, the total length of the delay line and the net power required are found to be 26 kV, 100 m and 100 MW, respectively, by fundamental calculation.

1. INTRODUCTION

The beam of cluster ions is expected to give several advantages to the additional injection heating of hot plasmas.[1)] Cluster ions are aggregates of 10 to 10^6 constituent atoms. They should have energies more than 100 keV/atom in order to achieve the efficient heating of hot plasmas. This needs 1 to 10^5 MeV/cluster ion. The high voltages generated and useful for this aim are restricted to several MV due to the limits of engineering ability. Hence it is very difficult to accelerate cluster ions to such high energies as above by conventional methods. Several methods for the injection have been carried out [2)] and proposed,[3)] but they are not considered to utilize the characteristics of cluster ions satisfactorily. In the case of this acceleration, the extremely high voltages are not necessary.

The beam of cluster ions contains those of various specific sizes and so the usual methods for the acceleration of charged particles can not be used. In order to overcome these difficulties, the collective acceleration of cluster ions by trapping in a moving potential well has been investigated theoretically[4)] and the conceptual design of the delay line which produces moving potential wells(troughs of potential waves) has been carried out.

The moving potential well must increase its velocity(phase velocity of the potential wave), when it proceeds on the delay line to accelerate the cluster ions trapped in itself. The phase velocity depends upon the inductance and the capacitance per unit length of the delay line. If the inductance and the capacitance per unit length decrease continuously with increasing distance from the sending end of the delay line, the phase velocity will increase as the well proceeds and the acceleration of the trapped cluster ions will be achieved. All the cluster ions trapped in the well get almost the same final velocity, and the energies given to all the constituent atoms are nearly the same.

The deuterium cluster ions are initially accelerated in a DC potential drop of 1 MV and then injected into the cluster ion accelerator. The

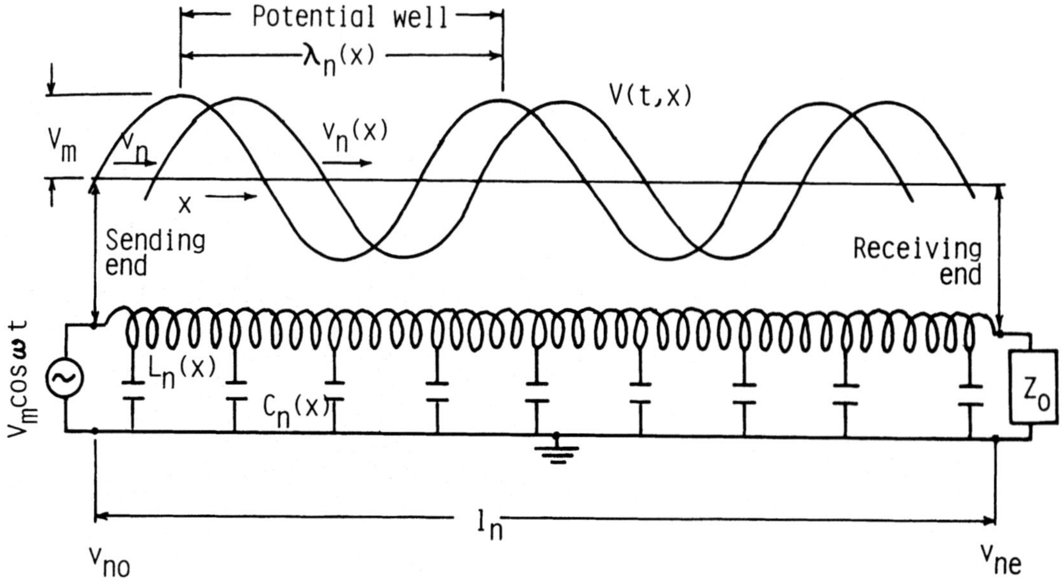

Fig.1 Schematic diagram of the n-th section of delay line

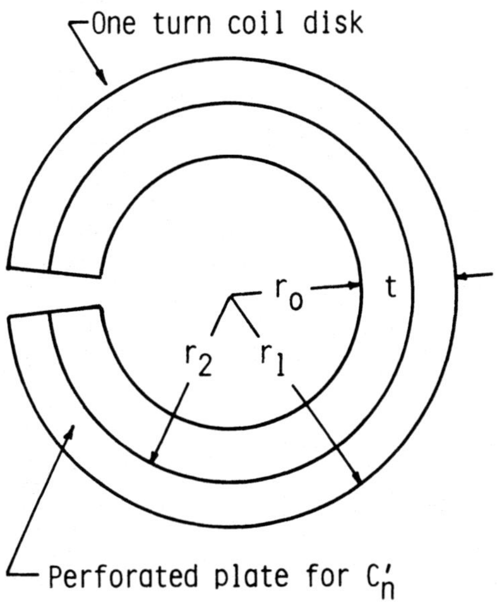

Fig.2 One turn coil disk and its parameters

accelerator is divided into nine sections, each of which has a reionizer which ionizes neutral fragments separated from their respective parent cluster ions by mutual collisions and collisions with residual gas molecules, a focusing device, a group of the accelerating disk electrodes perforated at their centers and connected with the respective turns of the coil of the delay line, a large pumping system and a high-frequency oscillator with large output power. In this paper the conceptual design of the delay line system is described.

If this acceleration is successful and used for heating the hot plasmas in large fusion reactors, the number of the beams injected will be decreased extremely compared with the present case of neutral beam injection.

2. FUNDAMENTAL CALCULATION

After the acceleration by the DC potential drop of 1 MV, the initial velocity of the cluster ions of specific size N/Z=1000 atoms/unit charge v_o reaches

$$v_o=3.10 \times 10^7 \text{ cm/s (1 keV/atom)}.$$

By the acceleration with moving potential wells, their final velocity v_f is increased ten times as large as v_o, that is

$$v_f=3.10 \times 10^8 \text{ cm/s (100 keV/atom)}.$$

Therefore, the phase velocity of the moving potential well v(x) also must increase in the same way as above.

The electric power to accelerate per unit length of the delay line p is made constant with respect to the distance from the sending end of the delay line x,

$$p=d[(1/2)I_{oo}m_o v^2(x)]/dx=I_{oo}m_o dv(x)/dt=I_{oo}m_o\alpha=\text{const.},\quad(2.1)$$

where I_{oo} is the equivalent current of the cluster ion beam, m_o the mass of a deuterium atom and α =dv(x)/dt. The acceleration α also is constant. From eq.(2.1), v(x) can be derived as $v^2(x)=v_o^2 + 2\alpha x$, and v(x) is approximately written as

$$v(x)\fallingdotseq v_o + \alpha x/v_o,\quad (v_o^2 \gg 2\alpha x).\quad(2.2)$$

If it is specified that $\alpha l/v_o=0.3v_o$, the last velocity v_e at the receiving end of the delay line becomes 1.3 times as large as v_o, where l is the length of the delay line. In order to reach $v_{ne}= v_f = 3.10 \times 10^8$cm/s, the delay line must be divided into 9 sections and in the n-th section,

$$\left.\begin{array}{l} v_{ne}= 1.3v_{no}=v_{(n+1)o} \\ v_{no}= 1.3^{n-1}v_{1o}= 1.3^{n-1}v_o\ ;\ v_{ne}= 1.3^n v_{1o}= 1.3^n v_o, \end{array}\right\}\quad(2.3)$$

and

$$v_f= v_{9e}= 1.3^9 v_{1o}= 10.60v_o,\quad(2.4)$$

where v_n is the phase velocity of the n-th section and $v_{1o}= v_o$. All these relations are shown in Fig.1.

The wave length $\lambda(x)$ increases with x, because v = fλ, where f is the frequency of the potential wave. In all the sections, their initial and last wave lengths have the same values and are specified to be λ_o= 6.20 cm and λ_e= 8.06 cm in view of electrical breakdowns between the turns of coil of the delay line and between the accelerating electrodes.

In each section the frequency of the potential wave has the following relation

$$f_n= 1.3^{n-1}f_1,\quad(2.5)$$

by which the electric field E_m is maintained to remain in a small range in each section through the relation $E_m= V_m\lambda(x)/(2\pi)$, where E_m and V_m are the amplitudes of electric field and potential, respectively. The value of f_1 is determined to be 5 x 10^6Hz by v_o and λ_o.

The net power needed in the n-th section P_n is

$$P_n=(1/2)I_{oo}m_o(v_{ne}^2 - v_{no}^2) = 1.69^{n-1}P_1, \quad (2.6)$$

where

$$P_1=(1/2)I_{oo}m_o(v_{1e}^2 - v_{1o}^2) =(1/2)I_{oo}m_o(v_e^2 - v_o^2). \quad (2.7)$$

The length of the n-th section is obtained as

$$l_n= P_n/p = 1.69^{n-1}P_1/p = 1.69^{n-1}l_1 , \quad (2.8)$$

where $l_1= P_1/p$.

The accelerating electric field E_m has the following relation,

$$Nm_o\alpha = ZeE_m , \quad (2.9)$$

where Z is the number of unit charges for a cluster ion. If the value of E_m is specified, the other parameters are evaluated. In view of high-frequency electrical breakdowns in vacuum and the rational structure of the coil of the delay line, the value of E_m is selected to be 1×10^4V/cm. Then

$$\alpha = 4.79 \times 10^{12} \text{ cm/s}^2,$$

where N/Z = 1000 atoms/unit charge.

As already specified, I_{oo}= 1000 A_{eq}= 6.25×10^{21} atoms/s. Hence, by eq.(2.1) $p = 1 \times 10^4$ W. By using the value of v_o and by eq.(2.3), the initial and last velocities of each section can be obtained. P_n and l_n are evaluated from the value of p and by eqs.(2.6), (2.7) and (2.8). The results are shown in Table I. In this case the inductance and the capacitance per unit length of the delay line $L_n(x)$ and $C_n(x)$ are obtained by using the following relations,

$$v_n(x) = [L(x)C(x)]^{-1/2} = f_n(\lambda_o + c_nx), \quad (2.10)$$

$$L_n(x) = [\xi_n(\lambda_o + c_nx)]^{-1} \quad ; \quad C_n(x) = [\eta_n(\lambda_o + c_nx)]^{-1}, \quad (2.11)$$

$$\xi_n\eta_n = f_n^2 \quad ; \quad (\eta_n/\xi_n)^{1/2} = R_o, \quad (2.12)$$

where $c_n=\alpha/(f_nv_{no})$ by eq.(2.2), R_o the common characteristic impedance through all the sections of the delay line and the value of the smallest inductance $L_9(l_9)$ is specified to be 1 μH/cm in view of the rational structure of the coil. In order to make E_m to be useful at the longer wave length λ_e= 8.06 cm, the amplitude of the potential wave is selected to be

$$V_m= E_m[\lambda_e/(2\pi)] = 1.28 \times 10^4 \text{ (V)}, \quad (2.13)$$

and the depth of the well is

$$V_d= 2V_m= 2.56 \times 10^4 \text{ (V)}.$$

3. CONCEPTUAL DESIGN OF DELAY LINE SYSTEM

From a practical standpoint, the design method for the delay line system has been considered. The delay line system is defined as a series connection of all the sections of the delay line and some of the sections are composed of the element delay lines connected in series-parallel. The description of this chapter is mainly of the element delay lines. One turn of the coil of element delay line is made of a circular disk perforated at its center. The coil pitch is one turn per cm, and the capacitance is formed by inserting a perforated circular plate in the gap between the coil disks. The radii of the hole of the disk, the disk itself and the plate hole are denoted by r_o, r_1 and r_2. The coil width t is defined as

$$t = r_1 - r_o , \quad (3.1)$$

as shown in Fig.2. The insulation between the disk and the plate is performed by teflon sheet of thickness d=0.2 cm.

The upper limit of electric field is determined to be $E_d = 5 \times 10^4$ V/cm by using SF_6 gas($\geq$1 atm) from the viewpoint of high-frequency electrical breakdowns. All the element delay lines are set in SF_6 gas. The skin depth is 10 μ at f=40 MHz, and the conducting material for the coils and the electrodes are made of laminated copper foils of 10μ in thickness.

If the current density in the coil and the output power of a unit high-frequency oscillator are 2 A/mm^2 and 2 MW, respectively, the current through element delay line is I'_m=200 A and the cross-sectional area of conductor of the disk is 100 mm^2. The apparent thickness of the disk q is 0.5 cm including the insulating films between the foils, and the net thickness q' is specified to be 0.25 cm. Hence, t becomes 4 cm.

If the parameters P'_n, V_m, v_{no}, t and d for an element delay line in the n-th section are given, the current I'_{mn} , the characteristic impedance R'_{no} , $L'_n(0)$ and $C'_n(0)$, r_o, r_1 and r_2 are obtained for the element delay line by the following relations, (g:power magnification)

$$P'_n = gP_n/(uw) \qquad (3.2) \qquad\qquad I'_{mn} = 2gP_n/(uw)/V_m \qquad (3.3)$$

$$R'_{no} = V_m/I'_{mn} \qquad (3.4) \qquad\qquad L'_n(0) = R'_{no}/v_{no} \qquad (3.5)$$

$$C'_n(0) = (R'_{no}v_{no})^{-1} \qquad (3.6)$$

$$r_o = \left\{-2.14t + [(2.14t)^2 - 4\pi(0.285t^2 - 8 \times 10^7 L'_n(0))]^{1/2}\right\}\frac{1}{2\pi} \qquad (3.7)$$

$$r_1 = r_o + t \qquad (3.8) \qquad\qquad r_2 = [r_1^2 - 4d \cdot C'_n(0)/2/\varepsilon]^{1/2} , \qquad (3.9)$$

where (uw) is the number of the element delay lines connected in series-parallel. Equation(3.7) is deduced approximately from the inductance of a single-layered coil of which the cross-section of conductor is rectangular. The structure of the coil of element delay line is shown in Fig.3. The parameters of an element delay line in each section are shown in Table II . The values of the inductance and the capacitance per unit length of each section calculated from $L'_n(x)$ and $C'_n(x)$ may be different from those in Table I, but the phase velocities in the corresponding sections are quite the same.

4. CONCLUSION

This method is suited to the acceleration of cluster ions and can utilize fully the characteristics of them and a strong injection heating will be carried out by a large equivalent current of the cluster ion beam and sufficient energies of constituent atoms. This accelerator will be practically constructed by using technologies developed so far in the fields of electrical , electronical, vacuum and accelerator engineerings. The electron ring device [5] for this aim also will be a good accelerator for cluster ions, however, its realization should be studied from a practical point of view.

5. REFERENCES

[1] BURROWS,B.J.C., DAWSON,P.G., MOSSON,G.A.G., TAY,E.S. and WATSON,H.H.H., 5th Symposium on Fusion Technology, Oxford, 2-5 July, 1968.(Paper 51)

[2] HAGENA,O.F. and HENKES,W., 9th Symposium on Fusion Technology, Garmisch-Partenkirchen, 14-18 June, 1976. Proceedings pp 885-890

[3] BECKER,E.W., FALTER,H., HAGENA,O.F., HENKES,W., KLINGELHÖFER,R., MOSER,M. O., OBERT,W. and POTH,I., Nucl.Fus. 17, 617-619 (1977)

[4] ENJOJI,H., International Conference on Plasma physics, Nagoya, 7-11 April, 1980. Proceedings Vol.1 p.291

[5] BOTTIGLIONI,F., COUTANT,J. and FOIS,M., Instr.& Method 106, 127-133 (1973)

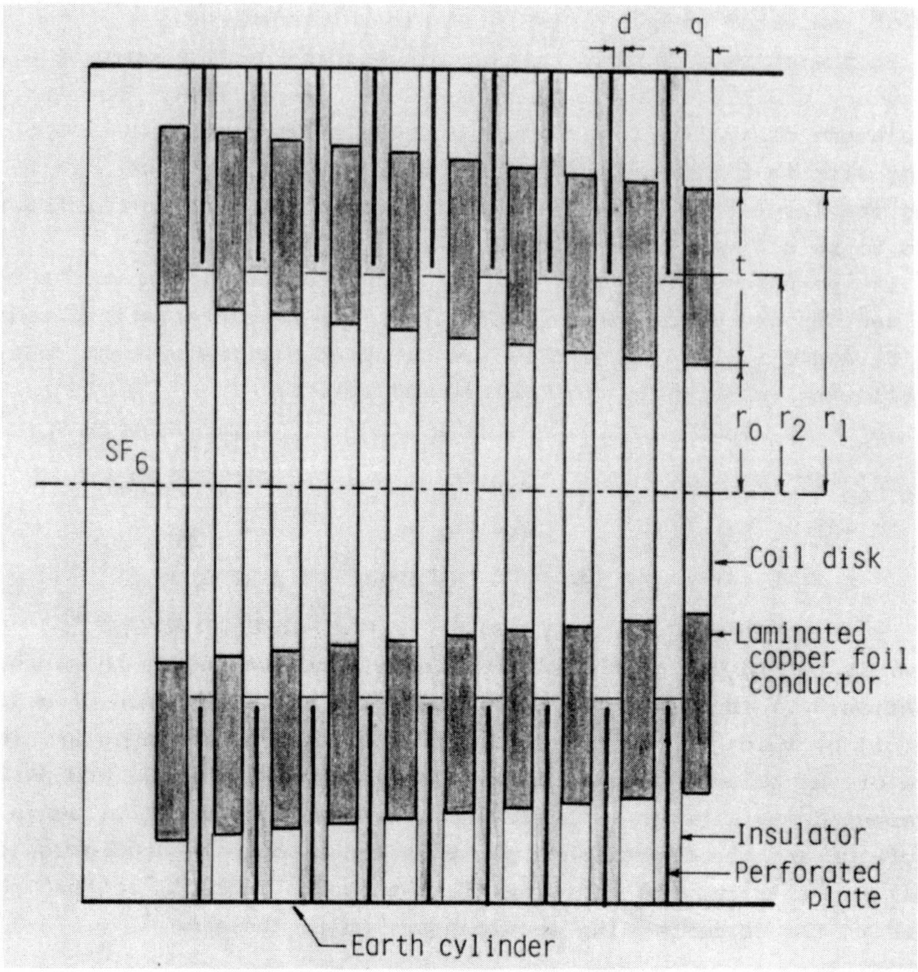

Fig.3 Conceptual drawing of a part of element delay line

TABLE I

Characteristics of Delay Line Obtained by Fundamental Calculation

(Σl_n=112 m)

Section No.n	x=0 x=l_n (cm)	$v_n(x)$ (cm/s)	f_n (MHz)	$\lambda(x)$ (cm)	$L_n(x)$ (μH/cm)	$C_n(x)$ (pF/cm)	P_n(net) (MW)
1	0 69.3	3.1 x 10^7 4.0 "	5.0	6.20 8.06	10.60 8.16	98.0 75.4	0.7
2	0 116.7	4.0 " 5.2 "	6.5	6.20 8.06	8.16 6.28	75.4 58.0	1.2
3	0 198.0	5.2 " 6.8 "	8.5	6.20 8.06	6.28 4.83	58.0 44.6	2.0
4	0 334.5	6.8 " 8.9 "	11.0	6.20 8.06	4.83 3.71	44.6 34.3	3.4
6	0 955.4	11.5 " 15.0	18.6	6.20 8.06	2.85 2.19	26.4 20.3	9.6
9	0 4611.3	25.3 32.9	40.8	6.20 8.06	1.30 1.00	12.0 9.3	46.3

TABLE II

Characteristics of Element Delay Line in Each Section

Section No.n	u;w	P'_n (MW)	v (10^7 cm/s)	t (cm)	I'_{mn} (A)	R'_{no} (Ω)	$L'_n(0)$ (μH/cm)	$C'_n(0)$ (pF/cm)	r_o (cm)	r_1 (cm)	r_2 (cm)	l_e (cm)
1	1;1	1.4	3.10	4	140	143	4.60	226	9.48	13.48	11.88	69.3
2	1;1	2.36	4.03	7	236	85	2.10	293	5.02	12.00	9.55	116.7
3	1;2	2.00	5.24	6	200	100	1.91	191	4.99	11.00	9.31	198.0
4	2;2	1.68	6.81	4	168	119	1.75	123	5.34	9.34	8.07	167.3
"	"	"	7.83	"	"	"	1.52	107	4.89	8.89	7.73	"
5	2;3	1.89	8.85	"	189	106	1.19	107	4.19	8.19	6.91	282.7
"	"	"	10.20	"	"	"	1.04	93	3.81	7.81	6.65	"
6	3;3	2.13	11.50	"	213	94	0.82	93	3.24	7.24	5.97	318.5
"	"	"	12.65	"	"	"	0.74	84	3.03	7.03	5.86	"
"	"	"	13.80	"	"	"	0.68	77	2.85	6.85	5.75	"
7	4;4	2.03	15.00	"	203	99	0.66	68	2.78	6.78	5.81	403.7
"	"	"	18.40	"	"	"	0.54	55	2.39	6.39	5.56	"
8	7;4	1.96	19.50	"	196	102	0.52	50	2.34	6.34	5.59	389.8
"	"	"	24.50	"	"	"	0.42	40	1.96	5.96	5.32	"
9	12;4	1.93	25.30	"	193	104	0.41	38	1.93	5.93	5.32	384.3
"	"	"	31.00	"	"	"	0.33	31	1.62	5.62	5.10	"

V_m= 2 x 10^4 V, d = 0.2 cm , l_e: length of element delay line, g = 2

ION BEAM GENERATION WITH AN INVERSE REFLEX TETRODE

K.Kasuya, K.Horioka, K.Kamada, S.Kato,
T.Takahashi, H.Kanazawa and K.Niu
Department of Energy Sciences
Tokyo Institute of Technology
Nagatsuta, Midori-ku, Yokohama, Japan 227

ABSTRACT

Characteristics of an inverse reflex tetrode were investigated to get high voltage and high power pulsed ion beams. The tetrode was fed by a coaxial blumlein line, which was charged by a marx generator. Charge neutral proton beams of 150 KeV were extracted with a propagation angle of about 1~1.5 degree. The time of flight and the energy of the beams were measured with a modified Faraday cup and a calorimeter.

1. INTRODUCTION

Reflex triodes provide us high voltage and high power ion beams to make or heat plasmas [1],[2] in the various fields of controlled thermonuclear fusion research. Because the conventional reflex triodes extract ion beams through the virtual cathodes which fluctuate in space and with time, the beam quality is not so good for the plasma applications. On the other hand, an inverse reflex tetrode [3] extracts ion beams through a real mesh cathode, and a favorable beam quality is expected, as follows.

(A) The zero potential surface for reference is fixed in space.

(B) The ion beams are charge-neutralized easily.

(C) Comparatively cold ion beams are obtained.

These characteristics are favorable for the beam transport (propagation and focussing). As the working impedance of the inverse reflex tetrodes is also very low and constant with time, a large current can be obtained with a low impedance power suppling line.

Fig. 1 shows the difference between the conventional reflex triode and the inverse reflex tetrode. The anodes (A.) are plastic thin films and are applied positive high voltage pulses. Electrons emitted from the cathodes (K.) are accelerated toward the anodes, and make reflexes via the anodes. Virtual cathodes (V.K.) are formed at the places near to the mean turning point of the reflex electrons.

This report is only a preliminary one to clarify the above characteristics in future in more detail. Our experimental facility and our preliminary experimental results of an inverse reflex tetrode are described shortly.

2. EXPERIMENTAL FACILITIES

A schematic diagram of the experimental facilities is shown in fig. 2. From the left to right in the same figure, there are a marx generator, a coaxial water blumlein line with a water sparc gap switch, and an inverse reflex tetrode (anode and cathode). The output voltage of the blumlein line is measured with a $CuSO_4$ voltage divider, and the tetrode total current is measured with a Rogowski coil. An axial magnetic field is applied along the axis of the vacuum vessel near to the tetrode with the pulsed external magnetic field coils, if needed.

Though the open-circuit maximum output voltage and energy of the marx generator are 500 KV and 5.2 KJ, experiments were performed with the line charging voltage of 150 KV. The open-circuit output voltage of the blumlein line in this case is about 300 KV. The output impedance of the line is 16 Ω.

The anode stalk is a thin aluminum cylinder with a diameter of 50 mm and a length of 115 mm. The cathode is made with a stainless steel mesh of 65% transparency, and the diameter is 30 mm. At the open end of the aluminum stalk, a polyester thin films of 4∿188 μm thickness are attached as anodes. Ion beams are extracted through the cathode mesh into the drift tube, in which a modified Faraday cup (50 mm in diameter) is located to measure the ion current along the axis. Instead of this modified Faraday cup, a calorimeter (50 mm in diameter) is used to measure the total beam energy. The vacuum vessel is evacuated with a vacuum pump to about 5×10^{-5} torr. Earlier detection of ion beams before the use of the modified Faraday cup or the calorimeter was performed with a paper for the blue copy, developing with a liquid after an ion beam irradiation.

3. RESULTS AND DISCUSSIONS

Typical CRT waveforms of the tetrode are shown in fig. 3. Fig. 3(A), (B) and (C) are the voltage measured with the $CuSO_4$ divider in fig. 2, the total tetrode current with the Rogowski coil and the ion beam current measured with the modified Faraday cup. The vias voltage of the Faraday cup is -450 V. The distance between the anode foil and the Faraday cup is 150 mm. The sensitivity of the Faraday cup is about 30 A/V. When the vias voltage is not applied to the Faraday cup, the beam current is not detected. This means that the beam is almost charge-neutral.

The same kinds of waveforms in fig. 3(C) are taken at three different positions (15, 35 and 55 cm from the anode foil) of the same Faraday cup. The time of flight of the ion beams versus the distance from the anode is shown in fig. 4. One is the plot for the wave fronts of the beam currents, and the other for the beam peak currents. The mean energy of the ion beam is estimated to be about 150 KeV for the both cases. This value corresponds to the proton speed of 5.5×10^6 m/sec.

The peak beam currents along the propagation axis measured with the Faraday cup and the calorimeter are shown in fig. 5 and 6. As the diameters of the beam detectors are 50 mm, the detectors can not cover the whole area of the beam at the second and third positions (35 and 55 cm). At the first position (15 cm) they can cover the whole area of the beam, and the values shown in fig. 5 and 6 are the total beam current and energy. The beam divergence angle is about 1∿1.5 degree with these figures.

Virtual cathode formation is very important for the inverse reflex tetrode to work. To check the formation of the virtual cathode, an additional metal plate is inserted into the anode stalk and the anode depth for the electron reflex is changed. The beam current is measured also with the

Faraday cup. Fig. 7 is the result of the peak current versus anode depth. As may be seen in this figure, an anode depth of 5 cm or so is necessary to form a virtual cathode. This is the case with no external magnetic field. The same kind of result with the external magnetic field of 4 KG is shown in fig. 8. This figure resembles qualitatively to fig. 7, while the current value is about two times larger in fig. 8 than in fig. 7.

The experimental results mentioned above are obtained with the anode foil thickness of 4 μm. Next, we change the thickness of the anode foil to see the reflex effect of the electrons. The ion beam current measured with the Faraday cup versus the foil thickness is shown in fig. 9. When the foil is thicker than 20 μm, the beam current does not change with the thickness or the external magnetic field. Electrons seem to penetrate through the foil once at most, in this case. On the other hand, the beam current increases, when the foil thickness is reduced with the external magnetic field. The reflex effect seems to be remarkable in this case. The maximum ion beam intensity is about 10 A/cm^2. The beam current decreases without the B field, in the thinner case.

4. CONCLUSIONS

Preliminary experiments to extract ion beams from an inverse reflex tetrode are performed with a pulse beam technology. Charge-neutral proton beam can be produced with a small divergence angle of 1∿1.5 degree. Although the peak beam current and the beam intensity (∿10 A/cm^2) is small until now, the virtual cathode formation is assured.

Each anode of the tetrode must be replaced with a new one after an ion beam extraction, if a thin plastic film is used as the plasma source. Although another type of beam source, the magnetically insulated diode can produce ion beams repetitively with one thick solid plastic anode, the charge neutrality of beams is not enough to be propagated along a long distance, compared with the reflex triodes. On the other hand, the efficiency of the reflex triodes for repetitive ion beam extraction with plastic meshes becomes smaller than the case with thin plastic films, because the electron reflex is more disturbed and the plasma ion source is not spatially uniform by the meshes, themselves.

Above considerations induce us to develop new anodes for reflex triodes, in order to extract high voltage and high power ion beams efficiently and repetitively with sufficient charge neutrality. For this purpose, a new type of anode plasma formation must be devised in future. We are now trying to test such a new idea.

5. ACKNOWLEDGEMENTS

A part of the blumlein line and an oscilloscope were lent by the Institute of the Laser Engineering, Osaka University and the Institute of Plasma Physics, Nagoya University. Authors thank to them, and also to the technical assistances of Messrs. K.Shioda, J.Yoshida, M.Okamoto, K.Shiomi, Y.Moriwaki, Y.Miyata, R.Okumura and Miss M.Sugawara.

6. REFERENCES

[1] KAMADA, K. et al., J. Phys. Soc. Japan 44, 1963 (1979)

[2] KAPETANAKOS, C.A. et al., Phys. Rev. Lett. 44, 1218 (1980)

[3] PASOUR, J.A. et al., Appl. Phys. Lett. 36, 646 (1980)

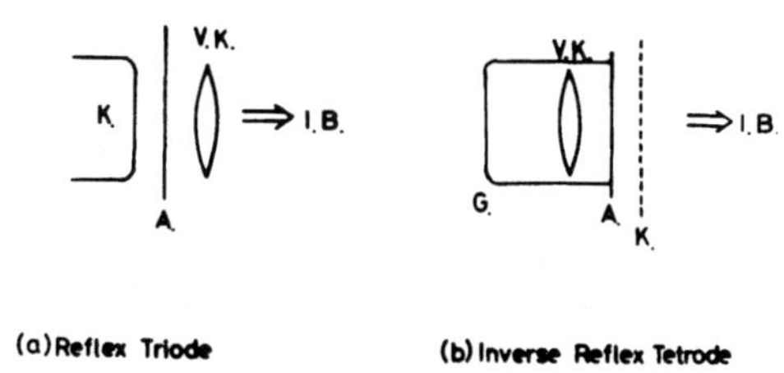

Fig.1. Comparison between two kinds of reflex ion beam sources

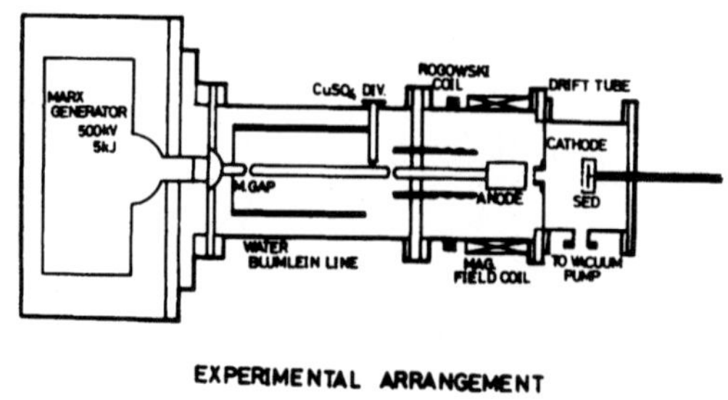

Fig.2A. Schematic diagram of experimental facilities

Fig.2B. Photograph of experimental facilities

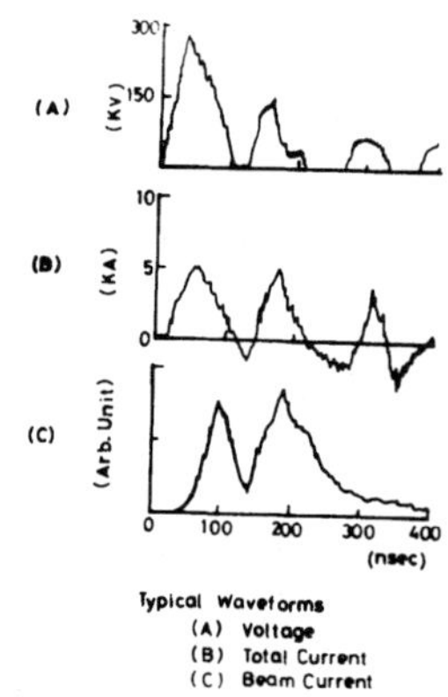

Fig.3. Typical CRT waveforms of IRT, (A) Voltage, (B) Total current, (C) Ion beam current

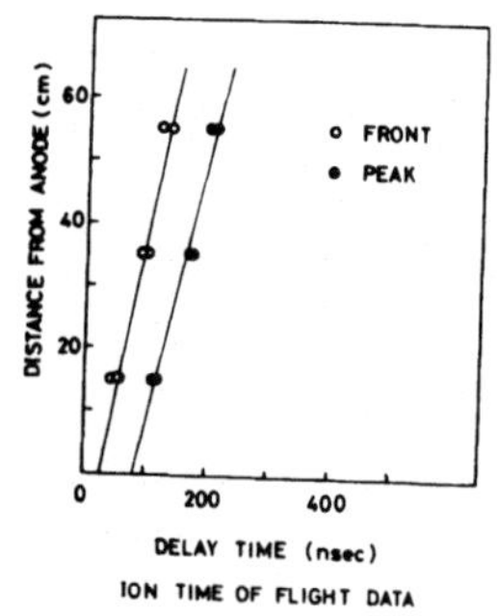

Fig.4. Time of flight of proton beams

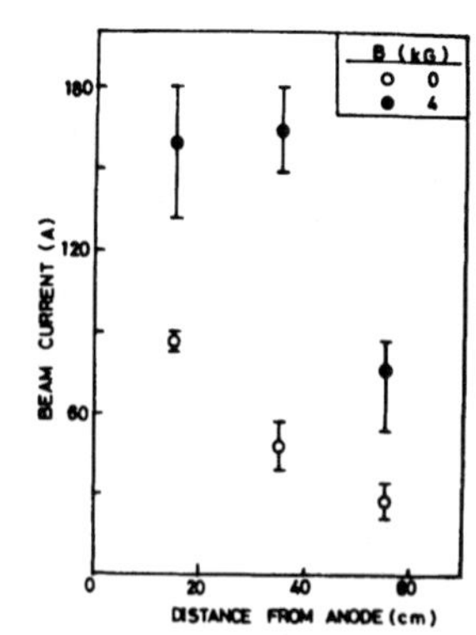

Fig.5. Peak current with a Faraday cup along the propagation axis with and without external magnetic field

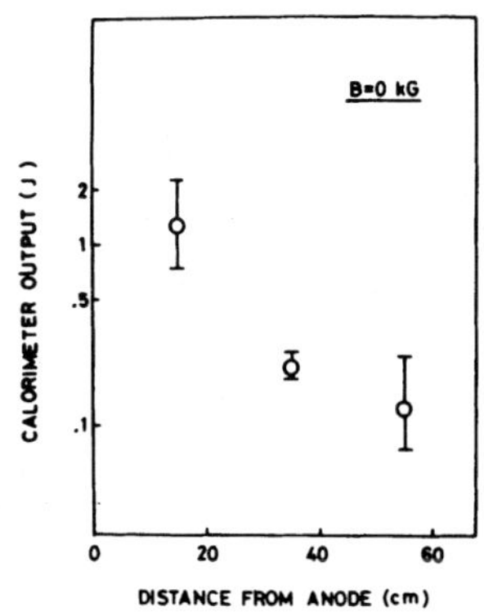

Fig.6. Energy measured with a calorimeter along the propagation axis

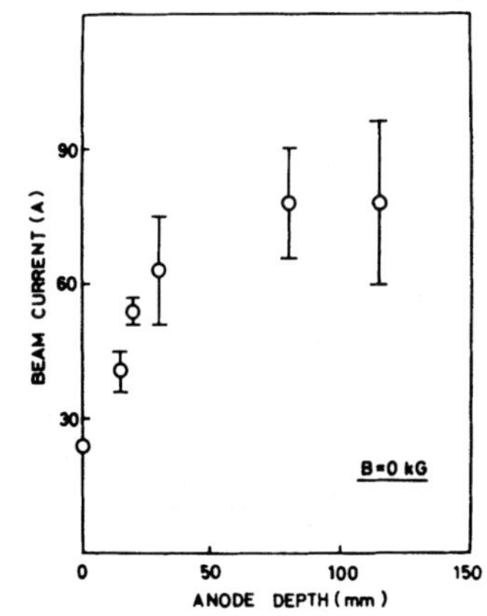

Fig.7. Peak ion current versus anode depth without axial magnetic field

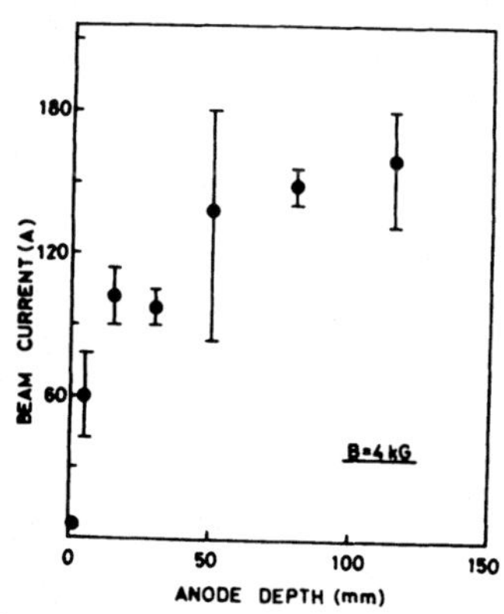

Fig.8. Peak ion current versus anode depth with axial magnetic field of 4KG

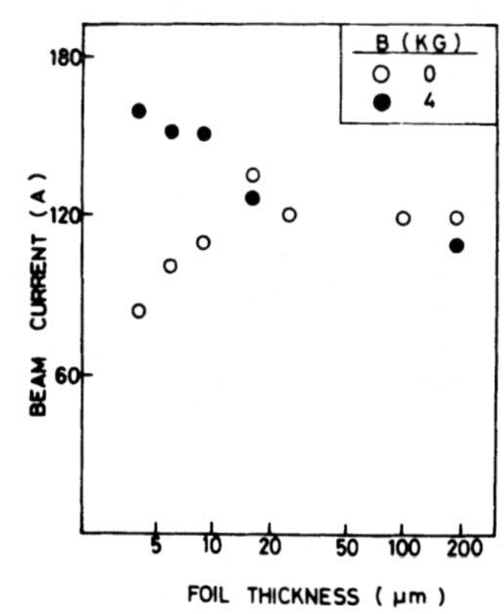

Fig.9. Peak ion current versus thickness of polyester anode foil

Reactor Studies

A Tandem Mirror Reactor With
An Inherent Thermal Barrier*

T. Kammash
D. L. Galbraith
University of Michigan
Ann Arbor, Michigan 48109 USA

ABSTRACT

The Basic Tandem Mirror Configuration is evaluated as a reactor allowing for unequal plug and central-cell electron temperatures as means of simulating the effect of a "thermal barrier". It is shown that the ratio of fusion power density to the power loss per unit volume in the plugs does increase when the temperatures are allowed to be unequal but simultaneously the central cell to plug volume ratio decreases sufficiently to result in a reduction in the Q-value of the system as compared to the case when the two temperatures are equal.

1. INTRODUCTION

The Tandem Mirror Concept has recently been advanced[1,2] as a promising scheme for enhancing the Q (ratio of fusion power to injected power) value of mirror machines which normally suffer from low Q-values as a result of excessive end losses. In the early assessment of these devices as potential power reactors[3,4] the plasma parameters that characterize the system were deduced on the assumption that the electron temperatures in the plugs and the central cell are the same. The electrostatic confinement for the fuel ions is maintained in the basic Tandem configuration by keeping the plug density well above that in the central cell. If the electron temperature is the same in both regions then the confining potential difference $e\,\Delta\varphi$ is just

$$e\,\Delta\varphi = T_e \ln\left(\frac{n_p}{n_e}\right) \tag{1}$$

where T_e is the (uniform electron temperature, n_e is the electron density in the central cell, and n_p is the electron density in the plugs. Since in the central cell the electron and ion temperatures are comparable, to within a factor of roughly two, and since good electrostatic confinement is possible only if $e\Delta\varphi$ is considerably larger than the fuel ion temperature T_c , it follows that n_p must be considerably larger than n_e . Unfortunately the mass and energy loss rates from the plugs scale almost exactly proportionally to n_p^2 , while the value of n_e is fairly well fixed by the required power density. Therefore good fuel ion confinement means a large energy loss rate from the plugs which keeps Q fairly low.

It has been pointed out[5] that this limitation can be overcome by maintaining the plug electrons at a hotter temperature than the central cell electrons. The resulting "thermal barrier", however, adds complexity (and possibly severe MHD instabilities) to the magnetic field configuration and requires "pumping out" of ions in a region of space between the central cell and the plugs. It does, nevertheless, result in a very considerable

*Work supported by US DOE

decoupling of the confining potential difference and the plug-to-central-cell density ratio.

It is natural, therefore, to inquire whether it might be possible to retain the relatively simple magnetic configuration of the "basic" Tandem Mirror Reactor while still enhancing its Q-value somewhat by allowing the plug electron temperature to rise higher than that of the central cell electrons. We address this question in this paper.

Analysis and Results

In carrying out the reactor analysis of a Tandem Mirror it has been customary to assume that the fusion reactions take place only in the central cell so that the particle and energy balance equations in this region include those of the thermal alpha particles in addition to the electrons and fuel ions. These equations along with those of the electrons and ions in the plugs are given elsewhere[3] and will not be reproduced here; only the modifications resulting from unequal plug and central cell electron temperatures will be presented here. Since it has also been shown previously[3,6] that thermal alphas tend to degrade the Q-value we will ignore them in this analysis and thus keep in mind that the results we obtain are generally better than they actually are.

If the central cell electron temperature is T_{ce} and the plug electron temperature is T_{pe} then the counterpart of Eq. (1) for the confining potential has been shown[7] to be

$$e\Delta\varphi \simeq T_{pe}\ln\left[\frac{n_p}{n_e}\left(\frac{T_{ce}}{T_{pe}}\right)^{1/2}\right] \tag{2}$$

With separate densities and temperatures for the plug and central cell electrons the appropriate electron balance equations in the steady state become:

a) Plug mass balance equation

$$\frac{n_p^2}{(n\tau)_p} - \frac{n_p^2}{(n\tau)_{pe}} - \frac{n_p^2}{(n\tau)_{pce}} + \frac{V_c}{V_p}\frac{n_e^2}{(n\tau)_{cpe}} = 0 \tag{3}$$

b) Central cell mass balance equation

$$\frac{n_c^2}{(n\tau)_c} + \frac{1}{2}n_c^2\langle\sigma v\rangle - \frac{n_e^2}{(n\tau)_{ce}} - \frac{n_e^2}{(n\tau)_{cpe}} + \frac{V_p}{V_c}\frac{n_p^2}{(n\tau)_{pce}} = 0 \tag{4}$$

c) Plug energy balance equation

$$\frac{n_p^2\left(E_p - \frac{3}{2}T_{pe}\right)}{(n\tau)_{iep}} + \frac{n_p^2 E_{inj}F_{eaux}}{(n\tau)_p} - \frac{n_p^2\bar{E}_{Lpe}}{(n\tau)_{pe}} - R_{EP}$$
$$- \frac{n_p^2\bar{E}_{Lpce}}{(n\tau)_{pce}} + \frac{V_c}{V_p}\frac{n_e^2\left(\bar{E}_{Lcpe} + e\Delta\varphi\right)}{(n\tau)_{cpe}} = 0 \tag{5}$$

d) Central cell energy balance equation

$$\frac{\frac{3}{2} n_c n_e (T_c - T_{ce})}{(n\tau)_{iec}} + \frac{1}{4} n_c^2 \langle \sigma v \rangle f_\alpha Q_e - \frac{n_e^2 \bar{E}_{Lce}}{(n\tau)_{ce}}$$

$$- R_{RC} - \frac{n_e^2 \bar{E}_{LCpe}}{(n\tau)_{cpe}} + \frac{V_p}{V_c} \frac{n_p^2 (\bar{E}_{Lpce} - e\Delta\varphi)}{(n\tau)_{pce}} = 0 \qquad (6)$$

In Eq. (3) the first term represents the fueling rate since it is equal to that of the ion. The second term represents the electron confinement loss to the outside while the third term represents the confinement loss to the central cell. The last term in that equation denotes a source of particles that reflects the rate of electron escape from the central cell to the plugs. Since the fueling rate in the central cell (which is equal to that of the ions) must accomodate the confinement loss as well as the fusion burn up it is given by the first two terms in Eq. (5). The remaining terms in that equation are identical to their counterparts in Eq. (4).

In the plug energy balance equation given by (5) the first term represents the energy exchange between the electrons and ions in that region with the denominator showing the time constant for that exchange. The term " F_{eaux}" denotes the ratio of radiative to neutral injection heating so that the third term in (5) represents the radiative heating of the electrons in the plugs. With $\bar{E}_{Lpce}$ representing the average energy of an electron escaping from the plugs into the central cell the last term in (5) denotes the energy supplied to the plug electrons by the central cell electrons which must overcome the potential $e\Delta\varphi$ in order to enter the plugs. The radiative loss term R_{RP} in this (and in the subsequent) equation contains both the bremsstrablung and synchrotron radiation losses for which the standard[6] formulas are used. The terms in Eq. (6) are identical to those in Eq. (5) with the exception of the second term in which f_α represents the fraction of the fast (fusion) alpha particles that do not escape from the system at birth, and Q_e denotes the fraction of the fast alpha energy that is transferred to the electrons as it thermalizes[3]. It should be noted that, since for the electrons the plug is a potential well, we can use the standard Pastukhov[8] formulas for $(n\tau)_{pce}$ and $\bar{E}_{Lpce}$; unfortunately we have no formulas for $(n\tau)_{cpe}$ and $\bar{E}_{Lcpe}$.

The four balance equations should (since the densities are determined elsewhere) be used to calculate the two electron temperatures, T_{pe} and T_{ce} plus the two electrostatic potentials $e\varphi_c$ and $e\Delta\varphi$ (or alternatively $e\varphi_p$). However, since we have no formulas for $(n\tau)_{cpe}$ and $\bar{E}_{Lcpe}$ we shall use Eq. (2) to calculate $e\Delta\varphi$, use Eq. (3) to calculate $(n\tau)_{cpe}$ and use Eqs. (4), (5) and (6) to calculate T_{pe}, T_{ce} and $e\varphi_c$. We shall further assume that

$$\bar{E}_{Lcpe} = \kappa T_{ce} \qquad (7)$$

where κ is an arbitrary parameter.

We should also note here that there should be one additional term in Eqs. (5) and (6). The central cell electrons pass fairly freely through the plug region with only a small fraction of these "passing" electrons being scattered sufficiently to become trapped in the plugs and thus contribute to $(n\tau)_{cpe}$ and $\bar{E}_{Lcpe}$. However, the remaining "passing" electrons act as a high energy beam, and like such a beam, they do lose some energy to the ambient (plug) electrons. If we denote by W_{cp} the rate at which energy is transferred from this "beam" to the plug electrons (per unit volume in the plug) then the right-hand side of Eq. (5) should be $-W_{cp}$ instead of zero, and the right-hand side of (6) should be $V_p/V_c\,(W_{cp})$. For the purposes of this calculation, however, we shall not try to evaluate W_{cp}, but shall set it equal to zero and use the parameter K in (7) to provide a rough approximation (qualitatively, though not quantitatively) to this effect.

As noted earlier, we solve Eq. (3) for $(n\tau)_{cpe}$ and when the result is substituted in Eq. (4) a combined plug-central cell mass balance equation results, namely

$$\frac{n_c^2}{(n\tau)_c} + \frac{1}{2} n_c^2 \langle \sigma v \rangle - \frac{n_e^2}{(n\tau)_{ce}} + \frac{V_p}{V_c}\left\{ \frac{n_p^2}{(n\tau)_p} - \frac{n_p^2}{(n\tau)_{pe}} \right\} = 0 \tag{8}$$

It is equally useful to have a combined plug-central cell energy balance equation. After some manipulation the equation of interest can be put in the form

$$\frac{3}{2}\frac{n_c n_e (T_c - T_{ce})}{(n\tau)_{iec}} + \frac{1}{4} n_c^2 \langle \sigma v \rangle f_\alpha Q_e - \frac{n_e^2 \bar{E}_{Lce}}{(n\tau)_{ce}} - P_{RC}$$
$$+ \frac{V_p}{V_c}\left\{ \frac{n_p^2 (E_p - \frac{3}{2} T_{pe})}{(n\tau)_{iep}} + \frac{n_p^2 (E_{inj} F_{eaux} - e\Delta\varphi)}{(n\tau)_p} - P_{RP} \right.$$
$$\left. - \frac{n_p^2 (\bar{E}_{Lpe} - e\Delta\varphi)}{(n\tau)_{pe}} \right\} = 0 \tag{9}$$

It is interesting to note at this point that the above energy balance equation contains the additional factor

$$\frac{V_p}{V_c} n_p^2 \left[\frac{1}{(n\tau)_{pe}} - \frac{1}{(n\tau)_p} \right] e\Delta\varphi$$

when compared to the same equation for the equal electron temperature case[6]. This term accounts for the fact that electrons are introduced by neutral injection into the plugs at the rate $n_p^2/(n\tau)_p$ but lost directly to the outside world at the different rate $n_p^2/(n\tau)_{pe}$. This difference is made up by electrons flowing, between plugs and central cell, over the potential difference $e\Delta\varphi$. Since normally $(n\tau)_{pe}$ is larger than $(n\tau)_p$ there must be a net flow from plugs into the central cell, and since these electrons must climb a potential wall to get there, there is a net energy transfer from the electrons to the electrostatic potential.

In order to perform the desired calculations we shall assume that the average escape energies that appear in the balance equations are given by

$$\bar{E}_{Lpe} = e\varphi_p + T_{pe} = e\varphi_c + e\Delta\varphi + T_{pe} \tag{10}$$

$$\bar{E}_{Lce} = e\varphi_c + T_{ce}$$
$$\bar{E}_{Lpce} = e\Delta\varphi + T_{pe} \tag{11}$$

recalling in the meantime that $\bar{E}_{Lcpe}$ is given by Eq. (7). With these assumptions the system of equations is solved numerically and selected results of these calculations are shown in table 1 for two injection energies, 1200 keV and 800 keV. For purposes of comparison the table also shows the results for equal electron temperatures i.e., for T_{pe} equal to T_{ce}, for two values of K, and for several values of electron auxiliary heating ratio, F_{eaux}.

We see from table 1 that not only is "Q" less when the two electron temperatures are allowed to vary separately, but also Q decreases as the temperature difference increases. The Q-value increases, however, as F_{eaux} increases. Further examination of this surprising result reveals that although the ratio of fusion power per unit volume produced in the central cell to the loss power per unit volume from the plugs does indeed increase when the two electron temperatures are allowed to be unequal; and further this ratio increases as K, and thus T_{pe}/T_{ce} increases, the volume ratio V_c/V_p seems to decrease dramatically precipitating a reduction in the Q-value.

When numerically investigating the cause of reduction of Q by examining the expression for V_c/V_p obtained from Eq. (9) it is seen that the most dramatic change can be traced to the term $(e\varphi_p + T_{pe})/(n\tau)_{pe}$. The confinement parameter $(n\tau)_{pe}$ decreases exponentially with T_{pe} and as this temperature is increased (for K values above unity which are physically meaningful) the dramatic change in this term accounts for the significant decrease in the volume ratio V_c/V_p. It appears possible, but by no means certain, that the volume ratio would increase if both T_{ce} and T_{pe} were reduced in such a manner that their ratio remained roughly constant. One way in which this might be accomplished is by introducing an impurity (high Z) ion in the central cell so as to enhance the bremsstrahlung losses in this region. Preliminary calculations based on this approach showed that increasing the bremsstrahlung by a factor of ten did result, in one instance, in mildly reducing the two temperatures at a roughly constant ratio, but this effect was far outweighed by changes in the other factors contained in the volume ratio with the end result that a significant drop in V_c/V_p, and correspondingly in the Q-value, took place.

References

[1] Fowler, T. K., Logan, B. G., Comments Plasma Phys. Controlled Fusion 2, 167 (1977)

[2] Dimov, G. I., et al, Plasma Physics and Controlled Nuclear Fusion Research 3, 177 (1977) IAEA, Vienna

[3] Kammash, T., Galbraith, D. L., Tenth Symposium on Fusion Technology, Padua, 1978. Proceedings P. 141.

[4] Moir, R. W., et al Lawrence Livermore Laboratory Report UCRL-52302, July 15, 1977

[5] Baldwin, D. E., Logan, B. G., Phys. Rev. Letters, 43, 1318 (1979)

[6] Galbraith, D. L., Kammash, T., Equilibrium Conditions for A Tandem Mirror Reactor, EPRI Report (to be published)

[7] Cohen, R. H. et al, Lawrence Livermore Laboratory Report UCRL-84147, March 27, 1980

[8] Pastukhov, V. P., Nucl. Fusion 14, 3 (1974)

TABLE I

Plasma and Reactor Parameters

For T_c = 30.0 kev

E_{inj} (kev)	F_{eaux}	$T_{pe}=T_{ce}$	K	T_{ce} (kev)	T_{pe} (kev)	E_p (kev)	$(n\tau)_p$ (sec/cm^3)	V_c/V_p	Q
1200	0.00	Yes	-	53.46	53.46	1121.96	2.458×10^{14}	184.17	2.176
		No	1.00	53.65	58.31	1156.64	2.656×10^{14}	91.55	1.243
			1.50	52.49	66.15	1117.26	3.915×10^{14}	-4.00	-0.078
	0.25	Yes	-	53.57	53.57	1128.65	2.416×10^{14}	385.18	3.609
		No	1.00	54.03	61.91	1195.04	2.684×10^{14}	228.57	2.661
			1.50	55.75	86.69	1387.28	3.265×10^{14}	80.65	
	0.50	Yes	-	53.61	53.61	1131.34	2.399×10^{14}	591.56	4.603
		No	1.00	54.29	65.18	1224.28	2.748×10^{14}	354.93	3.686
			1.50	56.01	89.65	1411.88	3.292×10^{14}	166.25	2.692
	1.00	Yes	-	53.64	53.64	1133.68	2.384×10^{14}	1008.85	5.869
		No	1.00	54.72	71.18	1273.56	2.881×10^{14}	567.31	4.988
			1.50	56.43	94.98	1452.11	3.372×10^{14}	317.85	4.157
800	0.00	Yes	-	49.61	49.61	882.44	1.529×10^{14}	65.87	0.506
		No	1.00	49.29	51.19	871.99	1.750×10^{14}	6.77	0.059
			1.50	-	-	-	-	-	-
	0.25	Yes	-	49.99	49.99	905.17	1.375×10^{14}	377.81	2.171
		No	1.00	50.24	57.25	950.64	1.559×10^{14}	119.07	0.859
			1.50	-	-	-	-	-	-
	0.50	Yes	-	50.10	50.10	911.62	1.332×10^{14}	779.32	3.654
		No	1.00	50.73	62.17	998.06	1.519×10^{14}	338.15	2.163
			1.50	52.14	83.64	1147.34	1.728×10^{14}	123.55	1.170
	1.00	Yes	-	50.17	50.17	915.92	1.303×10^{14}	1625.96	5.634
		No	1.00	51.40	70.96	1067.12	1.548×10^{14}	748.76	4.148
			1.50	52.89	92.36	1213.81	1.692×10^{14}	426.58	3.277

ENERGY ACCOUNTING FOR HEAVY ION FUSION REACTOR WITH VARIOUS DRIVER ACCELERATORS IN COMPARISON WITH REFERENCE POWER PLANT OF TOKAMAK TYPE

Akira Miyahara
Institute of Plasma physics, Nagoya University
Nagoya, Japan

Sunao Kawasaki
Kanazawa University, Kanazawa, Ishikawa
Japan
and
Susumu Ito and Kikuo Nakajima
Toshiba Corporation, Yokohama, Japan

ABSTRACT

Energy analysis is applied to inertial fusion reactors with various kinds of heavy ion accelerator drivers. Energy costs for the fusion power plants are calculated directly from the estimated weights of the component materials and the production energies of the raw materials per unit mass. The results are compared with the energy cost for the construction and the maintenance of the magnetically confined fusion plant of Tokamak type, to show that the heavy ion fusion (HIF) systems are comparable with the Tokamak and not far from being practical in the energy term.

1. INTRODUCTION

Only recently recognized were the importance and the utility of the energy analysis, particularly in the evaluation of a new future energy source which would replace the present fossil fuel. The net energy produced by any power plant in its lifetime, could be calculated and compared with, at least in principle, with the energy investment, which should be put into for the construction and the maintenance of the plant through the life, whether it consumes a fissionable material or some other renewable resources. Chapman[1] applied the energy analysis to estimate the future outlook of the nuclear fuel as an alternative to the crude oil, in deep doubt if the nuclear power plant could eventually pay from an energy standpoint. The concept of the energy analysis is so fundamental and could be so crucial in the energy strategy for the coming century, that it should be made clear for any possible future power plant, prior to the detailed argument on its physical structure. Rombough and Koen developed a model of the energy analysis applicable to energy systems and compared various nuclear power stations[2]. Magnetically confined fusion program was also checked in a similar way to reveal that the situation might be more unfavourable than that of the fission power[3]. Although still immature, the heavy ion fusion scenario[4] could be investigated in the energy term in our previous paper[5].

The study, however, has to necessarily involve a considerably wide range of statistical works covering the nearly whole field of the productive activity of the heavy industry. It highly depends on the energy flow model chosen in the analysis and it may suffer from a rather large uncertainty in the results. We here elaborate the previous work, assuming an improved and self-consistent design of the accelerator-reactor composite, to compare with the energy cost newly evaluated for a reference design of the commercial Tokamak reactor.

2. RECIPE OF ENERGY ANALYSIS

Once anyhow the energy flow diagram is fixed for the energy investment in the assumed power production system there are two approaches for calculating total energy investment in the system. One is used in the reference[2], where the fundamental assumption is a linear dependence of the output energy in any sector of the manufacturing processes to the monetary sale, i.e. total monetary output of the sector. The co-efficients of the conversion from the monetary language to the energy one are obtained and all sector's energy inputs are summed up to get the total energy investment, using the energy input-output matrix technique compiled by Herendeen. The details should be referred for to his original paper[6]. For the analysis available are the statistical data tables of the commercial and the industrial activities, published regulaly in many advanced countries[7].

The total energy investment thus calculated would be much more stable while the co-efficients of the monetary and the energy output to the ones are time dependent as well as the monetary sales. Monetary cost for any item would certainly fluctuate sporadically and continue to be raised up in the long-term. So far most energy analysis adopted this approach on account of the convenience of the statistical works, as done in the study for the fusion reactor[3].

The other way is called a direct approach where the direct energy consumptions for all phases of the production process are summed up and added to the indirect energy costs related to the process. Any productive process includes numerous indirect energy consumptions and the indirect cost could override the direct one. The direct approach may involve intricate energy flows and many technical difficulties, which would lead to large errors in the calculation. Our previous work[5] is based on the direct approach because of its simplicity of the principle. In this report, we make two major improvements in our analysis, in assuming more detailed designs of the accelerator-reactor assembly, especially in the inertial fusion reactor composite, and in making an erergy analysis for a reference fusion reactor of Tokamak type by the same direct approach technique as used for HIF concept, to compare with the HIF analysis for one side with the Tokamak energy study by the Input/Output matrix method[3] for the other side.

In the direct approach the energy plant is decomposed into all component materials with the estimated weights of which the unit respective production energies are multiplied to give the total direct energy investment. The energy related to the manufacturing and the assembling porcesses should be added to thereafter. The unit invested energies for the production of main raw materials are listed in table I.

TABLE I

Invested energy of raw materials

Item	Energy (x 10^7 J/kg)
Hot roll steel	2.18
Cold roll steel	1.17
Copper	7.50
Stainless steel	8.37
Superconductor	16.75
Li_2O	41.87
Cement	0.34
Alminium	12.98

The main parameters of the three chosen types of HIF scenarios are based on the designs presented in the reference[5]. Schematic diagrams of the accelerators are shown in fig.1. Some remarks on the design policies and the parameter considerations are given in the following.

2.1 RF linac based system

All the three systems are assumed to produce the same beam power of 30MJ/sec, corresponding to the thermal output of 3000 MW_t (with a moderate energy multiplication factor of 100), in accordance with a set of three conceptual reactor vessels anticipated to work, each of which is rated to have a capacity of 1000 MW_t. RF linac and storage ring combinations are due to designs presented by D.E. Young at '77 HIF workshop and by Storage Ring Group in '79 workshop. Repetition cycle is increased to be 10 Hz. No major change of other parameters were not supposed.

2.2 Synchrotron based system

The parameters of the synchrotron based system are modified considerably. The use of the superconductive storage ring is newly adopted to reduce the resistive loss in the coils producing the guide field, lest the circulating power should override the nominal output electric power. Increase of the repetition cycle up to 30 Hz may cause serious technical difficulties in the synchrotron operation, we, however, pass over the details and leave the problem to future investigation. Basic concept of the scenario is due to Hearthfire Reference Design by ANL.

If we choose a lower repetition rate than 30 Hz, keeping the output energy and the parameters of the synchrotrons and the storage rings unchanged, their numbers have to be increased because the allowable particle number accelerated in such circular machine is tightly limited with space charge effect on their stability characters, in contrast with the cases of the ordinary RF linacs and the induction linacs, where the increased repetition would give little influence on the machine operation.

2.3 Induction linac based system

The repetition cycle of 30 Hz is also assumed for the induction linac operation. It is not expected to meet any particular obstacle. It has not been demonstrated up to how many amperes the heavy ion particles could be transported stably in the induction linac The beam intensity assumed in the reference design, however, is supposed to be far below the critical value.

2.4 HIF reactor structure

No systematic investigation on the optimum reactor structure for HIF concept was not planned so far. A model similar to laser driven inertial fusion reactor is used here to calculate the energy investment. The cavity has a lithium wetted wall and an inner wall diameter of 9 m, which is surrounded by the blanket and the shield composite. Thermal output of the fusion reaction in the vessel is rated to be 1000 MW_t and the use of a set of three cavities is considered to operate at 3000 MW_t.

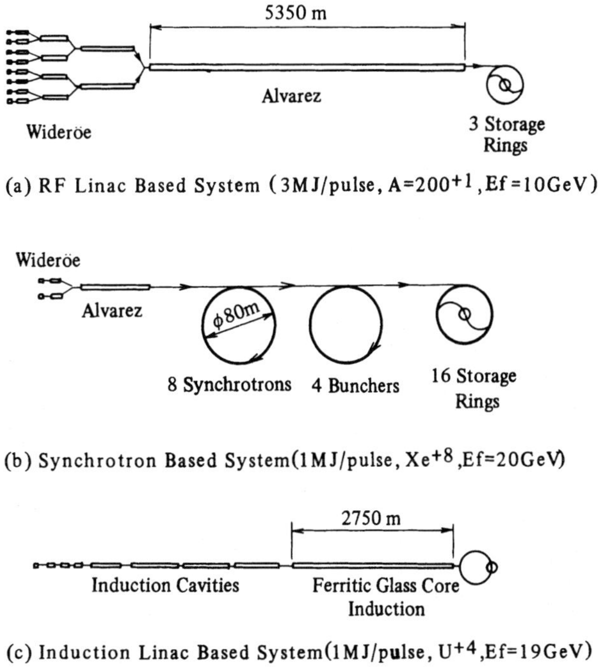

Fig. 1 Schematic diagrams of the Accelerators

TABLE II

Major parameters of Tokamak

Electrical power	1600 MW_e
Thermal power	3800 MW_t
Av. neutron wall load	3.5 MW/m^2
Major radius	7.0 m
Plasma half-width	1.94 m
Plasma elongation	1.6
Plasma current	11 MA
Safety factor (center)	1.4
Av. DT ion density	1.0×10^{20}
Av. ion temperature	17 keV
Current driven method	RF
Plasma heating method	RF
RF power to plasma	120 MW
Bt on axis	5.62 T
Bt max	11.0 T
No. of TF–coil	12
TF–coil material	Nb_3Sn/NbTi /Cu/Sus
Av. toroidal beta	0.067
Wall/Blânket structual material	Ferritic steel
Tritium breeding material	Li_2O
Net tritium breeding ratio	1.05
Blanket structure lifetime goal	8 yr

2.5 Tokamak reactor

The energy investment for the magnetically confined fusion reactor of Tokamak type using D-T mixture fuel, is calculated in order 1) to make compete HIF with Tokamak and 2) to compare the estimated energy inputs for Tokamak reactor with two approaches of the analysis stated before. The parameters on which the analysis is mainly based is due to "Starfire" Commercial Tokamak Reactor Concept, recently presented by ANL[7]. The electric power output has been selected to be about 1600 MW_e and the thermal output about 3800 MW_t, comparable with those assumed for HIF schemes. A continuous operation with the toroidal current drive by RF heating and a 75% plant availability is planned over the 30 years plant life. Major parameters are listed in table II. The recipe of the energy analysis for the Tokamak is the direct approach as used for the HIF programs and it makes possible comparison with the results for Tokamaks calculated previously, where used is the Input/Output Matrix method[3].

3. RESULTS AND DISCUSSION

Energy analysis is carried out over various stages of the accelerator assembly, which are devided into the essential elements such as the structures, vacuum systems, power supplies for the accelerating sections and the magnetic guides, housing and shield and so on.

Input energy costs including the direct and the indirect energy investments for the production of respective parts of these HIF systems are itemized and listed in table III, IV and V. The costs for reactor-generator parts are in table VI. Energy cost for the reference Tokamak reactor is shown in table VII.

In the energy term, RF linac and synchrotron depend strongly on how high repetition may be possible.

TABLE VII

Energy cost for the Tokamak reactor

Item		Weight (ton)	Energy (x 10^{15} J)
Tokamak reactor	TF–coil	25600	4.38
	EF–coil	1087	0.18
	Shield	24130	3.30
	Blanket	900 x 3	0.37
	Others	15455	2.26
Housing		360 x 10^4	3.01
Power supplies		–	7.72
Cryogenic system		10372	1.52
Vacuum system		962	0.08
Heat transfer system		46756	6.43
Turbine-generator system		43944	6.05
Total		377.1 x 10^4	35.31

TABLE VI

Energy cost for the reactor-generator system

(Unit 10^{12} J)

	Energy
Reactor	1143
Heat transfer system	4724
Turbine-generator system	4449
Housing	1879
Total	12195

TABLE III

Energy cost for the RF linac based system

(Unit 10^{12} J)

	Wideröe	Alvarez	Storage ring	Total
Accelerator structures	234	1830	834	2935
Vacuum system	0.7	3.8	5.4	9.9
RF power supplies	26.7	465	–	491.7
Magnet power supplies	–	2.5	50	52.5
Housing	177	1047	243	1467
Total	438.4	3348.3	1169.4	4956.1

TABLE IV

Energy cost for the synchrotron based system

(Unit 10^{12} J)

	Linacs	Synchrotron Rebuncher	Storage ring	Total
Accelerator structures	19.5	270	181	470.5
Vacuum system	0.1	42	26	68.1
RF power supplies	7.5	363	–	370.5
Magnet power supplies	–	510	670	1180
Housing	14.8	462	184	650.8
Total	41.9	1637	1061	2739.9

TABLE V

Energy cost for the induction linac based system

(Unit 10^{12} J)

	Pulsed drift tube	Low voltage induction	Fe-core induction	Ferritic glass induction	Total
Accelerator Structures	29	49	1968	2709	4755
Vacuum system	0.1	–	0.4	0.4	0.9
Power supplies	–	8	314	440	762
Housing	33	10	398	544	985
Total	72.1	67	2680.4	3693.4	6502.9

Fig. 2 shows the total energy input and the energy pay back time in years (defined by the total energy investment devided by net energy produced by the plant annually) as functions of the repetition frequency of the system operation. Remarkable improvements with increasing repetition are apparent for above two cases.
Fig. 3 shows the relative ratio of the energy costs of the respective components to the total energy.

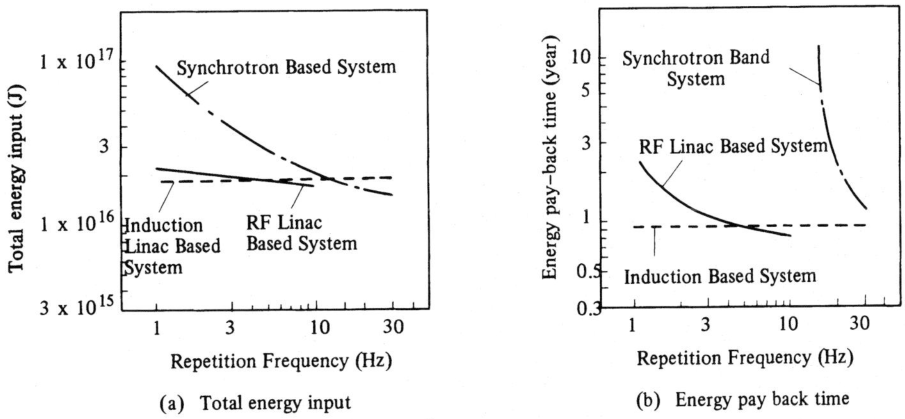

Fig. 2 Total energy input and energy pay back time for HIF systems

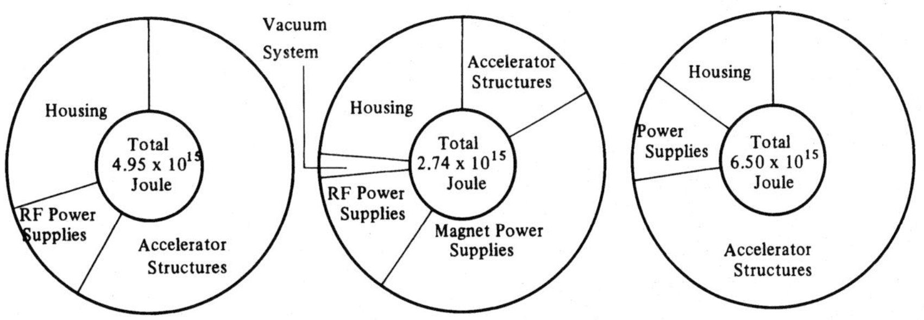

(a) RF Linac Based System (10Hz) (b) Synchrotron Based System (30Hz) (c) Induction Linac Based System (30Hz)

Fig. 3 Relative ratio of the energy costs of the respective components

We can make some conclusive remarks from these results :

1) Conceptual designs of HIF power plants having an energy pay back time of a few years are possible.
2) The situations for HIF and Tokamak are comparable in the energy term, instead of the apparent gross structure of the accelerators.
3) The HIF reactor cavity has the largest portion in the whole energy costs of HIF scenarios. The fact suggests an idea that the energy accounting of the inertial fusion program might not be sensible to the choice of the driver.
4) High repetition cycle is essential to get HIF concept practical, essentially in the synchrotron and the RF linac based systems, due to the circular beam accumulating element.
5) The energy cost of the Tokamak reactor is consistent with the EPRI's corresponding figure[3] within a factor of two. Then the difference between two approaches of the energy analysis would not make a great error.
6) The energy cost of HIF systems as well as the Tokamak is higher than that of the fission reactor of the comparable size. It, however, may not be far from being practical.
7) Our research leaves many problems unsolved, such as the fuel handling, environmental effects and some other subsidiary energy costs. The accuracy of the cost estimation, also has to be improved.

REFERENCES

[1] CHAPMAN, P., New Scientist 19, 866 (1974)

[2] ROMBOUGH, C.T. and KOEN, B.V., Technical Report ESL-31, Energy Systems Laboratories, College of Engineering, The Univ. of Texas, Austin (1974)

[3] EPRI-919 (1978)

[4] For the newest information of the heavy ion fusion concept and its hardware components, see the following. Proceedings of HIF workshops, held at Lawrence Berkeley Laboratory, Brookhaven and Argonne National Laboratories in 1977-79

[5] KAWASAKI, S. and MIYAHARA, A., IPPJ-454, Nagoya Univ. (1980) and Technical Digest of ICF/80 at San Diego (1980)

[6] For example see:
HERENDEEN, R.A., "An Energy Input-Output Matrix for the United States, 1963: User's Guide" Center for Advanced Computation, Univ. of Illinois, Urbana (1973)

[7] "Starfire" A Commercial Tokamak Reactor ANL/EPP/TM-125 Argonne National Laboratory (1979)

CULHAM CONCEPTUAL TOKAMAK REACTOR MkII
CONCEPTUAL LAYOUT OF BUILDINGS FOR A TWIN REACTOR POWER STATION

J.A.S. Guthrie and N.H. Harding
Associated Nuclear Services, 123 High Street, Epsom, Surrey, England.

ABSTRACT

This paper discusses the conceptual design of the nuclear complex of a 2400 MWe twin fusion reactor power station utilising common services and a single containment building. The design is based upon environmental and maintenance logistical requirements, the provision of adequate storage, workshop and construction facilities and the constraints imposed by the geometry of the main and auxiliary reactor coolant systems.

1. INTRODUCTION

Technical and economic evaluations of magnetic confinement fusion reactors indicate that, at economic power ratings, the first wall material will last only a few years and must be replaced several times during the overall reactor life.

In the Culham Conceptual Tokamak Reactor MkII (CCTRII) [1], fig. 1, the blanket structure is divided into 20 segments and because of the radiation hazard, special machines are provided for rapidly accessing and removing the segments [2-4]. The servicing machines include general purpose remote operated manipulators [5] and special purpose manipulator transporters (fig. 2) in addition to overhead cranes - also equipped with general purpose manipulators.

Clearly the access required by the servicing machines places a considerable design restraint on the location of services, coolant and power conversion equipment and has an overriding effect on the size and layout of the reactor hall. This paper describes the 'nuclear complex' of a fusion reactor power station containing two CCTRII reactors, each of 1200 MWe output and is a continuation of the work described in [6].

2. DESIGN CONSTRAINTS

2.1 Coolant systems

CCTRII produces 3400 MWt, 99% being deposited in the first wall/blanket/inner shield segment and preliminary studies [3,7] suggest that this heat can be recovered by a manifolded helium gas circuit (pressure 60 bar, inlet and outlet temperatures 350°C and 750°C respectively). The remainder (≅1%) passes to the outer shield which is cooled by water at atmospheric pressure [4]. In addition it is necessary to provide auxiliary cooling for the removal of afterheat from the segments (≅34 MW total) when the reactor is shut down [3], and when they are awaiting servicing.

2.2 Reactor services

Each reactor requires numerous service connections, for example: fuel gases, electrical power to injectors, vacuum pumps etc., connections between superconducting magnets and coils and energy storage devices, cryogenic liquids for the superconducting systems, gas return lines, vacuum pump exhaust lines, shield cooling water and control telemetry. The presence of these service connections must not inhibit essential access to any of the reactor components and systems.

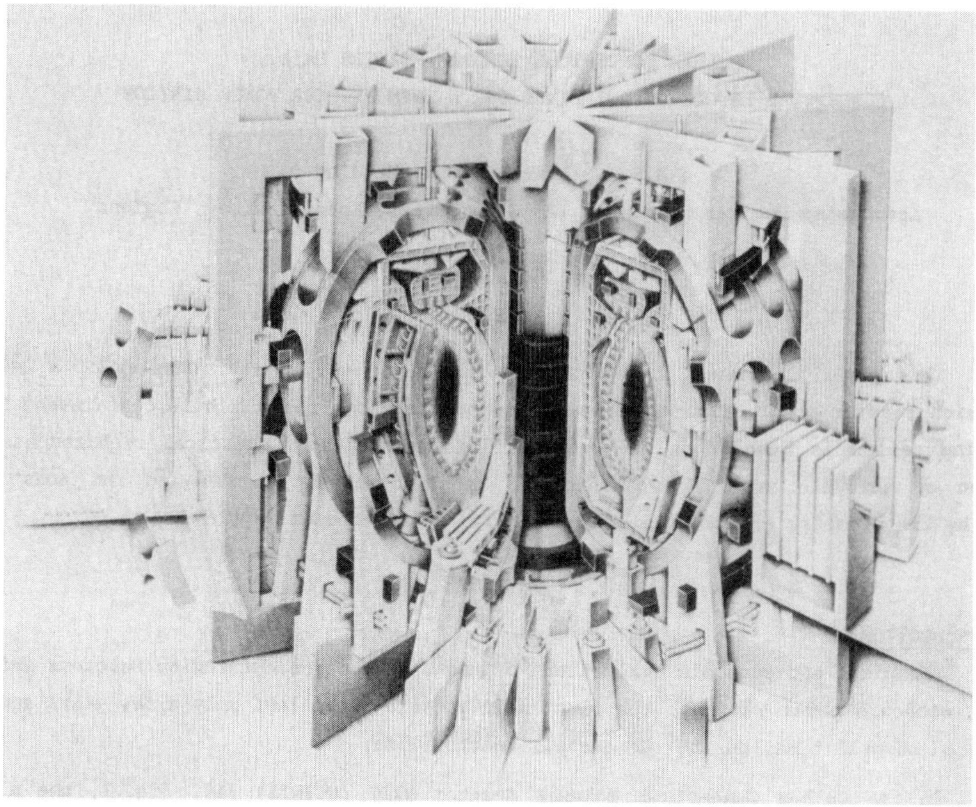

Fig. 1 Culham Conceptual Tokamak Reactor MkII

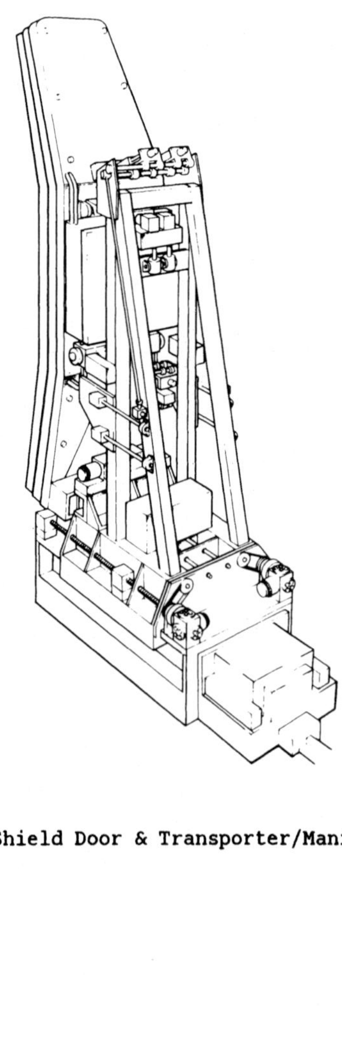

Fig. 2 Shield Door & Transporter/Manipulator

2.3 Maintenance and support logistics

Blanket segments will be removed at regular intervals via doorways in the fixed shield and replaced with new or refurbished segments. In order to minimise down time, it is essential that the servicing machines should have adequate access to the reactor.

The overall design should allow for recovery from major, as well as minor, component failures. An examination of failure recovery scenarios has identified one of the most difficult operations as the replacement of a failed toroidal field (TF) coil, the most significant features being the large amounts of radiologically protected storage space and access which are required.

2.4 Construction

On site fabrication, typically of 250 te segments, and coil winding of 20 m diameter coils, will be required. It is assumed that the permanent segment test facilities will be programmed to be available at an early date.

2.5 Radiological protection

Carruthers et al [8] examined fusion reactor power stations and concluded that the principal hazards are:

(a) to the general public - a sudden large release of tritium, the probability and severity of which is dependent upon whether or not a lithium fire is involved;

(b) to the operating staff - continuous small tritium release and the neutron activation of structural materials.

Assuming that the lithium is in a relatively non-combustible form, the specific radiological hazards are:

- radiation levels of ≅4000 R/hr at the fixed shield surface when the reactor is operating and considerably higher levels when a shield door has been removed;
- high radiation levels from in-transit irradiated blanket shield segments;
- radiation from auxiliary plant during operation and maintenance;
- tritium release into the containment buildings;
- tritium diffusion into the primary and secondary cooling circuits;
- tritium and particulate contamination released from components/sub-systems in the workshops.

2.6 Plant safety and reliability

Conventional practices to ensure the safety and reliability of nuclear facilities must of course be applied, typically: guaranteed power supplies, seismic design criteria, segregation of services etc.

2.7 Decommissioning

Ideally, the system should be designed so that, as far as is economically practicable, it will be possible to remove activated and tritium contaminated structural elements for storage or disposal; the remainder being decontaminated or entombed.

3. POWER STATION CONCEPT

3.1 General

The principal elements in the nuclear complex are:

(a) a reactor hall containing: two CCTRII reactors, two sets of steam raising plant and common cranage which is accessed by two pairs of air locks;

(b) two facilities for accommodating and cooling segments withdrawn from the reactors; each facility can contain up to ten segments and has direct access to the reactor hall;

(c) a high activity, remotely operated, workshop with facilities for servicing segments, breeder cells, injectors and diverters and containing a scrap processing and 'flasking out' facility;

(d) a fluid clean-up and tritium purification plant to extract tritium from all gas and liquid streams from the reactor halls and maintenance facilities; in the same room will be the tritium and deuterium storage facilities and the main ventilation systems;

(e) a parking and servicing area for the servicing machines.

Non-active areas which must be in close proximity to the 'nuclear island' are: the helium storage area, the helium liquefaction plant room, the electrical plant room and some control

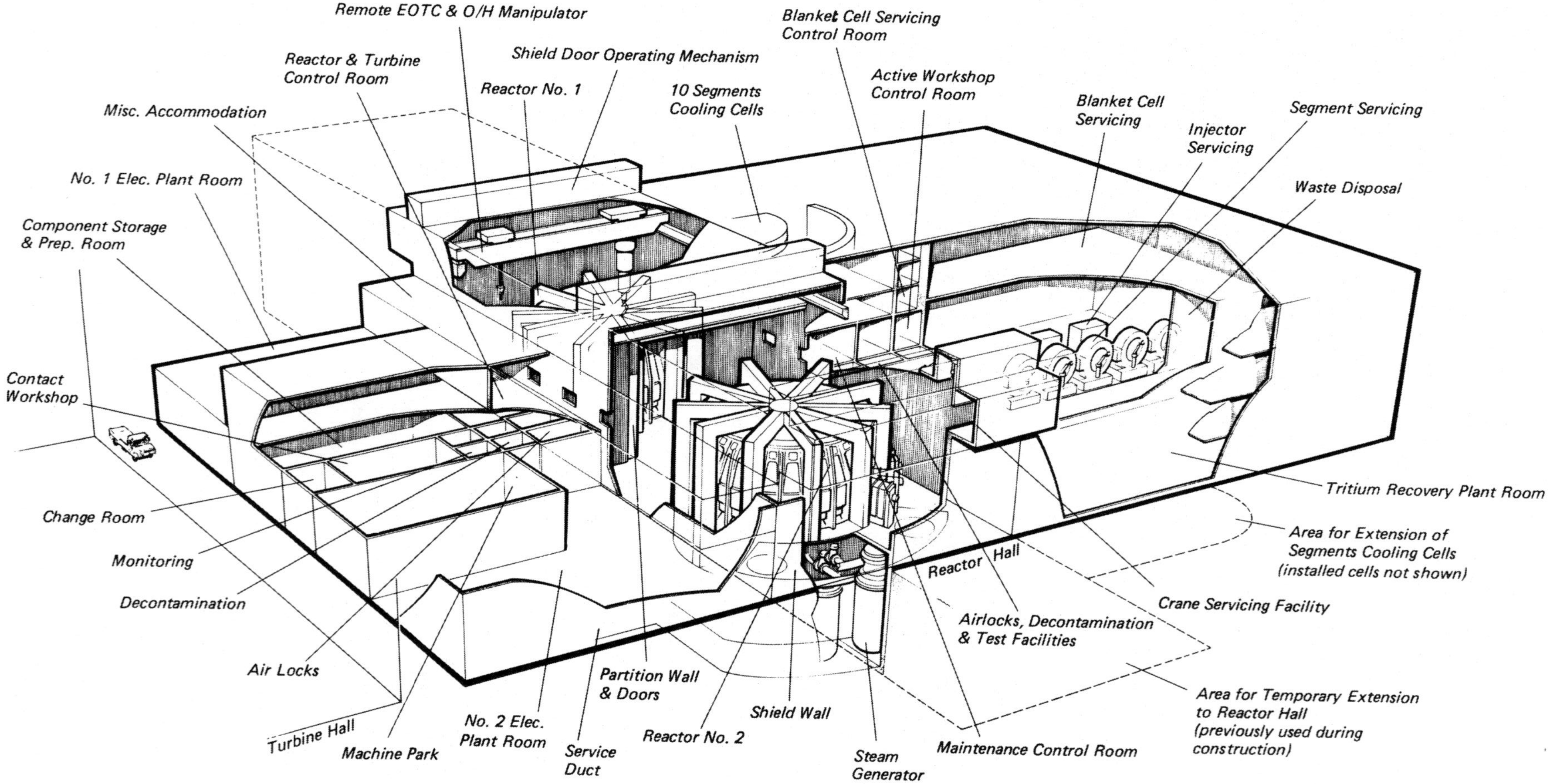

Fig. 3 Layout of Twin Reactor Power Station

rooms. All these elements are arranged as shown in fig. 3 and the most significant discussed below.

3.2 Reactor cooling and steam raising

CCTRII blanket cooling helium temperatures were chosen to enable conventional power plant steam conditions (565°C, 160 bar) to be achieved while maintaining adequate temperature differences in the steam generators. Steam reheat has been incorporated to minimise pipework complexity and to simplify the steam generator design. It is probable that proven 1200 MW turbogenerators will be available within the relevant timescales and one such unit per reactor is proposed although detailed consideration of availability aspects might favour the use of 600 MW machines with steam and feed range facilities. Current High Temperature Gas-cooled Reactor (HTGR) designs are for helium at conditions very similar to those quoted for CCTRII and HTGR data have been used to predict steam generator sizes, typically 0.25-0.5 m^3/MWt, and configuration. The steam generators are cylindrical steel vessels containing helical tubes attached to bottom mounted cylindrical steam and feed headers. Hot gas enters the top of the vessel and flows down over the counter flow heat transfer surface and the return cool gas flows up through the annular void between the helical tubes and the wall, so that the vessel is subjected to the cool gas temperature only. Gas flow control will probably be required for start up and during part load operation and therefore a synchronous or induction drive motor, supplied by a static frequency changer, is used to drive the helium circulator mounted at the top of each steam generator vessel.

Past experience suggests that the reliability of each steam generator/circulator unit is unlikely to exceed 99%. In order to achieve satisfactory overall power plant availability it must be possible to cool all blanket segments in spite of limited steam generator/circulator outages. Segments and steam generators are therefore interconnected via hot and cold gas manifolds. The connections from the manifolds to the blanket segments include isolating valves for use during segment exchanges and flow control valves for control of the segment structural temperature under load. The use of eight steam generators each rated at 430 MWt results in unit sizes which are current state-of-the-art for both steam generators and circulators and also meet road transport limitations. Steam generators are connected to gas manifolds via isolating valves. The proposed configuration, shown in fig. 4, provides short gas ducting length, no access constraint to the reactor, retention of any tritium leakages within the containment building and simplified pipework layouts.

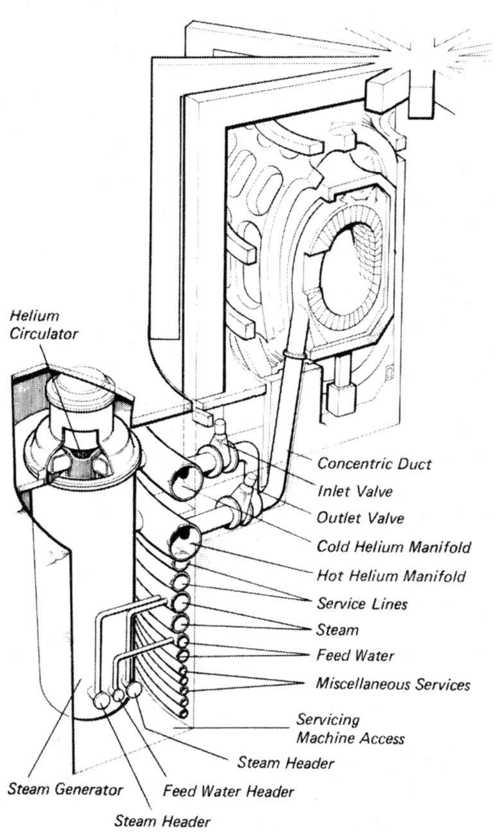

Fig. 4 Configuration of Coolant Circuit & Steam Generators

Removal of after-heat involves complex considerations with respect to temperature control and will not be easy to achieve with the main steam generator/circulator units. The layout therefore includes two auxiliary heat exchangers per reactor which will have guaranteed power supplies to the circulators and controls. Each unit will have the capacity to cool 30 segments (20 in the reactor plus ten in the cooling cells) and cross connections will be provided. These auxiliary heat exchangers are located in the steam generator trench thus utilising common servicing machines and provision for control of tritium leakage.

3.3 Reactor hall

The building structure is required to provide tritium containment and radiation shielding. The internal surfaces are clad with aluminium, selected on the basis of low tritium permeation, formed into a continuous membrane to maintain the internal atmosphere at slightly sub-atmospheric pressure. Oxygen should be excluded from the reactor hall on technical, safety and radiological grounds and an inert atmosphere, such as nitrogen, is preferred. The walls and roof will be ≅1.5 m thick for shielding purposes and designed to carry the differential pressure forces.

An aluminium clad concrete dividing wall, fitted with shield doors, is provided to separate the two reactors to limit the spread of tritium. Shield doors, which will be normally closed, will be opened to permit cranes and servicing machines to transfer directly between reactors. These penetrations would be sealed to allow the reactor hall to be opened for major servicing.

At the 27 m level, the structure is extended longitudinally at both ends to provide for crane and overhead manipulator servicing and radiation free parking areas.

The cost of the reactor hall foundations and structure will be a significant part of the total and it is desirable to minimise its dimensions - particularly the span. Dominant factors in size determination are:

- the size and layout of the reactor and steam generating plant;
- the access and storage space required during routine and non-routine maintenance.

It is not economically sensible to provide permanent facilities to cover unlikely maintenance or repair operations, such as a TF coil replacement and provision is therefore made only for routine servicing operations inside the permanent structures. Space outside and close to the reactor hall will be reserved for non-routine major servicing. As this space will be ideally located for reactor construction and coil winding, permanent foundations will be provided. The end walls of the reactor hall will therefore be designed to be taken down without affecting structural integrity and to allow load transfers between internal and external cranes. Temporary buildings and facilities will be provided only as required for specific maintenance operations within the constraint imposed by the existing foundations.

Examination of the storage and access requirements under various maintenance scenarios results in a reactor hall having the following internal dimensions: 100 m long x 50 m wide x 34 m high, which provides an annular roadway of 6 m width, radial access clearance of >7 m and the storage space required for routine maintenance. The layout of the steam generators and primary circuit provides ample space for ducts and cables under the floor and also allows the installation of auxiliary systems e.g. vacuum pumps. Service ducts which pass outside the containment building are located in underground passages fitted with sealed bulkheads. The steam generator trenches are connected by passages, fitted with isolating doors and served by lifts, to allow steam generator servicing machines to transfer between reactors. Two control rooms are provided, on opposite sides of each reactor, at the 13 m level viz:

(a) an operational control room for the reactor, steam generators, associated generating plant and atmosphere management;

(b) a control room to cover all routine and abnormal maintenance operations.

The control rooms are sited to have direct viewing of the reactor hall and are sized to provide space for control systems, computers and telemetry equipment. Nitrogen is extracted from the reactor hall via the steam generator trenches and service ducts to ensure that these areas, with potential for tritium and tritiated water vapour and water leakage, are purged. Gas is normally returned via high level ducts and the air locks.

3.4 Irradiated segment storage

Two storage facilities, each capable of holding 10 irradiated segments, are sited adjacent to each end of the reactor hall and are accessed from it. Each irradiated segment is connected to a concentric gas duct and cooled by helium circulated from the auxiliary heat exchangers. Conceptually the segment storage facilities are of the same type of construction as the reactor hall and can be extended to facilitate major reactor maintenance.

3.5 Remote workshop

After cooling, the segments are transferred via an airlock and decontamination unit to the high activity workshop where they are processed through a series of work stations in which irradiated blanket cells are removed, new cells fitted and a comprehensive test and inspection programme completed. The irradiated cells are transferred to the floor above, where they are dismantled, repaired, reassembled, tested and stored until required. Injectors, refuellers and diverters will also be serviced in the workshop area in special purpose facilities and a general purpose facility is also provided for non-routine maintenance. Permanently installed plant will be maintained by general purpose robot manipulators and overhead systems. This facility will generate quantities of active and/or tritiated scrap which will be compacted and packed into shielded containers prior to removal via a shielded airlock and monitoring complex. The building structure and finish is similar to that of the reactor hall and control rooms are provided for each of the automated process lines and general purpose facilities.

3.6 Fluid clean up and tritium purification plant

A similar plant for the fusion Engineering Test Facility (ETF) has been estimated [9] to require $\cong 3.6\times10^3$ m^2 area and 2.2×10^4 m^3 volume and as a first approximation the buildings shown in fig. 3 have been derived by proportionately scaling the ETF values. This type of plant should be very reliable and since activity levels will be low, contact maintenance should be feasible by personnel wearing protective clothing.

3.7 Machine park and contact workshop

The servicing machines are normally parked and maintained outside the reactor hall, entering and leaving through the airlock/decontamination and monitoring facility under the reactor operations control room.

4. CONCLUSIONS

The layout of the nuclear complex described in Section 3 and shown in fig. 4 has been developed to meet the design constraints defined in Section 2. This concept extends the system described in [6] by including: external major servicing and construction facilities, steam generation and provision for recent changes in routine maintenance concepts.

This work will continue with an examination of building construction and the preparation of cost estimates.

5. ACKNOWLEDGEMENTS

The authors would like to thank Mr. J.T.D. Mitchell, Culham Laboratory, Mr. C. Symonds of W.S. Atkins and Partners and their colleagues at ANS for their help and advice during the preparation of this paper.

6. REFERENCES

[1] Mitchell, J.T.D., ANS 3rd Topical Meeting on Technology of Controlled Nuclear Fusion, 1978. Proceedings p.954.

[2] Briaris, D.A., Stanbridge, J.R., Culham Laboratory Report CLM-R174 (1977).

[3] Briaris, D.A., Stanbridge, J.R., Culham Laboratory Report CLM-R184 (1978).

[4] Guthrie, J.A.S., Culham Laboratory Report CLM-R205 (1980).

[5] Horne, R.A., 26th Conference on Remote Systems Technology 1978, Proceedings p.55.

[6] Mitchell, J.T.D., Hollis, A., 9th Symposium on Fusion Technology, Proceedings p.429.

[7] Mitchell, J.T.D., Booth, J.A., Culham Laboratory Report CLM-R126 (1973).

[8] Carruthers, R. et al., Culham Laboratory Report CLM-R148 (1975).

[9] ETF Newsletter, Oak Ridge National Laboratory, Tennessee, p.2, May 1980.

THE EFFECT OF ENGINEERING CONSIDERATIONS BEING
ACCEPTED AS THE PRIME CONSTRAINT IN TOKAMAK
FUSION REACTOR DESIGN

R S Challender
UK Atomic Energy Authority
Risley, Warrington, England

ABSTRACT

Current Tokamak fusion reactor proposals have severe problems of maintenance and first wall replacement. An attempt is made here to approach the design function objectively, to accept from the outset constraining requirements relating to access and maintainability, and to assess the viability of the resulting embodiment. The general principles of design were originally developed jointly with Dr P Reynolds of the Culham Laboratory and presented at the Ninth Symposium on Fusion Technology.

1. INTRODUCTION

Most Tokamak fusion reactor design studies have led us to embodiments which are generally similar to each other and also similar to the early laboratory machines which were their progenitors. This traditional interlocking geometry of coils encircling a toroidal vacuum vessel has resulted in difficult and time-consuming maintenance procedures even on the early non-radioactive experimental machines. For a reactor plant which will become highly active, the problems of maintenance and first wall replacement in this geometry become formidable indeed. In most cases such problems have been met by proposing methods of remote maintenance using automated robot systems for servicing and repairing the plant; we, however, are highly sceptical of the ability of such robot equipment reliably to perform the duties required of it.

The efficiency and reliability of remote handling equipment have been demonstrated conclusively over many years in various industries, whether used in hazardous conditions or as aids to production, but in successful applications such as post-irradiation nuclear fuel handling or production processes in the car industry, the movements and operations required of the working elements of a particular robot system have been specialised and relatively simple. Each component of motion could be pre-planned and programmed either as a manual or automated operation and in the majority of successful applications the relatively simple cycle of operations is repetitive.

In our opinion this is a far cry from the duties required of some of the proposed remote maintenance equipment for Tokamaks which usually comprise general purpose machines which will have to remove considerable amounts of peripheral equipment and services in order to gain access to the area of plant to be repaired before undertaking cutting, removing, replacing and welding operations on any part of the plant requiring it. The extremely large number of potential fault conditions to be catered for, and in particular the considerable chance of unforeseen incidents preclude the detailed pre-planning and programming essential to the successful use of remote handling equipment. The gap between the duties performed by existing remote handling equipment and those proposed in some fusion reactor design studies is vast and unlikely to be bridged.

Another doubtful feature common to many of our current fusion reactor proposals is concerned with the sheer physical size of many of the components. Some studies have proposed toroidal coils 50 or 60 ft high with correspondingly large poloidal coils. These dimensions exceed by many times the maxima for transportation from a manufacturer's works, and the only available alternative would seem to be to manufacture the coils on the reactor site; but would we be prepared to set up a vast manufacturing complex involving such advanced high-technology industrial processes on every fusion reactor site?

Not only are component size and weight costly embarrassments as far as manufacture are concerned they also have adverse effects on the reliability and maintenance of the plant in that many proposals accept the fact that once placed in position during construction, the coils, for example, will have to remain there for the active life of the reactor. No maintenance or replacement will be possible.

2. THE CONSTRAINTS

The inherent problems outlined above emanate mainly from the traditional interlocking geometry of a toroidal vacuum vessel linked by surrounding coils and also from its large size. In an attempt to avoid or reduce them we have accepted a series of constraining requirements within which to develop a Tokamak embodiment. The constraints are as follows:

2.1 All super-conducting coils must be capable of being manufactured at works and transported to site.

2.2 All coils must be capable of removal and replacement.

2.3 First wall elements must be capable of periodic replacement.

2.4 Remote handling at the reactor must be limited to simple lifting and lowering operations with an overhead crane.

2.5 The mechanical disengagement of components before removal and the disconnection of services must be capable of being done manually in a non-radioactive area. The replacement of components and reinstatement of services must also be done manually.

2.6 The use of remote handling manipulators must be limited to pre-planned, repetitive operations in shielded cells away from the reactor.

3. THE DESIGN

3.1 Preamble

The reactor design presented here is not intended to be optimised. The reactor parameters (Table 2) have been derived to provide a compatible set of dimensions appropriate to the size of reactor we are considering in order to form a suitable framework on which to develop a novel concept of reactor layout. The parameters were derived by the methods described in Ref [1] with additional guidance and assistance from Refs [2], and [3].

3.2 Reactor size

In order to fulfil the requirement regarding transportability of toroidal coils regarding permissible height of load, their minor axis must be limited to about 6 metres. Making allowances for radial depth of coil, primary shield and blanket thickness we arrive at a figure of 2.4 metres for the width of the plasma volume at the first wall on the equator.

Assuming a figure of 0.85 for y (the ratio of plasma width to first wall width) we have a plasma semi-width of 1.02 m. In order to increase plasma volume and first wall area we have assumed an elliptical plasma with an axis ratio of 2 to 1. Using conservative values of first wall loading and power density this leads to a thermal power output of about 700 MW, and assuming an overall plant efficiency of 35% we could expect an electrical power generation of 250 MW; thus, a 1200 MW(e) station would have five reactors. Normally a number of small units of plant is considerably more costly than one large unit but a later section of the paper will show that other inherent advantages of the small units in this case reduce the cost difference considerably.

3.3 Plant layout (Fig 2)

The requirement for toroidal coils to be removeable and replaceable means that either each coil must be capable of being split into two or more parts, or the toroidal vacuum vessel must be capable of disconnection between each coil. Because the problems of manually accessible connections for super-conducting toroidal coil components appear to be more severe we have chosen the latter alternative.

Even so, the feasibility of making and breaking a bakeable vacuum seal in the torus manually in a highly radioactive environment is highly doubtful. We are, therefore, proposing that the toroidal vessel sectors are not sealed; each toroidal coil would be associated with its own section of blanket, primary shield and sector of toroidal shell and this would be handled as one complete assembly and moved vertically into position by crane. The edges of adjacent toroidal shell sectors would be close to each other but a gap would exist to provide a clearance when lifting in or out. The main vacuum boundary would be provided by a large cylindrical vessel with dished ends which enclosed the central reactor plant; it would have a removeable cover to gain access to the components inside it and number of standpipes would depend from the lower dished end through which would pass coolant ducts to the blanket sections and services to the coils. A secondary shield of boron steel against the inner surface of the vessel would surround the reactor plant and ensure that the main vacuum boundary (ie the vessel) would always be manually accessible for leak testing and repair if necessary. It would also enable manual disconnection of the services to any toroidal sector of the reactor at the standpipes below the vessel before removing and replacing the sector by means of the overhead crane. Horizontal, radially disposed standpipes at the mid-height of the vessel would be provided for plasma heating by neutral injection and for connection to the main vacuum pumps.

3.4 The vacuum system

The main outer vacuum vessel would first be evacuated to 10^{-5} or 10^{-6} torr. The surface of the segmented toroidal inner vessel would then be raised to baking temperature and the inner vessel reduced to a pressure of 10^{-8} or 10^{-9} torr. The relationship between the pressure inside the torus, the pressure outside, the pumping rate and the dimensions of the gaps between the toroidal vessel sectors is shown in Fig 1. It is clear that the pumping rate requirements are sensitive to the gaps between the torus sectors and considerable advantages would accrue if the gaps could be reduced after the sectors had been placed in position. To achieve this, the mating edge of each torus sector would be formed by a relatively narrow, electrically insulating, ceramic "joint" ring mounted on a bellows/diaphragm assembly around the edge of the elliptical cross-section of the torus. The space inside the bellows and diaphragm woud be maintained at a partial vacuum of about 0.5 ats so that when surrounded by normal atmospheric pressure the bellows/diaphragm assembly and joint ring would be retracted. The torus sectors would be loaded into position and as the main vessel was evacuated, the adjacent joint rings assemblies would extend generally to reduce the gap between them, closing the gap entirely in places.

3.5 The primary circuit

The first wall comprises overlapping ceramic tiles which can be removed and replaced after some years of service as a matter of routine in the maintenance area by pre-programmed remote handling methods. During operation, the heat generated in the tiles is radiated to the wall of a pressure vessel shaped like a torus sector which contains the primary shield and reflector in an outer annulus and the blanket in an inner annulus. Helium enters the outer annulus at 350°C, passes through the shield and reflector and then over the blanket. It leaves the inner annulus and is delivered at 750°C to the main heat exchangers, of which there are 4 associated with each reactor.

The primary shield comprises an assembly of boron carbide and steel plates. The reflector is graphite and the blanket is made up of an assembly of close-pitched tube assemblies containing static lithium. The helium coolant circulates outside the tubes, extracting the heat generated within them. The ends of the tube assembles emerge through the wall of the sector pressure vessel at various points. During the periodic rewalling procedure connection is made to these points in the maintenance cells to displace the irradiated lithium (for subsequent tritum extraction) with new lithium.

Superheated steam is generated in the heat exchangers and delivered to the turbine stop valve at 2300 psia and 1050°F.

4. COMPARATIVE COSTS

A detailed cost estimate for the Culham Mark IIb 1200 MW(e) proposal was prepared in 1977 and this has been used in a comparative cost study with the 5 x 250 MW reactor station proposed here. A summary of the detailed comparison is shown in Table 1. A number of smaller units is normally expected to cost more than one large unit of similar design. In this case, however, the smaller units allow, because of their smallness, considerable economical advantages in manufacture, construction and maintenance which go a long way to eliminating the cost differential. It is well understood that cost estimates made at such early stages in projects provide, at best, only very general indications, however, the comparison shows that the cost difference need not become an overwhelming disadvantage. Some of the inherent capital cost advantages of the small unit are:

4.1 Manufacturing super-conducting coils at works instead of at site.

4.2 Much lower requirement for complex remote handling equipment.

4.3 The effect of 4.2 on buildings and services.

4.4 Spare components are smaller and therefore cheaper.

In addition, operating cost advantages include:

4.5 Shorter shut down periods for rewalling means about 7% more mean exportd power output.

4.6 Baking of the torus surfaces for vacuum generation is readily achieved by opening the circuit to the main helium system serving the other 4 reactors.

Table 1 is a summary of a detailed cost assessment and the main factors used in the analysis were, (a) A 0.6 power relationship between output and cost and, (b) during the construction of the 5 identical reactor units an 85% improvement curve was assumed as recommended in Ref [4].

5. CONCLUSIONS

If the first generation of fusion reactors is based on the Tokamak principle we believe that a different layout from the Gordian knot perpetuated in the large experimental machines now being built will have to be developed. The embodiment described in this paper may not be the final answer, but if it stimulates some thought towards a more accessible arrangement it will have served its purpose.

A number of engineering drawings is available to accompany this paper. The author will be pleased to send copies to interested parties on request.

TABLE 1 Comparative Costs as % of Total Cost		
	1x1200MW	5x250MW
BUILDING & CIVIL WORK		
Reactor & Maintenance Bldgs	7.94	5.19
Bldg costs common to both	7.84	7.84
REACTOR PLANT & EQUIPMENT		
Reactor plant	51.15	56.57
Control	1.10	1.70
Instrumentation	2.65	3.02
Radiation monitoring	0.22	0.17
MECHANICAL & ELECTRICAL		
Mech plant & services	3.07	2.97
Elec plant & services	7.72	10.40
POWER HOUSE PLANT	8.27	8.27
STEAM GENERATORS	10.04	13.03
	100	109.16

TABLE 2 Plasma Parameters		
Mean thermal power	700	MW
Major radius	4.59	m
Aspect ratio	4.5	
Plasma elliptical ratio	2	
Plasma semi axes	1.02x2.04	m
Half width at first wall	1.2	m
Mean 1st wall power load	2.21	MW/m^2
Number of toroidal coils	16	
Mean plasma density	$1.18x10^{20}$	
Mean plasma temp	15	keV
Toroidal beta	0.085	
Poloidal beta	2.121	
Toroidal mag field (axis)	4.09	T
Toroidal mag field (max)	7.98	T
Plasma Current	4.6	MA

TABLE 3 Thermal Parameters	
Gross station output	1250 MW
Reactor electrical output	250 MW
Reactor thermal output	700 MW
Sector thermal output	43.75MW
He pressure at reactor inlet	20 ats
He temperature at reactor inlet	350° C
He temperature at reactor outlet	750° C
He mass flow per sector	21 kg/sec
He maximum velocity in ducts	40 m/sec
TSV pressure	2300 psi
TSV temperature	1050° F

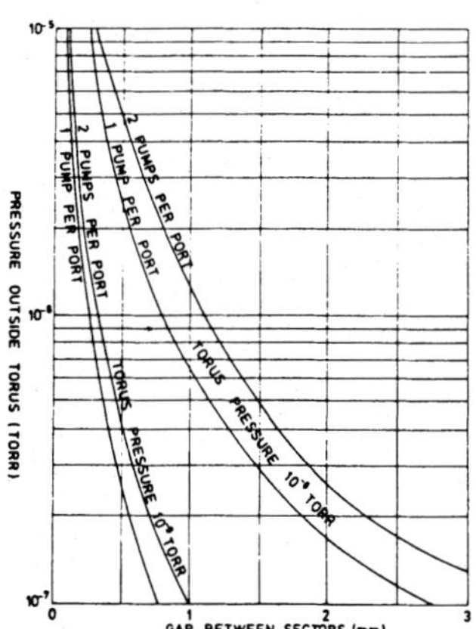

EFFECT OF GAPS ON VAC PUMPS

FIG I

6. REFERENCES

[1] GOLOVIN, T M et al, BNES Nuclear Fusion Reactor Conf, Culham, Sept 1969 Procs pp 194-221

[2] KAMMASH, T, Fusion Reactor Physics. Arbor Science Pubs Inc, Michigan (1975)

[3] BADGER, B et al, University of Wisconsin Report UWFDM-330 (NUMAK) 1979

[4] HIRSCHMANN, W B, AACE Bulletin-Vol 7, No 3 98-100 (1965)

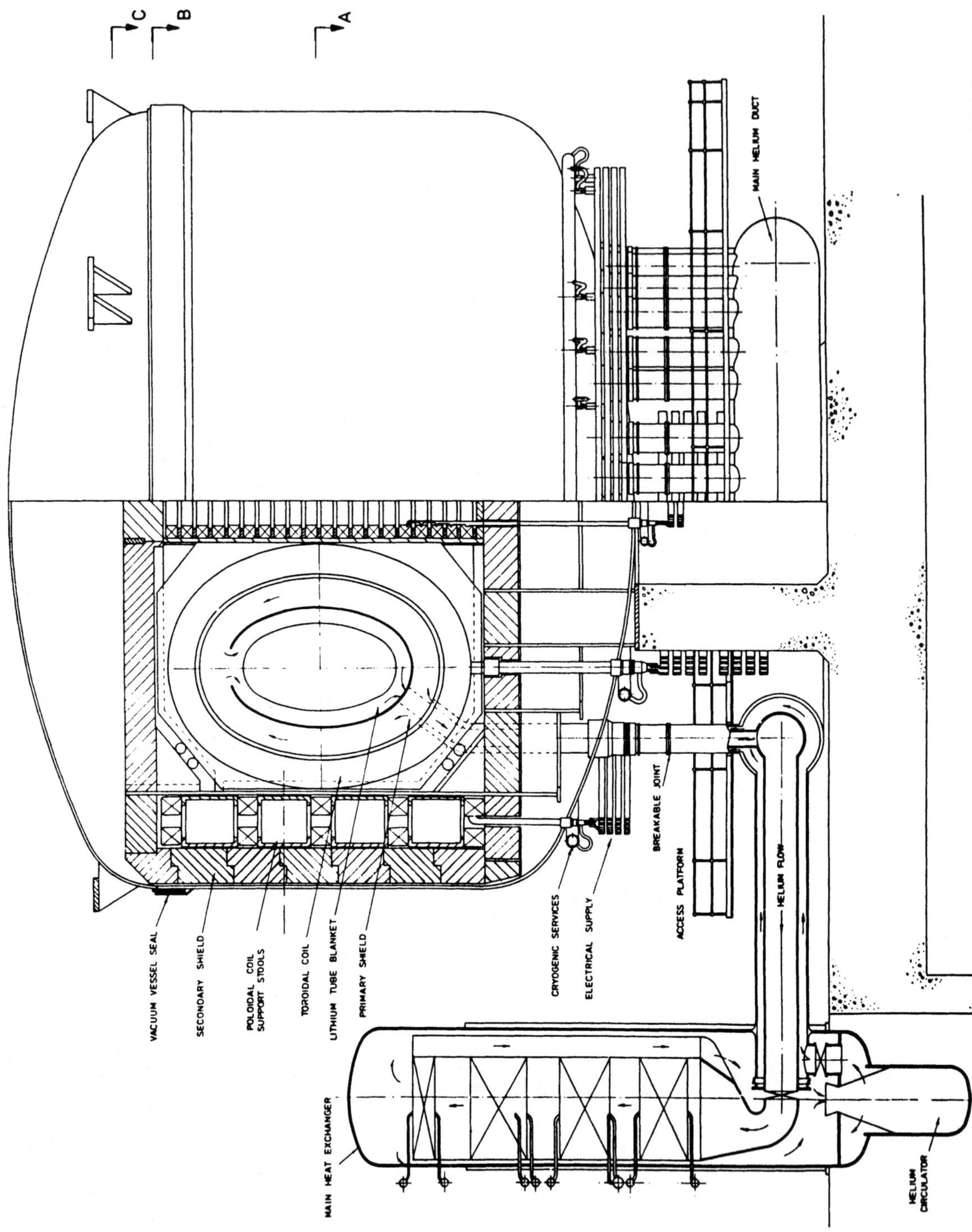

PLANT LAYOUT

FIG. 2

A STEADY-STATE TOKAMAK REACTOR WITH NON-DIVERTOR IMPURITY CONTROL - STARFIRE

Charles C. Baker
Argonne National Laboratory
Argonne, Illinois 60439, U.S.A.

ABSTRACT

STARFIRE is a conceptual design study of a commercial tokamak fusion electric power plant. Particular emphasis has been placed on simplifying the reactor concept by developing design concepts to produce a steady-state tokamak with non-divertor impurity control and helium ash removal. The concepts of plasma current drive using lower hybrid rf waves and a limiter/vacuum system for reactor applications are described.

1. OVERVIEW OF THE STARFIRE REACTOR

The purpose of the STARFIRE study [1-3] is to develop a design concept for a commercial tokamak fusion electric power plant. The major features of STARFIRE include a steady-state operating mode based on continuous rf lower-hybrid current drive and auxiliary plasma heating, solid tritium breeder material, pressurized water cooling, limiter/vacuum system for impurity control, all superconducting equilibrium field (EF) coils outside the superconducting toroidal field (TF) coils, fully remote maintenance, and a low-activation shield. The major parameters of STARFIRE are listed in Table I. An isometric view of the reference design is shown in Fig. 1.

The first wall/blanket (FW/B) is segmented into 24 sectors to permit removal between the 12 TF coils. The material of the FW/B is PCA (prime candidate alloy) stainless steel that operates at a maximum temperature of 425°C. The FW/B, limiter and rf waveguide are designed for a 16 MW-yr/m^2 integral neutron wall loading which results in a lifetime of about six years between changeouts. All components in contact with the plasma are coated with beryllium. The FW/B is cooled by pressurized water (15.4 MPa) with inlet and outlet temperatures of 280°C and 320°C, respectively, This permits the $LiAlO_2$ to operate within an appropriate temperature range to permit tritium release to a low-pressure helium purge stream without significant sintering of the solid tritium breeder material [4].

Table I. Major Parameters for STARFIRE

Net electrical power, MW	1200
Gross electrical power, MW	1440
Fusion power, MW	3510
Thermal power, MW	4000
Overall availability, %	75
Average neutron wall load, MW/m^2	3.7
Major radius, m	7.0
Plasma half-width, m	1.94
Plasma beta	0.067
Toroidal field on axis, T	5.8
Maximum toroidal field, T	11.1
Blanket structural material	Austenitic stainless steel
Tritium breeding medium	α-$LiAlO_2$
Neutron multiplier	Zr_5Pb_3 (solid) or Be
First wall/blanket coolant	Pressurized H_2O
Primary vacuum boundary	At inner edge of shield

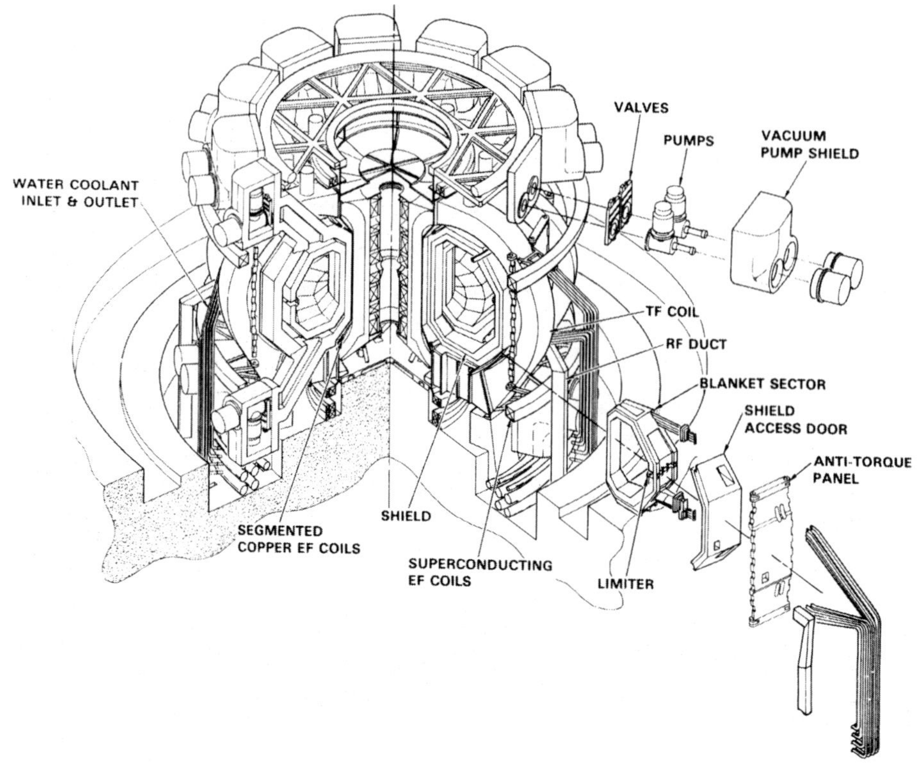

Fig. 1. STARFIRE reference design.

The shield serves as the vacuum boundary for the system. Twelve shield access doors with mechanical seals are provided between each pair of TF coils. Dielectric breaks are located in the outer regions of the shield where the maximum radiation dose is < 10^{10} rads. The shield will last the entire life of the reactor. The shield is 1.1 m thick and is composed of TiH_2, B_4C and a low-nickel steel. This composition and thickness will limit the radiation dose in the reactor building so that manned access is possible within 24 hours after shutdown. In addition, the activation of the shield and other components outside the blanket is low enough to permit recycling of the materials within 30-50 years after reactor decommissioning.

The maintenance concept for STARFIRE is similar to the approach developed by the Culham group [5]. The basic approach is to use modularized components with simplified "push/pull" procedures in the reactor building and to carry out almost all repairs in a hot cell away from the reactor. Fully remote maintenance is planned for all operations, although some manned access is possible as a backup. This will significantly reduce the radiation dose for plant workers. Availability goals of 75% for the overall plant result in permissible downtimes of 37 days/year for scheduled maintenance for the reactor and balance of plant (BOP), 34 days/year for unscheduled maintenance for the reactor, and 20 days/year for BOP unscheduled maintenance.

2. CURRENT DRIVE AND BURN CYCLE

There are several advantages that result from steady-state operation of a tokamak by non-inductive plasma current drive. These advantages include increased component and system reliability, higher neutron wall loadings, elimination of material fatigue of the first wall as a major life-limiting concern, no need for thermal and electrical energy storage, and elimination or significant reduction of the size of the ohmic heating (OH) coils.

Several current drive concepts were considered for STARFIRE [6]. Lower hybrid rf current drive was selected because its technological features of simple waveguides and existing power sources were compatible with reactor engineering considerations, and because the theory of rf current drive by lower hybrid waves is reasonably well established [7].

A cross-section of STARFIRE showing the rf system is illustrated in Fig. 2, and the rf system parameters are listed in Table II. The rf power is minimized by operating at low electron densities, radiating in a narrow band spectrum and minimizing the total plasma current. For STARFIRE, these considerations result in a plasma current of 10 MA (limiter q value of 5) with the current density peaked near the plasma edge where the density is low. Such a "hollow" current profile is consistent with a penetration depth by lower hybrid waves of about 15% of the plasma radius in the STARFIRE plasma. The wave frequency corresponds to twice the local lower hybrid frequency to avoid parametric instabilities. For an aspect ratio of 3.6 and a plasma elongation of 1.6, the plasma was found stable to betas of at least 6.7%, provided that the first wall provides some wall stabilization. The first wall has been designed with a current decay (L/R) time of 300 ms for this purpose.

The Brambilla theory [8] has been adapted to the STARFIRE conditions for design of the waveguides. Traveling waves are launched by phasing adjacent guides by $2\pi/3 = 120°$. The narrow opening in the toroidal direction (3 cm) and the spacing between openings (septum) define the toroidal wavelength. The spectral width is established by limiting the number of guides in the toroidal direction in one array to 18. The vertical dimension is 17 cm; four banks are stacked in the poloidal direction forming one module between each TF coil of size 0.66 m x 0.78 m. The average rf power density at the waveguide is 1.6 kW/cm^2 which should not

POWER BALANCE	MW
R F POWER TO PLASMA	90.4
(CURRENT DRIVE)	(66.5)
LOSSES	62.3
TOTAL R F ELECTRICAL POWER	152.7

POWER LOSSES		MW
A	GRILL	0.5
B	WAVEGUIDE	1.9
C	WINDOW (3)	0.2
D	WAVEGUIDE	18.2
E	CIRCULATOR	10.6
F	PHASE SHIFTER	1.5
G	C F A	21.8
ELECT. BLDG	POWER SUPPLIES	7.6
	TOTAL LOSSES	62.3

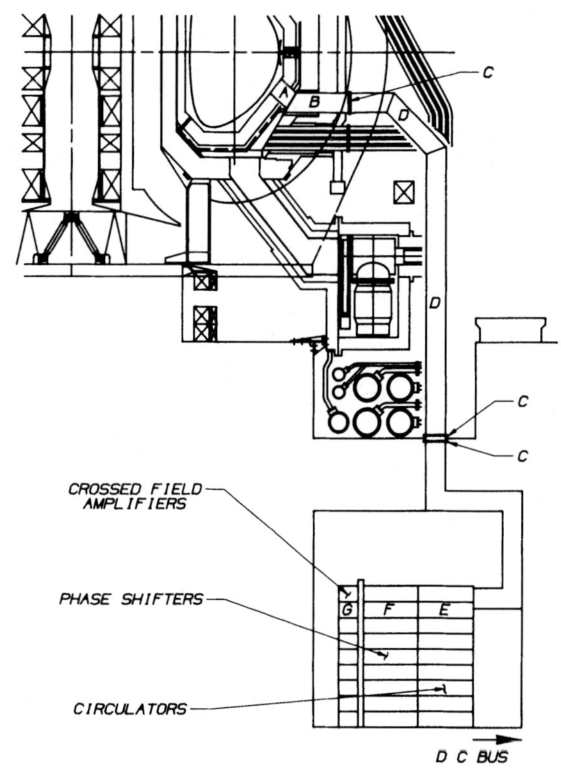

Fig. 2. STARFIRE rf system.

Table II. Lower Hybrid RF System Parameters for STARFIRE

1.67 GHz	wave frequency
1.40-1.86	spectrum required
10.9 cm	toroidal wave length
1.6 kW/cm^2	wave intensity at antenna
$2\pi/3$	phase difference
3 cm	guide opening in toroidal direction
0.70 cm	septum
17.0 cm	vertical (poloidal) guide opening
0.736	spectral fraction driving current
0.443	antenna reflection coefficient

result in any nonlinear plasma responses. The system will deliver ∿ 90 MW to the plasma of which ∿ 67 MW will drive the plasma current. The remaining power is in various side bands which is not useful for current drive, but is useful for plasma heating.

The rf duct assembly is mounted to the blanket and is constructed of PCA stainless steel. The interior of the duct is coated with copper to minimize power losses. Near the plasma, the waveguides are also coated with beryllium. A window of alumina is located near the TF coils to prevent electron cyclotron breakdown in the slots and to prevent backflow of tritium.

Figure 2 lists the power losses in the various components of the rf system. The rf system components are located in the basement of the reactor building and the power supplies are located in a separate building. A total electrical power of 153 MW is required for the rf system. Because substantial power is reflected by the array, it is important to recover this power. Crossed-field amplifiers appear to be well suited for this system.

For steady state operation, there is a broad range of startup, burn and shutdown scenarios. The STARFIRE burn sequence consists of the following phases:

- Electron cyclotron resonance heating power of 5 MW for about one second to break down and ionize the initial fill of DT gas (96% D) at about 10^{19} particles/m^3.
- Use of a small OH coil to produce an initial plasma current of ∿ 2 MA in 14 seconds.
- RF current drive at 45 MW for 250 seconds at low density which raises the plasma temperature to ∿ 6 keV, and increases the plasma current to the final value of 10 MA.
- Main heating phase with rf power at 90 MW for 200 seconds while the mainly deuterium plasma density is raised to its final value (0.8×10^{20} m^{-3}).
- A fusion ramp phase of 17 minutes when the fusion power is increased at a 5% per minute rate by slowly changing the tritium fraction from 4% to 50%. A small amount of impurities (0.1% of iodine) is added to maintain a stable plasma temperature.

The total startup time lasts about 24 minutes; the plasma will burn until a shutdown is required for maintenance or some other reason. Normal shutdown is essentially the reverse of the startup case described above. An abrupt shutdown capability is provided by injecting impurities into the plasma and creating a plasma disruption which will terminate the plasma in ∿ 100 ms. STARFIRE is designed to take several disruptions in a year. Rapid, but less severe, shutdowns of ∿ 2.5 s are also possible. It has also been shown that any hot spot formation on the first wall (∿ 1% of first wall area) will cause sufficient beryllium to ablate and extinguish the plasma in < 1 s, before any substantial damage can be done to the first wall. This illustrates one of the inherent safety features of fusion.

3. LIMITER/VACUUM SYSTEM

A limiter/vacuum system concept has been developed for STARFIRE for the purpose of impurity control and removal of the helium ash. The design utilizes a toroidal limiter, located at the outer side of the torus as shown in Figs. 1 and 3. For the STARFIRE design,

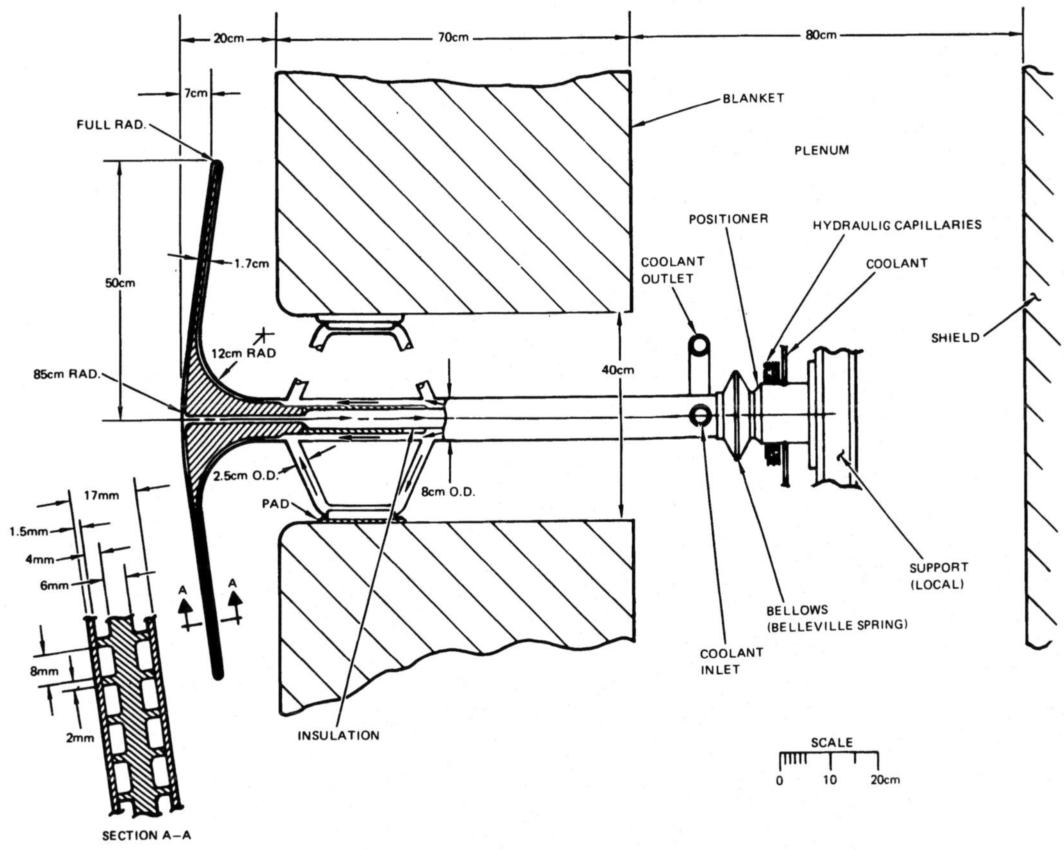

Fig. 3. STARFIRE limiter concept.

about 28% of the charged particle flux will be directed into the slot behind the limiter, where the charged particles will be neutralized. Some of the helium atoms will be directed into the 0.4 m high vacuum duct behind the limiter, which leads to a large vacuum plenum between the blanket and shield. There are 12 vacuum ducts at the top and bottom of the reactor with 1 m diameter penetrations through the shield (see Fig. 1). This arrangement significantly reduces neutron streaming. There are 48 compound cryopumps with 24 on-line at any given time, each with a helium pumping speed of 120 m^3/s. The pumps are rejuvenated every two hours.

Helium atoms streaming back out of the limiter slot have a very high probability of being ionized by the incoming plasma. The helium ions will be directed back to the interior surfaces of the limiter by electric fields set up by sheath effects at the limiter. Thus, there is a high probability that all of the α-particle flux entering the limiter slot will be removed by the limiter/vacuum system. This will result in about a 14% equilibrium α-particle concentration in the plasma, which is accommodated by providing an additional margin in the on-axis toroidal field of 0.85 T.

Because of significant charge-exchange reactions for the DT atoms in the limiter slot, only ∿ 10% of the DT flux will be pumped. This results in a high recycling of the DT particles and a high tritium burnup fraction (42%), with a resulting low inventory of tritium in the pumping and fueling system. This is one of the key features of the limiter/vacuum system.

Table III presents a summary of the major parameters of the limiter/vacuum system. In order to minimize the heat load to the limiter, most of the alpha-heating power to the plasma is radiated to the first wall by injecting a small amount of iodine along with the DT fuel stream. The total heat deposited on the limiter is 200 MW with a maximum heat flux of

Table III. Major Features of the Limiter/Vacuum System

Helium production rate, s^{-1}	1.24×10^{21}
Helium reflection coefficient, R_{α}	0.75
Hydrogen reflection coefficient, R_{DT}	0.90
Alpha particle concentration (n_{α}/n_{DT})	0.14
Beryllium (low-Z coating) concentration (n_{BE}/n_{DT})	0.04
Toroidal-field margin at plasma center, T	0.85
Fractional burnup, tritium	0.42
Tritium inventory in vacuum pumps and fueling system, g	200
Scrape-off region thickness, m	0.2
Limiter (one toroidal limiter centered at midplane)	
Structural material	Ta-5W, V-20Ti or FS-85
Low-Z coating material	Beryllium
Coolant	Water
Coolant inlet temperature, °C	115
Coolant outlet temperature (2 pass), °C	145
Maximum coolant pressure, MPa (psia)	4.2 (600)
Total heat removed from limiter, MW (90 MW transport, 56 MW radiation plus neutrals and 54 MW nuclear)	200
Maximum heat load (at leading edge), MW/m^2	4

4 MW/m^2. This heat is used for feedwater heating in the power conversion cycle. A detailed assessment of material candidates that included radiation effects, thermal-hydraulics, and stress analyses was performed. Tantalum, niobium, vanadium, and copper alloys appear to have the greatest potential as limiter materials. The limiter is designed to accommodate the electromagnetic forces induced under plasma transient conditions.

The limiter and first wall are coated with beryllium. The first wall will be eroded at a rate of 0.14 mm/year for STARFIRE's plasma conditions of a total first wall flux of 6.5×10^{22} particles/s. An initial coating of 1.2 mm on the first wall will have a lifetime of ∿ 7 yr. While the beryllium coating is also being sputtered off of the limiter, much of the beryllium from the limiter and first wall is being redeposited back on the limiter. Sufficient beryllium density (∿ 4%) is maintained in the plasma so that there is no net erosion of beryllium from the leading edge of the limiter.

4. SUMMARY

The STARFIRE design has incorporated a number of concepts which has resulted in a more simplified reactor with increased safety and environmental features. In particular, the limiter/vacuum system has the advantages that it does not require magnets, can have manageable heat loads, has a reasonable vacuum system which minimizes neutron streaming, has a high tritium burnup and minimizes tritium inventory in the fuel cycle, and it significantly reduces the complexity of the system leading to easier maintenance and assembly.

STARFIRE is designed to operate in a steady-state mode with the plasma current maintained by momentum transfer from lower-hybrid waves launched from end-fire waveguide arrays. The steady-state mode of operation significantly enhances the reactor potential of the tokamak with the specific advantages listed in Sec. 2. The lower-hybrid rf system appears adequate but its major disadvantage is the relatively large recirculating power requirements (∿ 10% of the plant gross electrical power output). Larger savings are potentially realizable with steady-state operation if the performance of the lower-hybrid current driver can be further improved or substantially better alternatives for the current driver are developed.

The results of the STARFIRE study indicate important directions for the development of tokamaks to significantly enhance their potential as power reactors. Simplifying the reactor design and improving component reliability must be important goals in developing fusion as a practical and economical energy source.

5. ACKNOWLEDGEMENT

The study reported in this paper represents the effort of many people at the Argonne National Laboratory, McDonnell Douglas Astronautics Company, General Atomic Company, and The Ralph M. Parsons Company. In particular, the author would like to express his appreciation to M. A. Abdou, J. N. Brooks, and D. A. Ehst at ANL for providing much of the input to this paper.

6. REFERENCES

[1] BAKER, C. C., et al., Argonne National Laboratory Report, ANL/FPP-80-1 (1980).

[2] BAKER, C. C., et al., Nucl. Eng. and Design (1980) to be published.

[3] ABDOU, M. A., et al., Eighth International Conference on Plasma Physics and Controlled Nuclear Fusion Research, Brussels, July 1980. Proceedings IAEA-CN-38/E-1.

[4] SMITH, D. L. and CLEMMER, R. G., Eighth Symposium on Engineering Problems of Fusion Research, San Francisco, November 1979. Proceedings, p. 433.

[5] MITCHELL, J. T. D., Third Topical Meeting on the Technology of Controlled Nuclear Fusion, Santa Fe, New Mexico, May 1978. Proceedings p. 954.

[6] STARFIRE Interim Report, Argonne National Laboratory, ANL/FPP/TM-125 (1979).

[7] EHST, D. A., Nucl. Fus. 19, 1369 (1979).

[8] BRAMBILLA, M., Nucl. Fus. 16, 47 (1976).

By acceptance of this article, the publisher or recipient acknowledges the U.S. Government's right to retain a nonexclusive, royalty-free license in and to any copyright covering the article.

THE BELT-SCREW-PINCH REACTOR AND OTHER HIGH-BETA SYSTEMS

M. Bustraan[1], B. Brandt[2], G.C. Damstra[3], J.A. Hoekzema[2],
H.Th. Klippel[1], W. Schuurman[2], H.J. Veringa[1], and K.A. Verschuur[1].

[1] ECN, Petten; [2] FOM-Rijnhuizen, Nieuwegein; [3] KEMA, Arnhem, The Netherlands.

ABSTRACT

In a screw-pinch reactor the expenditure for plasma implosion and compression can be reduced and the reacting volume and burn time can be enlarged. This is possible by pinch ignition of only a few percent of the fuel. Fusion energy then ignites injected fuel pellets and expands the plasma. The magnitude of the pulsed magnetic fields is such as to make the application of superconducting coils feasible. An economical reactor model is described. A comparison is made with tokamak and reversed field pinch reactor designs.

1. INTRODUCTION

High-beta plasma systems can be stabilized with special current and pressure profiles, thus becoming possible low-field alternatives to tokamak reactors. In Nieuwegein the screw pinch is studied (SPICA) [1]. Parameter studies of screw-pinch reactors with circular and elongated cross-section were presented at previous SOFTS [2,3]. The efficiency and the economics were poor. The main reason was that the volume of the thermonuclear plasma after ignition by compression heating was very small, so that a high confining magnetic field was necessary. Moreover, since this high field could not be generated on a short enough timescale by superconducting coils, normal conducting coils were adopted, giving rise to high ohmic losses. In order to circumvent these drawbacks an other approach to obtain a burning plasma should be employed. A promising heating method is the use of thermonuclear energy. This may be achieved by the injection of fuel after ignition of a low-density plasma in the usual screw pinch way. It is assumed that the injection leaves the screw-pinch configuration intact. The bulk of the plasma in this case stems from injection and is heated by thermonuclear alpha-particles. This method can only work if the plasma keeps itself at a stability limit for beta by means of mild instabilities. These instabilities enhance particle and energy transport inside the plasma column, creating a plasma with almost homogeneous density and temperature. During the injection phase the plasma is held at about 10 keV by controlled injection of fuel. The ignited plasma which initially has a small volume, then expands to fill a substantial part of the available volume. The injection phase ends with the combined injection of fuel and impurities and the burn phase starts with a plasma that is in thermal equilibrium at a temperature around 7 keV and at the stability limit for beta. For a screw-pinch reactor with optimal geometry (belt-shape) this limit is assumed to be 0.50. The thermal equilibrium may be stable due to heat conduction to the wall or it may be maintained by slow feedback control of the confining magnetic field. During the burn phase the temperature of the plasma increases gradually as the concentration of alpha-particles in the plasma increases, see Fig. 5. Compared to a previous design of a BSPR [3], the required implosion and compression energies are reduced, and the relative burnup is decelerated. As a result the efficiency of the reactor is much higher.

Following the historical trend in other fusion reactor designs, the working principle described above brings a reactor system with reduced dimensions and power production within

TABLE I. Parameters of the BSPRIL

input parameters		Cu-coils	S.C.-coils	output parameters		Cu-coils	S.C.-coils
aspect ratio	A =	5	5	half width of the torus	b =	1,44 m	1,44 m
maximum β	β_{max} =	0,5	0,5	half height of the torus	h =	4,32 m	4,32 m
thermal output power	P_{th} =	2,5 GW	2,5 GW	large radius of the torus	R =	7,2 m	7,2 m
average wall loading	P_w =	2 MW/m^2	2 MW/m^2	implosion energy	W_i =	15 MJ	15 MJ
max. E-field at wall	E_{wm} =	4×10^4 V/m	4×10^4 V/m	compression energy	W_a =	19,5 GJ	13,5 GJ
thermal efficiency	η_{th} =	0,42	0,42	burn time	t_b =	19 s	51 s
ETS efficiency	η_{ETS} =	0,95	0,95	cycle time	t_c =	33 s	65 s
torus elongation	e =	3	3	burnup		13,6 %	27 %
thickness blanket + shield	d_b =	0,8 m	1,2 m	bias field	B_o =	0,04 T	0,04 T
thickness TF coils	d_{TF} =	1,4 m	0,7 m	B after implosion	B_e =	0,11 T	0,11 T
voltage ETS	V =	21 kV	21 kV	B during burn	$\bar{B}$ =	2,8 T	2,8 T
safety factor at wall	q =	2	2	T during burn	$\overline{kT}$ =	7 keV	7 keV
energy multiplication of the blanket	Q_m =	1,2	1,2	relative plasma volume after compression	v_1 =	0,02	0,02
filling factor coils	f_c =	0,7	0,25	during the burn	v_2 =	0,5	0,5
initial filling density	n_o =	7×10^{18} m^{-3}	7×10^{18} m^{-3}	fuel injection	m =	1,4 gr	1,4 gr
				recirculating power	P_{circ} =	404 MW_e	200 MW_e
				net electric power	P_{net} =	676 MW_e	880 MW_e
				net efficiency	η_{net} =	27.0 %	35,2 %

Table II. Economic parameters of the BSPRIL

	unit cost estimate	Cu-coils		S.C.-coils	
		volume/ weight	cost (M$)	volume/ weight	cost (M$)
blanket	165 k$/$m^3$	446 m^3	74	446 m^3	74
shield	10 k$/$m^3$	500 m^3	5	1300 m^3	13
TF coils Cu	20 $/kg	9040 tons	181		
S.C.	10 $/$m^1$/kA			101 MA	34
PF coils Cu	20 $/kg	3050 tons	61		
S.C.	10 $/$m^1$/kA				11
Implosion bank	0.5 $/J	15 MJ	8	15 MJ	8
ETS system	0.011 $/J	19.5 GJ	215	13.5 GJ	148
cryogenics	6000 $P^{0.7}$ ($/kW)			48 MW_e	10
pellet acc. system	1 $/J	140 MJ	140	140 MJ	140
acc. power supply	0.02 $/J	140 MJ	3	140 MJ	3
conventional plant	150 $/$kW_t$	2.5 GW_t	375	2.5 GW_t	375
total direct costs			1062 M$		816 M$
indirect costs	150%		1593 M$		1224 M$
total capital costs			2655 M$		2040 M$
P_{net}		676 MW_e		880 MW_e	
rel. direct costs			1571 $/$kW_e$		927 $/$kW_e$
rel. capital costs			3927 $/$kW_e$		2318 $/$kW_e$

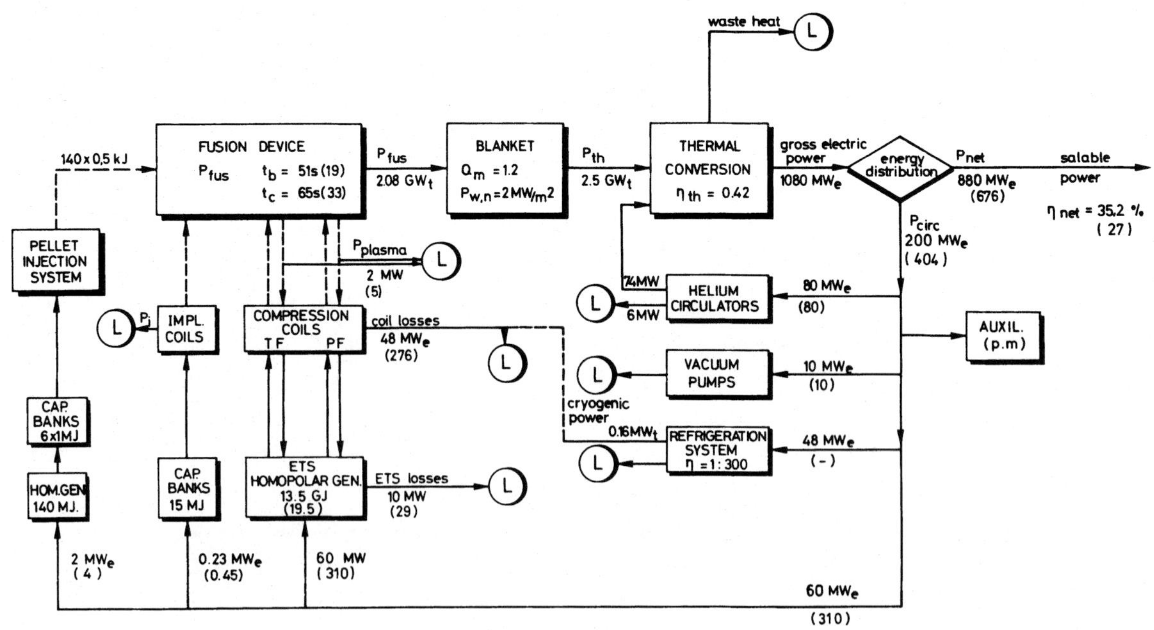

REFERENCE DATA ARE GIVEN FOR THE DESIGN WITH S.C. COILS.
DATA IN ( ) ARE FOR THE DESIGN WITH NORMAL CONDUCTING Cu-COILS.

Fig. 1. Energy flow diagram of the BSPRIL.

reach. Furthermore, because the magnetic field is now comparatively low, it also becomes possible to use superconducting coils. For the screw-pinch reactor with injection and thermal power of 2.5 GW_t, the main parameters of an optimized reactor (BSPRIL) are shown in table I. In section 2 the technical aspects of the BSPRIL are discussed, both for the cases of resistive and superconducting coils. In section 3 both screw-pinch reactor concepts are compared with the related tokamak and reversed field pinch concepts.

2. TECHNICAL ASPECTS OF THE BSPRIL

2.1. Blanket and shield system

As for the preceding BSPR, blanket modules with lateral dimensions of 0.5 x 0.5 m^2 are chosen. Again the breeder material in the modules is lithium, the reflector and shield are made of stainless steel, and the coolant is helium. For an acceptable tritium breeding ratio a minimum thickness of 40 cm for the breeder zone is required. Using normal conducting Cu-coils with inorganic insulators, a minimum thickness of 40 cm for reflector + shield (75% Fe, 25% B_4C) is required. The radiation heating in the coils is then negligible compared to the ohmic heating. If superconducting coils with epoxy insulation are used, the reflector + shield thickness has to be increased to 80 cm. Here the radiation heating of the coils is small compared to the heat dissipated in the coils due to the pulsed operation. Annealing will not be necessary during the life of the plant. Based on an average wall loading of 2 MW/m^2 and an energy multiplication factor in the blanket of 1.2 the average power density in the blanket will be about 2.6 MW/m^3. Calculations with the code FURNACE [4] show that the poloidal variation of the fast neutron flux in the first wall is (+25%, -20%). The power density and the breeding rate vary between (+15%, -20%). These variations are less than those for the preceding BSPR, which is due to the race track geometry of the vessel and the much larger relative plasma volume of the BSPRIL. The 4000 modules each have to be covered with an insulating layer to withstand the induced voltage generated in the wall during the implosion phase. Therefore, the modules have to be connected to the support structure and coolant ducts by ceramic joints.

2.2. Power conversion system

A power conversion system consisting of a primary helium coolant circuit with a modest helium outlet temperature (650 °C) combined with a steam turbine generator circuit is considered. In this way an economically attractive conversion system is obtained [5]. A thermal efficiency of 42% is expected here. The inlet temperature of the helium is 250 °C and the pressure is 50 bar. Assuming a relative pressure loss of 5% the ideal pumping power will be 80 MW. Subdividing the blanket into 24 toroidal sections, each section needs coolant ducts of about 30 cm diameter and the main duct to the heat exchanger will be about 1.5 m in diameter. Two steam turbine generator units of nominal 540 MW each are envisaged. The power supply to the electric systems of the reactor itself (the recirculating power) accounts in the overall energy balance to a total amount of 200 MW_e for the reactor with s.c. coils and 400 MW_e if Cu-coils are used. The energy-flow diagram of the BSPRIL is depicted in fig. 1.

2.3. Implosion coil system

The implosion system must deliver a fast rising magnetic field at the start of each cycle. During the implosion phase an electric field strength of 40 kV/m is generated at the first wall. As was suggested in the former BSPR design the blanket modules will be used directly as the implosion circuit, resulting in a voltage of 20 kV between the modules. Each of the modules is connected to a storage capacitor system, which includes a power crowbar. Consequently, the total storage bank with a capacity of 15 MJ is divided in 4000 units of 3.75

kJ, 30 kV each. Uniform distribution of these capacitors on one gallery around the reactor is possible. The required screw-pinch configuration of the magnetic field is obtained by connecting the modules in an appropriate pattern. The insulating layer covering each blanket module has to be about 2 mm thick. The average power needed to recharge the capacitors is less than 0.5 MW.

2.4. Compression coils

Separate innerlying toroidal-field (TF)-coils and outerlying poloidal-field (PF)-coils are considered for the compression system. The TF-coil system is subdivided into 24 sections and the PF-coils into 4 sections. The rectangular shape of the coils with demountable upper yokes as assumed in the previous designs is abandoned in favour of a more realistic shape of the coils which are not demountable. For the elongated first wall we have chosen, a race-track shape of the coils was selected.

2.4.1. Design with normal conducting Cu-coils

A parametric study leads to an optimal coil thickness for the Cu-coils of $d_{TF} = 1.40$ m where only the inner legs of the coils are tapered. The cross-section of the outer legs of each coil section is 1.40 x 1.23 m^2. The weight of each coil section is 376 tons. The distance between the outer legs (1.3 m) is sufficient for the passage of the vacuum ducts, for manipulation equipment, and probably sufficient for vertical removal of the blanket modules. The magnetic field of 2.8 T at the torus axis requires a total current of 101 MA-turns in the coil system and 4.2 MA-turns per coil section. Taking 21 kV as available voltage supply from the storage system, 5 turns per section have to be taken, carrying 0.84 MA. The current density in the Cu-windings will be about 4 MA/m^2. The ohmic heating in these coils averaged over the cycle time of 35 s amounts to about 276 MW for which a water flow of 1.6 m^3/s is needed if water cooling at 60 °C is chosen. The average distance between the coils and the ETS system is 8 m. 28 water-cooled copper busbars carrying 0.84 MA each between the ETS system and the coil sections are envisaged. The lateral dimensions are such that the ohmic losses in the busbars are 5% of those in the coils. To compensate for all ohmic losses in the coils and transmission lines an average power of 310 MW is supplied to the ETS system.

2.4.2. Superconducting coils

With superconducting coils a considerable reduction of the losses will be expected, although the superconducting coils are pulsed. For the present design, for which a field rise of 50 T/s is needed, the following estimates are made for the cryogenic losses arising from the pulsed operation of the superconducting coils:

- coupling losses during the periods of transient of the magnets account for 4.3 MJ per cycle;
- relaxation effects due to non-instantaneous homogeneization of the current in the s.c. composite during energizing and de-energizing of the magnet accounts for 5.3 MJ per pulse;
- losses due to non-perfect transposition of the parallel subsections in the coils are taken into account to an amount of 0.7 MJ per pulse.

Averaged over the cycle time (65 s for the reactor with s.c.-coils) the cryogenic losses are 160 kW. Using an efficiency of 1:300 for the helium refrigeration system, a continuous electric power supply of 48 MW_e is required, this is 2 MW_e per coil section. Assuming an allowable current density of 5.10^8 A/m^2 in the s.c. material, a superconducting to matrix ratio of 1:10, and a filling factor of 25%, each coil section (including insulation, support, and dewars) has a cross-section of about 0.7 x 0.5 m^2 and a weight of 30 tons for the windings. The significantly reduced size of these s.c. coils compared to the normal conducting

Cu-coils implies both easier access to the blanket and cost savings of the coil system. Based on unit costs of 10 \$/m/kA each coil section will cost 1.4 M\$ compared to 7.5 M\$ for a Cu-coil section. The costs of the refrigeration system are estimated [6] to be about 10 M\$. An estimate of the required cooling capacity due to heat inleak and to ohmic losses for optimized s.c. current leads shows that the mass flow of the cold helium to the current leads (if allowed to heat up to room temperature) does not exceed the amount of off-gas due to the losses in the coils. Consequently, this off-gas can be used for cooling of the leads.

2.5. ETS system

The reactor with Cu-coils needs a compression energy of 16.9 GJ in the TF-coil system and 2.6 GJ in the PF system. For the reactor with s.c. coils these values are 12.4 GJ and 1.15 GJ respectively. These energies are supplied and stored periodically in homopolar machines with an efficiency of 95%. The charging voltage of the ETS system is 21 kV (20 kV on the windings of the coils and 1 kV on the transmission lines). Each of the TF-coils has to be connected to a 0.7 GJ storage unit delivering 0.84 MA. Contrary to the conceptual design of a 1.3 GJ, 11 kV homopolar machine [7], consisting of a drum-type rotor with diameter of 2.7 m and total length of 20 m, it is expected that with a disk-type rotor a more compact homopolar machine of the same energy and voltage can be realized. Proceeding from this idea 2 homopolar machines connected in series are envisaged for each TF-coil. The design parameters of each machine are 0.35 GJ, 11 kV, 0.84 MA, outer diameter 3 m, and length 5 m. The total number of 24 x 2 homopolar generators for the TF-coil system and 4 x 2 generators for the PF-coil system can be installed on one gallery around the reactor. For the reactor with superconducting coils the storage energy in the ETS system is 30% smaller because less energy has to be supplied during the pulse to compensate for the losses. This leads to smaller dimensions and lower costs for this ETS system.

2.6. Pellet injection system

In the BSPRIL with Cu-coils 0.20 g DT will be burned per burn cycle and 1.47 g DT has to be supplied to the plasma vessel during each cycle. Starting each cycle with the implosion and compression of a filling mass of only 2% of the final mass during the burn, 1.40 g DT has to be injected after the compression phase. Injection of fuel by means of pellets was found to be a promising fuelling technique. The injection of 140 pellets of 10 mg each is considered. The pellets have to be deposited in the thermonuclear plasma, where they are evaporated and heated. The plasma expands either by the pressure increase if the pellet reaches the center, or by a burning front if it is ablated at the plasma border. The pellets are assumed not to alter the low-density outer layer of the plasma and the force-free screw-pinch currents. A velocity of 10^4 m/s of the pellets turns out to fulfil these requirements.

A promising method for pellet acceleration is laser propulsion [8]. To accelerate pellets up to 10^4 m/s the necessary ablation pressure can probably be achieved with existing laser technology.

The kinetic energy of one 10 mg pellet with a velocity of 10^4 m/s is 500 J. Assuming a pessimistic efficiency of the total acceleration system of 0.05%, the energy supply to the laser gun is about 1 MJ per pellet. Although the efficiency is low, this has a minor influence on the energy balance and on the net efficiency of the reactor. The total power supply of 140 MJ can be extracted from a homopolar generator. Assuming a repetition rate of 6 Hz for a laser gun, a total of 6 laser guns is needed to inject all pellets within a period of 4 s. The straight acceleration tubes of these 6 guns may be incorporated with 6 alternative vacuum ducts, thus providing a uniform fuel supply and a reduction of the number of blanket

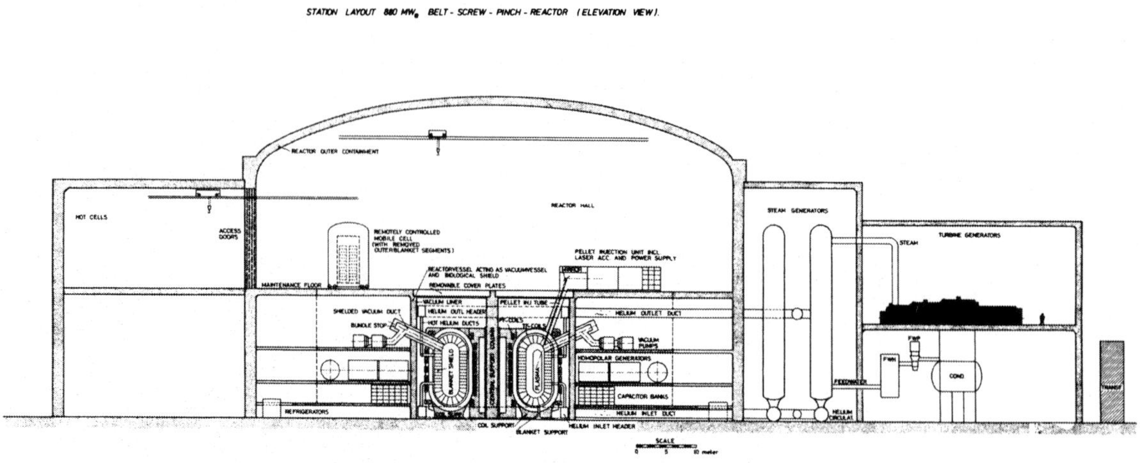

Fig. 2. Station layout of the BSPRIL.

TABLE III. Comparison between Tokamak, Reversed-Field Pinch, and Screw-Pinch Reactor Designs.

	TOKAMAK	RFP	SP
aspect ratio	(-) low high primary to plasma current ratio?	(+) <u>high</u> → easy access low prim. to plasma current ratio	(-) intermediate
elongation	difficult to maintain high	not necessary	high
stability	(-) max. β uncertain (-) risk of disruption	stability at changing reversal uncertain: passive stabilization by <u>shell</u> → segmentation difficult?	(-) stability at β-limit uncertain, mild instability required
plasma current	risetime uncertain	risetime uncertain	(-) fast rise with toroidal field
toroidal field	(-) high, but <u>stationary</u> (+)	(+) low, but pulsed (-)	(+) low, but pulsed (-)
beta	(-) low	(+) <u>high</u>	(+) high
pulse time	long or short if long, refuelling and ash removal needed	(-) short $t_{pulse} > t_{diff}$ possible? fuelling and ash not necessary	(-) short force-free region maintainable at $t > t_{diff}$? Ash removal?
heating to ignition	ohmic + additional heating → (-) costly	<u>ohmic</u>, perhaps additional in case of impurities	implosion + adiabatic compression (-) high voltage → <u>insulated wall</u> → coating necessary
formation phase	(-) turbulent → wall contact (+) low losses	(-) turbulent (-) possibly high losses	(+) <u>stable</u> (+) no plasma losses
run down	(+) low energy losses	(-) high losses	intermediate
fuel injection	possible	possible?	<u>essential for economics</u>
active feedback	(-) vertical field for equilibrium	(-) long-period stabilization	(+) first wall remains flux surface
impurities	problematic, through plasma-wall interaction, additional heating and long pulse	turbulent formation; little heating power reserve, but short pulse	no turbulence, short pulse but insulators at first-wall difficult to clean
coils	superconducting	superconducting, but normal possible	superconducting, but normal possible
ETS system	PF and additional heating 8 GJ	PF + TF circulating 8.5 GJ, 21 kV	(-) PF + TF circulating 14 GJ, 21 kV
blanket	(+) <u>stationary operation</u> allows higher mean wall loads	(-) shell implies lower T-breeding, and lower thermal efficiency	(-) insulated wall at pulsed operation
coil design	(-) high mechanical stresses for high fields	(-) high $\dot{B}$ questionable with supercond. coils	(-) high $\dot{B}$ questionable with supercond. coils
general design	large ducts for divertor and neutral injectors	shell elements difficult to disassemble	short transmission lines → difficult access to blanket
capital costs	highest (3500 $/kW_e$)	lower (3000 $/kW_e$)	lower (2500 $/kW_e$)

penetrations. With a laser efficiency of 1%, the laser has to deliver 10 kJ light to each pellet during the 1 ms time of flight of the pellets in the gun tube. Alternatively 10 successive pulses of 1 kJ on each pellet may be considered.

2.7. General design and cost evaluation

A general impression of a possible station layout of the reactor with s.c.-coils is given in figure 2. Due to the implosion heating mechanism a closed conducting first wall is not permitted. Therefore a closed metallic vacuum-tight wall has to be situated outside the coils supported by a thick concrete wall. These walls act as reactor vessel and biological shield. The energy storage systems are located outside the reactor vessel as close as possible to the coils, to minimize the length of the transmission lines. This tightly packed structure around the nuclear reactor makes horizontal access to the blanket impossible. Therefore access to the blanket is supposed to be vertical after removal of the demountable cover plates of the reactor vessel.

To remove the unburned fuel and the ash during the down time of the reactor, a pumping rate of 800 m^3/s at 10^{-6} torr requires 10 MW_e. To minimize the number of ducts through the blanket, one duct of 1.5 m diameter per 2 toroidal sections, each leading to 2 parallel series of vacuum pumps, is sufficient. The ducts enter the torus in the upper corner of the outer blanket, where access is easy. For protection of the vacuum pumps to neutron streaming effects, the vacuum pumps are connected perpendicular to the ducts.

A cost evaluation for the BSPRIL reactor has been performed on unit cost estimates derived form other system studies, e.g. [9], see table II. The main cost components of the nuclear part of the reactor are the blanket (including the primary heat transfer system), the coils, the ETS system, and the fuel injection system. The non-nuclear parts of the reactor: turbine plant, electric plant, and the civil structures are more or less independent of the type of reactor and can be indicated as "convential plant", accounting for about 150 $/kW$_t$.

For the BSPRIL reactor the costs of the nuclear part of the reactor are about 1.5 to 2 times the costs of the conventional plant. The total direct costs of the reactors are 1060 M$ (1570 $/kW$_e$) for the Cu-coil design and 820 M$ (930 $/kW$_e$) for the design with superconducting coils. The additional cumulative indirect costs and interest during construction, which are about 1.5 times the direct costs, have also to be taken into account. This leads to total capital costs of 2660 M$ (3930 $/kW$_e$) and 2040 M$ (2320 $/kW$_e$), respectively.

3. COMPARISON

In this section the physical, technological, and economical aspects of the BSPRIL are compared with those of the high-β, pulsed reversed-field pinch (RFP) reactors, and of the low-β tokamak reactors. The net electric power output of all reactors lie in the range of 600 MW_e. The comparative high-β concepts are the RFP-reactors from Los Alamos [10] and Culham [11]. The related tokamaks are the Pulsed-Tokamak of Culham [12] and the NUWMAK of Wisconsin [13].

The characteristic similarities and differences of the various reactor concepts are given in condensed form in table III. Figures 3 and 4 show the power comparison and the direct costs comparison, respectively.

The magnetic field profiles of the screw pinch and the tokamak are similar but while the toroidal-field strength in the screw pinch is lower, the plasma pressure is considerably higher. The highest beta can be utilized if fuel injection is applied, so that the plasma volumes also become comparable. The RFP also has a high value of β, and it is interesting to note the rather low values of β close to the coil. Other favourable points of the RFP are the possibility to use a slender torus with circular cross-section, and the ohmic heating to

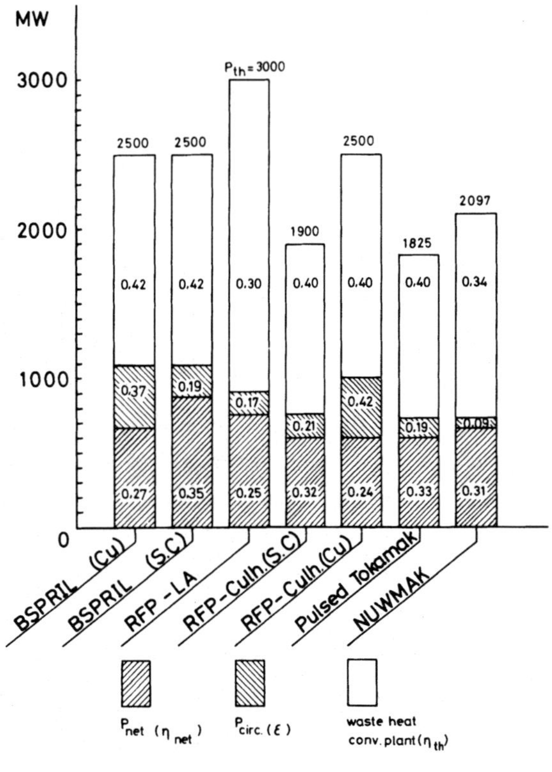

Fig. 3. Comparison of power balances.

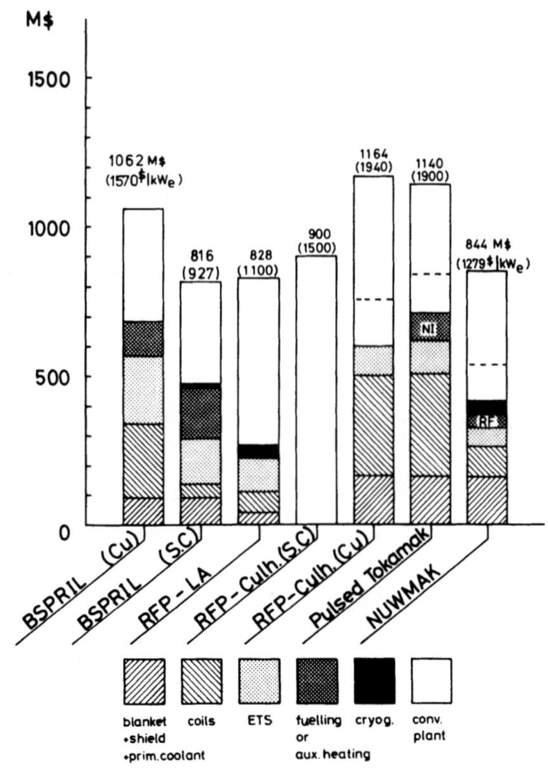

Fig. 4. Comparison of direct costs.

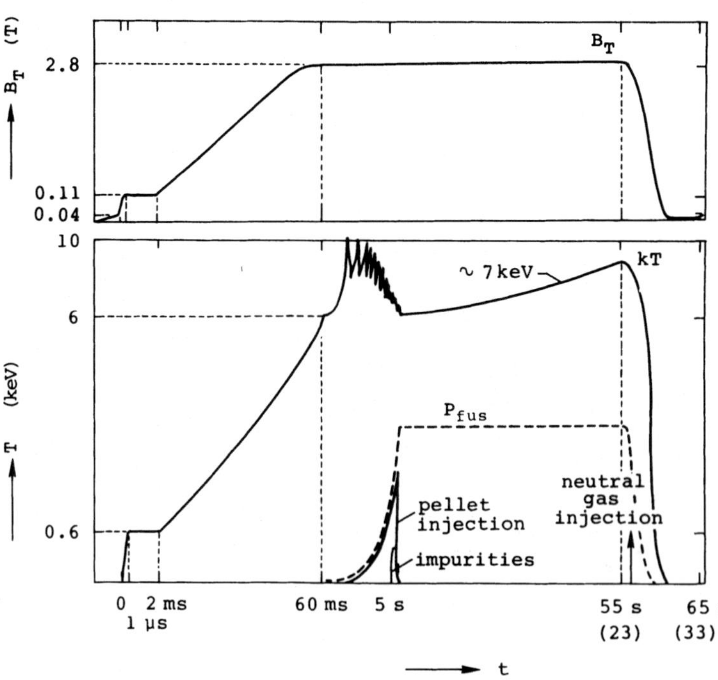

Fig. 5. Time history in the BSPRIL.

ignition (still unproven). The copper shell for plasma stabilization of the RFP degrades the neutron flux and the tritium production, and complicates the energy conversion. The tokamak profits from its long pulse if the ash and impurities can be removed. The screw pinch claims a stable formation phase as the toroidal field rises together with the plasma current. Due to the high electric fields an insulated plasma vessel is required, its integrity under pulsed operation and neutron flux is a specific problem. A high thermal efficiency, e.g. by helium heat transfer, is as essential as a low circulating power fraction. Superconducting coils are necessary for a positive power balance in low-beta systems. For high-beta systems they are necessary to make them competitive.

4. CONCLUSION

With the use of pellet injection and s.c.-coils, the screw pinch has evolved into an economically competitive system.

References

[1] Bobeldijk, C., et al., "Current decay and stability in SPICA", Proc. 6th Int. Conf. on Plasma Phys. and Contr. Nucl. Fusion Res., Berchtesgaden 1976, vol I, p. 493.

[2] Bobeldijk, C., et al., "Parameter study of a screw-pinch reactor", Proc. 9th Symp. on Fusion Techn., Garmisch-Partenkirchen 1976, p. 367.

[3] Bustraan, M., et al., "A reactor study on a belt-shaped screw-pinch", Proc. 10th Symp. on Fusion Techn., Padua 1978, p. 107.

[4] Verschuur, K.A., "FURNACE. A program system for neutron and photon transport calculations in large toriodal geometries", ECN report to be published.

[5] Aronstein, R., Ghose, S.K., "Cost components of fusion power plants", Trans. ANS, 26 (1977) 45.

[6] Strobridge, T.R., "Feasibility study of a multipurpose refrigerator for a superconducting cable test facility", EPRI-282, July 1975.

[7] Thomassen, K.I., "Conceptual engineering design of a one-GJ fast discharging homopolar machine", EPRI-ER-246, (1976).

[8] Burgess, M.D., et al., "Acceleration of macro particles by laser light", CLM-P546, Culham, June 1978.

[9] Badger, B., et al., "UWMAK-III, A non-circular tokamak power reactor design", UWFDM-150, Wisconsin, 1976.

[10] Hagenson, R.L., et al., "The reversed-field-pinch reactor (RFPR) concept", LA-7973-MS, Los Alamos, Aug. 1979.

[11a] Hancox, R., et al., "A 600 MW(e) reversed-field-pinch reactor study", CLM-P501, 1977.

[11b] "A reversed-field-pinch reactor with superconducting windings", Proc. 7th Int. Conf. on Plasma Phys. and Contr. Nucl. Fusion Res., Innsbruck 1978, p. 323.

[12] Spears, W.R., Hancox, R., "A pulsed tokamak reactor study", CLM-R197, Culham, 1979.

[13] Badger, B., et al., "NUWMAK, A tokamak reactor design study", UWFDM-330, Wisconsin, 1979.

This work was performed as part of the research programme of the association agreement of Euratom and FOM with financial support from ZWO and Euratom.

INERTIAL CONFINEMENT FUSION REACTOR SYSTEMS

T. G. Frank, I. O. Bohachevsky, J. H. Pendergrass
Los Alamos Scientific Laboratory
P.O. Box 1663
Los Alamos, NM 87545

ABSTRACT

A variety of reactor cavity concepts, drivers, and energy conversion mechanisms are being considered to realize commercial applications of ICF. Presented in this paper are: (1) a review of reactor concepts with estimates of practically achievable pulse repetition rates, (2) a survey of drivers with estimates of the requirements on reactor conditions imposed by beam propagation characteristics, and (3) an assessment of compatible driver-reactor combinations.

1. INTRODUCTION

Since about 1969, research has been carried out to develop an alternative to magentically confined controlled thermonuclear fusion -- that of compressing, heating and confining thermonuclear fuel by inertial forces generated by the interaction of an intense, pulsed beam energy source (or driver) with a pellet containing the fuel. The outer region of a fusion pellet consists of an absorber/ablator material in which energy from a driver source is deposited. This material is blown off creating a recoil impulse which, together with plasma pressure, heats and compresses the fuel. Thermonuclear ignition occurs at the center of the fuel and propagates radially outward in a time that is short compared to the time required for the pellet to disassemble, resulting in fusion of an appreciable fraction of the fuel. Understanding of the fundamental physics of driver-pellet interactions and pellet dynamics is being developed through combined theoretical and experimental investigations.

Inertial confinement fusion (ICF) experimental programs have, thus far, relied principally on the use of short-pulse lasers in pellet implosion experiments, and the construction and use of lasers as research tools will be emphasized during the next several years to establish the technical feasibility of ICF. Ion beam accelerators are promising alternatives to lasers for driving fusion pellet microexplosions, and programs for development of accelerators with acceptable characteristics are being conducted.

For commercial applications, fusion pellet microexplosions must be repeatedly contained in reactor cavities in a manner that prevents severe damage to reactor components and permits convenient, economic rècovery of the energy for conversion to electricity or some other usable form. First-generation ICF applications will be based on the tritium-deuterium fuel cycle. Reactor cavities must be surrounded by blanket regions containing lithium because tritium for the fuel cycle must be produced in interactions between fusion neutrons and lithium. Different driver types impose different conditions on reactor cavity environments so that facility design for commercial applications must be done for integrated systems.

In the remainder of this paper, reactor concepts now being studied and acceptable driver-reactor combinations are discussed.

2. REACTOR CONCEPTS

Fusion pellet microexplosions release energy as x rays, energetic pellet debris, and high-energy neutrons. Reactor components must be protected from excessive damage by x rays and pellet debris, which may cause material loss by evaporation and/or sputtering of exposed surfaces. Several different approaches to protecting cavity walls from x rays and pellet debris are being studied to assess their feasibility, to identify technology requirements, and to determine their acceptability for use in combination with various drivers. Serious damage to reactor structures also results from exposure to high fluences of high-energy neutrons. Provision for moderating neutron energies to minimize damage to reactor structures are included in some reactor concepts.

Inertial-confinement-fusion reactor concepts can be divided into two major categories with regard to accommodating the photon and debris energy released by pellet microexplosions: (1) concepts in which the energy absorbing surfaces are significantly perturbed by photon and debris energy deposition but are regenerated between pellet microexplosions and (2) concepts in which photon and debris energy are either directly absorbed in solid reactor cavity components or in which the debris may be diverted from the reactor cavity.

For concepts in the first category, the cavity wall is protected against damage from photons and pellet debris by either a gas or a protective liquid metal layer. Soft x-ray and pellet-debris energy is deposited in the protective material. For some designs, there are restrictions on pellet microexplosion repetition rate due to the time required to restore the cavity after a pellet microexplosion to conditions necessary for pellet injection and beam transport.

In the second category of reactor concepts, photons and pellet debris are deposited in semi-permanent cavity liners. Near-surface energy deposition can cause evaporation and the pellet debris can cause sputtering of the liner surface. A variant in this category is the use of magnetic fields to divert the ionized pellet debris out the ends of a cylindrical cavity leaving only the x rays to be accommodated by the cavity wall. There are tradeoffs for minimum damage to material surfaces between relative x ray and pellet debris energy yields and energy spectra. There are generally no practical constraints on maximum pellet microexplosion repetition rate for reactor designs in this category.

There have been a large number of ICF reactor concepts proposed during the past decade. Through analytic evaluations of performance, studies of interface conditions in integrated systems, and comparative economic assessments, the most attractive features of these concepts are being incorporated into a few designs that are compatible with the different driver types being developed.

2.1 Lithium Wetted-Wall Concepts

The wetted-wall reactor concept was originally proposed in 1971 [1]. It has been studied and occasionally modified throughout the past decade. A schematic of this concept is shown in fig. 1. The reactor chamber is spherical and is surrounded by a blanket region containing liquid lithium and structural components. The cavity wall is formed by a porous metal through which coolant lithium flows to form a protective layer on the inside surface. The soft x-ray and pellet-debris energy is deposited in the protective lithium layer resulting in partial evaporation and ablation. The lithium vapor is subsequently exhausted through a supersonic nozzle at the bottom of the reactor into a condensor. The protective layer is restored between pellet microexplosions by radial inflow of lithium from the blanket region.

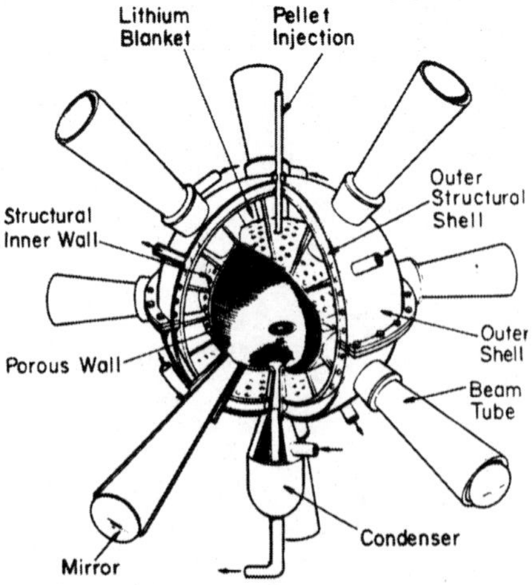

Fig. 1. Wetted-wall reactor concept.

The dominant stresses induced in wetted-wall reactor concepts result from the recoil impulse of the lithium ablated from the interior surface of the cavity wall and from thermal expansion of the lithium blanket due to neutron energy deposition. Stainless steel reactor cavity and blanket structural walls ~ 1-cm thick provide adequate strength to accommodate the generated pressures and impulses.

2.2 High Yield Lithium Injection Fusion Energy (HYLIFE)

The HYLIFE reactor concept [3] was developed to satisfy several specific requirements and objectives, including: modification of the flux and energy spectra of radiation emitted by pellet microexplosions with flowing fluids that can be reestablished after each microexplosion, 30-year operational lifetime without replacement of damaged or radioactive structure, and minimization of development time.

The reference HYLIFE concept, shown schematically in fig. 3, consists of an 8-m-high, 10-m-diameter chamber in which a blanket of liquid lithium shields the steel wall from x rays, pellet debris and high-energy neutrons. The liquid lithium blanket is composed of a dense hexagonal array (0.5 packing fraction) of 20-cm-diameter jets. A 300-jet array provides an effective blanket thickness of 1 m between the pellet microexplosion and the first structural steel wall. The pellet and the driver beams are injected horizontally through specially arranged corridors in the array of jets as shown in fig. 4.

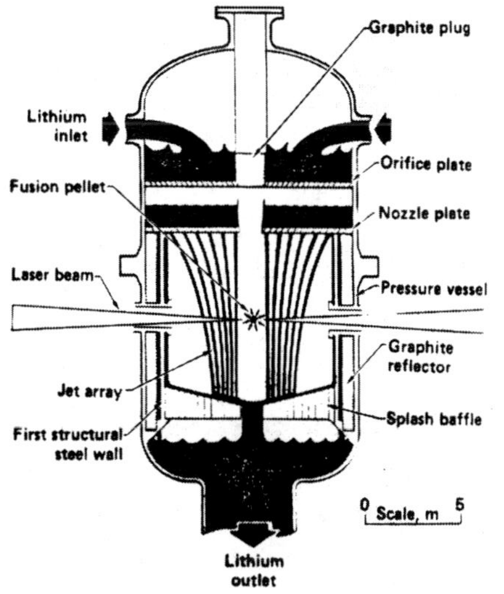

Fig. 3. HYLIFE reactor concept.

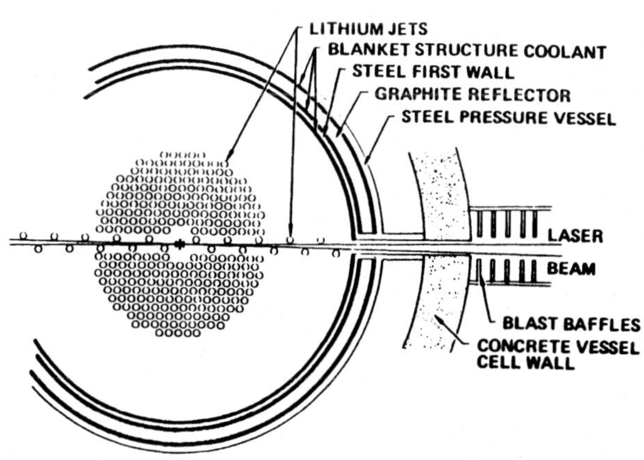

Fig. 4. Beam transport corridor in HYLIFE reactor concept.

The energy of the volumetric expansion of lithium, which results from neutron absorption, is primarily deposited in liquid-liquid interactions of colliding jets. The pressure of the lithium vaporized from the inner surface of the blanket by soft x rays and pellet debris exerts an outward force on the blanket, after coalescence of the jets, causing the lithium to expand outward and collide with the pressure vessel wall; however, the resulting stress is estimated to be acceptable. The large surface area of flowing lithium acts as a condensation pump on which the vaporized lithium is condensed between pellet microexplosions. It is estimated that the jet array will be reestablished and the cavity pressure reduced to that corresponding to the lithium vapor pressure in ~ 1 s following a pellet microexplosion. An attractive feature of the HYLIFE concept is that the neutron energy wall loading is reduced by a factor of ~ 20 by the lithium blanket leading to an anticipated chamber lifetime equal to plant lifetime.

Lithium vapor densities at the time of pellet injection and beam transport can be predetermined by limiting the maximum lithium temperature in the cavity. The nominal maximum

The wetted-wall reactor is proposed for use with a laser driver. The vapor density in the cavity must be reduced after a pellet microexplosion to 10^{15} to 10^{16} atoms/cm^3 for efficient transport of laser beams. The exhaust nozzle is appropriately sized to evacuate the cavity to this lithium vapor density in ~ 0.8 s. From this and other considerations the pellet microexplosion repetition rate is constrained to ~ 1 Hz or less.

The wetted-wall reactor concept suffers from two potential disadvantages: (1) a perceived difficulty in monitoring the reestablishment of the protective lithium layer on the cavity interior surface and (2) the limitation on pellet microexplosion repetition frequency. In addition, this concept is not readily adapted to very low cavity vapor density operation. These disadvantages are circumvented by a modified concept [2] shown in fig. 2. In the modified concept, liquid lithium is injected tangentially through a circular slit nozzle at the top of the spherical reactor cavity at a rate sufficient to remove the x-ray and pellet-debris energy as sensible heat of the liquid with only a modest increase in temperature. Positive coverage of the cavity interior surface is assured by centrifugal forces. Vaporized lithium from pellet microexplosions recondenses on the surface of the injected lithium stream in time intervals much less than required for exhaust through a nozzle. The exhaust nozzle is replaced by a simple drain at the bottom of the cavity. The injection nozzle and diverter vanes around beam ports are protected by lithium films that are maintained by forced flow through porous structures.

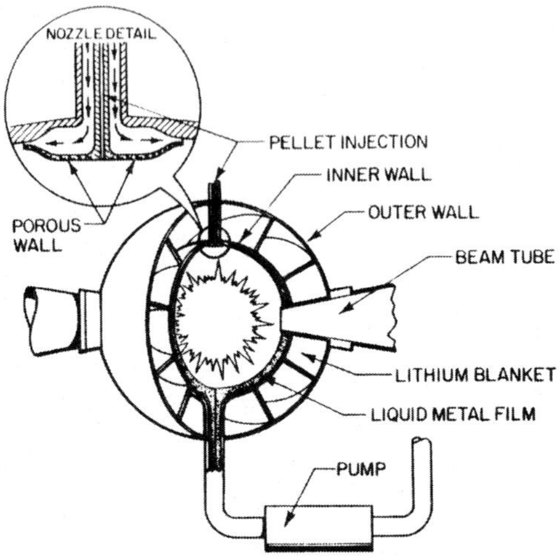

Fig. 2. Modified-wetted-wall reactor concept.

This reactor concept could be operated with very low cavity vapor densities by limiting the maximum temperature of the fluid inside the cavity to values corresponding to low vapor pressures. Separate coolant streams of different materials could be used in the reactor cavity and blanket to provide flexibility in choices of vapor pressure and temperature.

A large fraction (60 to 70%) of the energy release from pellet microexplosions is deposited directly in the blanket regions of wetted-wall reactors by high-energy neutrons. Blanket coolant is introduced near the cavity wall by structures concentric with the beam transport tubes. The lithium then flows radially outward through the 1-m-thick blanket. Uniform radial flow is achieved by including sufficient impedance to flow in successive structural shells.

Reference design studies have been done for 150-MJ pellet microexplosions at repetition rates of 1 and 10 Hz, respectively, for the wetted-wall and the modified-wetted-wall concepts. The cavity radii were chosen to be 2 m from neutron damage considerations. For the modified-wetted-wall concept, a lithium flow rate of 2.4 m^3/s is required to limit the temperature increase of the cavity coolant to an arbitrary 100 K. This flow rate could be provided by a circular nozzle with a 1-cm-wide slit with an injection velocity of 100 m/s. Analyses of flow profiles through a 2-m-radius cavity indicate that the lithium thickness increases from 1 cm at the top of the cavity to 7 cm at the bottom. Pumping requirements are less than 1% of the electric power produced.

lithium temperature for use with laser drivers is 770 K, corresponding to a lithium vapor density of ~ 3×10^{15} atom/cm^3. If a lithium vapor density limit of 10^{13} atom/cm^3 is required, the lithium temperature cannot exceed ~ 620 K.

The operating characteristics of the reference HYLIFE power plant includes a pellet yield of 2700 MJ at a repetition rate of 1 Hz. The circulating lithium flow rate is 140 m^3/s of which 8% is diverted to a heat exchanger. The temperature increase of the lithium as it flows through the reactor chamber is 13 K. The pumping power required for the primary lithium loop is 1.6% of the net electric power produced.

2.3 Gas Filled Reactor Cavities

The use of a noble gas in reactor cavities to minimize the damaging effects of x rays and pellet debris has been investigated in two regimes of gas pressure. Inclusion of a low-pressure gas in reactor cavities has been considered for use with laser drivers, whereas higher presures are appropriate for use with light particle beam drivers.

The most exhaustively studied gas-filled reactor concept for use with laser drivers is the SOLASE design [4]. The SOLASE cavity wall and blanket structure is made of graphite, and the reactor coolant is circulating lithium oxide particles. The cavity is filled with 10^{15} to 10^{16} atoms/cm^3 of xenon or neon. This gas stops the ion debris and attenuates the soft x rays. This energy deposition heats the gas to 1-3 eV. The gas then reradiates the absorbed energy in a time interval much longer than the pulse in which it was originally released. Residual heat is removed from the cavity by flowing the gas through the cavity.

The SOLASE reference cavity design has a 6-m radius. The pellet yield is 150 MJ and the repetition rate is 20 Hz. The anticipated lifetime of the structure is 1 yr. An advantage of the graphite structure is its low-induced radioactivity permitting limited hands-on maintenance two weeks after shutdown.

Light-ion-beam reactor concepts require a relatively high-density gas (10^{18} to 10^{19} atom/cm^3) in the cavity for beam propagation along ionized channels from the particle source to the pellet. For cavity gas density greater than 10^{18} atom/cm^3, all of the pellet debris energy and most of the x-ray energy released by the pellet microexplosion are deposited at relatively short ranges resulting in the formation of a hydrodynamic shock. Accommodation of the shock overpressure by the cavity structure leads to optimized cavity designs with relatively large radii. The energy deposited in the cavity gas is reradiated and conducted to the cavity wall in times very long compared to deposition times. Equilibrium gas temperatures are quite high which may pose a pumping problem if the cavity gas is continuously circulated.

Typical operating characteristics for a light-ion-beam driven reactor include 75-MJ pellet yields in 3×10^{18} atom/cm^3 gas with a repetition rate of 10 Hz and a cavity radius of 4 m [5].

2.4 Ablative Liners

For applications requiring a very good vacuum in the reactor cavity, the most attractive cavity concept may be a steel or refractory metal structure with protection from x rays and pellet debris provided by a liner that is allowed to evaporate and ablate at a controlled rate. Protective liners made of carbon have been investigated for this purpose. The cavity radius is made sufficiently large to limit surface erosion of the liner so that replacement is not required more often than once per year.

An example of a reactor with a carbon-lined cavity is shown in fig. 5. The cavity radius required to limit surface erosion to 2 to 3 cm in one year of operation with 150 MJ microexplosions at 10 Hz is 10.5 m.

2.5 Magnetic Deflection

Since the debris from fusion pellet microexplosions is ionized, it can be defelected away from sensitive reactor components by magnetic fields [6]. The use of magnetic deflection to protect cavity walls and optical components has been investigated. The essential features of such a reactor concept are shown schematically in fig. 6. The reactor cavity is cylindrical, with an impressed steady-state magnetic field produced by a solenoid located concentric with, and exterior to, a lithium blanket region. The ionized pellet debris are diverted by the magnetic fields either through magnetohydrodynamic ducts or to specially designed energy sinks in the ends of the cavity.

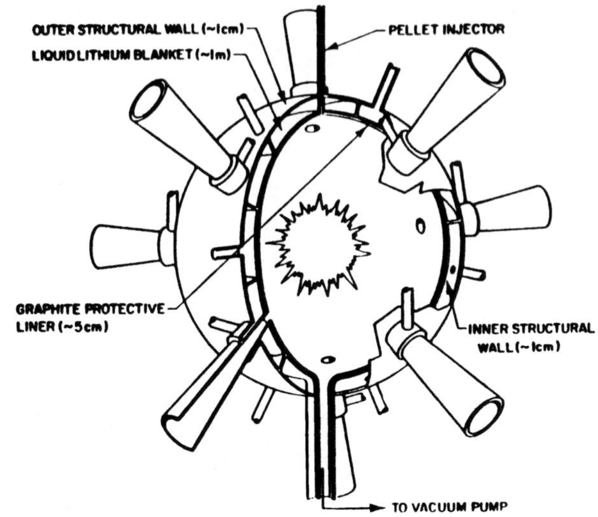

Fig. 5. Sacrificial wall reactor concept.

Conceptual designs of reactors protected by magnetic deflection are constructed of steel or a refractory metal with additional cavity wall protection provided by carbon liners. Designs have been evaluated for use with 150 MJ pellet yields with 10-Hz repetition rates. Cavity radii corresponding to 1-yr. carbon liner lifetimes are ~ 2.5 m for cavities filled with 10^{15} atoms/cm^3 of xenon and 7.5 m for high-vacuum cavities.

3. Driver-Reactor Compatibility

There are three classes of pulsed beam energy sources now being investigated and evaluated for possible use as drivers for ICF; they are lasers, heavy-ion beams, and light-ion beams. Each of these drivers imposes different conditions on reactor interfaces and on cavity conditions suitable for beam injection.

For laser drivers, it is necessary to have optical components (mirrors) in direct line of sight of the pellet microexplosion. These components are protected from damage from x rays and pellet debris by distance and by a tenuous gas in the beam transport tubes and/or magnetic fields to divert ionized debris.

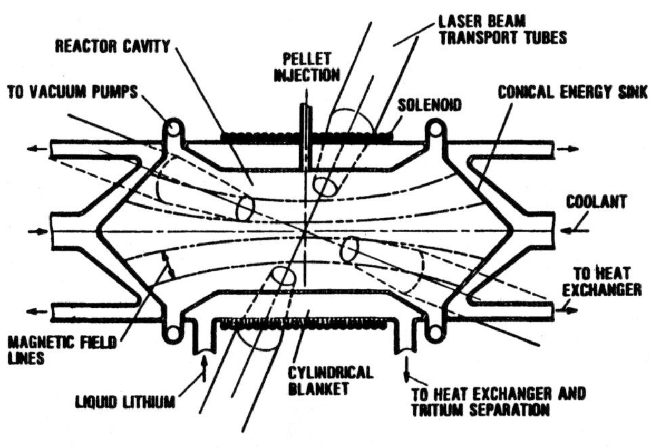

Fig. 6. Magnetic deflection of ion debris in ICF reactor

Laser beam transport inside reactor cavities can be accomplished through low-density gas; however, there are a great many processes that can result in scattering and defocusing of laser beams, and the upper limit on gas density for efficient transport of focused beams has not been accurately determined. A combination of experimental results and theoretical analyses indicate that gas densities less than ~ 5×10^{15} atom/cm^3 do not affect beam propagation significantly.

Driver-reactor interface conditions for heavy-ion beam drivers are currently being assessed. Important considerations will include pressure differences between the reactor cavity and the accelerator, protection of focusing and other magnets from damage by high-energy neutrons, beam propagation distances, and beam injection configurations. Although several approaches to beam transport and focusing inside the reactor cavity are being studied, the least uncertainty is associated with ballistic propagation of singly ionized atoms. For particle kinetic energies now being considered, an upper limit on reactor cavity density of 10^{12} to 10^{13} atoms/cm^3, where two-stream instability is not too serious, is expected to be the prevailing requirement. However, even gas densities in this range are several orders of magnitude higher than can be tolerated in the accelerator so that vigorous differential pumping at the interface will be required. Beam propagation and focusing in gas densities below 10^{13} atom/cm^3 for distances from the focusing magnets of several meters is straightforward; however, space-charge effects may require that the number of beams be large.

Light-ion beams impose entirely different conditions than lasers or heavy-ion beams on driver-reactor interface and cavity conditions. A relatively high-density gas is required in the cavity for beam propagation along ionized channels from the particle source to the pellet. Gas densities which satisfy requirements for beam propagation are in the range 10^{18} to 10^{19} atoms/cm^3. Preionization along beam paths in the cavity is provided by low-power lasers. Voltage is then applied to the channel electrodes establishing arcs along the preionized paths and creating low-density plasma channels. The ion beam is then magnetically confined to the channel as it propagates to the pellet.

Compatible driver-reactor combinations that have been identified in this discussion are indicated in Table I.

TABLE 1

Compatible Driver-Reactor Combinations for Inertial Confinement Fusion

Reactor Concept	Driver Type: Lasers	Driver Type: Heavy Ion Beams	Driver Type: Light Ion Beams
Wetted-Wall	X		
Modified-Wetted-Wall	X	X	
HYLIFE	X	X	
Low-pressure gas-filled	X		
High-pressure gas-filled			X
Sacrificial liner	X	X	

4. REFERENCES

[1] Booth, L. A., Los Alamos Scientific Laboratory Report LA-4858-MS, Vol. 1 (1972).

[2] Pendergrass, J. H. et al., American Nuclear Society Meeting, Las Vegas, NV, 8-13 June 1980. Transactions p. 40.

[3] Meier, H. R. et al., 2nd International Conference on Liquid Metal Technology in Energy Production, Richland, WA, 20-24 April 1980.

[4] Moses, G. A. et al., 3rd Topical Meeting on the Technology of Controlled Nuclear Fusion, Santa Fe, NM 9-11 May 1978. Proceedings pp. 448-457.

[5] Cook, D. L., Sandia Laboratories Report SAND 80-0466C (1980).

[6] Booth, L. A. Proceedings of the IEEE 64, 10, 1460-1482 (1976).

MODELLING OF DIVERTOR FLOW FOR A REACTOR REGIME

J G Morgan and P J Harbour

Culham Laboratory, Abingdon, Oxon OX14 3DB, UK.
(Euratom/UKAEA Fusion Association)

ABSTRACT

We describe the development of a one-dimensional model for the flow of plasma in the scrape-off layer and divertor of a reactor-scale tokamak. Some analytic results and preliminary numerical results are presented.

1. INTRODUCTION

There are two approaches to the problem of extracting power from a reactor by the use of a divertor. Either the scrape-off layer may be in a collisionless regime at low density and high temperature, transporting energy by convection, or it may operate at high density and a lower temperature, with collisions and thermal conduction dominant. Present small tokamaks operate at a density which implies near-collisionless conditions in the scrape-off layer. This low collisionality may result in a high temperature in a reactor divertor which is undesirable [1]. In some recent work by the authors and their colleagues [2-9] it has been proposed that the scrape-off plasma in a reactor should operate at high density and moderate temperature so that it can be in pressure balance with the plasma in the divertor which can consist of a dense, cool target of recombining, radiating plasma. With such a system the heat flux into the divertor is likely to be high; moreover, plasma-wall interactions (eg, sheath effects, arcing and sputtering) are likely to be minimised because the plasma is cold near the wall, both in the scrape-off region and in the divertor.

One proposal for achieving this high density, low temperature configuration is that a substantial fraction of the gas flowing out of the divertor chamber should be recirculated back into the scrape-off plasma of the main torus [9]. This recirculated gas is expected to be ionised in the scrape-off plasma and to augment the flow of plasma into the divertor. With such an approach it should be possible to increase the fractional burn-up of fuel from about one per cent without recirculation to 5% or more.

In reality, the scrape-off region of a tokamak is a very complex system with many interacting plasma and atomic processes occurring in a genuinely three-dimensional magnetic geometry. The slow radial diffusion of plasma across the separatrix turns into a rapid flow along the field lines into a divertor or limiter. This flow is augmented by the ionisation of neutrals recycled from the wall of the toroidal chamber. A fully 3-D computational model is near, or beyond, the limits of present-day techniques and machines, so simplifications are essential to achieve progress at modelling. Classical transport along field lines is large compared with even Bohm-like anomalous radial transport so a good first approximation is to consider a one-dimensional model describing the direction parallel to the field lines with transverse flow included as volume sources and sinks.

In this paper we describe the development of a model for the flow of plasma in the scrape-off region of a tokamak with a divertor.

2. THE MODEL

In a reactor regime we expect a collisional scrape-off layer with mean free paths shorter than, or at most comparable with, the scale lengths of the flow. Hence it is generally appropriate to use a fluid model to describe the plasma. We follow the usual prescription of taking moments of the electron and ion Boltzmann equations, closing the system of equations by the use of transport coefficients [10]. We use a two-fluid model with an average ion species, at present consisting of pure D-T fuel without impurities.

It is to be expected that the longitudinal flow will adjust on timescales very short compared with rates of change of the reactor plasma so a steady-state description of the flow is appropriate and the problem reduces to the following system of ordinary differential equations:

Continuity: $$\frac{d}{dz}(n_{i,e}v_{i,e}) = S_{i,e} \qquad \ldots (1)$$

Momentum: $$m_{i,e}\, n_{i,e}\, v_{i,e}\frac{dv_{i,e}}{dz} = -\frac{dp_{i,e}}{dz} + \frac{d}{dz}\left(\eta_{i,e}\, n_{i,e}\, T_{i,e}\, \tau_{i,e}\frac{dv_{i,e}}{dz}\right) \pm en_{i,e}E \pm R^{ei} + R_{i,e} - m_{i,e}\, v_{i,e}S \qquad \ldots (2)$$

Energy: $$\frac{3}{2}n_{i,e}\, v_{i,e}\frac{dT_{i,e}}{dz} + p_{i,e}\frac{dv_{i,e}}{dz} = \frac{d}{dz}\left(\kappa_{i,e}\frac{n_{i,e}T_{i,e}\tau_{i,e}}{m_{i,e}}\frac{dT_{i,e}}{dz}\right) + \eta_{i,e}\, n_{i,e}\, T_{i,e}\, \tau_{i,e}\left(\frac{dv_{i,e}}{dz}\right)^2 \pm Q^{ei} + Q_{i,e}$$
$$- v_{i,e}R_{i,e} - \frac{3}{2}T_{i,e}S_{i,e} + \frac{1}{2}m_{i,e}v_{i,e}^2 S_{i,e} \qquad \ldots (3)$$

where:

z is the position co-ordinate and a uniform magnetic field has been assumed so that the flow is "constant area",

$m_{i,e}, n_{i,e}, v_{i,e}, T_{i,e}, p_{i,e}$ are the mass, number density, velocity, temperature and pressure of ions (i) and electrons (e),

$\eta_{i,e}, \kappa_{i,e}$ are numerical coefficients of viscosity and thermal conductivity [10]
$\eta_e = 0.9733$, $\eta_i = 1.28$, $\kappa_e = 3.16$ and $\kappa_i = 3.9$,

$\tau_{i,e}$ are collision times [10]

$$\tau_e = \frac{3\sqrt{m_e}}{4\sqrt{2\pi}}\cdot\left(\frac{4\pi\varepsilon_o}{e^2}\right)^2\cdot\frac{1}{\log\Lambda_{ei}}\cdot\frac{1}{n_i}\cdot T_e^{3/2} = \frac{5.366\times 10^{39}}{\log\Lambda_{ei}}\left(\frac{T_e}{J}\right)^{3/2}\Big/\left(\frac{n_i}{m^{-3}}\right) \quad s$$

and $$\tau_i = \frac{3\sqrt{m_i}}{4\sqrt{\pi}}\cdot\left(\frac{4\pi\varepsilon_o}{e^2}\right)^2\cdot\frac{1}{\log\Lambda_{ii}}\cdot\frac{1}{n_i}\cdot T_i^{3/2} = \frac{3.240\times 10^{41}\sqrt{A}}{\log\Lambda_{ii}}\left(\frac{T_i}{J}\right)^{3/2}\Big/\left(\frac{n_i}{m^{-3}}\right) \quad s$$

(where A is the atomic weight, 2.5 here, and $\log\Lambda_{ei}$ and $\log\Lambda_{ii}$ are Coulomb logarithms),

R^{ei} is the force due to electron-ion collisions
$$R^{ei} = 0.51\frac{m_e n_e}{\tau_e}(v_e - v_i) + 0.71\, n_e\frac{dT_e}{dz},$$

Q^{ei} is the equipartition heat transfer rate due to electron-ion collisions
$$Q^{ei} = \frac{3m_e n_e}{m_i\tau_e}(T_e - T_i),$$

e and E are the magnitude of the electron charge and the electric field, and, finally,
$S_{i,e}, R_{i,e}, Q_{i,e}$ represent sources of particles, momentum and energy for ions and electrons.
When a choice of sign is involved the ion equations have + and the electron equations have −.

The equations can be simplified by the assumption of quasi-neutrality, $n_e = n_i$, which applies at every position; hence $dn_e/dz = dn_i/dz$. Also, if we exclude the possibility of external sources of one species only, such as ion beams, then $S_e = S_i$. So the current density, $j = e(n_i v_i - n_e v_e)$, is independent of z and

$$\frac{dv_e}{dz} = \frac{1}{n_i}\left(1 - \frac{v_e}{v_i}\right) S_i + \frac{v_e}{v_i}\frac{dv_i}{dz} .$$

Thus we can use these charge-balance assumptions to express v_e and dv_e/dz in terms of v_i and dv_i/dz so the two momentum equations can be summed to eliminate the electric field (and R^{ei}). Now the system consists of one first-order and three second-order differential equations.

Firstly we shall describe further simplifications which permit analytic solutions of the equations in a variety of circumstances. Secondly we describe a numerical approach to the solution of the full equations (1-3).

3. ANALYTIC RESULTS

Some analytic results can be readily obtained, even for flows with varying area as in a bundle divertor. If there is no electric current ($v_e = v_i$) or viscosity, the equations of continuity and momentum may be written

$$\frac{d}{dz}(nvA) = SA, \qquad \text{... (4)}$$

$$(m_i + m_e)\, nv\frac{dv}{dz} = mnv\frac{dv}{dz} = -\frac{d}{dz}[n(T_e + T_i)] - mvS. \qquad \text{... (5)}$$

Here A(z) is the area of the flow channel. It has been assumed that the volumetric particle source S contributes no momentum to the flow. Equations (4) and (5) combine to yield the gradient in the Mach number, $M = v/\sqrt{(T_e + T_i)/m}$,

$$\frac{dM}{dz} = \left(\frac{M}{1-M^2}\right)\left[-\frac{1}{A}\cdot\frac{dA}{dz} + \frac{(1+M^2)S}{nv} + \frac{(1+M^2)}{2(T_e+T_i)}\cdot\frac{d}{dz}(T_e+T_i)\right] \qquad \text{... (6)}$$

This expression, a generalisation of that used by Harbour [8] shows how the gradient in flow speed (or Mach number) is enhanced or diminished by changes in area or by particle sources and temperature gradients. In a scrape-off plasma with $dT/dz = 0$; $dA/dz = 0$; $S = \text{const}(> 0)$ the flow must start from rest, so is initially subsonic ($M > 1$) and eq (6) shows that M increases, provided $M < 1$. At the entrance to a bundle divertor there is a rapid convergence and subsequent divergence of the flow. Neglecting terms in dT/dz and S, eq (6) shows that M increases towards unity. Downstream of the throat M will decrease unless it has reached one at the throat which will permit a supersonic flow in this expanding region. The behaviour here is similar to that in a Laval nozzle [11]. The maximum Mach number predicted for the uniform part of the scrape-off plasma for the DITE bundle divertor was 0.166 [12] and 0.32 for INTOR [8].

In a poloidal divertor the area does not vary much but eq (6) shows that a decreasing temperature will drive a supersonic flow to higher Mach numbers and a subsonic flow to lower Mach numbers. For example, if $dA/dz = 0$ and $S = 0$ the solution is

$$\left(\frac{M_o^2+1}{M^2+1}\right)\frac{M}{M_o} = \sqrt{(T_e+T_i)/(T_{eo}+T_{io})}, \qquad \text{... (7)}$$

where the subscript o refers to values at the upstream end of the flow.
This solution has subsonic and supersonic branches depending on whether M_o is above or below unity, and it is readily seen that decreasing temperatures move M away from one on either side.

The above relates speed and temperature without reference to position; the energy equation is needed to make this connection. A complete solution is possible in the case of an idealised, constant-area exhaust channel with no sources and sinks such as ionisation, recombination, charge exchange or radiation. Further assumptions are that there is no electric current so $v_e = v_i$, there is no viscosity, the electron and ion temperatures are the same, removing equipartition, and the Coulomb logarithms are taken as constant, so that the thermal conductivity is proportional to $T^{5/2}$. Then eqs (1-3) become

$$\frac{d}{dz}(nv) = 0 \qquad \ldots (8)$$

$$(m_i + m_e)\, nv \frac{dv}{dz} = -\frac{d}{dz}(2nT) \qquad \ldots (9)$$

$$\frac{d}{dz}\left(\kappa_o T^{5/2} \frac{dT}{dz}\right) = 3\, nv \frac{dT}{dz} + 2nT \frac{dv}{dz} \qquad \ldots (10)$$

Equations (8) and (9) integrate trivially to yield v in terms of T and then eq (10) can be integrated once to give a first order separable equation for T in terms of z which can be solved in closed form, although T(z) is a lengthy expression which will not be given here!

However, one conclusion can be stated simply. Provided that the density is high enough for eq (10) to be valid, electron thermal conduction is a very powerful process. The zero-velocity $T \propto z^{2/7}$ solution is perturbed by only a few per cent by the right-hand side of eq (10) even when v approaches the ion-acoustic speed, which the above scrape-off layer models indicate is of the order of the highest speed to be expected in the exhaust channel.

Such analyses can provide a physical understanding of the solutions to the full equations. These can exhibit subsonic/supersonic transitions near the divertor entrance and elsewhere and it is important to identify the cause of the transition to see if it is physically reasonable.

4. NUMERICAL MODEL

One benefit of using a computer to solve the eqs (1-3) is that realistic expressions can be used for the sources and sinks; we are not restricted by the requirement of analytic tractability.

We choose, at this stage, to include in the model an externally imposed density of cold, neutral, atomic D-T gas with functional form

$$n_o = n_{ob} + n_{ot} \exp(-z/\ell_o)$$

which represents a (small) constant background, n_{ob}, and a release of neutrals at the target, n_{ot}, which are attenuated by the plasma with scale length ℓ_o. Here z measures the distance upstream from the target. These neutrals interact with the plasma via various atomic processes. We use expressions for rate coefficients from Freeman and Jones [13] for charge exchange and electron ionisation, from Tarter [14] for line radiation and from Seaton [15] for radiative recombination to give the following sources and sinks in eqs (1-3):

$$S_i = S_e = S_{ionis.} + S_{recomb.} = n_e n_o \langle\sigma v\rangle_{ionis.} - n_e n_i \alpha_{recomb.}$$

$$R_e = R_{e,recomb.} = -m_e v_e n_e n_i \alpha_{recomb.}$$

$$R_i = R_{i,cx} + R_{i,recomb.} = -m_i v_i n_i n_o \langle\sigma v\rangle_{cx} - m_i v_i n_e n_i \alpha_{recomb.}$$

$$Q_e = Q_{e,ohmic} + Q_{e,ionis.} + Q_{e,recomb.} + Q_{e,rad.}$$

$$= 0.51 \frac{m_e n_e}{\tau_e}(v_e - v_i)^2 - \chi\, n_e n_o \langle\sigma v\rangle_{ionis.}$$

$$- (\chi + \tfrac{3}{2} T_e + \tfrac{1}{2} m_e v_e^2) n_e n_i \alpha_{recomb.} - n_e(n_i + n_o) P(T_e)$$

$$Q_i = Q_{i,cx} + Q_{i,recomb.} = -\left(\frac{3}{2} T_i + \frac{1}{2} m_i v_i^2\right)\left(n_i n_o \langle\sigma v\rangle_{cx} + n_e n_i \alpha_{recomb.}\right)$$

where the notation for the rate coefficients follows the usual conventions, χ is the ionisation energy (2.18×10^{-18} J) and $P(T_e)$ is the specific radiated power. Note that for $P(T_e)$

Tarter [14] uses the total density of hydrogen nuclei, both atoms and ions, and he includes continuum radiation as well as line radiation.

We model the flow from the scrape-off layer into a poloidal divertor (which is reasonably well represented by our "constant area" assumption) by considering a flux bundle that intersects the divertor target and following it back through the divertor entrance and around the torus. This flux bundle can be considered to be "unwound" from the torus and straightened to match our one-dimensional longitudinal model [2]. Outward radial diffusion from the reacting core plasma across the separatrix is modelled by additional volume sources in the portion of the model which represents the scrape-off layer. The radially outflowing plasma is here considered to join the flow without longitudinal momentum so there is no addition to the R terms in eq (2) but particles contribute to S in eq (1) and convected and conducted energy contribute to Q_e and Q_i in eq (3). We consider these contributions to be externally specified constants at present.

Natural symmetry boundary conditions can be applied at the upstream end of the scrape-off layer where there is a "watershed" in the flow to the two divertor entrances. These boundary conditions are

$$dn_i/dz = 0,\ dT_e/dz = 0,\ dT_i/dz = 0 \text{ and } v_i = 0.$$

Note that the condition $dv/dz = 0$ does not apply because of the non-zero source term in the continuity equation.

The situation at the downstream (target) end is not so simple. We wish to model two possibilities: firstly the case of surface recombination of the plasma on a metal target, and secondly the case of a gas target in which atomic and radiation processes take out all the plasma energy leading to a cool, recombining plasma surrounded by dense, cool neutral gas [4,16].

A sheath must insulate a metal target from the plasma; conditions in the sheath can be modelled under a collisionless assumption [1,17] but then there is a complex intermediate region between the sheath and the bulk plasma on the scale of a mean free path. In this intermediate region collisions become dominant over electrostatic space charge effects and the distribution functions become near-Maxwellian. Deciding what are the appropriate boundary conditions for the bulk plasma is an area of current work. In the case of a gas target the downstream boundary conditions on the plasma should have little influence on the majority of the flow. At present in the model we impose specified values on n, T_e and T_i, generally low temperatures and a high density. With the upstream conditions these give seven boundary conditions which complete the system to be solved.

The two-point boundary value problem is solved by discretising the equations and using a variable-step, variable-order solution package from the Harwell subroutine library implementing the algorithm of Lentini and Pereyra [18]. Of course, the variables are non-dimensionalised before handing over to the solution package and they are also transformed. Logarithmic variables are used for the ion particle flux, $n_i v_i$, the ion velocity and the two temperatures. The other three variables are linear in:

i. the total viscous momentum flux $\eta_e n_e T_e \tau_e (dv_e/dz) + \eta_i n_i T_i \tau_i (dv_i/dz)$
ii. the conducted electron heat flux $(\kappa_e n_e T_e \tau_e/m_e)\ (dT_e/dz)$
iii. the conducted ion heat flux $(\kappa_i n_i T_i \tau_i/m_i)\ (dT_i/dz)$.

Because of the logarithmic variable, the $v_i = 0$ upstream boundary condition cannot be imposed exactly and a small, but non-zero, fixed value of v_i is used instead. At the divertor entrance between the scrape-off layer and the exhaust channel the extra sources to model radial diffusion are not switched off abruptly, rather they are ramped down to zero over a short distance.

5. NUMERICAL RESULTS

There is a large parameter space to explore, even within the constraints of a one-dimensional model. We have so far restricted attention to one geometry of reactor-like

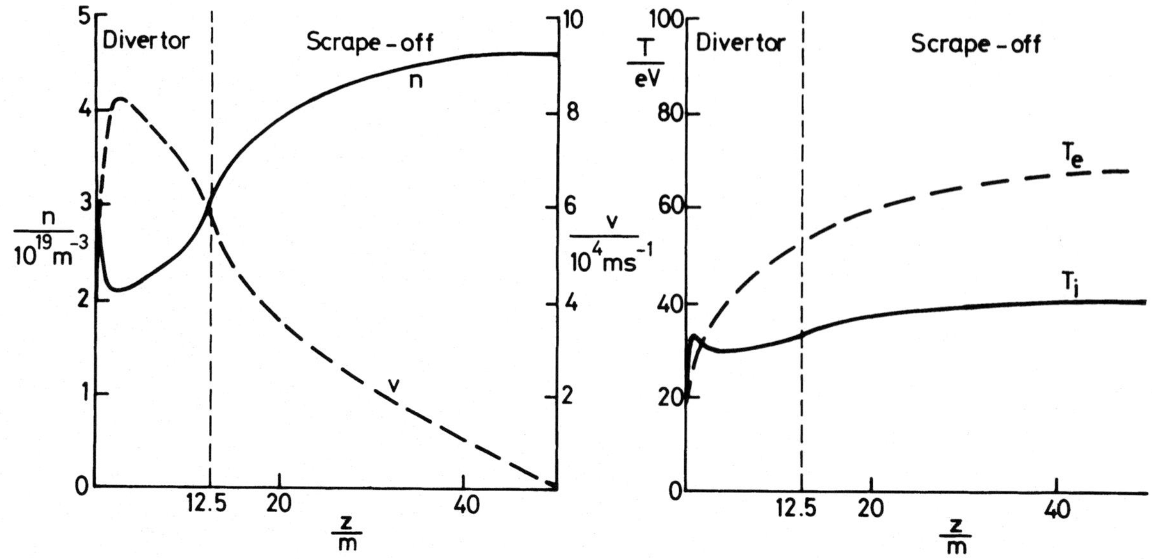

Fig. 1 n, v, T_e and T_i with little influence from neutrals, $n_{ob} = 10^{11}$, $n_{ot} = 10^{16}$ m^{-3} and $l_o = 1$m. Scrape-off layer radial sources are $S_r = 4.5 \times 10^{22}$ m^{-3} s^{-1} and $Q_{er} = 3 \times 10^6$, $Q_{ir} = 3 \times 10^5$ Wm^{-3}

dimensions. All the results reported below are for a scrape-off channel of half-length 37.5 m (ie, from the upstream "watershed" to each divertor entrance) followed by a divertor channel of length 12.5 m. The results are also all for ambipolar flow.

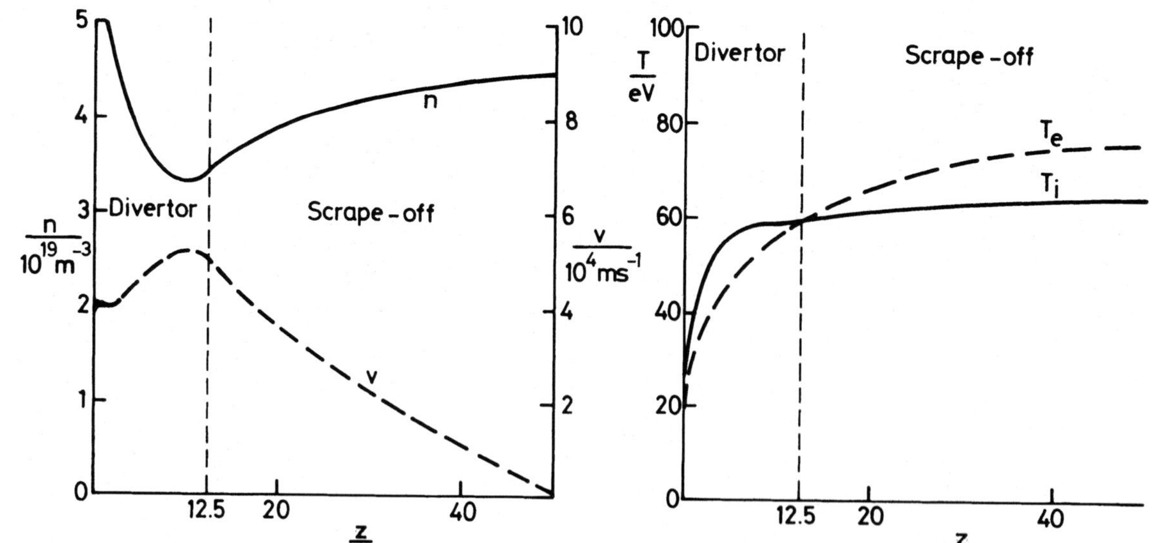

Fig. 2 n, v, T_e and T_i with more neutrals and larger energy sources from radial diffusion. $n_{ob} = 10^{13}$, $n_{ot} = 4 \times 10^{17}$, $l_o = 1.5$, $S_r = 4.5 \times 10^{22}$, $Q_{er} = 3.5 \times 10^6$ and $Q_{ir} = 10^6$.

Even with these restrictions we have to examine the effects of varying the radial diffusion input to the scrape-off layer, the amount of neutral gas in the exhaust chamber and the artificial target boundary conditions. This is still work in progress and here we can present only a very short selection of the latest results.

We illustrate the longitudinal profiles of density, speed, electron and ion temperatures for two cases in Figs.1 and 2. For Fig.1 we have chosen a low neutral density which has little influence on the plasma via ionisation, etc, and the profiles are determined solely by the plasma effects described in eqs (1), (2) and (3) and the imposed source terms in the scrape-off layer. Some points to note are that the flow in the scrape-off layer is subsonic and appears to be limited to $M = 1$ at the divertor entrance. The removal of the particle source at the divertor entrance allows a smooth transition to supersonic flow in the first part of the exhaust channel. The electron temperature profile is dominated by conduction but the ion temperature profile is moulded by flow terms.

In Fig.2 the larger neutral density increases the plasma density in the exhaust channel by ionisation. This throttles back the plasma flow making it entirely subsonic. Also, the increased ion radial energy source raises the ion temperature and makes ion thermal conduction more important in determining the temperature profile, although flow effects are still noticeable. Charge exchange on the cold neutrals takes out a significant amount of energy from the ions but a higher neutral density would be needed to remove all of the ion energy; we have not yet achieved the conditions for a "gas target".

The choice of arbitrary target boundary conditions can cause a rapid readjustment of the flow within the last metre or so of the exhaust channel. However, experimentation with the values used for these boundary conditions shows that their influence upstream decays rapidly with distance.

6. CONCLUSION

We have described a one-dimensional, two-fluid model for the flow of a divertor plasma. This model is being developed as part of a broader programme of reactor exhaust studies. The eventual aim of this work is to be able to define the regime in which a radiative gas target might be able to operate and the range of recirculation parameters which might be required in order both to maintain the radiative target and to achieve a reasonable burn-up fraction. Radiation is present in the target region in the results presented here, but a fully self-consistent solution including a radiative gas target has yet to be demonstrated.

7. REFERENCES

[1] HARBOUR, P.J. and HARRISON, M.F.A., Nucl. Fus. 19, 695-701 (1979).

[2] HARRISON, M.F.A., HARBOUR, P.J., HOTSTON, E.S. and MORGAN, J.G., 4th International Conf. on Plasma-Surface Interactions, Garmisch 21-25 April 1980, to be published.

[3] HARRISON, M.F.A., HARBOUR, P.J., HOTSTON, E.S. and MORGAN, J.G., 8th International Conf. on Plasma Physics and Controlled Nuclear Fusion Research, Brussels 1-10 July 1980, IAEA-CN-38/V-2-2.

[4] HARRISON, M.F.A., "Considerations of the Physical Problems of a Self-consistent Exhaust and Divertor System for a Long Burn Tokamak", Culham Laboratory, unpublished (1980).

[5] HOTSTON, E.S., "Refuelling and the Scrape-off Layer of INTOR", Culham Laboratory, unpublished (1980).

[6] MORGAN, J.G., "Simple Modelling of the Scrape-off Layer", Culham Laboratory, unpublished (1980).

[7] HOTSTON, E.S., "Power Flow in the Scrape-off Layer", Culham Laboratory, unpublished (1980).

[8] HARBOUR, P.J., "A Model of Particle Transport within the Scrape-off Plasma and Divertor", Culham Laboratory, unpublished (1980).

[9] HARBOUR, P.J., "Controlled Recirculation of Exhaust Gas from the Divertor into the Scrape-off Plasma", Culham Laboratory, unpublished (1980).

[10] BRAGINSKII, S.I., Reviews of Plasma Physics 1, 205-311 (1965).

[11] LIEPMANN, H.W. and ROSHKO, A., "Elements of Gas Dynamics", Wiley, New York (1957).

[12] PROUDFOOT, G. and HARBOUR, P.J., 4th International Conference on Plasma-Surface Interactions, Garmisch 21-25 April 1980, to be published.

[13] FREEMAN, R.L. and JONES, E.M., Culham Laboratory Report CLM-R137 (1974).

[14] TARTER, C.B., J.Quant.Spectrosc.Radiat.Transfer 17, 531-536 (1977).

[15] SEATON, M.J., Mon.Not.Roy.astr.Soc. 119, 81-89 (1959).

[16] BRUNELLI, B. and HARBOUR, P.J., Informal meeting on Long Pulse Tokamak Operation, *in* Plasma Physics and Controlled Nuclear Fusion Research 1978, Vol.1, Nucl.Fus.suppl., 823-828 (1979).

[17] EMMERT, G.A., WIELAND, R.M., MENSE, A.T. and DAVIDSON, J.N., Phys.Fluids 23, 803-812 (1980).

[18] LENTINI, M. and PEREYRA, V., SIAM J. Numer. Anal. 14, 91-111 (1977).

ELECTROMAGNETIC TRANSIENT EFFECTS IN THE STRUCTURE OF EXPERIMENTAL FUSION REACTORS

E. Coccorese * - R. Martone **

* Istituto di Elettrotecnica, Univ. di Napoli, Via Claudio 21, Naples (Italy)
** Dipartimento Elettrico, Univ. della Calabria, Arcavacata, Cosenza (Italy)

ABSTRACT

With reference to the proposed parameters for the NET/INTOR experiment, a possible Poloidal Field configuration is proposed, including the transformer and the equilibrium winding.

A transient analysis of the P.F. system is then performed, aiming to evaluate, by means of a computer code developed for this purpose, the eddy currents in the vacuum vessel structure and the resulting electromagnetic volume forces.

Finally, by means of an analytical method, eddy currents and electromagnetic forces in the blanket structure are evaluated, under specified assumptions and approximations.

1. INTRODUCTION

Heavy eddy currents are expected in the blanket and in the vacuum vessel of the experimental fusion reactors, due to time varying magnetic fields; therefore their thermal and mechanical effects have to be carefully considered when developing the design.

Assuming to have a steady-state Toroidal Field, time varying fields are due to the Poloidal System (including the plasma, the transformer and the equilibrium winding) in particular during plasma start-up and shut-down phases. In addition, in case of a plasma disruption, the plasma field goes rapidly to zero and again the appearance of eddy currents is expected.

Major radius (m)	5.2
Minor radius (m)	1.2
Elongation factor	1.6
Plasma current (MA)	6.4
Toroidal Field (T)	5.5

Table I - NET/INTOR parameters

In the present paper we refer, for the calculations, to the proposed parameters of NET/INTOR experiment [1] (see Tab. I).

In sect. 2, after an equilibrium analysis, a possible P.F. configuration is proposed, including the transformer and the equilibrium winding.

In sect. 3 the eddy currents and their effects on the vacuum vessel are evaluated, both during plasma start-up and in the event of a plasma disruption.

Sect. 4 shows the calculation of eddy currents effects in the blanket.

Finally, sect. 5 shows the main conclusions arising from the calculations performed.

2. POLOIDAL FIELD CONFIGURATION

2.1 Equilibrium analysis

An equilibrium analysis has been performed within the limits of MHD theory [2]. In particular a new procedure (of the fixed-boundary type) has been used [3] which allows the evaluation of an equilibrium current-density profile starting from an arbitrarily assigned one.

Among the various cases examined, we report here in Figs. 1 and 2 the results achieved when a "flat" profile is chosen.

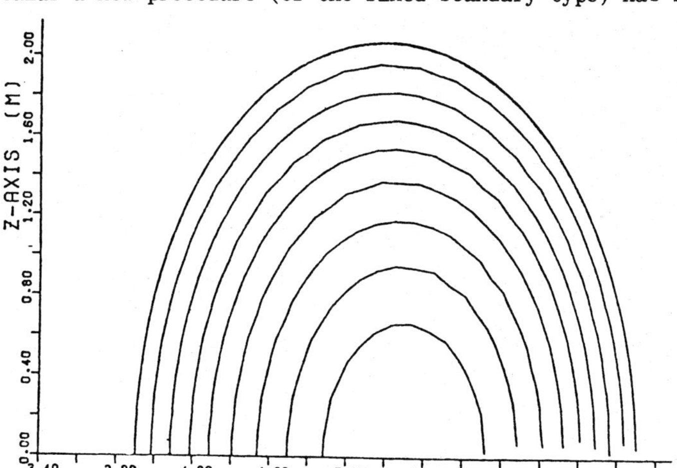

Fig. 1 - Equilibrium Field Map

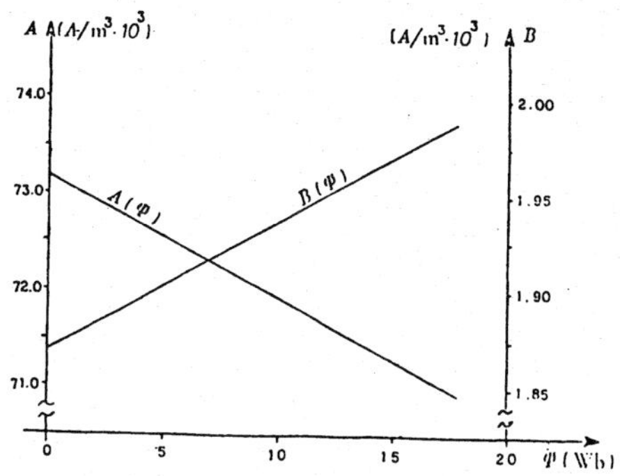

Fig. 2 - Computed A(ψ) and B(ψ) functions

All the following calculations will refer to this plasma current-density profile.

2.2 Equilibrium winding

The Poloidal Field winding has to be placed outside the Toroidal Field coils (to avoid mechanical linking); so, due to the weak magnetic linkage with the plasma ring, a large amount of ampere-turns (with respect to the plasma current) is required.

On the other hand the choice of the position of the coils is quite wide because the only additional constraint is to allow the dismounting of the blanket modules [4]. Therefore, when designing the winding, one has to evaluate both position of the coils and their currents in optimal way (minimizing, for example, the total ampere-turns or the amount of conducting material).

As a consequence, to design the equilibrium winding, the allowed area has been subdivided in N_E elementary conductors: the equilibrium condition has been imposed on N_B points of the plasma boundary as follows :

$$(1) \quad \sum_{1}^{N_E} {}_j \, M_{ij} I_j + \psi_i^{(PL)} = \psi_B \pm \varepsilon_E \; ; \quad i = 1, \ldots, N_B \; ; \quad j = 1, \ldots, N_E$$

where:

I_j is the generic current of the equilibrium winding;

$\psi_i^{(PL)}$ is the flux contribution of the plasma current at the ith point of the plasma boundary;

ψ_B is the value of the flux at the plasma boundary;

ε_E is an imposed tolerance;

M_{ij} is a mutual coupling coefficient.

It should be noticed that, being the plasma current-density distribution an equilibrium one, eq. (1) is sufficient to ensure equilibrium.

A set of additional constraints has then been added :

$$(2) \qquad |I_i| \leq G_{max} S_i$$

where:

S_i is the cross-section of the ith conductor;

G_{max} is the max. current-density allowed in the conducting material.

Having chosen N_E and N_B in such a way that $N_E \gg N_B$, the system (1),(2) generally admits an infinity of solutions for the $N_E + 1$ unknowns

$$I_1, \ldots, I_{N_E}, \psi_B ,$$

among which one can choose the one which optimizes a certain quality index, such as $\sum_i |I_i|$ if one wants to minimize the total ampere-turns.

The relative calculations were carried out by means of a computer code already developped [5].

The so-evaluated equilibrium winding configuration is shown in Fig. 3, while its main characteristics are summarized in Tab. II.

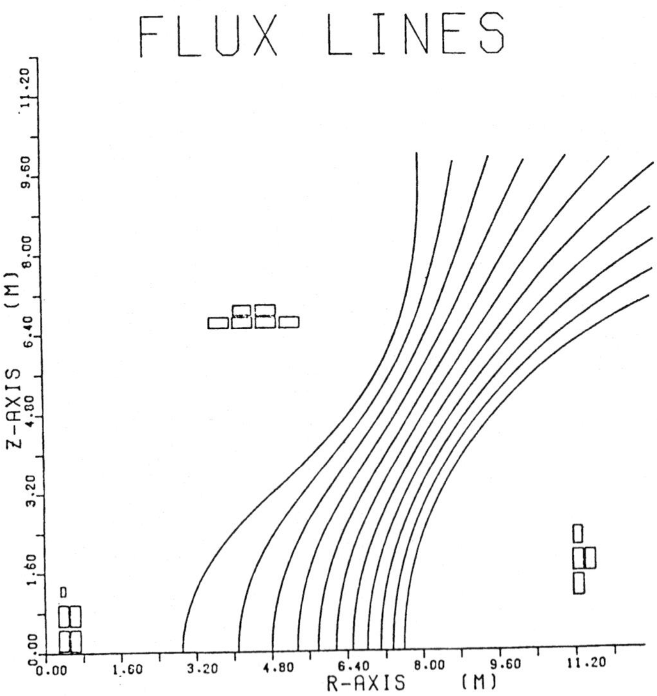

Fig. 3 - Equilibrium Winding Configuration

Material	Copper
Weight (T)	693
Current (MA)	24.0
Max losses (MW)	127
Bound Cond. Accuracy (%)	0.5

Table II - Main Charact. of the Equil. Winding

2.3 Transformer winding

The above procedure can be used also for the transformer winding. In fact, as the transformer has to produce zero-field within the plasma region, another system, similar to (1),(2) can be written, where eqs. (1) are substituted by :

$$(3) \quad \sum_{1}^{N_T} {}_j M_{ij} I_j = \psi_T \pm \varepsilon_T \quad ; \quad i = 1,\ldots,N_B \quad ; \quad j = 1,\ldots,N_T$$

where ψ_T is the flux that the transformer links with the plasma.

The only difference is that ψ_T is not arbitrary but is determined by the following relationship :

$$(4) \quad (1+p)\psi_T = L_P I_P + \phi_J - \phi_E$$

where :

p is the biasing factor of the transformer;

L_P is the self-inductance of the plasma ring;

I_P is the total plasma current;

ϕ_J is the additional (not-inductive) flux required;

ϕ_E is the flux that the equilibrium winding links with the plasma.

Although the parameters of the reactor suggest the use of superconducting material for the transformer, a copper solution is still feasible; therefore both solutions have been studied.

The relative configurations are shown in Figs. 4 and 5, while Tab. III shows the main characteristics of the winding in both cases.

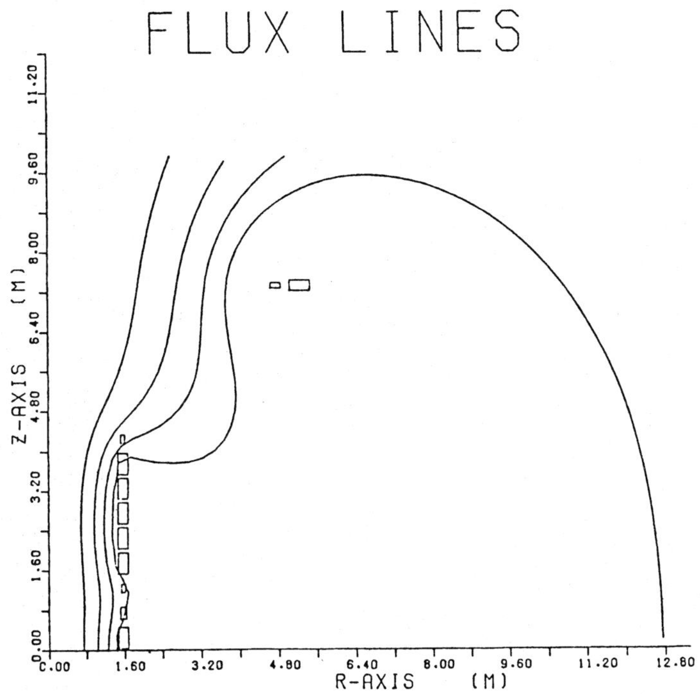

Fig. 4 - Superconduct. Transformer Configuration

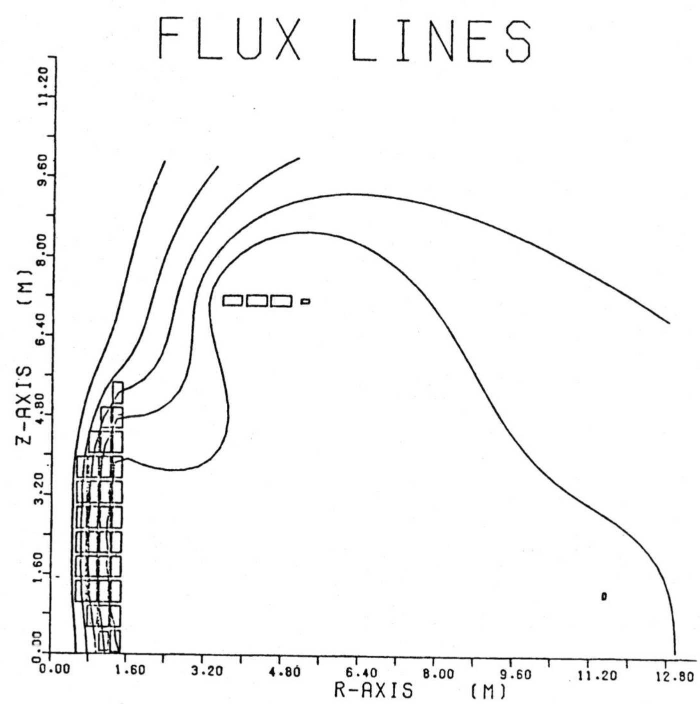

Fig.5 - Copper Transformer Configuration

Material		Supercond	Copper
Weight	(T)	166	497
Current	(MA)	27.7	66.3
Max losses	(MW)	-	99.9
Max disturb. field	(G)	25	30

Table III - Main Characteristics of the Transformer winding

3. EDDY CURRENTS IN THE VACUUM VESSEL

3.1 General

Time-varying fields (occurring especially during start-up, shut-down phases and in the event of a plasma disruption) induce eddy currents in the metallic structure of the vacuum vessel. The resulting Joule losses and electromagnetic stress could be dangerous for the structure, especially in its weakest parts, such as the bellows. As other undesirable effect there is a possible interference with the equilibrium field.

All above is of particular concern when, as at the present stage of NET design, no flux-breakers are foreseen.

3.2 Calculation method

The main assumption is to consider the vacuum vessel as an axisymmetric metallic structure which behaves like another "secondary winding" for the transformer. Due to the non-axisymmetry of the vessel (rigid sectors and bellows) not only toroidal but also saddle currents are expected. However, at this preliminary stage, it seemed useful to have a simple axisymmetric model; the evaluation of the saddle currents can then be easily performed by means of an already developed method [6].

The analysis of the plasma start-up has been performed by means of a circuital model, whose main elements are :

a) transformer winding :the already calculated configuration and final current distribution allow its representation in terms of an ohmic-inductive circuit;

b) equilibrium winding : similar considerations apply and in addition it is assumed that the power supply system is able to drive the current with the same dynamics of the plasma current;

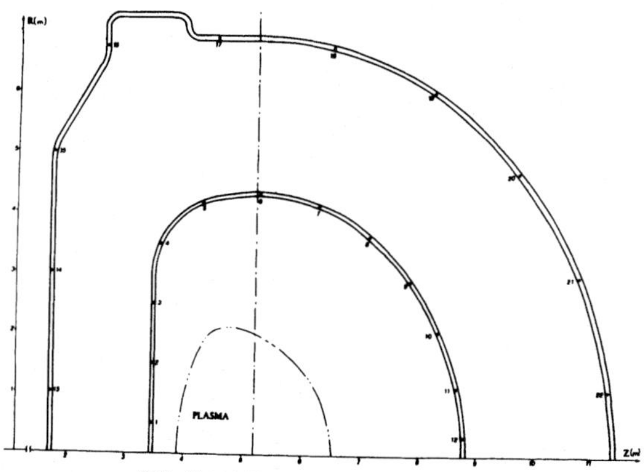

Fig.6 - Schematic View of the vacum vessel

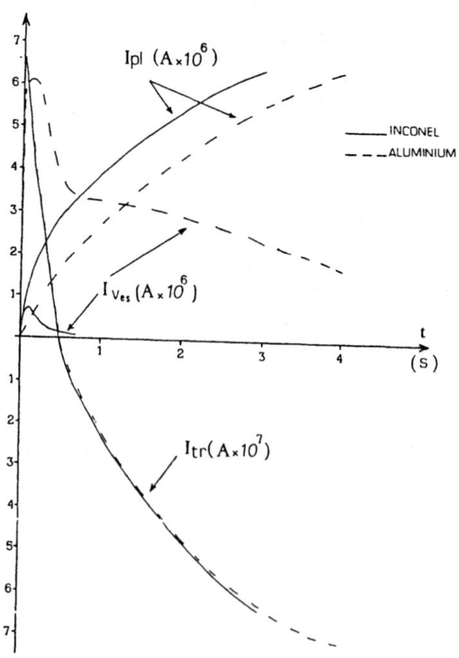

Fig.7 - Current Waveforms during start-up

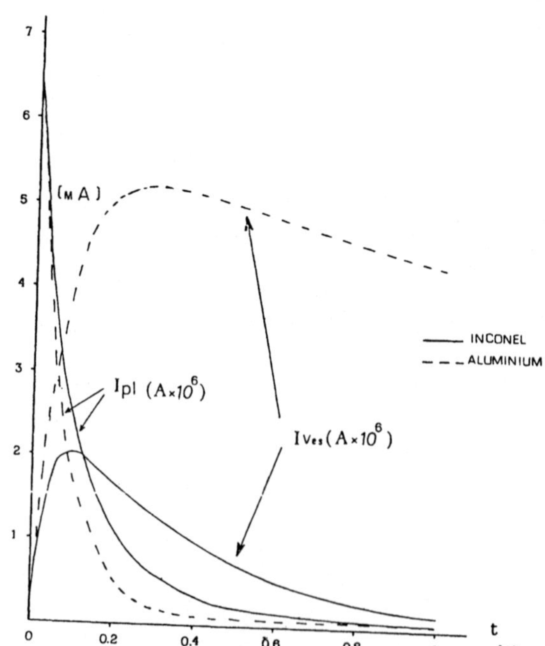

Fig.8 - Current Waveforms during Plasma Disruption

	G_{max} (A/mm²)	
No.	A	B
1	4.55	62.5
2	4.72	66.5
3	4.97	69.5
4	4.74	64.5
5	3.96	48.4
6	3.64	33.0
7	3.65	24.6
8	4.06	20.6
9	5.10	24.5
10	6.20	31.2
11	7.10	36.8
12	7.48	39.3
13	5.60	141.0
14	5.60	145.0
15	3.63	90.5
16	2.78	45.0
17	6.03	95.3
18	3.70	28.5
19	4.38	20.9
20	6.60	34.0
21	11.20	61.8
22	16.70	96.8

Table IV - Max. Curr.-dens. in the bellows (start-up)
A = Inconel
B = Aluminium

No.	J_{max} (A/mm²)	
	A	B
1	55.0	282.
2	45.6	211.
3	32.3	124.
4	22.4	65.5
5	22.2	52.5
6	25.5	54.5
7	28.3	55.5
8	32.7	66.6
9	35.5	69.5
10	37.5	72.0
11	39.1	73.0
12	42.0	97.5
13	10.4	26.8
14	4.6	9.4
15	1.5	1.0
16	2.1	1.4
17	5.6	3.4
18	9.3	6.0
19	11.7	7.8
20	13.4	8.7
21	13.1	9.8
22	11.3	11.0

Table V - Max. Curr.-dens. in the bellows (disrupt.)
A = Inconel
B = Aluminium

c) transformer power supply : a discharge resistor and an energy storage are assumed to force down the current and provide its inversion;

d) plasma : it is simulated by an inductance in series with a suitable non-linear resistor evaluated on the basis of the additional Volt-sec required;

e) vacuum vessel : by ideally subdividing it in elementary toroidal rings it is represented by an ohmic-inductive circuit.

Therefore the model assumes the form of a set of differential equations, whose solution gives information on the following points:

a) currents of the poloidal system, i.e. transformer, equilibrium coils, plasma and vacuum-vessel;

b) power and energy demand from the power supply system;

c) power and energy dissipated in the coils and in the vacuum vessel.

In addition, by evaluating the fields involved, one can calculate the electromagnetic forces and the disturbance field produced by the eddy currents in the vacuum vessel. In particular, the electromagnetic forces in the vacuum vessel are calculated within the bellows, where also a torque effect due to the interaction with the toroidal field is expected.

A similar model is used to calculate the effects of a plasma disruption.

3.3 Results of the calculations

The calculations were performed by means of a computer code developed on the basis of the above method. The structure studied [4] (sketched in Fig.6) is made of an inner and an outer containment realized with 12 rigid sectors connected by bellows 2 mm thick and 1.4 m long (when stretched). Reported results (see Figs.7,8 and Tabs.IV,V) refer to quantities evaluated in 22 points (Fig.6) of the bellows. The copper solution was considered for the transformer: two cases were examined, corresponding to a vacuum vessel made of inconel or aluminium.

4. EDDY CURRENTS IN THE BLANKET STRUCTURE

4.1 General

As in the vacuum vessel, the eddy currents in the blanket structure produce the same undesirable effects. However, in the NET case, an axisymmetric approach by no means can be used, being the structure made up of several modules electrically insulated in the toroidal sense. In addition, the geometry is so complicated that an accurate analysis should require a 3-D code. Nevertheless, at the present preliminary phase of the design, it seemed useful to use a quasi-analytical approach, that under strongly simplifying assumptions, is able to give both rough estimates and scaling laws for the effects of the eddy currents produced.

4.2 Calculation method

The blanket structure is assumed to be a linear and homogeneous medium of conductivity σ. From Maxwell equations, neglecting displacement currents and assuming that the field $\bar{B}_i$ produced by the eddy currents is much smaller than the inducing field $\bar{B}_e$, we have :

$$\nabla^2 \bar{B}_i = \mu\sigma\, \partial \bar{B}_e/\partial t \qquad (5)$$

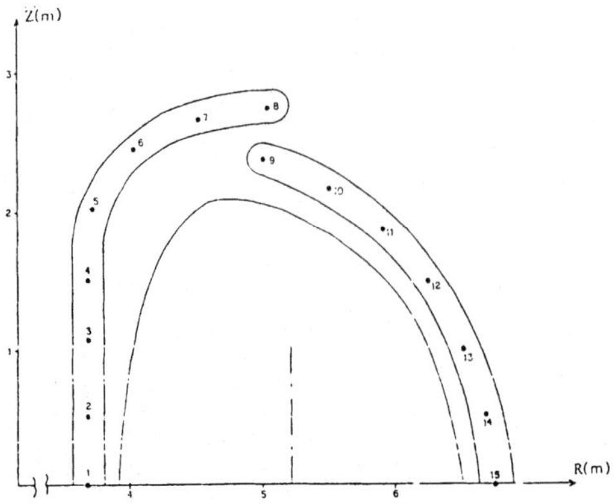

Fig.9 - Schematic View of the Blanket

No.	$J_{\rho\,max}$ (A/mm²)		$J_{z\,max}$ (A/mm²)		$f_{\rho\,max}$ (MN/m³)		$f_{z\,max}$ (MN/m³)		$f_{\tau\,max}$ (MN/m³)	
	A	B	A	B	A	B	A	B	A	B
1	.35	19.5	2.3	125	.84	102	.13	15.8	2.8	155
2	.33	18.4	2.1	118	.75	91	.12	14.1	2.6	146
3	.30	16.3	1.9	104	.59	71	.09	11.3	2.4	133
4	.25	13.9	1.6	89	.42	51	.07	8.2	2.0	111
5	.21	11.8	1.4	76	.32	38	.05	5.9	1.7	92
6	.18	9.8	1.3	69	.26	32	.04	5.4	1.3	72
7	.16	8.7	1.3	70	.26	32	.04	-4.8	1.0	56
8	.14	7.9	1.3	70	.27	33	.03	4.1	0.8	44
9	.17	[illegible]	[illegible]	[illegible]	.36	43	.03	3.6	1.0	53
10	.17	9.4	1.6	90	.44	54	.04	5.0	0.9	49
11	.17	9.4	1.8	98	.51	63	.05	6.0	0.8	46
12	.18	9.8	1.9	108	.62	76	.06	7.1	0.8	46
13	.19	10.3	2.1	118	.75	91	.07	8.1	0.8	47
14	.20	10.9	2.3	127	.85	103	.07	9.0	0.9	48
15	.19	10.7	2.3	127	.85	104	.07	8.6	0.8	45

Table VI - Max. Curr.-dens. and Volume Forces in the Blanket structure (Aluminium)

Assuming in addition that the inducing poloidal field is everywhere tangential to the blanket (τ-axis), a 2-D rectangular geometry can be introduced, where the eddy currents have components only along ρ (i.e. perpendicular to the blanket surface) z and (i.e. toroidal) directions (Fig.9). Considering the generic cross-section of the blanket in the (ρ,z) plane, the current density components (limiting only to the first term of the series involved) are :

$$J_\rho = -\pi/b\; K\, \dot{B}_\tau \sin(\pi/\Delta\,\rho)\cos(\pi/b\,z) \qquad J_z = \pi/\Delta\; K\, \dot{B}_\tau \cos(\pi/\Delta\,\rho)\sin(\pi/b\,z)$$

where Δ is the thickness of the blanket, b is the length (in z direction) of the module, $\dot{B}_\tau$ is the time-derivative of the poloidal inducing field and

$$(6) \qquad K = (4/\pi)^2 \cdot 6/\pi^2 \cdot 1/[(1/\Delta)^2 + (1/b)^2]$$

Eqs. (6) allow the evaluation of the electromagnetic volume forces in the blanket, by means of the following relations (where B_T is the toroidal field):

$$(7) \qquad f_\rho = J_z B_\tau \qquad f_z = J_\rho B_\tau \qquad f_\tau = J_\rho B_T \quad .$$

The ohmic power dissipated in the blanket is given by (where V is the volume of the blanket):

$$P_{tot} = \frac{1}{\sigma}\int_V (J_\rho^2 + J_z^2)\, dV$$

Assuming an uniform $\dot{B}_\tau$ and an average $b = 2\pi R_O/N_B$ (where R_O is the major radius of the plasma and N_B the number of blanket sectors) one has a simple scaling law for the ohmic power:

$$P_{tot} = 2N_B\left(\frac{4}{\pi}\right)^5 \sigma\, \frac{(\Delta/2)^3(\pi R_o/N_B)^3}{(\Delta/2)^2 + (\pi R_o/N_B)^2}\, l_B\, \dot{B}_\tau^2$$

where l_B is the toroidal length of the blanket.

4.3 Results of the calculations

The blanket structure foreseen so far for NET is made of N_B = 18 sectors, 20 cm thick. The inducing field B_τ has been already evaluated in sect. 3 (case of inconel). As indicated in Fig. 9, 15 blanket sections in the (ρ,z) plane have been considered; the results are shown in Tab. VI.

5. CONCLUSIONS

The calculations performed for P.F. configuration show that a copper solution for the transformer is still possible, even if very costly in terms of power-supply and heating removal.

The comparison between inconel and aluminium as structural material for the vacuum vessel shows that intorelable eddy currents are expected in the latter case: therefore the employment of aluminium should be excluded unless flux breakers are foreseen.

A good evaluation of the eddy currents in the blanket require both a detailed knowledge of the blanket structure and of the fields dynamics, however a rough estimate can be made by means of the approximate model proposed.

ACKNOWLEDGEMENTS

The authors wish to thank Mr. A. Di Chiano (CCR EURATOM-Ispra) for the help received in carrying out the numerical calculations relative to the blanket.

REFERENCES

[1] INTOR Design specifications for use in preconceptual design phase, June 1980

[2] SHAFRANOV V.D.: "Plasma equilibrium in a magnetic field", Reviews of Plasm.Phys.,2 (1966)

[3] BOBBIO S., COCCORESE E., MARTONE R.,: "A new procedure for the design of the equilibrium-field currents for non-circular plasmas", 11° SOFT Oxford, 15-19 Sept 1980

[4] FARFALLETTI-CASALI F. et al.: "Preliminary studies of the systems integration of NET/INTOR" 11° SOFT, Oxford, 15-19 Sept 1980

[5] COCCORESE E., MARTONE R. :"A design procedure for air core transformers in toroidal axisimmetrical geometry" 10° SOFT, Padua, 4-9 Sept 1978

[6] RUBINACCI G.: "The thermo-magnetic transient in the vacuum vessel of high field compact experiment", 11° SOFT, Oxford, 15-19 Sept 1980

AN ANALYSIS OF THE ESTIMATED CAPITAL COST OF A FUSION REACTOR

A A Hollis and L S Evans
Research Reactors Division, AERE Harwell, United Kingdom

ABSTRACT

The cost of building a fusion reactor similar to the Culham Conceptual Tokamak reactor Mark IIB is assessed and compared with other published capital costs of fusion and fission reactors. It is concluded that capital-investment and structure-renewal costs for a typical fusion reactor as presently conceived are likely to be higher than for thermal-fission reactors.

1. INTRODUCTION

Some estimates of the likely capital cost of a fusion reactor have been prepared [1,2]. The work reported here includes additionally operational and maintenance features, allowing somewhat more realistic estimates to be made of the capital cost of a power station. This has been considered in more detail in report [9].

2. THE REACTOR CONCEPT

The 1200 MW(e) Culham Conceptual Tokamak Reactor Mk IIB [10] incorporates an elliptical plasma cross-section and poloidal field windings with the inside horizontal dimension across the D-shaped toroidal field coils increased to allow space for coolant ducts and access for servicing. The high wall loading requires blanket replacement as a routine operation, and the extensive provisions [3] needed for the maintenance of the reactor are featured in the concept. Further aspects include (a) superconducting toroidal field coils and water-cooled copper poloidal field coils, (b) a helium-cooled blanket of cellular structure with lithium moderator and breeder, an outer shield of boron carbide situated within a fixed water-cooled shield and support structure which forms the vacuum boundary, (c) a blanket in 20 sections with horizontal access between the magnets through a removable door in the fixed shield, (d) neutral-beam-injection heating, (e) vacuum-tight shielding buildings to house the nuclear island and maintenance facilities, (f) helium-to-steam heat exchangers and conventional turbogenerators. A diverter is included in the concept but has not been costed due to inadequate data at this time.

3. ACCOUNTING METHOD

The total capital cost includes building the nuclear-power station and bringing it into commercial operation. In addition to the 'direct costs' and those directly associated on an item-by-item basis with the equipment structures and materials that comprise the complete power plant, there are also numerous other costs that may be grouped under the two headings of 'indirect costs' and 'special costs'. In this study the capital cost has been built up, generally using the account numbers defined by the IAEA [4] as a guide in the economic evaluation of bids for nuclear-power plants in which a uniform numerical system is based on a physical subdivision of the project. The account numbers and physical subdivision of the plant are:

Account Number	Description
	Direct Costs
20	Land and land rights
21	Structures and site facilities
22	Reactor-plant equipment
23	Turbine-plant equipment
24	Electric-plant equipment
25	Miscellaneous-plant equipment
26	Cooling towers
	Indirect Costs
91	Construction-facilities plant and equipment
92	Engineering & management services
93	Other costs
94	Interest during construction
	Special Costs
110	Recommended spare parts
120	Installation labour
130	Contingencies
140	Warranties and fees
150	Escalation
160	Freight

Within Account 22, Reactor-Plant Equipment (the Nuclear Island), are included a number of new subdivisions to suit the special case of the fusion reactor; these are Primary Toroidal Structures, the main features of the nuclear island, Reactor Ancillary Equipment, plant situated within the reactor hall, and Blanket- and Injector-Handling Equipment, plant situated outside the reactor hall necessary for the operation of the reactor equipment. Also Power Supplies to the Field Coils has been introduced into Account 24. However, since this equipment is vital to the operation of the nuclear island, it is better for comparison with other systems that the cost of power supplies is included in the nuclear island of CCTR II.

4. DIRECT COSTS

Only general cost estimates were made. Schematic and diagrammatic layouts were used with sizes and throughputs to suit the reactor output and dimensions, and adjustments were made to cater for the translation of these layouts into real physical configurations. Most of the costs were calculated by estimating the quantities of materials required and multiplying by a standard unit of cost for a similar structure or piece of equipment, e.g. £13K/tonne for blanket segment structure. Notional amounts were included for the identifiable small support systems, large errors in these costs having an insignificant effect on the total capital costs, and any overestimation compensating for the likelihood that not all support systems have been included.

The figures for the conventional power-house plant and equipment have been derived from other studies of nuclear projects of a comparable size, where necessary with an adjustment to cater for the increased complexity. The initial supply of special, i.e. non-fuel and non-structural, materials and heat-transfer fluids under Account 26 has not been included.

The direct capital costs of a single 1200 MW(e) reactor at 1977 prices is estimated to be £910M. A breakdown of the costs is shown in Table I.

5. INDIRECT COSTS

The indirect costs associated with recent nuclear-plant cost estimates have been reported [5]. For the BWR and PWR reactors examined, engineering and management services (Accounts 91 and 92 together) varied between 20% and 44%, other costs between 14% and 19%, contingencies between 10% and 14%, and interest between 50% and 56% of the direct costs. In this study the figures chosen are:

Account 91 Construction-facilities plant and equipment	35%	of the direct costs
Account 92 Engineering & management services		
Account 93 Other costs	15%	
Account 94 Interest during construction	50%	

The indirect costs are estimated to be £913M. The breakdown of these costs is shown in Table II.

6. SPECIAL COSTS

Because of the form of the available information, installation labour is part of the estimated costs of the individual items and is included as part of the items making up the direct costs. No entries are made for spare parts, warranties and fees, escalation, and freight. A nominal sum has been included to allow for unexpected costs.

7. TOTAL COSTS

The gross capital cost of a 1200 MW(e) fusion reactor, at 1977 prices, is estimated to be £2,000M or £1,666/kW(e), made up from:

	Total £M	£/kW(e)
Direct costs	900	758
Indirect costs	913	761
Special costs (contingency ≈10%)	177	147
TOTAL	2,000	1,666

8. DISCUSSION

This estimate, although speculative, provides a basis for discussion on the capital-cost aspects of the economics of providing electricity by Tokamak fusion reactors.

Direct comparisons with the costs of other reactor systems can be misleading because of difficulties which arise about the units of currency and conversion rates, the date for which the estimate is applicable and any inflationary factors to be applied, and the items of cost that may or may not be included. However, as the turbogenerators and associated plant (Accounts 23 and 24) can be assumed to be fairly similar for selected systems it is possible to study the costs under the various account numbers relative to the cost of this conventional plant.

The estimates for or costs of three other systems have been used:

(a) UWMAK III [2], a 5,000 MW(t) Tokamak reactor,

(b) a 1,200 MW(e) PWR [4],

(c) a 1,190 MW(e) BWR [6]

In order to make the data comparable some small adjustments have been made; for Culham IIB Power Supplies to the Field Coils has been transferred to Account 22, and for UWMAK II a number of items under Accounts 24 and 25 have been transferred to Account 22.

Account 22, Reactor Plant and Equipment (the Nuclear Island), contributes some 72% towards the direct costs of the power station. Within this account, the primary toroidal structures, i.e. coils, blanket, shield, core etc., contribute £282M, 30% of the direct costs, and the appurtenant ancillary equipment contributes £228M, 25% of the direct costs.

Although the reactor blanket is accredited with only £60.7M, 6.5% of the direct costs, its impact is worth noting. The need to replace the blanket periodically results in the requirement for:

(a) a containment building to shield against high-level radiation,

(b) the remote operation and maintenance of equipment within the containment building,

(c) equipment within the containment building for storing and cooling used segments,

(d) extensive shield workshops with remotely controlled equipment for repairing used blanket segments and other items,

(e) facilities for the storage, transport and disposal of used blanket segments and other equipment,

(f) facilities for the fabrication, testing, and commissioning of the large intricate blanket segments from say 20 t factory-made sub-assemblies.

Provision of equipment and facilities for repair and replacement of blanket segments has a significant influence on the direct costs.

It should be noted that these features will influence costs throughout the life of the reactor. The periodic replacement of the blanket will contribute directly to the generation costs. The capital cost of the blanket will therefore be a recurring charge in a structure-replacement cycle similar to the fuel-cycle costs of a fission reactor. A further consequence is the need for operating/maintenance staff over and above that found in a conventional power station. Staff will be required to operate and service the complex remotely controlled equipment provided to exchange blanket segments in the reactor hall and dismantle and break up the used segments, and to fabricate, repair, test and commission replacement segments and dispose of any radioactive waste structural material. Economy in the overall scheme of blanket renewal is likely to be a most fruitful area for savings in running cost.

Delays during construction leading to cost escalation are a feature of large-plant construction [7], and those at the job site make the largest contribution to these increases. With a large proportion of the cost due to the largely in-situ fabricated primary toroidal structure, fusion reactor costs will be very sensitive to escalation during construction.

The study indicates at Table III that the capital cost of a fusion reactor is likely to be higher than its thermal-fission counterpart. Because of the nature of the information available the estimate contains large elements of judgement as to what can be assumed to be probable; nevertheless it draws attention to those features which have a large influence on costs and are therefore worthy of more detailed study. Aronstein and Ghose [8] report that the reactor per se promises not to be the major cost element of a fusion-power plant and assume that the reactor accounts for only 20% of the total plant costs and is not different from LWR experience; however the work reported here suggests that the nuclear island is likely to account for some 72% of the total capital costs.

As discussed earlier, difficulties can arise in making comparisons with costs of other reactors and systems, and until a common system is used these difficulties will remain. However the use of the IAEA guide at least ensures that all the items leading up to the total cost are considered for inclusion, and allows broad analysis and comparisons on the lines presented above.

9. CONCLUSIONS

While accurate quantitative conclusions cannot be drawn, this study has indicated that capital-investment and structure renewal costs for a typical fusion reactor as presently conceived are likely to be higher than those for thermal-fission reactors. In particular the cost of the nuclear island based on CCTR Mk IIB is shown to be some 5 to 9 times that of the nuclear island of a thermal-fission power station.

10. ACKNOWLEDGEMENT

The rates and unit costs which form the basis of Table I and hence of this estimate were provided by A. Burman of the Estimations and Sanctions Section, Risley Nuclear Power Development Establishment. The authors gratefully acknowledge this major contribution.

11. REFERENCES

[1] HANCOX, R. and MITCHELL, J.T.D., Reactor costs and maintenance with reference to the Culham Mark II conceptual Tokamak reactor design. IAEA-CN-35/11-1.

[2] UNIVERSITY OF WISCONSIN. UWMAK III. UWFDM-150 (1976)

[3] HOLLIS, A.A. and MITCHELL, J.T.D., A Tokamak reactor with servicing capability. 9th SOFT garmisch-partenkirken, June 1976.

[4] UNITED ENGINEERS AND CONSTRUCTORS INC., Capital cost of PWR reactor plant. COO-2477-5 (1977).

[5] WOITE, GEORG., Capital investment costs of nuclear-power plants. IAEA Bulletin - Vol. 20, No. 1.

[6] UNITED ENGINEERS AND CONSTRUCTORS INC., Capital cost of Boiling-Water Reactor plant. COO-2477-6 (Vol. 1). June 1977.

[7] RICHARDSON, JOHN, A., Construction costs and manpower performances for current U.S. nuclear plants. J. Inst. Nuc. Eng. Vol. 19, No. 3.

[8] ARONSTEIN, R.E. and GHOSE, S.K., Cost components of fusion-power plant. J. Inst. Nuc. Eng., Vol. 19, No. 5.

[9] HOLLIS, A. and EVANS, L.S., An analysis of the estimated capital cost of a fusion reactor. AERE Report 9933.

[10] MITCHELL, J.T.D., Blanket replacement in toroidal fusion reactors, Proc. A.N.S. 3rd topical meeting on technology of controlled nuclear fusion Vol. 2, p.954, 1978.

TABLE I

Estimated DIRECT Capital Investment Costs for the Mk IIB Reactor

Account No.	Description	£M	£/kW(e)	% of total direct costs
21	Structure & site facilities	113.7	94.8	12.5
22	Reactor - plant equipment	656.1	546.75	72.0
23	Turbine - plant equipment	77.2	64.3	8.5
24	Electrical - plant equipment	30	25	3.3
25	Miscellaneous - plant equipment	11.7	9.8	1.3
26	Cooling towers	22	18.33	2.4
	TOTAL	910.7	759	100

TABLE II

Estimated INDIRECT Costs for the Culham

Mk IIB Reactor

Account No.	Description	£M	£/kW(e)	% of total indirect costs
91	Construction-facilities, plant & equipment)			
92	Engineering & management services)	320	267	35
93	Other costs	137	114	15
94	Interest during construction	456	380	50
	TOTAL	913	761	100

TABLE III

Comparison of Fusion- and Fission-Reactor Direct Costs

Estimated Costs, £M

Account No.	Description	Culham IIB	UWMAK III	BWR COO 2477/6	PWR COO 2477/5
21	Structure & site facilities	113.7	284	56.66	50.68
22	Reactor - plant equipment	656.1	460.2	63	66.74
23	Turbine - plant equipment	77.2	261	58.34	55.64
24	Electrical - plant equipment	30	112	20.38	19.72
25	Miscellaneous - plant equipment	11.7	30	-	-
26	Cooling towers	22	20	11	10.79
	Total direct costs £M	910.7	1167	*312	*306.6
	Reactor power MW(e)	1,200	1,985	1,190	1,200
	Total direct costs £/kW(e)	759	588	262	256
	Year of estimate	1977	1976	1976	1976

* Includes £103M for:- main transformer, switch gear & transmission facility costs and also initial fuel costs.

FACTORS AFFECTING THE MINIMUM CAPITAL COST
OF A TOKAMAK REACTOR

R. Hancox
Culham Laboratory, Abingdon, Oxon. OX14 3DB UK
(Euratom/UKAEA Fusion Association)

ABSTRACT The Mk IIA Culham conceptual tokamak reactor design is a 2500 MWe steady-state reactor developed on the basis of a cost optimisation. A revised 1200 MWe conceptual design, the Mk IIB, used a lower wall loading and lower thermodynamic efficiency. A detailed costing of the Mk IIB design, however, showed it to have an unacceptably high capital cost. Since this high cost is a common characteristic of many fusion reactor designs, the cost optimisation of the Mk II design has been reconsidered.

1. REACTOR DESCRIPTION

The Culham MkII conceptual tokamak reactor design was first published in 1976 [1]. It incorporated several features currently being discussed in relation to future large tokamak experiments, such as an elliptically shaped plasma minor cross-section and an iron transformer with saturated centre core to minimise the plasma aspect ratio, and therefore has a superficial similarily to the JET experiment.

A revised design was published in 1977 [2] with more conservative parameters and the addition of a single-null poloidal divertor, and these two designs are now designated Mk IIA and IIB. Both assume quasi-steady operation with refuelling and a means of exhausting reaction products, although these features were not considered in detail. Other features which were implied but not specified in detail included the means of initiating the plasma current and heating the plasma to ignition and means for controlling the plasma temperature and hence the reactor power output. Some of these aspects were discussed in a more recent study [3] of a pulsed tokamak reactor, the Mk IIC. The initial assumptions on which these designs were based and the resulting parameters are shown in tables I and II.

TABLE I

Mk II reactors, initial assumptions

	IIA	IIB	IIC
Net power output, MWe	2500	1200	600
Poloidal beta - K_β	1.0	1.0	1.0
- E_β	0.5	0.5	0.66
Safety factor (q_0)	1.0	1.0	1.0
Plasma ellipticity	1.75	1.75	1.68
Burn time/off time	∞	20	1.93
Max. toroidal mag. field, T	≤ 8.0	≤ 8.0	~8
Max. core magnetic field, T	≤ 6.0	≤ 6.0	~8
Blanket & shield thickness, m	1.5	1.5	2.1
Thermal efficiency, %	45	40	40
Recirculating power, %	5	8	19
Max. neutron dose, x 10^{25} /m^2	100	20	40

TABLE II

Mk II reactors, parameters

	IIA	IIB	IIC
Net power output, MWe	2500	1200	600
Gross thermal power, MW	5830	3240	1825
Major radius, m	7.4	6.7	7.8
Minor radius, m	2.1	1.9	2.0
Aspect ratio	3.5	3.5	3.9
Plasma current, MA	11.7	10.2	11.0
Toroidal beta($\bar{\beta}_t$), %	9.3	9.2	7.7
Tor. mag. field (on axis), T	4.0	4.0	3.9
Peak core magnetic field	4.2	5.6	7.8
Wall power loading, MW/m^2	6.7	4.5	2.1
Unit cost (£/kWe in 1976)	156	254	510

1.1 Tokamak Reactor Mk IIA

The parameters of the Mk IIA reactor were deliberately chosen to obtain a low capital cost. Thus a high output power was assumed and a higher wall loading than was used in similar designs at that time. The maximum values allowed for plasma parameters such as the plasma pressure ratio β and safety factor q or for engineering parameters such as the maximum level of toroidal magnetic field B_m, although optimistic, were not too unrealistic.

The blanket of the Mk IIA reactor is constructed in a cellular form with liquid lithium breeder and helium cooling. The choice of structural material was not discussed in detail, but it was assumed that operation would be possible at a sufficiently high temperature to give 45% efficiency for the conversion of heat to electricity.Because of the high wall loading it was accepted that the first wall structure would suffer extensive radiation damage and therefore have to be replaced during the life of the reactor. Thus the Mk II designs have incorporated the means of maintaining and repairing the internal reactor structure by remotely operated machines, and the development of these maintenance concepts has been a predominant feature of the Culham reactor studies.

1.2 Tokamak Reactor Mk IIB

The Mk IIB reactor represents a development of the IIA design to include a polidal divertor and further detail in the construction and maintenance of the blanket and shield. The engineering assumptions on which the design was based were more conservative with reductions in the net power output, wall loading, and thermal efficiency, and an increase in the recirculating power for auxiliary equipment.

To obtain a more realistic structure in the 1.5m space allowed for the blanket and shield the breeding blanket consisted of an inner zone of lithium-lead and an outer zone of lithium orthosilicate enclosed in an inconel cell. The inner section of shield was mounted within the blanket cell to maximise the energy recovery, both being cooled by helium supplied through a co-axial duct.

1.3 Tokamak Reactor Mk IIC

The Mk IIC design was developed specifically to allow a comparison of the tokamak reactor with a Reversed Field Pinch reactor. In both cases pulsed operation was assumed without refuelling during the burn. Because of its pulsed operation the wall loading was again reduced. The net power output was also reduced to 600 MWe, with the result that the physical size of the reactor was similar to that of the Mk IIA and IIB designs.

Further consideration was given in the Mk IIC study to the requirements for initiating and heating the plasma and to the poloidal magnetic field system required to maintain and control the plasma. Twenty primary coils and fourteen vertical field coils were required to give the necessary magnetic field distribution. Filamentary niobium-titanium conductors were assumed for the poloidal field windings, whereas niobium-tin strip was proposed for the toroidal field windings. For the pulsed reactor an air cored transformer appears preferable.

2. COST STUDY

The parameters of the designs described above were all determined by a cost optimisation procedure using a computer program TCOST [4] developed at Culham. The program includes the equations linking the physical and technical parameters of the reactor, and calculates a cost based on the physical size or power rating of each component in a possible design. The program then adjusts parameters within given limits to find either the minimum total capital cost of a reactor or the minimum unit cost at a given power output. The quoted cost refers to the reactor and its building and does not include conventional generating plant.

The cost data used as input to the program were generated from published costs of fission reactors or engineering structures of a similar size and complexity. The input and output data has also been checked against information given in earlier published reactor designs and against data used in the optimisation of the JET experiment [5]. The main components of the cost input are given in table III. The physical relationships and cost data incorporated in the program were updated several times until 1978, but costs have been kept at 1976 values to facilitate comparisons between new designs and the original Mk IIA costing.A limitation of the input data, inherant in any program of this type, is that it cannot take account of the influence of minor changes in designs or materials. Whilst adjustments have been made for major changes, for example the change from niobium-titanium to niobium-tin superconductor, it has not been possible to reflect changes in the structural design except through gross changes in size or weight.

The capital costs of the three reactor designs calculated with the program TCOST are shown in table II. It is clear that as the physical and engineering assumptions have become more conservative the unit cost has incresed. An independent, and more accurate, costing of the Mk IIB design [6] has indicated that the later costs are unacceptably high, and that every effort must be made to reduce the costs if a tokamak reactor is to be competitive compared to alternative power sources. For this reason the influence on the capital cost of several of the initial assumptions and limitations has been investigated.

As a starting point most of the assumptions corresponding to the Mk IIB design have been used. However, in order to leave as many free parameters as possible the maximum toroidal magnetic field, the maximum magnetic field in the core and the maximum wall loading have been increased. Assumptions and optimum parameters for the standard case are shown in table IV.

2.1 Plasma Beta and Safety Factor

The poloidal plasma pressure ratio βp was assumed to be related to plasma aspect ratio R/a through the constants K_β and E_β so that $\beta_p = K_\beta (R/a)^{E_\beta}$. In the standard design the values of K_β and E_β were taken to be 1.0 and 0.5 which is currently thought to be a reasonable assumption, but the effect on unit capital costs of values of K_β up to 2.0 are

TABLE III

Input cost data, at 1976 values

Superconductor	7.5×10^{-2}	£/kA.m.T
Magnet stabilizer (copper)	5.6×10^{4}	£/m^3
Magnet manufacture	2.8×10^{4}	£/m^3
Magnet structure (steel)	5.0×10^{3}	£/m^{2}T^2
Cryogenic insulation	2.0×10^{3}	£/m^2
Refrigeration	1.1×10^{2}	£/MW
Blanket	1.7×10^{4}	£/m^2
Shield	5.5×10^{4}	£/m^2
Primary circuit	9.0×10^{3}	£/MW
Reactor auxiliaries	6.5×10^{3}	£/MW
Iron transformer core	1.1×10^{4}	£/m^3
Poloidal power supply	2.0×10^{4}	£/MJ
Plasma heating	3.5×10^{5}	£/MW
Reactor building	4.5×10^{2}	£/m^3

TABLE IV

Initial Parameters for Study, Optimum Parameters and Sensitivities

	IP	OP	S
Net power output, MWe	1200		-0.44
Poloidal beta - K_β	1.0		-0.31
Safety factor $(q_0)^\beta$	1.0		0.66
Burn time/off time	20		-0.02
Max. toroidal mag. field, T	≤15	9.5	0
Max. core magnetic field, T	≤15	15.0	-0.02
Blanket & shield thickness, m	1.5		0.56
Thermal efficiency, %	40		-0.57
Recirculating power, %	8		0.05
Wall power loading, MW/m^2	≤20	6.1	0
Major radius		6.0	
Minor radius		1.6	
Unit cost (£/kWe in 1976)		194	

shown in fig. 1 for three sizes of reactor. Clearly values below unity lead to high costs, but the reductions in unit cost for values much above unity are not great. The effects of some more restricted engineering parameters are indicated.

The safety factor q_0 on the plasma axis was assumed to be unity in the standard design. For values in the range 1 to 2 the unit costs are roughly proportional to the safety factor, emphasizing the importance of reaching in practice the theoretical limit. The effect of ellipticity in the plasma minor cross-section has not been studied, since the theoretical model in the program linking the current in the poloidal field windings with the plasma shape was not considered sufficiently accurate to give reliable results.

2.2 Fixed Engineering Parameters

The effect of changing the reactor power output is shown in fig. 2. In the central range of sizes the variation is close to the 2/3 power law common in engineering, and it is only at sizes below 600 MWe that the unit cost rises more rapidly. Thus, whilst there is an advantage in constructing larger units, the cost of fusion reactors will not change strongly in relation to other power sources if comparisons are made between systems of the same size. The question of the optimum wall loading is considered later, but it should be noted from fig. 2 that if the wall loading is limited the advantage of large systems is reduced.

It is well known that the space required for the blanket and shield on the inboard side is an important parameter in determining the cost of a reactor. In the standard design it was fixed at 1.5 m, and increasing it to 2.0 m increases the unit cost by 20%. The corresponding space on the outboard side is 4.0 m to allow full access for maintenance and repair of the reactor, and here a 0.5 m increase only raises the cost by 2%.

The efficiency of conversion of heat to electricity is also important in determining the unit cost. The sensitivity to this assumption is included in the table of results on the over-simple assumption that the higher temperature operation required for higher efficiency does not imply higher material prices. A more detailed investigation of this relationship would be valuable, but is beyond the scope of this study.

2.3 Wall Loading Limit

In this study the limit on the wall power loading (defined as the total power deposited on, or passing through, the wall) was set at the unrealistically high value of 20 MW/m^2. In most cases, however, the minimum costs were obtained with lower levels of wall loading, and

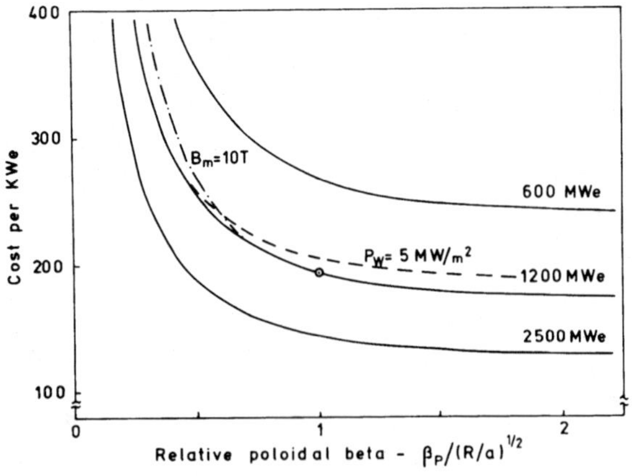

Fig.1 Effect of poloidal beta on cost

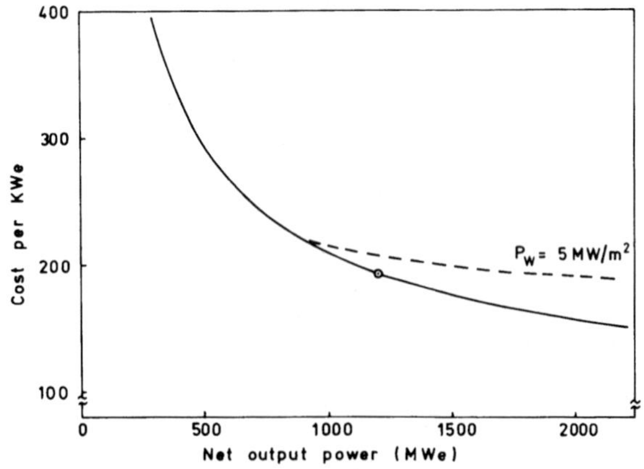

Fig.2 Effect of power output on cost

only for reactors of output power exceeding 1000 MW_e does the optimum level exceed a value of 5 MW/m^2 which has been considered practical in several reactor designs. As already noted, the cost advantage of large reactors is closely associated with the possibility of operating with high wall loadings, and with wall loadings below about 3 MW/m^2 the advantage of reactors with power outputs above 600 MWe almost disappears.

Since one of the least certain factors in the physics assumptions is the value of poloidal beta, it is of interest to investigate how the optimum wall loading changes with beta. The variation of the optimum wall loading is shown in fig. 3, together with the optimum maximum toroidal magnetic field. The optimum wall loading increases with increasing beta, but saturates until the maximum magnetic field in the core reaches its limit and then increases again. Decreasing the limiting value of the maximum core magnetic field increases the optimum wall loading, whereas at lower values of beta reducing the maximum toroidal magnetic field decreases the optimum wall loading. It is concluded that improvements in the achievable value of beta, which are desirable to reduce capital costs, also require an increase in the wall loading. However, as seen from fig. 1, the effect on the cost of limiting the wall loading at higher values of beta is not strong, which shows that the optimum is rather broad.

2.4 Magnetic Field Limits

In the present study the upper limits for the maximum toroidal magnetic field (i.e. the magnetic field at the inboard surface of the toroidal field coil) and the maximum core magnetic field were set at the high value of 15 T. In all cases of interest the minimum costs are obtained with values of the maximum toroidal magnetic field the below 15 T, and the cost penalty of using lower levels of maximum core magnetic field is quite small unless values below about 5 T are specified.

The optimum value of maximum toroidal magnetic field is around 10 T, and does not depend strongly on the reactor size. It is also seen from fig. 3 that it does not depend strongly on the poloidal beta assumed. Thus under the conditions assumed in this study a value in the range 8 to 10 T gives costs close to the minimum, and either niobium-titanium or niobium-tin superconductor could be used, depending on the margin of safety and reliability required.

2.5 Sensitivities

To judge the relative importance of the variables considered in the previous sections, the sensitivity of the cost to changes in each variable are shown in table IV. The sensitivity, S, is the slope of the curve of unit cost, C, as a function of a variable V, i.e. S = (V/C) (dC/dV). In some cases the sensitivity is zero, because the optimum value

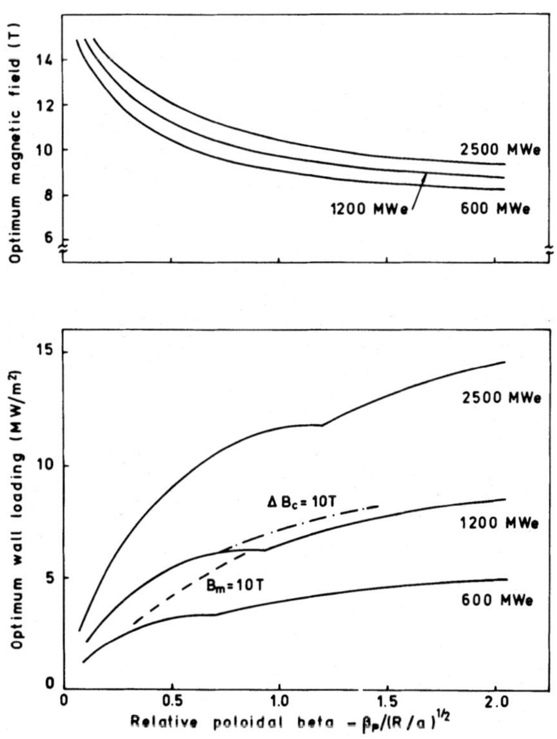

Fig.3 Optimum wall loading and toroidal field

of the variable is less than the higher limit used for the standard design. The highest values of the sensitivity are obtained for the plasma safety factor q_0 and the blanket and shield thickness, but as indicated before the chosen values for these parameters are already optimistic. Nevertheless, means of controlling the plasma pressure profile to reduce the safety factor throughout the plasma might be effective in minimising costs, and a redistribution of the breeding blanket with a low level of tritium breeding in a thin blanket on the inboard side of the reactor and a high level over the larger outboard area could also be effective.

3. CONCLUSIONS

The effects of several changes in design parameters on the unit cost of a tokamak reactor have been studied. No single change leads to a substantial cost reduction, and the parameters which most stronly influence the cost such as the plasma safety factor q_0 or the blanket and shield thickness have already been set at levels which are unlikely to be significantly improved. Thus reducing tokamak reactor costs must depend on improving all physics and engineering parameters or finding cheaper means of achieving the same parameters.

The optimum level of wall loading increases with the power output of the reactor and the value of poloidal beta obtainable. Reductions in unit costs can be achieved if these three parameters can all be increased, but neither increases in beta or power output are effective if the wall loading is limited to values below the optimum. For values of poloidal beta greater than the square-root of the aspect ratio and values of output power above 1200 MWe the optimum wall loading exceeds values used in many tokamak reactor designs, and thus the limitation on wall loading emerges as a major factor in raising the unit costs.

The optimum level of the maximum toroidal magnetic field is in the range 8 to 10 T for a range of other variables, and the sensitivity to the chosen value of maximum core magnetic field is low so that most reactor designs are already close to optimum in these respects. For this reason, and because the optimum wall loading is roughly proportional to the power output, the physical sizes of most optimised designs are similar. Thus the maintenance concepts developed with the Mk II design will remain valid for a range of optimum designs.

The analysis undertaken here is intended to illustrate the relative importance of various parameters in determining the capital cost of a tokamak reactor, and to indicate the ways in which this cost might be reduced. The study is far from complete and does not include several effects such as changes in ellipticity of the plasma cross-section, alternative structural materials or operating temperatures or the implications of different divertor systems, and much remains to be done in reducing the cost of fusion reactors to an economically attractive level.

4. REFERENCES

[1] Hancox, R. and Mitchell J.T.D., Proc. 6th Int. Conf. on Plasma Physics and Controlled Nuclear Fusion, Berchtesgaden, October 1976. Vol. 3, p. 193.

[2] Mitchell, J.T.D., Proc. 3rd Topical meeting on Technology of Controlled Nuclear Fusion, Santa Fe, May 1978. Vol. 2, p. 954.

[3] Spears W.R. and Hancox R., Culham Laboratory Report CLM-R197, July 1979.

[4] Tam, K.N., M.Sc. Thesis, University of Surrey, 1973.

[5] Sheffield J. and Gibson A., Nuclear Fusion, Vol. 15, 677, 1975.

[6] Evans L.S. and Hollis A.A. To be published in Proc. 11th Symp on Fusion Technology.

HYFIRE: A TOKAMAK--HIGH-TEMPERATURE ELECTROLYSIS SYSTEM*†

J.A. Fillo, J.R. Powell, M. Steinberg, R. Benenati, F. Horn, H. Isaacs, O.W. Lazareth, H. Makowitz, and J. Usher

Brookhaven National Laboratory
Upton, NY 11973

ABSTRACT

Brookhaven National Laboratory (BNL) is carrying out a comprehensive conceptual design study called HYFIRE of a commercial fusion Tokamak reactor, high-temperature electrolysis system. The study is placing particular emphasis on the adaptability of the STARFIRE power reactor to a synfuel application. The HYFIRE blanket must perform three functions: a) provide high-temperature (~1400°C) process steam at moderate pressures (in the range of 10 to 30 atm) to the high-temperature electrolysis (HTE) units; b) provide high-temperature (~700° to 800°C) heat to a thermal power cycle for generation of electricity to the HTE units; and c) breed enough tritium to sustain the D-T fuel cycle. In addition to thermal energy for the decomposition of steam into its constituents, H_2 and O_2, electrical input is required. Fourteen hundred degree steam coupled with 40% power cycle efficiency results in a process efficiency (conversion of fusion energy to hydrogen chemical energy) of 50%.

1. INTRODUCTION

Brookhaven National Laboratory is carrying out a comprehensive conceptual design study called HYFIRE of a commercial fusion Tokamak reactor, high-temperature electrolysis (HTE) system. The purpose of the study is to provide a mechanism to further assess the commercial potential of fusion via a Tokamak reactor for the production of synthetic fuel. The HYFIRE reactor design is based on the Tokamak commercial power reactor, STARFIRE [1], the primary difference residing in the type of blanket between the two reactors as well as power cycle design. In addition to exploring a range of blanket and power cycle options to determine those best suited for H_2 production, the study is placing particular emphasis on the adaptability of a Tokamak power reactor to a synfuel application.

Details of the STARFIRE reactor study are documented in ref. 2. The key technical objective of the STARFIRE study has been to develop an attractive embodiment of the Tokamak as a commercial power reactor consistent with credible engineering solutions to design problems. This same philosophy is carried over to the HYFIRE study with an eye towards assessing what major changes are required for the synfuel reactor. HYFIRE is based on the deuterium/tritium/lithium fuel cycle.

The primary criteria for commercial attractiveness emphasized in the STARFIRE study are economic, safety, and environmental impact. These criteria are, of course, of equal concern for HYFIRE where economics must include the economics of producing hydrogen and safety and environmental impact include the production of hydrogen and oxygen.

Section 2 is an overview of the reactor concept. Section 3 covers blanket design of the synfuel process modules as well as tritium breeding/power cycle modules. Section 4 is an overview of the HTE process and design of the electrolyzer, including some aspects of product conditioning. Section 5 covers the overall process design including the thermal power cycle for electrical power generation and coupling the HTE process. Section 6 presents the key conclusions based on preliminary analysis.

*Work performed under the auspices of the U.S. Department of Energy, Washington, D.C. 20545.

†By acceptance of this article, the publisher and/or recipient acknowledge the U.S. Government's right to retain a nonexclusive, royalty-free license in and to any copyright covering this paper.

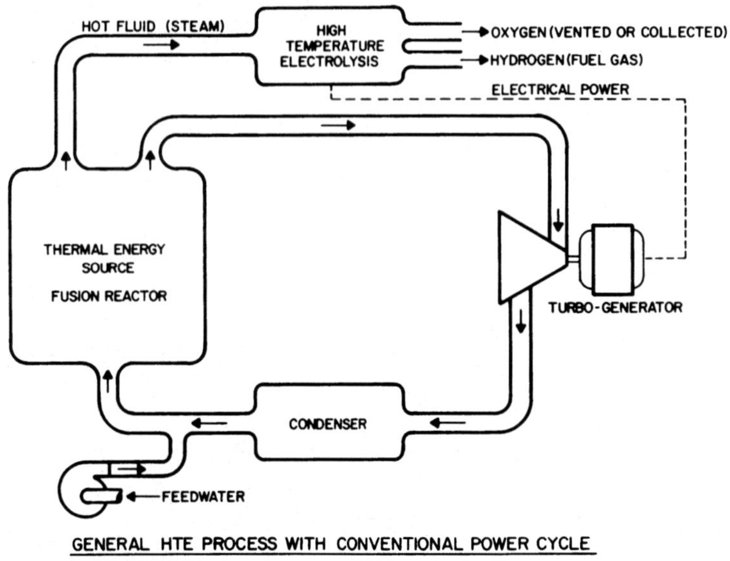

Fig. 1.

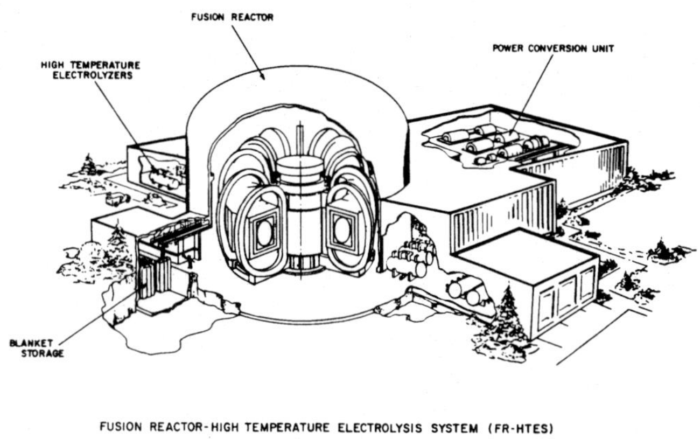

Fig. 2.

2. OVERVIEW OF REACTOR CONCEPT

Figure 1 shows a simplified flow sheet for an HTE/fusion synthetic fuel plant. All electrical production goes to the HTE cells (and to operation of the fusion reactor) to make hydrogen/electricity for sale, depending on market demand. Two blanket types are inferred; the first type heats steam to high temperatures ($T \geq 1000°C$) for delivery to the HTE cells, while the second heats a working fluid for the thermal power cycle and electricity generation as well as for tritium breeding.

Major systems of HYFIRE are shown in fig. 2. As in STARFIRE, HYFIRE operates in a steady-state mode with the plasma current driven by lower hybrid rf. The plasma impurity control and exhaust system is based on the limiter/vacuum concept. The reactor has a 7-m major radius and produces 4000 MW of thermal power with an average neutron wall load of 3.6 MW/m^2. This translates into ~1050 MT of hydrogen produced per day.

All superconducting equilibrium field coils are located outside the 12 toroidal-field coils and 4 small segmented copper coils are located inside for plasma stability control. The shield provides neutron- and gamma-ray attenuation and serves as the primary vacuum boundary for the plasma. Twelve shield access doors are provided to permit removal of 24 toroidal blanket sections. Twelve electrolyzer units and associated heat exchangers are housed in a building circumferentially surrounding the reactor. In addition, the power generating units are shown.

Steady-state operation of the Tokamak is assumed. As the STARFIRE study indicates, and which is equally applicable to HYFIRE, there are a number of technological and engineering benefits for a commercial reactor that would be derived from steady-state operation. Among these are (1) component and system reliability is increased; (2) material fatigue is eliminated as a serious concern; (3) higher neutron wall load is acceptable; (4) thermal energy storage is not required; (5) the need for an intermediate coolant loop is reduced; (6) electrical energy storage is significantly reduced or eliminated; and (7) an ohmic heating solenoid is not needed, and external placement of the equilibrium-field coils is simplified. It has been estimated that the combined benefits of steady state can result in a saving in the cost of energy as large as 25 to 30%.

All fusion applications will probably require reasonably high plant availabilities (fraction of the time the plant is on-line), on the order of 0.5 to 0.8, due to the high-capital investment for the reactor. As with any electric generation system, fusion reactors connected to the grid will have to have high reliability, with relatively few outages per year. Reliability requirements for a synfuel plant will be less demanding, though, since the product can readily be stored off-line. Fluctuations in plant output can thus be readily smoothed out by using available storage to meet demand requirements if the plant shuts down. While high-capital cost reactors will have to have high plant factors for economic reasons, they could be allowed to shut down unexpectedly fairly often for short periods (i.e., to start up the plasma), if they were not connected to an electrical grid.

3. BLANKET DESIGN

The HYFIRE blanket must perform three functions: a) provide high-temperature ($\geq$1000°C) process steam at moderate pressures (in the range of 10 to 30 atm) to the high-temperature electrolysis units; b) provide high-temperature (~700° to 800°C) heat to a thermal power cycle for generation of electricity to the HTE units; and c) breed enough tritium to sustain the D-T fuel cycle. The dual requirements, generation of high-temperature process steam for the HTE's and high-temperature heat for the thermal power cycle, differentiates the HYFIRE and STARFIRE blanket system.

Setting the requirements that the global breeding ratio equal 1.1 to allow for doubling time requirements, perturbations, etc., for HYFIRE places a premium on space, i.e., it will probably be necessary to breed tritium in regions of the process steam blanket modules. Tritium from the power cycle part of the blanket must make up the tritium deficiency.

The two-temperature-zone blanket [3] approach is mandatory for the process steam portion of the energy supply. The modules will have relatively cool shells (~300°C) with thermal insulation between the shell and the high-temperature (~1400°C) interior. The two-temperature design concept is also carried over for the power cycle modules. As in STARFIRE, the first wall/blanket structure or shell is PCA stainless steel.

Three blanket options are under study for HYFIRE. Each option has an HTE steam module region and a power cycle module region with tritium breeding in each region. Tritium breeding is to be accomplished with solid breeders, and tritium inventory in the blanket should be minimized. A possible problem with tritium holdup in Li_2O has been raised by the STARFIRE study. This can be circumvented either by using neutron multipliers (Be, Pb) and a solid breeder (either Li_2O or $LiAlO_2$), by scavenging with D_2 or H_2 in the He purge circuit, or by using a liquid breeder material (e.g., PbBiLi mixture). Tritium will be released to He purge streams, and not to the main circuit. Module arrangement along toroidal field lines, as in STARFIRE, is preferred, since this minimizes differences in maintenance procedures between HYFIRE and STARFIRE. The inboard blanket-shield region will probably be used for HTE steam modules, with a thin secondary zone behind for tritium breeding.

For the HTE modules refractory oxides, e.g., ZrO_2 or Al_2O_3, from the high-temperature region of the blanket must be stable under exposure to the steam or steam/hydrogen process stream under radiation and thermal cycling conditions, although the latter may be mitigrated due to steady-state performance by HYFIRE. Such materials will fill the interior of the blanket as solid rods or balls, and will also be used as a low-density solid block or fibrous thermal insulation between the high-temperature interior and the structural shell. Materials compatibility tests [4] in steam and steam/hydrogen indicate that ZrO_2 and Al_2O_3 are suitable for long-term service up to ~1500°C (the present testing limit at BNL). Tests with SiC and MgO indicate these materials are restricted to somewhat lower temperatures.

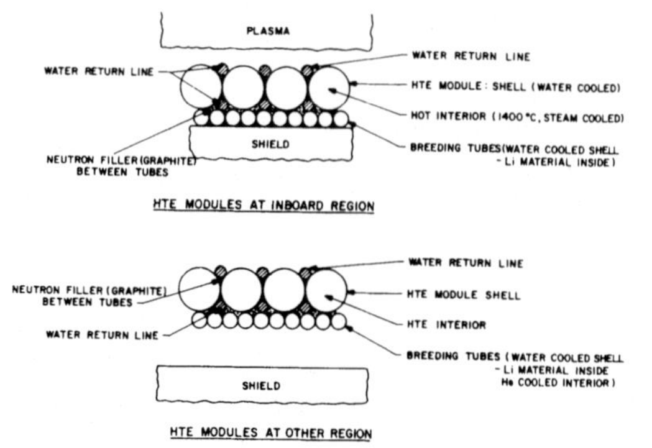

MARK I HTE PROCESS STEAM BLANKET

Fig. 3a.

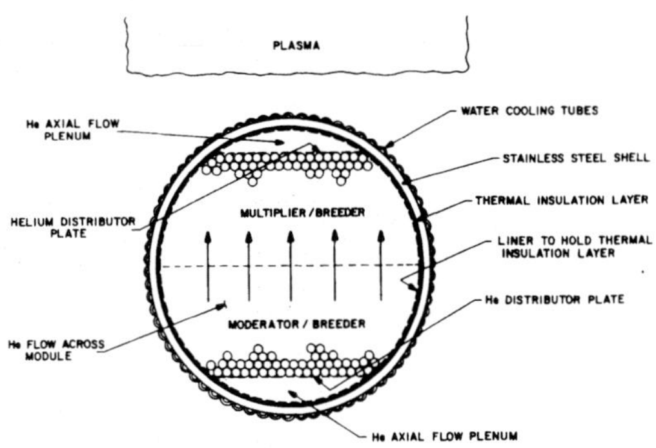

CROSS SECTION OF MARK I BREEDING BLANKET MODULE

Fig. 3b.

TABLE I

Representative Tritium Breeding Results

MARK I	MARK III	MARK II
Breeding blanket		Breeding blanket
1.2 to 1.6	1.0 to 1.4	1.1 to 1.3
HTE breeding		HTE breeding
.3 to .5 (no front breeding)		.3 to .5 (no front breeding)

Representative process (HTE) and power/tritium blanket modules are shown in figs. 3a and 3b. All of the larger modules (∿40 to 50 cm diameter) are of the two-temperature type with steam cooling of the hot interior for HTE modules, primary He cooling (∿20 atm) of the hot interior for tritium breeding modules with heat exchange to a secondary He (∿70 atm) stream in small heat exchangers directly behind the blanket and water cooling of the module shells.

Representative tritium breeding ratios for the three blanket options are shown in table I. All options are viable from the standpoint of neutronics and thermal hydraulics. There is not adequate energy deposition in the hot blanket interior for the HTE's and power cycle in MARK II. MARK I is a preferred design since tritium is radioactively isolated from the main He coolant stream, stress on the module structure is low due to low primary coolant pressure, and no leakage of steam from the steam generator into the blanket is possible because of the intermediate He-to-He heat exchanger.

4. HTE PROCESS

The electrochemical decomposition of water into hydrogen and oxygen is an endothermic reaction requiring both heat and electricity. The efficiency of production of electricity from fusion reactor heat is limited by the Carnot relationship and various irreversibilities in the power cycle. With conventional steam power cycles, electrical generation efficiency will be on the order of 40%. Since the heat input component for water decomposition is used directly at essentially 100% efficiency, there is a definite advantage to make the ratio of the direct heat input to the electrical energy input as large as possible. At a temperature, T, the input thermal energy equals $T\Delta S$, where ΔS is the entropy change for the reaction. The electrical energy input equals the Gibb's free energy change, ΔF, for the reaction, and the sum of these energy changes equals the reaction enthalpy, ΔH.

As temperature increases the reaction enthalpy remains virtually constant. The Gibb's free energy or electrical energy input, however, decreases with increasing temperature and the thermal energy input, $T\Delta S$, increases. The ratio of thermal energy ($T\Delta S$) to electrical energy (ΔF) increases with electrolysis temperature; this results in higher process efficiency so that more hydrogen production can be generated for a given fusion energy input. For HYFIRE,

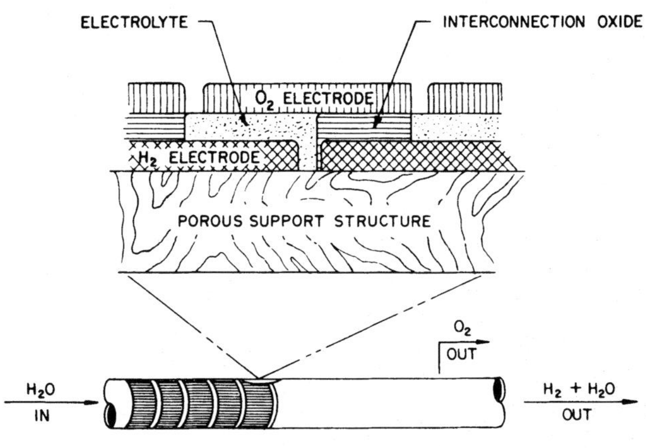

Fig. 4

the design temperature of interest is 1400°C.

The heat input, TΔS, absorbed by the HTE cells during electrolysis is supplied from the sensible heat content of the process streams. For practical electrolyzer designs, the steam/H_2 stream will cool by 100° to 200°C as it proceeds through the electrolyzer.

A schematic of the Westinghouse fuel cell is shown in fig. 4. This design also serves as the basis for the high-temperature electrolyzer since an electrolyzer is a fuel cell in reverse. High-temperature electrolysis uses arrays of tubes of relatively small diameter (∿1 cm), thick-walled porous ceramic (e.g., stabilized ZrO_2) on which a succession of thin electrode layers of suitably-doped ceramics are deposited. The H_2 and O_2 ceramic electrodes are separated by a thin (several mils) electrolyte layer of yttria-stabilized ZrO_2. Electrodes are electrically connected in series along each tube to minimize IR losses. A large number (∿10^5) of electrolyzer tubes are then connected in parallel in a large pressure vessel. Typical steam pressures in a high-temperature electrolyzer are on the order of 10 to 20 atm.

Previous studies [5] of HTE processes have assumed that steam is directly heated in the hot interior of the HTE process heat modules, and then passes into the HTE electrolyzer. The steam is cooled as it passes through the electrolyzer by the endothermic electrolysis process. In order to keep the electrolyzer temperature at a high average value, it is necessary to electrolyze only a small fraction (∿10%) of the steam during its passage and return the steam-H_2 mixture for reheat to another section of the blanket. The optimum number of series of reheats and electrolyzers depend on various parameters. For reasons of ducting and connections to the 24-blanket sectors of HYFIRE, the number of electrolyzers are fixed at 12, one for each two blanket sectors, and process parameters are adjusted to reflect the fixed numbers.

This type of electrolysis process arrangement is characterized by: 1. Steam-H_2 mixtures flow through the blanket, with the H_2/steam ratio varying from 0 for the first electrolyzers to ∿10/1 at the exit of the electrolyzer string. The refractory in the hot blanket interior must thus withstand steam/H_2 mixtures at temperatures of ∿1400°C. 2. Radioactive isotopes picked up by the steam-H_2 stream will go along with the H_2 product, necessitating cleanup by filtration or absorption (e.g., in ion exchange resins).

Other types of HTE process arrangements are possible to mitigate activation of the H_2 product. Rather than circulating steam through the blanket to remove heat, instead, it passes straight through the electrolyzers, either in series or parallel flow, exiting as almost pure H_2. Heat is provided to the electrolyzers (and removed from the blanket) by circulating O_2 plus inert gas (e.g., He) from the shell side of the electrolyzer. Oxygen would be separated from the inert gas at the end of the process and discharged to the atmosphere or whatever market was available.

Another design approach would be to make the HTE electrolyzer slightly longer (e.g., about 10% longer), with a separate shell-side zone to transfer heat from the He blanket coolant to the steam-H_2 mixture flowing inside the nonporous ZrO_2 HTE tube. A separation partition between the O_2 shell-side zone and the He shell-side zone is required, with a flowing gas sweep to prevent slight mixing of gases in the two zones.

4.1 Hydrogen recovery

If the conversion of steam to hydrogen in the high-temperature electrolyzers is to be limited to 10% (to minimize electrical requirements) steps must be taken to separate the product hydrogen from the waste steam. One way to accomplish this separation is to cool the process stream and condense out the water. A product hydrogen stream in excess of 99% purity can be achieved in this way based on recouperative heat exchange at the high-temperature end of the process and normal cooling tower water at the low-temperature end of the process.

Since the conversion per pass is so low, the recycle ratio is quite high, i.e., it takes 10 passes through the HTE system before a mole of steam is completely converted to hydrogen, and since that steam must be brought up to process conditions then completely condensed in each pass through the system, much attention must be paid to energy conservation if the process is to be a viable one. One way to conserve energy would be to use the high-temperature steam leaving the HTE system to drive a turbine thereby producing electricity which can be recycled to the process. In this scheme, the process stream leaving the HTE would first be cooled recouperatively to about 538°C, a temperature compatible with steam turbine blade materials. It would then pass through the turbine (which would not have any extraction streams) exiting at about 1.5 psia and approximately saturated. Based on a turbine efficiency of 81%, 19 kWh of electricity can be recovered per lb-mole of hydrogen. Only about 70% of the steam can be condensed out of this stream using cooling tower water. The resulting low pressure wet hydrogen steam must be compressed to some suitable product discharge pressure (the actual pressure will depend on the end use intended for the hydrogen). This compression would be done in several stages with intercooling between stages. Each intercooler will remove some of the remaining water vapor thus making a product stream of 98 to 99% dry hydrogen possible. Based on assumed compressor efficiencies of 80% and a final discharge pressure slightly above 1 atm, approximately 3 kWh of compressor work are required leaving a net of about 16 kWh/mole of hydrogen product. It should be noted that the above is based on a reasonable selection of process parameters with virtually no effort as yet to optimize these process conditions.

5. BLANKET/POWER CYCLE/PROCESS COUPLING

Depending on whether steam is directly transported to the electrolyzer or an intermediate heat exchanger is introduced between the electrolyzer and blanket, or whether only a small fraction (∿10%) of the steam is converted to hydrogen in a single pass and extracted or a string of electrolyzers are placed in series, coupling between the blanket, electrolyzer and power cycle can be quite different. Each case must be analyzed separately.

As an example of a directly heated electrolyzer, the blanket (already discussed) has three distinctly different heat generation zones, each with its own process fluid and with different process parameters.

One zone which involves the front wall of all modules operates at the relatively modest temperature of 320°C and is kept cool by the circulation of high pressure (1000 to 2000 psi) water. The heat generation in this zone is about 43.5% of the total reactor heat output which amounts to 1740 MW(t). The pressurized water coolant from this zone is circulated through a steam generator in which 1000 psia saturated steam is generated on the secondary side.

Another zone, which will lie behind the front wall of about 50% of all blanket modules, is optimized for tritium breeding. This zone will operate at 800°C and is cooled by pressurized (70 atm) helium gas which will be used to superheat the steam referred to above. The heat generated in this zone is about 35% of the total reactor heat output which amounts to 1400 MW(t).

The third and final zone, which will lie behind the front wall of the remaining 50% of all blanket modules, is designed with the needs of the high-temperature electrolyzers in mind. This zone will operate in excess of 1400°C and serves to resuperheat relatively low pressure (10 atm) steam to 1400°C before it goes on to the electrolyzers. The heat generation in this zone is 21.5% of the total reactor heat output which amounts to 860 MW(t).

Looking at the steam circuit, make up water is added to recycle water and together they enter a steam boiler not unlike that of a PWR-type fission reactor. The 1000 psia saturated steam so produced is superheated to about 760°C, by means of the hot helium mentioned above, at which condition it enters the high-pressure steam turbine, exiting at 10 atm and about 482°C at which conditions it enters the third zone of the blanket mentioned above. It leaves the blanket at 1400°C and goes directly to the high-temperature electrolyzers where approximately 10% is converted to hydrogen and oxygen. The oxygen is wasted after passing through heat exchangers to recover its heat content. The hydrogen together with the unreacted steam is passed on to the hydrogen recovery portion of the flow sheet which is described in Section 4.

Preliminary studies and calculations indicate that gross power cycle efficiency in the 40 to 45% range appear achievable in HYFIRE using STARFIRE power recirculating parameters and He power requirements. Corresponding H_2 production efficiency (total fusion energy to the chemical energy of the hydrogen produced) is in the 50 to 55% range with the potential of reaching 60%.

6. CONCLUSIONS

Based on HYFIRE studies to date, the following observations are made: a) the MARK I blanket appears more attractive (simultaneously meets global tritium breeding requirements and required energy splits between process steam and helium, potential high-thermal efficiency); b) attractive tritium breeders such as $LiAlO_2$ and liquid lead with dissolved lithium have been identified; c) gross power cycle efficiencies in the 40 to 45% range appear achievable; and d) high H_2 production efficiencies in the 50 to 55% range appear achievable.

7. REFERENCES

[1] BAKER, C.C. et al., 8th Symposium on Engineering Problems of Fusion Research, San Francisco, 13-16 November 1979. Proceedings pp 1614-19.

[2] BAKER, C.C. et al., Argonne National Laboratory Report, to be published.

[3] POWELL, J.R. et al., 7th Symposium on Engineering Problems of Fusion Research, Knoxville, 25-28 October 1977. Proceedings pp 1732-35.

[4] HORN, F.L. et al., First Topical Meeting on Fusion Reactor Materials, Miami Beach, 29-31 January 1979. Proceedings pp 439-43.

[5] FILLO, J.A., Editor, Department of Energy, DOE Report HCP/T0016-01 (1978).

A HIGH FLUX ENERGY AND PARTICLE COLLECTION SCHEME FOR DIVERTORS

K. A. Werley, J. G. Gilligan, and G. H. Miley
Fusion Studies Laboratory, Nuclear Engineering Program
University of Illinois, Urbana, Illinois 61801 U.S.A.

ABSTRACT

A lithium rain approach is described for handling the large divertor ion and energy fluxes (10^{25} ions/m^{2}s and 10 MW/m^2) inherent to a power producing reactor. Previous concepts, which use a solid wall or gravity driven liquid metal flow, face serious feasibility questions dealing with thermo-mechanical wall material limitations or MHD retardation forces on a moving conducting fluid. The new scheme uses a rain of high speed (1-166 m/s) lithium droplets, formed by nozzles outside of the magnetic field, to collect diverted ions and their energy.

1. INTRODUCTION

Managing the high particle and energy fluxes associated with divertors of reactor grade plasmas is a difficult technological challenge. For example, the Oak Ridge National Laboratory (ORNL) bundle divertor requirements for The Next Step (TNS) reactor concept are 226 MW and 3.5×10^{23} ion/s.[1] A stationary, cooled, solid plate collector is limited by stress, erosion, embrittlement, fatigue and creep considerations to a surface load of about 0.5 MW/m^2. Also, collecting the particles with conventional vacuum pumps at $\sim 3 \times 10^{-6}$ torr requires a throughput of 3×10^9 ℓ/s compared to the largest present day pumps which handle 5×10^4 ℓ/s. Thus, the requirements can be quite severe.

Suggestions to alleviate these problems include protecting the wall with a liquid film and using hydride-forming surfaces (e.g., titanium, zirconium) to remove particles. Both of these ideas have drawbacks. Liquids evaporate easily, so a flowing design is needed in which a liquid element spends only a short time under a heat load. Also, once the pumping surfaces are saturated they must be replaced and regenerated at 800-1000°C for several hours. A suggestion which ostensibly combines the best features of the above two ideas while eliminating their drawbacks is that of a gravity-driven lithium flow down a screen (Westinghouse design) or flat plate (Wisconsin UWMAK-I design). Lithium acts as a getter material, forming lithium-hydride, and presumably would flow fast enough for heat removal. However, a moving conducting fluid in a magnetic field induces electric currents and a force resisting motion. As a result, Wells [1] has concluded that gravity-driven flow cannot move nearly fast enough to remove the heat in the presence of the MHD force. However, he suggests that by breaking up the lithium into small droplets, so there is no return path for the electrons to flow, MHD effects are greatly reduced. This paper examines this lithium droplet concept and points out some significant advantages over conventional collection schemes.

This scheme consists of guiding the diverted plasma beam into a region of lithium rain created by forcing lithium through nozzles (see fig. 1). The lithium rain absorbs the beam kinetic energy and simultaneously "getters" hydrogen isotopes. The droplets then fall into a collection tank located in a region of low magnetic field where MHD effects are minor and from there are directed to a heat exchanger, lithium hydride separator, and subsequently, back through the system.

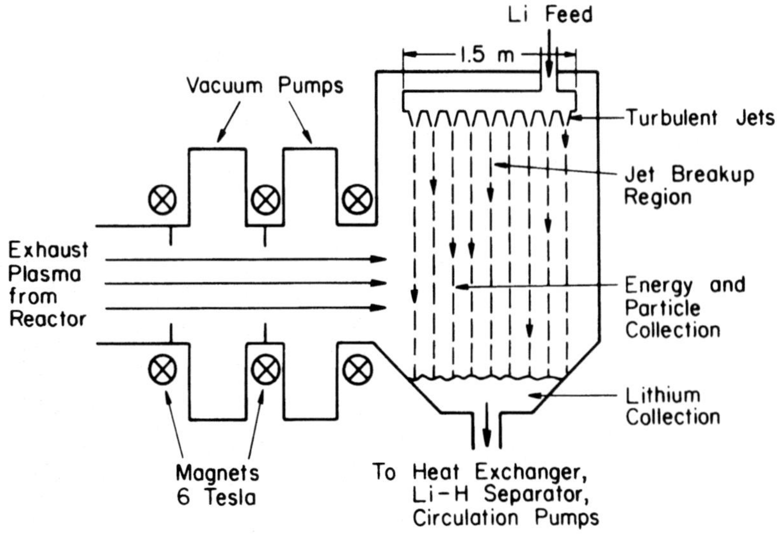

Fig. 1. Lithium rain energy-particle collection.

2. NOMENCLATURE

For clarity and length considerations we define the following notation.

b = breakup length	L = zone length	α = thermal diffus. $= k/\rho c_p$
B = magnetic field	t = dimen. time $= 4\alpha\tau/d^2$	η = resistivity
c_p = specific heat	v = droplet speed	ρ = density
d = droplet diameter	W = zone width	σ = surface tension
F = heat flux	We = Weber no. $= d\rho v^2/\sigma$	Σ = atten. cross sect.
H = zone height		τ = time
k = thermal conduct.		

3. DROPLET DESCRIPTION

Before discussing the overall system we examine the life of an individual droplet. We model the plasma-rain interaction region as a rectangular parallel piped with sides L, W, and H. WH equals the beam area and L is the distance an ion has to travel to cross the rain region. Assuming the droplets to be equally sized and evenly spaced, we can divide the interaction zone into tiny cells of height h and lateral spacing ℓ (see fig. 2).

3.1 Droplet motion

The first step in the lithium rain scheme is the formation of droplets in a region outside the magnetic field. This is accomplished by turbulent flow through a nozzle to form a liquid jet which breaks up into droplets. Chen and Davis [2] present photographs of how turbulent jets form a string of evenly spaced droplets with a size about equal to the nozzle bore. Phinney [3] gives an empirical formula which correlates the breakup length, b, to the Weber number.

$$b = d(55 + 1.085\ We^{1/2}) = d(55 + 40.2d^{1/2}v) \qquad (1)$$

In this and subsequent formulae the numerical constants are evaluated using mks units and liquid lithium (350°C) properties of ρ = 500 kg/m^3, c_p = 4190 joules/kg°K, k = 46 W/m°K, α = 2.2×10^{-5} m^2/s, σ = 0.365 Nt/m and η = 3.5×10^{-7} ohm-m.

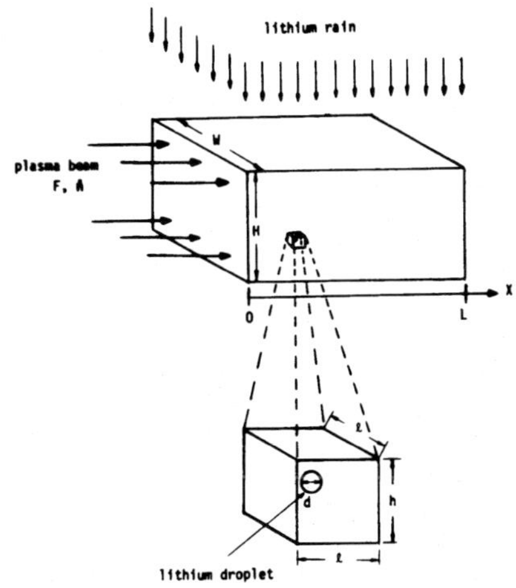

Fig. 2. Model of the beam-rain interaction zone and a volume element containing a single droplet.

The minimum acceptable droplet speed is determined by the time spent in the interaction zone. A droplet should neither overheat nor accelerate (due to gravity) and change its spacing. To hold the velocity change across the zone to less than 10% requires v^2/H be greater than 98.1 m/s^2. The maximum speed is set by an acceptable supply pressure which, following Wells [1], is taken as 6.89 MPa (1000 psi) which corresponds to a maximum speed of 166 m/s.

The droplet size is also limited. Once the droplet has formed it enters a region of increasing magnetic field which induces electric currents in the lithium causing droplet deformation, deceleration and lateral displacement. Walker [4] considers the deformation assuming the field to vary as half a sine wave. He estimates a maximum droplet size d_{max} in terms of the maximum field B and the length of the gradient region Z.

$$d_{max} = 8(2\eta\sigma Z/\pi B v)^{1/3}/3^{1/2} = 0.02\ (Z/vB)^{1/3} \tag{2}$$

Walker and Wells [5] show that the deceleration and displacement for the droplets are negligible for parameter ranges of interest. Thus, a stable spherical droplet can be formed over a distance between several meters and a few centimeters which moves at a speed between several and 166 meters per second. A droplet with a diameter less than a millimeter to a few centimeters will cross the field gradients and interaction zone and fall into the collector.

3.2 Droplet temperature and particle trapping

If we assume the field gradient to be small in the interaction zone, then the deformation will be small and the droplet can be approximated by a sphere. The heat conduction problem of a sphere in a directed heat flux (see fig. 3) has been solved.[6] The temperature evolution along a diameter of the sphere is described by fig. 4. The dimensionless time variable, t, is defined in the nomenclature, and t=0 corresponds to the time when the sphere of initial temperature, T_0, enters the heat flux. Eq. 3 defines the droplet average temperature. Eq. 4 gives the maximum temperature. The maximum time spent under the heat flux is H/v.

$$T_{AVG} = T_0 + 3F\tau/2\rho c_p vd = T_0 + 71.6\times10^{-6}\ FH/v^2d \tag{3}$$

$$T_{MAX} = T_{AVG} + f(1,0,t)\ Fd/2k \tag{4}$$

Note that for $t > 0.5$, f (1,0,t) equals 0.689.

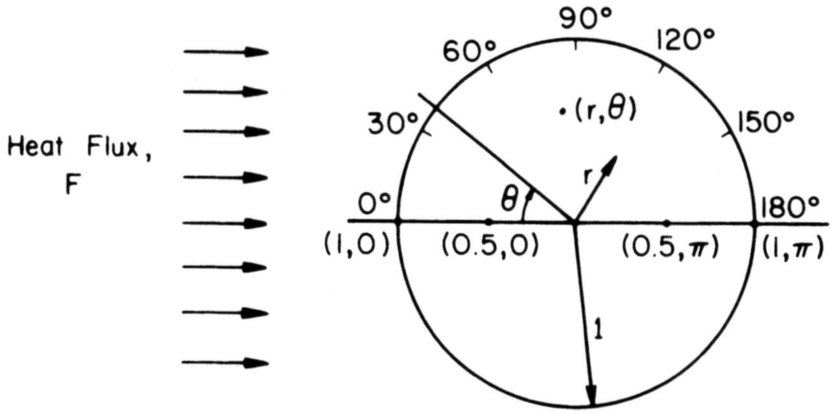

Fig. 3. Dimensionless geometry.

Erents, McCracken and Goldsmith [7] give experimental results for the trapping of deuterium ions in lithium. They report a 97% trapping efficiency for doses up to $5\text{x}10^{23}\text{m}^{-2}$ over a temperature range from the melting point (459°K) to 650°K. Above 700°K the efficiency falls rapidly towards zero. Thus, the lithium rain can potentially remove tremendous amounts of hydrogen. For instance, if a droplet spends a tenth of a second in the interaction zone, dose rates of $5\text{x}10^{24}\text{m}^{-2}\text{s}^{-1}$ can be removed. Also, the rain represents an extended lithium surface so that untrapped hydrogen, after the first encounter, still stand an excellent chance of being trapped. Since helium will not chemically bond to lithium, it is unknown whether the rain will remove helium. High energy helium ions may embed deep enough into a droplet that the droplet can be removed from the chamber before the helium can diffuse out. Since, in general, much more hydrogen enters the divertor than helium, even if the rain does not pump helium the conventional vacuum pump requirements will be greatly reduced.

The minimum allowed temperature is chosen to be 490°K (the melting point plus a 31° safety factor). Assuming that we want to pump hydrogen sets the maximum temperature to about 700°K. This corresponds to a lithium vapor pressure of $2.5\text{x}10^{-4}$ torr, which is higher than the reactor neutral gas pressure of 10^{-5}-10^{-6} torr. Thus, a differential pumping system is needed which may also be necessary for helium removal. If conventional pumps are used to remove all the particles, pressures as high as 10^{-1}-10^{-2} torr may be needed to handle the throughputs.

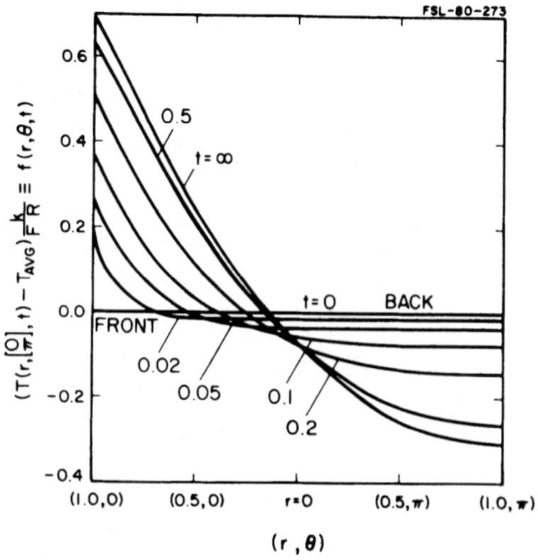

Fig. 4. Temperature profile in sphere (along the diameter parallel to the external heat flux).

4. SYSTEM DESCRIPTION

4.1 Droplet spacing and interaction zone length

From the photographs of ref. [6] we estimate the vertical droplet spacing to be $h = 1.5d$. The lateral spacing is set by how close the nozzles can be packed without the droplets colliding. We choose this distance to be $\ell = 4d$. Thus, the lithium volume fraction is 2.18% and the probability of the plasma beam to strike the first column of droplets is the area ratio, $a = \pi d^2/4h\ell = 0.131$. Since this is the probability of attenuation in a distance ℓ, the attenuation can be characterized by $e^{-\Sigma x}$ where the macroscopic cross section is $\Sigma = 0.0327/d$. So, to intercept 99% of the flux requires an interaction zone length of $L = 141d$, which will be between 0.1 and 3m.

4.2 Average outlet temperature

We have seen that the droplet's average temperature is given in eq (4), while the flux across the interaction zone drops as $e^{-\Sigma x}$. Thus if all the droplets move at the same speed then the droplets at the back of the zone (large x) will reach a lower temperature, so when collected, the lithium mixing cup temperature rise will be $\Delta T_{mc} = \Delta T_{AVG}/4.6$ for a 99% flux collection. However, if $v = v_0 e^{-\Sigma x}$, then the slower droplets spend more time in the zone so F/v remains constant and $\Delta T_{mc} = \Delta T_{AVG}$. This exponential velocity profile is quite advantageous as it permits a higher grade heat source for generating electricity and it also reduces the lithium flow rate by a factor of 4.6. The ORNL TNS bundle divertor design [1] has the flux striking opposite sides of the interaction zone, in that case $\Delta T_{mc} = \Delta T_{AVG}/2.3$.

4.3 Mass flow rate and recirculation power

The mass flow rate is given by $\dot{m} = \rho v A$, where the effective flow area is $A = LW\pi d^3/6\ell^2 h = \pi LW/144$. In this section, we consider pipe flow where v is the average fluid velocity. The potential energy loss rate which must be restored to raise the fluid is given by $\dot{m}gX$, where $g = 9.81 m/s^2$ and X is the height the lithium must be raised. Frictional pumping losses are given by $\Delta p A v$, $\Delta p = f' Y \rho v^2/2D_e$ where f' is the Darcy-Weisbach friction factor, D_e is the equivalent diameter and Y is the piping length.

5. SUMMARY

We used the descriptions and limitations of the previous sections to calculate and list, in table I, representative parameters for lithium rain collector designs. Case 1 utilizes the maximum droplet velocity and a small interaction zone height to maximize the heat and particle flow removed. Cases 1a and 1b are for flat velocity profiles and compare the effects of different droplet diameters while case 1c demonstrates the profit of using an exponential velocity profile. The lithium rain design handles a maximum heat flux of 21 MW/m^2 (higher if H is lowered below 1m), however, the design pays the price in that the mixing cup temperature rise is only a few degrees and frictional losses are in the 10 MW range. Using an exponential profile increases T_{mc} and decreases the potential energy losses by a factor of 4.6 and reduces frictional losses by two orders of magnitude. Case 2 presents 1.0 and 2.5 MW/m^2 flux designs. These designs are quite impressive as they allow outlet temperatures 100°K above present light water reactor coolant temperatures. Thus conversion efficiencies approaching 35% can be expected for large systems. Also, coolant circulation powers are small. Small diameter droplets are preferred since they minimize the temperature difference across a droplet and maximize the average temperature rise. However, it is not clear exactly what the lower size limit is. Case 3 is a 10 MW/m^2 design which is about the largest flux which permits acceptable temperatures for thermal-to-electrical conversion systems.

TABLE I

Lithium Rain Collector Parameters

	Case 1			Case 2		Case 3	
	a	b	c	a	b	a	b
Droplet velocity, m/s	166	166	166	13.5	22.8	90	90
Velocity profile	flat	flat	expon.	expon.	expon.	flat	expon.
Droplet diameter, m	.0086	0.0036	0.0036	0.002	0.002	0.002	0.002
Interaction zone height, m	1	1	1	1	1	1.8	1.8
Heat flux, MW/m^2	21.7	21.4	21.4	1.0	2.5	10.0	10.0
Ion flux, 10^{24}ions/m^2s	83	83	83	6	11	45	45
Outlet temperature, °K	491.4	4903.3	505.3	686.0	662.0	507.0	570.0
Breakup length, m	5.81	1.7	1.7	0.16	0.19	0.43	0.43
Pot. energy loss, kW	235	61	13	0.5	0.8	16.0	3.7
Frictional loss, kW	14,900	10,700	110	0.09	0.40	1,740	18
Mass flow rate, kg/sec	2210	932	200	9	15	227	60

For the above cases, T_{in} = 490°K, T_{MAX} = 700°K, and zone width W = 1m.

In conclusion, the rain represents an extended free lithium surface which can getter large throughputs of hydrogen ions ($5x10^{23}$ ions/m^2) and can simultaneously remove surface heat loads in excess of 10 MW/m^2 (over a magnitude larger than a solid wall). Other features of the rain include (1) the elimination of hot spot, stress, erosion, fatigue and hydrogen embrittlement problems, and (2) the reduction of MHD effects and coolant circulation powers. Any one of these features would represent a marked improvement to most designs. The fact that they occur simultaneously makes the rain scheme quite remarkable. Problems and limitations include (1) safety requirements for handling liquid metals, (2) temperature limitations by vapor pressures and hydrogen trapping, (3) the need for experimental data on helium pumping ability, and (4) the feasibility of lithium-hydride separation.

This approach leads to a compact and simple design which can be applied to widely varying conditions by adjusting the droplet speed, size and spacing. It appears to have the best technical feasibility of any scheme proposed to date, and while more data and a "proof of principle" experiment is needed, the concept is near term in the sense that such an experiment could be built today.

6. REFERENCES

[1] WELLS, W. M., Oak Ridge National Lab. Report ORNL/TM-6727 (1979).

[2] CHEN, T and DAVIS, J. R., ASCE J. Hyd. Div. 90, 182-3 (1964).

[3] PHINNEY, R. E., AIChE J. 21, 996-999 (1975).

[4] WALKER, J. S., "Deformation and Break-Up of a Liquid-Metal Drop Moving Across a Non-Uniform Magnetic Field," Univ. of Ill. Dept. of Theoretical and Applied Mechanics.

[5] WALKER, J. S. and WELLS, W M., Oak Ridge National Lab. Report ORNL/TM-6976 (1979).

[6] WERLEY, K. A. and GILLIGAN, J. G., Univ. of Ill. Fusion Studies Lab. Report FSL-12 (1980).

[7] ERENTS, S. K., McCRACKEN, G. M., and GOLDSMITH, P. J., Phys. D: Appl. Phys. 4 (1971).

[8] LIPOV, M. I.V. and MURAV'EV, E. V., The I. B. Kurchatov Order of Lenin Institute of Atomic Energy Report OLS-79-442 (1979).

A "COMPACT" REVERSED-FIELD PINCH (CRFP) REACTOR DESIGN

George H. Miley and Richard A. Nebel
Fusion Studies Laboratory
Nuclear Engineering Program
University of Illinois
Urbana, Illinois 61801

1. INTRODUCTION

Many fusion reactor design studies have resulted in large and complex nuclear "islands", making maintenance difficult, integration into smaller electrical networks impractical, and posing financial problems due to the large capital costs involved. Consequently, a new class of concepts, termed compact tori (CT), have evolved that could potentially lead to smaller, less complex systems[1]. These concepts include Field Reversed Mirrors, Field Reversed Theta Pinches, and Spheromaks. They all share, to a varying extent, the common features of a closed toroidal configuration and a high-β field that is largely created by internal plasma currents. The Reversed Field Pinch (RFP) has a magnetic configuration that is similar in many respects to these CT. Consequently it too can be viewed as a candidate for a "compact" system. While a relatively large major radius (and large aspect ratio) will be retained, the plasma chamber itself and the surrounding structure can enjoy a "compactness" and still provide easy access.

2. PRIOR RFP CONCEPTUAL REACTOR CHARACTERISTICS

RFP conceptual reactor studies have been carried out at both the Culham Laboratory and the Los Alamos Scientific Laboratory[2-5]. These reactor concepts typically have a major radius of ~ 12-15m and a minor radius of ~ 1.5-1.75m. Thermal power levels are of the order of 2000-3000 MW. As such, these reactors are relatively large. This does not infer that they may not ultimately be attractive, however. Indeed, the RFP has a series of notable features that provide important advantages. Thus, the ability to heat to ignition with ohmic heating results both in a cost savings and simplicity compared to approaches requiring auxiliary heating via neutral beams or RF. The relatively high β provides a good power density and offers the possibility of eventually extending the concept to advanced fuels like catalyzed-deuterium. The lack of an aspect ratio constraint on stability allows a large aspect ratio design which in turn, by virtue of the open central region, simplifies maintenance and allows modular construction.

The goal set for the present study then was to examine the feasibility of a smaller RFP reactor in the 100 MWe range that would still retain many of the advantages cited above for the larger units. Such a reactor, labeled a Compact RFP (CRFP), would provide a flexibility in size that may be essential for early exploration of fusion power, by providing a route to a lower capital cost device suitable for a near term demonstration plant.

3. CRFP STUDIES

Scaling down the dimensions of the large RFP designs by roughly a factor of two provides a plasma volume compatible with the goal of a 100 MWe output. In order to examine the transport and stability properties of an RFP burn in a plasma of this size, we employed the 1-D dynamic burn code developed earlier for RFP studies and described elsewhere[6]. This code includes both global and local stability checks. Violation of the former results in burn termination while local violations are assumed to result in turbulent transport.

Continuity balances and alpha heating are treated in a conventional manner. Results from these calculations are presented in table I along with comparison parameters for the LASL RFP reactor (RFPR) reference design. It is observed that, while the volume of the CRFP is decreased by a factor of 8 compared to the RFPR, its net electrical output is only reduced by a factor of 7.5. Consequently, the CRFP enjoys a slight advantage in power density although its Q_p-value has dropped from 14.4 to below 10, indicating an undesirable increase in recirculating energy. The output power of 100 MWe satisfies the original goal of a "small" reactor, however. Indeed, a further decrease in size causes a rapid decrease in Q_p due to radial energy losses via thermal conduction. (As in the RFPR, an energy confinement time of $\tau_E \sim 200\ \tau_{Bohm}$ was employed.) Indeed, to avoid excessive losses in the CRFP, it was necessary to increase magnetic field by a factor of 1.2 in order to increase the density.

Table I

Comparison of CRFP Parameters with RFPR

PARAMETER	RFPR	CRFP
First wall radius, r_w(m)	1.5	0.75
Major radius, R(m)	12.7	6.35
Burn time, τ_B(s)	21.6	10.3
Cycle time, τ_C(s)	26.6	15.3
Toroidal field on centerline, $B_{\phi o}$(T)	6.93	8.5
Toroidal current, I_ϕ(MA)	20.0	12.0
Average plasma density, n(1/m^3)	$2.0(10)^{20}$	$3.8(10)^{20}$
14.1 MeV fusion neutron energy, W_N(MJ/m)	676.5	173.0
Total thermal power, P_{TH}(MWt)	3000	628
Net electrical power, P_E(MWe)	750	100
Q_p = (fusion energy)/(ohmic & field energy)	14.4	9.7

The CRFP burn time was shorter than for the RFPR as indicated by the respective burn-up fractions. Gross MHD stability was maintained throughout the burn, but Suydam's criterion was violated on centerline, leading to a turbulent core region that results in enhanced transport. However, as found in the RFPR[6], the turbulence does not result in a major deterioration in the energy balance since the outer radius zone remains quiescent.

An indication of the relative size of the CRFP can be gained from fig. 1. Its relative location between the RFPR and Zeta reinforces the earlier suggestion that a device of this type would be potentially attractive as an early demonstration device.

4. PROBLEMS AND IMPROVEMENTS

Due to the capital intensive character of fusion devices, to be competitive economically, any reactor in the 100 MWe range must, of necessity, offer high performace and modular construction (i.e., mass production of key components). It appears that the same

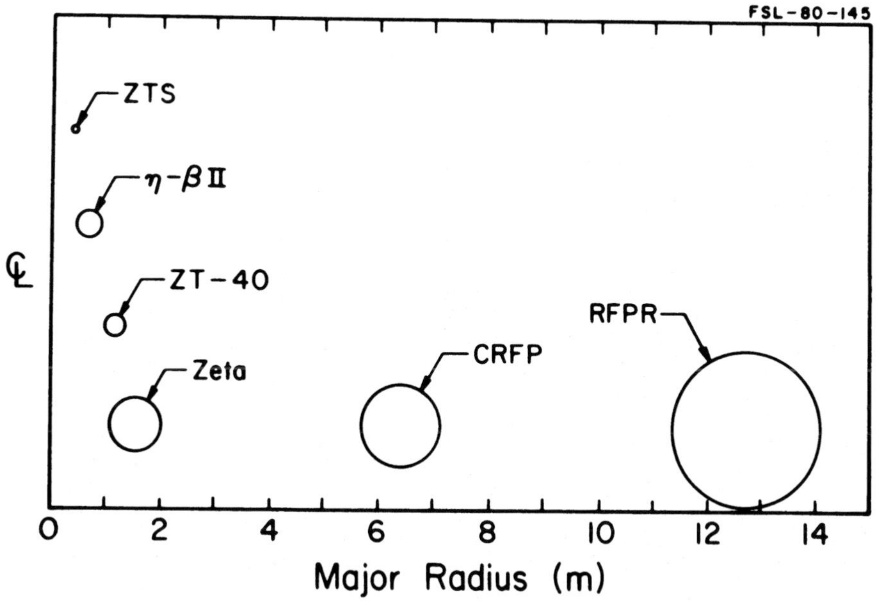

Fig. 1. Relative sizes of RFP experiments and reactors.

design considerations employed in the RFPR can lead to satisfaction of the latter requirements. On the other hand, the low energy multiplication (Q_p) and associated large recirculation power requirement suggests an undesirable expense. Previous work[7] has indicated that significant improvements in Q_p might be obtained by refueling. Since batch burns (non-refueled) are extinguished by fuel depletion, refueling the plasma increases the burn time, the amount of fusion energy produced per pulse, and consequently Q_p. However, refueling introduces new problems of fuel penetration and stability. Since the degree of anomalous mixing between the refuel ions and the plasma is unknown, two extremes, representing complete mixing (gas refueling) and no mixing (pellet injection with in-situ particle deposition), are considered for the CRFP.

4.1 Gas Refueling

Tokamak experiments[8] indicate that anomalous mixing between the cold refuel gas and the plasma occurs on a fast (μsec) time scale. Consequently, for the gas refueling simulations the refuel ions were assumed to immediately redistribute themselves along the plasma density profile. This is done by introducing a fuel source rate of the form:

$$S(r) = S_0 \, J_0(\alpha r), \qquad \alpha = 2.405/r_w. \tag{1}$$

Results indicate that the burn time is approximately doubled over the batch burn CRFP and Q_p is increased by about 75% (see table II, Case A). Once again, gross MHD stability is maintained through the entire burn.

4.2 Pellet Refueling

If the refuel ions do not anomalously mix with the background plasma, it is necessary to inject the refuel ions in order to refuel the interior of the plasma. Pellet injection appears to be the most energy economic manner to do this. For these simulations, the ablation model of Parks, et al.[9] was used to determine the refuel deposition profile. The refuel ions and electrons were assumed to immediately equilibrate with the background where they are deposited.

Table II

Summary Description of Design Parameters

PARAMETER	GAS REFUELED CRFP (CASE A)	PELLET REFUELED CRFP CRFP (Case B)
First wall radius, r_w(m)	0.75	0.75
Major radius, R(m)	6.35	6.35
Burn time, τ_B(s)	21.0	18.1
Cycle time τ_C(s)	26.0	23.1
Toroidal field on centerline, $B_{\phi o}$(T)	8.4	8.4
Toroidal current, I_ϕ(MA)	13.2	13.2
Average plasma density, N($1/m^3$)	$2.7(10^{20})$	$2.7(10)^{20}$
14.1 MeV fusion neutron energy, W_N(MJ/m)	333.0	256.5
Total thermal power, P_{TH}(MWt)	820	706.0
Net electrical power, P_E(MWe)	240	160.0
Q_p=(fusion energy)/(ohmic & field energy)	17.2	13.2
Pellet radius (cm)	-	0.195
Pellet velocity (m/sec)	-	$6.2(10^4)$
Pellet injection rate (sec^{-1})	-	0.5
Effective refuel rate $(m\text{-}sec)^{-1}$	$3.1(10^{19})$	$2.8(10^{19})$

Results indicate that the pellet-refueled CRFP mode performs better than the batch-burn mode but not as well as the gas-refueled mode. The difference is principally due to the pellets not reaching the plasma axis where the temperature is highest. Pellet refueling increased Q_p by about 30% over the batch burn mode (see table II, Case B).

5. CONCLUSION

These results indicate that a CRFP operating in the range of 100 MWe should be feasible. While the large aspect ratio leads to a 6.4-m major radius, the design still retains many attributes of conventional compact tori, such as simplicity and ease of access. The relatively low Q_p with batch burns might be satisfactory for a demonstration unit. However, to obtain power-plant "grade" Q_p, either refueling (possibly by pellets) or larger size plasmas appear necessary.

6. REFERENCES

[1] MILEY, G. H., "Theoretical Considerations and Applications of Compact Torus Concepts," Japanese-US Exchange Workshop on Compact Tori, Princeton University, Dec. 1979.

[2] LAWSON, J. G., "Reversed Field Pinch Reactor Study, I. Physical Principles," CLM-R 171 (1977).

[3] HANCOX, R. and SPEARS, W. R., "Reversed Field Pinch Reactor Study, II. Choice of Parameters," CLM-R 172 (1977).

[4] HOLLIS, A. A. and MITCHELL, J. T. D., "Reversed Field Pinch Reactor Study, III. Preliminary Engineering Design," CLM-R 173 (1977).

[5] HAGENSON, R. L., KRAKOWSKI, R. A., and CORT, G. E., "The Reversed-Field Pinch Reactor," Los Alamos report LA-7973-MS (1979).

[6] NEBEL, R. A., et al., "Comparison of Zero-Dimensional and One-Dimensional Thermonuclear Burn Computations for the Reversed-Field Pinch Reactor (RFPR)," Los Alamos report LA-8185-MS (1980).

[7] NEBEL, R. and MILEY, G. H., "Refueling and Control of RFP Burns," Third ANS Topical Meeting on the Technology of Controlled Nuclear Fusion, Santa Fe, NM (1978) pp. 15-18.

[8] APGAR, E., et al., in Plasma Physics and Controlled Nuclear Fusion Research (Proc. 6th Int. Conf., Berchtesgaden 1976) 1, 274 (1977).

[9] PARKS, P. B. and TURNBULL, R. J., Phys. Fluids, 21, 1735 (1978).

CONCEPTUAL STUDIES
ON HIGH GAIN LASER FUSION REACTOR
WITH MAGNETICALLY GUIDED L_I FLOW

S. Ido, K. Imasaki, Y. Izawa, Y. Kato, Y. Kitagawa,
M. Matoba, K. Mima, T. Mochizuki, S. Nakai, K. Nishihara,
H. Nishimura, T. Norimatsu, T. Sasaki, T. Yabe, T. Yamanaka
and C. Yamanaka

Institute of Laser Engineering, Osaka University,
Yamada-Kami, Suita, Osaka, 565, JAPAN

ABSTRACT

A concept of a D-T fusion laser reactor (SENRI-I) with magnetically guided inner Li flow is presented. It is assumed that input laser energy is 1∿5 MJ and pellet gain is 1000∿200. The scaling law of pellet gain is examined. The inner Li flow is used as a first wall protector, a coolant and a tritium breeder. It is found that flow shape and velocity are well controlled and the disassembling is suppressed by magnetic field pressure. One dimensional neutron transport calculations show that tritium breeding ratio is ∿ 1.6.

1. INTRODUCTION

This paper presents a design of a D-T target laser fusion reactor[1.2.3.4.] to clarify technical and physical problems on the way to a commercial fusion reactor. We investigate an original design of a D-T target ICF reactor driven by the multi-MJ laser system which utilizes thick Li flow (64 cm) inside a spherical cavity to moderate the first wall loading and neutron fluence. The Li flow is also used as an 'inner blanket' and it is guided by the steady magnetic field. Overview of a reactor cavity and a blanket is shown in fig. 1, and parameters are shown in table 1. Pellet gain optimized for burning processes is presented. Neutronics in a core and a blanket is also presented.

2. BURNING PLASMA ——— SCALING LAW OF PELLET GAIN

The pellet gain Q (= fusion power output E_f/input laser energy E_L) should be greater than 100 to achieve a system efficiency η_s of 30% for a power reactor as discussed in §3. Figure 2 shows that the pellet gain Q becomes of the order of 100∿1000 for a laser of multi-megajoules[4,5], where $Q = E_f / E_L = \varepsilon_{Lp} g_f$, and $\varepsilon_{Lp} = \varepsilon_a \varepsilon_c$ is the coupling efficiency between the laser output energy and D-T fuel, $g_f = E_f/E_p$ is the fuel gain, $\varepsilon_a = E_a/E_L$ is the absorption efficiency and $\varepsilon_c = E_p/E_a$ is the hydrodynamic efficiency, where E_a, E_p are absorbed energy and plasma energy. It is assumed that the ignitor plasma is required to be hot enough (T_I>5keV) and large enough ($\rho_I R_I$>0.8gr/cm^2) to ignite the thermonuclear reaction wave[7,8] and ε_{Lp} is 0.07, 0.05 and 0.03 respectively. The scaling law obtained from fig. 2 can be written as

$$Q \simeq 10^4 \, E_L^{3/8} \, \varepsilon_{Lp}^{11/8} \qquad (1)$$

where E_L is in MJ. It is clear that required laser energy for over 1000MJ thermal energy generation is 1.1MJ for a optimistic case (ε_{Lp}=0.07) and 5MJ for a pessimistic case (ε_{Lp}=0.03). Q is 1000 for ε_{Lp}=0.07, E_L=1.1MJ and it is 230 for ε_{Lp}=0.03, E_L=5MJ. Exact value for ε_{Lp} is not determined experimentally so far. It must be investigated in breakeven experiments which will be performed by laser facilities with $E_L \gtrsim 100$kJ, such as NOVA(LLL), ANTARES(LASL) and KONGOH PROJECT (OSAKA).

3. ENERGY BALANCE OF THE TOTAL SYSTEM

A system efficiency η_s of a laser fusion reactor is defined to be a ratio of electric output energy to total thermal output energy including thermonuclear fusion and heat generated in a blanket. The system efficiency is required to be larger than 30%, since the waste energy may cause thermal polution and reduce energy economy.

The gross electric output power E_g is estimated to be

$$E_g = \bar{\eta}\,\{\,(M_bQ + 1)E_L + \gamma(\eta_L^{-1} - 1)E_L\,\}\ , \qquad (2)$$

for a given Q, where $\bar{\eta}$ is the conversion efficiency from thermal to electric energy, M_b is the blanket gain, η_L is the laser efficiency, and γ is the recovery efficiency of waste thermal energy. On the other hand, the electrical power recycled to maintain the laser operation (E_L/η_L) and to evacuate a reactor cavity ($E_v=\alpha_pE_f$), is assumed to be given by

$$\varepsilon E_g = E_L/\eta_L + \alpha_p(M_bQ + 1)\,E_L\ , \qquad (3)$$

where ε is the recirculating power fraction. The system efficiency η_s is then determined by $\eta_s = E_g(1-\varepsilon)/(M_bE_f)$.

When we set $\bar{\eta}$=40%, γ=0, α_p=5%, M_b=1.2 (a pure fusion system) as standard values, the system efficiency η_s is shown by solid lines in fig.3. If the laser efficiency η_L is 10%, the practical power plant is able to be designed for the pellet gain greater than 200. For a fission-fusion hybrid reactor, the blanket gain, M_b is $\sim$4(^{232}Th) and $\sim$18 (Natural Uranium). When natural uranium is used for a blanket, the necessary pellet gain is about 20 for η_L= 5% as shown by a broken line in fig.3.

4. INNER BLANKET (Li FLOW) DESIGN

4.1 Overview

A concept of magnetically guided Li flow reactor is shown in fig. 1. The inner wall of SUS is covered by a thick Li layer which flows down from top to bottom in a reactor cavity. SUS-316 is used as structural materials because its material characteristics is well known and reliable. On the other hand, the temperature is kept to be low ($\leq$500°C)[6] because of compatibility between SUS and Li. Thick Li flow is guided by static magnetic field. The magnetic field stabilizes fluctuations of the Li-flow and decelerates the flow throuth the Lorentz force $J\times B/c$. The flow velocity and thickness can be then controlled. We put four types of coils surrounding the reactor cavity as shown in fig. 1. The design of coil shape, current strength and coil positions are investigated in another article. The total thickness of inner and outer Li layer is designed to be 70 cm to convert the neutron energy to thermal energy sufficiently and to breed sufficient tritium. This ensures a low level of neutron fluence to a structural wall and a long life of the reactor cavity. The magnetic field is also useful to protect the Li layer from α-particles and other charged particles and to weaken the heat pulse generated in a very thin surface (∿10 μm^t) of the Li layer. Only X-ray heats the surface layer up to much more than the boiling temperature. Although Li evaporates from the surface, the surface temperature is cooled down quickly by the evaporation and the shock wave formation. The Li surface may then act as a cryopump.

4.2 Energy Deposition and Structural Problems in Li Layer

Neutrons and X-rays generated by the microexplosion deposit their energy in the Li layer. Li layer pressure is about 10 MPa for neutron heating and 10 GPa for X-ray heating. These pressure pulses generate hydrodynamic motions and disassembling of the Li flow. The disassembling and shock waves may destruct the Li layer and give damages on the reactor first wall[9,10].

In fig. 4, we show results of simulations on the hydrodynamic motion of the Li layer due to neutron bulk heating, where an inner surface of Li flow is at (a) R=2.8m (f_n=8.2MW/m^2) and at (b) R= 5.0m (f_n=2.55MW/m^2), when inner radius of SUS first wall is 5.64m. The profile of neutron energy deposition is presented by neutron transport calculations as shown in fig. 7. We see that the magnetic field of several KG decelerates the two phase Li layer expansion and disassembling is suppressed. It is obvious that Li is pushed back by magnetic field and that charateristic oscillation time is about Alfvén transit time. We conclude that the magnetic field may protect the Li layer from plasma and α-particle and prevent the disassembling.

5. NEUTRONICS IN A PELLET AND A BLANKET

5.1 Neutronics in a high density D-T plasma

In high pellet gain fusion, fuel is compressed to ∿ 1000ρ_0 where ρ_0=0.2g/cc is solid D-T density, and ρR becomes 1∿6g/cm^2. For these values of ρR, elastic and inelastic collisions of neutrons with D and T ions cannot be neglected. Figure 5 shows the neutron energy spectrum for various values of ρR, which is calculated by the one dimensional neutron transport code 'ANISN'[11,12,13]. In a typical implosion with burning, ρR is ∿3g/cm^2 in our burning calculations for Q∿100∿400. It is clear in fig. 5 that the softening of neutron spectrum is evident for $\rho R \geq 1$. Number of 14 MeV neutrons without scattering or only with self-scattering is 0.908, 0.680 and 0.494 per source neutron for ρR=0.7, 3.0 and 6.0g/cm^2 respectively, where it is assumed that density distribution of DT fuel is uniform. Fusion neutrons not only deposit their energy to plasma by scattering but also reproduce neutrons by D(n, 2n)H reations. Number of neutron per D-T reaction is 1.011, 1.043 and 1.061 for ρR=0.7, 3.0 and 6.0g/cm^2.

5.2 Neutronics in a blanket

Neutron transport and tritium breeding ratio are calculated by the 1-dimensional neutron transport code 'ANISN'. The structure of the calculated blanket in a spherical geometry is shown in fig. 6. Most energy from a microexplosion is deposited in an inner Li blanket (64 cm). An outer Li blanket (6 cm) cools structural materials (SUS-316). There is a graphite layer (20 cm thickness) as a refletor which is cooled by a gas coolant. More than 80% of total energy is deposited in an inner Li blanket. Deposited energy is larger than injected neutron energy, where $\Delta E \simeq [\ {}^{6}Li(n,\alpha)T$, heating] - [${}^{7}Li(n, n'+\alpha)T$, cooling] $\simeq$ 4 MeV. The profile of temperature increase per 1000 MJ microexplosion in a core is shown in fig. 7, where number of D-T reactions is 3.55×10^{20}. A solid line and a broken line correspond to 14 MeV monochromatic neutrons and 1.043 $\times(\rho R=3$ softened neutrons) respectively. The 64 cm thick inner Li blanket is heated up to $\sim1\sim3$ deg for a 1000MJ microexplosion and sufficiently reduces thermal loading of structural materials. The first wall (SUS-316) is heated up to ~1.5 deg/(1000MJ microexplosion) because specific heat ratio of SUS-316 is much smaller, where SUS-316 is heated mainly by γ-ray.

Tritium breeding ratio is 1.678, 1.596 and 1.492 for 14.1 MeV, $\rho R=3$ and $\rho R=6$ softened neutron respectively. In each case it is expected that sufficient tritium breeding ratio is obtained in the blanket shown in fig. 6.

6. TRITIUM AND LITHIUM CYCLE

A concept of cooling system consists of 3 cycles through the heat exchangers. In the first cycle, liquid Li flow cools the reactor through in the inner blanket and the outer blanket. As an intermadiate cooling cycle (the secondary system), liquid Na flow is used. In the third cycle, steam is produced in the steam generators and it goes into the turbines.

Tritium recovery system is under examination. Because 1000MJ microexplosion/sec corresponds to 3.55×10^{20} DT reactions/sec and typical burn ratio Y is ~0.5, tritium consumption rate is 3.42 mg/sec, and 34.4 Ci/sec. As described in §5, tritium breeding ratio is ~1.6. Therefore 27.5 Ci of T is bred in liquid lithium and 17.7 Ci of unburned tritium is absorbed in an inner Li flow. This fact implies that activation rate of the Li flow blanket is 44.7 Ci/sec and consumption rate of natural Li is 1.4 ton/year.

7. CONCLUSION

Our studies show that the value of Q can attain $100\sim1000$ by several megajoules lasers and scaling law is $Q\sim(100\sim1000)E_L^{3/8}$ (E_L in MJ) depending on ε_{Lp} for optimized ρR. In our concept, thick Li flow inside a spherical reactor container is used. Thermal, neutronic and mechanical analysis shows that inner Li flow sufficiently protects the SUS spherical vessel. It is shown by 1-dimensional neutron transport calculations that tritium breeding ratio in $64cm^t$ inner Li flow and $6cm^t$ outer Li flow is ~1.6, depending on the structure of the blanket and neutron spectrum. Detail designs are under examination.

R E F E R E N C E S

[1] Annual Progress Report on Laser Fusion Program, p.193, Institute of Laser Engineering, Osaka University, ILE-APR-78 (1978).

[2] Report of System Design of ICF Reactor, Institute of Laser Engineering, Osaka University (1979), in Japanese.

[3] YAMANAKA, C., et al., 'Conceptual Design of 1240 MW_{th} Laser Fusion Reactor in ILE OSAKA', Research Report of ILE, ILE-7913P (1979).

[4] IDO, S., et al., 8th Int. Conf. on Plasma Phys. and Contr. Nucl. Fusion Res., IAEA-CN-38/E-3 (1980).

[5] KIDDER, R. E., Nucle. Fusion 16, 405(1976).

[6] FURUKAWA, K., private communication.

[7] ASANO, N., NISHIHARA, K., NOZAKI, K. and TANIUTI, T., J. Phys. Soc. Japan 41, 1774(1976).

[8] NOZAKI, K. and NISHIHARA, K., J. Phys. Soc. Japan 43, 1393 (1977).

[9] GLENN, L. A., and YOUNG, D. A., Nucle. Engineering and Design, 54(1979) 1.

[10] GREEN, L. A., Preprint UCRL-83051.

[11] HATTORI, K. and KONDO, S., private communication.

[12] SEKI, Y. and IIDA, H., JAERI-M 8818 (1980).

[13] BERANEK, F., CONN, R. W., Nucle. Technol., 47, 406(1980).

Table 1 Parameters of Laser Fusion Reactor Concept

(A)	Laser	Gas phase (CO_2 etc.)	
		Number of beams	8
		Output energy	1 ∿ 5 MJ
		Shot cycle	1 Hz
		Peak power density	200 TW
		Pulse duration	100 nsec
		Efficiency	10 %
(B)	Pellet	Multi-shell	
		Fuel DT	
		Radius	∿ 3 mm
(C)	Output power		1240 MW (thermal)
			496 MW (gross electric)
			426 MW (net electric)
(D)	Reactor vessel with magnetically guided inner Li flow		
		SUS container	R = 5.64 m (spherical)
		Cavity	R ∿ 5.00 m
		Inner Li flow	d ∿ 0.64 m

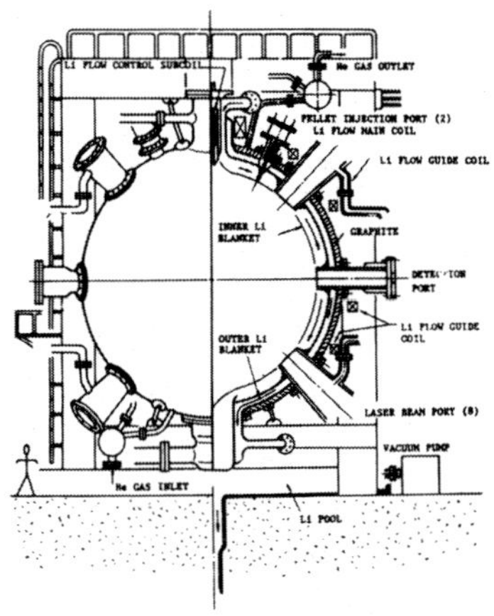

Fig. 1 Overview of SENRI-I, a magnetically guided Li flow laser fusion reactor.

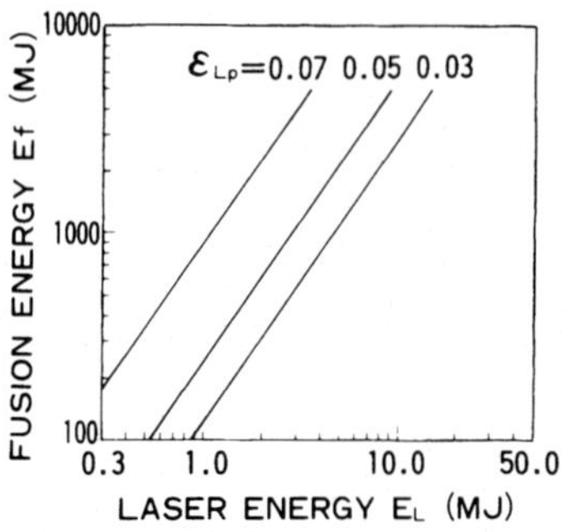

Fig. 2 Scaling law of pellet gain Q vs. laser energy E_L. ε_{Lp} is the corona-core coupling efficiency.

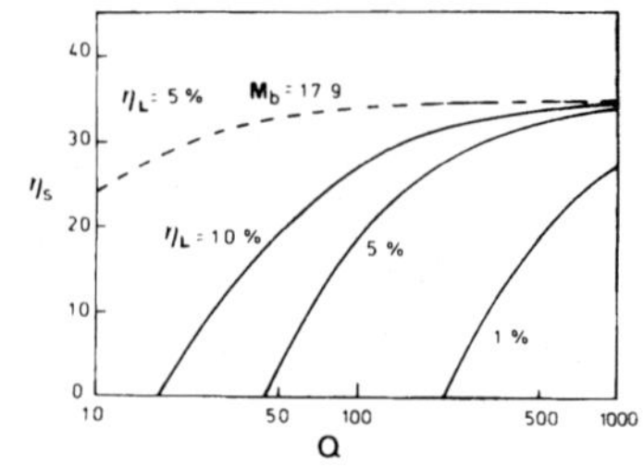

Fig. 3 System efficiency η_s vs. pellet gain Q for various values of η_L.

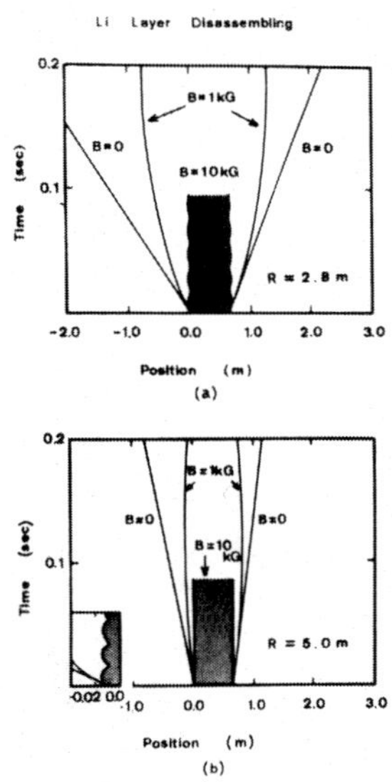

Fig. 4 Space-time diagram of Li layer surface motion. Three lines are for B=0, B=1KG and B=10KG. Inner radius of Li is (a)·2.8m and (b) 5.0m.

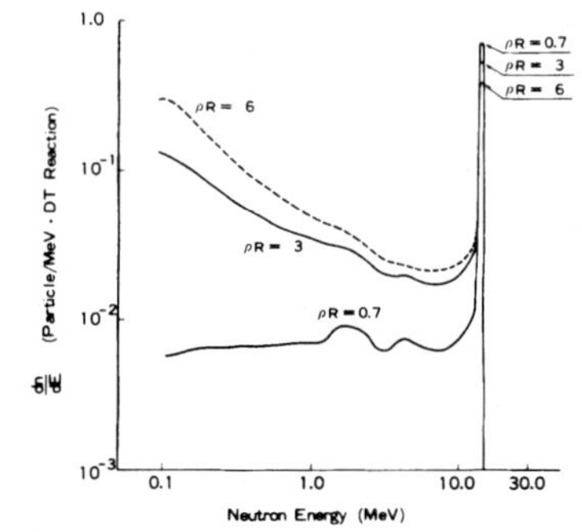

Fig. 5 Neutron distributions scattered in a core plasma for ρR = 0.7, 3.0 and 6.0 g/cm^2.

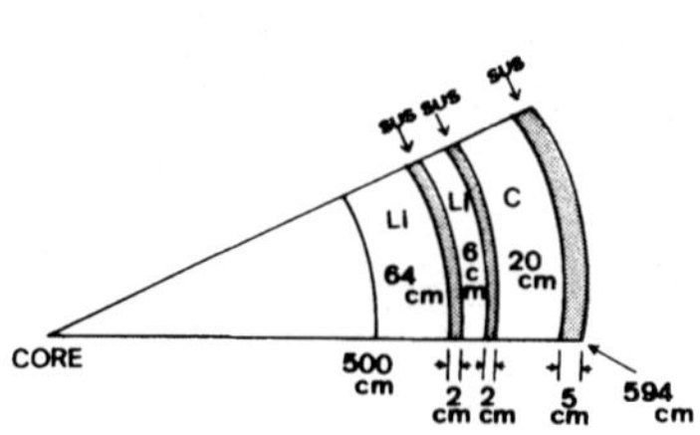

Fig. 6 Structure of a blanket.

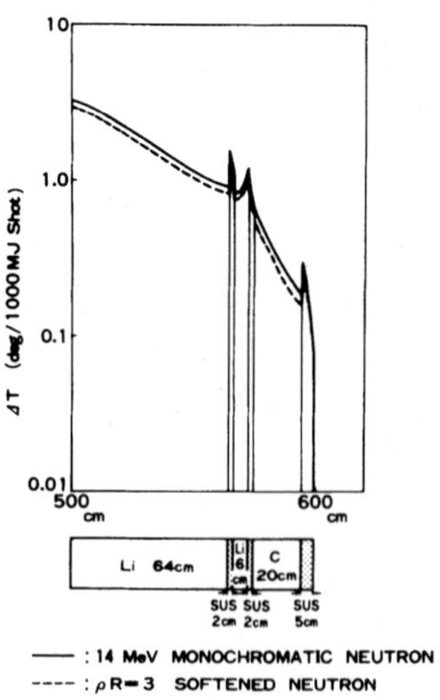

Fig. 7 Temperature increase of a blanket for a 1000 MJ microexplosion.

PRELIMINARY STUDIES OF THE SYSTEMS INTEGRATION OF NET-INTOR

F Farfaletti-Casali, G P Casini, M Rieger, U Buzzi and P Gritzmann

Commission of the European Communities
Joint Research Centre - Ispra Establishment
I-21020 Ispra (Va), Italy

A van Melle

T.N.O., 7334 DT Apeldoorn-Zuid, The Netherlands

ABSTRACT

The paper presents the first preliminary studies carried out at the JRC-Ispra in the field of mechanical configuration and systems integration of an Experimental Power Tokamak-type Reactor, to be built after JET, with emphasis on assembly and disassembly problems.

Two assembly and disassembly schemes have been studied:

- in the first, the handling of the internal renewable components is performed through access ports located on the top of the Reactor,
- in the second, the handling of these components is performed horizontally through access ports in the mid-plane of the Reactor.

It is shown that, with these two mechanical configurations, it is possible to satisfy the requirements of a machine of the NET-INTOR generation, in the case of a reactor concept without divertor.

1. INTRODUCTION

This paper presents the first preliminary studies carried out at the JRC-Ispra, in the field of mechanical configuràtion and of systems integration of the nuclear island of an Experimental Power Tokamak-type Reactor of the next generation, to be built after JET. Such a reactor concept, as it will be presented in this paper, coincides with that of NET (the Next European Torus, according to the definition of the European Community) and corresponds to that of INTOR (the International Tokamak Reactor, as it is being defined at the IAEA workshops for the preliminary definition of the machine).

For some basic features of the reactor, reference has been made to the main INTOR guideline parameters (major radius R = 5.2 m; minor radius a = 1.3 m; elongation factor 1.6; thermal power about 620 MW; 12 toroidal field coils, etc.).

According to the European point of view that such a device should be as simple as possible and should only rely on proven techniques, the basic studies presented here contemplate only the major, already defined, sub-systems and not those sub-systems, such as the divertor, for which there is, for the moment, insufficient information to define them.

Because of the impossibility to choose, in this preliminary phase, the best mechanical configuration, due to the lack of some judgement elements which can be determined only after the insertion of some not well defined sub-systems, two alternative solutions - A1 and A2 - were studied, in the most flexible way, to cover, as far as possible, all the general layouts and all the basic schemes of the machine.

Four schemes have been identified for the general arrangement of a machine like NET-INTOR; 2 schemes for the so-called A solutions and 2 schemes for the so-called B solutions (see fig. 1). The A solutions can avoid the opening of the torus during all routine scheduled interventions, while the B solutions require the opening of the torus for some scheduled maintenance interventions.

We decided to concentrate our studies on the two solutions A, A1 (renewal of the blanket segments from the top of the reactor) and A2 (renewal of the blanket segments horizontally), but in such a way as to also cover schemes B1 and B2. This is obtained by separating, inside the reactor, the renewable parts from the permanent ones.

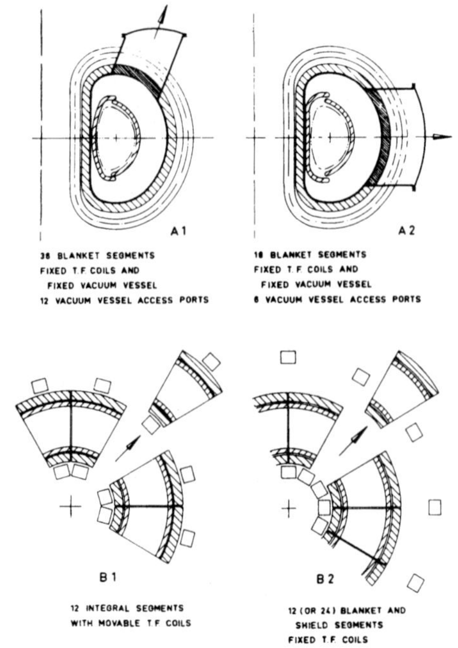

Fig. 1

The first are the internal components, or "internals", i.e. those parts facing the plasma, like the blanket segments, with a life-time which can be less than that of the plant, whose substitution can be periodically scheduled. The second ones are the parts which, in principle, have the same life-time as the plant and which only have to be substituted due to unscheduled damage, hopefully never. It is clear that the "internals" have to assure sufficient protection against radiation damage to the permanent parts; for example the blanket segments will have a minimum thickness sufficient to guarantee that the vessel has the same life-time as the plant. In fact, in the two solutions studied, the disassembly schemes A are usually followed to remove the "internals", while schemes B can be used during the assembly and the decommissioning of the machine and for the unscheduled interventions on the permanent parts.

2. AIM OF THE STUDIES

The mechanical configurations, A1 and A2, have been expressly developed to satisfy the requirements of a machine of the NET-INTOR generation, according to the European point of view of an experimental staged device capable of easily achieving subsequent objectives in subsequent operational phases.

The specific objectives of these studies were mainly that of:

- minimizing the constraints imposed by interference between the different components, in particular avoiding any interaction between the internal components and the toroidal field coil system, during the scheduled maintenance operations;
- avoiding the opening of the torus during all the operations on the internal components, i.e. the parts which have been strongly activated and which have a tritium inventory, thus allowing a certain amount of access into the reactor hall;
- allowing for the substitution of the internal components during the various operational phases, without modifying the toroidal vessel and the other permanent parts; this is obtained by increasing the dimensions of the vessel in order to allow the substitution of thicker blanket segments to the original ones.

3. MAIN COMMON FEATURES OF THE MECHANICAL CONFIGURATIONS STUDIED

a) Separation between the basic permanent structures (including the toroidal vessel and the magnet shield) and the internal components, (or the "internals", the non-permanent parts inside the toroidal vessel), which have to be substituted for the change of the operational or testing phases of the reactor, or for the end of their life. The "internals" include the first wall and blanket segments (non-breeding or breeding, at low or high temperature), the blanket test sections for the future power reactors, the limiters, the divertor plates, etc.

b) The inner (or primary) containment is the inner vacuum-tight surface of the structure constituting the toroidal vessel and the magnet shield; this surface is the separation element between the permanent structures and the "internals". It is conceived in such a way to allow all the scheduled interventions on the "internals" to be carried out remotely through access ports, without opening the torus and without dismounting the permanent parts.

c) The complete segmentation of the toroidal reactor is independent for the permanent structures and for the "internals". The toroidal vessel and magnet shield (as an integrated structure) are divided into 24 toroidal segments, 12 under the toroidal field coils and 12 in between them, rigidly clamped together and welded in correspondence to their inner surface (inner containment); the segments between the toroidal field coils and provided with the access ports for the "internals".

The cutting and the rewelding of the inner containment, between the adjacent segments, are done from the inside of the toroidal chamber by a remotely operated tool. The withdrawal of the toroidal segments of vessel and shield, is done in a radial direction opening the torus, according to the disassembly schemes B of fig. 1.

The first wall and blanket (as an integrated structure), are divided into 36 toroidal segments in solution A1 and into 18 segments in solution A2; they are supported inside the toroidal vessel and are provided in correspondence to the access ports with shielding plugs, coolant supply tubes and other servicing tubes and cables.

The withdrawal of the toroidal segment of blanket is done through the access ports, without opening the torus, according to the disassembly schemes A of fig. 1. The blanket test sections are incorporated inside the blanket segments more easily withdrawable, directly in correspondence to the access ports.

d) The scheduled interventions on the "internals" are carried out under a tight containment, using a transfer flask, while the unscheduled operations on the permanent structures can be done opening the torus.

The loading-unloading equipment for renewing the blanket segments and the other internals is contained inside a transfer flask, which can be connected in a tight way on the access ports, before their opening. In such a way it is possible to carry out all the operations inside the toroidal vessel (disconnection or cutting of the coolant supply tubes and subsequent connection or rewelding, withdrawal of the blanket segments and of their shielding plugs and subsequent reinsertion, etc.) under a controlled atmosphere (in principle, also under vacuum) avoiding any contact and communication between the "internals" and the external components and any tritium contamination outside the toroidal vessel.

e) The toroidal field coils system is independent from the other systems. The 12 toroidal field coils are arranged around a central core which acts as central guiding structure and are provided with lateral supports which act as anti-shear panels. The assembly appears as indicated schematically for solutions A1 and A2 respectively. This toroidal field coil system constitutes a rigid self-sustaining assembly at cryogenic temperature which, in principle, needs neither support from the other systems nor any external anti-torque structure, but only the support of its weight on the basement. (See fig. 2 and 3.) The poloidal field coils are not linked with the toroidal field coils and can be removed without dismantling the toroidal field coils system. For this purpose the outer poloidal field coils have a diameter greater than the maximum diameter of the toroidal field coils system in order to be easily removable by crane. The lowest poloidal field coils have to be lowered under the basement and laterally shifted for removal.

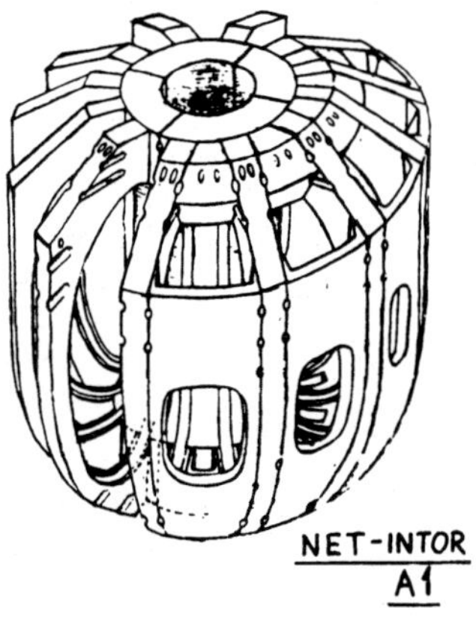

Fig. 2

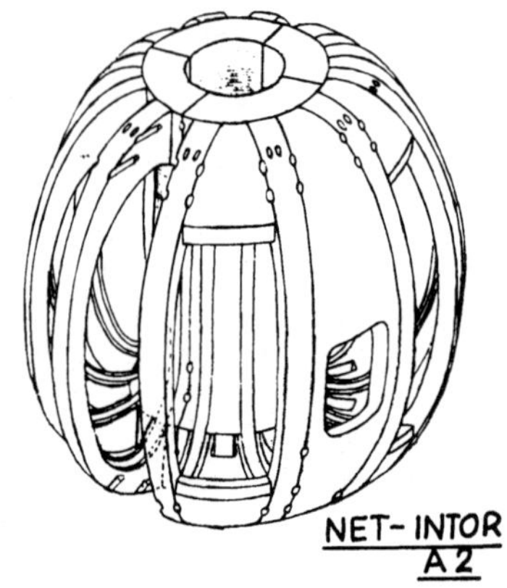

Fig. 3

f) An outer (or secondary) containment is arranged around the reactor. As far as the position of this containment is concerned, two variants are possible, indicated in fig. 4 with reference to the configuration A2. In the first one which is considered the basic solution, the outer containment is located outside the toroidal field coils system, but inside the poloidal field coils; it constitutes, together with the inner containment, the vacuum boundary of the common cryostat of the bellows, to reduce the induced currents and it is toroidal of shape. Each poloidal field coil has an independent cryostat and can be removed without warming the toroidal field coils system.

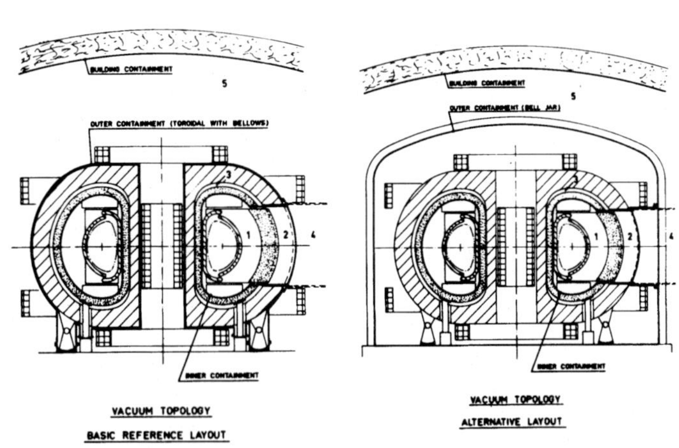

Fig. 4

In the alternative solution the outer containment which has the shape of a bell-jar, is located outside both the toroidal and poloidal field coils and constitutes, together with the inner containment, the common cryostat of both systems. In this case the poloidal field coils can be removed only dismantling the bell-jar and warming the toroidal field coils system; the main advantage of this solution is that of avoiding the presence of large bellows on the outer containment. In both variants the access ports pass through the inner and outer containment, tightly joining them.

4. CHARACTERISTICS OF SOLUTIONS A1 AND A2

The basic philosophy and the main lines of the design, as indicated in the previous sections, are the same for configurations A1 and A2. The two configurations are mainly characterized by geometrical differences:

- in solution A1 the renewal of the internal components is carried out through 12 access ports on the top of the reactor and the number of blanket segments is 36;
- in solution A2 the renewal of the internal components is carried out horizontally through 6 access ports in the equatorial plane of the reactor and the number of blanket segments is 18.

Both solutions A1 and A2 were studied for the case of a non-breeding blanket, water-cooled at low temperature [1] and for a reactor scheme without a poloidal divertor.

The mechanical configuration A1 is shown in fig. 5. In this solution each section of the blanket segment hangs under the segment plug and at the lower side it has a nose engaging in a guiding structure of the toroidal vessel, which is made of stainless steel and is also water-cooled.

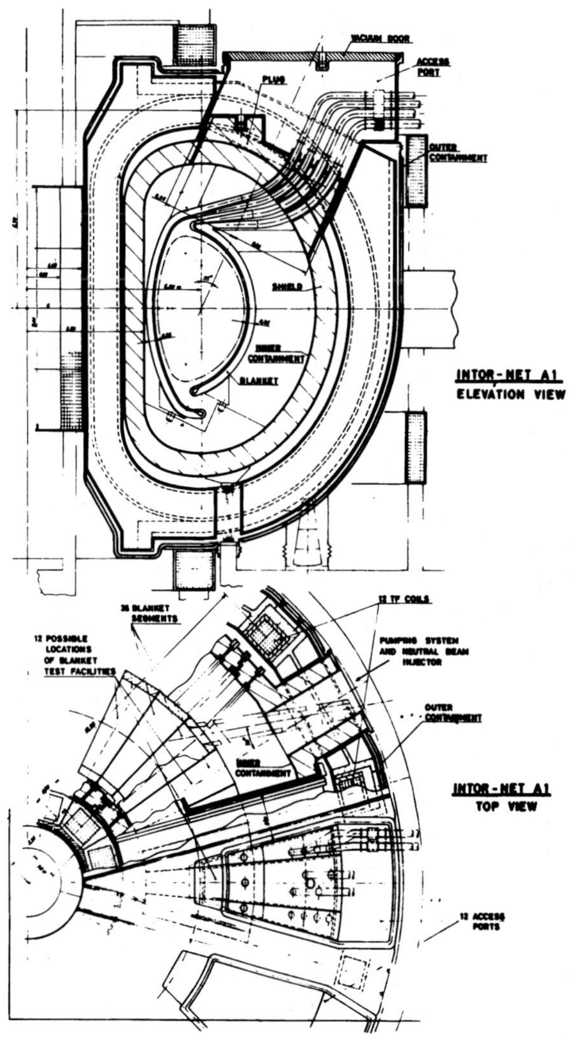

Fig. 5

The segments of the toroidal vessel located under the toroidal field coils incorporate bellows and current breakers to give the required electrical resistance.

The segments of the toroidal vessel located between the toroidal field coils are provided on the upper side with an access port and with a demountable tight door located inside a flange, which can be connected to the nose of the transfer flask (fig. 6). These segments are also provided with the shielded holes for the vacuum pumps and for the neutral injectors, arranged in the equatorial plane, without any interference with the access ports. The procedure for the renewal of the blanket segments can be identified looking at fig. 5.

After the opening of the access port and the disconnection of the cooling tubes and of the other servicing tubes and cables, done by a remotely operated manipulator located inside the transfer flask, a special lifting tool can be engaged on the upper side of the segment plugs. Obviously, the two lateral segments can be withdrawn only after the removal of the central one, which can be directly lifted with translation in sloping direction inside the transfer flask.

The power blanket test modules can be incorporated inside the 12 segments which are in the central position under the access ports and which are the more easily changeable ones.

Fig. 7 shows an example of test modules incorporated inside the outboard section of one of the 12 segments which can be used as test facilities.

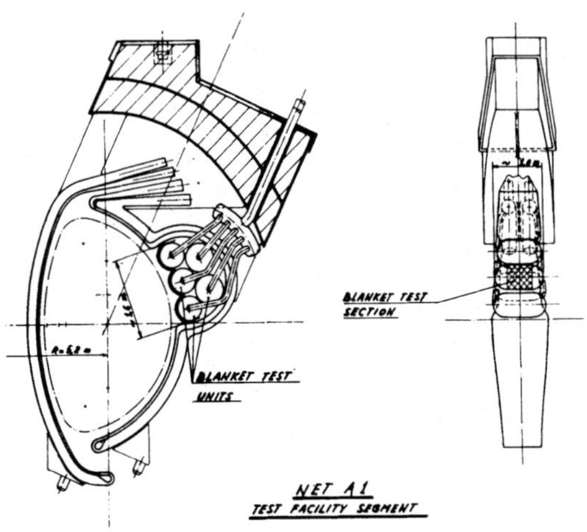

Fig. 7

provided with only 6 access ports, between alternate pairs of toroidal field coils, the remaining 6 spaces being used for strong shear-panels to support the coils and for neutral injection and pumping ports.

The blanket segments and the respective plugs are sustained by a horizontal basement located inside the toroidal vessel and the access ports and their tilting forces are supported by proper guides engaging into the structure of the vessel.

The 6 access ports can be connected to the nose of the transfer flask, as indicated in fig. 9. As far as the renewal of the blanket segments is concerned, we can operate horizontally in a similar way to that which is done vertically in solution A1.

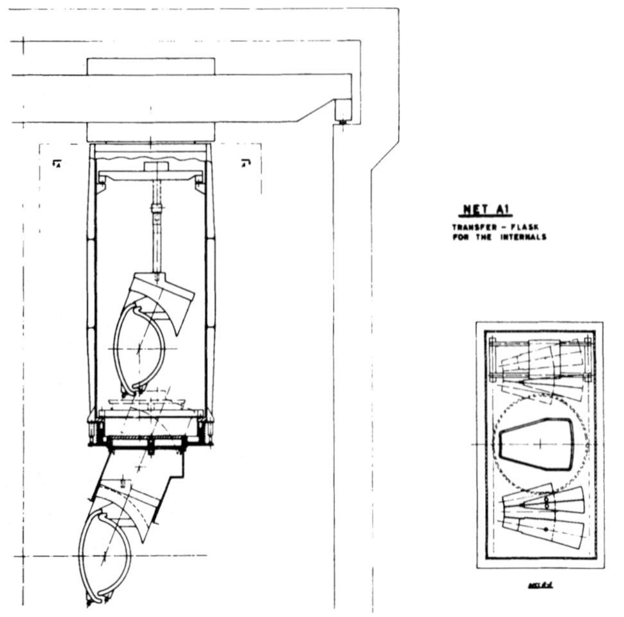

Fig. 6

The mechanical configuration A2 is indicated in fig. 8. In this solution the toroidal vessel is

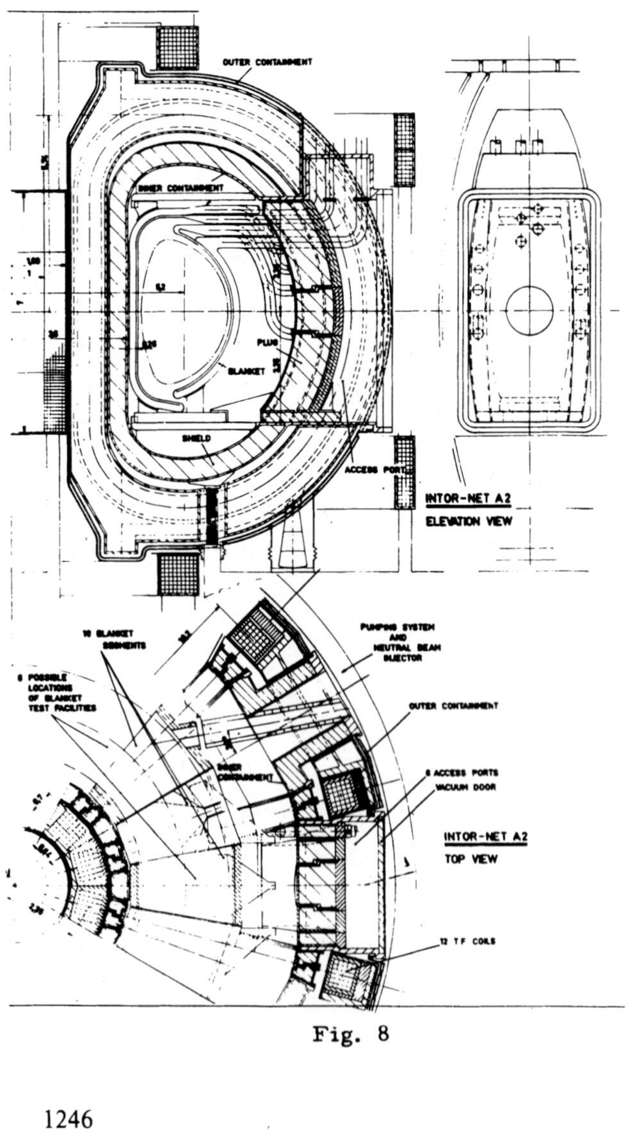

Fig. 8

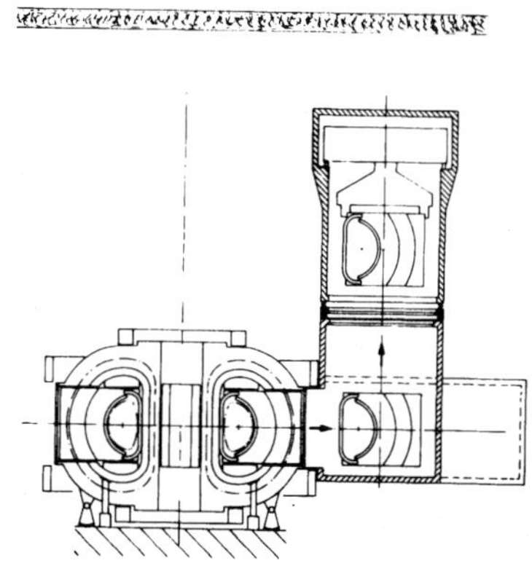

NET A2
TRANSFER-FLASK FOR THE INTERNALS

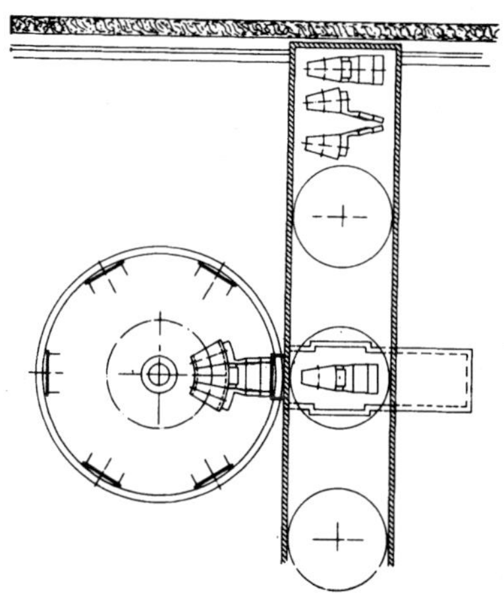

Fig. 9

After the opening of the access port and the disconnection of the cooling tubes by a remotely operated manipulator located inside the transfer flask, it is possible to withdraw, in a radial direction, the central blanket segment and its plug inside the transfer flask. The two lateral segments and their plugs can be withdrawn only after the removal of the central one, shifting them laterally and radially on the supporting basement.

The power blanket test modules can be incorporated inside the 6 segments in the central position of each access port, which are the more easily changeable.

Fig. 10 shows an example of test modules incorporated inside the outboard section of one of the 6 segments which can be used as test facilities. The test modules can be horizontally withdrawn independently from the blanket segments.

5. CONCLUSIONS

It is too early to choose the best general layout and the most convenient scheme for the assembly, disassembly and maintenance of NET-INTOR.

We will simply try to give, as an example, a first comparative analysis of solutions A1 and A2 and to identify some advantages and disadvantages of these solutions, on the basis of what has already been studied, which mainly concerns the renewal of the blanket segments and the more general aspects of the layout.

- Advantages

A1: - the extended use of cranes and other lifting equipment,

- a building of small diameter,
- a simpler structure of the transfer flask.

A2: - more space for the withdrawal of the blanket segments and consequent fewer segments,

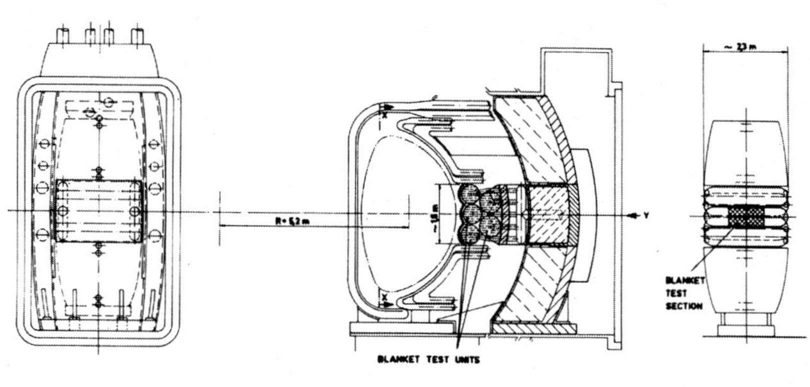

NET A 2
TEST FACILITY SEGMENT

Fig. 10

- more effective control of the vertical stability of the plasma.

- Disadvantages

A1: - electrical asymmetry of the toroidal vessel,

- absence of shear-panels between each pair of toroidal field coils in correspondence to the 12 access ports,
- more constraints on the location of the poloidal field coils above the reactor.

A2: - some difficulties for the horizontal displacement of blanket segments and plugs.

It is clear that a comparative analysis of the different solutions and their possible variants has to be repeated after the addition of each new system to the reactor. For the moment we can simply say that both the mechanical configurations A1 and A2 can satisfy the requirements of a simple machine of the NET-INTOR generation, in the case of a reactor concept without divertor. In the case of a reactor with poloidal divertor, where the life-time of the targets is expected to be shorter than that of the other internal renewable components, a new optimized mechanical configuration should be studied (for instance, providing for the withdrawal of the divertor targets segments independently from the blanket segments, through access ports at the bottom of the toroidal chamber).

6. ACKNOWLEDGEMENTS

The studies presented in this paper benefited by valuable contributions from Messrs. Aymar, Leloup, Torossian and Faivre of the CEA, Fontenay-aux-Roses, and from Messrs. Komarek, Erb and Jesske of KFK-Karlsruhe.

7. REFERENCES

[1] CASINI, G., BIGGIO, M., FARFALETTI-CASALI, F., FENICI, P., GAUTSCH, O., PONTI, C., RIEGER, M. and SCHILLER, P. (CEC, JRC-Ispra) and KLIPPEL, M.Th. (ECN, Petten), Main Features of an Aluminium Non-Breeding Blanket for Experimental Fusion Reactors, presented at this conference (1980).

PRELIMINARY INVESTIGATION OF BURN CONTROL IN NET-INTOR

A. Sestero, B. Brunelli, M. Gasparotto, R. Guidarelli-Mattioli★

Associazione EURATOM-CNEN sulla Fusione, Centro di Frascati,
C.P. 65 - 00044 Frascati, Rome, Italy

ABSTRACT

For the temperature stabilization of a Tokamak reactor, after the working temperature is choosen by means of the impurity control, a combination of two methods is proposed, namely, modulation of a fraction of the additional heating plus modulation of the equilibrium vertical field. With reference to the modulation of the additional heating the control aspects are studied in terms of the modern control theory, showing that the conditions are verified for complete observability and complete controllability.

1. INTRODUCTION

In the authors' opinion the most promising options for feed-back stabilization of the equilibrium temperature (if unstable) in a Tokamak reactor, at this moment, appear to be control via *adiabatic compression-expansion and* control via *modulation of a fraction the additional heating power.*

This opinion is rooted in the authors' preference for avoiding the risks tied to not well proven methods (e.g. magnetic field ripple or β-limit controls) and, on the other hand, in the conviction that the working temperature can be selected at will (within a reasonably extended range) by selecting and maintaining the level of impurity concentration in the plasma.

Notice that within the envisaged burn control scenario one does rely on apparati (vertical field coil system, additional heating system and exhaust system) that must be anyhow already available in the reactor for meeting other basic objectives. Notice also that the power levels required for the control task would be substantially less than the nominal levels needed respectively by the vertical field modulation system and additional heating system for performing their primary duties, to such an extent that the considered option turns out to be compatible with the requirements of reactor economics.

Greater details on the subject of this paper can be found in [1]; relevant references can be found there as well as in [2] and [3].

2. MATHEMATICAL MODEL OF THE BURNING PLASMA

A characteristic difficulty of the burn control lies with the markedly space-dependent features of the thermal instability, namely, with the tendency of the temperature to peak strongly at the plasma centre (rather than changing uniformy everywhere). Certainly, either the method that is chosen for the temperature stabilization of the reactor is capable of preventing an excess peak-

★ Guest of the Frascati Laboratories for his Doctorate work in Nuclear Engineering.

ing of the temperature profile, or this end must be separately achieved by other means. In any case, however, as a result of all the actions that are to be undertaken on the plasma, one will have to ensure an approximate stabilization of the shape of the temperature profile. In view of this circumstance, it appears thus meaningful to carry out a preliminary investigation of the control problem on the basis of a non space-dependent calculation, whereby reference is made only to global quantities, obtained by integration over the volume of the plasma, taking into account assigned density and temperature radial profiles.

In view of the above, the plasma model with which we shall be dealing in the present note will consist only, essentially, of the global balance equations for the energy densities $E(J/m^3)$ of each of three particles species, alpha particle (α), electrons (e) and ions (i):

$$\begin{cases} \dfrac{dE_\alpha}{dt} = G(E_i) - \dfrac{E_\alpha}{\bar{\tau}_\alpha(E_e)}, \\ \dfrac{dE_e}{dt} = \dfrac{E_\alpha}{\bar{\tau}_\alpha(E_e)} - \dfrac{E_e - E_i}{\bar{\tau}_{eq}(E_e)} - L(E_e), \\ \dfrac{dE_i}{dt} = \dfrac{E_e - E_i}{\bar{\tau}_{eq}(E_e)} \end{cases} \tag{2.1}$$

where G and L are the gain term and loss term, respectively; the times $\bar{\tau}_\alpha$ and $\bar{\tau}_{eq}$ are energy transfer times calculated as suitable averages over density and temperature profiles of the folloing expressions:

$$\tau_\alpha = - \frac{W}{dW/dt} = - \frac{1}{2}\frac{w}{dw/dt} = \frac{0.99 \times 10^{19}}{\ell n\, \Lambda} \frac{T_e^{3/2}}{n_e} \text{ sec}, \tag{2.2}$$

$$\tau_{eq} = \frac{2.45 \times 10^{19}}{\ell n\, \Lambda} \frac{T_e^{3/2}}{n_e} \text{ sec} \tag{2.3}$$

where for the ion species an effective mass $(2 \times 3)^{\frac{1}{2}} \simeq 2.45$ in proton units has been assumed, as appropriate for a 50 - 50 DT mixture; (W is here the α-particle energy, w its velocity, $\ell n\, \Lambda$ is Coulomb's logarithm, T_e is the electron temperature, n_e the electron density (see, e.g. [4])). If we choose as reference case $n_e = 10^{20}$ m^{-3}, T_e = 13.3 keV, Coulomb's logarithm has about the value 17; hence from Eq. (2.2) one obtains τ_α = 0.28 sec. This value is by no means negligible compared to the e-folding time (of the order of one second) for ther thermal instability; thus it appears appropriate to introduce the associated retarding effect into our model.

By the same token, it takes time for the energy to be transferred from the electrons to the ions: $\tau_{eq} = 2.45 \cdot \tau_\alpha = 0.69$ sec.

For the generation term G we have assumed the approximate expression [5]:

$$G = 1.4\ 10^{-13}\ \overline{n_i^2 \langle\sigma v\rangle} = 6.99 \cdot 10^{-7}\ \overline{(n_i T_i)^2} = 1.6 \cdot 10^{-6} \left(\overline{n_i T_i}\right)^2 = 1.6 \cdot 10^{-6}\ E_i^{2\,(\dagger)} \tag{2.4}$$

(†) $\overline{(n_i T_i)^2} = 2.29 \left(\overline{n_i T_i}\right)^2$ if averaged over the plasma cross section, given the profiles $n_i(r) \sim 1 - r^2/a^2$ and $T_i(r) \sim (1 - r^2/a^2)^2$.

Furthermore we have assumed that all losses are dominated by an electron anomalous loss term, which is taken to be of the form [6]:

$$L \sim E_e^{\frac{1}{2}} \tag{2.5}$$

For the equilibrium case, referred to by a subscript "o" in the various quantities, and defined by taking d/dt = 0 in Eqs (2.1), one easily obtains the following relations (with $\tau_{\alpha o} = \tau_\alpha(E_{eo})$):

$$\begin{aligned} & E_{eo} = E_{io}, \\ & \frac{E_{\alpha o}}{\tau_{\alpha o}} = G_o = L_o \end{aligned} \tag{2.6}$$

Next, introducing:

$$X_1 = E_\alpha - E_{\alpha o}, \quad X_2 = E_e - E_{eo}, \quad X_3 = E_i - E_{io}, \tag{2.7}$$

we proceed to the linearization of Eqs (2.1) with respect to the $X_j (j = 1, 2, 3)$:

$$\begin{cases} \dfrac{dX_1}{dt} = - \dfrac{X_1}{\tau_{\alpha o}} + \dfrac{X_2}{\tau_{\alpha e}} + \dfrac{X_3}{\tau_G}, \\[2ex] \dfrac{dX_2}{dt} = + \dfrac{X_1}{\tau_{\alpha o}} - \dfrac{X_2}{\tau_{\alpha e}} - \dfrac{X_2 - X_3}{\tau_{eqo}} - \dfrac{X_2}{\tau_L}, \\[2ex] \dfrac{dX_3}{dt} = + \dfrac{X_2 - X_3}{\tau_{eqo}} \end{cases} \tag{2.8}$$

We shall observe that other new terms would appear in Eqs (2.8), if the plasma were permitted to change its major radius as a consequence of the variation of its internal energy: in other words, in deriving our models we have been actually assuming throughout that a very effective plasma position control system is separately operating, with the capability of maintaining the plasma major radius rigorously constant.

In conclusion, for the reference case $n_e = n_i = 10^{20}\ m^{-3}$, $T_e = T_i = 13.3$ keV, that we are considering, the numerical values of the coefficients appearing in Eqs (2.8) turn out to be the following:

$$\frac{d}{dt}\begin{pmatrix} X_1 \\ X_2 \\ X_3 \end{pmatrix} = \begin{pmatrix} -3.57 & +3.06 & +4.08 \\ +3.57 & -5.53 & +1.45 \\ 0 & +1.45 & -1.45 \end{pmatrix} \begin{pmatrix} X_1 \\ X_2 \\ X_3 \end{pmatrix} \tag{2.9}$$

with eigen-values $\lambda_1 = +0.60$, $\lambda_2 = -3.45$, $\lambda_3 = -7.70$ and eigenvectors (arranged column-wise):

$$\Lambda_{jK} \equiv \begin{Vmatrix} 0.76 & -0.58 & 0.45 \\ 0.53 & -0.66 & -0.87 \\ 0.38 & 0.48 & 0.20 \end{Vmatrix} \tag{2.10}$$

The first mode is thus thermally unstable with growth time $\lambda_1^{-1} = 1/0.60 = 1.67$ sec.

3. FEED BACK TEMPERATURE STABILIZATION

For the control of our system we have added to the equations (2.9)a single *entry* u in the form ★:

$$\dot{X}_j = A_{jk} X_k + b_j u \qquad (3.1)$$

and prescribed a single *exit* y of the form:

$$y = c_k X_k + d \cdot u \qquad (3.2)$$

Choosing a particular *entry* means specifying the vector (b_1, b_2, b_3). To what extent the three eigenmodes of the system are *controllable*, with our choice of b_j, can be evaluated by comparing the magnitudes of the components b'_k of the vector in the frame of the eigenvectors:

$$b'_k = \Lambda^{-1}_{kJ} b_j \qquad (3.3)$$

For instance, if we decide to stabilize the equilibrium temperatures of the tokamak reactor by modulating the injection of RF, which would heat directly the electrons, it means that $(b_1, b_2, b_3) \equiv (0, 1, 0)$. The transformation (3.3) using (2.10) gives: $b'_1, b'_2, b'_3) \equiv (+0.48, -0.02, -0.84)$: the controllability of the second mode is worst than that of the two others(fortunately the second mode is already stable due to the corresponding eigenvalue $\lambda_2 = -3.45$ being of negative signe).

A particular choice of the *exit* y defines the vector (c_1, c_2, c_3) in (3.2). The consequent relative *observability* of the three eigenmodes can be estimated considering the vector components $c'_k = c_J \Lambda_{JK}$.

Two interesting examples of possible exits are:

a) $y \sim \dot{E}$, ($\dot{E}$ being the time derivative of the plasma pressure $E = E_\alpha + E_e + E_i$. It follows immediately that $y = \dot{X}_1 + \dot{X}_2 + \dot{X}_3$ ‡).Summing up the three equations (2.9) term by term we obtain:

$$(C_1, C_2, C_3) \equiv (0, -1.02, +4.08) \text{ and } (C'_1, C'_2, C'_3) \equiv (+1.0, +2.6, +1.7) \qquad (3.4)$$

hence we can conclude that the observability of the three eigenmodes is equally good.

b) $y \sim \dot{X}_3$ ‡ ($\dot{X}_3$ being derived from the time derivative of the neutron flux ϕ from the plasma, since ϕ is a known function of $E_i = E_{io} + X_3$. In the linear approximation around the equilibrium point, it is $\dot{\phi} \sim \dot{X}_3$).

★ As usual, a sum over repeated indices is understood.

‡ The justification for preferring the measure of dX_j/dt, instead of X_j is based on these considerations: in the expression $X_j = E_j - E_{io}$ the two quantities on the R.H.S., to a large extent, cancel each other out, thus producing a large relative error in X_j. This is not the case instead for the time-derivative expression $\dot{X}_j$.

From the third of equations (2.9) it follows:

$(C_1, C_2, C_3) \equiv (0, +1.45, -1.45)$ and $(C_1', C_2', C_3') \equiv (+0.22, -1.65, -1.56)$

It appears that the first mode (the unstable one) is less observable than the others.

As regards to the measure of the time derivative of the neutron flux from the plasma, the statistical fluctuations impose rather stringent conditions, if the measure has to come out of them.

For instance a single fast detector of the Lithium-glass type (dead time = 0.2 μs) is marginally adequate for a $\dot{\phi}$ measurement with 5% relative error. A 10 times finer control (implying roughly 10 times less circulating control power) could however in principle be achieved by employing an array of 100 detectors, still manageable in a fusion reactor.

On the basis of modern linear control theory (see e.g. [7]), we shall now make a few statements, illustrating the control possibilities that are open to us. Under conditions of good observability of all eigenmodes (as, e.g., in the case (3.4)), an *asymptotic state estimator* can be built, which, starting from the single (scalar) exit, is capable of reconstructing the complete (vector) state of the system, in principle with an arbitrarily high accuracy in an arbitrarily short time (see, e.g., [7]).Hence, when the complete state of the system is known, one can feed back from it trough the available scalar entry; in so doing, under conditions of good controllability of all eigenmodes, a true *mode synthesis* can be achieved, whereby the new eigenvalues of the system are in principle arbitrarily prescriptible [7].

On a less sophisticated level, one may also try feeding back directly from the exit (rather than from the complete vector state of the system, or from a part of it). One may, e.g., consider the so-called *proportional control*, obtained by relating the entry to the exit via simple proportionality (after subtracting the term du in (3.2)):

$$u = \eta \cdot c_K X_K$$

where η is a (real) scalar, to be suitably chosen. For our reference case, with $c_K \equiv (0, 1, 0)$, it has found that stabilization can be achieved for negative values of η lying in the range $-1.00 > \eta > -6.86$, with pure damping occurring in the subrange $-1.00 > \eta > -2.11$, and damped oscillations in the complementary subrange $-2.11 > \eta > -6.86$. Optimal stabilization (that is, lowest value for real part of highest eigenvalue) occurs in the vicinity of the dividing value $\eta = -2.11$ (with the highest eigenvalue dropping to about the value -2.00, from the value +0.5964 it had in the absence of feed-back). Thus also simple proportional control from the exit could be satisfactory for the considered case, although it would not, of course, be so flexible as the previously outlined mode synthesis procedure.

REFERENCES

[1] A. Sestero, B. Brunelli, M. Gasparotto, R. Guidarelli-Mattioli, *A Preliminary Investigation of the Burn Control Problem in INTOR*, Report 80.34 Associazione EURATOM-CNEN sulla Fusione, Centro di Frascati, Frascati, Italy

[2] "International Tokamak Reactor, Zero Phase" - IAEA, Vienna (1980) pg.249--254

[3] Ya.I. Kolesnichenko, Nucl. Fusion 20, 727 (6) (1980)

[4] L. Spitzer, "Physics of Fully Ionized Gases", Interscience Publishers, New York, 1962, Chapter 5

[5] B. Brunelli, Il Nuovo Cimento, 55B, 264 (2) (1980)

[6] L. Pieroni, F. Santini and S.E. Segre, Il Nuovo Cimento 56B, N. 2, 210 (1980)

[7] C.T. Chen, "Introduction to Linear System Theory",Holt, Rivehart and Wiston, Inc., New York (1970), Chapter 7

NEUTRON WALL LOAD, POWER DENSITY AND PAY-BACK TIME

K.H. Schmitter
Max-Planck-Institut fuer Plasmaphysik
Euratom-Association, D-8046 Garching

ABSTRACT

It is attempted to assess the impact of the power density in future Tokamak reactor nuclear islands on the economical competitiveness with today's pressurized water fission reactors in a global manner. Lower limits for the first wall neutron loading and lifetime are estimated considering economical aspects and energy pay-back time constraints as well.

1. INTRODUCTION

It is doubtful whether investigations about the economic aspects of fusion on the basis of a detailed costing of components for conceptual reactor designs can already yield sensible results in the present phase of transition from problem identification to problem solution. The essential technical systems and components of a fusion reactor and their manufacture cannot yet be specified with sufficient exactness. This paper describes an attempt to estimate data on the economy of tokamak power plants relative to that of pressurized water reactor (PWR) power plants by means of simple, global considerations of some important relevant features. The objective is to answer the question of the economically imposed, approximate minimum load capacity and lifetime of the structural material of the first wall and blanket.

2. FACTORS FOR ECONOMIC SUITABILITY ASSESSMENT

Important features for assessing the technical and economic suitability of energy systems for power plants are [1]:

- the mean volumetric power density
- system complexity
- availability.

Most fusion reactor design studies are concerned with system optimization in respect of these features.

The mean volumetric power density of a D-T fusion reactor depends on the plasma power density ($\sim \beta^2 B^4$) and the sum of the volumes of reactor systems located outside the plasma, such as blanket and shielding system, magnet system, start-up heating system, fuel supply system, divertor system, vacuum system, etc. Its value is limited by the neutron load capacity and lifetime of the first wall and blanket structures. These depend not only on the material properties, but also on the reactor design and operation parameters [2].

First wall and blanket components of D-T reactors must be periodically replaced before their lifetimes expire. Scheduled maintenance, requiring extraordinarily complicated remote operations, will, under realistic assumptions, also have a significant impact on the plant economy. In order to achieve competitive overall availability ($\gtrsim$ 80 %), long lifetimes of the first wall and blanket structural materials must be achieved. Considering scheduled and unscheduled outages as well, a tokamak first wall lifetime of at least 7 years is assumed to be the lower limit for reaching competitive availabilities with any maintenance concept [3].

3. STRUCTURAL MATERIAL

Steel of various types is the only structural material available at present for fusion reactor construction. With austenitic steels at operating temperatures of < 500^0 C it is hoped to make integrated neutron wall loadings up to 6 MWa/m^2 possible. Ferritic steels whose applicability is being investigated at present with allowance for their magnetic properties, will perhaps allow this value to be doubled or even trebled [4].

The question now arises whether ferritic steel might already be sufficient to provide economically acceptable power densities in fusion reactors or whether new materials will have to be developed for this purpose. The time required to develop a new material until it is ready for industrial application, is estimated by experts today to be at least 20 years. Thus not only is the answer to the materials question extremely important, this "activity" also very probably lies on a critical path in the development of D-T fusion into a competitive energy source.

4. NEUTRON WALL LOADING AND POWER DENSITY

If the net electric power density p_{el} of the tokamak reactor is defined as the quotient of the plant net electric power P_{el} and the volume V_c which is bounded by the surface of a cylinder enclosing the toroidal tokamak magnet

$$p_{el} = P_{el}/V_c \tag{1}$$

one obtains for STARFIRE [5] (Fig. 1) and NUWMAK [6] (Fig. 2) the values in Table I. The table also gives the neutron wall loadings p_w and for NUWMAK the value of the power per unit weight p^*_{el} of the nuclear island.

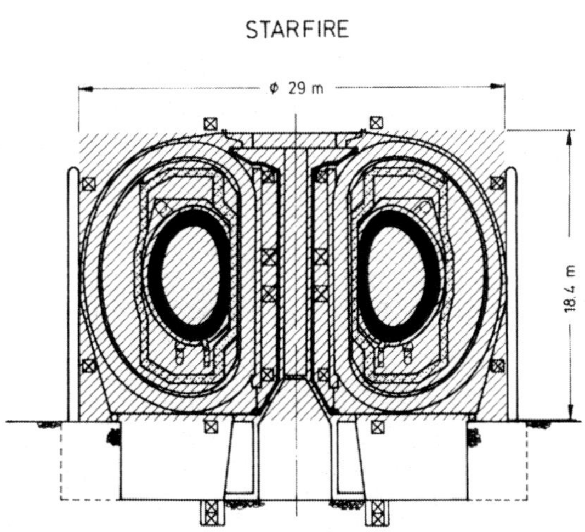

Fig. 1 STARFIRE cross section [5]. Reference volume indicated.

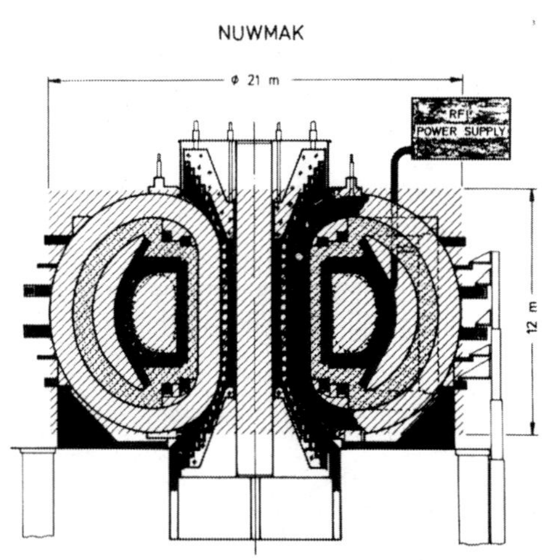

Fig. 2 NUWMAK cross section [6]. Reference volume indicated.

If a linear relation between the volumetric power density and neutron wall load

$$p_{el} = k_1 \, p_w \qquad [W_{el}/cm^3] \tag{2}$$

is assumed, extrapolation of the data from the two fusion reactor designs with

TABLE I

Net electric power, neutron wall loading and power density

	STARFIRE	NUWMAK	
P_{el}	1150	660	MW
V_c	$12.15x10^3$	$4.16x10^3$	m^3
pw_o	350	400	W/cm^2
pel_o	0.095	0.159	W/cm^3
$p^*_{el_o}$		0.086	W/g

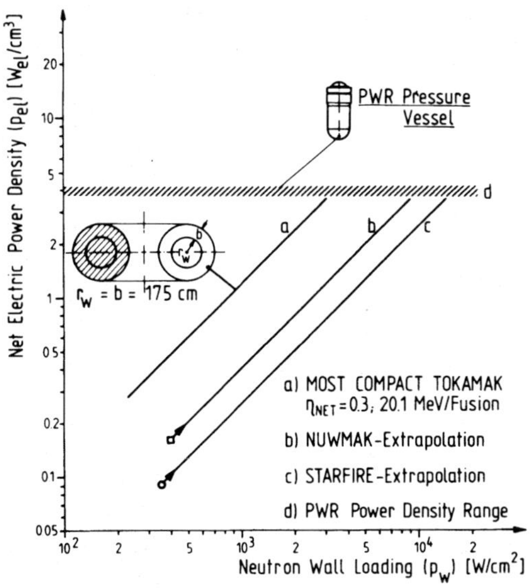

Fig. 3 Neutron wall loading p_W and net electric power per unit volume of nuclear island p_{el}

$$k_1 = \frac{P_{el}}{V_c \cdot pw_o} \tag{3}$$

yields the curves shown in Fig. 3, where the dependence of the blanket volume on the power density is ignored (optimistic extrapolation!). Compacting the construction beyond a certain limit is achieved at the expense of complexity and availability. It cannot yet be stated whether the compact NUWMAK architecture here is optimum. The upper compacting limit is characterized in Fig. 3 by the line of the so-called "most compact" tokamak reactor ($A = 3$; $r_W = b = 1,75$ m), whose power refers to the sum of the net volumes of the plasma vessel (A, r_W) and the outer system (b), which comprise the blanket, shielding and magnet only. The reference value taken for the power density is that prevailing in the pressure vessel of pressurized water reactors (PWR) [7]. The structure of a light-water reactor is less complex than that of a future D-T tokamak reactor would be and the materials required for its construction will, with all probability, entail lower specific energy costs than tokamak materials. In addition, the reference volumes chosen here for the tokamak reactors do not include essential sub-systems of the nuclear island (e.g. start-up heating, fuel injection, selective vacuum pumps) because too little is as yet known about these. Power density comparisons made on this basis should therefore hardly lead to a pessimistic assessment of the economic chances of the tokamak as a power reactor principle.

From Fig. 3 it can be deduced that the power density of the PWR could only be attained in a tokamak reactor of NUWMAK-type design, if structural material is available, which permits a neutron wall loading of about 90 MW/m^2. For a lifetime of 7 years this corresponds to an integrated wall loading of 630 MWa/m^2.

Assuming also a linear relation between the power density per unit weight p^*_{el} and the neutron wall loading

$$p^*_{el} = k_2 \cdot p_W \quad [W_{el}/g] \tag{4}$$

the extrapolation of NUWMAK's $p^*_{el_o} = 0.086$ W/g (Fig. 4) yields a somewhat lower breakeven value: The same net electric power per unit weight as in the PWR (1.32 W/g) is already reached at a neutron wall loading of about 70 MW/m^2. The corresponding integrated neutron wall loading is 490 MWa/m^2.

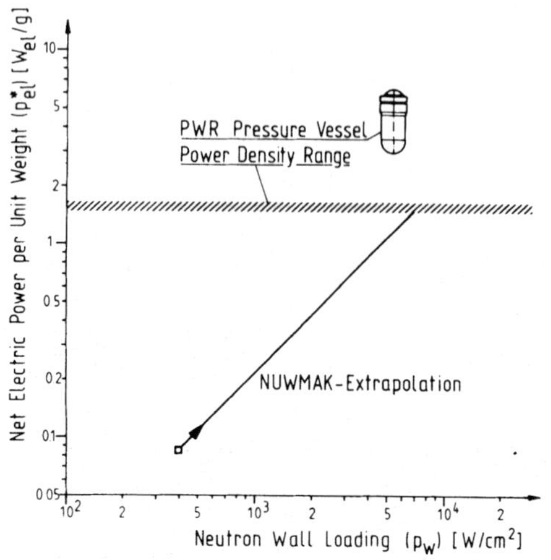

Fig. 4 Neutron wall loading p_w and net electric power per unit weight of nuclear island p^*_{el}

5. ENERGY PAY-BACK TIME

The energy pay-back time t_{pb} is the quotient of the energy required for producing a power plant and its net power output. This time might yield more meaningful criteria for assessing energy systems. Pay-back times for light water power plants between about 0.9 years and 2.3 years have been published [8], [9]. Pay-back time and power density are linked to each other. They may be linked in the following form:

$$t_{pb} = t_{pb_0} \cdot \{1 + \alpha \cdot [(\frac{p_{el_0}}{p_{el}})^{\varkappa} - 1]\} \qquad (5)$$

where:

α = the fraction of the nuclear island pay-back time to the total plant pay-back time,

$\varkappa$ reflects the ratio of average specific material energy cost of the tokamak reactor as compared to those of the PWR ($\varkappa \geq 1$), unity has been chosen in the following as a conservative assumption.

Taking as reference value a mean pay-back time t_{pb_0} of 1.5 years for PWR power plants, one can draw curves for the pay-back time of tokamak power plants as a function of the net electric power per unit volume Fig. 5, and as a function of the net electric power per unit weight Fig. 6.

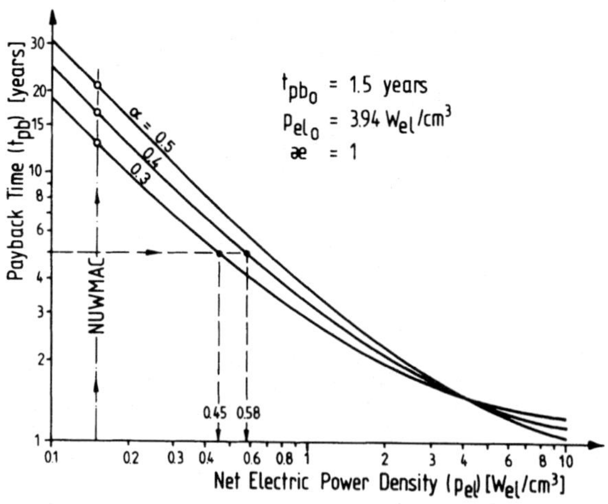

Fig. 5 Net electric power per unit volume of nuclear island p_{el} and plant pay-back time t_{pb}

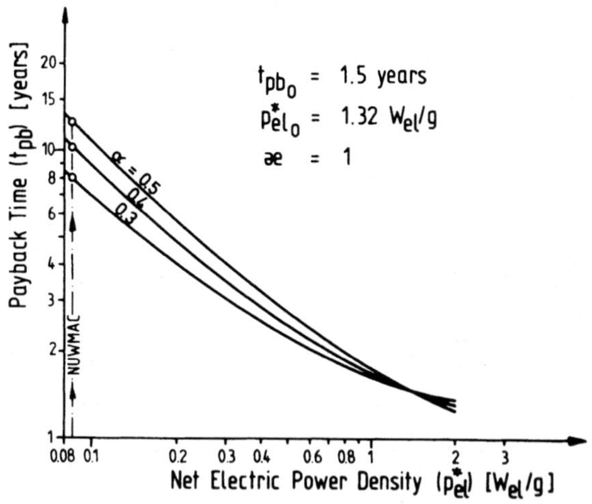

Fig. 6 Net electric power per unit weight of nuclear island p^*_{el} and plant pay-back time t_{pb}

Assuming that a pay-back time of 5 years can still be tolerated and α = 0.37 (i.e. the present PWR-plant value), one gets from Fig. 5: p_{el} = 0.54 W/cm^3 as a lower limit for the required power per unit volume of the tokamak reactor and from Fig. 3, curve b, the corresponding neutron wall loading $p_w \geq$ 13.5 MW/m^2. Thus, a pay-back time of 5 years for a NUWMAK-type fusion power plant, being designed for a first wall replacement time of 7 years, would require structural materials with an integrated wall load capacity of $Q_w \geq 95$ MWa/m^2 in such a reactor. Corresponding values taken from the power density per unit weight extrapolation, (Fig. 6, Fig. 4) are about 35 % lower, that is $Q_w \geq 60$ MWa/m^2.

6. CONCLUSIONS

It was shown that future tokamak reactors would suffer from extremely high neutron wall loadings when operated at values of power density which correspond to those in present day PWR's. Even the relatively low neutron wall loading calculated for power densities that would yield an energy pay-back time of about 5 years, would probably not lead to reasonable lifetimes (> 7 years) if steel is used as structural material.

ACKNOWLEDGEMENT

I wish to express my sincere appreciation to Prof. Pfirsch for useful discussions and to Prof. Eckhartt and Dr. Brinkschulte for their help and support in this work.

REFERENCES

[1] SCHMITTER, K.H., "Fragen zur Kraftwerkseignung des Tokamaks", IPP 4/170, June 1978

[2] DAENNER, W., J. RAEDER, "First wall life prediction by the FWLTB computer program", 11th SOFT, Oxford, September 1980

[3] "Fusion reactor remote maintenance study", EPRI ER-1046, Final Report April 1979

[4] RAWLS, J.M. et al., "Assessment of martensitic steels as structural materials in magnetic fusion devices", GA-A 15749 UC-20d, January 1980

[5] BAKER, C.C. et al., "STARFIRE - A Commercial Tokamak Reactor", Proceedings of 8th Symp. on Engineering Problems of Fusion Research, San Francisco 1979

[6] BADGER, B. et al., "NUWMAK a tokamak reactor design study", UWFDM-330, Madison, March 1979

[7] ALTVATER, W., KWU Erlangen, private communication

[8] CHAPMAN, P.F., "Energy analysis of nuclear power stations", Energy Policy, Dec. 1975

[9] MORAW, G. et al., "Energiebilanz von Kraftwerken und Ausbauprogrammen", Atomwirtschaft-Atomtechnik XXII, Nr. 1 1977

This work was performed under the terms of the agreement of association between the Max-Planck-Institut fuer Plasmaphysik and EURATOM.

WITAMIR-I, A TANDEM MIRROR REACTOR WITH NON-ZERO NU

J. Santarius, K. Audenaerde, J. Beyer,* G. Emmert, J. Kesner, G. Kulcinski, D. Larbalestier, W. Maurer,** C. Maynard, R. Perry, J. Scharer,* I. Sviatoslavsky, D.-K. Sze
University of Wisconsin, Nuclear Engineering Department
1500 Johnson Drive, Madison, Wis., 53706, USA

ABSTRACT

Witamir-I, a tandem mirror reactor, is a study by the University of Wisconsin fusion engineering design group. Various engineering aspects and physics considerations are presented. Witamir-I is a high Q device with particularly attractive maintenance features and blanket characteristics.

1. INTRODUCTION

Witamir-I [1] is a tandem mirror reactor (TMR) with thermal barriers. The original tandem mirror concept of plugging a solenoid with electrostatic potentials created by high density in end plug cells was invented by Fowler and Logan [2], and independently by Dimov, et al [3]. It required large amounts of neutral beam power to maintain the plug density, motivating Baldwin and Logan [4] to invent the thermal barrier concept—a potential dip between the central cell and plug which thermally insulates the plug electrons, allowing them to be efficiently heated with radio frequency (RF) power. Studies of a thermal barrier TMR then began, [5], [6] using a Boltzmann relation between density, potential, and electron temperature. Subsequent work by Cohen, et al [7] found that the Boltzmann relation for the plug potential, $\Delta\phi$, must be replaced by

$$e\Delta\phi = T_{ep} \ln\left[\frac{n_p}{n_b}\left(\frac{T_{ec}}{T_{ep}}\right)^{\nu_c}\right] \tag{1}$$

where ν_c ranges from 0.5 to 1.5 and $\nu_c \approx 0.5$ for the Witamir-I magnetic field configuration; $\nu_c = 0$ is the Boltzmann case; n_p and n_b are the plug and barrier densities; and T_{ec} and T_{ep} are the central cell and plug electron temperatures. Non-zero ν_c makes high $\Delta\phi$ more difficult to achieve, and increases power flow between plug and central cell electrons which decreases Q, the ratio of fusion power out to plasma input power. Witamir-I is the first complete reactor study using non-zero ν_c and including the effect of using RF power to create a hot mirror-trapped electron population in the barrier to enhance ϕ_b, the thermal barrier potential.

The following sections discuss some features of the Witamir-I parameter set, engineering aspects of the main reactor regions, and plasma physics considerations.

2. PARAMETERS

Magnetic field, potential, and density profiles are shown in Fig. 1 for one end of Witamir-I. ϕ_b is created by the density drop from neutral beam charge exchange pumping and flux tube expansion as the magnetic field falls from 14 T to 1.4 T. ϕ_c, which confines central cell ions, is created by RF heating of plug electrons and by neutral beam injection. ϕ_e is the ambipolar potential occurring because electrons are more collisional than ions and scatter more quickly into the loss cone. The hot electron density, n_{eh}, is created by RF heating at the barrier center. Power, machine, and plasma parameters are listed in Table 1. The pumping parameter, g_b, is the ratio of total barrer ion to passing ion density.

---

*University of Wisconsin, Department of Electrical and Computer Engineering.

**Kernforschungszentrum Karlsruhe, Federal Republic of Germany.

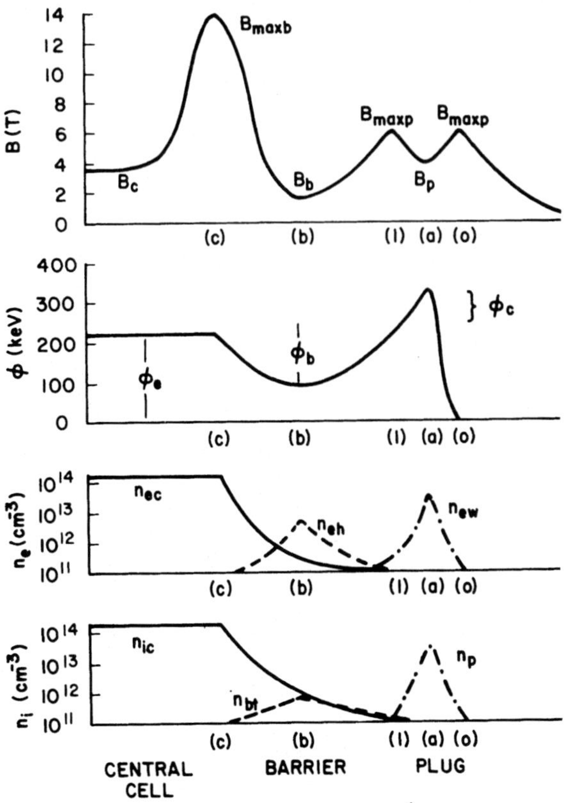

Fig. 1. Magnetic field, potential, and density for one end of the machine.

A 3000 MW fusion power plant is chosen in order to give 1200 to 1500 MWe of gross electric power. Then, Q = 30 gives a reasonable value of about 100 MW of input power to the plasma. In particular, total RF power is less than 50 MW. High Q requires a central cell with short radius and long length because input power is proportional to barrier and plug volumes which are tied to central cell radius through magnetic flux conservation, while fusion power output is proportional to central cell volume = $\pi r_c^2 L_c$. A small radius allows smaller magnetic field coils for a given field strength, giving less expensive coils.

3. CENTRAL CELL

An inherent advantage of a TMR is the accessibility of the blanket and shield for maintenance. Figure 2 shows a blanket module being removed from the reactor. Accessing the various coils, blanket modules, and shield

TABLE 1

Power and Machine Parameters

Powers are given as absorbed plasma power without efficiencies folded in.

Parameter	Value	
Q	28.0	
Fusion power	3000.	MW
Neutron wall loading	2.4	MW/m^2
Central cell power density	11.3	MW/m^3
Plug ECRF power	16.5	MW
Plug neutral beam power at 500 keV injection	2.4	MW
Plug trapping fraction	0.13	MW
Barrier ECRF power	33.3	MW
Barrier neutral beam power		
9.6 keV injection	12.7	MW
190. keV injection	42.5	MW
Central cell surface heat load	2.75	W/cm^2
Plug surface heat load	3.43	W/cm^2
Barrier surface heat load	80.0	W/cm^2
Central cell wall radius	0.97	m
Central cell length	165.	m
Barrier length	10.	m
Plug length	5.5	m
Central cell magnetic field	3.6	T
Barrier maximum field	14.0	T
Barrier minimum field	1.4	T
Plug maximum field	6.0	T
Plug minimum field	4.0	T

Plasma Parameters

Central Cell		
Density	1.51×10^{14}	cm^{-3}
Ion temperature	32.5	keV
Electron temperature	32.8	keV
Potential, ϕ_c	102.	keV
Beta, β_c	0.40	
Plasma radius	0.72	m
$(n\tau)_{ic}$	7.8×10^{14}	sec cm^{-3}
Barrier		
Density average	6.9×12	cm^{-3}
Mean hot electron energy, E_{eh}	270.	keV
Passing electron fraction, F_{ec}	0.27	
Pumping parameter, g_b	2.0	
Pumping fraction at low energy	0.95	
Pumping fraction at high energy	0.05	
Potential, ϕ_b	141.	keV
Beta, β_b	0.235	
Plasma radius average	0.59	m
Plug		
Density average	2.7×10^{13}	cm^{-3}
Mean ion energy	905.	keV
Electron temperature	123.	keV
Potential, $\phi_c+\phi_e$	326.	keV
Cohen parameter, ν_c	0.5	
Beta, β_p	0.64	
Plasma radius	0.77	m
$(n\tau)_{ip}$	9.8×10^{13}	sec cm^{-3}

modules simply requires hydraulically moving an element to the most convenient hatch and moving others out of the way.

A cross-section of the central cell is shown in Fig. 3. The support scheme is simple and practical, and the blanket design, which is easy to fabricate and may be factory assembled, is particularly appealing. There is negligible thermal stress, and the part of the blanket nearest the plasma is made of extruded, seamless HT9 tubes with no welds exposed to the plasma.

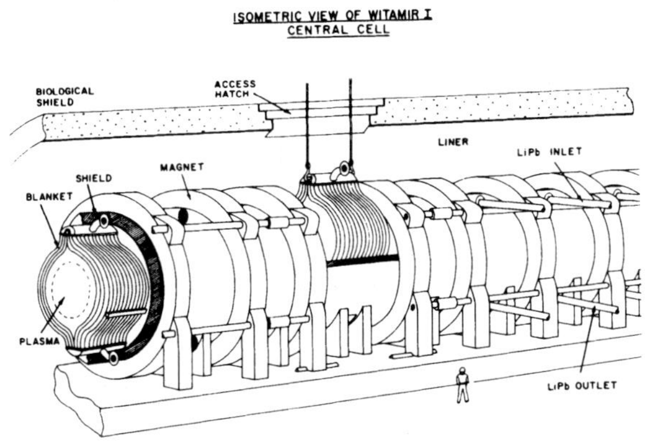

Fig. 2. Blanket module removal.

As the tritium breeder and coolant, $Li_{17}Pb_{83}$ was chosen since it has low tritium solubility, thus reducing the tritium inventory, and is relativity inert to water, thus minimizing the consequences of an accident. A detailed analysis of the blanket [8] yields an excellent energy multiplication of 1.37, and a tritium breeding ratio of 1.07.

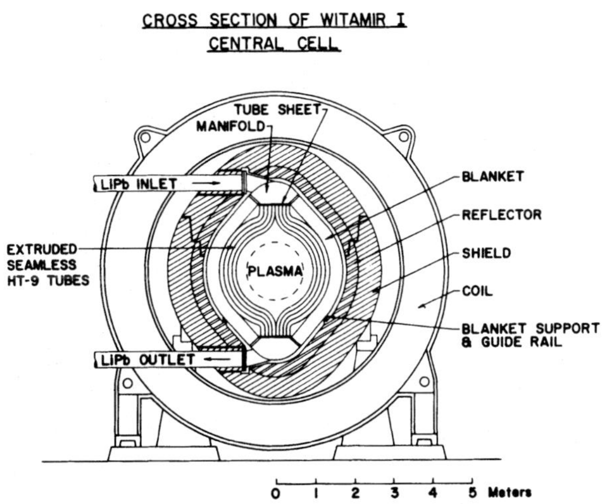

Fig. 3. Central cell cross-section.

The central cell magnets, which give 3.6 T on axis from slightly over 6 T on the conductor, are made of NbTi and Al, and present no difficulties. Ripple is less than 5%. Figure 4 shows the magnetic field coils for one end of the machine. The oversized central cell coil will be discussed in sec. 4.

4. BARRIER

Because the thermal barrier concept is a very recent addition to TMR design, experimental verification of some plasma physics questions remains. In particular, maintenance of a large electron temperature differential must be verified, and details of the pumping process must be explored.

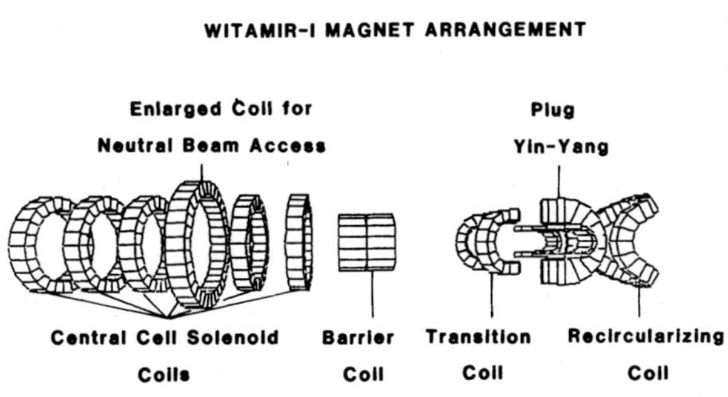

Fig. 4. Magnets for one end.

The barrier field coil consists of three concentric cylinders, giving 14 T on axis and a maximum of 15 T on conductor. The inner, middle, and outer coil superconductors are Nb_3Sn, NbTiTa, and NbTi, all stabilized by Cu and operated at 1.8°K.

Ions scattering into the barrier must be "pumped" out to maintain a density dip and thus ϕ_b. Witamir-I implements this by charge exchange off of neutral beams injected at approximately 10°. Insertions of these requires the enlargement of one central cell coil on each end (see Figs. 4 and 5). The beams are focused near the central cell end of the barrier coil.

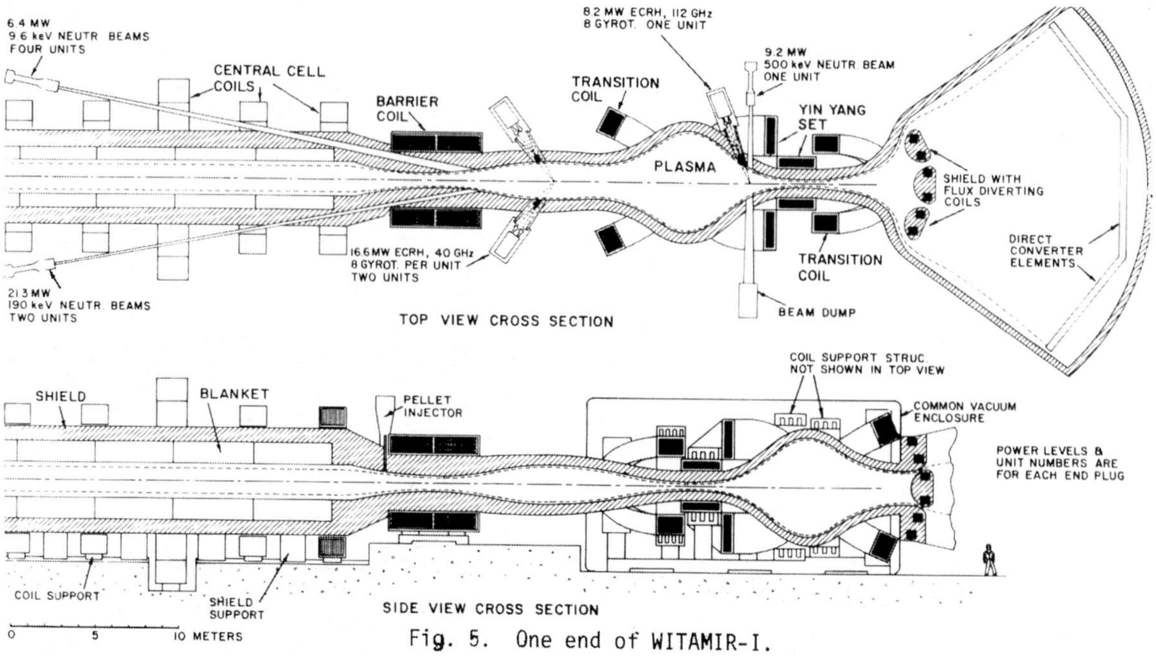

Fig. 5. One end of WITAMIR-I.

Electron cyclotron range of frequencies (ECRF) heating will be used at the bottom of the barrier to create a hot, mirror-trapped electron population [4]. This will reduce the number of electrons passing through both central cell and plug, thus giving a larger ϕ_b because the central cell electrons have approximately a Boltzmann distribution. The transport system designed for this purpose is shown in Fig. 6.

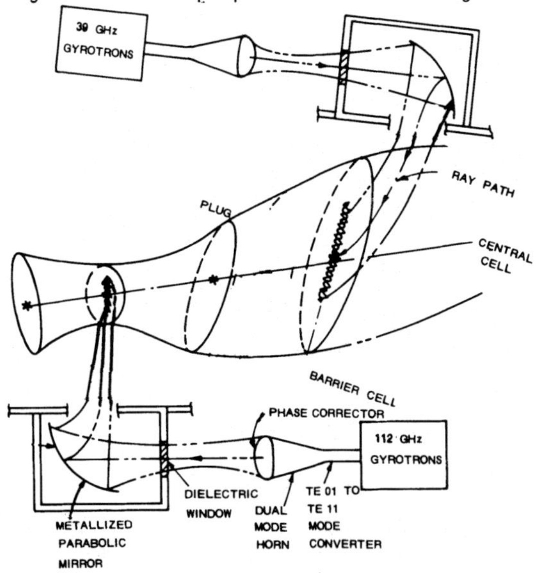

Fig. 6. ECRH transport system (not to scale).

ECRF heating in the barrier complicates the physics and machine while allowing higher Q and some lower technology requirements. Although the barrier ECRF option is chosen here, it should be stressed that a tandem mirror reactor without barrier ECRF seems viable, if not quite as attractive, in terms of Q.

5. PLUG AND DIRECT CONVERTOR

The plug region is essentially a minimum-B mirror machine, for which much experimental work exists, so theory is generally well-based. The Witamir-I magnet design (see Fig. 5) has a maximum of about 8 T at the yin-yang conductors, which are NbTi cooled to 4.2°K; stabilization is by Cu, and forces are manageable.

Neutral beam injection at 500 keV requires negative ion technology, which is in a developmental stage. It presently appears difficult to build a TMR without neutral beam energy $\gtrsim$ 150 keV, where negative ion technology is required.

Two options were considered for heating plug electrons: 1) ECRF heating in the plug at about the 4 T point, and 2) ion cyclotron range of frequencies (ICRF) heating in the middle of the plug. The first, chosen for Witamir-I, is being developed. The second, while technologically preferable, poses physics questions. Figure 6 shows the ECRF transport system.

The field from the recircularizing coil shown in Fig. 4 goes into a direct convertor which electrostatically drains energy from ions that escape over ϕ_c. By maintaining ϕ_c slightly higher on one end of the machine, essentially all ions will escape out the other end, thus necessitating a direct convertor on only one end of the mahcine. It will handle approximately 502 MW of power at an efficiency of 67%.

6. PLASMA PHYSICS

A detailed description of the plasma physics model is contained in [1]. Much of the physics, especially of the barrier, is based on work in [9], with suitable changes for differences in magnetic field configuration, etc. The main terms in the power balance are heating by fusion alphas, neutral beams, or ECRF power, balanced by end loss, radiative processes, and energy carried out by charge exchange neutrals from the pumping process. Internally, collisional energy conduction and convection are also important.

The central cell ions are deuterium and tritium. Protons are used in the plugs because the consequent absence of fusion neutrons from plug fusions allows the yin-yang coils to be designed with only central cell streaming neutrons in mind. Better microstability is also expected because protons have smaller gyroradii than deuterium or tritium.

MHD stability questions remain with regard to the magnetic field design, with theory indicating a maximum central cell beta, the ratio of plasma pressure to magnetic field pressure, of about 25% and experiments exceeding theoretical beta limits. The value 40% was chosen for central cell beta here, and seems reasonable in light of probable future improvements in the field design and the fact that the MHD theory is somewhat tentative.

7. CONCLUSIONS

The Witamir-I study shows that high Q operation appears feasible with all magnets except possibly the barrier coil within extrapolatable technology. Neutron streaming has a minimal effect on the end plugs. A TMR is easier to maintain than a tokamak because of the linear geometry and wide coil spacing, and the excellent energy multiplication of 1.37 makes the $L_{17}Pb_{83}$ blanket look very attractive.

8. REFERENCES

[1] B. Badger, et al, University of Wisconsin Report UWFDM-400 (1980).

[2] T.K. Fowler and B.G. Logan, Comm. on Pl. Ph. 2, 167 (1977).

[3] G. Dimov, et al, Fiz Pl. 2, 597 (1976).

[4] D.E. Baldwin and B.G. Logan, Phys. Rev. Lett. 43, 1318 (1979).

[5] J. Santarius and R. Conn, University of Wisconsin Report UWFDM-340 (1980).

[6] G. Carlson, et al, Lawrence Livermore Laboratory Report UCRL-52836 (1979).

[7] R. Cohen, et al, Lawrence Livermore Laboratory Report UCRL-84147 (1980) (to be published in Nuc. Fus.).

[8] R.T. Perry and C.W. Maynard, "Neutronic-Design Study of Witamir-I," (these proceedings).

[9] D. Baldwin, et al, Lawrence Livermore Laboratory Report UCID-18496 (1980).

Fusion Materials

RADIATION DAMAGE SIMULATION EXPERIMENTS ON A Mn-Cr AUSTENITIC STAINLESS STEEL IN A HVEM.

M.Snykers
SCK-CEN Boeretang 2400 MOL, Belgium

E.Ruedl
J.R.C. ISPRA,Italy

ABSTRACT

Manganese-chromium austenitic stainless steel specimens (without helium) are irradiated in a HVEM in the temperature range 400-650°C. Void swelling is observed at all temperatures. The peak swelling temperature is higher than 650°C. The total swelling obtained is slightly lower than for nickel-chromium austenitic steels such as FV 548 and an AISI 316 type (the latter with the same carbon content as the manganese-chromium steel). Irradiation induced phase instability is observed at temperatures > 500°C, together with an increase of the swelling rate with increasing irradiation dose. The second phase is analysed by means of electron diffraction. It corresponds to α-Fe and it has a well defined orientation relationship with the matrix. It is formed by a process of nucleation and diffusional growth.

1. INTRODUCTION

The economic aspects of the future fusion reactor is to a large extent determined by the lifetime of the first wall. From the present potential first wall materials no definitive choice can be made as yet. Instead, alternative choices are proposed such as the Mn-Cr steels for the Ni-Cr austenitic steels [1]. In these steels the nickel is partially or completely replaced by manganese.

The influence on the phase stability has been studied by the manufacturer; no thermal instability was observed at temperatures lower than 600°C [2].

The influence on other properties such as compatibility with liquid lithium, void swelling and activation are being studied. Some results concerning the corrosion behaviour in static lithium are published [1]; it is still too early to make an evaluation as compared with Ni-Cr austenitic steels. No results concerning void swelling are published as yet. Some work is performed in Harwell on a Ni-Mn-Cr austenitic steel [3].

In order to obtain rapidly information concerning the swelling behaviour of Mn-Cr austenitic steels, HVEM irradiations are performed. The present paper deals with the swelling results obtained on a heliumfree Mn-Cr steel of the type AMCR, with a very low nickel content, delivered by Creusot-Loire. The aim of this study is to evaluate this Mn-Cr steel as compared with the more common Ni-Cr austenitic stainless steels.

2. EXPERIMENTAL

2.1. Specimen preparation

Electron microscopy specimens are prepared from AMCR bulk material with a composition as given in table I and with a heat treatment SA/1000°C - water cooled. Discs with a diameter of 3 mm are mechanically grinded to a thickness of 0.15 mm and further electro-polished by a double jet in a 10 % perchloric - 90 % acetic acid solution at room temperature.

TABLE 1

Chemical Composition Of The AMCR Steel

Fe	Mn	Cr	Ni	C	N	O_2
Bal	17.5	10	0.67	0.2	570 ppm	40 ppm

Electron irradiations are performed in the HVEM of the University of Antwerp (RUCA)equipped with a side entry goniometer stage in the temperature range 400-650°C at an irradiation dose rate of about 5×10^{-3}dpa s^{-1}. The temperature increase due to beam heating is taken into account [4].

2.2. Evaluation of the results

Micrographs of the irradiated area are magnified to a standard magnification of 90.000X and histograms are made of the void size distributions for a standard area by means of a Zeiss particle size analyser. The number of the voids per histogram varies between 30 and 400. The diameter of the mean void $d_{\bar{v}}$ and the total swelling S are given by :

$$d_{\bar{v}} = \left(\frac{1}{N} \sum_{i=o}^{N} d_i^3 \right)^{1/3}$$

and

$$S = \Delta V / (V - \Delta V) = N\bar{v} / (V - N\bar{v})$$

where N is the total number of voids considered, d_i is the measured void diameter, V is the volume of the specimen considered, $\bar{v}$ is the mean void volume (= $1/6\ \pi d_{\bar{v}}^{3}$).

The thickness of the void layer is measured by means of a stereographic method [4].

3. EXPERIMENTAL RESULTS

3.1. Swelling as a function of dose

Graphs of the diameter of the mean void volume, $d_{\bar{v}}$, the void density, n, and the swelling, S, as a function of irradiation dose are shown in Figs. 1,2 and 3 for temperatures from 400 to 650°C. Results are shown for doses < 40 dpa. Those obtained for higher doses are difficult to analyse especially at temperatures > 500°C, where a second phase formation occurs, in the irradiated areas.

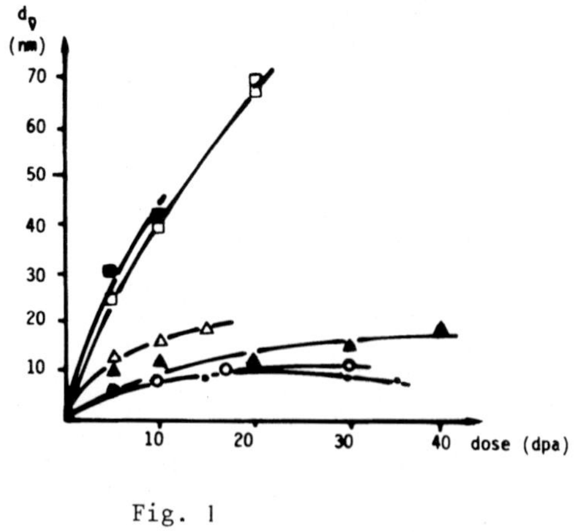

Fig. 1

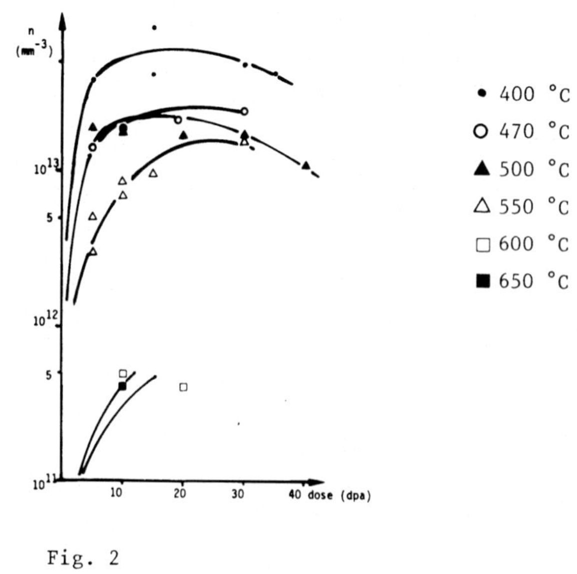

Fig. 2

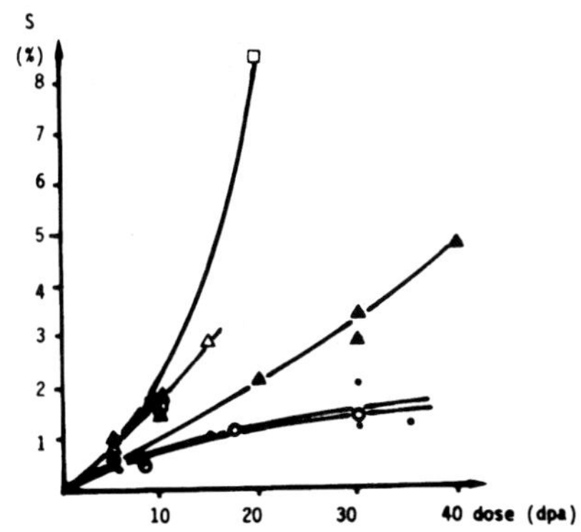

Fig. 3

Fig. 1, 2 and 3

Diameter of the mean void volume (Fig.1) void density (Fig.2) and percentage void swelling (Fig.3) as a function of irradiation dose for temperatures ranging from 400 to 650°C.

The graphs of the void density versus dose show maxima, for temperatures < 500°C. They are shifted to higher doses for higher temperatures. They are due to increasing void coalescence and disappearance at higher irradiation doses.

The total swelling increases nearly linearly with increasing irradiation dose for temperatures lower than 500°C. For higher temperatures the swelling rate increases with increasing dose, a phenomenon which is probably due to the phase instability.

The high swelling rate at 600°C is due to the fast increase of the void diameter.

3.2. Swelling as a function of temperature

Graphs of the diamater of the mean void volume $d_{\bar{v}}$, the void density, n, and the total swelling, S, as a function of the irradiation temperature are shown in Figs. 4, 5 and 6.

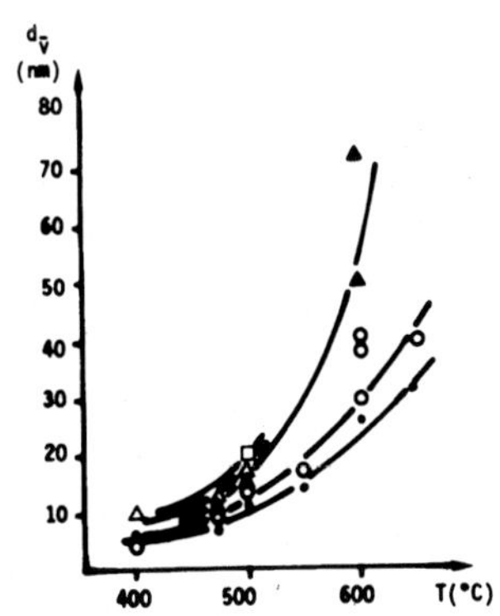

Fig. 4

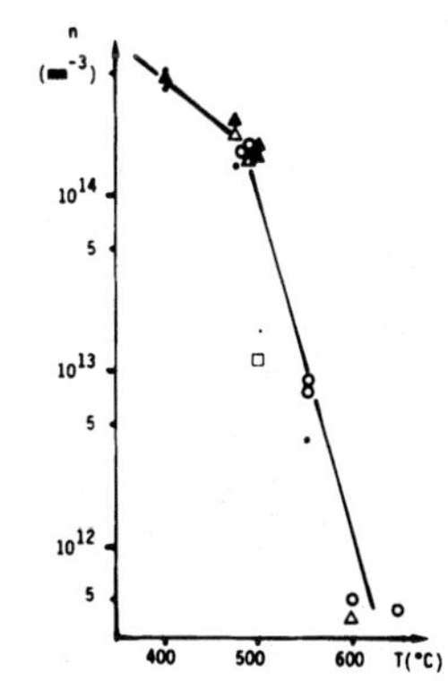

Fig. 5

• 5 dpa
○ 10 dpa
▲ 20 dpa
△ 30 dpa
□ 40 dpa

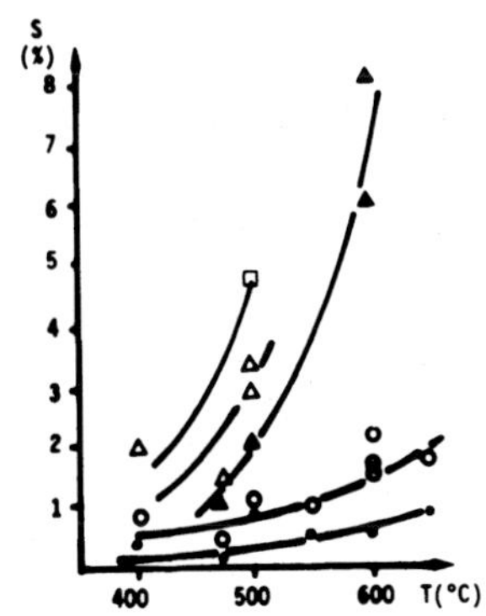

Fig. 6

Fig. 4, 5 and 6

Diameter of the mean void volume (Fig.4) void density (Fig.5) and percentage void swelling (Fig.6) as a function of temperature for different doses.

The diameter of the mean void volume increases rapidly with increasing temperature for temperatures higher than 500°C. This can be seen as well on the micrographs shown in Fig.7. The void density decreases nearly linearly with increasing temperature. The total swelling increases with increasing temperature. The peak swelling temperature is not reached yet at 650°C. It is difficult to determine the peak swelling temperature because of the poor visibility of the voids at high temperatures due to the second phase formation.

3.3. Structural stability

3.3.1. Irradiation stability

The AMCR steel is structurally unstable during electron irradiation at temperatures > 500°C. A second phase forms in the irradiated areas. It nucleates both in the grains and at grain boundaries and grows in the form of lamellas, as is shown in Fig.7. For comparable irradiation doses, the amount of new phase increases with increasing irradiation temperature. No phase transformation is observed outside the irradiated areas, not even at the higher temperatures. No interaction between this second phase and the voids can be observed. The formation of this phase reduces considerably the visibility of the voids.

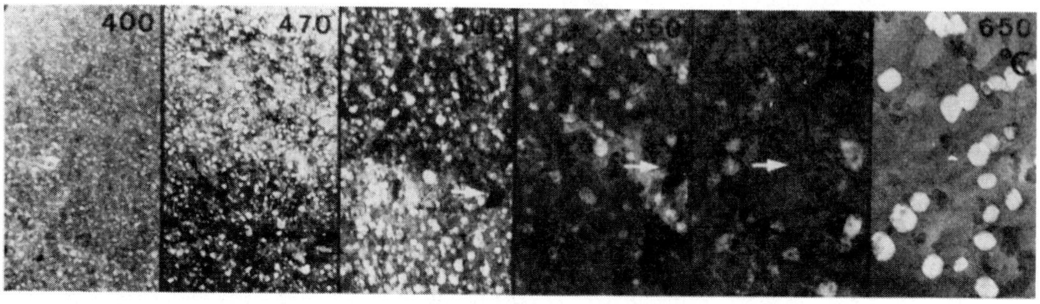

The arrows indicate the precipitated phase.

Fig. 7. Micrographs of areas irradiated to a dose of 10 dpa at different temperatures. Notice the second phase formation at high temperatures.

A micro diffraction analysis indicates that the new phase has a bcc structure with a lattice parameter similar to that of α-Fe. The structural relationship with the fcc-matrix is $(111)_{bcc}//(110)_{fcc}$ with $[0\bar{1}1]_{fcc}$ $//[1\bar{1}1]_{fcc}$ and $[2\bar{1}\bar{1}]_{bcc}//[21\bar{1}]_{fcc}$. Some reflections not belonging to the bcc phase are present; their origin could not be determined yet.

3.3.2. Thermal stability

In order to investigate the thermal stability of this type of material, differential thermal analysis and heat treatments are performed. The DTA measurements show an endothermal peak at about 325°C and an exothermal peak at about 600°C. Bulk specimens, heat treated at 600°C for 1500hr, and thinned for electron microscopy observation show the formation of carbides with an fcc structure, presumably of the $M_{23}C_6$ type. The carbides are formed at grain boundaries and within the grains, sometimes at planar defects such as microtwins present in the original material. The peak at 600°C observed during DTA measurements corresponds with the carbide precipitation observed after heat treatment. The absence of carbide precipitation observed during electron irradiation may be due to the short time at temperature and/or to a thin foil effect.

4. DISCUSSION

The present work is part of a complete evaluation of the Mn-Cr austenitic steels as an alternative choice for the Ni-Cr austenitic steels.

In this discussion our results will be compared with those obtained from HVEM irradiations of solution annealed Ni-Cr austenitic steels not doped with helium. Such results are available from commercial FV548SS [5] and from AISI 316SS with various carbon contents [6]. For the last one only the variety with 0.2 % carbon concentration, similar to the AMCR steel, will be taken into consideration.

The swelling results for the three steels are quite similar. The total swelling at 500 and 600°C of the AMCR steel is 20-50 % lower than for the 0.2 % C-AISI 316SS and the FV 548SS. The void density is about the same as for the FV 548SS and about 30 % higher than for the AISI 316SS. The increase of the swelling rate with increasing dose for temperatures higher than 500°C, which occurs in the AMCR steel is not observed in the AISI 316SS and the FV 548SS. This phenomenon is observed together with the formation of a second phase. According to our experiments carried out earlier, the AISI 316SS remains stable under HVEM irradiations at temperatures < 650°C [7]. Micrographs from the paper of Williams and Eyre [5] show that the FV 548SS is stable under HVEM irradiations at temperatures < 600°C. The increase of the swelling rate in the AMCR steel may be due to the phase instability as in predicted by theory of void swelling [8,9]. However, no interaction is observed between the voids and the newly formed phase. The exact mechanism of swelling enhancement is not clearly determined as yet.

The irradiation induced phase observed here has a lamellar form and a bcc structure. It has a martensitic type relationship with the matrix. It is not formed by a shear process but by a diffusional process. The fact that it is not found after heat treatment on bulk material at temperatures higher than 500°C suggests that it may be due to a thin foil effect. Other kinds of phase transformations are observed after heat treatments, where carbides are formed and after exposure to liquid lithium, where a hexagonal ε-phase is formed [1]. These phases, are not present in the electron irradiated specimens.

5. CONCLUSIONS

HVEM irradiation results of a manganese-chromium austenitic stainless steel of the type AMCR in the temperature rang 400-650°C are available. Compared with HVEM irradiation results obtained on some nickel-chromium stainless steels, the replacement of nickel by manganese has no adverse effect on the void swelling properties in the absence of helium. The peak swelling temperature is higher than 650°C as is the case for the nickel-chromium stainless steels.

An irradiation induced phase instability is observed at temperatures > 500°C. The second phase is α-Fe type; it is formed by a nucleation and diffusional growth process. This phase instability is probably a thin foil effect. No thermal phase instability is observed at temperatures < 600°C.

6. ACKNOWLEDGEMENTS

The authors wish to thank Mr. W. Timmermans from the S.C.K./C.E.N. for the DTA measurements and the team from the HVEM laboratory from the University of Antwerp for their technical assistance.

7. REFERENCES

[1] COEN, V., KOLBE, H., ORECCHIA, L., SASAKI, T., J.NUCL. MAT 85 and 86, 271 (1979)

[2] CREUSOT - LOIRE RESEARCH TEAM : private communication

[3] HARRIES, D.R. PRIVATE communication

[4] SNYKERS, M., JANSSENS, C., S.C.K./C.E.N. REPORT BLG 521

[5] WILLIAMS, T.M., EYRE, B.L.,J.NUCL.MAT 59,18 (1976)

[6] MAKIN, M.J., WALTERS, G.P.,AERE-R 7934 pp 299-312

[7] SNYKERS, M.,unpublished results

[8] BRAILSFORD, A.D.,J.NUCL.MAT., 56,7 (1975)

[9] MANSUR, L.L.,NUCL.TECHN. 40, MID-AUG (1978)

RANDOM WALK OF HELIUM BUBBLES IN VANADIUM

P.J. Goodhew and S.K. Tyler

Department of Metallurgy and Materials Technology,
University of Surrey,
Guildford,
England

ABSTRACT

Helium bubbles will be generated in the first wall of any fusion reactor. These bubbles will either cause swelling and embrittlement in their own right or act as precursors to a population of voids, depending on the temperature and various irradiation and material parameters.

Bubbles can, in principle, grow either by the re-solution and re-precipitation of the inert gas (ripening) or by migration and coalescence. We have studied the growth of bubbles arising from the room temperature implantation of helium to a concentration of $5 \times 10^{26} He/m^3$ in Vanadium. We demonstrate in this paper that during isothermal annealing at 950^oC ($0.55\ T_m$) faceted helium bubbles undergo Brownian motion and grow only by coalescence. No bubble was observed to shrink and hence ripening can be ruled out. The rate controlling process which limits the migration and hence growth rate at this temperature could be surface diffusion or the nucleation of ledges on bubble facets. We conclude both from our hot stage electron microscope observations and from isothermal annealing of bulk specimens that ledge nucleation is rate limiting even for bubbles less than 10nm in size. The energy of a monatomic ledge on a {100} surface in Vanadium is deduced to lie between 3 and 5×10^{-11} J/m. We also conclude that bubbles undergoing Brownian motion at $> 0.5\ T_m$ need not be spherical, as has frequently been assumed in the past.

1. INTRODUCTION

Populations of helium bubbles will be generated in the first wall of any fusion reactor. These bubbles will either cause swelling and embrittlement in their own right or will act as precursors to a population of voids, depending on the temperature and irradiation conditions. It is therefore essential that we understand the mechanisms of growth of the bubbles in order to analyse and eventually to predict their behaviour.

There are three mechanisms by which gas bubbles in a solid may grow: a) ripening, which would result from the re-solution of gas in small bubbles followed by its re-precipitation in large bubbles; b) Migration and coalescence or c) Injection of gas atoms and vacancies. Most bubble growth experiments to date have been carried out in the absence of displacement damage and because of the low measured solubility of helium in metals (1) it has been widely assumed that migration and coalescence is the only significant mechanism (e.g. 2,3). Only two groups of workers have observed bubble migration in metals directly (4,5) and in both cases an f.c.c. metal was studied. In Barnes and Mazey's experiments a temperature gradient was imposed, up which there was a directed flow of bubbles (4). Ruedl and Kelly (5) studied the movement of Krypton bubbles in platinum under isothermal conditions but presented no direct evidence of random migration and coalescence.

The mobility of gas bubbles is expected to drop rapidly as their size increases beyond a few nanometers diameter whatever rate-controlling mechanism is postulated (6). It is not clear therefore whether in isothermal conditions there will be sufficient mobility (in the absence of any local driving force) for Brownian motion of bubbles to lead to coalescence and hence growth of the mean size of the bubble population. Since our experiments (7), and those of others (3,8) indicate that the mean bubble size in metals does continue to increase at least up to bubble diameters of tens of nanometers it is necessary to discover whether the assumed growth mechanism does actually operate at the temperatures used. We therefore carried out some in-situ heating stage experiments to study bubble migration in Vanadium at 950°C within the electron microscope.

2. EXPERIMENTAL

Vanadium sheets (thickness 100μm) were initially annealed for one hour at 1050°C in a vacuum of 5×10^{-5} Pa. They were then implanted at room temperature with He^+ ions at four energies, resulting in a uniform concentration of helium $5 \times 10^{26} He/m^3$ in a zone between 200 and 650nm from the implanted surface. Disc specimens for electron microscopy were prepared by anodizing followed by back-thinning as described in detail elsewhere (9). The specimens were examined in the single-tilt heating holder of a JEM 200CX TEMSCAN microscope operating at 200 KeV.

The temperature of the specimen was raised rapidly to 950°C and the stage was allowed to stabilize for one hour. It was then possible to record micrographs at half-hourly intervals while the specimen remained at temperature. The bubble migration rate was slow enough to make microscopy possible at 200,000x but fast enough for measurable displacements to occur in half an hour. All micrographs were recorded in multi-beam kinematical conditions with approximately 0.5μm of underfocus.

3. RESULTS AND DISCUSSION

3.1 Bubble Shape

We observed that all bubbles which were large enough for their shape to be resolved (say > 3nm) adopted, at 950°C, a cuboidal shape bounded by {100} facets but with rounded edges. This can be interpreted as follows: If helium is effectively insoluble in Vanadium, even at 950°C, then after coalescence the resultant bubble will contain an excess pressure until sufficient vacancies can diffuse to the bubble to relieve it. This will take a long time at $0.55\ T_m$ (10). Bubbles seen in our experiments should thus not be equilibrium bubbles, in which the pressure exactly balances the surface energy, but will contain an excess pressure which will result in strain energy being stored in the lattice around the bubble. We can write this as

$$\text{Strain energy} \propto P - \sum \gamma_i A_i$$

where P is the actual gas pressure and γ_i is the surface energy of the i face with area A_i. Clearly in order to minimize the strain energy it is necessary to increase the sum of the surface energy terms. Now in the absence of a significant vacancy supply the bubble is of constant volume and hence can often maximise $\sum \gamma_i A_i$ by increasing A_i even if γ_i is thus reduced, in other words by forming faces of larger area but slightly lower energy per unit area. Thus instead of the expected sphere (maximum γ_i) we may see the development of lower energy planes of greater area. We have already shown that this can result in fully faceted cuboidal bubbles at room temperature in niobium alloys [10]. In the present case we clearly have the partial development of {100} facets either in response to a modest excess pressure or (perhaps more likely) because at this elevated temperature the energy minimum at the {100} orientation is much reduced, so that complete faceting is not possible.

3.2 Brownian Migration and Coalescence

The sequence of micrographs in Fig. 1 shows the truly random walk nature of bubble migration in the microscope at 950°C. The migration to and fro of an individual bubble was observed many times, as was the coalescence of two bubbles to form a single larger bubble [11]. There is no evidence from our experiments of the directed flow of bubbles which would be driven by a local gradient of stress or temperature. It therefore appears that random Brownian motion is sufficient to cause bubble coalescence at the bubble densities observed here (~ 10^{22} bubbles/m^3). We shall now examine this conclusion quantitatively to see if the likely rate-controlling processes for bubble migration could permit such migration distances in half an hour.

The two most likely rate-controlling processes are surface diffusion around the inside of the bubble or the nucleation of ledges on faceted bubble surfaces. We have previously derived expressions for the diffusion coefficients under zero net driving force for a cubic bubble of side a [7]. These are

$$D_{b_s} = \frac{4 D_s \Omega^2}{\alpha_s^2 \, a^4} \qquad .....(1)$$

if surface diffusion is rate-controlling and

$$D_{b_f} = \frac{\pi \, a \, D_s}{6 \, \alpha_s} \exp\left(\frac{-\pi \, a \, \varepsilon}{2 \, k \, T} \right) \qquad .....(2)$$

if the nucleation of monatomic ledges limits the bubble migration. Reasonable values of the parameters in these equations are : surface diffusion coefficient, $D_s = 1.14 \times 10^{-15}$ m^2 /sec [3]; surface jump distance, α_s = 1nm; atomic volume, $\Omega = 1.38 \times 10^{-29} m^3$. Figure 2 shows D_b as a function of a from the two equations with, in the ledge nucleation case, several possible values of ε, the ledge energy per unit length. It can be seen that, provided that $\varepsilon < 5 \times 10^{-11}$ J/m, the slowest and hence rate-controlling mechanism will be surface diffusion for small bubble sizes and ledge nucleation for larger bubbles. For bubbles of the average size observed in our hot stage experiments, about 2nm, a bubble diffusion coefficient of about 3×10^{-20} m^2/sec, limited by surface diffusion, is indicated. From this we can deduce the average distance migrated, $\bar{x}$, using the standard one-dimensional random walk expression

$$\bar{x} = \sqrt{2 D_b t}$$

For the conditions already quoted this expression gives a mean distance migrated of 10nm in half an hour. Since the mean separation of bubbles is about 50nm this in very good agreement with our observation of occasional coalescences. There can be little doubt that Brownian migration leading to random coalescences does take place at 0.55 Tm in Vanadium.

3.3 Rate Controlling Process

We have also conducted more extensive isothermal heat treatments outside the microscope and have analysed the growth of the average bubble size with time. The results, shown in Fig. 3, can best be fitted using growth models limited by ledge nucleation rather than surface diffusion [12]. From such experiments we deduce that the ledge energy ε on a {100} facet in Vanadium must lie in the range 3-5 x 10^{-11} J/m. The conclusion that nucleation of ledges is rate-controlling, even in a partially faceted bubble with rounded corners, is not surprising when we consider that the many steps and ledges implicitly present at the corners provides no nuclei for the removal of atoms from a face of the bubble (Fig. 4.) This is probably the true rate-controlling step for migration.

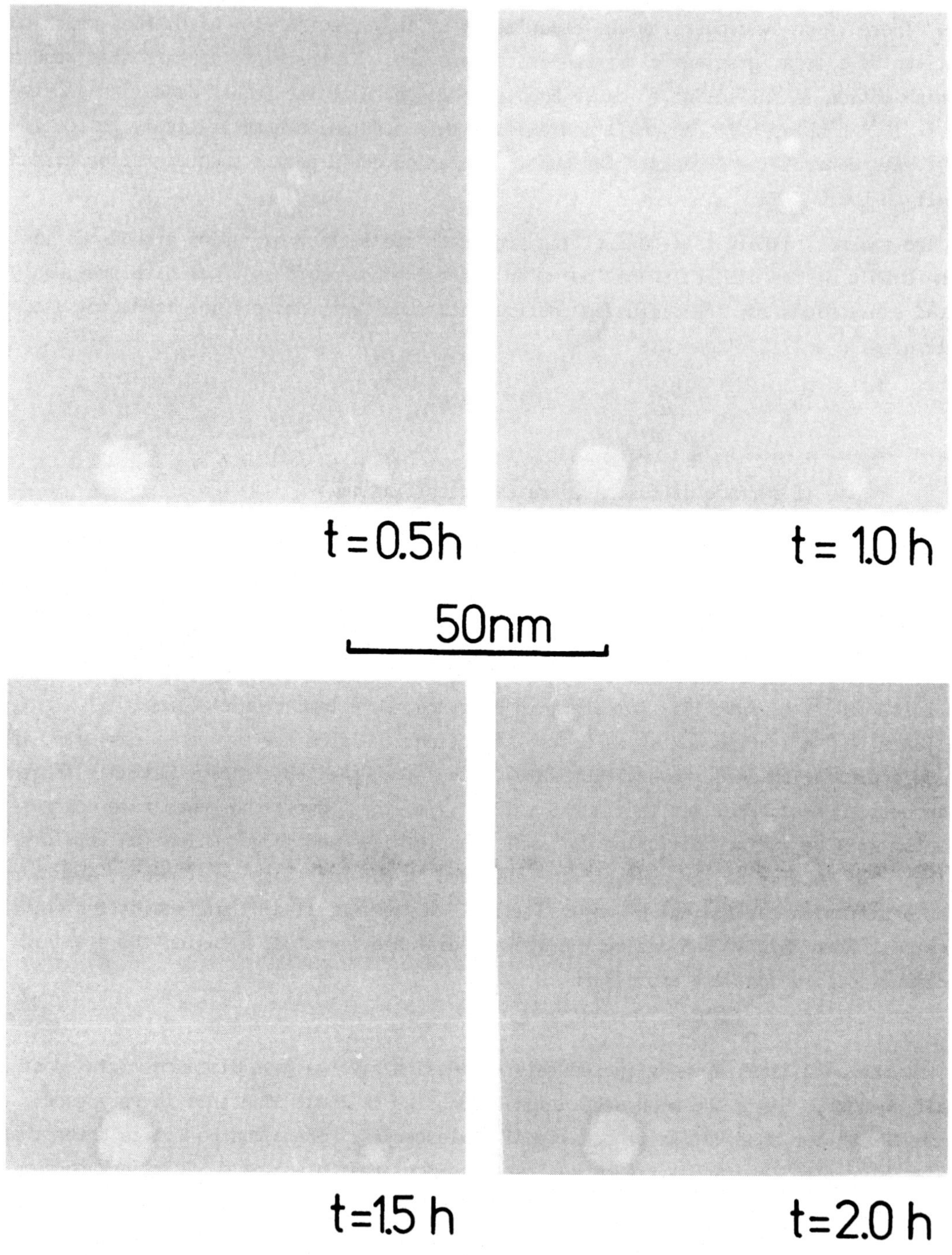

FIGURE 1 : A sequence of micrographs taken in the microscope heating stage at 950°C. Several of the smaller bubbles can be seen to have migrated at random with respect tothe largest bubble.

5. REFERENCES

[1] RIMMER, D.F., and COTTRELL, A.H., Phil. Mag. 2 1345-1353 (1957)

[2] CHEN, K.Y., and COST, J.R., J. Nucl. Mat. 52 59-74 (1974)

[3] SMIDT, F.A. and PIEPER, A.G., J.Nucl. Mat. 51 361-365 (1974)

[4] BARNES, R.S. and MAZEY D.J., Proc. Roy. Soc., A 275 47-57 (1963)

[5] RUEDL, E. and KELLY, R. J. Nucl. Mat. 16 89-102 (1965)

[6] GOODHEW, P.J., TYLER, S.K. and WALDRON, M.B. In Preparation

[7] TYLER S.K. and GOODHEW P.J. J. Nucl. Mat. 74 27-33 (1978)

[8] BRASKI, D.N., SCHROEDER, H and ULLMAIER, H. J. Nucl. Mats 83 265-277 (1979)

[9] AITKEN, D, TYLER, S.K. and GOODHEW, P.J. J. Phys. E. 11 511-513 (1978)

[10] TYLER, S.K. and GOODHEW P.J., J. Microscopy 116 55-63 (1979)

[11] TYLER, S.K. and GOODHEW, P.J. Proc. 10th Int. Symp. "Effects of Radiation on Materials" Savannah 1980. In press.

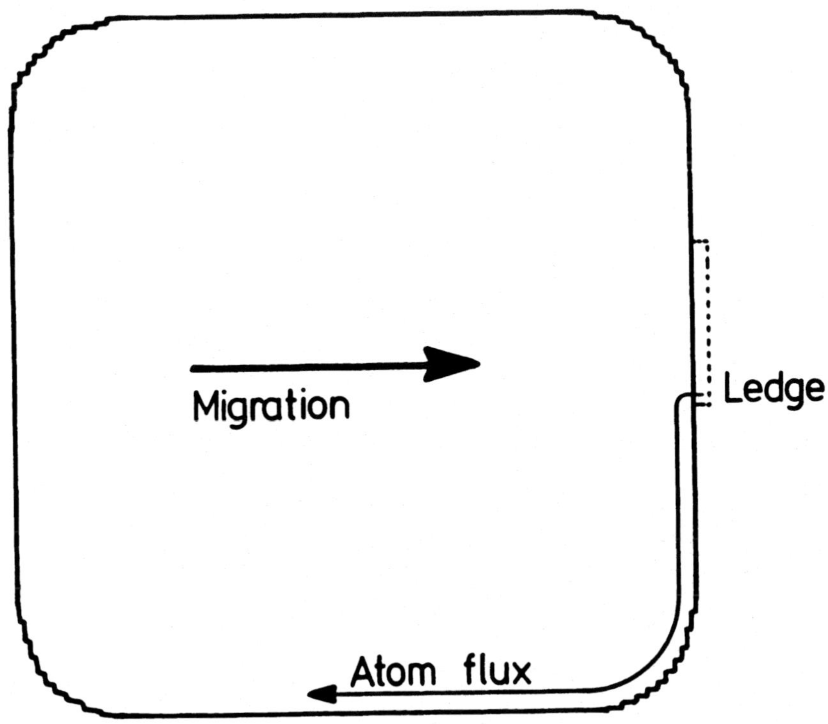

FIGURE 4 : Bubble migration occurs by diffusion of atoms around the internal surfaces of the bubble. Despite any rounding of the corners it is necessary to nucleate a ledge on the leading face before each monatomic layer can be removed.

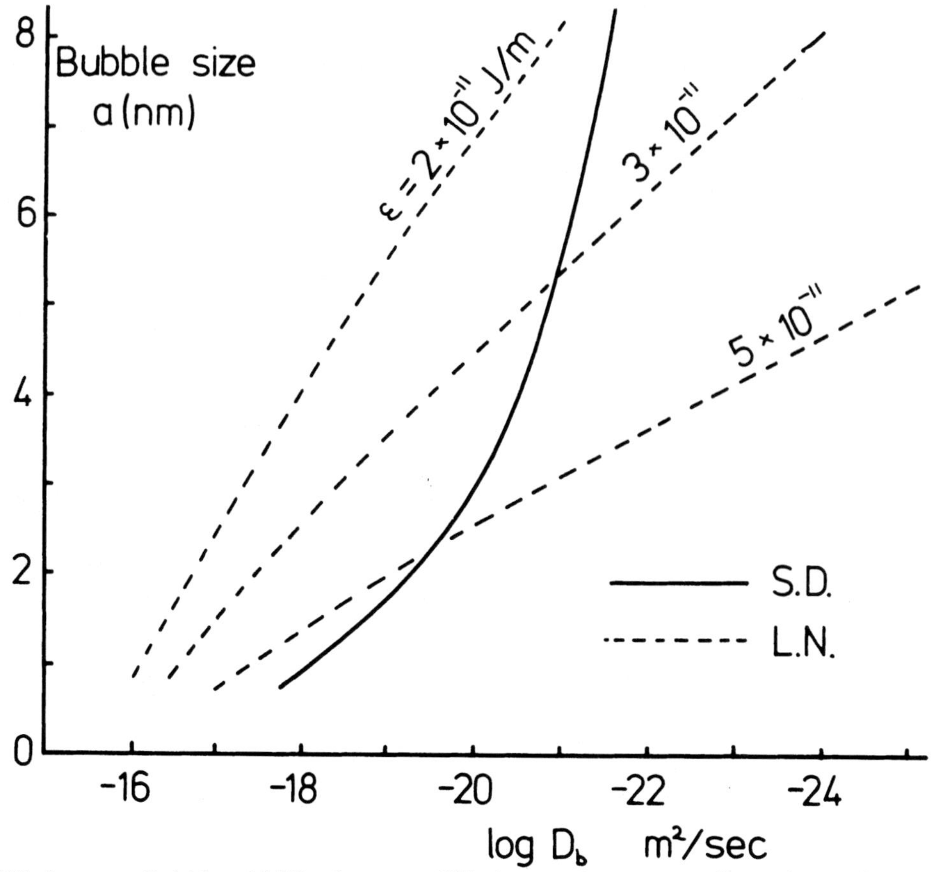

FIGURE 2 : Bubble diffusion coefficient, D_b, as a function of bubble size according to equation (1), for surface diffusion control (SD) and equation (2), for ledge nucleation control (LN). Three values for the ledge energy, ε, have been used for illustration.

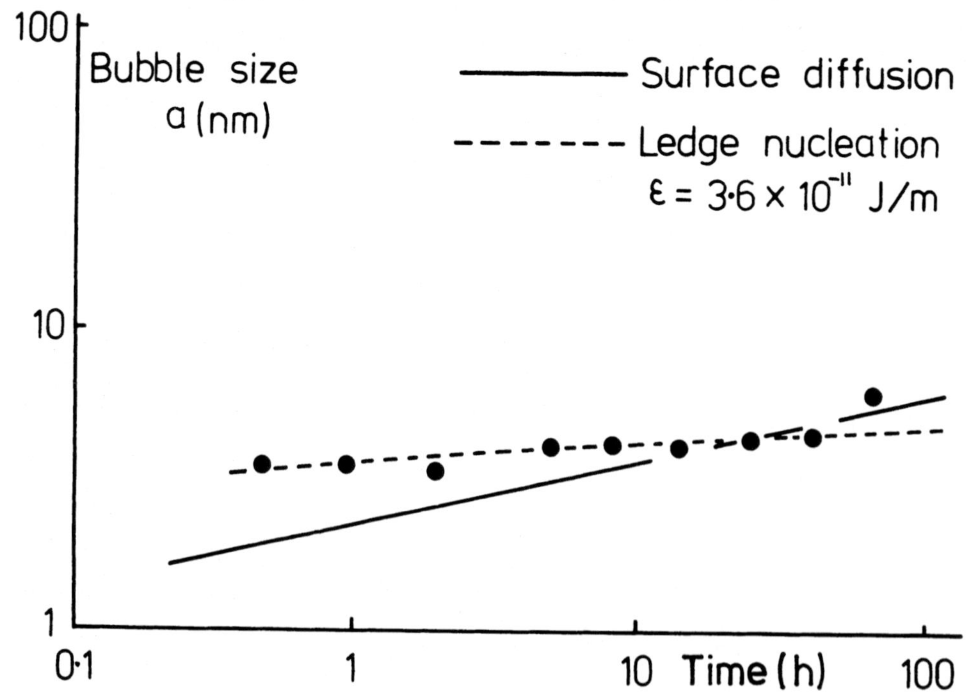

FIGURE 3 : Experimental results from isothermal anneals (full circles) compared with predicted bubble sizes using surface diffusion controlled kinetics (full line) and ledge nucleation controlled kinetics (dotted line). Clearly ledge nucleation gives the better fit.

ELECTRON BEAM WELDABILITY OF NITROGEN STRENGTHENED AUSTENITIC STRUCTURAL MATERIALS FOR FUSION MAGNETS

G. Friesinger, K.P. Jüngst, A. Nyilas
Kernforschungszentrum Karlsruhe
Institut für Technische Physik, Postfach 340, D-7500 Karlsruhe, FRG
and
G. Kuhnen
Brown, Boverie & Cie, TGF, CH-5242 Birr, Switzerland

ABSTRACT

In this work the application of the electron-beam (EB) technique for the joining of structural materials used in fusion technology is reported. The selected structural material for the casings of the superconducting torus TESPE is a fully austenitic AISI 316 LN type stainless steel (SS). EB-welding of this type of steel brings serious problems because of the depletion of the dissolved nitrogen, resulting in metallurgical damage of the joints. The variation of the welding parameters and their effect on weld quality are studied. The major weld parameter for avoiding weld pool porosity is the heat input. For a 15 mm thick butt joint (L-type) the welding parameter is optimized by different EB-weld tests. It is demonstrated that a 0.3 mm foil, used as a filler material, was capable of dissolving excess nitrogen during the welding period resulting in a significant reduction of nitrogen depletion in the weld metal.

1. INTRODUCTION

As part of the development work on fusion reactor systems, the testing of large superconducting magnets under different load conditions is necessary. Therefore a small compact toroidal magnet system (TESPE |1| with six D-shaped superconducting coils is under construction at present. The mean torus diameter of this system is approximately 1.2 m. Fig. 1 represents the main dimensions of the coil casings. This device will be capable of storing a maximum energy of ~ 9 MJ. The superconducting NbTi coils are enclosed in fully welded SS casings to withstand the high mechanical forces transmitted by the magnetic field of ~ 7 Tesla during the operation at liquid helium environment. The SS must have the following mechanical properties at the operating temperature of 4 K, a yield strength of > 1000 MPa, tensile strength > 1500 MPa, and a plane strain fracture toughness of > 100 $MPa\sqrt{m}$. The casings, however, have not only to take the magnetomechanical forces of the coils but are also designed to act as helium vessels for the bath cooled coils, and therefore they have to be helium tight.

To prevent superconductor movement the coils have to fit into the casings with a minimum tolerance. Thus a minimum distortion of the geometry after the welding process is essential. Especially this requirement led to the application of the EB-welding technique for the joining process of the SS used for the coil casings. EB-welding is capable of producing single pass weld joints with a very narrow fusion zone. In comparison to conventional arc-welding processes, the distortion of the EB-welded construction is minimal. In addition, the weld metal contamination during the melting period of the weld pool is ne-

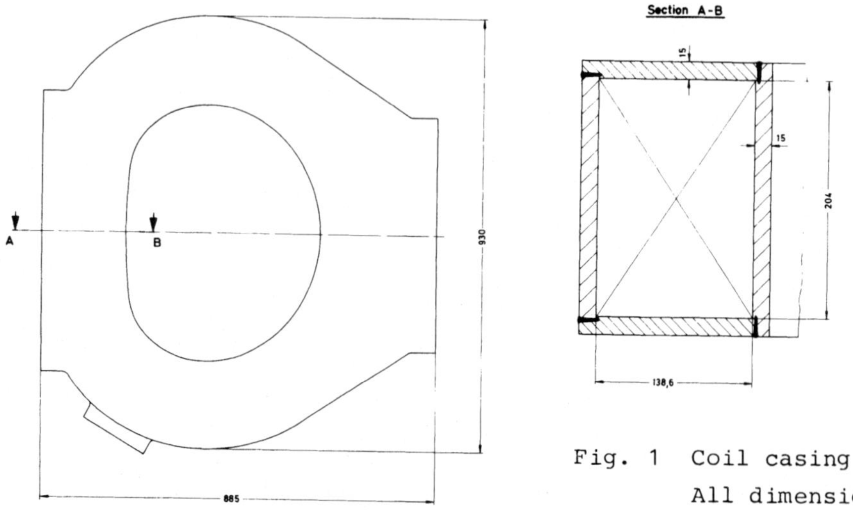

Fig. 1 Coil casing of TESPE system.
All dimensions in mm.

gligible, as the entire procedure takes place in a vacuum chamber.

The main objective of the present work is to represent the EB-weldability of the selected SS. The careful selection of the EB-welding parameters is described. Procedures to overcome the obvious difficulties arising from nitrogen depletion are discussed in detail and a test programme for verification of the EB-weldability for the coil casing joints is outlined.

2. MATERIAL

The material for the coil casing is a nitrogen strengthened austenitic chromium nickel SS. Nitrogen as alloying element in SS is a strong austenitic stabilizer and increases the yield point drastically because of its interstitial solubility in the austenitic phase. The selected material with the German specification number 1.4429 (~ AISI 316 LN) has superior cryogenic mechanical properties and it fulfils the main requirements for the use as coil casing for the fusion magnets. Material chemistry and tensile data are presented in table I. The alloy was supplied by the manufacturer in a commercially fully softened condition (heat treated at 1050°C for 1/2 hour and water quenched). Apart from the good mechanical and metallurgical behaviour of the selected base metal, a good joining characterization is an essential requirement for the construction of the designed coil casings. Reported data |3, 4, 5| on the EB-weldability of nitrogen strengthened austenitic SS suggest serious problems with weld pool eruptions during the EB-joining process that results in weld

TABLE I

BULK MATERIAL PROPERTIES (MATERIAL 1.4429)

Chemistry, wt %

C	Si	Mn	P	S	Cr	Mo	Ni	N	Fe
0.032	0.41	1.26	0.016	0.016	16.7	2.68	13.7	0.16	bal.

Mechanical properties [2]

Temperature K	Yield strength MPa	Tensile strength MPa	Elongation %	Reduction of area %
295	431	709	39	70
4	1160	1580	38	44

splatter and porosity. Poor weld quality of this type of SS has often limited its application in the past.

3. THERMODYNAMICS OF NITROGEN AND EB-WELDING

The main cause of the weld porosity in EB-welds of nitrogen bearing SS is the rapid evolution of dissolved nitrogen. The outgassing phenomena of nitrogen in liquid metal are known and can be thoroughly explained by the thermodynamics. The solubility of nitrogen in liquid pure iron obeys Sievert's Law at pressures up to 1 bar. In addition, nitrogen was found [6] to obey Sievert's Law if iron is alloyed with various elements, too. According to Sievert's Law

$$\%\ [N]_{Fe} = K_N \cdot P_{N_2}^{1/2} \qquad (1)$$

where K_N = equilibrium constant; P_{N_2} = nitrogen partial pressure in bar; $\%\ [N]_{Fe}$ = nitrogen content in liquid iron in % weight. The solubility of nitrogen in pure iron at 1873 K is 0.045% under 1 bar nitrogen partial pressure. The influence of additional alloying elements can be readily given by interaction coefficients of each individual element according to the extended eq. (1)

$$\%\ [N]_{Fe} = \frac{K_N}{f_N^{Fe-i}} \cdot P_{N_2}^{1/2} \qquad (2)$$

where $$\log f_N^{Fe-i} = \sum_{i=1}^{n} e_N^i \cdot \%\ [i]_{Fe}$$

is the sum of all interaction coefficients in the alloy. Table II represents the interaction coefficients e_N^i of major alloying elements in the SS.

Taking the given interaction coefficients of table II and the concentrations of alloying elements in table I one calculates a nitrogen solubility at 1873 K of SS 1.4429 of 0.17% at 0.79 bar nitrogen partial pressure. This

TABLE II

Interaction Coefficients, e_N^x, in Liquid Fe-N-X Alloys at 1873 [6]

Element, x	C	Si	Mn	P	S	Cr	Mo	Ni
Interaction coeff. e_N^x	0.25	0.047	-0.020	0.051	0.013	-0.045	-0.011	0.010

calculated value is in excellent agreement with the analysed nominal value of 0.16% nitrogen (table I). Reducing the partial pressure of nitrogen to the approximate value in an EB-vacuum chamber of ~ 10^{-4} mbar results, according to Sievert's rule, in a nitrogen content in the liquid alloy of ~ 7.10^{-5}%, which is nearly 2500 times lower than the equilibrium content of nitrogen in this alloy under normal conditions. However, because of the high density and high surface tension of liquid steel, the growth of nitrogen bubbles under equilibrium conditions is limited to a nitrogen partial pressure > 10 mbar [7], which corresponds to a nitrogen content of ~ 0.023%. This value must not be exceeded, if nitrogen release has to be prevented. Therefore the high difference in solubility for vacuum and atmospheric conditions, respectively, is responsible for nitrogen gas evolution. When a nitrogen saturated melt is brought rapidly from 1 bar to vacuum environment, which in reality happens during EB-welding, the result is vigorous bubble evolution and weld damage. One way to overcome this difficulty is to find optimum values for the weld parameters which influence the kinetics of the process. If nitrogen release occurs, the melt must be carefully controlled by heat input parameters to prevent the solidification of the weld pool with the existing bubbles, moreover the entrapped gas bubbles must be outgassed before solidification starts.

4. EB-WELD TESTS

A series of tests was conducted with two high voltage (150 KV) EB-equipments: a small Heraeus 3 KW (at KfK) and a Heraeus 30 KW EB-welding facility [8]. The first tests were made with the smaller equipment at constant beam current level (10 mA) to observe the penetration of the weld zone into the material 1.4429. Fig. 2 represents the results of this investigation. The welding speed was varied from 40 mm/s to 5 mm/s (lower plot). The EB-gun was placed in this case in a vertical position and the maximum depth achieved was ~ 8 mm. It was observed that an increased penetration depth decreased the weld zone quality with respect to porosity due to nitrogen evolution. The joint configurations for this series of the EB-weld tests were all butt welds on plate and were prepared using well machined workpieces.

For the tests with the 30 kW EB-equipment the joint configurations were L-type butt welds which allowed the simulation of the coil casing joint. Rings of ~ 300 mm diameter were radially welded with round plates. The EB-gun was placed in a horizontal position. The upper plot in fig. 2 shows the penetration results for various speeds at 18 mA current level. By further increasing the beam current up to 120 mA, the penetration could be increased up to 16 mm at the high welding speed of 30 mm/s. It was detected, however, that for constant penetration depth of 16 mm the increase of the welding speed deteriorates the weld zone quality drastically. Figure 3 represents the surface of longitudinally machined weld sections resulting from this series of the tests. Tests with radial beam rotation at 50 Hz level showed no improvements of weld quality, moreover a slight increase of individual pore sizes was detected, which is a contradiction to previous reports on similar tests |9|.

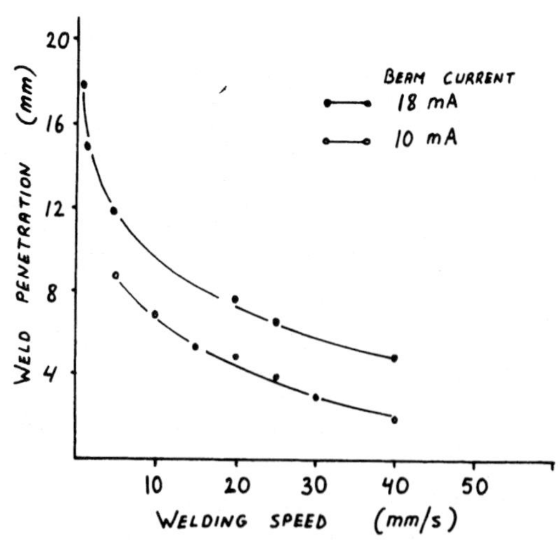

Fig. 2 Weld penetration of material 1.4429 at various speeds

ELECTRON BEAM WELD TESTS OF MATERIAL
1.4429 (~ AISI 316 LN)
BUTT JOINT (L-TYPE) 15 MM THICKNESS

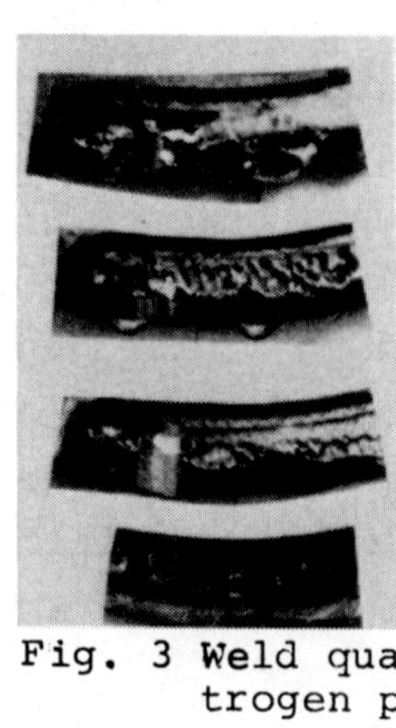

Beam Current mA	Welding Speed mm/s	Beam Rotation circular 50Hz
120	30	yes
70	20	no
35	10	no
20	5	yes

Fig. 3 Weld quality with reference to nitrogen porosity in the fusion zone

Results of nitrogen analysis by the inert gas fusion technique revealed a maximum decrease of nitrogen content in the vicinity of the EB-fusion zone of approximatly 15% in the worst case (with vigorous bubble evolution). This means, the stabilizing of the austenitic phase and also the strengthening effect of the nitrogen will be decreased.

In the next series of weld tests, the welding speed was decreased down to 1 mm/s. The observations showed that such a low welding speed results in a heating-up of the workpiece and gives rise to shrinkage of the joint. Therefore the welding speed was limited to 2 mm/s for ~ 16 mm weld penetration and a beam current of 15 mA. At this optimum welding speed, a slight improvement of the weld porosity was detected.

A significant further improvement of weld quality was achieved by inserting a nitrogen-free, thin filler material in form of a foil (0.3 mm thickness). The various tests with different filler materials revealed an optimum of weld quality for the joint configuration of 1.4429 (AISI 316 LN) bulk material and 1.4401 (AISI 316) filler material. The beam was focussed a few mm above the top of the filler material. Fig. 4 represents the micrograph of the

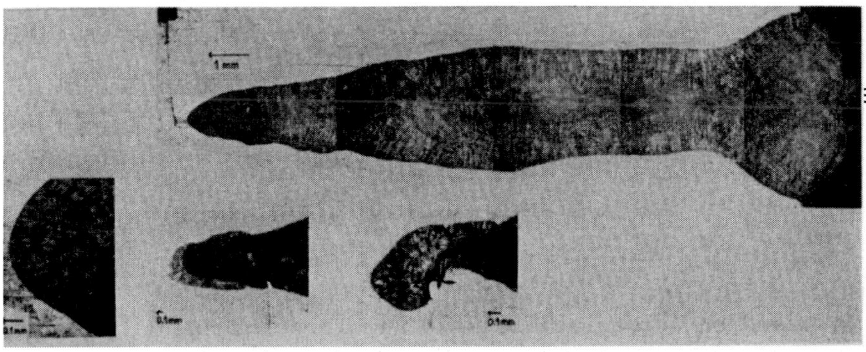

Fig. 4 Micrograph of weld metal with filler material in the midplane

weld zone in transverse section. No nitrogen porosity was detected at the zone where the filler material was inserted. A slight porosity in the root section was detected below 15 mm depth due to the absence of filler material in this portion of the joint. Fig. 5 represents the X-ray photograph of the weld zones

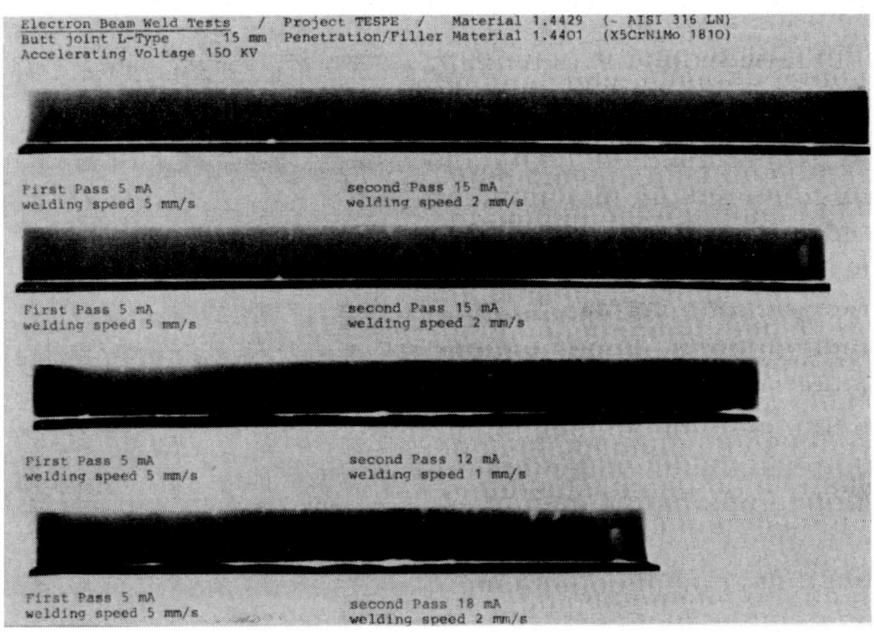

Fig. 5 X-ray photographs of different EB-weld tests

with no porosity in the region of filler material and with apparent porosity at the root. The top bead appearances were in all cases of excellent quality. These EB-weld tests were all performed on horizontally positioned workpieces. Further tests will give more information about the weld quality in case of 90° tilt of the weld joint. The capillary effect could reveal different welding behaviour in this case.

5. CONCLUSION

For the nitrogen strengthened austenitic stainless steel 1.4429 (~ AISI 316 LN) the influence of EB-welding parameters on weld quality was investigated by various EB-weld tests. It was observed that nitrogen bubble evolution, resulting in porosity, can be controlled by careful selection of EB-welding parameters.

In case of a butt joint (L-type) of 15 mm depth the use of a 150 kV gun, a beam current of 15 mA, and a welding speed of 2 mm/s resulted in sound welds.

A significant further decrease of porosity was achieved by insertion of a 0.3 mm thick filler into the midplane of the butt joint. The most suitable filler material is the nitrogen-free 1.4401 (AISI 316). Micrographs show no porosity in that weld zone where the filler foil was located.

6. ACKNOWLEDGEMENT

This work was based on a contract between Kernforschungszentrum Karlsruhe and Max-Planck-Institut für Plasmaphysik, Garching, concerning the cooperation in the field of superconductivity. It was supported by EURATOM.

7. REFERENCES

|1| Dustmann, C.H., Friesinger,G. and Jentzsch,K., SOFT 10, Padua, Sept. 1978

|2| Albrecht, C., LCT Final Report (Phase 1), Siemens AG, Erlangen 1980, unpublished

|3| Brooks, J.A., Welding Journal, 54 (1976), Res. Suppl. 189-195 s

|4| Casey, H., Metal Construction, July 1976, 296-299

|5| Messler, R.W., Journal of Nuclear Materials 85 & 86 (1979) 361-365

|6| Pehlke, R.I., Elliott, J.F., Trans. AIME 218 (1960) 1088-1101

|7| Kraus, Th., Schweizer Arch. für ang. Wiss. und Tech., 28 (1962) 452-471

|8| Kuhnen, G., Technical Report, Brown, Boverie & Cie (1979) TGF-3-TN 79/1

|9| Fletcher, M.J., Electron Beam Welding, Mills & Boon Limited, London 1971

IRRADIATION CREEP AND MICROSTRUCTURAL CHANGES OF STAINLESS STEEL UNDER LIGHT ION IRRADIATION

P. Jung, T.C. Reiley, W. Kesternich
Institut für Festkörperforschung der Kernforschungsanlage
Jülich, D-5170 Jülich, Assoziation-Euratom-KFA

ABSTRACT

The effect of impurity concentration and dislocation structure on the creep rate of type 316 stainless steel has been investigated. A 33% cold worked, low-impurity version showed higher creep rates and an extension of quadratic stress dependence to lower stresses compared to commercial 20% cold worked material. A dislocation structure much more homogeneous than by cold working, was obtained by prestraining solution-annealed 316 at high temperatures and high stresses. This homogeneous structure allowed a more detailed investigation of microstructural development under irradiation. The irradiation creep rates were about a factor of two lower than those of the cold worked material. Irradiation creep rates and their dependence on stress, temperature and composition are compared to fast reactor results.

1. INTRODUCTION

In addition to void swelling and high temperature embrittlement, irradiation creep can pose serious limitations to the lifetimes of structural components in advanced fission and future fusion reactors. Critical parameters for designing a tokamac's first wall like dimensional stability or strain-to-failure depend on irradiation creep processes. Since in-pile creep experiments are difficult and costly, simulation experiments by light ion irradiation offer a promising alternative for mechanistic studies of these processes. Advantages over in-reactor experiments are significant in the areas of higher defect production rates, lower radioactivity and better experimental versatility. A further advantage in simulating a fusion reactor environment is the adjustable ratio of helium to defect production rates by changing the energy of the light ions [1].

Important parameters in any irradiation creep theory are dislocation structure, dislocation-defect interaction and irradiation-induced defect structure. These parameters are studied in type 316 stainless steel by varying pretreatment, composition and irradiation dose.

The materials examined were:

1. Commercial type-316 stainless steel, 20% cold worked (316CW).
2. The same material, solution annealed for 1 hour at 1050% and crept thermally for 24 hours at 600°C under a stress of 275 MPa (316PC). After this treatment the material showed a uniaxial strain of about 10%.

3. A Fe-Cr-Ni-Mo alloy with type 316 composition, but with a very low level of C, Si, Mn etc. This material was used in 33% cold worked condition (316L-CW).

The irradiation creep experiments were performed at the Jülich Compact Cyclotron using 7 Mev protons and 10 Mev deuterons, respectively. Details of the apparatus are described elsewhere [2]. Microstructural analysis of the specimens was performed in a Philips EM 400, transmission electron microscope (TEM).

2. IRRADIATION CREEP

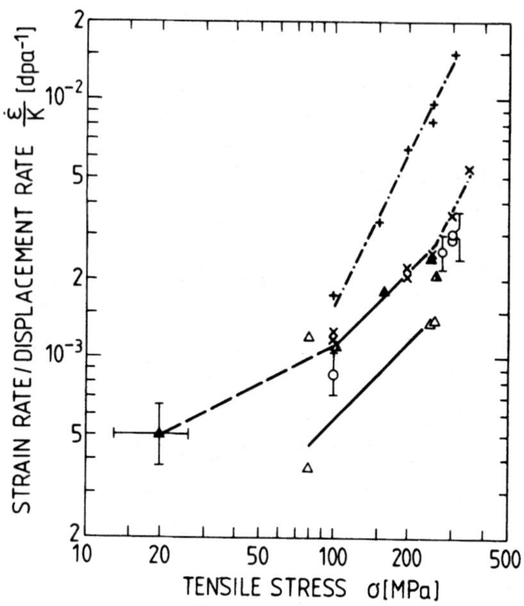

Fig. 1: Irradiation creep rate as a function of stress. The data points correspond to the following materials, irradiation temperatures and irradiation particles:

+: SS316L - 33% c.w., 300°C, 9.2 MeV-d

x: SS316 - 20% c.w., 300°C, 9.2 MeV-d

o: SS316 - 20% c.w., 300°C, 6.2 MeV-p

△: SS316 - 10% precrept, 300°C, 6.2 MeV-p

◭: SS316 -10% precrept, 375°C, 6.2 MeV-p

▲: SS316 - 10% precrept, 420°C, 6.2 MeV-p

The paricle energy is given at the front surface of the 50 μm specimens.

In fig. 1 the irradiation creep rates are given as a function of tensile stress. The low-carbon alloy (316L-CW) shows the highest creep rates and a quadratic dependence on stress from 100 to 300 MPa. Commercial 316CW shows quadratic stress dependence only in the high stress range (above 250 MPa) with a creep rate which is then by a factor of almost four lower than that of the low-carbon alloy. At lower stresses from 250 to 100 MPa a linear dependence on stress is observed. The data for the commercial steel are taken at an irradiation temperature of 300°C for proton and deuteron irradiations, respectively. The two sets of data agree quite well within the error bars of typically 20%, indicating that the displacement rates for the two types of particels are calculated consistently [1].

Recent creep rate data on 20% cold worked 316 from other authors using 13.2 Mev protons [3] are 40% lower than our data. Dpa calculation in Ref. [3] agree within 30% with those given in Ref. [1]. Experiments using 55 Mev [4] and 20.7 Mev [5] α-particles, respectively, employed dpa-calculations which differed by factors of 1.8 and 3.6 from the results in Ref. [1]. If the dpa-calculations of Ref. [1] are used, creep rates are obtained which are by about 50% higher [4], respectively by about 50% lower [5] than the present data. Measurements of the temperature dependence [1] showed, that the creep rate of SS316 increases by a factor of about two by changing the irradiation temperature from 300°C to 400°C. The same dependence on temperature was found for the precrept specimens which, at identical temperatures, showed by about a factor of two lower creep rates then the cold worked material. The precrept material also shows linear stress dependence between 100 and 300 MPa. A single data point at 20 MPa indicates a tendency to an even lower stress dependence below 100 MPa. As the very low creep rates in this regime may cause large experimental errors, (compare the data points at 80 MPa) further investigations are necessary to substantiate this result.

3. MICROSTRUCTURE

Fig. 2: SS316 CW, unirradiated, showing very inhomogeneous dislocation structure

After 20% cold working the commercial material shows the well known structure of slip bands and deformation faults, which separate areas of much lower dislocation density (fig. 2). This very inhomogeneous distribution of dislocations precludes the determination of reasonable numbers for dislocation density.

On the other hand the precrept specimens show a fairly homogeneous dislocation structure (fig. 3) in agreement with Refs. [6,7]. Individual slip planes can be distinguished

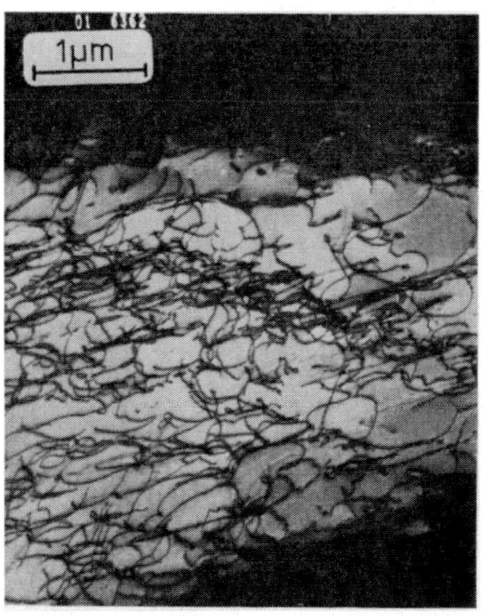

Fig. 3: SS316 PC, unirradiated, showing homogeneous dislocation structure

Fig. 4: SS316 PC, unirradiated, looking edge on at a slip system

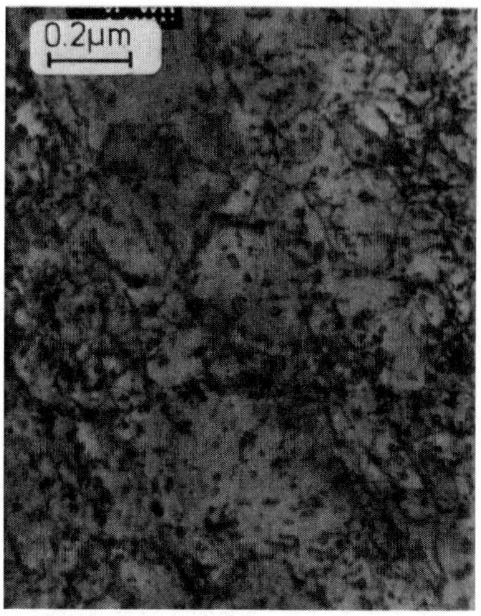

Fig. 5: SS316 PC, irradiated at 300°C to a dose of 0.2 dpa, showing a dense population of loops

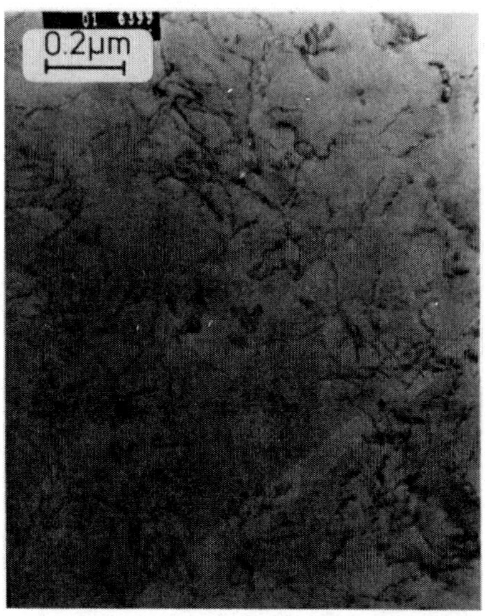

Fig. 6: SS316 PC, irradiated at 375°C to a dose of 0.28 dpa, showing few large loops, partially "granulated".

by looking edge on at a slip system (fig. 4). Microstructural parameters, derived from unirradiated material and after irradiation are listed in table I.

Irradiation at 300°C to a dose of 0.11 dpa creates small loops, mostly visible as black-white contrasts. This structure is easily distinguished from the few deformation loops observed in the unirradiated material. At doses around 0.2 dpa the loop population is further increased (fig. 5), at the expense of the network dislocation density. As stress was frequently changed during the irradiation creep experiments, only limited information on the stress dependence of microstructure was obtained. For example at 300°C a specimen which was irradiated at stresses between 80 and 250 MPa to a dose of 0.21 dpa (see table I) showed network dislocation density (ρ_N), loop number density (n_L) and typical (d_L) and maximum ($d_{L,max}$) loop diameters well within the range of the results of an unstressed specimen of similar dose.

At 375°C (0.28 dpa) the loop number density dropped drastically with only few large loops observable. Some of them have a granulated shape and clearly show stacking fault contrast in fig. 6.

At 420°C, most loops are already entangled into the network and only occasionally isolated loops can be distinguished.

With respect to total dislocation density, no significant change by irradiation is observed. At 300°C the total dislocation density may be slightly increased due to the dense loop population which overcompensation the reduction in network density. But spatial variations preclude a safe statement. At 420°C, the dislocation density is probably slightly reduced by irradiation Furtheron the more or less planar arrangements of dislocations on the different slip planes (fig. 4) is gradually transformed to a true three dimensional network, and the dislocation structure is coarsened by the appearance of long

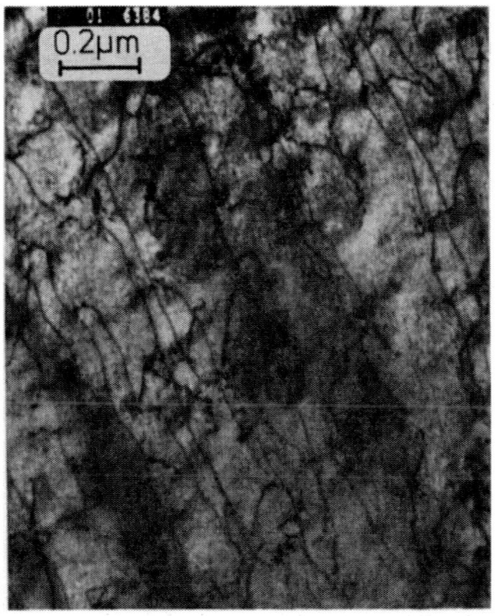

Fig. 7: SS316 PC, irradiated at 420°C to a dose of 0.26 dpa, showing long dislocations.

TABLE I

Microstructural parameters of 316 PC

T_{irr} [°C]	Ø [dpa]	ρ_N [$10^{14}/m^2$]	n_L [$10^{20}/m^3$]	d_L ($d_{L,max}$) [10^{-9}m]	ρ_{total} [$10^{14}/m^2$]
unirradiated	0	1.8±0.6	0	0	1.8±0.6
300°C	0.19	1.7±0.3	22.±3.	17(50)	2.9±0.5
σ=0MPa	"	0.6±0.1	38.±4.	12(50)	2.0±0.3
300°C σ=80-250	0.21	0.6±0.2	22.±6.	12(50)	1.4±0.5
375°	0.28	-	-	-(70)	1.6±0.2
420°	0.09	-	-	-	1.7±0.3
"	0.26	1.0±0.2	0.5±0.2	50(100)	1.0±0.2

dislocation segments (fig. 7).

The 316L-cw specimens show still higher concentration of slip bands and deformation faults than the 20% cold worked commercial material. Only in a few areas distinguishable dislocations are observed. Therefore no detailed analysis of microstructural development was possible. The general behaviour was similar as described above: A close loop population at 300°C at doses above 0.1 dpa and larger loops at higher temperatures. Again at intermediate temperatures (350°C) the faulted, "granulated" loops are observed occasionally. The deformation structure (fig. 2) was not significantly reduced up to the highest doses (0.3 dpa) used.

4. COMPARISON TO FAST NEUTRON RESULTS

From fast reactor irradiation it is known that solution annealed and 20% cold worked 316 approach at high doses similar dislocation structures. Around 420°C, the saturation density was determined to $3.10^{14}m^{-2}$ [8] and about $4.10^{14}m^{-2}$ [9], respectively. That means, the dislocation structure of 316 PC is similar in homogeneity and density to the structure of high-dose-irradiated 316.

On the other hand relatively large changes of microstructure seem to be only of little influence on the creep rate of stainless steel, as the very different dislocation structures of the 20% cold worked and the precrept material result only in a factor two difference in creep rate. Furtheron nucleation and growth of dislocation loops, especially during the 300°C irradiation cause no perceivable change in creep rate. Creep rates become measurable at doses around 0.02 dpa and remain constant up to the highest doses obtained (0.3 dpa). Transient effects have been observed at low temperatures in nickel [10,11] but in stainless steels only at high stresses and temperatures [4,12] where also transient thermal creep rates are appreciable [13].

To avoid systematic errors in comparing irradiation creep rates of light ion and fast neutron experiments, the following points must be considered:

a) Dose rates in terms of atomic displacement rates (K) must be calculated consistently. This was done in Ref. [1] for light ions. There it was shown, that displacement rates for fast neutron irradiation as calculated by the NRT-formalism [14] must be reduced by a factor of about 0.35 for stainless steels to account for spontaneous recombination of defects in large cascades. In the following this "efficiency" correction is applied to the data.
b) The stress state (σ) of the specimens will be transferred to tensile stress by using the Soderberg [15] formalism.

For linear stress and dose rate dependence steady state irradiation creep rate $\dot{\varepsilon}$ can be characterized by the so-called creep coefficient $C = \dot{\varepsilon}/K\sigma$ (in units of $[10^{-6}$/dpa MPa$]$). Dose rate effects are difficult to measure in fast reactor experiments as different dose rates obtained at different reactor positions are inevitably coupled with a change in neutron spectrum [16,17]. On the other hand in light ion simulation experiments a linear dose rate dependence of irradiation creep rate has been savely established for cold worked stainless steels between 300°C and 500°C for dose rates from about 1 to 5.10^{-6} dpa/sec [5,12,18].

The present data give at 300°C, C=5. for 316 PC and C=10. for 316 CW. By about a factor of two higher values are obtained at 420°C. Recent evaluation of DFR data [16] give C=10. for 316 CW in the temperature regime 247-360°C. ANL data [19] at 400°C give a creep rate of 316 CW which increases with dose, approaching C=5. at high doses. That means both pretreatments of the commercial 316 used in the present investigations give creep rates which are within the scatterband of fast neutron data. Considering the microstructure, it is felt that the precrept material may even give a better approximation of the high dose neutron data.

A further important point is, that for both materials the same dependence on stress (linear) and temperature is observed as for SS316 CW in fast reactor experiments [20,21].

Type 316 stainless steel in which carbon was precipitated by aging showed in fast reactor experiments 2.5 times higher creep rates than unaged material [19]. This factor is consistent with the present result of a four times higher creep rate in carbon free material. The in-reactor experiments also give a quadratic stress dependence [21] for the aged material between 100 and 200 MPa in accordance with our results.

A final comparison of light ion to fast neutron results can be made by looking at the relative creep strength of type 316 versus 321 stainless steels. Light ion results on 321-60% c.w. at 500°C give C=50. [12, 1] as compared to C= 20. for 316-20% c.w. at 420°C [1]. If the difference in temperature is considered (the difference in cold working may be disregarded [20]), irradiation creep strength of 321 is obtained to be by about a factor of two lower than that of 316. This agrees with in-reactor results [22] where factors between 1 and 2 are observed.

5. CONCLUSIONS

1. Type 316 stainless steel precrept to 10% strain at 275 MPA and 600°C shows a homogeneous microstructure in contrast to cold worked material. This structure allows observation of microstructural development under irradiation already at low doses.

2. Irradiation creep rates of this material are by a factor of two lower compared to 20% cold worked 316.

3. Irradiation creep rates in light ion simulation experiments show identical dependence on stress and temperature, on composition and on reduction in carbon content as data from fast reactors.

4. Absolute creep rates for type 316 stainless steel derived from different light ion simulation experiments agree within a factor of two and are within the range of fast reactor data if displacement doses are calculated correctly.

5. Light ion experiments are therefore equivalent to fast reactor experiments for simulating irradiation creep in a fusion reactor environment.

6. REFERENCES

[1] P. Jung, C. Schwaiger, H. Ullmaier, J. Nucl.Mater. 85+86, 867-871 (1979)

[2] P. Jung, J. Viehweg, C. Schwaiger, Nucl. Instr. and Methods 154,207-212 (1978)

[3] E.K. Oppermann, J.L. Straalsund, G.L. Wire, Nucl. Techn. 42,71-81 (1979)

[4] T.C. Reiley, R.L. Auble, R.H. Shannon, J. Nucl. Mater. 90,271-281 (1980)

[5] P. Blanchard, P. Jary, J. Delaplace, J. Nucl. Mater. 74,267-276 (1978)

[6] K.D. Challenger, J. Moteff, Met. Trans. 4, 749-755 (1973)

[7] H.J. Kestenbach, T. Liuz da Silveira, S.N. Monteiro, Met. Trans. A. 7A, 155-158 (1976)

[8] H.R. Brager, J. Nucl. Mater. 57, 103-118 (1975)

[9] P.J. Barton, B.L. Eyre, D.A. Stow, J. Nucl. Mater. 67,181-197 (1977)

[10] D.J. Michel, P.L. Hendrick, A.G. Pieper, J. Nucl. Mater. 75,1-6 (1978)

[11] C.H. Henager, J.L. Brimhall, E.P. Simonen, J.Nucl.Mater. 90,290-296(1980)

[12] J.A. Hudson, R.S. Nelson, R.J. McElroy, J.Nucl.Mater. 65, 279-294(1977)

[13] P. Jung, A.A. Sagués, H. Schroeder, H.J. Auer, J.Nucl.Mater. 74, 348-350 (1978)

[14] IAEA Specialists Meeting, Nuclear Engg. and Design 33, 91 (1975)

[15] I.Finnie, W.R. Heller, Creep of Engineering Materials (McGraw Hill, New York 1959) p. 172

[16] J.P. Foster, A. Boltax, J. Nucl. Mater. 89, 331-337 (1980)

[17] G.W. Lewthwaite, D. Mosedale, J. Nucl. Mater. 90,205-215 (1980)

[18] C. Schwaiger, P. Jung, H. Ullmaier, J. Nucl. Mater. 90, 268-270 (1980)

[19] G.L. McVay, R.E. Einzinger, G.L. Hofmann, L.C. Walter, J. Nucl.Mater. 78, 201-209 (1978)

[20] J.L. Straalsund, Radiation Effects in Breeder Reactor Structural Materials, Scottsdale 1977, p. 191

[21] L.C. Walters, G.L. McVay, G.D. Hudman in [10], p. 277

[22] D. Mosedale, G.W. Lewthwaite, Creep Strength in Steels and High Temperature Alloys, Sheffield 1972 (The Metals Soc., London 1974) p.169

AN ALUMINIUM ALLOY FOR A NON BREEDING BLANKET NET/INTOR

F. Brossa, G. Musso and G. Piatti
Commission of the European Communities
Joint Research Centre - Ispra Establishment
Materials Science Division
21020 Ispra - Italy

ABSTRACT

All existing commercially available Al alloys contain a sufficient quantity of impurities, such as Fe, Ni, Mn, Zn..., so that their activity is important. These materials should not be used in NET/INTOR first wall/blanket structures. In order to achieve low residual radioactivity a new alloy, based on a very high purity aluminium (99.995) and high purity alloying elements selected from the few low activity elements such as Mg, Si, V, C, N and O, has been developed by the authors. The major features (property area: mechanical, physical, chemical compatibility) of this material have been experimentally determined in order to evaluate it as a first wall material for NET/INTOR.

1. INTRODUCTION

One of the options for the NET/INTOR design is that of considering the machine as composed of a basic structure and test facilities. The basic structure has to be designed as much as possible to be reliable for the all reactor lifetime. Low temperature operation for the first wall and blanket (non breeding), which implies no electricity production, appears attractive because one can expect that a number of the problems related to the thermodynamical design and radiation damage will be severe, as shown by the Ispra group [1]. In this framework the Al has been proposed and studied as an interesting competitor of steel [1]. Previously, suggestions concerning the use of Al in fusion reactor were done by the Brookhaven group [2].

The main interest in Al comes from its activation decay schemes which are constituted by short lived isotopes. Other advantages are relatively low cost, well established industry and plentiful resources. However it is recognized that it is not profitable to use pure Al because on INTOR/NET long term service at relatively high temperature (~120°C) is required. [1]. On the other hand it is not allowed to employ a commercially available Al alloy because all existing Al alloys contain a sufficient quantity of impurities, such as Fe, Ni, Mn, Zn..., to make their activity important [3]. In order to achieve low residual radioactivity new materials based on a very high purity Al (99.995) alloyed with high purity elements selected from the few low activity elements such as Mg, Si, V, C, N and O, have to be developed.

Taking into account these considerations the authors have developed an Al-0.7 wt% Mg - 0.4 % Si alloy starting from very pure Al (99.995) and alloying Mg 99.99 and Si 99.99. The material then has been characterized by an experimental evaluation of the microstructure and some selected properties particularly sensitive to impurities such as tensile properties, corrosion resistance, sputtering yield, plasma disruption, total infrared emissivity and electrical resistivity. The results are presented in the present work and compared with the analogous properties of the commercially available 6063 alloy, which was the model for the design of the present experimental alloy.

2. EXPERIMENTAL AND RESULTS

2.1. Alloy preparation

Different batches of the alloy have been prepared following the conventional metal-forming methods employed in the industry of aluminium: casting of ingots and then hot extrusion in order to produce shapes of circular cross section. The material has shown an excellent formability. The composition range of the alloys determined by chemical analysis is shown in the table I. The impurity level has been maintained below 100 ppm which is very low for an Al alloy if we consider that the average amount of impurities in the commercial

6063 alloy is of the order of 5000 ppm. In particular in our alloys some impurities as Ni and Mn (high activity elements) are at the level of 2 ppm which is acceptable [3].

To harden the alloy, the following precipitation hardening procedure has been adopted:

- Solution heat treating (510°C - ½ hr) of the extruded bars, in order to dissolve a maximum amount of the second phases in the α Al solid solution, and then water quenching in order to retain this solution down to room temperature.
- Precipitation treatment (175°C-24 hrs). At this temperature the super-saturated solid solution undergoes changes which lead to the formation of very fine precipitates (Guinier-Preston zones) dispersed in the Al matrix.
- In the end a full hardness is developed in the alloy.

TABLE I

Composition (by weight) of the Experimental and 6063 Alloys

Element	Experimental alloys	6063
Mg (%)	0.65 - 0.75	0.15 - 0.90
Si (%)	0.35 - 0.45	0.20 - 0.60
Fe (ppm)	30 - 40	3500 (max)
Cu (ppm)	20 - 30	1000 (max)
Ni (ppm)	1 - 2	500 (max)
Zn (ppm)	1 - 2	1000 (max)
Mn (ppm)	1 - 2	1000 (max)
Ti (ppm)	<1	1000 (max)
Cr (ppm)	<1	1000 (max)
Al -	balance	balance

This type of heat treatment, designated by T6 Temper symbol, is the same as that used in the 6063 alloy when a strong mechanical strength is requested It should be observed that in the commercial alloy the presence of the impurities give rise to other precipitates ($FeAl_3$, $FeAl_6$, Fe_6SiAl_8) probably in sufficient quantities to contribute to the alloy hardening [4].

A second age hardening, designated by T5, has also been performed. It corresponds to an artificial ageing treatment (180°C-24 hrs) after the extrusion operation without solution heat treatment. This type of temper is less hard then T6 type, but more useful from a practical point of view during fabrication in industrial processes.

Finally a third temper, designated O, and corresponding to a full annealing, has been considered to obtain a soft and more stable structure from a thermodynamic point of view. The use of the annealed treatment after fabrication is recommanded for various Al alloys for service in corrosion environment. The temper scheme is the follows: recrystallisation annealing (420°C-2 hrs) of the extruded bars followed by a very slow furnace cooling (10°C max per hr. to 150°C). This results in a two-phase structure ($Al+MgSi_2$) in which coarse particles of the second phase are dispersed in the Al (particle size 0.5 - 5 μm) due to coalescence process which occurs during the slow cooling in the furnace of the alloy bars.

The material as extruded and heat treated has been examined from microstructural point of view. Optical microscopy, transmission electron microscopy (TEM) and scanning electron microscopy (SEM) observations were employed for the investigation. Examples of typical microstructures observed of the alloys in the full annealed and age hardened conditions are given in Figs. 1 and 2 respectively. In the first case Keller'etch has revealed the coarse precipitates of Mg_2Si that formed during annealing for 2 hrs at 420°C (see the optical micrograph of the Fig. 1). In the second case the fineness of the precipitates (zones Guinier-Preston) is revealed only by TEM analysis as clearly shown in Fig. 2. Also note precipitation of some Mg_2Si particles along grain boundaries.

2.2. Mechanical Properties

Mechanical properties were evaluated for the heat-treated alloys (O Temper, T5 Temper and T6 Temper) by tensile tests at 20°, 100° and 120°C on round samples (diameter 8 mm and gauge length 40 mm). Figs 3-4-5 show respectively tensile strength, yield strength and elongation at rupture. The data concerning the 6063 alloy are taken from the literature [5]. A slight decrease is observed in the ultimate tensile strength of the experimental alloy (T6 Temper) relative to the commercial 6063 alloy (T6 Temper), while the differences in yield strength and elongation are very small. The results show that the effect of additional strengthening in 6063 alloy due to Fe impurities is less than expected and practically negligible [4]. From a practical point of view and for design purposes the experimental alloy (low impurity level) can be considered equally resistant as commercially available 6063. As concerning the artificially aged only treatment (T5), the hardening is confirmed less strong than T6. Finally in the full annealed material (O Temper) it is observed a strong softening effect (resistance decreasing and elongation increasing) because during the annealing treatment the density of dislocations decreases, the strain hardening effects are cancelled and the final microstructure is completely recrystallized. In fact the dislocation density is dependent on the impurity amount and, when this amount is reduced to a minimum value, as in the case of our alloy, the softening effect is more strong. Consequently in the high purity alloy there are minimum values for the resistance and maximum values for the ductility.

After rupture the specimens have been examined by SEM analysis of the fracture surface. In the age hardened samples(T5 and T6 Temper) the rupture is of intergranular type at room temperature (Fig. 6) whereas at 120°C there is a strong change (Fig. 7) in the sense that the rupture is characterized by many dimples which occurs near the Mg_2Si particles precipited. This type of rupture is based on void nucleation at particles or inclusions, growth of the voids by plastic deformation and subsequent coalescence of the voids untill the final fracture. This ductile rupture (dimple formation and so on) is observed also at room (Fig. 8) and at high temperature in the annealed broken samples.

From the previous remarks it is clear that the T6 condition gives the best combination of resistance and ductility at room and at 120°C of the three Tempers considered. However to make a conclusion for selection and design considerations it is necessary to dispose of creep data. These tests have been planned.

An alternative to the Al-0.7 Mg - 0.4 Si (impurity level $<$ 100 ppm) which comes from the 6000 series alloys, is Al-3 wt% Mg (impurity level $<$ 150 ppm) now under development at ISML-Alumetal (Novara). In this case the alloy is hardened by solid solution and cold working (not hardenable by heat treating). This fact can be present some practical advantage, less costs for example. Some data of this alloy [6] are presented for this alloy.

2.3. Corrosion tests

The assessment of the corrosion resistance of the alloy has been performed on samples (diameter 16 mm, thickness 1.5 mm) obtained from the extruded bar not heat treated) in an autoclave at 120°C in high purity (deionized) water up to 4000 hrs. This test has been planned following the general observation that a drastic effect of pressure and temperature is observed in the corrosion of Al by water [4]. The results do not confirm the previous observation as can be seen from Table II. After an initial very low weight change due to formation of a protective layer of oxide the corrosion is stopped. Examination of the samples after the test by optical micrography reveals the presence of some pittings very localized, but not exfoliation or intergranular corrosion or other indications of corrosion.

2.4. Physical properties

2.4.1. Physical sputtering

In order to evaluate the erosion rate of the first wall of NET/INTOR it is necessary to know the sputtering coefficients. The experimental values have been measured on samples of extruded Al-0.7 Mg-0.4 Si alloy for energies between 0.1 and 10 KeV. The sputtering yield values obtained are presented in Fig. 9. They are nearly identical, other conditions being equal, to the corresponding Al data previously measured [7], which have been utilized for the wall erosion calculation [1] taking into account 10^6 pulses of 100 sec. as INTOR life and a particle flux 47.5% D, 47.5 T, 5% He. The erosion value obtained ($\sim$ 1.5 cm) can be considered valid also for the our experimental alloy.

TABLE II

Corrosion Resistance to Water at 120°C

Time (hrs)	Weight Variation (mg/cm^2)
800	+ 1.4
1200	+ 1.2
4000	+ 1.2

2.4.2. Plasma disruptions vaporization

Different evaluations have been given in order to estimate the rate of evaporation for a disruption and with results often desagreeing [8]. With the purpose of giving an experimental contribution to this problem, the thermochemical behaviour of the first wall has been simulated by a thermal shock test, generated by an electron beam of up to 6 kW focused on a surface of 10 mm^2 (pressure 10^{-5} Torr) with a heat load of 60 kW/cm^2 for 40 m sec has been applied to different specimens of our Al-0.7 Mg-0.4 Si alloy (as extruded material). Disruption parameters similar to those of INTOR have been performed: 15-20 kW/cm^2 as power density focused on the sample surface and time disruption slightly more than 20 m sec calculate for the INTOR [1]. The damage produced is shown in the optical micrograph of Fig. 10. A rough damage calculation based on experimental value confirms that the erosion of first wall due to plasma disruption vaporization is marginal compared with sputtering [1].

2.4.3. Total optical emissivity

The total hemispherical emissivity in the infrared region at 85°C has been measured on discs of the our extruded alloy of 50 mm diameter mechanically polished using an apparatus constructed in Ispra Laboratories and based on the calorimetric method [9]. The obtained values: 3.2% and 3.1% show a very low thermal emissivity in the infrared region similar to pure Al.

2.4.4. Electrical resistivity

Samples 5 mm in diameter and 100 mm in length, prepared by turning the extruded quenched and Temper (T6) bars, were used for resistance measurements. These were performed at -196°C by the potentiometrical method using a Honeywell Rubicon potentiometer. The following values has been obtained:

ρ = 727 nΩ-cm (as extruded and quenched from solution treatment condition

ρ = 530 nΩ-cm (as extruded and Temper T6 condition)

2.4.5. Other physical properties

As concerning the other physical properties necessary for a complete evaluation of a material candidate as NET/INTOR structural members, the authors did not perform measurements. In effect it is a matter of properties which are practically independent of the impurities and in good approximation we can use the 6063 alloys values [10]. A summary of the more important properties is tabulated in Table III.

2.4.6. Technological properties

The alloy possesses excellent fabricability and weldability characteristics as usual for the alloys of the 6000 series. The choice for the final selection or design of the T6 condition is no problem from the point of view of the fabbrication of complex shapes.

Table III

SELECTED PROPERTIES OF EXPERIMENTAL Al-0.7Mg-0.4 Si Alloys (Impurity level < 100 ppm)

Density at 20°C Kg/m^3	Liquidus Temperature °C	Solidus Temperature °C	Thermal expansion µm/m/°C	Specific Heat J/g	Thermal Conductivity J/cm^2/cm/°C sec	Electrical Resistivity nΩ-cm
2.70	651	616	21.8 (-50°C to +20°C) 23.4 (20°C to 100°C) 24.5 (20°C to 200°C)	1.00 (at 100°C)	2.185 (0 Temper) 2.017 (T_6 Temper)	530 (at -196°C) (T_6 Temper)

Total infrared Emissivity %	Sputtering yield atoms/ion	Typical Mechanical Properties (T6 Temper)					Weight Variation (water at 120°C) after 4000 hrs mg/cm^2
		Tensile Strength MPa	Yield Strength MPa	Elongation %	Stress to Rupture in 100 hrs at 120°C MPa	Modulus of Elasticity GPa	
3.1 (at 85°C)	1.5×10^{-2} (10^{-1} - 2×10^{-1} KeV)	220 (24°C) 190 (100°C) 180 (120°C)	195 (24°C) 175 (100°C) 165 (120°C)	12 (24°C) 15 (100°C) 15 (120°C)	150	68.9	+ 1.2

3. CONCLUSIONS

The mechanical, chemical and physical data (experimental and compiled data) shown in this paper in order to evaluate the main features of an Al alloy first wall and non breeding blanket for NET/INTOR lead to following conclusions:

1) The experimental alloy Al-0.7 Mg - 0.4 Si (total impurity level < 100 ppm) shows good characteristics: good corrosion resistance, moderate mechanical resistance, light weight, excellent working properties, improved finish and responds to age hardening treatment.

2) Under normal particle and heat fluxes NET/INTOR conditions the extruded experimental alloy (T6 Temper) could be considered as a potentially viable material (if the maximum temperature can be hold below 150°C).

3) In order to conclude it is necessary to know the radiation damage behaviour of the alloy. For that reason an experimental program is planned.

ACKNOWLEDGEMENTS

The ideas presented here have evolved over the last two years, with the help of numerous discussions with colleagues and many experts. Particularly we wish to thank Prof. F. Gatto and Dr. P. Fiorini (ISML-Alumetal, Novara) for many suggestions, the help on experimental problem and the permission to report some unpublished data; Dr. J. Bohdansky (Max-Planck-Institut für Plasmaphysik, Garching) for the sputtering measurement; Prof. M. Farag (American University in Cairo, Egypt) for criticism in reading the paper; Dr. P. Schiller and Prof. G.P. Casini for many suggestions. The experimental work was carried out by M. Airola, E. Haine, A. Misirocchi, C. Ronsecco, E. Weir, R. Baranzini and G.P. Ossola.

REFERENCES

[1] CASINI, G.P., SCHILLER P., and al. 11th SOFT, Oxford, 15-19 September 1980.
[2] POWELL J.R., MILES F.T., ARONSON A, and WINSCHE W.E., Brookhaven National Laboratory Report BNL - 18236 (June 1973).
[3] ROCCO P, PONTI C., and CARETTA A., 11th SOFT, Oxford, 15-19 September 1980.
[4] MONDOLFO L.F., Aluminium Alloys: Structure and Properties, Butterworths, London (1976).
[5] ASTM, Special Technical Publication, Nr. 291
[6] GATTO G., FIORINI P., ISML-Alumetal, Novara, to be published (1981).
[7] ROTH J., BOHDANSKY J., and ORRENBERGER W., Max-Planck-Institute für Plasma Physick Report, IPP 9/26 (1979).
[8] WOLFER W.G., GOLD R.E., SMITH D.L., and KULCINSKY G.L., USA-Contribution to the 3rd Meeting of the INTOR Workshop (16-28 June 1980), EUR FU BRU/XII 501/80/EDV 42 Report (1980).
[9] QUATAERT D., Joint Research Centre, Ispra Establishment, Materials Division, to be published (1981).
[10] Metals Handbook, Vol. 1, Properties and Selection of Metals, 8th edn., Metals Park, Ohio (1961) pp. 946-947.

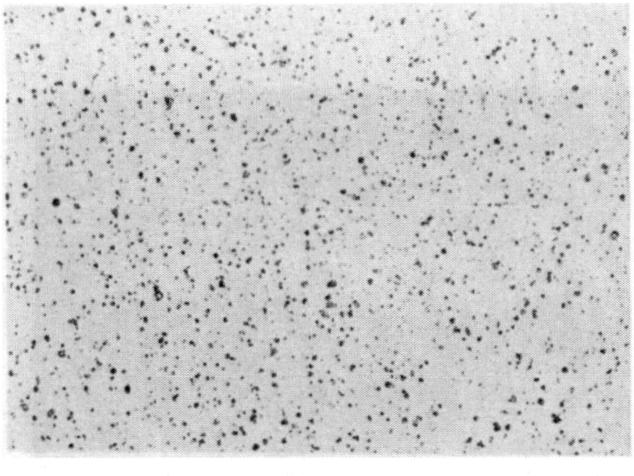

Fig. 1. Microstructure of an extruded and full annealed (O-Temper) Al-0.7 Mg-0.4 Si alloy (impurities level < 100 ppm) (magnification 100:1).

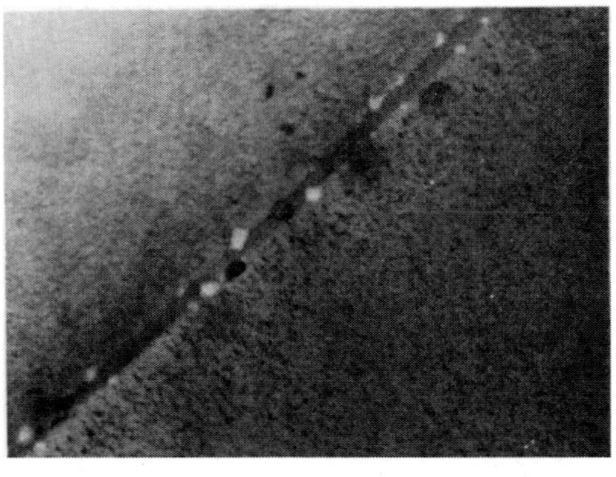

Fig. 2. Electron micrograph (TEM) showing precipitation (Guinier-Preston zones) of an extruded and age hardened (T6 Temper) Al-0.7 Mg-0.4 Si alloy (impurity level < 100 ppm) (magnification 30.000:1).

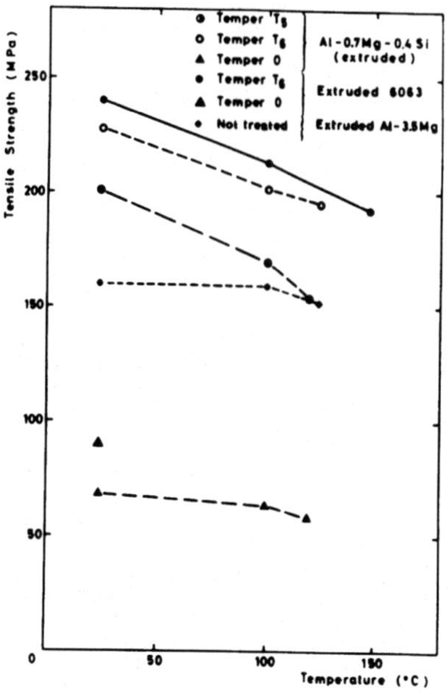

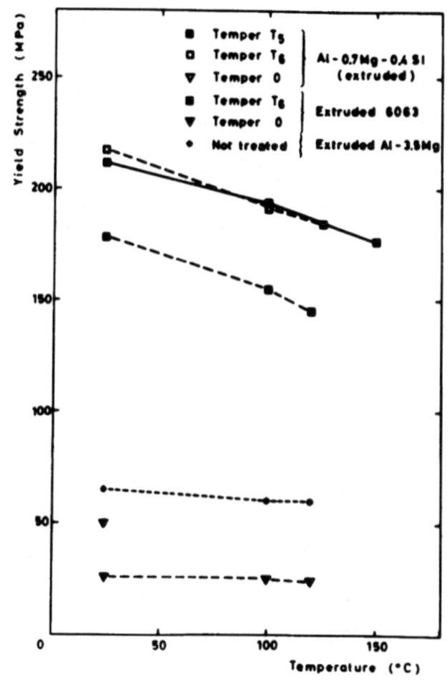

Fig. 3. Tensile properties (ultimate tensile strength) of different Al extruded alloys: a) heat treated (O, T5 and T6 Temper) Al-0.7 Mg - 0.4 Si (impurities level < 100 ppm); b) heat treated (O and T6 Temper) 6063 /5/; c) as extruded Al-3.0 Mg impurity level < 150 ppm /6/

Fig. 4. Tensile properties (Yield strength) of different Al extruded alloys: a) heat treated (O,T5 and T6 Temper) Al-0.7 Mg - 0.4 Si (impurity level < 100 ppm); b) heat treated (O and T6 Temper) 6063 /5/; as extruded Al-3.0 Mg (impurity level < 150 ppm) /6/.

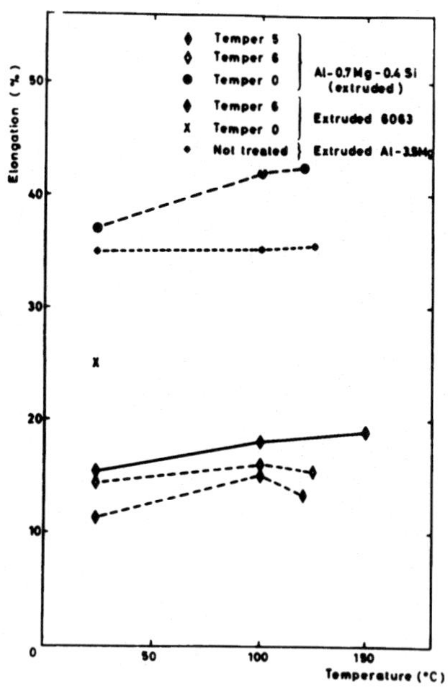

Fig. 5. Tensile elongation at rupture of different Al extruded alloys: a) heat treated (O,T5 and T6 Temper) Al-0.7 Mg - 0.4 Si (impurities level < 100 ppm); b) heat treated (O and T6 Temper) 6063 /5/; c) as extruded Al-3.0 Mg (impurity level < 150 ppm) /6/.

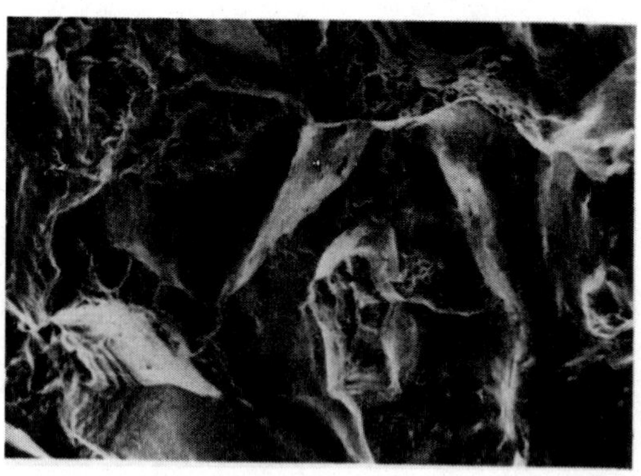

Fig. 6. Intergranular rupture (SEM) micrograph in extruded and heat treated (T5 Temper) Al-0.7 Mg - 0.4 Si (impurity level < 100 ppm). Alloy deformed in tensile test at 20°C. (magnification 30:1).

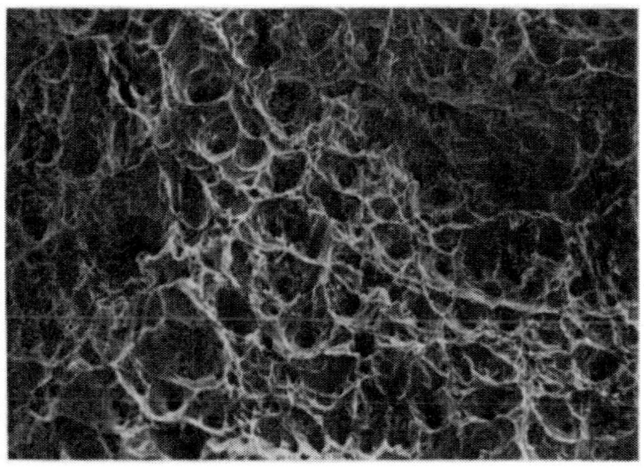

Fig. 7. Features of a ductile rupture (SEM micrograph obtained by traction at 120°C in a extruded and heat treated (T5 Temper) Al-0.7 Mg - 0.4 Si (impurity level < 100 ppm) alloy (magnification 200:1).

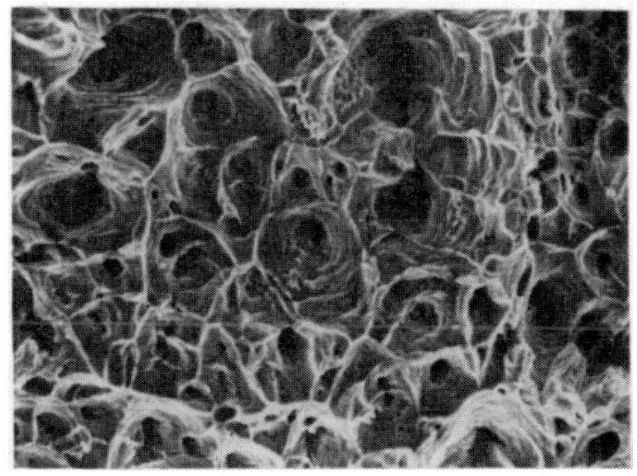

Fig. 8. Dimples associates with precipitates (Mg_2Si) of rupture zone of a sample of a extruded and full annealed (O Temper) Al-0.7 Mg - 0.4 Si (impurity level < 100 ppm) alloy deformed by tensile test at room temperature (magnification 100:1).

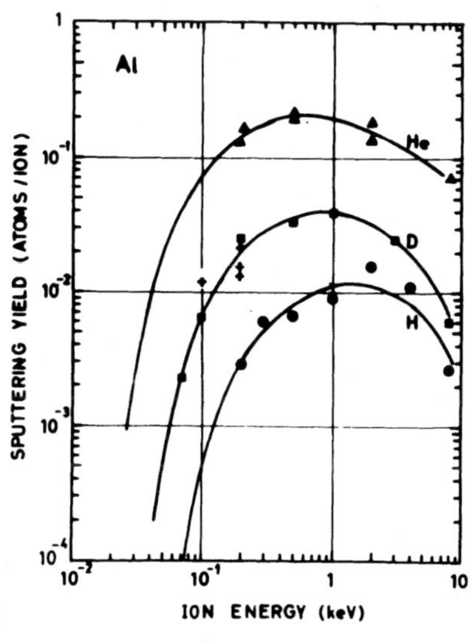

Fig. 9. Energy dependence of the sputtering yield of Al (● ■ ▲) /7/ and of the extruded Al-0.7 Mg - 0.4 Si (impurity level < 100 ppm) alloy (+) with H, D and He ions at low temperature.

Fig. 10. Damage produce on extruded Al-0.7 Mg - 0.4 Si (impurity level < 100 ppm) alloy be an electron beam pulse (power density ~ 20 kW/cm^2) seen by optical microscopy. (magnification 50:1). The molten layer reaches a maximum of 0.4 mm.

FIRST WALL DESIGN AND LOW-Z MATERIAL DEVELOPMENT IN JT-60

H. Nakamura, M. Masuda, Y. Miyauchi, M. Shimizu,
N. Ogiwara, M. Yamamoto, M. Ohta

Japan Atomic Energy Research Institute
Tokai-Mura, Naka-gun, Ibaraki-ken, Japan

K.Owada

Hitachi Ltd.
Saiwai-cho, Hitachi-city, Ibaraki-ken, Japan

ABSTRACT

The present status of JT-60 first wall design and low-Z material development are described. Thermal analysis shows that in case of radiation cooling the maximum temperatures are 2000°C for 10 MW-10 sec/10 min deposition to innermost 3 toroidal fixed limiters and 1300°C for 20 MW-5 sec/10 min deposition to magnetic limiter plates, respectively. Therefore, some considerations to increase thermal conduction between first walls and support structures are necessary for long pulse operartion. Stress analysis shows that mechanical stresses of molybdenum fixed limiters and magnetic limiter plates are lower than the permissible ones. In case of liners, the bending stress of molybdenum liners around magnetic limiter coil exceed the premissible ones. However, Inconel 625 liners have no problem.

Low-Z material development of pre-coating, in-situ coating and bulk methods are now under way. In this paper, the recent results of pre-coating are described. Up to now, small size sample coating of TiC and C and its evaluation tests have been performed. Coating thickness is 20 μm. Thermal fatigue test was carried out. Temperature rise is 1000°C/10 sec. Repetition is 100 cycles. TiC samples survived up to 50∿100 cycles. However, C samples flaked away after one cycle. Detailed investigation of these results are now under way.

1. Introduction

JT-60 is a large tokamak with a poloidal magnetic limiter and its completion is expected in September 1984. The main machine parameters are as follows; R=3.0 m, a=0.95 m, I_p=2.7 MA and B_t=45 kG. The target parameters of the plasma are in the range of $\bar{T}$=5∿10 keV and $\bar{n}\tau$=2∿6× $10^{-19}m^{-3}$sec. The main objects of JT-60 are to obtain a long pulse, reactor grade plasma and to investigate its physical and technological aspects relevant to a fusion reactor. In order to perform the above objectives, the design of JT-60 first walls must fulfill the special requirements from the view points of technology (thermal and mechanical design)and physics(impurity control). The thermal design of JT-60 first walls is so much difficult due to the larger stored energy of the plasma and longer duration than those of currently operating tokamaks. For example, the stored energies of the plasma and discharge durations are 100 kJ/shot and 100 msec in JFT-2, and 200 MJ/shot and 10 sec in JT-60, respectively. The mechanical design of JT-60 first walls is also so much difficult due to a large electromagnetic force at major disruption. The design effort is necessitated so that the first walls are able to withstand the above electromagnetic force. Control of the first wall material impurity in large tokamaks is essentially important to achieve a long pulse and reactor grade plasma. Therefore, the effectiveness of the low-Z first wall is pointed out to reduce the deterimental effect of the wall material impurity without the magnetic limiter and/or the edge cooling.

In the present paper, the present status of the design of the JT-60 first wall, which is now under way, is discussed in Chaps.2, 3 and 4. And in Chap.5 described is the present status of the low-Z material development.

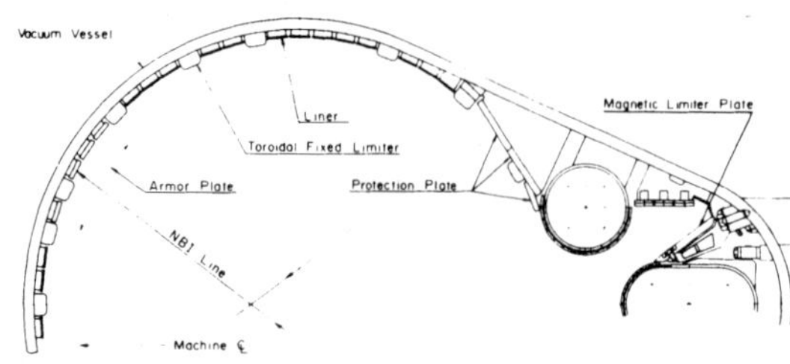

Fig.1 Cross sectional view of JT-60 vacuum vessel.

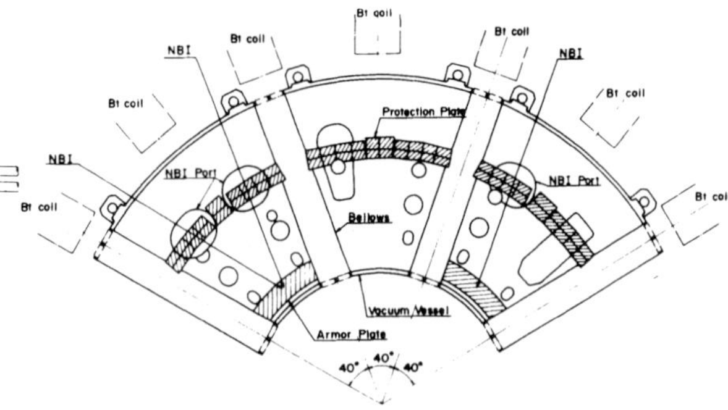

Fig. 2 Plane view of a part of JT-60 vacuum vessel.

2. First wall description

The cross sectional view and plane view of JT-60 vacuum vessel are shown in figs.1 and 2, respectively. JT-60 first walls are classified into liners, toroidal fixed limiters, magnetic limiter plates, armor plates and protection plates. Except for liners, molybdenum is selected as first wall materials keeping surface temperature rise as low as possible during long pulse operation. As candidate material of liners, molybdenum and Inconel 625 are under examination from the points of mechanical strength, thermal properties, manufacturing and cost. Liners are classified into bellows covering ones, rigid ring ones and magnetic limiter coil ones depending on their roles. To receive large heat load, seventeen rows of toroidal fixed limiters with elliptical cross section are placed inside the vacuum vessel, as shown in fig.1. Magnetic limiter plates are installed in burial chamber and cooled by N_2 gas. The armor plates and the protection plates are attached inside the vacuum vessel for protection against the neutral beam strikes and localized high heat flux due to ripple and orbit losses, respectively. JT-60 first walls are designed to be easily exchangeable to study variuos kinds of wall materials. Surface modification plan of low-Z material coating onto molybdenum first walls is now being put into practice.

3. Design requirement

3.1 Thermal load

JT-60 operations are divided into two modes, fixed limiter one and magnetic limiter one. Total thermal load to first walls, duration time and repetition time are 22 MW, 5∿10 sec and 600 sec, respectively. At present, exact estimation of the thermal load to first walls in large tokamaks is difficult, which strongly depends upon impurity contents and/or first wall design. Therefore, the design conditions of the thermal loads to first walls in normal operation are based on the assumption of the following four categories, THERMAL LOAD-I,II,III and IV as shown in table I. THERMAL LOAD-I is caused by synchrotron radiation loss and deposited to all the surfaces of first walls. THERMAL LOAD-II is caused by line radiation and charge exchange loss and deposited to the first walls which face main plasma. The heat flux due to THERMAL LOAD-I and II corresponds to 10 W/cm^2 for the liners. THERMAL LOAD-III and IV are caused by conduction and convection loss to fixed limiters and magnetic limiter plates, respectively. In the present design, THERMAL LOAD-IV is 5 MW for guaranteed value and 20 MW for target value. The latter value corresponds to 30 MW operation in JT-60.

Besides the above THERMAL LOAD-I∿IV, the following thermal loads related to neutral beam injection are taken into account for, that is, RIPPLE LOSS, ORBIT LOSS, NBI SHINE-THROUGH and NBI UNTRAPPED LOSS[1]. The design conditions of these thermal loads are also shown in table I. RIPPLE LOSS implies the fast ion loss that is trapped in toroidal ripple wells. The total loss is estimated to be 0.4 MW for 0.35 % ripple rate and $10^{14} cm^{-3}$ plasma density. Peak and average heat fluxes are 60∿80 W/cm^2 and 20∿30 W/cm^2, respectively. ORBIT LOSS implies ripple induced banana drift loss and localized to outermost limiters and protection plates. The total loss is estimated to be 1.8 MW for the same conditions as RIPPLE LOSS. Heat fluxes are 1 kW/cm for the limiters and 20 W/cm^2 for protection plates. NBI SHINE-THROUGH implies the direct bombardment of neutral beam. This case occures mainly at major disruption during NBI heating. The maximum heat flux to the armor plates is 2000 W/cm^2 for 2 beam superposition[2]. NBI UNTRAPPED LOSS implies the unattenuated loss of injected neutral beams and heat flux to the armor plates is 200 W/cm^2 for 5×10^{13} cm^{-3} average plasma density.

TABLE I Design conditions of the thermal load to first walls

	FIXED LIMITER OPERATION	MAGNETIC LIMITER OPERATION	DEPOSITION AREA
DURATION TIME	10 sec	5 sec	
THERMAL LOAD-I	2 MW	2 MW	160 m^2
THERMAL LOAD-II	10 MW	10 MW	114 m^2
THERMAL LOAD-III	10 MW	5 MW	2.3~12.8 m^2
THERMAL LOAD-IV	——	5/20 MW	5.7 m^2
RIPPLE LOSS	0.4 MW	TBD	11 m^2
ORBIT LOSS	1.8 MW	TBD	TBD
NBI-SHINE THROUGH	2000 W/cm^2	2000 W/cm^2	TBD

3.2 Disruption

The mechanical design condition of electromagnetic force is based on the following assumption of plasma disruption. Plasma disruption is divided into minor disruption and major disruption.

(1) Minor disruption ; The time dependent behavior of loop voltage is assumed as follows.

$$V = \begin{cases} -V_o \sin^2[(t-t_o)/\tau] & t_o \leqq t \leqq t_o + \tau \\ 0 & t_o \leqq t \end{cases}$$

where V_o = 3 kV, 100 μsec $\leqq \tau \leqq$ 400 μsec. Minor disruption of 10^5 cycles are expected during the life of JT-60, 10 years.

(2) Major disruption ; The major disruption of plasma current is assumed as follows.

$$I_p = I_{po}\exp[-(t-t_o)/\tau] \qquad t_o \leqq t$$

where I_p ; plasma current, I_{po}; I_p at $t = t_o$ and τ; time constant of major disruption. The shortest time constant of major disruption is assumed to be 1 msec, of which discharge is expected 10^3 shots during the life of JT-60.

4. Analysis

In this section, the results of thermal and mechnical stress analyses of liners, fixed limiters and magnetic limiter plates are described. The analyses of armor plates and protection plates are now under way.

4.1 Thermal analysis

In this present design, thermal contact between first walls and support structures is not clear. Therefore, the analysis was performed under the conditions of radiation cooling and of radiation and thermal conduction cooling.

(1) Liners

To estimate equilibrium base temperature, steady state thermal analysis was perfomed by the model which contained fixed limiters to consider the heat flux due to its thermal radiation. Heat flux from plasma is 10 W/cm^2. In case of only radiation cooling, the equilibrium base temperature are 490°C for Inconel 625 liners and the temperature several tens degrees low for molybdenum ones, respectively. However, in case of radiation and thermal conduction cooling, these temperatures decrease by 100~200°C.

(2) Fixed limiters

Thermal analysis was performed in the case of 10 MW deposition to all limiters, 3 outermost limiters and 3 innermost limiters, respectively. The maximum heat flux and the calculated temperatures are shown in table II for radiation cooling. This table shows that all limiters deposition case is permitted. However, 3 limiters deposition case is not permitted because the equilibrium base temperatures exceed the usable temperature of bolt, 600∿700°C, and the maximum temperatures exceed the recrystallization temperature of molybdenum.

(3) Magnetic limiter plates

Thermal analysis was performed in the case of 5 MW and 20 MW depositions. In this calculation, the irradiation width is assumed to be 10 cm with separatrix swinging. The calculated values for radiation cooling are shown in table III. This table shows that 5 MW-5 sec operation is permitted and 20 MW-5 sec operation is not permitted in case of radiation cooling. If thermal conduction between magnetic limiter plates and support structures is expected, the equilibrium base temperature for 20 MW deposition is reduced from 840°C to 600°C and 20 MW-5 sec operation becomes to be possible.

Therefore, some considerations to increase the thermal conduction between first walls and support structures are necessary for long pulse operation with large additional heating power.

TABLE II Calculated temperatures of toroidal fixed limiters in case of radiation cooling.

Deposition Area	Max. Heat Flux	Equilibrium Base Temp.	Transient Temp. Rise
All limiters	175 W/cm^2	600 °C	120 °C
Outermost 3 limiters	774 W/cm^2	1000 °C	465 °C
Innermost 3 limiters	1330 W/cm^2	1200 °C	782 °C

TABLE III Calculated temperatures of magnetic limiter plates in case of radiation cooling

Total Power to MLP	Max.Heat Flux (10 cm width)	Equilibrium Base Temp.	Transient Temp. Rise
5 MW	88 W/cm^2	600 °C	110 °C
20 MW	350 W/cm^2	840 °C	460 °C

4.2 Stress analysis

(1) Mechanical properties of first wall materials

(a) Molybdenum

Tension test was performed at the temperatures from a room temperature to 1400°C. Observed wide scattering of the data results from the difference of manufacturing method among some manufacturers. The stress intensity of 20 kg/mm^2 at a room temperature and 12 kg/mm^2 at 500°C is adopted in our design. The fatigue test results of candidate molybdenum at 550°C are shown by the curve 1 in fig.3 together with the other molybdenum test results [3,4]. The design curve is obtained based upon ASME code Sec.III. The correction by the mean (residual) stress reduces considerably the permissible stress level in the range of cycle more than 10^5 bacause the ratio of 0.2 % proof stress to ultimate stress exceeds 0.8. Design has pay attention to the brittleness in the range lower than the transition temperature around 300°C or higher than the recrystallization temperature around 1100°C.

(b) Inconel 625

The permissible stress intensity is 19.4 kg/mm^2 at 550°C. Fatigue design curve is decided through many tests.

(2) Stress analysis

Though the detailed stress analysis of all first walls has not been completed, the rough evaluation of stress has been done for liners, fixed limiters and magnetic limiter plates.

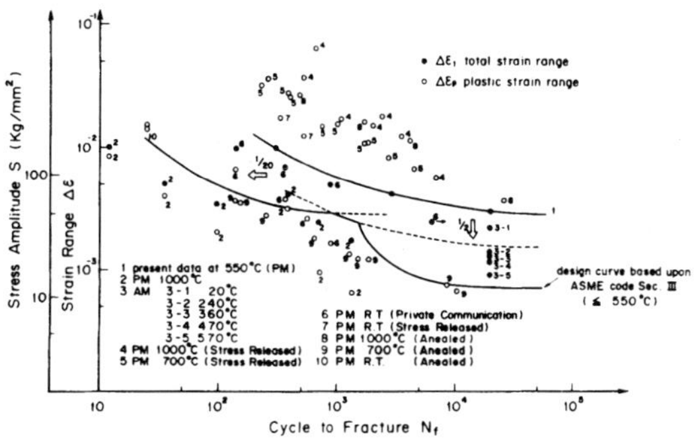

Fig.3 Fatigue design curve of molybdenum

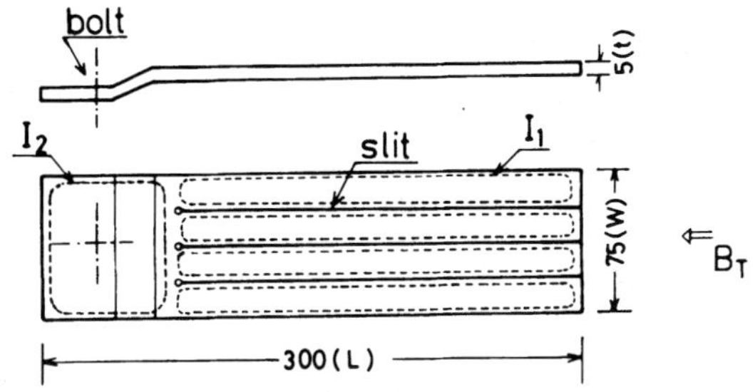

Fig.4 Induced eddy current flow of liner with slits. Dimensions are in case of rigid ring molybdenum liners.

(a) Liners

In case of 1 msec major disruption, the variation amplitude of magnetic field perpendicular to liners are 3.5 kG for bellows covering liners and magnetic limiter coil liners and 6.5 kG for rigid ring liners. Induced eddy current flows near the liner edge, as shown in fig.4. The results of analysis shows that the liners where the stress is the largest is the magnetic limiter coil molybdenum liners. In this case, the maximum eddy currents are I_1=2 kA and I_2=5.7 kA. This currents correspond to bending stress of 50 kg/mm^2. Therefore, in case of 1 msec major disruption, molybdenum is not applicable to magnetic limiter coil liners. The exact evaluation of dynamic magnification factor is under way. On the other hand, Inconel 625 liner has no problem from the view point of mechanical strength.

(b) Fixed limiters

The mecahnical stress due to electromagnetic force is not a serious problem. Rough evaluation indicates that thermal stress is considered to be the most serious problems. This analysis is under way.

(c) Magnetic limiter plates

Magnetic limiter plates have slits to reduce the electromagnetic force. In case of 1 msec major disruption, induced eddy current and stress are 25 kA and 8.9 kg/mm^2, respectively. This value is lower than the permissible stress of 21.6 kg/mm^2 (=1.8 Sm) even if the dynamic magnification factor of 2.0 is taken into account.

5. Low-Z material development

The details and strategy of JT-60 low-Z material development are described in ref.[5]. At the present situation, selection of the first wall materials suitable for fusion reactor is difficult because the details of reactor plasma conditions are not clear. Therefore, JT-60 low-Z material development program is divided into the following two stages; first stage development intending to improve the plasma parameters by the start of JT-60 operation and second stage development applicable to fusion reactors. Application methods of low-Z material in the first stage development are classified into pre-coating, in-situ coating and bulk method. As the pre-coating method, physical vapor deposition was selected considering its good adhesion and low deposition temperature. As one of the in-situ coating methods, co-axial type magnetron sputter one was selected. This method was applied to DIVA and JFT-2 tokamaks. As the bulk method, the lamination structure by pyrolytic graphite was selected considering its good thermal properties. Now, the development of these three methods are under way. In this chapter, we describe the recent results of the pre-coating development. Up to now, small size sample coating of TiC and C and their evaluation tests have been performed. Coating thickness is 20 μm taking into account for the erosion lifetime of limiters.

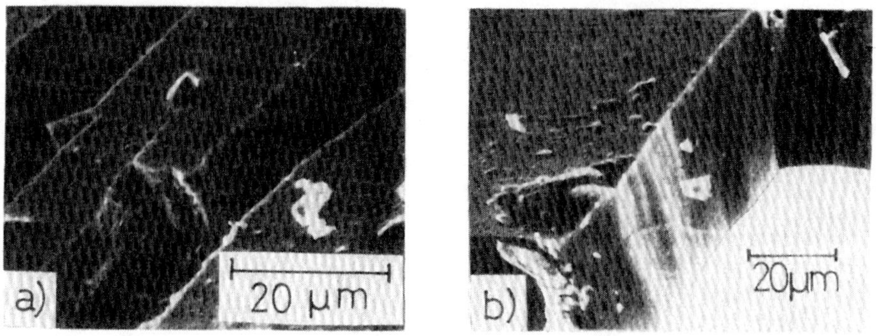

Fig.5 SEM pictures of cross section of coating layer
(a) TiC/Mo (b) C/Mo

5.1 TiC coating

Hollow cathode type ion plating was selected as TiC coating method. Deposition temperatures are 300°C, 500°C and 700°C. The result of X-ray diffraction measurement exhibits crystallization of TiC coating layer and small effect of the deposition temperature. A SEM picture of cross section in case of 700°C deposition temperature is shown in fig.5-a.

5.2 C coating

DC glow discharge type ion plating was selected as C coating method. Deposition temperatures are ambient temperature, 300°C, 500°C and 700°C. The result of X-ray diffraction measurement exhibits that C coating layer is amorphous carbon. In case of C coating by DC glow discharge, the effect of the deposition temperature is very large. At the deposition temperatures above 500°C, coating layer flakes away gradually without external stress. A SEM picture of cross section in case of ambient deposition temperature is shown in fig.5-b.

5.3 Thermal fatigue test

To evaluate the properties under high heat load deposition of limiters, the thermal fatigue test was carried out by infra-red lamp up to 100 cycles. Duration and repetition time are 10 sec and 300 sec, respectively. Temperature rise was 1000 degrees per 10 sec.

In the case of TiC coating, flaking of coating layer was observed between 50∿100 cycles for most samples. However, one of the samples fabricated at the deposition temperature of 700 °C survived up to 100 cycles. In the case of C coating, the failure of samples was observed after only one cycle. At present, detailed investigation of these results are under way.

ACKNOWLEDGEMENT

We would like to thank Drs. Y.Iso, M.Yoshikawa, T.Iijima and H.Tomioka for their continuous guidance and encouragement. We also appreciate to ULVAC corporation for preparing low-Z coating samples.

REFERENCES

[1] TANI, K. et al., submitted to Nucl. Fusion.
[2] NAKAMURA, H. et al., JAERI-M 8359 (1979).
[3] NAGATA, N. et al., J. Nucl. Mater. 85&86, 839-843 (1979).
[4] Molybdenum Metal, Climax Molybdenum (1960).
[5] NAKAMURA, H., Proc. US Japan Workshop on Divertors, First Wall Materials and Impurity Control (Tokai, 1980), JAERI-M 8971 (1980) paper 10-2.

List of Poster Sessions

LIST OF DISCUSSION SESSIONS

Tuesday Morning

Blanket and Shielding

Leader: G.P. Casini
Chairman: A. Malein

Tuesday Afternoon

Magnet Technology

Leader: J.E.C. Williams
Chairman: B. Taquet

Vacuum Systems

Leader: P. Reynolds
Chairman: H. Conrads

Wednesday Morning

Data Acquisition and Control

Leader: A.H. Read
Chairman: A.M. van Ingen

Fusion Technology in Universities

Leaders: A. Shercliff,
G.H. Miley and J. Salge
Chairman: R. Hancox

Thursday Morning

Neutral Injection

Leader: T.S. Green
Chairman: A. Torossian

Power Supplies

Leader: K. Thomassen
Chairman: A.M. van Ingen

Thursday Afternoon

Reactor Studies

Leader: C. Baker
Chairman: K.H. Schmitter

Fusion Materials

Leader: R. Bullough
Chairman: P. Schiller

List of Participants

Name	Address	Country
Ford,G.W.K.	Australian AEC	Australia
Voss, M.F.	Innsbruck University	Austria
Bosia,G.	Ecole Royale Militaire	Belgium
Darvas,J.	CEC, Brussels	Belgium
Malein,A.	CEC, Brussels	Belgium
Nihoul,J.	SCK-CEN,Mol	Belgium
Noterdaeme,J.M.	Ecole Royale Militaire	Belgium
Snykers,M.	SCK-CEN, Mol	Belgium
Verbeek,R.J.	CEC, Brussels	Belgium
Gordon,C.	Ontario Hydro	Canada
Stangeby,P.	Toronto University	Canada
Jensen,P.B.	Risø National Laboratory	Denmark
Sorensen,H.	Risø National Laboratory	Denmark
Warming, (Mrs.) L.	Risø National Laboratory	Denmark
Julian,F.A.	Dunmanway, Cork	Eire
Albrecht,C.	Siemens AG	FRG
Bieder,H-E.	BBC, Mannheim	FRG
Bloemer,B.	Braunschweig Technical Univ.	FRG
Bogensberger,H.G.	NUKEM GmbH Hanau	FRG
Bogusch,E.	Interatom GmbH	FRG
Brockmann,H.J.	KFA Julich	FRG
Brossman,U.B.	IPP, Garching	FRG
Buende,R.	IPP, Garching	FRG
Cloth,P.	KFA Julich	FRG
Conrads,H.	KFA Julich	FRG
Cooper,D.	IPP, Garching	FRG
Cosler,A.	KFA Julich	FRG
Daenner,W.	IPP, Garching	FRG
Deimel,P.	Stuttgart University	FRG
Deksnis,E.	KFA Julich	FRG
Dorst,D.	IPP, Garching	FRG
Dring	KFA, Julich	FRG
Dustmann,C-H.	BBC, Mannheim	FRG
Eidens,J.	KFA, Julich	FRG
Falter,H.D.	JET,Culham	FRG
Forsyth,N.	HT Reaktorbau, Mannheim	FRG
Freisinger,J.	1st Inst. Physics, Giessen	FRG
Fricke,K.G.	Braunschweig Technical Univ.	FRG
Friesinger,G.	KFZ Karlsruhe	FRG
Fuchs,G.	KFA Julich	FRG
Gernhardt,J.	IPP, Garching	FRG
Giesen,B.	KFA Julich	FRG
Gorg,B.	KFA Julich	FRG
Grieger,G.	IPP, Garching	FRG
Guggi, D.	KFA, Julich	FRG
Haubenberger,W.D.	IPP, Garching	FRG
Heinz,W.	KFZ Karlsruhe	FRG
Hemmerich,J.L.	KFA Julich	FRG
Hofmann,G.A.	Maxwell Laboratories	FRG
Hornschemeyer,W.	Kabelmetal Osnabruck	FRG
Ihle,H.R.	KFA, Julich	FRG
Jeske,U.	KFZ Karlsruhe	FRG
Joneja,O.P.	KFA, Julich	FRG
Jung,P.	KFA.Julich	FRG
Kaletta,D.	KFZ Karlsruhe	FRG
Kesternich,W.	KFA Julich	FRG
Knobloch,A.	IPP, Garching	FRG
Kohlhaas,W.	KFA Julich	FRG
Koppendorfer,W.	IPP, Garching	FRG
Korten,M.	KFA, Julich	FRG

Kotzlowski,H.	IPP, Garching	FRG
Krause,H.	IPP, Garching	FRG
Krauth,H.	KFZ Karlsruhe	FRG
Lehr,K-J.	BBC, Mannheim	FRG
Liu,C.C.	IPP, Garching	FRG
Mast,K.F.	IPP, Garching	FRG
Migge,H.	Hahn-Meitner-Inst. Berlin	FRG
Mioduszewski,P.	KFA Julich	FRG
Nathrath,N.	Messerschmitt	FRG
Niewiera,W.	Kabelmetal Osnabruck	FRG
Nyilas,A.	KFZ Karlsruhe	FRG
Perchermeier,J.	IPP, Garching	FRG
Pillsticker,M.	IPP, Garching	FRG
Rapp,H.	IPP, Garching	FRG
Roehrig,D.	KFA Julich	FRG
Saffert,J.	JET,Culham	FRG
Salge,J.	Braunschweig Technical Univ.	FRG
Schmitter,K-H.	IPP, Garching	FRG
Schneider,F.	IPP, Garching	FRG
Schramm,J.	Braunschweig Technical Univ.	FRG
Schroeder,H.	KFA Julich	FRG
Schwarz,U.	KFA Julich	FRG
Seiler,H.	BBC, Mannheim	FRG
Streibl,B.	IPP, Garching	FRG
Tenten,W.	KFA Julich	FRG
Theenhaus,R.	KFA Julich	FRG
Thueringer,R.	1st Inst. Physics, Giessen	FRG
Wedler,H.	IPP, Garching	FRG
Wesner,F.	IPP, Garching	FRG
Zehlein,H.	KFZ Karlsruhe	FRG
Baget,R.	Paris University	France
Bareyt,B.	CEN Fontenay-aux-Roses	France
Bochirol,L.	CEN, Grenoble	France
Bottereau,J.M.	CEN Fontenay-aux-Roses	France
Claudet,G.	CEN Grenoble	France
Cohen,A.	CEN Fontenay-aux-Roses	France
Delaplace,J.	CEN Saclay	France
Druaux,J.	CEN Fontenay-aux-Roses	France
Gauthier,A.	CEN, Grenoble	France
Guini,M.	Cogelex (Cegaltom)	France
Kuus,H.	CEN Fontenay-aux-Roses	France
Lebars, (Mrs.) J.	CEN Saclay	France
Leloup,C.	CEN Fontenay-aux-Roses	France
Libeyre,P.	CEN Fontenay-aux-Roses	France
Lipa,M.	CEN, Grenoble	France
Magne,R.	CEN, Grenoble	France
Perin,J.P.	CEN, Grenoble	France
Porrot,E.	CEN, Grenoble	France
Removille,J.R.P.	Luxatom	France
Rey,G.	CEN, Grenoble	France
Rijnoudt,E.	CEN Fontenay-aux-Roses	France
Taquet,B.	CEN, Grenoble	France
Torossian,A.	CEN Fontenay-aux-Roses	France
Touche,J.	CEN Fontenay-aux-Roses	France
Dutta Gupta,P.B.	ITT Kharagpur	India
Bagatin,M.	Padua University	Italy
Bishop,G.R.	JRC, Ispra	Italy
Bobbio,S.	University of Naples	Italy
Brunelli,B.	CNEN, Frascati	Italy
Carrara,V.	NIRA SpA Genova	Italy
Casini,G.	JRC, Ispra	Italy
Cenacchi,G.	CNEN, Bologna	Italy
Coletti,A.	CNEN, Frascati	Italy
Della Porta,P.	SAES Getters SpA Milan	Italy
Esposito,F.	University of Naples	Italy
Farfaletti Casali,F.	JRC, Ispra	Italy
Ferrario,B.	SAES Getters spA Milan	Italy
Ferro,C.	CNEN, Frascati	Italy
Gnesotto,F.	CNR, Padua	Italy

Guarnieri,M.	Padua University	Italy
Martone,R.	University of Calabria	Italy
Maschio,A.	Padua University	Italy
Materassi,A.	Varese	Italy
Piatti,G.	JRC, Ispra	Italy
Rocco,P.	JRC, Ispra	Italy
Rubinacci,G.	University of Calabria	Italy
Sambo,C.	Officine Galileo, Florence	Italy
Sassi,M.	CNEN, Frascati	Italy
Schiller,P.	JRC, Ispra	Italy
Schrefler,B.	Padua University	Italy
Stella,A.	Padua University	Italy
Taroni,A.	CNEN, Bologna	Italy
Tenti,P.	Padua University	Italy
Van Rutten,F.	JRC, Ispra	Italy
De Menna, L.	University of Naples	Italy
Aoki,T.	Nagoya University	Japan
Enjoji,H.	Mie University	Japan
Goto,S.	Osaka University	Japan
Ido,S.	Osaka University	Japan
Ioka,S.	Toshiba Corporation	Japan
Itou,Y.	Hitachi Limited	Japan
Kasuya,K.	Tokyo Inst. of Technology	Japan
Maekawa,A.	Hitachi Limited	Japan
Miura,A.	Toshiba Corporation	Japan
Murata,T.	Toshiba Corporation	Japan
Nabae,A.	Nagaoka Technical University	Japan
Nakajima,K.	Toshiba Corporation	Japan
Ogata,A.	JAERI, Ibaraki	Japan
Ohta,M.	JAERI, Ibaraki	Japan
Sato,Y.	Electrotechnical Lab. Ibaraki	Japan
Takahashi,I.	Nagaoka Technical University	Japan
Tanahashi,S.	IPP, Nagoya,	Japan
Uemura,A.	Hitachi	Japan
Ueyama,J.	Hitachi	Japan
Yamato,H.	Toshiba Corporation	Japan
Yasosima,Y.	Toshiba Corporation	Japan
Damstra,G.C.	Kema Arnhem	Netherlands
Elen,J.D.	ECN Petten	Netherlands
Felkers,H.F.	Meratoom, The Hague	Netherlands
Griffioen,A.C.	FOM, Rijnhuizen	Netherlands
Hellingman,P.	FOM, Rijnhuizen	Netherlands
Hoekzema,J.A.	FOM, Rijnhuizen	Netherlands
Hogewey,G.M.D.	FOM, Rijnhuizen	Netherlands
Klippel,H.Th.	ECN Petten	Netherlands
Kooijman,W.	FOM, Rijnhuizen	Netherlands
Lisser,J.	Holec Machines	Netherlands
Manintveld,P.	FOM, Rijnhuizen	Netherlands
Sterk,A.B.	FOM, Rijnhuizen	Netherlands
Van Donselaar,M.F.	FOM, Rijnhuizen	Netherlands
Van Ingen,A.M.	FOM, Rijnhuizen	Netherlands
Van Melle,A.	TNO Apeldoorn	Netherlands
Verschuur,K.A.	ECN Petten	Netherlands
Nowak,A.	S.Kaliski Inst.	Poland
Taczanowski,S.	Inst of Physics & Nuc Tech	Poland
Manero-Amoros,F.	Junta de Energia Nuclear	Spain
Bergstroem,J.B.	Royal Inst. Tech. Stockholm	Sweden
Stenquist,C.	Studsvik Energiteknik	Sweden
Cheetham,A.	CRPP, Lausanne	Switzerland
Hofmann,F.	CRPP, Lausanne	Switzerland
Jaeger,J-F.	EIR, Wurenlingen	Switzerland
Keller,R.	CRPP, Lausanne	Switzerland
Kuse,D.	BBC, Baden	Switzerland
Lietti,A.	CRPP, Lausanne	Switzerland
Maix,R.K.	BBC Zurich	Switzerland
Schmid,T.	BBC Zurich	Switzerland

Simik,A.	CRPP, Lausanne	Switzerland
Richardson,D.	Culham Laboratory	U.K,
Baker,L.J.	AERE Harwell	U.K.
Banford,H.M.	SURRC, Glasgow	U.K.
Barnes,P.M.	Culham Laboratory	U.K.
Bayes,D.V.	Culham Laboratory	U.K.
Beedham,E.	GEC Machines Limited	U.K.
Bombi,F.	JET, CEC	U.K.
Bond,A.	Culham Laboratory	U.K.
Booth,J.	JET, CEC	U.K.
Boschi,A.	JET, CEC	U.K.
Brown,F.	Avica Equipment Limited	U.K.
Bullough,R.	UKAEA, Harwell	U.K.
Butterworth,S.C.	Culham Laboratory	U.K.
Calderbank, (Mrs.) V.	Culham Laboratory	U.K.
Celantano,G.	JET, CEC	U.K.
Challender,R.S.	UKAEA, Risley	U.K.
Ciscato,D.	JET, CEC	U.K.
Cole, H.C.	Culham Laboratory	U.K.
Constantine,G.	AERE Harwell	U.K.
Corbyn,D.B.	JET, CEC	U.K.
Coupland,J.R.	Culham Laboratory	U.K.
Crawley,H.J.	Culham Laboratory	U.K.
Davenport,P.A.	Culham Laboratory	U.K.
De Kock,L.	JET, CEC	U.K.
De Vere,A.P.C.	Culham Laboratory	U.K.
Dietz,K.J.	JET, CEC	U.K.
Duquenoy,H.R.	JET, CEC	U.K.
English,C.A.	UKAEA, Harwell	U.K.
Firth,L.	Culham Laboratory	U.K.
Fry,M.G.J.	Brentford Electric	U.K.
Fry,T.M.	Associated Nuclear Services	U.K.
Fullard,K.	JET, CEC	U.K.
George,M.W.	UKAEA, Harwell	U.K.
Goede,A.P.H.	JET, CEC	U.K.
Goodhew,P.J.	University of Surrey	U.K.
Gott,H.	Associated Nuclear Services	U.K.
Gray,J.W.	Culham Laboratory	U.K.
Green,T.S.	Culham Laboratory	U.K.
Guthrie,J.A.S.	Associated Nuclear Services	U.K.
Hammond,D.P.	Culham Laboratory	U.K.
Hancox,R.	Culham Laboratory	U.K.
Harbour,P.J.	Culham Laboratory	U.K.
Harding,N.H.	Associated Nuclear Services	U.K.
Harries,D.R.	UKAEA, Harwell	U.K.
Harrison,M.F.A.	Culham Laboratory	U.K.
Hatfield,J.	Culham Laboratory	U.K.
Hayward,R.J.	Culham Laboratory	U.K.
Hemsworth,R.S.	JET,Culham	U.K.
Heywood,G.C.H.	Culham Laboratory	U.K.
Hildreth,G.H.	Brentford Electric	U.K.
Hoskins,A.J.	Culham Laboratory	U.K.
Huart,M.	JET, CEC	U.K.
Huguet,M.	JET, CEC	U.K.
James,T.E.	Culham Laboratory	U.K.
Jarvis,O.N.	UKAEA, Harwell	U.K.
Kear,M.J.	Culham Laboratory	U.K.
Lavender,K.E.	Culham Laboratory	U.K.
Lawson,J.D.	Rutherford Laboratory	U.K.
Lomer,W.M.	UKAEA, Harwell	U.K.
Martin,A.R.	Culham Laboratory	U.K.
Mason,L.A.	UKAEA, Harwell	U.K.
Mazey,D.J.	UKAEA, Harwell	U.K.
Mitchell,J.T.D.	Culham Laboratory	U.K.
Mondino,P.L.	JET, CEC	U.K.
Morgan,J.G.	Culham Laboratory	U.K.
Morrison,C.A.	Imperial College	U.K.
Murphy,E.	Culham Laboratory	U.K.
Noll,P.	JET, CEC	U.K.
Obert,W.	JET, CEC	U.K.
Pease, R.S.	Culham Laboratory	U.K.

Phillpott,J.	Culham Laboratory	U.K.
Pierce, (Miss) S.	UKAEA, Harwell	U.K.
Pohlchen,R.	JET, CEC	U.K.
Quigley,T.M.	Oxford University	U.K.
Raynor,F.	Culham Laboratory	U.K.
Read,A.H.	Culham Laboratory	U.K.
Rebut,P-H.	JET, CEC	U.K.
Reynolds,P.	Culham Laboratory	U.K.
Salpietro,E.	JET, CEC	U.K.
Sanders,B.C.	Culham Laboratory	U.K.
Shercliff,J.A.	Warwick University	U.K.
Sinton,E.K.	Culham Laboratory	U.K.
Sonnerup,L.	JET, CEC	U.K.
Spears,W.R.	Culham Laboratory	U.K.
Stork,D.	JET, CEC	U.K.
Thompson,E.	JET, CEC	U.K.
Tozer,B.A.	CEGB, Marchwood	U.K.
Trotman,D.L.	Culham Laboratory	U.K.
Usselmann,E.A.	JET, CEC	U.K.
Varley,G.L.	Culham Laboratory	U.K.
Vickers,A.J.	Culham Laboratory	U.K.
Walters,C.R.	Rutherford Laboratory	U.K.
Weaver,J.	Culham Laboratory	U.K.
Wilcock,P.D.	Culham Laboratory	U.K.
Wood,M.H.	UKAEA, Harwell	U.K.
Worsnop,P.A.	Culham Laboratory	U.K.
Baker,C.C.	Argonne National Laboratory	U.S.A.
Barkley,W.R.	EIMAC Varian, San Carlos	U.S.A.
Berkner,K.H.	Lawrence Berkeley Laboratory	U.S.A.
Budny,R.	PPL Princeton	U.S.A.
Carter,D.R.	RCA Pennsylvania	U.S.A.
Cecchi,J.L.	PPL Princeton	U.S.A.
Cummins,J.D.	Aydin Energy Division	U.S.A.
Down, M.G.	Westinghouse R & D Centre	U.S.A.
Dry,D.N.	Aydin Electronics	U.S.A.
Eckard,R.	Lawrence Livermore Laboratory	U.S.A.
Fillo,J.A.	Brookhaven National Laboratory	U.S.A.
Frank,T.G.	LASL, Los Alamos	U.S.A.
Glad,A.S.	General Atomic, San Diego	U.S.A.
Hesselgrave,D.H.	Aydin Energy Division	U.S.A.
Hurley,C.A.	Lawrence Livermore Laboratory	U.S.A.
Jordan,C.	Lecroy, California	U.S.A.
Kammash,T.	University of Michigan	U.S.A.
Miley,G.H.	University of Illinois	U.S.A.
Moir,R.W.	Lawrence Livermore Laboratory	U.S.A.
Petree,F.	Ebasco, New York	U.S.A.
Rappe,G.H.	PPL Princeton	U.S.A.
Rawls,J.M.	General Atomic, San Diego	U.S.A.
Reardon,P.J.	PPL Princeton	U.S.A.
Sager,P.H.	General Atomic, San Diego	U.S.A.
Santarius,J.F.	Wisconsin University	U.S.A.
Santoro,R.T.	Oak Ridge National Laboratory	U.S.A.
Schilling,G.	PPL, Princeton	U.S.A.
Stambaugh,R.	General Atomic, San Diego	U.S.A.
Theil,E.	Lawrence Berkeley Laboratory	U.S.A.
Thomassen,K.	Lawrence Livermore Laboratory	U.S.A.
Williams,J.	MIT, Massachusetts	U.S.A.